国家出版基金项目
NATIONAL PUBLICATION FOUNDATION

猪解剖学与组织学彩色图谱

COLOR ATLAS OF ANATOMY AND HISTOLOGY OF THE PIG

陈耀星　王子旭◎著

北京科学技术出版社

图书在版编目（CIP）数据

猪解剖学与组织学彩色图谱 / 陈耀星，王子旭著. —北京：北京科学技术出版社, 2018.5
ISBN 978-7-5304-8753-2

Ⅰ. ①猪… Ⅱ. ①陈… ②王… Ⅲ. ①猪—动物解剖学—动物组织学—图谱 Ⅳ. ①S828.1-64

中国版本图书馆CIP数据核字（2016）第309058号

猪解剖学与组织学彩色图谱

作　　者：陈耀星　王子旭
策划编辑：王　晖
责任编辑：王　晖　于庆兰
责任校对：贾　荣
责任印制：李　茗
封面设计：异一设计
出 版 人：曾庆宇
出版发行：北京科学技术出版社
社　　址：北京西直门南大街16号
邮政编码：100035
电话传真：0086-10-66135495（总编室）
0086-10-66113227（发行部）　0086-10-66161952（发行部传真）
电子信箱：bjkj@bjkjpress.com
网　　址：www.bkydw.cn
经　　销：新华书店
印　　刷：北京盛通印刷股份有限公司
开　　本：889mm × 1194mm　1/16
字　　数：435千
印　　张：34.25
版　　次：2018年5月第1版
印　　次：2018年5月第1次印刷
ISBN 978-7-5304-8753-2/S · 212

定　　价：430.00元

编委会名单

主　著　陈耀星　王子旭

副主著　曹　静　李　健　董玉兰　邱利伟

参　著（以姓氏笔画为序）

丁家波　万华云　万建青　马保臣

马淑慧　王　铁　王文利　王团结

王利永　艾克拜尔·热合曼　卢嘉茵

白欣洁　宁淑杰　边　疆　刘　磊

刘嘉静　李　梅　李　婕　杨昆鹏

吴思捷　张玉仙　张利卫　张莹辉

陈付菊　陈福宁　林如涛　岳　亮

赵俊杰　姚文生　贾六军　原展航

高　婷　郭青云　康　凯　董彦君

蒋　南　程君生　熊娟娟

前 言

我国目前已是世界养猪大国，但生猪生产效率还低于世界先进水平，中国养猪业大而不强。为了实现从养猪大国走向养猪强国，保障养猪业的健康与可持续发展，就必须加强对猪的营养健康、繁殖改良和疾病防治等开展深入研究，这迫切需要有详细的猪体结构资料。另一方面，由于猪和人在生理解剖、营养代谢、生化指标等特征上有较大的相似性，尤其是心血管系统结构与人更为相似，猪是最接近人类的模式动物之一，是研究人类疾病防治的理想实验动物，其研究和开发利用也需要猪体的解剖学与组织学基础资料。然而有关猪解剖学与组织学彩色图谱的专著还相当匮乏。鉴此，我们在北京科学技术出版社的支持下，获得国家出版基金资助，完成了这本《猪解剖学与组织学彩色图谱》。

本图谱内容包括体表结构与被皮系统、骨骼系统、肌肉系统、消化系统、呼吸系统、泌尿系统、雄性生殖系统、雌性生殖系统、心血管系统、淋巴系统、神经系统、内分泌系统和感觉器官等13章。共收集实图643幅，选图考究，包括大体解剖图、组织显微照片、组织化学与免疫细胞化学染色照片、电镜照片和先进的3D成像技术图片、细胞内染色照片，还有对各知识点的简要文字说明，较全面覆盖了猪的解剖学和组织学知识。这些标本图片不仅帮助读者认识猪体解剖学和组织学结构，也方便了他们了解临床运用的图像科技。为了便于读者阅读和使用，本图谱采用了中英文注释的方式，并在书后附加了名词索引，方便参考查阅。

参加本图谱编著工作的有中国农业大学的陈耀星、王子旭、曹静、董玉兰、董彦君、张利卫、蒋南、白欣洁、马淑慧、边疆、刘磊、宁淑杰、熊娟娟、原展航、岳亮、杨昆鹏、林如涛、高婷、李梅、万华云、卢嘉茵、刘嘉静、王铁和李婕，河南科技大学的李健，中国兽医药品监察所的王团结、赵俊杰、张莹辉、程君生、吴思捷、姚文生、万建青、康凯、丁家波和王利永，北京麋鹿生态实验中心的郭青云，塔里木大学的艾克拜尔·热合曼，中国牧工商（集团）总公司的马保臣，中国医学科学院阜外医院的贾六军，青海大学的陈付菊，北京农业职业学院的张玉仙和王文利，北京昌平中西医结合医院的陈福宁。本图谱的编著工作自2014年正式启动，经过3年的努力，包括选购动物、制作和挑选标本、拍摄、精选图片、图片加工、标注、中英文注释等过程，凝聚了全体参著人员的心血和汗水。全书最后由中国农业大学陈耀星教授、王子旭高级实验师和曹静副教授统稿。

在付梓之际，衷心地感谢为本图谱编著努力工作的所有人员。衷心地感谢北京农学院胡格教授无私提供了部分塑化标本。感谢中国农业大学动物医学与动物科学专业本科生蒋无砚、陈顺琪、逄金吉、薛晓柳、陈登金、吕巧琳、关芷玲、肖志斌、张羽飞、李建波、徐义涵等同学帮助对解剖标本图片加工处理，由于你们的辛勤劳动，使本图谱的图版更加精美。感谢国家出版基金委员会、北京科学技术出版社和北京市高等学校教育教学改革项目（2015-ms048）的资助。感谢北京科学技术出版社各位编辑的精心编校，谢谢你们同我们之间默契和成功的合作。

图谱编著工作是一项浩瀚的工程。鉴于我们的经验不足和水平有限，加之时间仓促，如果存在错误和欠妥之处，竭诚希望广大读者和同行批评指正。

陈耀星　王子旭

2016年12月于北京

目　录 | CONTENTS

第一章　体表结构与被皮系统................................1
第二章　骨骼系统................................27
第三章　肌肉系统................................79
第四章　消化系统................................113
第五章　呼吸系统................................205
第六章　泌尿系统................................239
第七章　雄性生殖系统................................267
第八章　雌性生殖系统................................297
第九章　心血管系统................................327
第十章　淋巴系统................................369
第十一章　神经系统................................399
第十二章　内分泌系统................................441
第十三章　感觉器官................................467
参考文献................................481
中英索引................................482
英中索引................................510

第一章

体表结构与被皮系统

Chapter 1

Structure of body surface and integumentary system

一、体表结构

猪体可分成运动系统、消化系统、呼吸系统、泌尿系统、生殖系统、心血管系统、淋巴系统、神经系统、内分泌系统、被皮系统和感觉器官，从外表可划分成头部、躯干、前肢和后肢。

1．头部（head）：包括颅部和面部。

（1）颅部（cranial part）位于颅腔周围，可分为枕部、顶部、额部、颞部、耳部和眼部。

（2）面部（facial part）位于口腔和鼻腔周围，可分为眶下部、鼻部、咬肌部、颊部、唇部、颏部和下颌间隙部。

2．躯干（trunk）：分为①颈部，包括颈背侧部、颈侧部和颈腹侧部；②背胸部，包括背部（分鬐甲部和背部）、胸侧部（肋部）和胸腹侧部（分胸前部和胸骨部）；③腰腹部，分为腰部和腹部；④荐臀部，包括荐部和臀部；⑤尾部。

3．前肢（forelimb）：包括肩部、臂部、前臂部和前脚部。前脚部又可分腕部、掌部和指部。

4．后肢（hindlimb）：分为臀部、股部、膝部、小腿部和后脚部。后脚部又可分跗部、跖部和趾部。

二、被皮系统

被皮系统（integumentary system）包括皮肤和皮肤的衍生物。皮肤被覆于猪体表面，由复层扁平上皮和结缔组织构成，内含大量血管、淋巴管、汗腺以及丰富的感受器（如痛、温、触、压觉感受器）。因此，皮肤是重要的感觉器官，并具有保护深部组织、防止体液蒸发、调节体温和排泄废物的功能。毛、汗腺、皮脂腺、乳腺、枕、蹄等都是由皮肤衍变而成，故称为皮肤的衍生物。

1．皮肤（skin）：覆盖于体表，在口裂、鼻孔、肛门和尿生殖道外口等处与黏膜相接。皮肤可分为表皮、真皮和皮下组织3层。皮肤颜色因品种而异，与皮肤细胞内所含黑色素颗粒和类胡萝卜素有关。

（1）表皮（epidermis）为皮肤最表面的结构，由复层扁平上皮构成，由浅向深依次为角质层（horny layer）、透明层（clear layer）、颗粒层（granular layer）、棘层（spinous cell layer，*stratum spinosum*）和基底层（basal layer）。基底层与真皮相连，其细胞增殖能力很强，可不断产生新的细胞，以补充表层角化脱落的细胞；角质层由大量角化的扁平细胞堆积而成，细胞死亡后即脱落。表皮内有丰富的游离神经末梢，有接受疼痛刺激的功能，但表皮内无血管。

（2）真皮（dermis）位于表皮深层，是皮肤最厚也是最主要的一层，由致密结缔组织构成，其中胶原纤维粗大、交织成网，坚韧且富有弹性。皮革就是由真皮鞣制而成的。真皮内有毛囊、皮脂腺、汗腺、血管、淋巴管、神经和感受器分布，在结构上可分为乳头层（papillary layer）和网状层（reticular layer）。临床上作皮内注射，就是把药物注入真皮内。

（3）皮下组织（subcutaneous tissue）位于真皮的深层，由疏松结缔组织构成，又称浅筋膜（superficial fascia）。皮下组织内有皮血管、皮神经和皮肌。在营养良好的猪皮下蓄积大量的脂肪，称猪膘。

2. 毛（hair）：是一种角化的表皮结构，坚韧而有弹性，是温度的不良导体，具有保温作用和一定的经济价值。猪的被毛（clothing hair）遍布全身，有粗毛与细毛之分。以短而直的粗毛为主，分布于头部和四肢。猪颈、背部有特殊的长毛，称为鬃；尾部有尾毛和系关节后部的距毛。

毛是表皮的衍生物，由角化的上皮细胞构成。毛露于皮肤表面的部分称毛干（hair shaft），埋在皮肤内的部分称毛根（hair root），毛根末端膨大呈球状为毛球（hair bulb）。毛球细胞分裂能力强，是毛的生长点。毛球的顶端内陷呈杯状，真皮结缔组织伸入其内形成毛乳头（hair papilla），相当于真皮的乳头层，含有丰富的血管和神经，毛球可通过毛乳头获得营养物质。毛囊（hair follicle）包围于毛根周围，可分成表皮层和真皮层。表皮层由皮肤表皮向真皮内陷入，包围于毛根之外，称根鞘（root sheath）；真皮层构成结缔组织鞘，包于根鞘之外。在毛囊的一侧有一束斜行的平滑肌，为竖毛肌（arrector muscle），受交感神经支配，收缩时使毛竖立。毛干组织结构由外至内依次为毛小皮、毛皮质和毛髓质。

毛有一定寿命，生长到一定时期就会衰老脱落，为新毛所代替，即换毛（molting）。猪换毛的方式为持续性换毛，即换毛不受季节和时间的限制，如猪鬃。

3. 皮肤腺（cutaneous gland）：由表皮陷入真皮内形成，包括乳腺、汗腺和皮脂腺。

（1）乳腺（mammary gland）属复管泡状腺，公猪和母猪虽都有乳腺，但只有母猪的乳腺能充分发育并具有泌乳功能，且形成较发达的乳房。乳房（mammae, breast）的最外面是薄而柔软的皮肤，其深面为一浅筋膜和一深筋膜。深筋膜的结缔组织伸入乳腺实质内，构成乳腺的间质，将腺实质分隔成许多腺叶和腺小叶。猪乳房成对排列于腹白线两侧，常有5～8对，每个乳房有1个乳头，每个乳头有2～3个乳头管。

（2）汗腺（sweat gland）为单管状腺，分泌部位于真皮，导管长而扭曲，多开口于毛囊，少数直接开口于皮肤表面。汗腺分泌汗液，起排泄废物和调节体温的作用。猪的面部与颈部汗腺发达。

（3）皮脂腺（sebaceous gland）为分枝泡状腺，位于真皮内，毛囊和立毛肌之间。在有毛的部位，其导管开口于毛囊；在无毛部位，则直接开口于皮肤表面。皮脂腺分泌脂肪，有润滑皮肤和被毛的作用。

（4）腕腺（carpal gland）为特化的皮肤腺，为前肢掌骨内侧表皮内陷的结构，其分泌方式为顶浆分泌，公、母猪均有。当公、母猪交配时，公猪利用此结构给母猪作上“标记”。

（5）耵聍腺（ceruminous gland）为耳道的腺体，顶浆分泌腺和皮脂腺生成耵聍。

4. 蹄（hoof）：是四肢的着地器官，位于指（趾）端。由皮肤演变而成，其结构似皮肤，也具有表皮、真皮和少量皮下组织。表皮因角质化而称角质层，构成蹄匣（hoof capsule），无血管和神经；真皮部含有丰富的血管和神经，呈鲜红色，感觉灵敏，通常称肉蹄（dermis of the hoof）。猪蹄为偶蹄，有两个直接与地面接触的主蹄（principal hoof）和两个不与地面接触的悬蹄（dewclaw）。

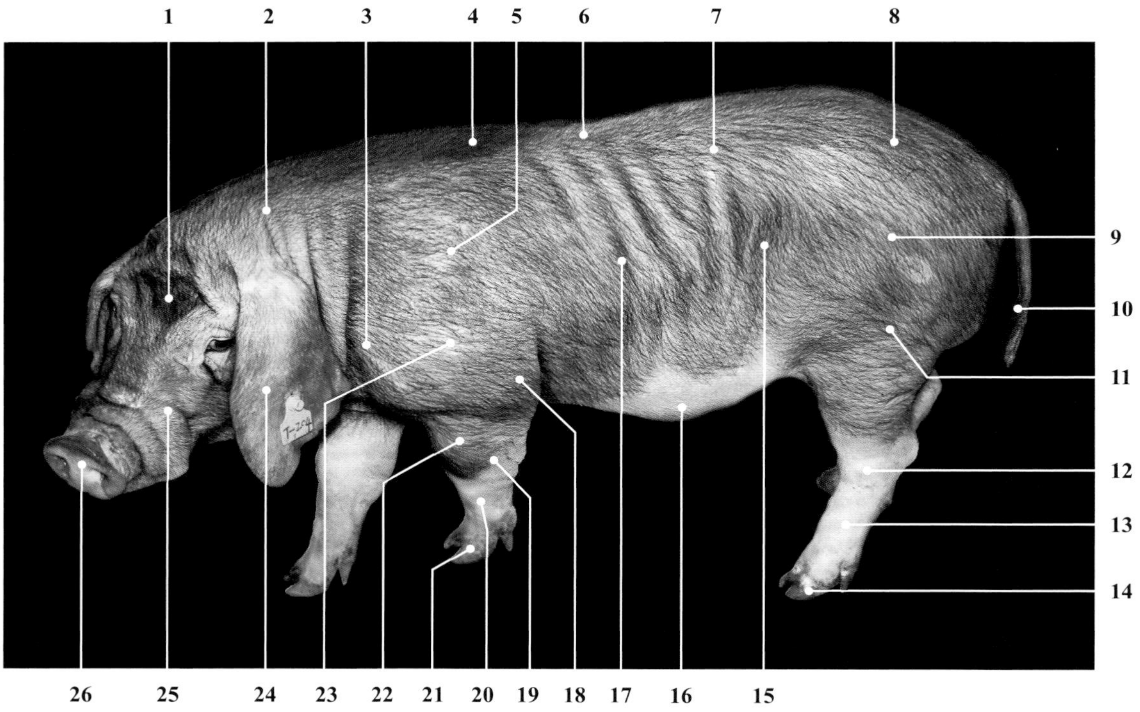

图 1-1　公猪各部解剖名称（左侧观）
Figure 1-1　Anatomical terms of the boar body (left view).

1- 颅部 cranial part
2- 颈部 cervical part
3- 肩部 shoulder joint
4- 鬐甲部 withers
5- 肩胛部 scapular region
6- 背部 back
7- 腰部 lumbar part
8- 荐臀部 sacral-gluteal region
9- 股部 thigh region
10- 尾部 tail region
11- 小腿部 crural region
12- 跗部 tarsal region
13- 跖部 metatarsal region
14- 趾部 digital region
15- 腹部 abdomen
16- 脐部 umbilical region
17- 胸部 thoracic part
18- 肘部 elbow region
19- 腕部 carpal region
20- 掌部 metacarpal region
21- 指部 digital region
22- 前臂部 antebrachial region
23- 臂部 brachial region
24- 耳部 aural region
25- 面部 facial part
26- 吻突 rostral disc

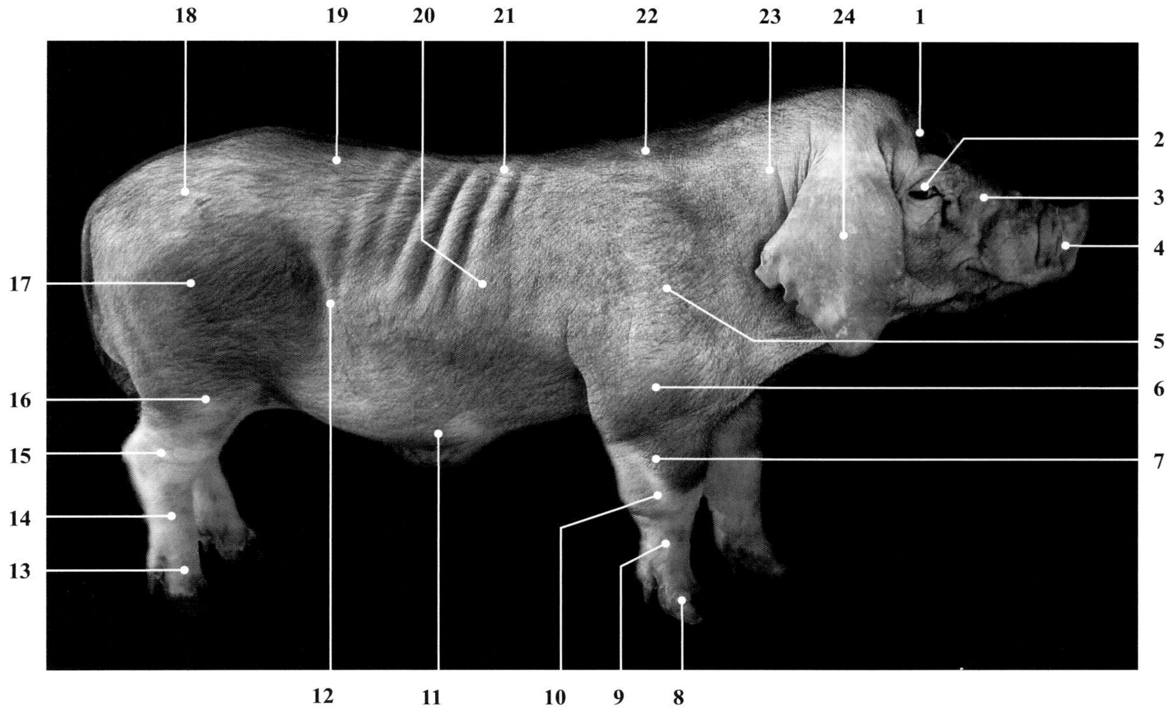

图 1-2　公猪各部解剖名称（右侧观）
Figure 1-2　Anatomical terms of the boar body (right view).

1- 颅部 cranial part
2- 眼 eye
3- 面部 facial part
4- 吻突 rostral disc
5- 肩胛部 scapular region
6- 臂部 brachial region
7- 前臂部 antebrachial region
8- 指部 digital region
9- 掌部 metacarpal region
10- 腕部 carpal region
11- 脐部 umbilical region
12- 腹部 abdomen
13- 趾部 digital region
14- 跖部 metatarsal region
15- 跗部 tarsal region
16- 小腿部 crural region
17- 股部 thigh region
18- 臀部 coxal region
19- 腰部 lumbar part
20- 胸部 thoracic part
21- 背部 back
22- 鬐甲部 withers
23- 颈部 cervical part
24- 耳部 aural region

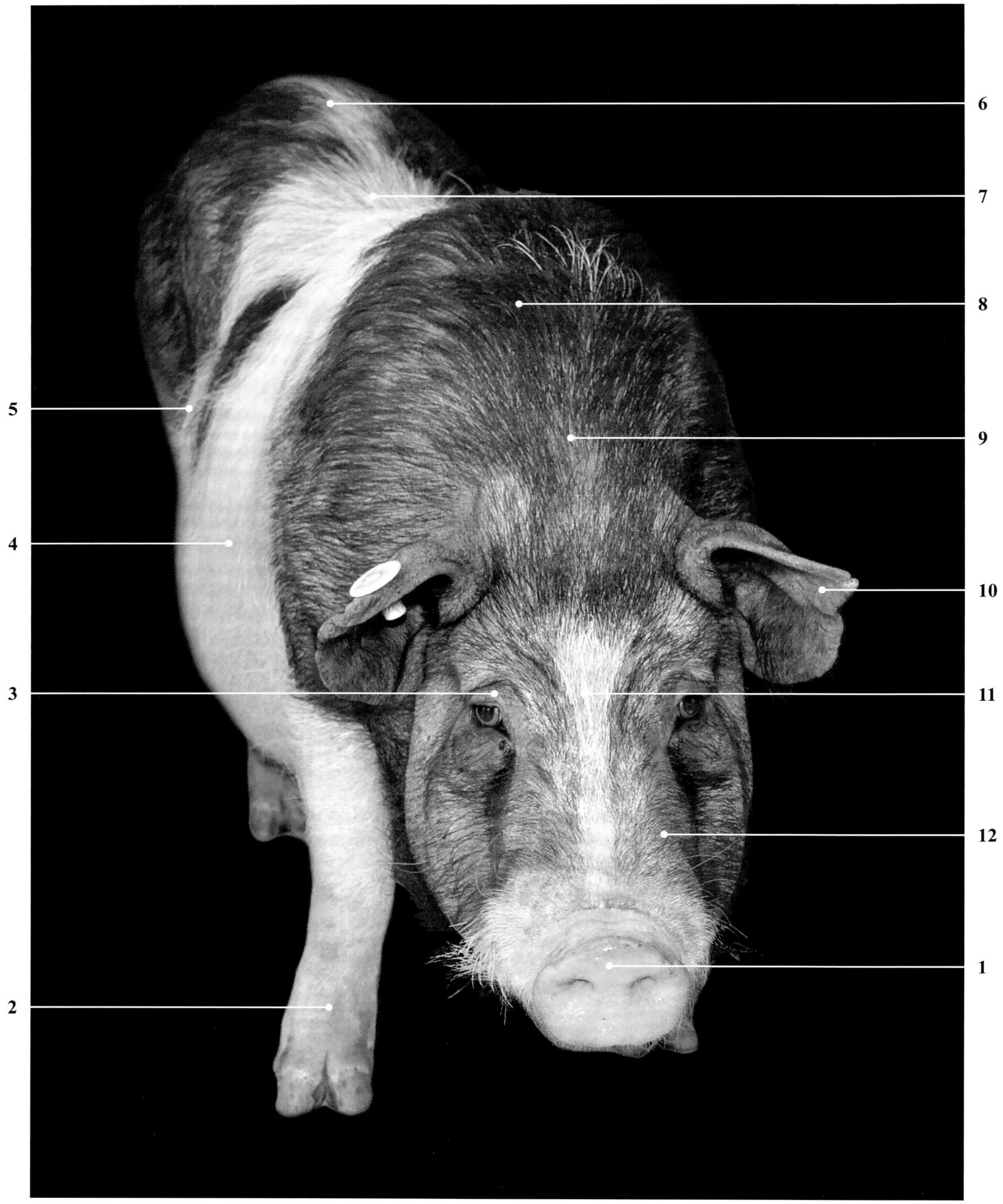

图 1-3　母猪各部解剖名称（前面观）
Figure 1-3　Anatomical terms of the sow body (cranial view).

1- 吻突 rostral disc
2- 前肢 forelimb
3- 眼部 ocular region
4- 胸部 thoracic part
5- 腹部 abdomen
6- 荐部 sacral part
7- 腰部 lumbar part
8- 鬐甲部 withers
9- 颈部 cervical part
10- 耳部 aural region
11- 颅部 cranial part
12- 面部 facial part

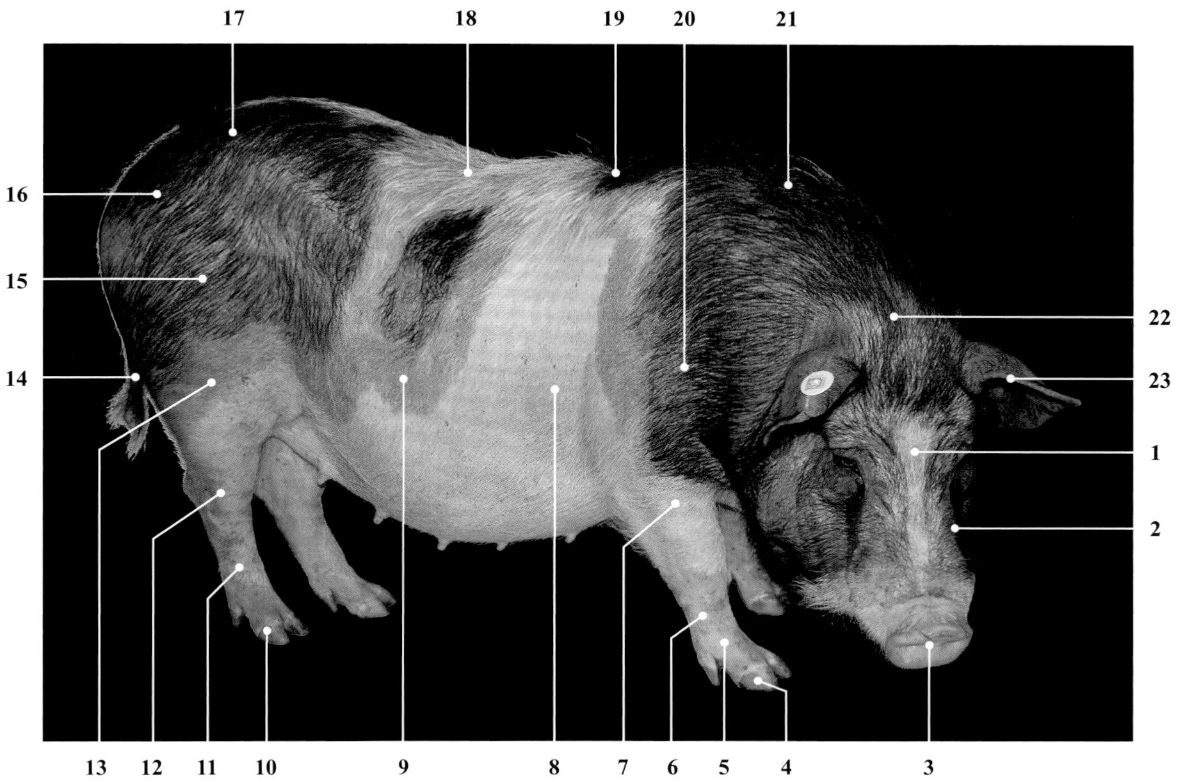

图 1-4 母猪各部解剖名称（右侧观）
Figure 1-4 Anatomical terms of the sow body (right view).

1- 颅部 cranial part
2- 面部 facial part
3- 吻突 rostral disc
4- 指部 digital region
5- 掌部 metacarpal region
6- 腕部 carpal region
7- 前臂部 antebrachial region
8- 胸部 thoracic part
9- 腹部 abdomen
10- 趾部 digital region
11- 跖部 metatarsal region
12- 跗部 tarsal region
13- 小腿部 crural region
14- 尾部 tail region
15- 股部 thigh region
16- 臀部 coxal region
17- 荐部 sacral part
18- 腰部 lumbar part
19- 背部 back
20- 肩胛部 scapular region
21- 鬐甲部 withers
22- 颈部 cervical part
23- 耳部 aural region

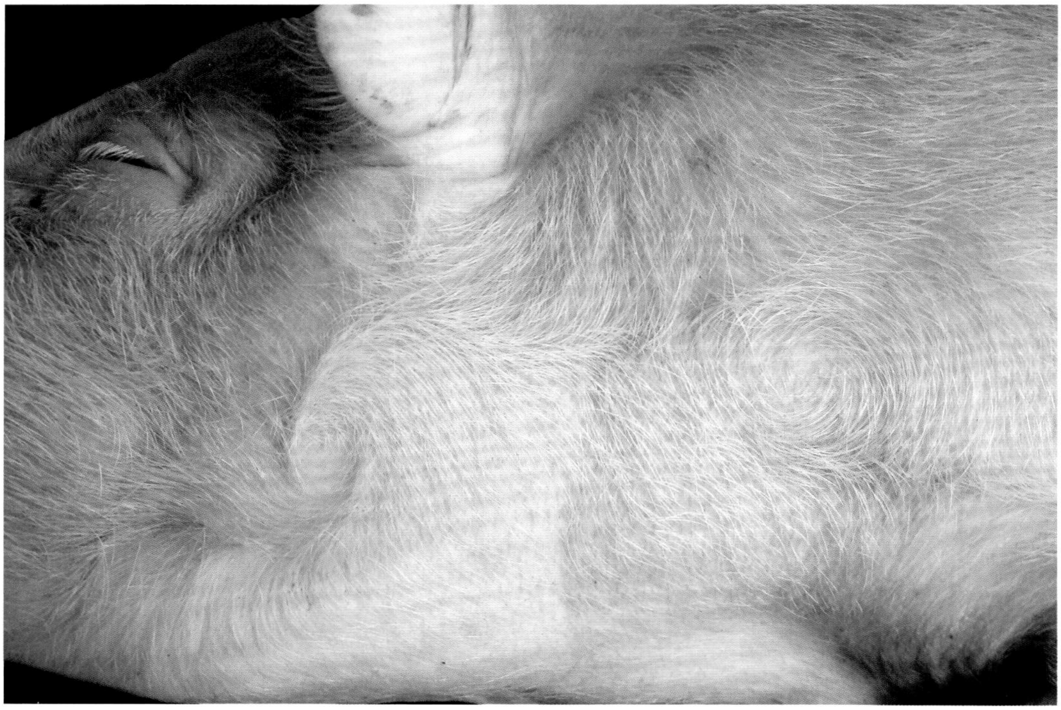

图 1-5　左侧颈部毛流（旋涡毛流）
Figure 1-5　Left cervical flumina pilorum (vortex flumina pilorum).

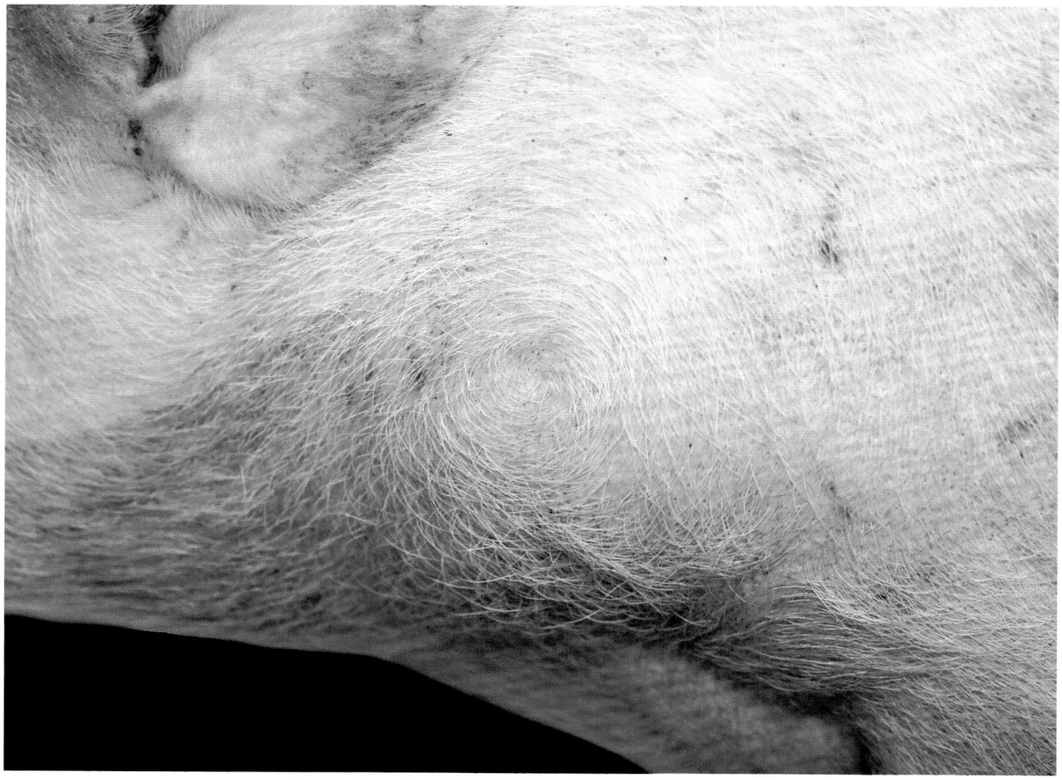

图 1-6　左侧颈部毛流（旋涡毛流）
Figure 1-6　Left cervical flumina pilorum (vortex flumina pilorum).

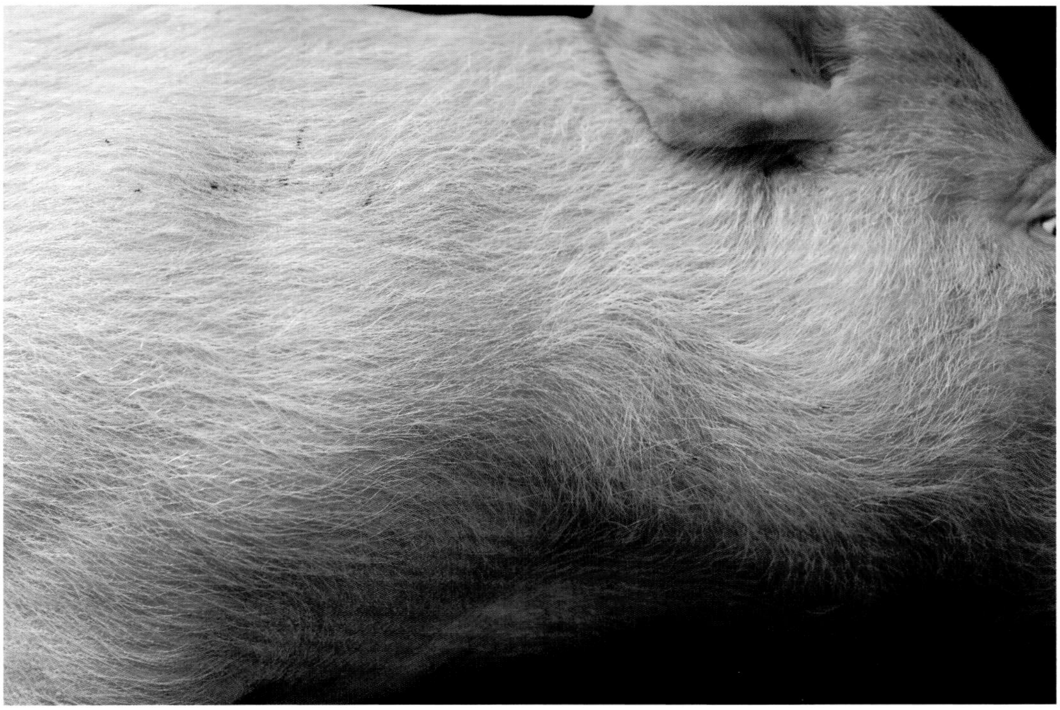

图 1-7 右侧颈部毛流
Figure 1-7 Right cervical flumina pilorum.

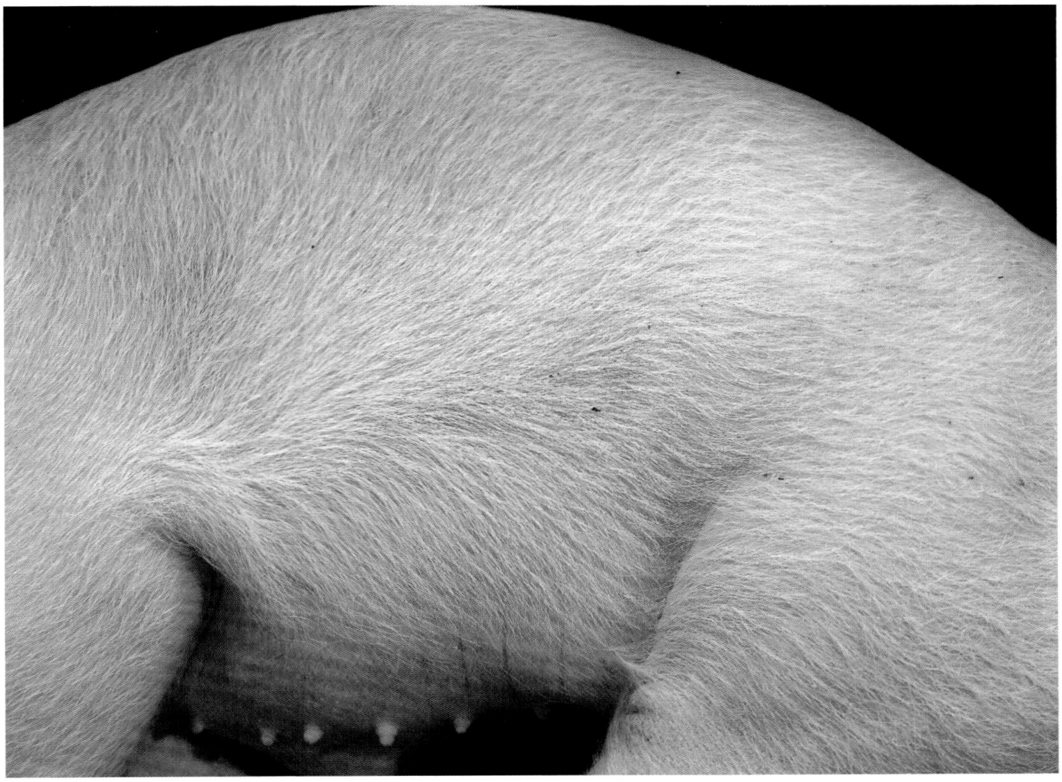

图 1-8 胸腹部毛流
Figure 1-8 Thoracoabdominal flumina pilorum.

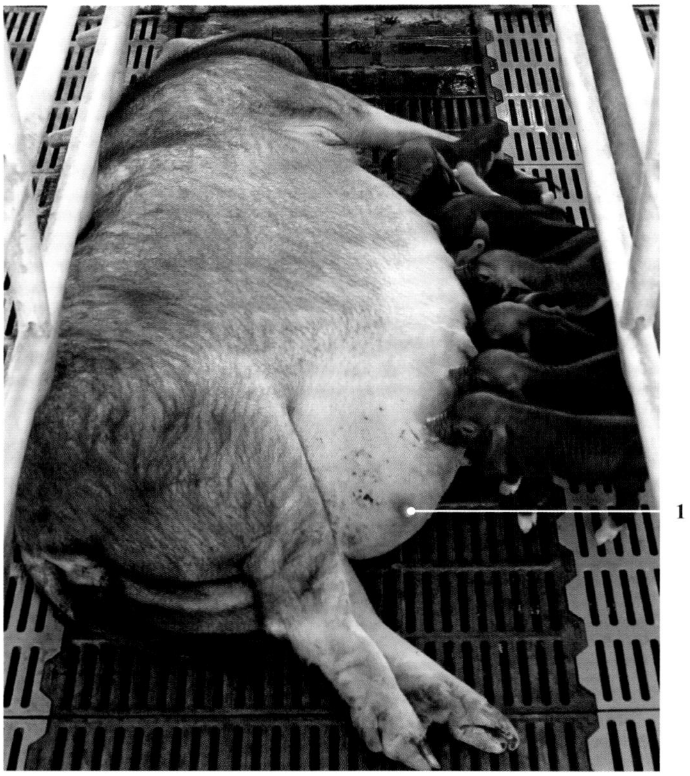

图 1-9　泌乳期乳房
Figure 1-9　Lactating breast.

1- 乳头 nipple

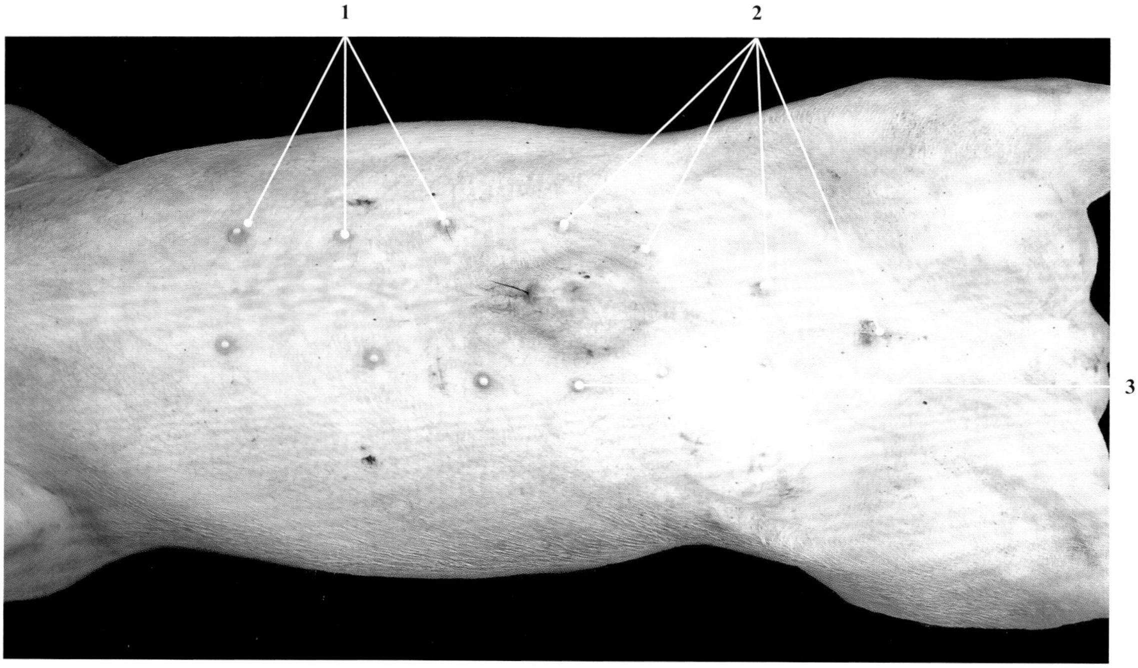

图 1-10　公猪胸腹部乳房
Figure 1-10　Thoracoabdominal breast of the boar.

1- 胸部乳房 chest breast　　**2-** 腹部乳房 abdomen breast　　**3-** 乳头 nipple

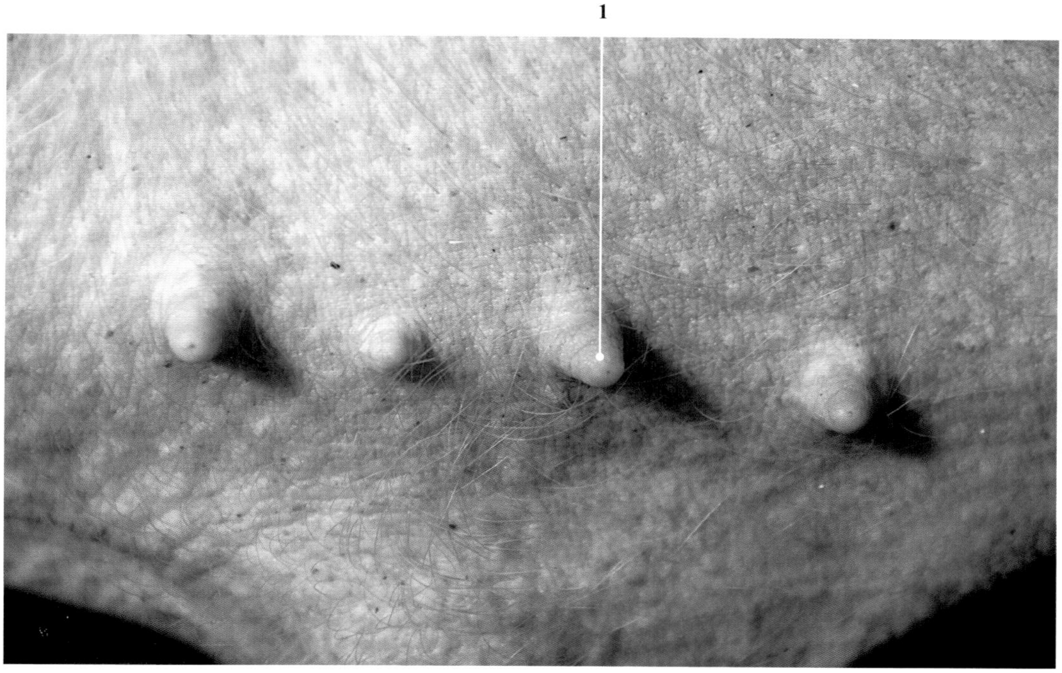

图 1-11 乳头
Figure 1-11 Nipple.

1- 乳头 nipple

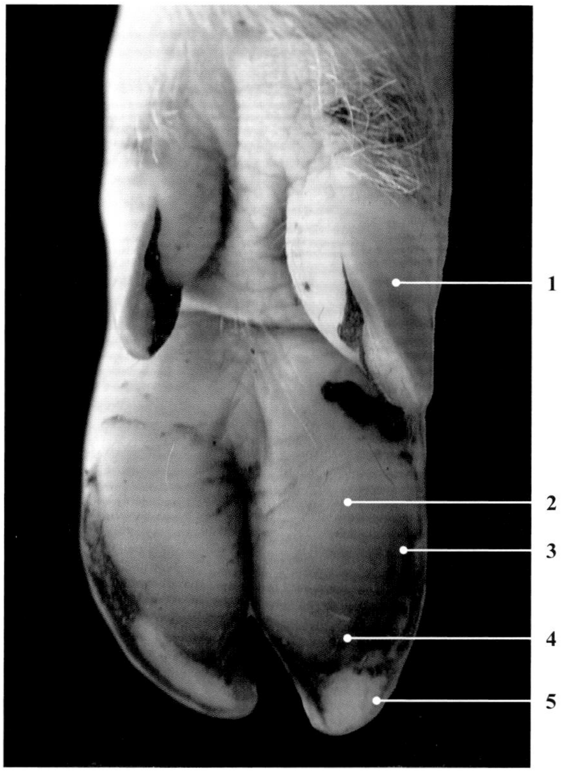

图 1-12 蹄
Figure 1-12 Hoof.

1- 悬蹄 dewclaw
2- 蹄球 bulb of hoof
3- 主蹄 principal hoof
4- 蹄底 sole of hoof
5- 蹄壁 wall of hoof

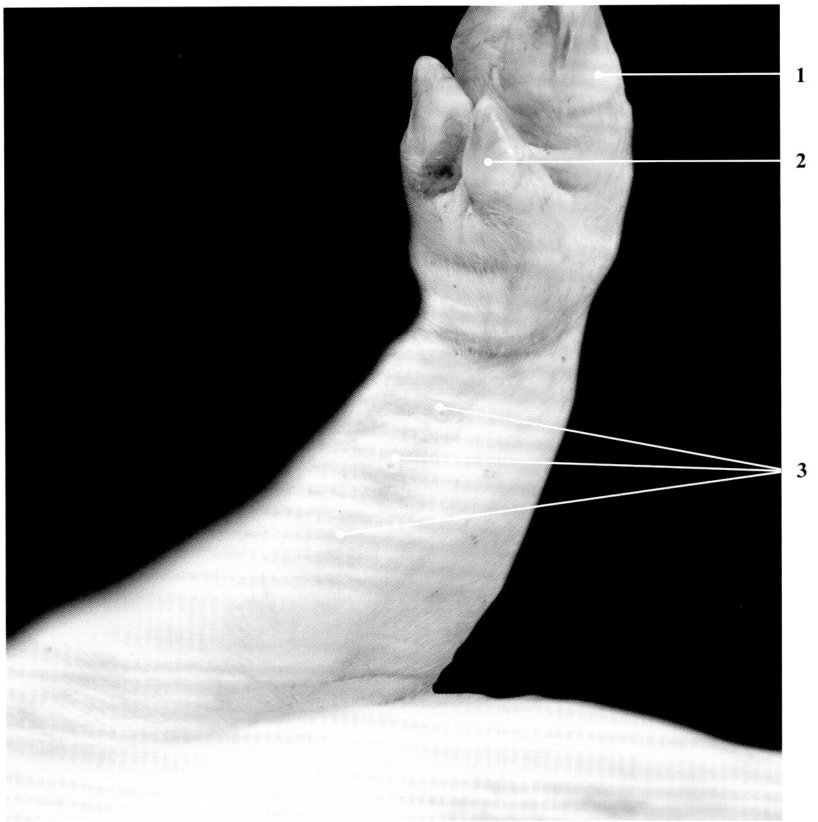

1- 主蹄 principal hoof
2- 悬蹄 dewclaw
3- 腕腺 carpal gland

图 1-13　腕腺
Figure 1-13　Carpal gland.

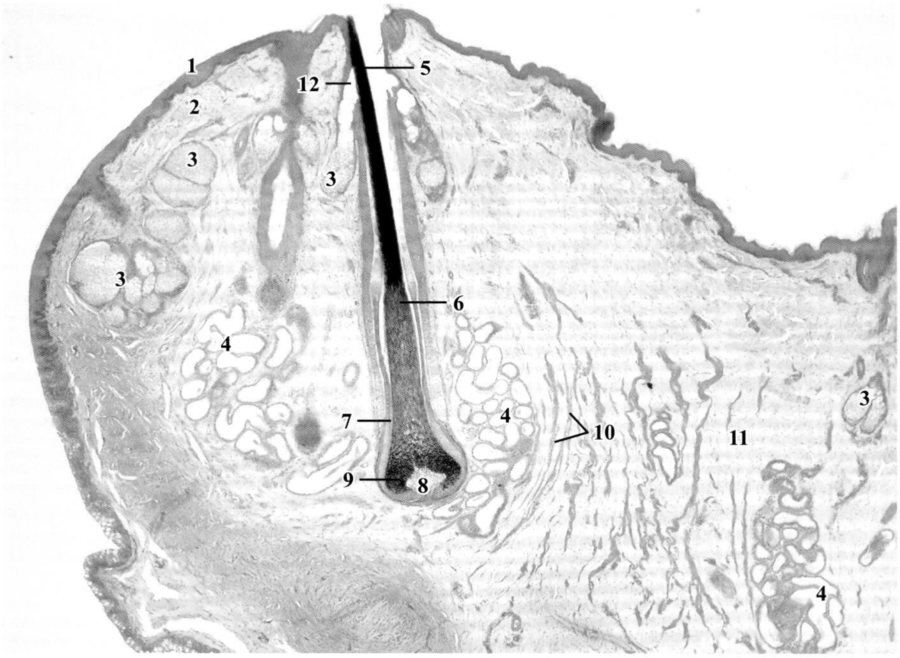

1- 表皮 epidermis
2- 真皮 dermis
3- 皮脂腺 sebaceous gland
4- 汗腺 sweat gland
5- 毛 hair
6- 毛根 hair root
7- 毛囊 hair follicle
8- 毛乳头 hair papilla
9- 毛球 hair bulb
10- 竖毛肌 arrector muscle
11- 皮下组织 subcutaneous tissue
12- 皮脂腺分泌导管 excretory duct of the sebaceous gland

图 1-14　有毛皮肤组织切片（HE 染色，低倍镜）
Figure 1-14　Histological section of a hairy skin (HE staining, lower power).

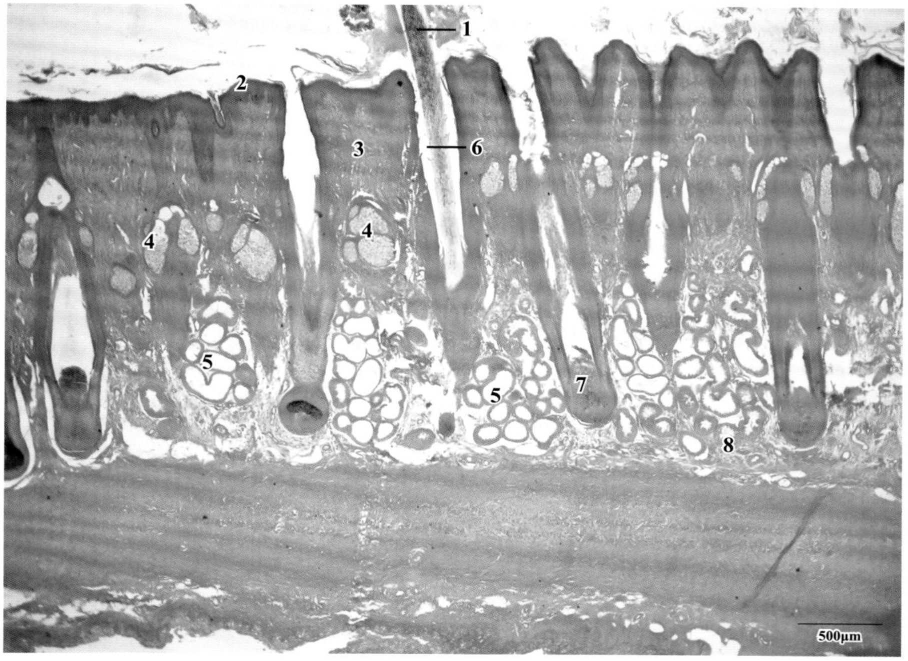

1- 毛干 hair shaft
2- 表皮 epidermis
3- 真皮 dermis
4- 皮脂腺 sebaceous gland
5- 汗腺 sweat gland
6- 毛根 hair root
7- 毛球 hair bulb
8- 皮下脂肪 subcutaneous fat

图 1-15　有毛皮肤组织切片（HE 染色，低倍镜）
Figure 1-15　Histological section of the hairy skin (HE staining, lower power).

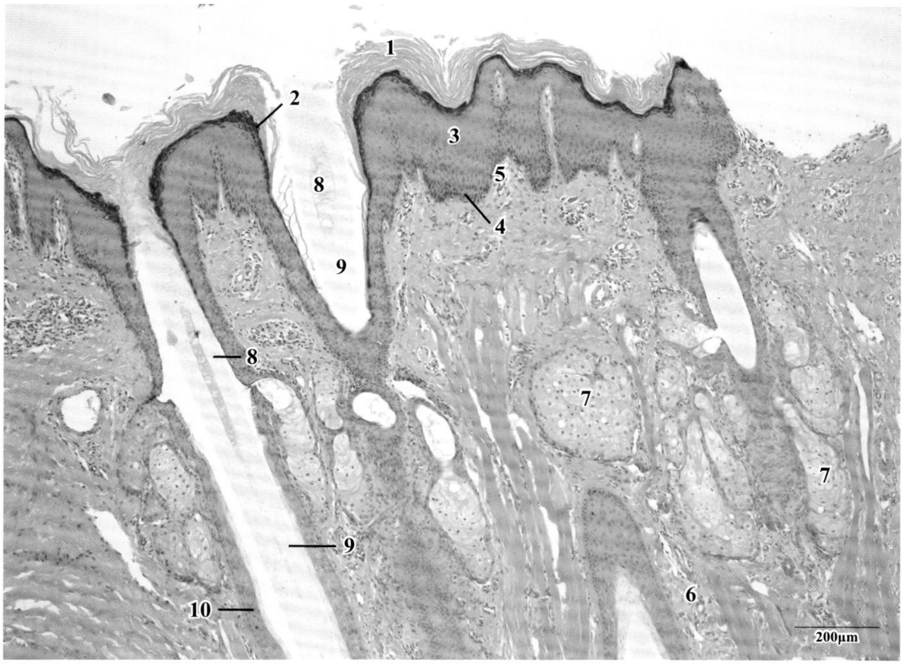

1- 角质层 horny layer
2- 颗粒层 granular layer
3- 棘层 spinous cell layer
4- 基底层 basal layer
5- 真皮乳头层 dermis, papillary layer
6- 真皮网状层 dermis, reticular layer
7- 皮脂腺 sebaceous gland
8- 毛髓质 hair medulla
9- 毛皮质 hair cortex
10- 毛囊 hair follicle

图 1-16　有毛皮肤组织切片（HE 染色，高倍镜）
Figure 1-16　Histological section of the hairy skin (HE staining, higher power).

1- 角质层 horny layer
2- 颗粒层 granular layer
3- 棘层 spinous cell layer
4- 基底层 basal layer
5- 真皮 dermis
6- 基底细胞 basal cell
7- 棘细胞 spinous cell
8- 颗粒细胞 granular cell
9- 角质细胞 keratinocyte

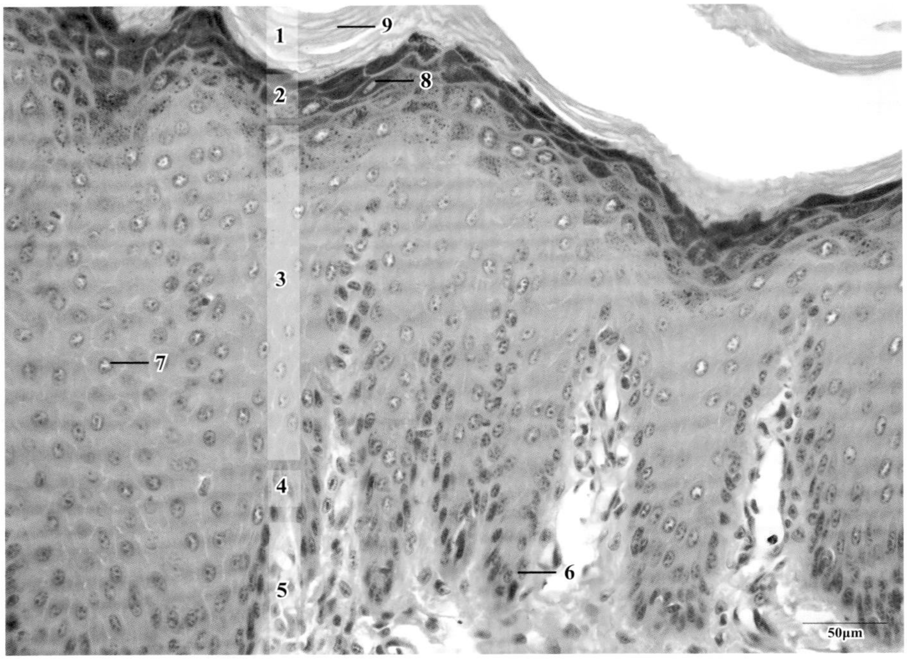

图 1-17 有毛皮肤组织切片，示表皮（HE 染色，高倍镜）
Figure 1-17 Histological section of the hairy skin showing the epidermis (HE staining, higher power).

1- 角质层 horny layer
2- 颗粒层 granular layer
3- 棘层 spinous cell layer
4- 基底层 basal layer
5- 真皮 dermis
6- 基底细胞 basal cell
7- 黑素细胞 melanocyte
8- 棘细胞 spinous cell
9- 颗粒细胞 granular cell
10- 毛细血管 blood capillary

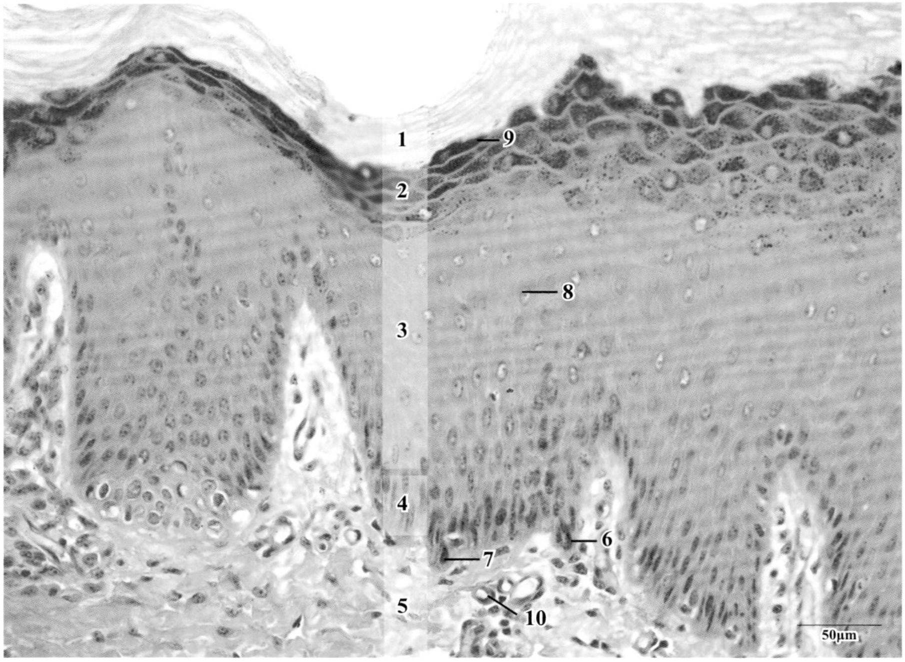

图 1-18 有毛皮肤组织切片，示表皮（HE 染色，高倍镜）
Figure 1-18 Histological section of the hairy skin showing the epidermis (HE staining, higher power).

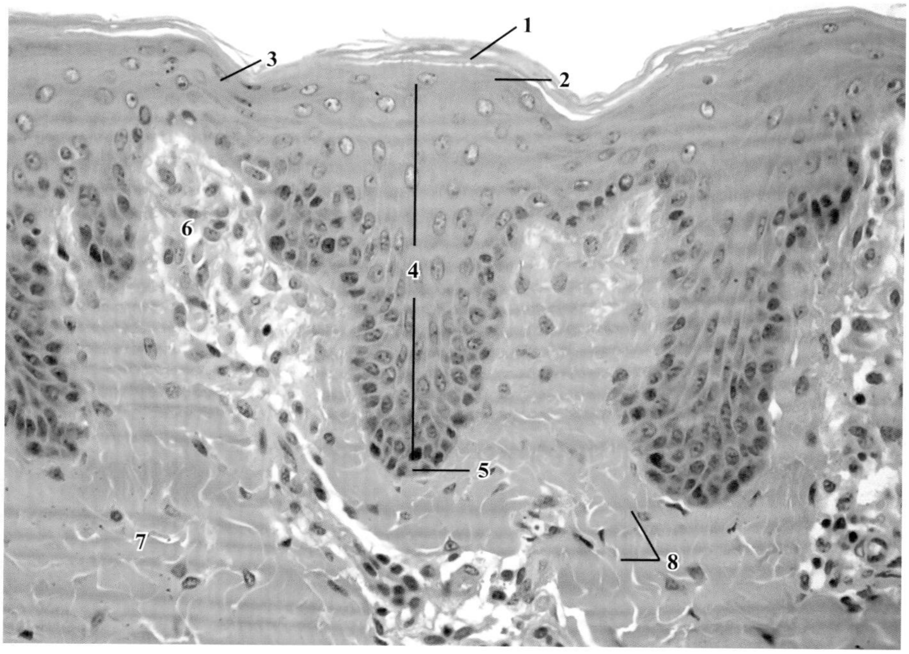

1- 角质层 horny layer
2- 透明层 clear layer
3- 颗粒层 granular layer
4- 棘层 spinous cell layer
5- 基底层 basal layer
6- 真皮乳头层 dermis, papillary layer
7- 真皮网状层 dermis, reticular layer
8- 胶原纤维束 collagen fiber bundle

图 1-19　蹄部皮肤组织切片（HE 染色，高倍镜）
Figure 1-19　Histological section of the hoof skin (HE staining, higher power).

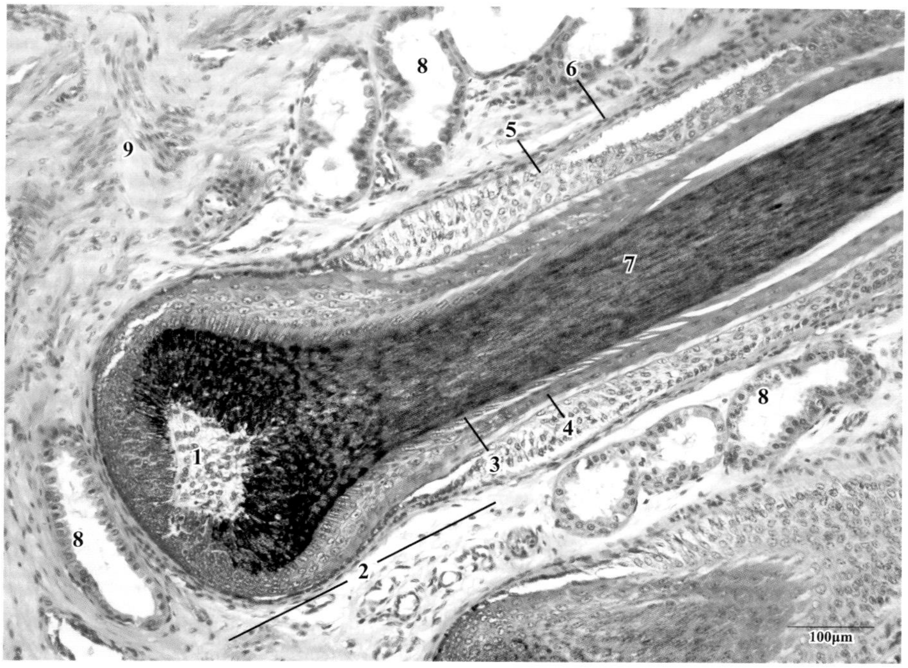

1- 毛乳头 hair papilla
2- 毛球 hair bulb
3- 毛小皮 hair cuticle
4- 内根鞘 inner root sheath
5- 外根鞘 outer root sheath
6- 真皮根鞘 dermis, root sheath
7- 毛髓质 hair medulla
8- 汗腺 sweat gland
9- 真皮 dermis

图 1-20　毛囊组织切片（HE 染色，高倍镜）
Figure 1-20　Histological section of the hair follicle (HE staining, higher power).

1- 毛乳头 hair papilla
2- 毛小皮 hair cuticle
3- 毛皮质 hair cortex
4- 内根鞘小皮 cuticle of internal root sheath
5- 内根鞘 inner root sheath
6- 外根鞘 outer root sheath
7- 汗腺 sweat gland

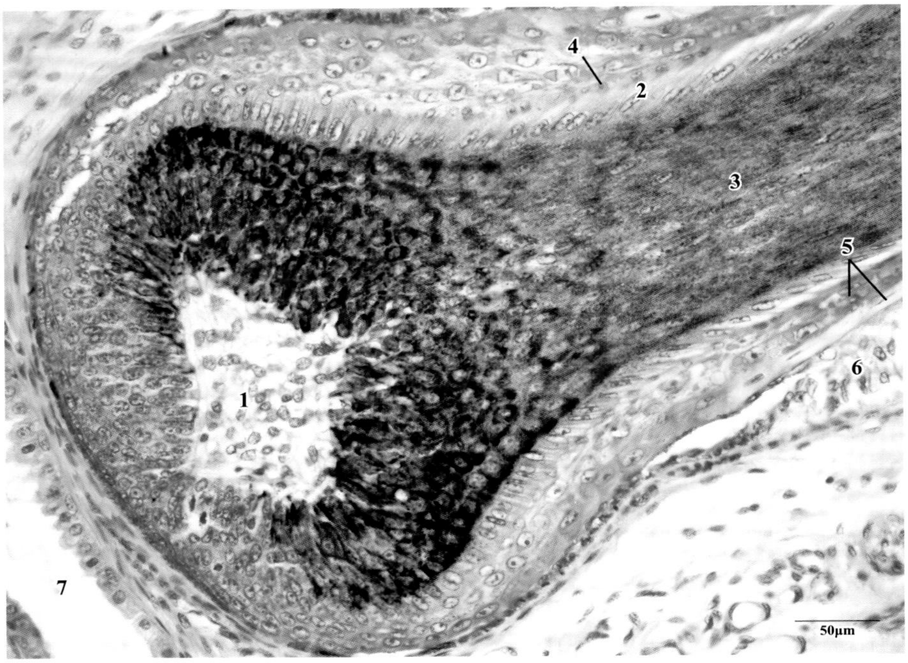

图 1-21 毛球组织切片（HE 染色，高倍镜）
Figure 1-21 Histological section of the hair bulb (HE staining, higher power).

1- 毛皮质 hair cortex
2- 毛小皮 hair cuticle
3- 内根鞘小皮 cuticle of internal root sheath
4- 内根鞘 inner root sheath
5- 外根鞘 outer root sheath
6- 真皮根鞘 dermis, root sheath
7- 汗腺 sweat gland

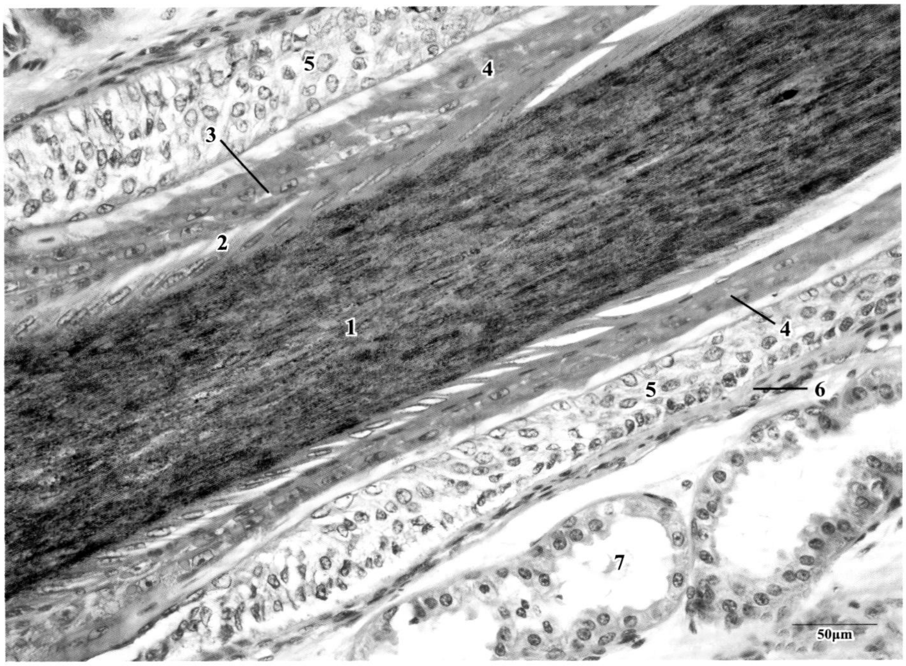

图 1-22 毛根组织切片（HE 染色，高倍镜）
Figure 1-22 Histological section of the hair root (HE staining, higher power).

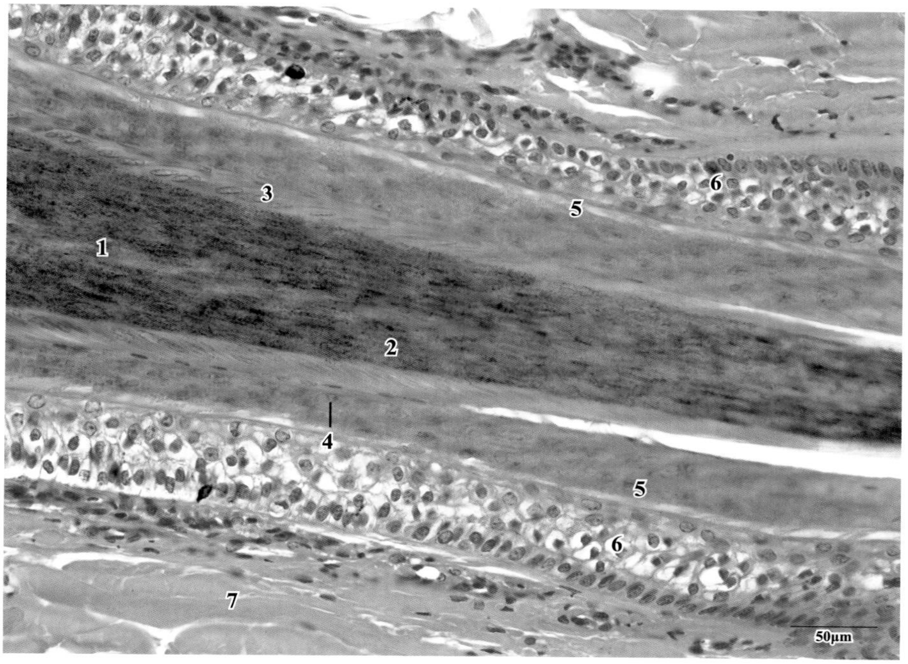

1- 毛髓质 hair medulla
2- 毛皮质 hair cortex
3- 毛小皮 hair cuticle
4- 内根鞘小皮 cuticle of internal root sheath
5- 内根鞘 inner root sheath
6- 外根鞘 outer root sheath
7- 胶原纤维束 collagen fiber bundle

图 1-23　毛根组织切片（HE 染色，高倍镜）
Figure 1-23　Histological section of the hair root (HE staining, higher power).

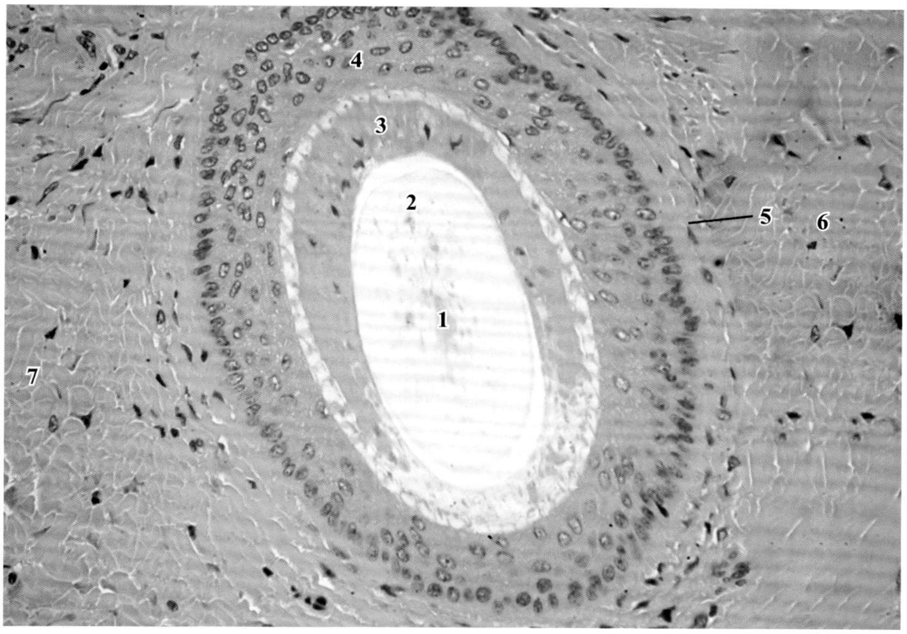

1- 毛髓质 hair medulla
2- 毛皮质 hair cortex
3- 内根鞘 inner root sheath
4- 外根鞘 outer root sheath
5- 真皮根鞘 dermis, root sheath
6- 真皮网状层 dermis, reticular layer
7- 胶原纤维束 collagen fiber bundle

图 1-24　毛囊组织横切片（HE 染色，高倍镜）
Figure 1-24　Histological cross section of the hair follicle (HE staining, higher power).

1- 毛根 hair root
2- 内根鞘 inner root sheath
3- 外根鞘 outer root sheath
4- 真皮根鞘 dermis, root sheath
5- 毛细血管 blood capillary
6- 胶原纤维束 collagen fiber bundle

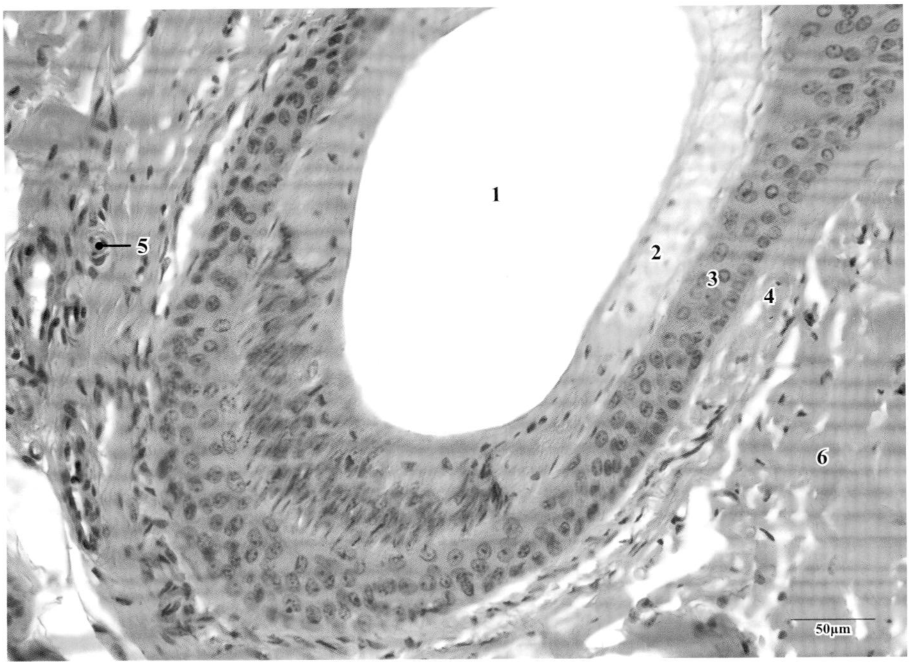

图 1-25　毛囊组织横切片（HE 染色，高倍镜）
Figure 1-25　Histological cross section of the hair follicle (HE staining, higher power).

1- 毛皮质 hair cortex
2- 毛小皮 hair cuticle
3- 内根鞘小皮 cuticle of internal root sheath
4- 内根鞘 inner root sheath
5- 外根鞘 outer root sheath
6- 真皮根鞘 dermis, root sheath
7- 汗腺 sweat gland
8- 微动脉 arteriole

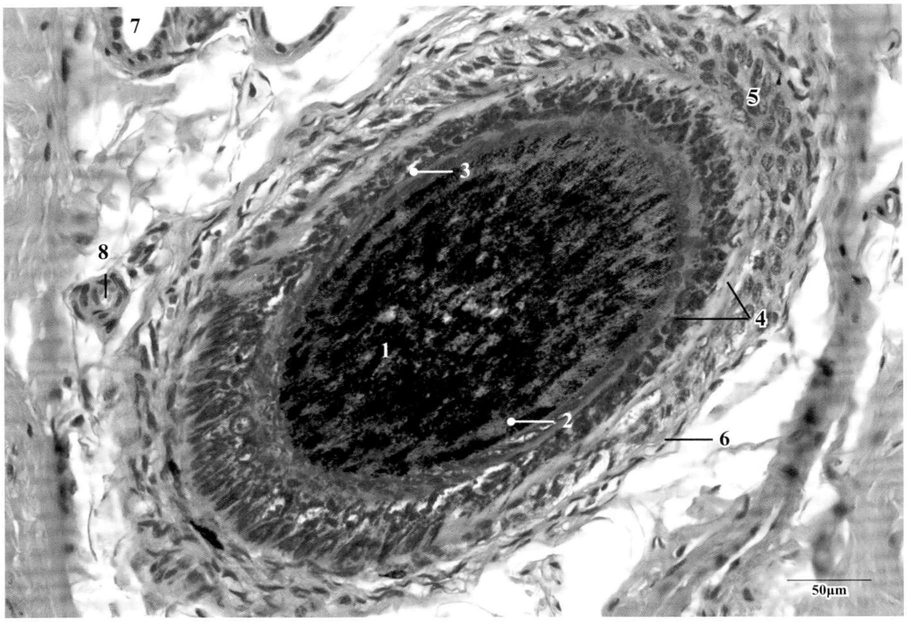

图 1-26　毛囊组织横切片（HE 染色，高倍镜）
Figure 1-26　Histological cross section of the hair follicle (HE staining, higher power).

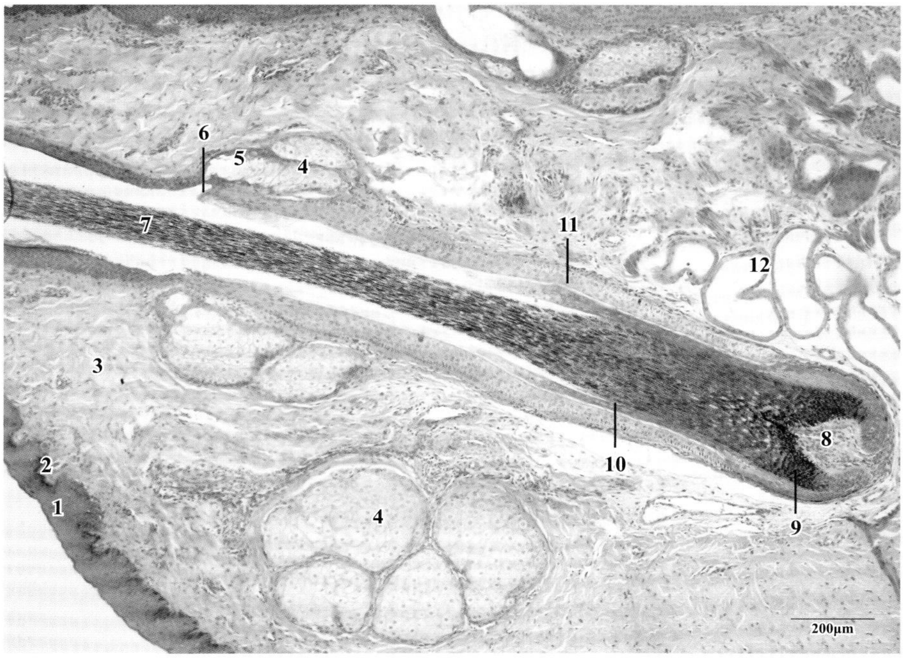

1- 表皮 epidermis
2- 真皮乳头层 dermis, papillary layer
3- 真皮网状层 dermis, reticular layer
4- 皮脂腺 sebaceous gland
5- 皮脂腺导管 duct of sebaceous gland
6- 导管口 opening of duct
7- 毛根 hair root
8- 毛乳头 hair papilla
9- 毛球 hair bulb
10- 内根鞘 inner root sheath
11- 外根鞘 outer root sheath
12- 汗腺 sweat gland

图 1-27　皮肤组织切片，示皮脂腺（HE 染色，低倍镜）
Figure 1-27　Histological section of a skin showing the sebaceous gland (HE staining, lower power).

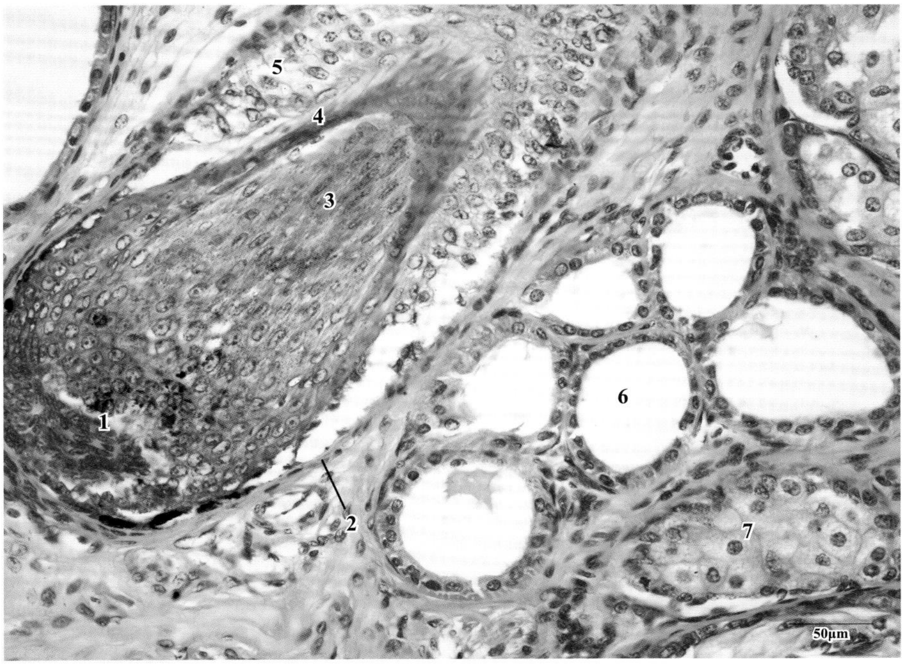

1- 毛乳头 hair papilla
2- 真皮根鞘 dermis, root sheath
3- 新生毛皮质 cortex of new hair
4- 内根鞘 inner root sheath
5- 外根鞘 outer root sheath
6- 汗腺 sweat gland
7- 皮脂腺 sebaceous gland

图 1-28　皮肤组织切片，示汗腺（HE 染色，高倍镜）
Figure 1-28　Histological section of a skin showing the sweat gland (HE staining, higher power).

1- 分泌导管 excretory duct
2- 腺细胞 gland cell
3- 肌上皮细胞核 nucleus of myoepithelial cell
4- 分泌小管 secretory tubulus

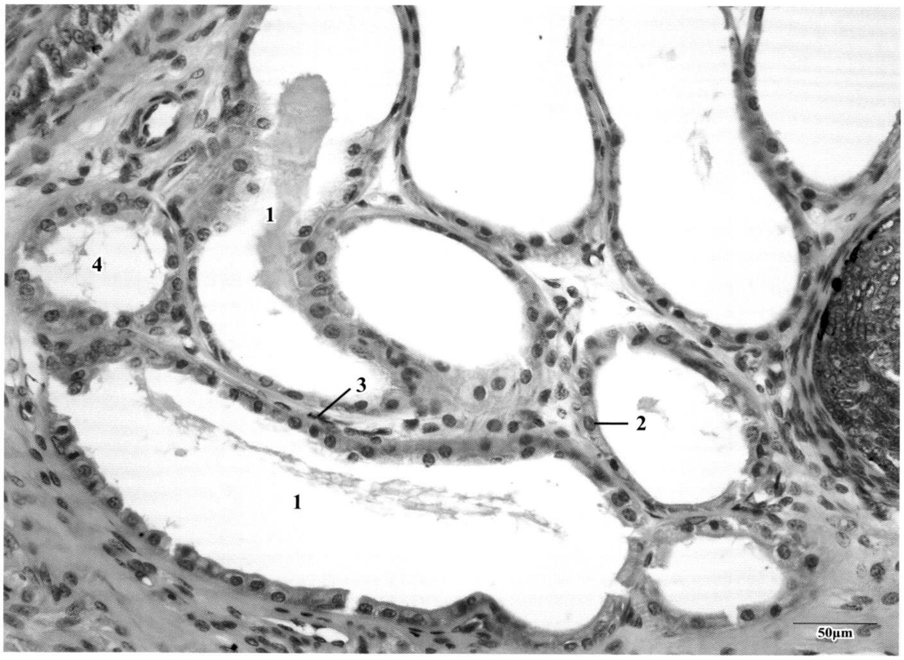

图 1-29　汗腺组织切片（HE 染色，高倍镜）
Figure 1-29　Histological section of the sweat gland (HE staining, higher power).

1- 分泌导管 excretory duct
2- 皮脂腺 sebaceous gland
3- 空泡细胞 vacuolated cell
4- 基底细胞 basal cell

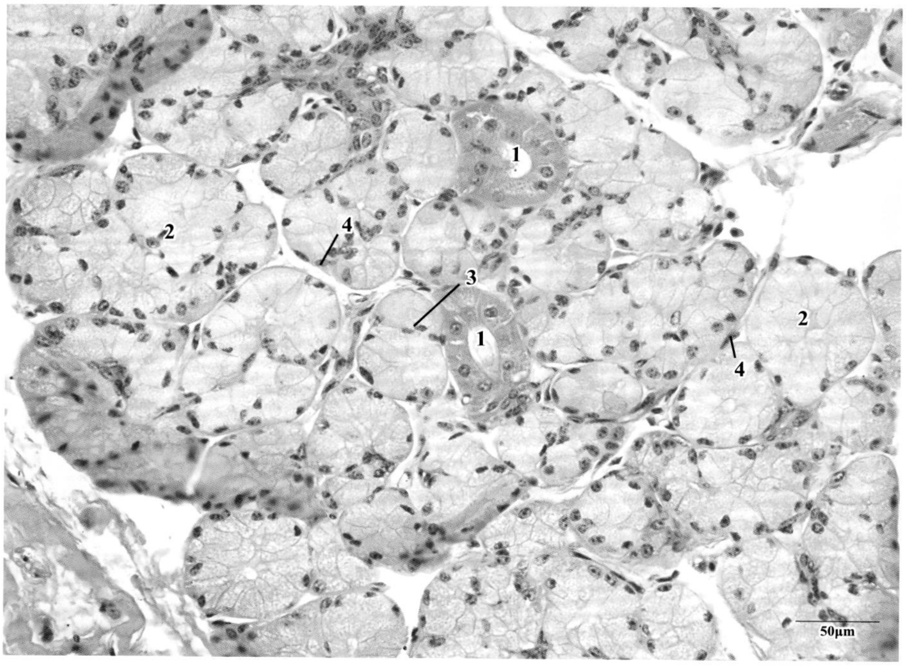

图 1-30　皮脂腺组织切片（HE 染色，高倍镜）
Figure 1-30　Histological of the sebaceous gland (HE staining, higher power).

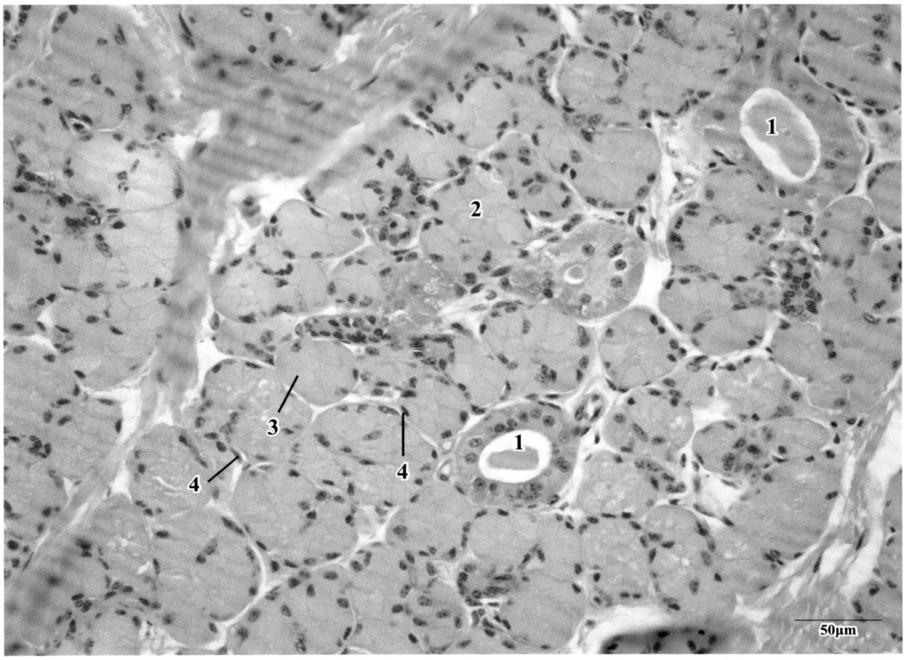

1- 分泌导管 excretory duct
2- 皮脂腺 sebaceous gland
3- 空泡细胞 vacuolated cell
4- 基底细胞 basal cell

图 1-31　皮脂腺组织切片（HE 染色，高倍镜）
Figure 1-31　Histological section of the sebaceous gland (HE staining, higher power).

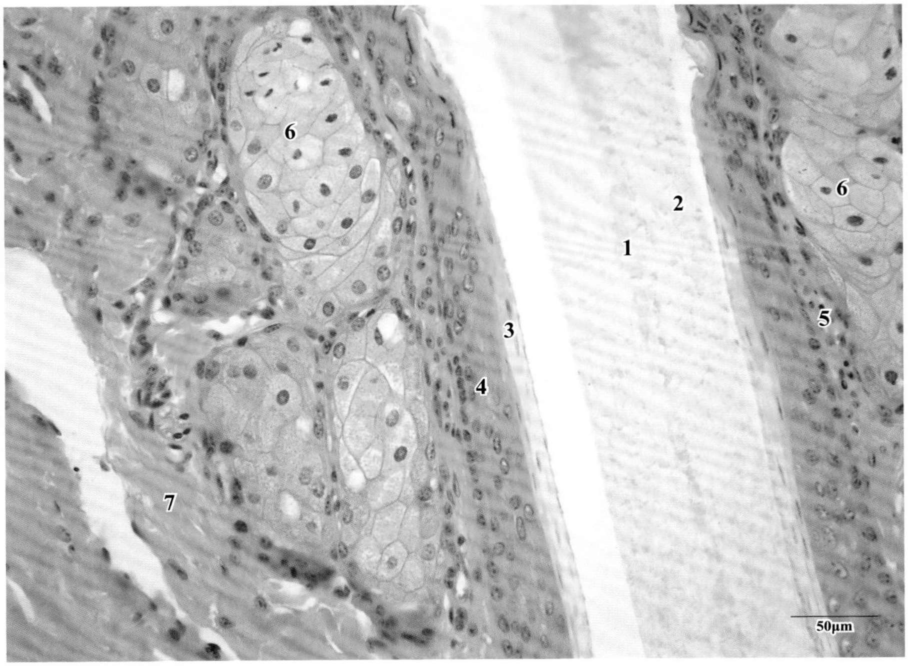

1- 毛髓质 hair medulla
2- 毛皮质 hair cortex
3- 内根鞘 inner root sheath
4- 外根鞘 outer root sheath
5- 真皮根鞘 dermis, root sheath
6- 皮脂腺 sebaceous gland
7- 真皮网状层的胶原纤维束 collagen fiber bundlem dermis reticular layer

图 1-32　有毛皮肤组织切片，示皮脂腺（HE 染色，高倍镜）
Figure 1-32　Histological section of a hairy skin showing sebaceous gland (HE staining, higher power).

1- 皮脂腺 sebaceous gland
2- 腺细胞 gland cell
3- 基底细胞 basal cell
4- 胶原纤维束 collagen fiber bundle

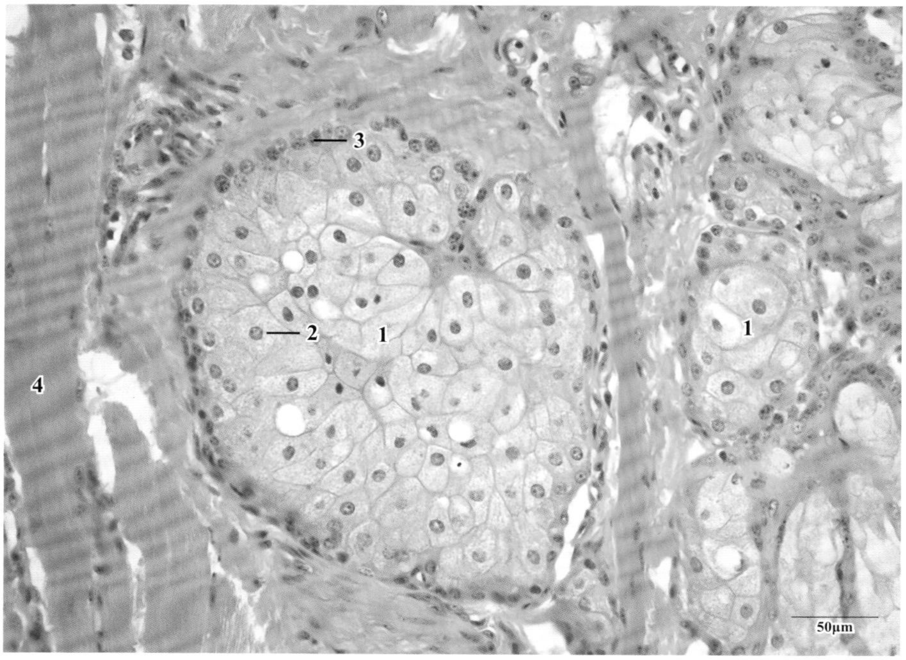

图 1-33　有毛皮肤组织切片，示皮脂腺（HE 染色，高倍镜）
Figure 1-33　Histological section of a hairy skin showing sebaceous gland (HE staining, higher power).

1- 腺小叶 glandular lobular
2- 小叶内结缔组织 innerlobular connective tissue
3- 腺泡 acinus
4- 毛囊 hair follicle

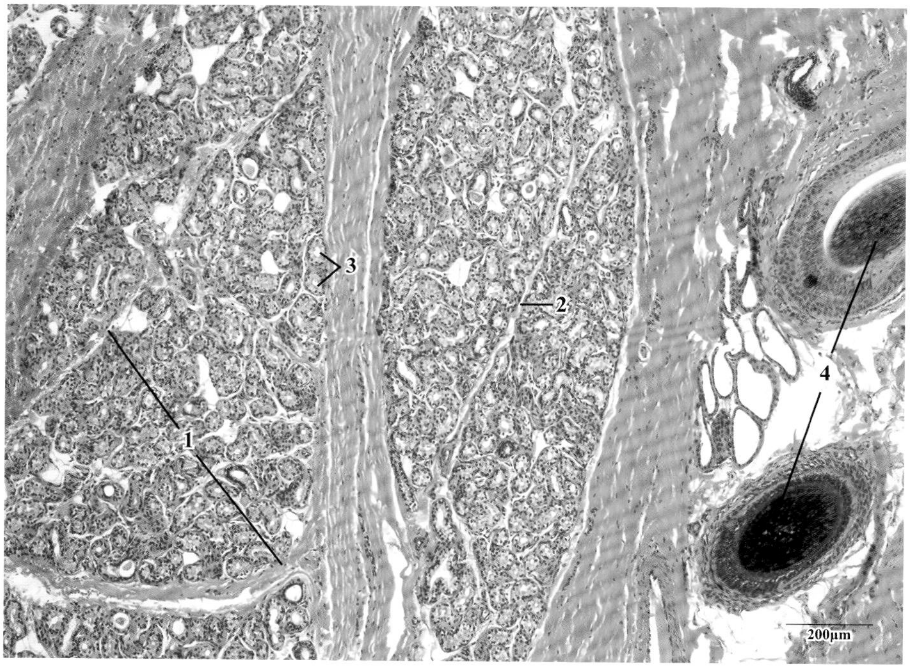

图 1-34　腕腺组织切片（HE 染色，低倍镜）
Figure 1-34　Histological section of the carpal gland (HE staining, lower power).

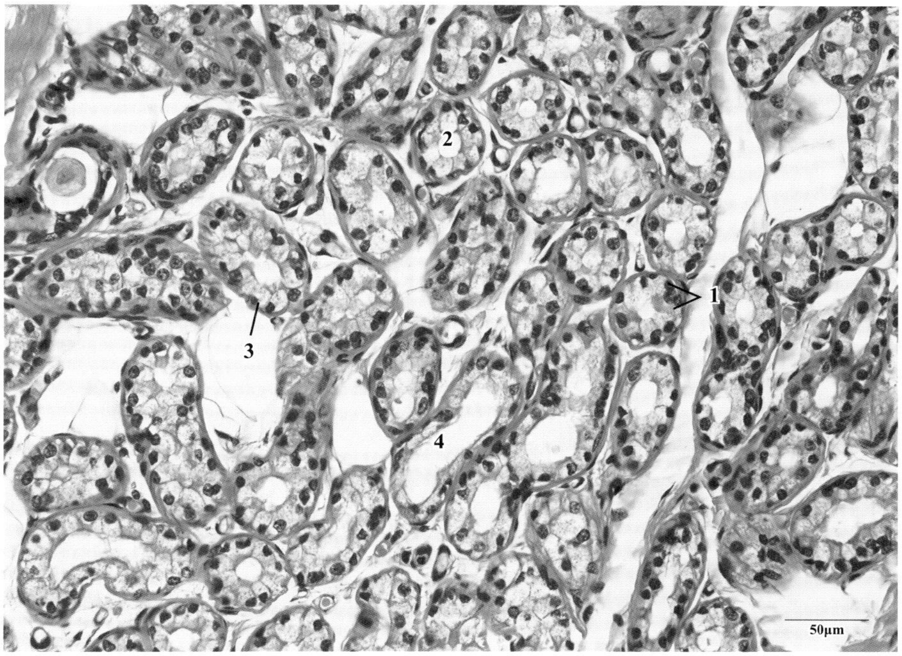

1- 腺泡 acinus
2- 腺泡腔 acinus cavity
3- 腺细胞 gland cell
4- 分泌导管 excretory duct

图 1-35　腕腺组织切片（HE 染色，高倍镜）
Figure 1-35　Histological section of the carpal gland (HE staining, higher power).

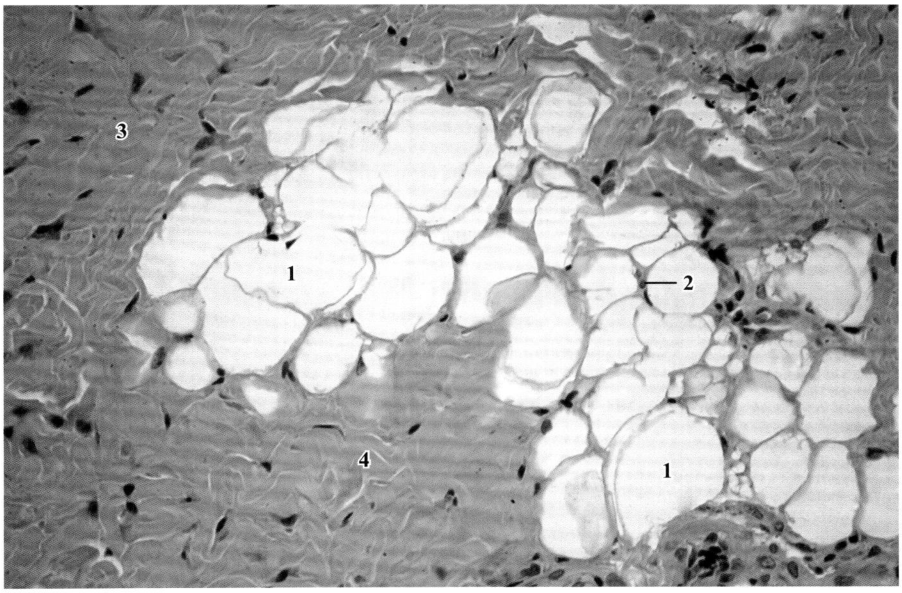

1- 脂肪细胞 adipocyte
2- 脂肪细胞核 nucleus of adipocyte
3- 真皮 dermis
4- 胶原纤维束 collagen fiber bundle

图 1-36　蹄部皮肤组织切片，示真皮（HE 染色，高倍镜）
Figure 1-36　Histological section of the hoof skin showing the dermis (HE staining, higher power).

1- 脂肪细胞 adipocyte
2- 红细胞 erythrocyte,red blood cell (RBC)
3- 脂肪细胞核 nucleus of adipocyte

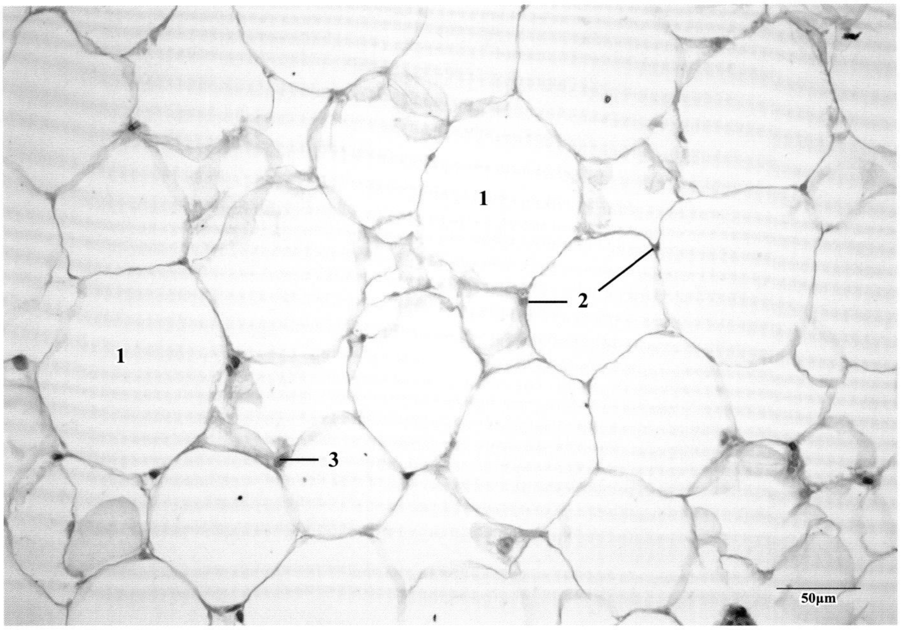

图 1-37　皮下脂肪组织切片（HE 染色，高倍镜）
Figure 1-37　Histological section of the subcutaneous fat (HE staining, higher power).

1- 表皮 epidermis
2- 角质层 horny layer
3- 透明层 clear layer
4- 真皮乳头层 dermis, papillary layer
5- 真皮网状层 dermis, reticular layer

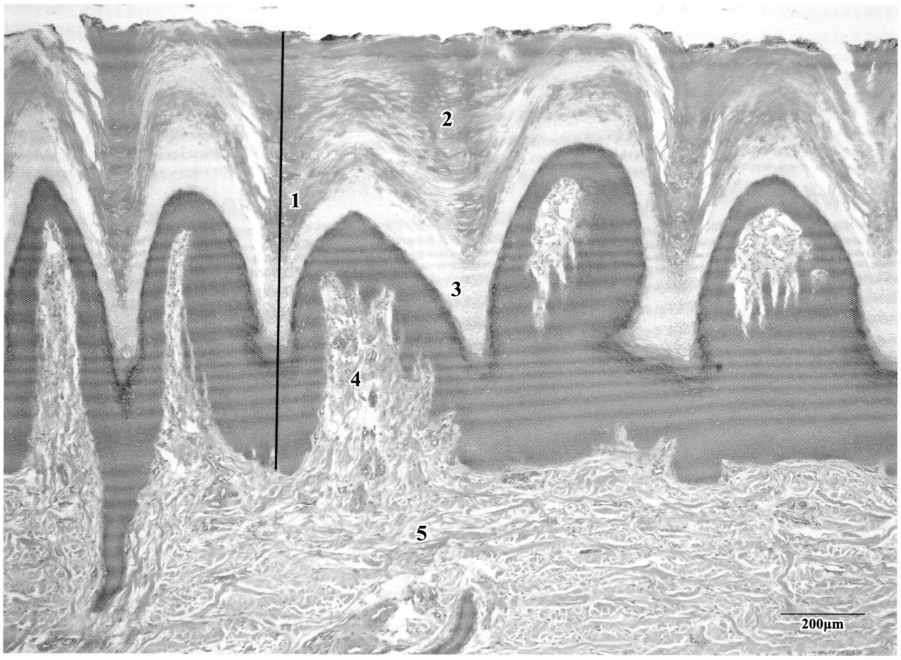

图 1-38　无毛皮肤组织切片（HE 染色，低倍镜）
Figure 1-38　Histological section of the nonhairy skin (HE staining, lower power).

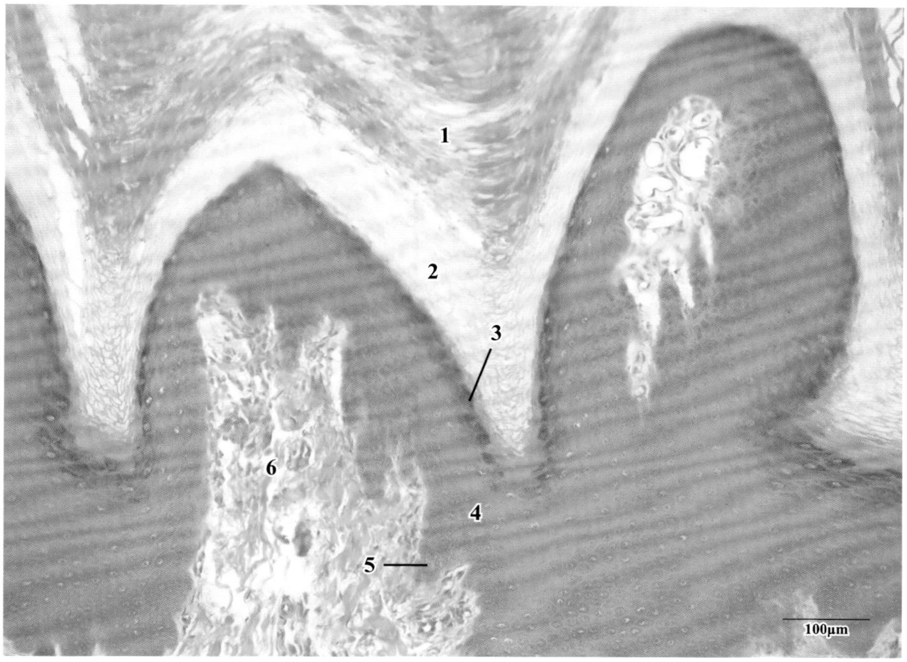

1- 角质层 horny layer
2- 透明层 clear layer
3- 颗粒层 granular layer
4- 棘层 spinous cell layer
5- 基底层 basal layer
6- 真皮乳头层 dermis, papillary layer

图 1-39　无毛皮肤组织切片，示表皮（HE 染色，高倍镜）
Figure 1-39　Histological section of a nonhairy skin showing the epidermis (HE staining, higher power).

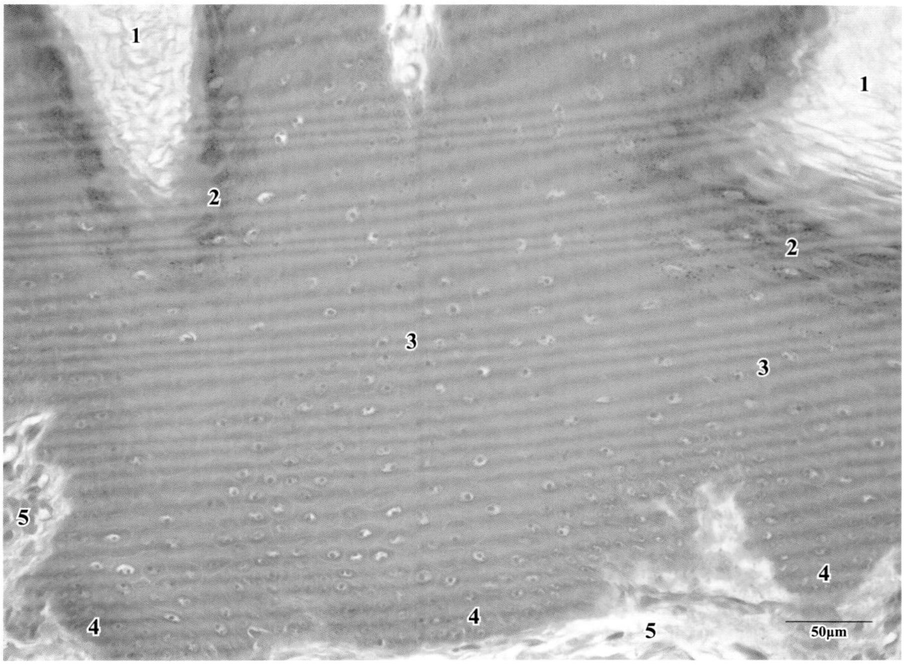

1- 透明层 clear layer
2- 颗粒层 granular layer
3- 棘层 spinous cell layer
4- 基底层 basal layer
5- 真皮 dermis

图 1-40　无毛皮肤组织切片，示颗粒细胞（HE 染色，高倍镜）
Figure 1-40　Histological section of a nonhairy skin showing the granulosa cell (HE staining, higher power).

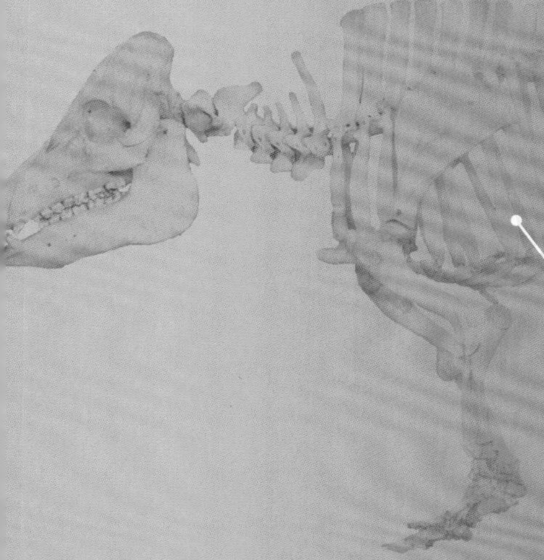

第二章

骨骼系统

Chapter 2
Skeletal system

一、骨

猪的全身骨由于品种不同和个体间差异，在骨的数量上有所不同，变动范围是281～288块，按存在部位分为头骨、躯干骨和四肢骨。全身骨借骨连结形成骨骼，构成猪体的支架，使猪体形成一定的形态，保护体内器官——脑、心、肺、胃、肠等，并充当肌肉收缩时的杠杆。

表2-1 猪全身骨骼的划分

- 猪全身骨骼
  - 中轴骨骼
    - 头骨
      - 颅骨：枕骨、额骨、顶骨、顶间骨、筛骨、颞骨、蝶骨
      - 面骨：上颌骨、切齿骨、鼻骨、吻骨、颧骨、泪骨、腭骨、翼骨、犁骨、鼻甲骨、下颌骨、舌骨
    - 躯干骨：椎骨、肋、胸骨
  - 四肢骨骼
    - 前肢骨：肩胛骨、肱骨、前臂骨(桡骨、尺骨)、腕骨、掌骨、指骨、籽骨
    - 后肢骨：髋骨（髂骨、坐骨、耻骨）、股骨、膝盖骨、小腿骨（胫骨、腓骨）、跗骨、跖骨、趾骨、籽骨

（一）头骨

头骨（skull）分为颅骨和面骨。

1．颅骨（cranial bone）：包括枕骨、蝶骨、筛骨、顶间骨、顶骨、额骨和颞骨。

（1）枕骨（occipital bone）构成颅腔的后壁和底部的一部分。枕骨的后上方有横向的枕嵴。猪的枕嵴特别高大。枕骨的后下方有枕骨大孔，后接椎管。枕骨大孔的两侧有枕骨髁和项结节（nuchal tubercle），枕骨髁与寰椎构成寰枕关节。髁的外侧有颈静脉突，髁与颈静脉突之间的窝内有舌下神经孔。

（2）蝶骨（sphenoid bone）构成颅腔下底的前部，位于枕骨底部前方。由蝶骨体和两对翼（眶翼、颞翼）以及一对翼突组成，形如蝴蝶。在幼龄猪，分为前蝶骨及后蝶骨。前蝶骨包括前部蝶骨体及两个眶翼。后蝶骨包括后部蝶骨体、两个颞翼及两个翼突。老龄猪的前、后蝶骨愈合。蝶骨体内的空洞为蝶窦。蝶骨的后缘与枕骨及颞骨形成不规则的破裂孔。其前缘与额骨及腭骨相连处有筛孔、视神经孔和眶圆孔与颅腔相通。这些孔、裂都是血管和神经的通路。

（3）筛骨（ethmoid bone）位于颅腔和鼻腔之间。由一个垂直板、一个筛板和两个筛骨迷路组成。垂直板位于正中，将鼻腔后部分为左、右两部。筛骨迷路向前突入鼻腔后部。筛骨迷路后方是多孔的筛板，构成颅腔的前壁。

（4）顶间骨（interparietal bone）为一小骨，位于枕嵴正中的前方。出生前与枕骨融合在一起。

（5）顶骨（parietal bone）为成对骨，构成颅腔的顶壁，其后面与枕骨相连，前面与额骨相接，两侧为颞骨，参与形成颞窝。

（6）额骨（frontal bone）为成对骨，位于顶骨的前方，鼻骨的后方，构成颅腔的前上壁和鼻腔的后上壁。额骨的外部有突出的眶上突，构成眼眶的上界。眶上突的基部有眶上孔。突的后方为颞窝；突的前方为眶窝，是容纳眼球的深窝。额骨的内、外板以及与筛骨之间，形成额窦。幼猪额窦仅位于前部，成年后额窦逐渐扩大到全部额骨及顶骨。

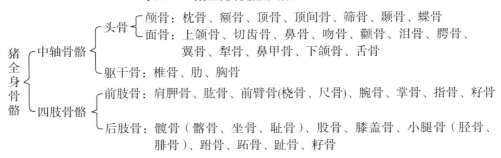

（7）颞骨（temporal bone）为成对骨，位于颅腔的侧壁，又分为鳞部和岩部。鳞部与顶骨、额骨及蝶骨相连。在外面有颧突伸出，并转而向前与颧骨的突起合成颧弓。颧突根部有髁状关节面，与下颌髁成关节。岩部位于鳞部与枕骨之间，是中耳和内耳的所在部位。

2．面骨（facial bone）：主要构成鼻腔、口腔和面部的支架，由成对的鼻骨、泪骨、颧骨、上颌骨、切齿骨、腭骨、翼骨、鼻甲骨和不成对的犁骨、下颌骨、舌骨和吻骨组成。

（1）上颌骨（maxillary bone）位于面部的两侧，构成鼻腔的侧壁、底壁和口腔的上壁，几乎与面部各骨均相接连。它向内侧伸出水平的腭突，将鼻腔与口腔分隔开。齿槽缘上具有臼齿齿槽，前方无齿槽的部分，称齿槽间缘。骨内有眶下管通过。骨的外面有面嵴和眶下孔。

（2）切齿骨（incisive bone）又称颌前骨（premaxillare bone），位于上颌骨前方，构成鼻腔的侧壁和下底及口腔上壁的前部。骨体上有切齿齿槽，骨体向后伸出腭突和鼻突。腭突向后接上颌骨的腭突，鼻突则与鼻骨之间形成鼻颌切迹。

（3）鼻骨（nasal bone）位于额骨的前方，构成鼻腔顶壁的大部。

（4）吻骨（rostral bone）位于鼻骨前方，构成吻突的基础。

（5）泪骨（lacrimal bone）位于上颌骨后背侧，眼眶底的内侧。其眶面有泪囊窝和鼻泪管的开口。

（6）颧骨（zygomatic bone）位于泪骨腹侧，构成眼眶的下界。前接上颌骨的后缘，下部有面嵴，并向后方伸出颞突，与颞骨的颧突结合形成颧弓。

（7）腭骨（palatine bone）位于上颌骨内侧的后方，形成鼻后孔的侧壁与硬腭的后部。可分为水平部和垂直部，分别构成硬腭和鼻后孔侧壁的骨质基础。

（8）翼骨（pterygoid bone）是成对的狭窄薄骨片，位于鼻后孔的两侧。

（9）犁骨（vomer）位于鼻腔底面的正中，背侧呈沟状，接鼻中隔软骨和筛骨垂直板。

（10）鼻甲骨（turbinal bone）是两对卷曲的薄骨片，附着在鼻腔的两侧壁上，并将每侧鼻腔分为上、中、下3个鼻道。

（11）下颌骨（mandible）是头骨中最大的骨，有齿槽的部分，称为下颌骨体，下颌骨体之后没有齿槽的部分，称下颌支。下颌骨体呈水平位，前部为切齿齿槽，后部为臼齿齿槽。切齿齿槽与臼齿齿槽之间为齿槽间隙。下颌骨体外侧前部有颏孔。下颌支呈垂直位，上部有下颌髁，与颞骨的髁状关节面构成关节。下颌髁之前有较高的冠状突。下颌支内侧面有下颌孔。两侧下颌骨体和下颌支之间，形成下颌间隙。

（12）舌骨（hyoid bone）位于下颌间隙后部，由几枚小骨片组成，即一个舌骨体以及成对的角舌骨、甲状舌骨、上舌骨、茎突舌骨和鼓舌骨构成。舌骨体有向前突出的舌突。鼓舌骨与两侧颞骨的岩部相连。舌骨有支持舌根、咽和喉的作用。

（二）躯干骨

躯干骨包括椎骨、肋和胸骨。

1．椎骨（vertebrae）：按其位置分为颈椎、胸椎、腰椎、荐椎和尾椎。通常颈椎7枚，胸椎13～16枚，腰椎5～7枚，荐椎4枚，尾椎20～23枚。所有的椎骨按从前到后的顺序排列，由软骨、关节和韧带连结在一起，形成身体的中轴，有保护脊髓、支持头部、悬挂内脏、传递冲力等作用，称为脊柱（vertebral column）。

（1）颈椎（cervical vertebrae）第1颈椎呈环形，又称寰椎（atlas）。寰椎由背侧弓和腹侧弓构成。前面有成对关节窝，与枕骨髁构成关节；后面有与第2颈椎构成关节的鞍状关节面。寰椎两侧的宽板为寰椎翼。第2颈椎又称枢椎（axis），椎体发达，前端突出，称为齿状突，与寰椎的鞍状关节面构成可转动的关节，棘突发达，呈板状，无前关节突。第3～6颈椎形态相似，椎体发达，椎头和椎窝明显；关节突发达，横突分前后两支。在横突基部有横突孔（transverse foramen），各颈椎横突孔连接在一起，形成横突管（transverse canal），供血管和神经通过。第7颈椎的椎体短而宽，椎窝两侧有与第1肋骨构成关节的关节面，棘突明显。

（2）胸椎（thoracic vertebrae）椎体大小较一致，在椎头和椎窝的两侧均有与肋骨头构成关节的前、后肋窝。棘突发达，横突短，有小关节面与肋骨结节构成关节。

（3）腰椎（lumbar vertebrae）椎体长度与胸椎相近；棘突较发达，其高度与后段胸椎相等；横突长，呈上下压扁的板状，伸向外侧，有利于扩大腹腔顶壁的横径。

（4）荐椎（sacral vertebrae） 是构成骨盆腔顶壁的基础。猪的荐椎愈合较晚，愈合后称为荐骨（sacrum）。荐骨前端两侧的突出部叫荐骨翼。第1荐椎椎体腹侧缘前端的突出部称为荐骨岬。荐骨的背面和盆面每侧各有4个孔，分别称为荐背侧孔和荐腹侧孔，是血管和神经的通路。棘突不发达。

（5）尾椎（coccygeal vertebrae）除前3或4枚尾椎具有椎骨的一般构造（椎体、椎弓、突起和椎孔等）外，其余尾椎椎弓、棘突和横突则逐渐退化，仅保留有椎体。

2. 肋（rib）：包括肋骨和肋软骨。肋骨（costal bone）构成胸廓的侧壁，左右成对。肋骨的椎骨端（近端）有肋骨小头和肋骨结节，分别与相应的胸椎椎体和横突构成关节。相邻肋骨间的空隙称为肋间隙。每一肋骨的下端接一肋软骨（costal cartilage）。经肋软骨与胸骨直接相接的肋骨为真肋。猪的真肋有7对。肋软骨不与胸骨直接相连，而是连于前一肋软骨上的肋骨称为假肋。最后肋骨与各假肋的肋软骨依次连接形成的弓形结构称为肋弓，是胸廓的后界。

3. 胸骨（breast bone, sternum）：位于胸底部，由6枚胸骨节片借软骨连接而成。其结构包括胸骨柄、胸骨体和剑状软骨，两侧有肋窝，与真肋的肋软骨相接。

（三）前肢骨

前肢骨包括肩胛骨、肱骨、前臂骨和前脚骨，其中前臂骨包括桡骨和尺骨，前脚骨包括腕骨、掌骨、指骨（又分为系骨、冠骨和蹄骨）和籽骨。

1. 肩胛骨（scapula）：为三角形扁骨，外侧面有一纵形隆起为肩胛冈。猪的冈结节发达，且弯向后方，肩峰不明显。肩胛冈前方为冈上窝，后方为冈下窝，供肌肉附着。肩胛骨内侧面的上部为锯肌面，中、下部凹窝为肩胛下窝。肩胛骨的上缘附有肩胛软骨，远端有一圆形浅凹为肩臼。肩臼前方突出部为肩胛结节。

2. 肱骨（humerus）：又称臂骨（bone of arm），分为骨干和两个骨端。近端后部球状关节面是肱骨头，前部内侧是小结节，外侧是大结节。两结节之间为臂二头肌沟。骨干呈不规则的圆柱状，形成一螺旋状沟为臂肌沟，外侧上部有三角肌粗隆，内侧中部有卵圆形的大圆肌粗隆。肱骨远端有内、外侧髁。髁间是肘窝。窝的两侧是内、外侧上髁。猪的三角肌粗隆不发达，但大结节粗大。

3. 前臂骨（skeleton of forearm）：包括桡骨和尺骨。桡骨（radius）在前内侧，尺骨（ulna）在后外侧。猪的尺骨比桡骨长，两骨之间有较大的前臂间隙。

4. 腕骨（carpal bone）：位于前臂骨与掌骨之间，为小的短骨，排成上下两列。近列腕骨4枚，自内向外为桡腕骨、中间腕骨、尺腕骨和副腕骨；远列腕骨4枚，自内向外依次为第1、第2、第3和第4腕骨。

5. 掌骨（metacarpal bone）：近端接腕骨，远端接指骨。猪的第3和第4掌骨发育良好，第2和第5掌骨则变小，第1掌骨缺失。

6. 指骨（digital bone）：指骨从上至下顺次包括系骨（近指节骨）、冠骨（中指节骨）和蹄骨（远指节骨）。蹄骨近端前缘突出，称伸腱突；底面凹且粗糙，称屈腱面。猪有4指。第3、第4指发达，第2、第5指小。

7. 籽骨（sesamoid bone）：包括近籽骨和远籽骨。

（四）后肢骨

后肢骨包括髋骨、股骨、膝盖骨（髌骨）、小腿骨和后脚骨。髋骨是髂骨、坐骨和耻骨的合称。小腿骨由胫骨和腓骨组成。后脚骨包括跗骨、跖骨、趾骨和籽骨。

1. 髋骨（hip bone）：由髂骨、坐骨和耻骨结合而成。三块骨在外侧中部结合处形成深杯状的关节窝，称为髋臼，与股骨头成关节。左、右侧髋骨在骨盆中线处以软骨连结形成骨盆联合。骨盆（pelvis）是指由两侧髋骨、背侧的荐骨和前4枚尾椎以及两侧的荐结节阔韧带共同围成的结构，呈前宽后窄的圆锥形腔。前口以荐骨岬、髂骨和耻骨为界；后口的背侧为尾椎，腹侧为坐骨，两侧为荐结节阔韧带后缘。母猪骨盆的底壁平而宽，公猪则较窄。

（1）髂骨（ilium）位于髋骨的外上方，髂骨翼宽大，髂骨体窄小。髂骨翼的外侧角粗大，为髋结节；内侧角为荐结节。

（2）坐骨（ischium）位于髋骨后下方，构成骨盆底的后部。坐骨前缘与耻骨围成闭孔；坐骨结节粗大。左、右侧坐骨的后缘连成坐骨弓。两侧坐骨内侧缘由软骨连结形成坐骨联合，形成骨盆联合的后部。

（3）耻骨（pubis）较小，位于前下方，构成骨盆底的前部。耻骨后缘与坐骨前缘共同围成闭孔。两侧耻骨内侧缘由软骨连结形成耻骨联合，构成骨盆联合的前部。

2. 股骨（femoral bone, femur）：近端粗大，股骨头的中央有一凹陷为头窝，供圆韧带附着，与髋臼构成关节；外侧有粗大的大转子；内侧有小转子。股骨远端粗大，前部是滑车关节面，与膝盖骨成关节；后部由股骨内、外侧髁构成，与胫骨构成关节。

3. 膝盖骨（kneecap）：或称髌骨（patella），呈顶端向下的楔形，位于股骨远端的前方。

4. 小腿骨（skeleton of leg）：包括胫骨和腓骨。胫骨（tibia）位于内侧；近端粗大，有胫骨内、外侧髁，与股骨髁成关节；远端有螺旋状滑车，与胫跗骨成关节。腓骨（fibula）发达，位于胫骨近端外侧。

5. 跗骨（tarsal bone）：共7枚，位于小腿骨与跖骨之间，分为近、中、远三列。近列有2枚，内侧是距骨（胫跗骨），外侧是跟骨（腓跗骨）。中列仅有1枚中央跗骨。远列由内向外依次是第1、第2、第3和第4跗骨。

6. 跖骨（metatarsal bone）：与前肢掌骨相似，但较细长。

7. 趾骨（digital bone）：分系骨、冠骨和蹄骨，与前肢指骨相似。

8. 籽骨（sesamoid bone）：位置、形态与前肢籽骨相似。

二、骨连结

1. 躯干骨的连结：包括脊柱连结和胸廓连结。其中脊柱连结包括椎体间连结、椎弓间连结（如项韧带nuchal ligament）、寰枕关节和寰枢关节；胸廓连结包括肋椎关节（costovertebral joint）和肋胸关节（costosternal joint）。

2. 头骨的连结：颅顶大部分形成骨缝，颅底各骨间为软骨连结或骨性连结，其特点是连合较为牢固，不能活动。颞下颌关节（temporomandibular joint）又称下颌关节，是颅骨间唯一一对滑膜连结。由颞骨腹侧关节结节与下颌骨的髁突构成。颞下颌关节属于联动关节，同时运动，可进行开口、闭口和侧运动。

3. 前肢骨的连结：前肢各骨之间自上向下依次形成肩关节（humeral joint）、肘关节（elbow joint）、腕关节（carpal joint）和指关节（finger joint）。指关节又包括掌指关节（metacarpophalangeal joint，又称系关节或球节fetlock joint）、近指节间关节（proximal interphalangeal joint，又称冠关节coronal joint）和远指节间关节（distal interphalangeal joint，又称蹄关节coffin joint）。

4. 后肢骨的连结：包括荐髂关节（sacroiliac joint）、髋关节（hip joint）、膝关节（stifle joint）、跗关节〔tarsal joint，又称飞节（hock）〕和趾关节。趾关节包括系关节（跖趾关节）、冠关节（近趾节间关节）和蹄关节（远趾节间关节）。

三、骨的基本结构

骨由骨膜、骨质、骨髓、血管和神经构成。

1. 骨膜（periosteum）：由致密结缔组织构成，被覆于除关节外的骨表面为骨外膜，被覆于骨髓腔面、骨小梁表面、中央管及穿通管内表面的薄层结缔组织为骨内膜。骨外膜分2层，外层较厚，纤维密集而粗大；内层较薄，结缔组织疏松，含骨原细胞、成骨细胞、小血管和神经。骨内膜纤维细而少，由一层扁平细胞构成，能分裂分化成骨细胞。

2. 骨质（bone substance）：是构成骨的主要成分，分骨密质（compact bone）和骨松质（spongy bone）。骨密质位于骨的外周，构成长骨的骨干和骺以及其他类型骨的外层，坚硬、致密。骨松质位于骨的深部，呈海绵状，由互相交错的骨小梁构成。骨松质小梁的排列方向与受力的作用方向一致。骨密质和骨松质的这种配合，使骨既坚固又轻便。

骨组织（bone tissue, osseous tissue）由骨基质和骨组织的细胞组成。骨基质为固体，由骨盐、骨胶纤维及少量黏蛋白组成。骨基质结构呈板层状，称为骨板。同一骨板内的骨胶纤维相互平行，相邻骨板的骨胶纤维相互垂直或呈一定角度。骨板的结构使骨具有很强的支撑能力。骨组织的细胞包括骨原细胞、成骨细胞、骨细胞和破骨细胞。前三者与骨基质的生成有关，最后一个与骨基质的溶解吸收有关。

根据骨板排列形式的不同，可分为环骨板、骨单位和间骨板。环绕在骨干外表的为外环骨板（outer circumferential lamella），较厚，有10～40层骨板；环绕在骨髓腔的为内环骨板（inner circumferential lamella），较薄，仅由数层骨板组成。骨外膜中的小血管横穿外环骨板深入骨质中，贯穿环骨板的横向小管为穿通管（perfoating canal），穿通管与纵行排列的骨单位中央管通联，是小血管、神经和骨内膜的通道，并含有组织液。骨单位（osteon）呈圆筒状，由4～20层同心圆排列的骨单位骨板围绕形成，介于内、外环骨板之间，是骨干中骨密质的主要成分。骨单位中心为一条纵行的中央管。各骨单位表面有一层含骨盐较多而胶原纤维较少的骨基质，为骨黏合线。骨单位最外层骨板内的骨小管均在骨黏合线处折返，不与相邻骨单位的骨小管相通。骨单位最内层骨板的骨小管与中央管相通，从而形成血管系统与骨细胞之间物质交换的通路。骨间板（interstitial lalmella）是填充在骨单位之间的一些不规则的骨板，是旧的骨单位被吸收后的残余部分。

3．骨髓（bone marrow）：分红骨髓和黄骨髓。红骨髓位于骨髓腔和所有骨松质的间隙内，具有造血功能。成年家畜长骨骨髓腔内的红骨髓被富含脂肪的黄骨髓代替，但长骨两端、短骨和扁骨的骨松质内终生保留红骨髓。当机体大量失血或贫血时，黄骨髓又能转化为红骨髓而恢复造血功能。骨松质中的红骨髓终身存在。

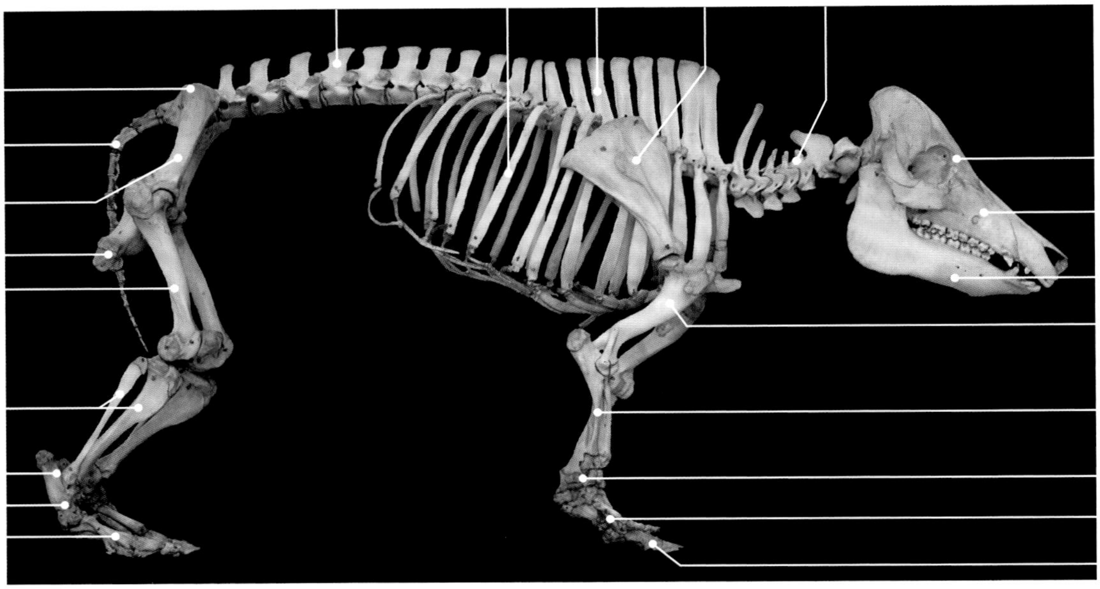

图 2-1　全身骨骼右侧观
Figure 2-1　Right lateral view of the skeleton.

1- 额骨 frontal bone
2- 上颌骨 maxillary bone
3- 下颌骨 mandible
4- 肱骨 humerus
5- 前臂骨 skeleton of forearm
6- 腕骨 carpal bone
7- 掌骨 metacarpal bone
8- 指骨 digital bone
9- 趾骨 digital bone
10- 跖骨 metatarsal bone
11- 跟骨 calcaneus
12- 胫骨和腓骨 tibia and fibula
13- 股骨 femoral bone, femur
14- 坐骨 ischium
15- 髂骨 illium
16- 尾椎 coccygeal vertebrae
17- 荐结节 sacral tuber
18- 腰椎 lumbar vertebrae
19- 肋骨 costal bone
20- 胸椎 thoracic vertebrae
21- 肩胛骨 scapula
22- 颈椎 cervical vertebrae

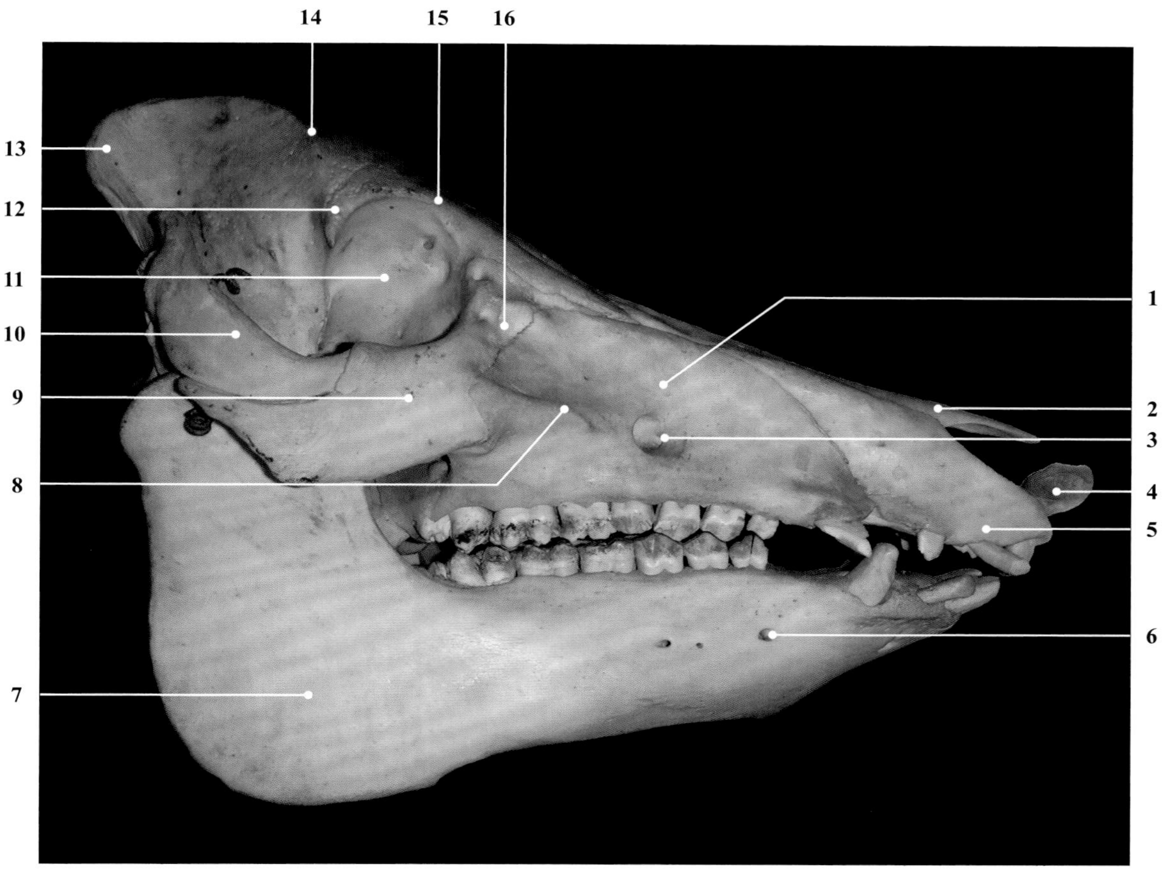

图 2-2 头骨右侧观
Figure 2-2 Right lateral view of the skull.

1- 上颌骨 maxillary bone
2- 鼻骨 nasal bone
3- 眶下孔 infraorbital foramen
4- 吻骨 rostral bone
5- 切齿骨 incisive bone
6- 颏孔 mental foramen
7- 下颌骨 mandible
8- 面嵴 facial crest
9- 颧骨 zygomatic bone
10- 颞骨颧突 zygomatic process of temporal bone
11- 眶窝 orbital fossa
12- 额骨颧突 zygomatic process of frontal bone
13- 枕骨 occipital bone
14- 顶骨 parietal bone
15- 额骨 frontal bone
16- 泪骨 lacrimal bone

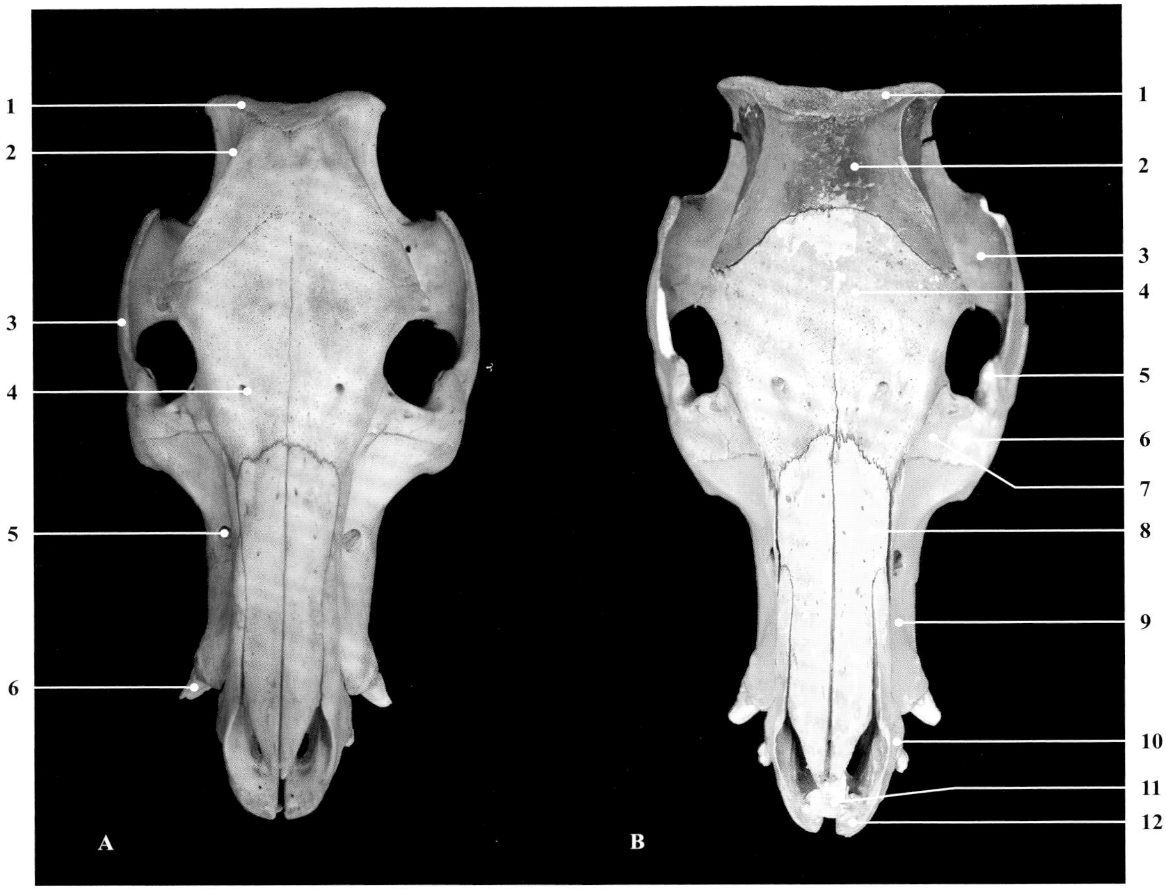

图 2-3　头骨背面观
Figure 2-3　Dorsal view of the skull.

A. 原色 Primary color

1- 枕嵴 occipital crest
2- 顶嵴 parietal crest
3- 颞骨颧突 zygomatic process of temporal bone
4- 眶上孔 supraorbital foramen
5- 眶下孔 infraorbital foramen
6- 犬齿 canine

B. 彩色 Color

1- 枕骨 occipital bone
2- 顶骨 parietal bone
3- 颞窝 temporal fossa
4- 额骨 frontal bone
5- 颧弓 zygomatic arch
6- 颧骨 zygomatic bone
7- 泪骨 lacrimal bone
8- 鼻骨 nasal bone
9- 上颌骨 maxillary bone
10- 切齿骨鼻突 nasal process of the incisive bone
11- 吻骨 rostral bone
12- 切齿骨骨体 body of the incisive bone

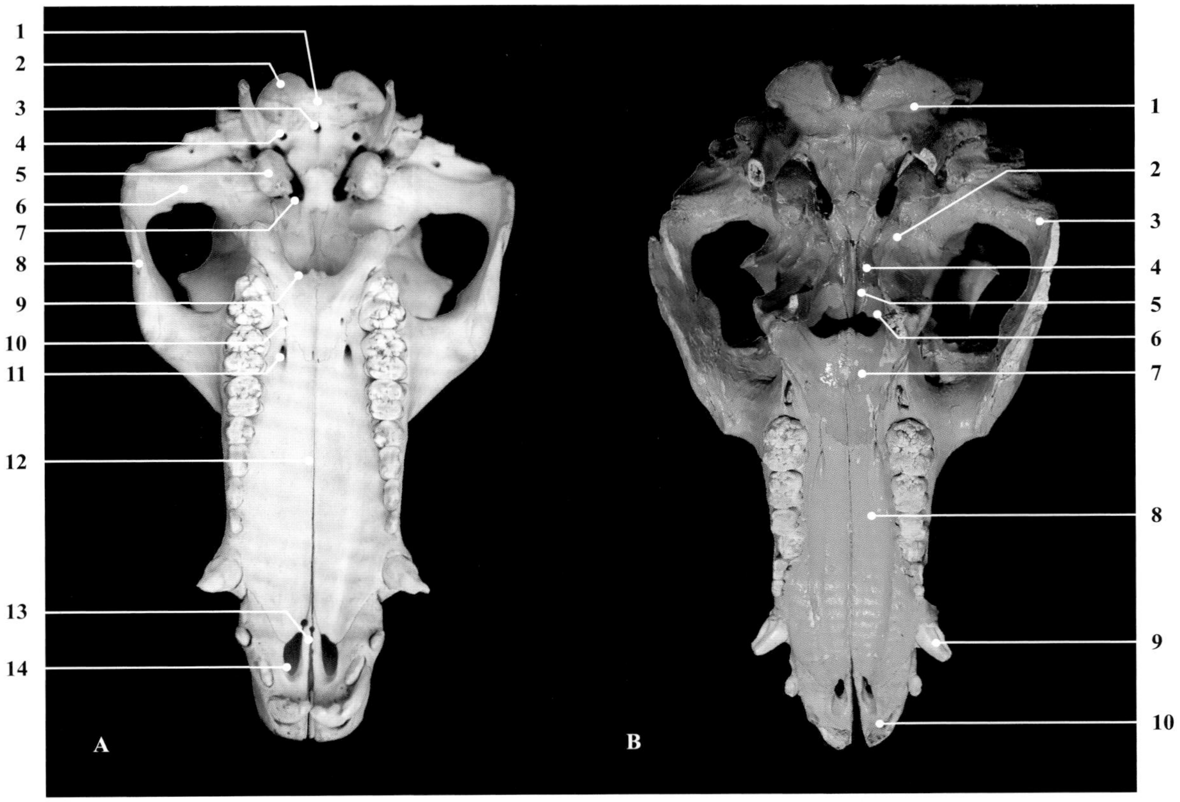

图 2-4　头骨腹侧观
Figure 2-4　Ventral view of the skull.

A. 原色 Primary color
1- 枕骨大孔 foramen magnum
2- 枕骨髁 occipital condyle
3- 枕骨基部 basioccipital bone
4- 舌下神经孔 hypoglossal foramen
5- 鼓泡 tympanic bulla
6- 下颌窝 mandibular fossa
7- 破裂孔 foramen lacerum
8- 颧弓 zygomatic arch
9- 鼻后孔 posterior nasal apertures
10- 腭小孔 lesser palatine foramen
11- 腭大孔 greater palatine foramen
12- 腭正中缝 median palatine suture
13- 切齿骨腭突 palatine process of the incisive bone
14- 腭裂 palatine fissure

B. 彩色 Color
1- 枕骨 occipital bone
2- 蝶骨 sphenoid bone
3- 颞骨 temporal bone
4- 翼骨 pterygoid bone
5- 犁骨 vomer
6- 腭骨垂直板 perpendicular plate of palatine bone
7- 腭骨水平板 horizontal plate of palatine bone
8- 上颌骨腭突 palatine process of the maxillary bone
9- 犬齿 canine
10- 切齿骨骨体 body of the incisive bone

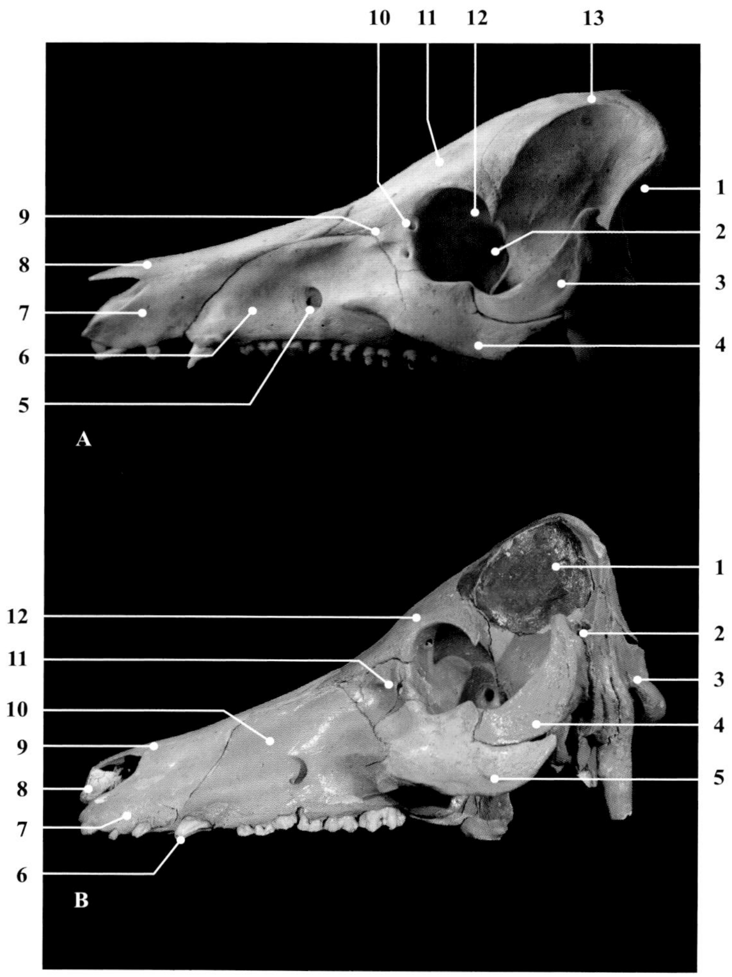

图 2-5　头骨左侧观
Figure 2-5　Left lateral view of the skull.

A. 原色 Primary color
1- 枕骨 occipital bone
2- 视神经孔 optic canal,optic foramen
3- 颞骨颧突 zygomatic process of temporal bone
4- 颧骨 zygomatic bone
5- 眶下孔 infraorbital foramen
6- 上颌骨 maxillary bone
7- 切齿骨 incisive bone
8- 鼻骨 nasal bone
9- 泪骨 lacrimal bone
10- 泪孔 lacrimal foramen
11- 额骨 frontal bone
12- 眶窝 orbital fossa
13- 顶骨 parietal bone

B. 彩色 Color
1- 顶骨 parietal bone
2- 外耳道 external auditory meatus
3- 枕骨 occipital bone
4- 颞骨颧突 zygomatic process of temporal bone
5- 颧骨 zygomatic bone
6- 犬齿 canine
7- 切齿骨 incisive bone
8- 吻骨 rostral bone
9- 鼻骨 nasal bone
10- 上颌骨 maxillary bone
11- 泪骨 lacrimal bone
12- 额骨 frontal bone

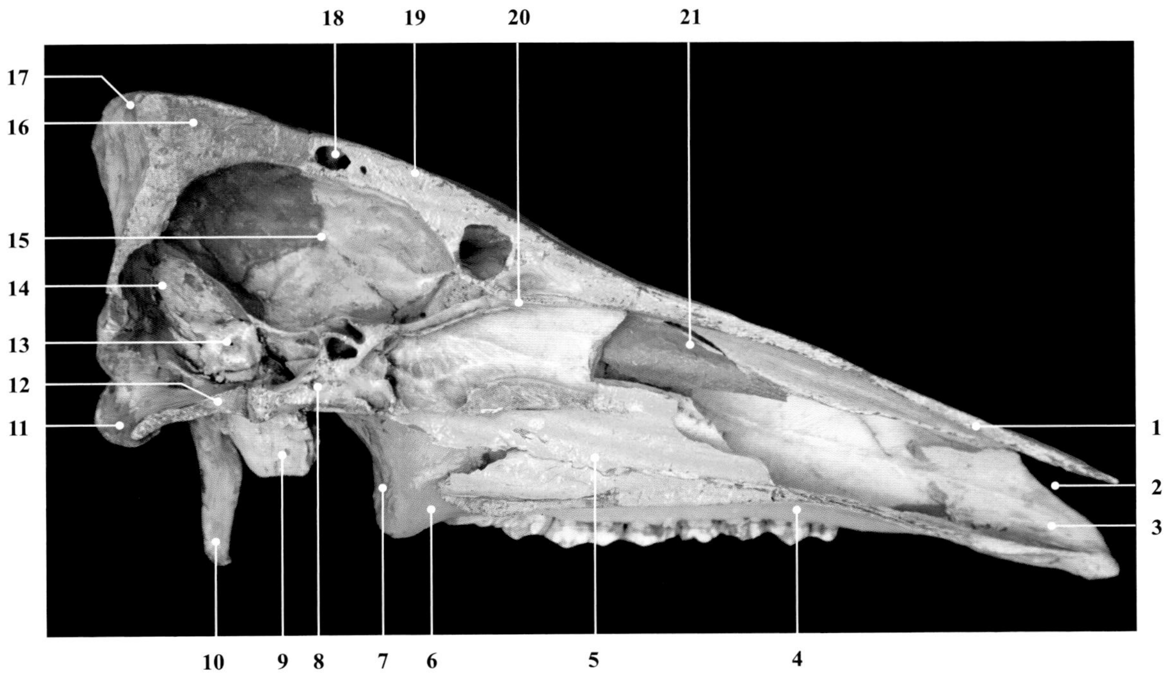

图 2-6 头骨正中矢状面
Figure 2-6 Paramedial section of the skull.

1- 鼻骨 nasal bone
2- 鼻颌切迹 nasomaxillary notch
3- 切齿骨 incisive bone
4- 上颌骨 maxillary bone
5- 下鼻甲骨 ventral turbinal bone
6- 腭骨 palatine bone
7- 翼骨 pterygoid bone
8- 蝶骨体 basisphenoid
9- 鼓泡 tympanic bulla
10- 枕骨颈静脉突 jugular process of the occipital bone
11- 枕骨大孔 foramen magnum
12- 枕骨基部 basioccipital bone
13- 内耳道 internal acoustic meatus
14- 颞骨岩部 petrous part of temporal bone
15- 颅腔 cranial cavity
16- 顶骨 parietal bone
17- 枕骨 occipital bone
18- 额窦 frontal sinus
19- 额骨 frontal bone
20- 筛骨板 ethmoidal plate
21- 上鼻甲骨 dorsal turbinal bone

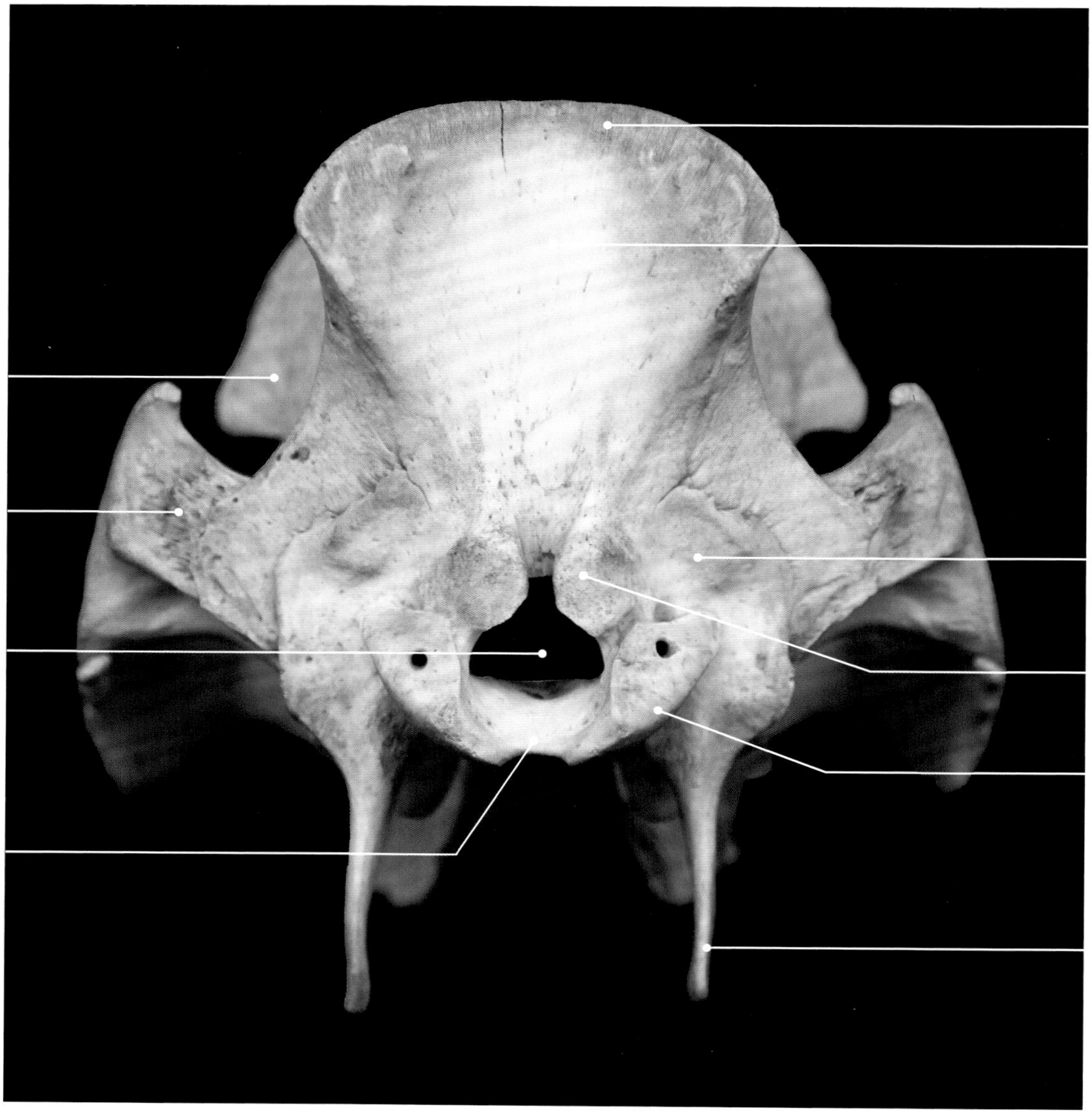

图 2-7　头骨后面观
Figure 2-7　Caudal view of the skull.

1- 枕嵴（项嵴） occipital crest (nuchal crest)
2- 枕骨鳞部 squamous part of occipital bone
3- 枕骨外侧部 lateral part of occipital bone
4- 项结节 nuchal tubercle
5- 枕骨髁 occipital condyle
6- 枕骨颈静脉突（髁旁突）jugular process of the occipital bone (paracondylar process)
7- 枕骨基部 basioccipital bone
8- 枕骨大孔 foramen magnum
9- 颞骨 temporal bone
10- 顶骨 parietal bone

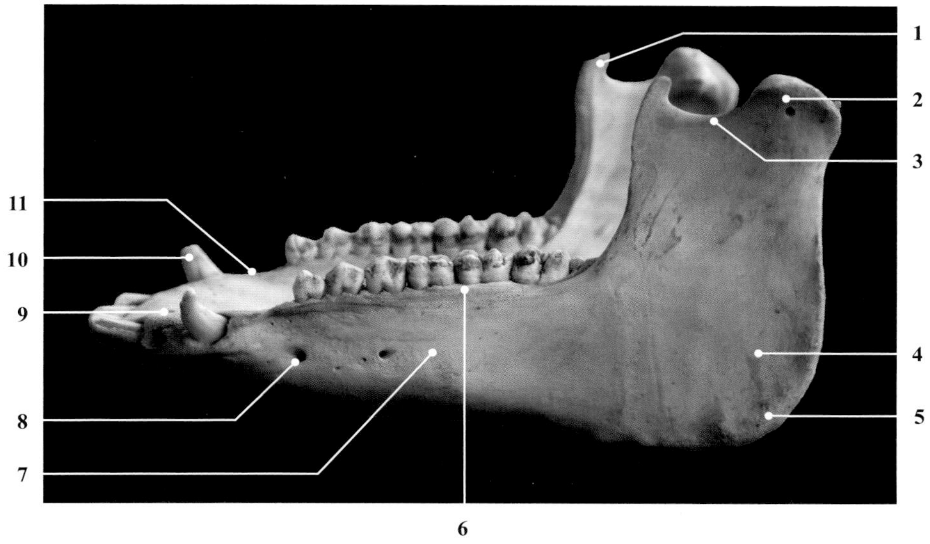

图 2-8 下颌骨左外侧观
Figure 2-8 Left lateral view of the mandible.

1- 冠状突 coronoid process
2- 髁状突 condylar process
3- 下颌骨切迹 mandible notch
4- 咬肌窝 masseteric fossa
5- 下颌角 angle of mandible
6- 齿槽缘 alveolar margin
7- 下颌骨体臼齿部 molar part of mandible body
8- 颏孔 mental foramen
9- 下颌骨体切齿部 incisive part of mandible body
10- 犬齿 canine
11- 齿槽间缘 interalveolar margin

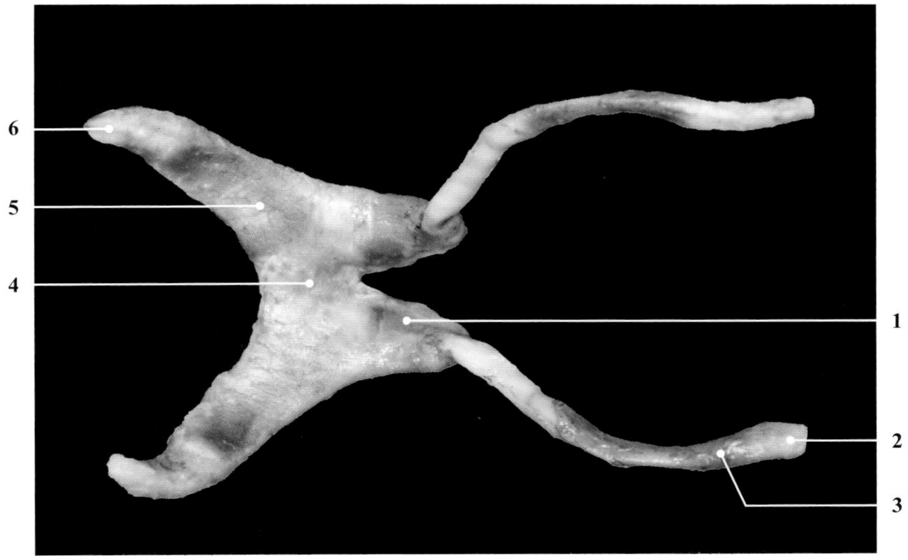

图 2-9 舌骨背面观
Figure 2-9 Dorsal view of the hyoid bone.

1- 上舌骨 epihyoid
2- 鼓舌骨 tympanohyoid
3- 茎突舌骨 stylohyoid
4- 角舌骨 ceratohyoid
5- 甲状舌骨 thyrohyoid
6- 甲状舌骨软骨 cartilage of the thyrohyoid bone

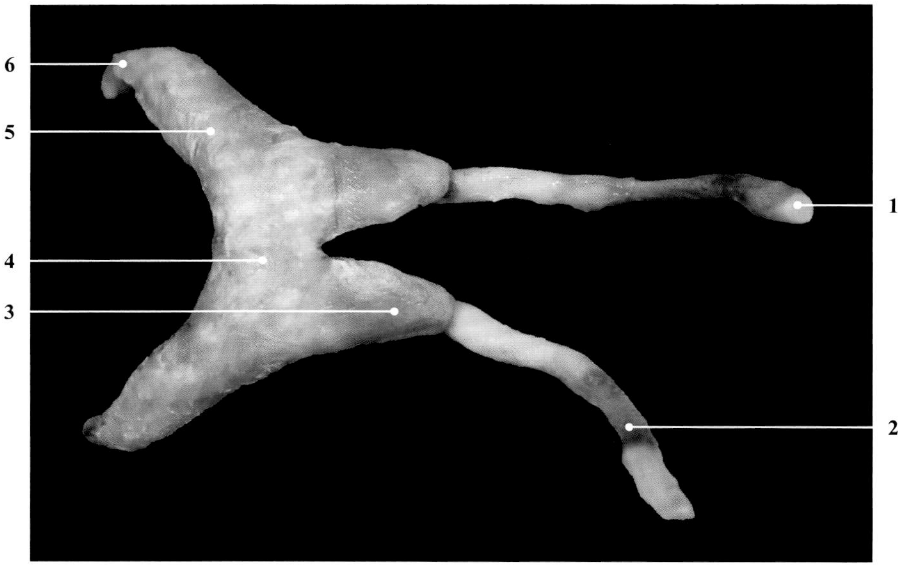

图 2-10　舌骨腹面观
Figure 2-10　Ventral view of the hyoid bone.

1- 鼓舌骨 tympanohyoid
2- 茎突舌骨 stylohyoid
3- 上舌骨 epihyoid
4- 角舌骨 ceratohyoid
5- 甲状舌骨 thyrohyoid
6- 甲状舌骨软骨 cartilage of the thyrohyoid bone

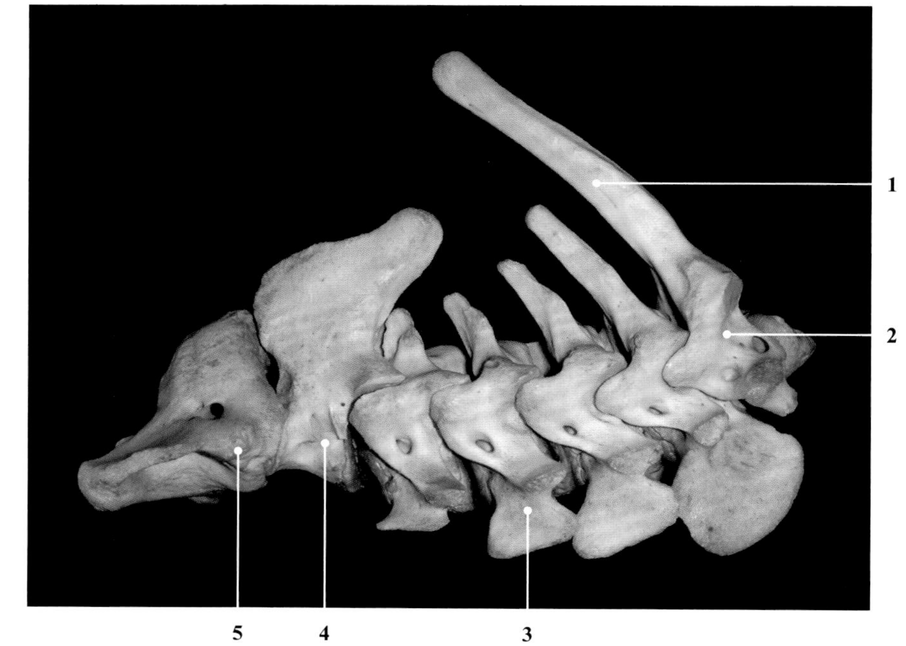

图 2-11　颈椎左外侧观
Figure 2-11　Left lateral view of the cervical vertebrae.

1- 第 7 颈椎棘突 spinous process of the 7th cervical vertebra
2- 第 7 颈椎 7th cervical vertebra
3- 第 4 颈椎 4th cervical vertebra
4- 枢椎 axis
5- 寰椎 atlas

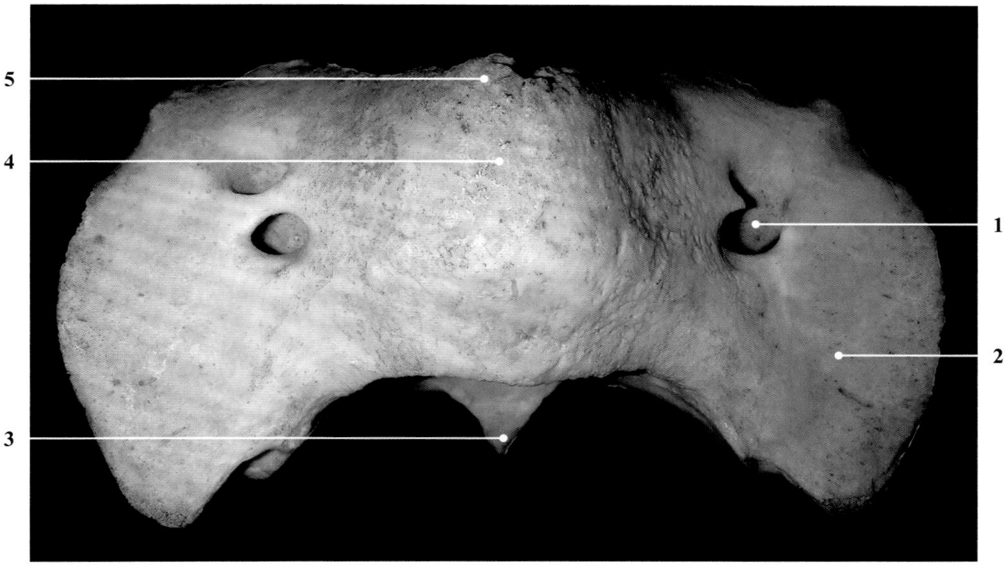

图 2-12 第 1 颈椎（寰椎）背侧观
Figure 2-12 Dorsal view of the first cervical vertebra (atlas).

1- 翼孔和椎外侧孔 alar foramen and lateral vertebral foramen
2- 寰椎翼 wing of atlas
3- 腹侧结节 ventral tubercle
4- 背侧弓 dorsal arch
5- 背侧结节 dorsal tubercle

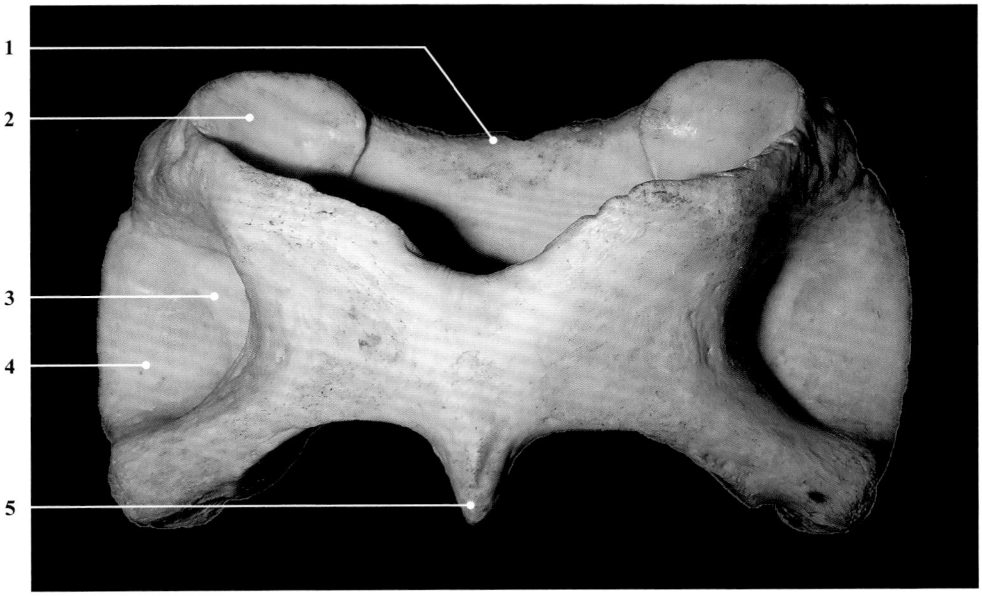

图 2-13 第 1 颈椎（寰椎）腹侧观
Figure 2-13 Ventral view of the first cervical vertebra (atlas).

1- 背侧弓 dorsal arch
2- 关节窝 articular fossa
3- 翼窝 alar fossa
4- 寰椎翼 wing of atlas
5- 腹侧结节 ventral tubercle

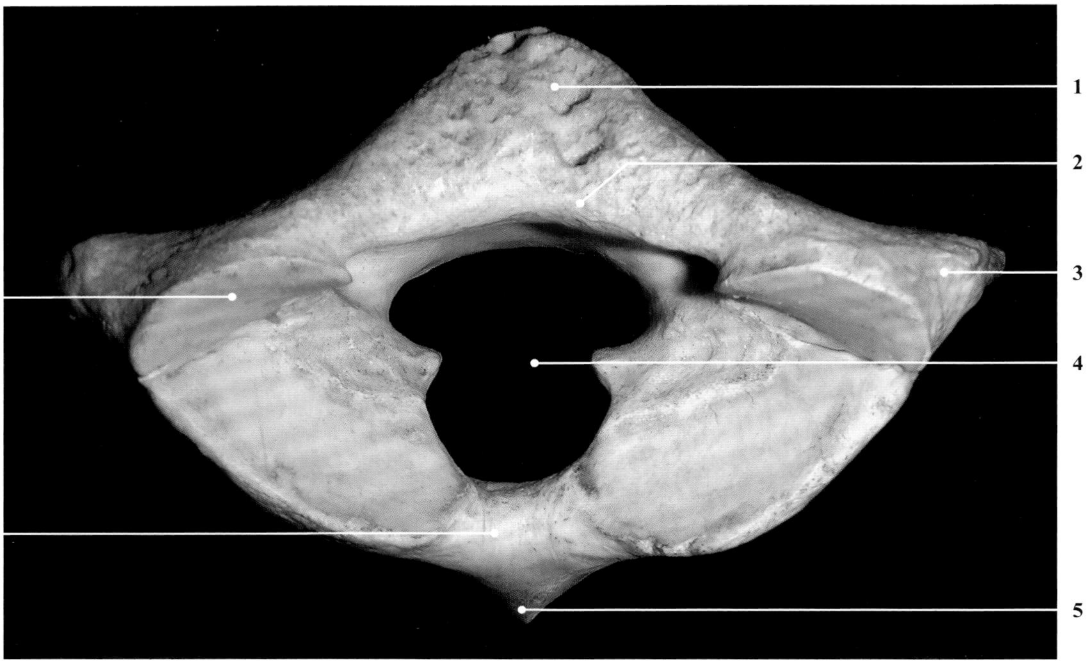

图 2-14　第 1 颈椎（寰椎）前面观
Figure 2-14　Cranial view of the first cervical vertebra (atlas).

1- 背侧结节 dorsal tubercle
2- 背侧弓 dorsal arch
3- 寰椎翼 wing of atlas
4- 椎孔 vertebral foramen
5- 腹侧结节 ventral tubercle
6- 腹侧弓 ventral arch
7- 关节凹 articular fovea

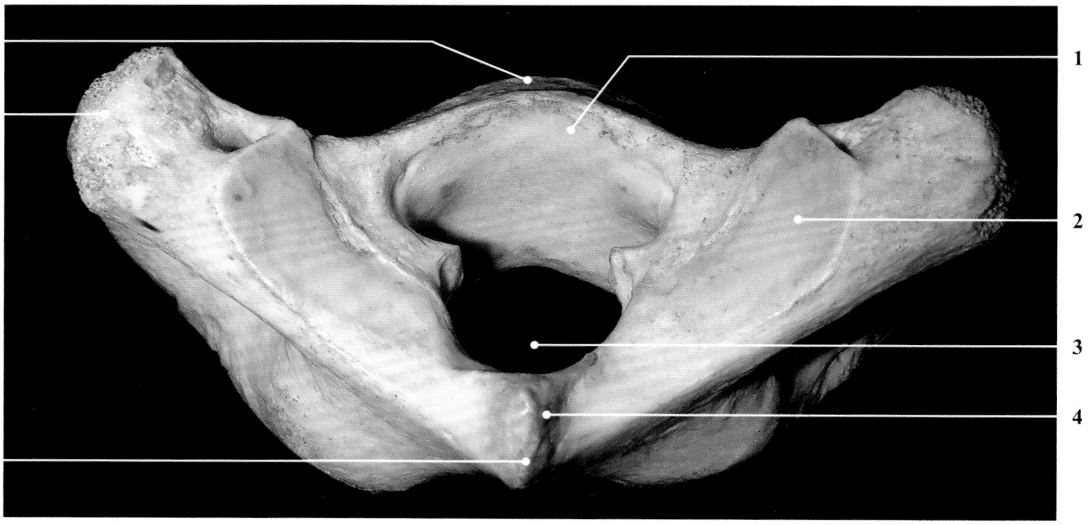

图 2-15　第 1 颈椎（寰椎）后面观
Figure 2-15　Caudal view of the first cervical vertebra (atlas).

1- 背侧弓 dorsal arch
2- 后关节凹 caudal articular fovea
3- 椎孔 vertebral foramen
4- 腹侧弓 ventral arch
5- 腹侧结节 ventral tubercle
6- 寰椎翼 wing of atlas
7- 背侧结节 dorsal tubercle

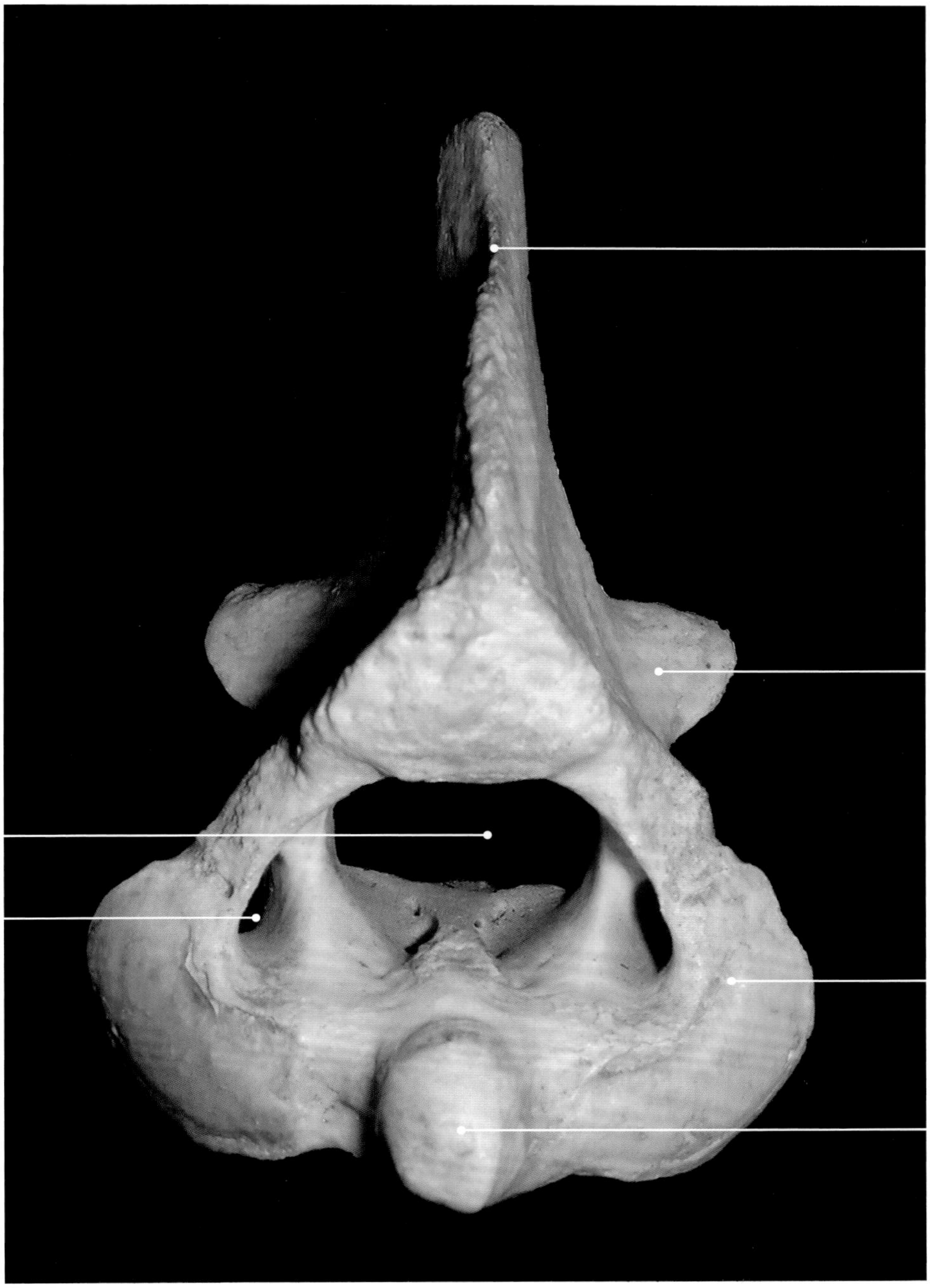

图 2-16　第 2 颈椎（枢椎）前面观
Figure 2-16　Cranial view of the second cervical vertebra (axis).

1- 棘突 spinous process
2- 后关节突 caudal articular process
3- 前关节突 cranial articular process
4- 齿突 dens
5- 横突孔 transverse
6- 椎孔 vertebral foramen

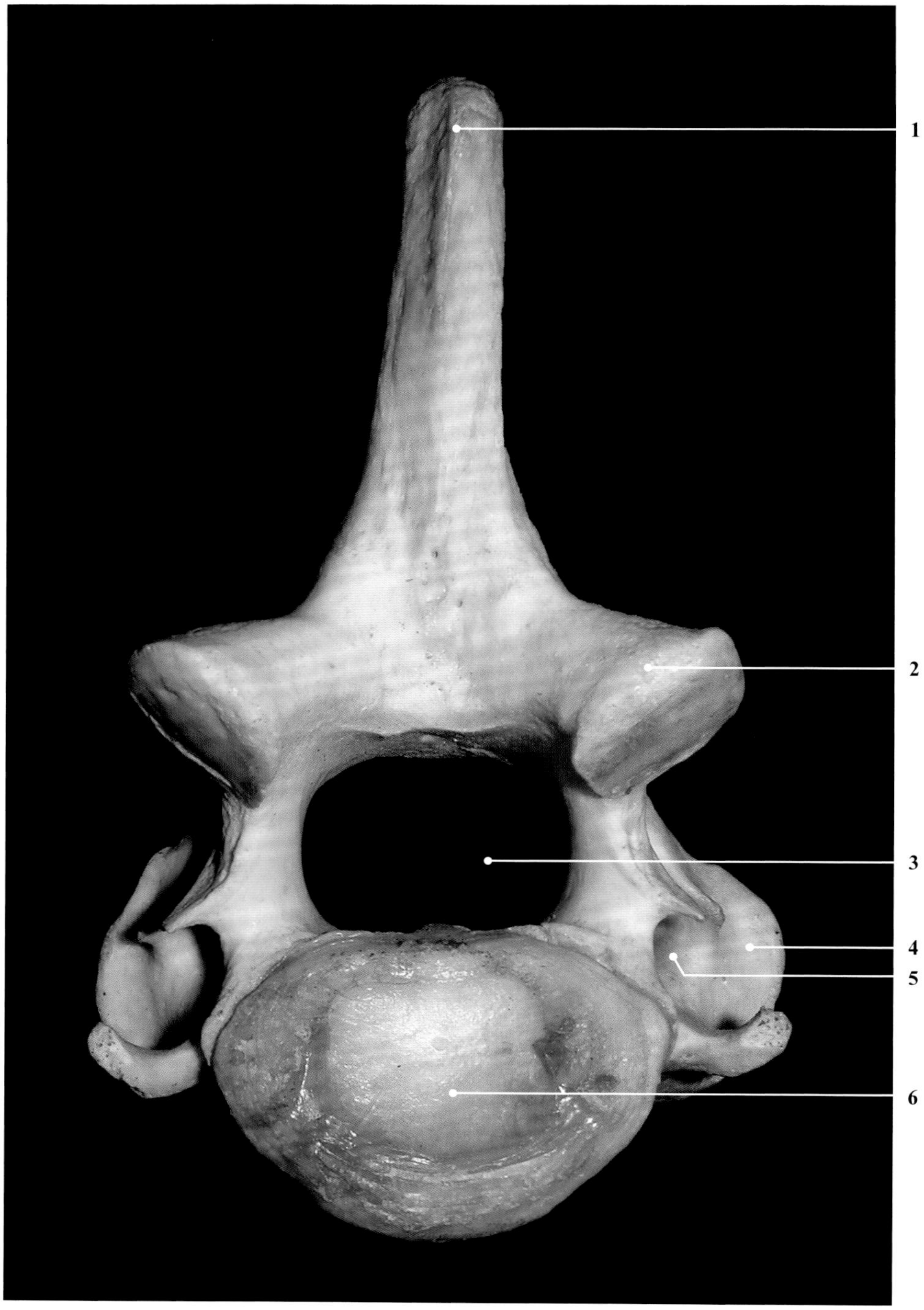

图 2-17　第 2 颈椎（枢椎）后面观
Figure 2-17　Caudal view of the second cervical vertebra (axis).

1- 棘突 spinous process
2- 后关节突 caudal articular process
3- 椎孔 vertebral foramen
4- 横突 transverse process
5- 横突孔 transverse foramen
6- 椎窝 vertebral fossa

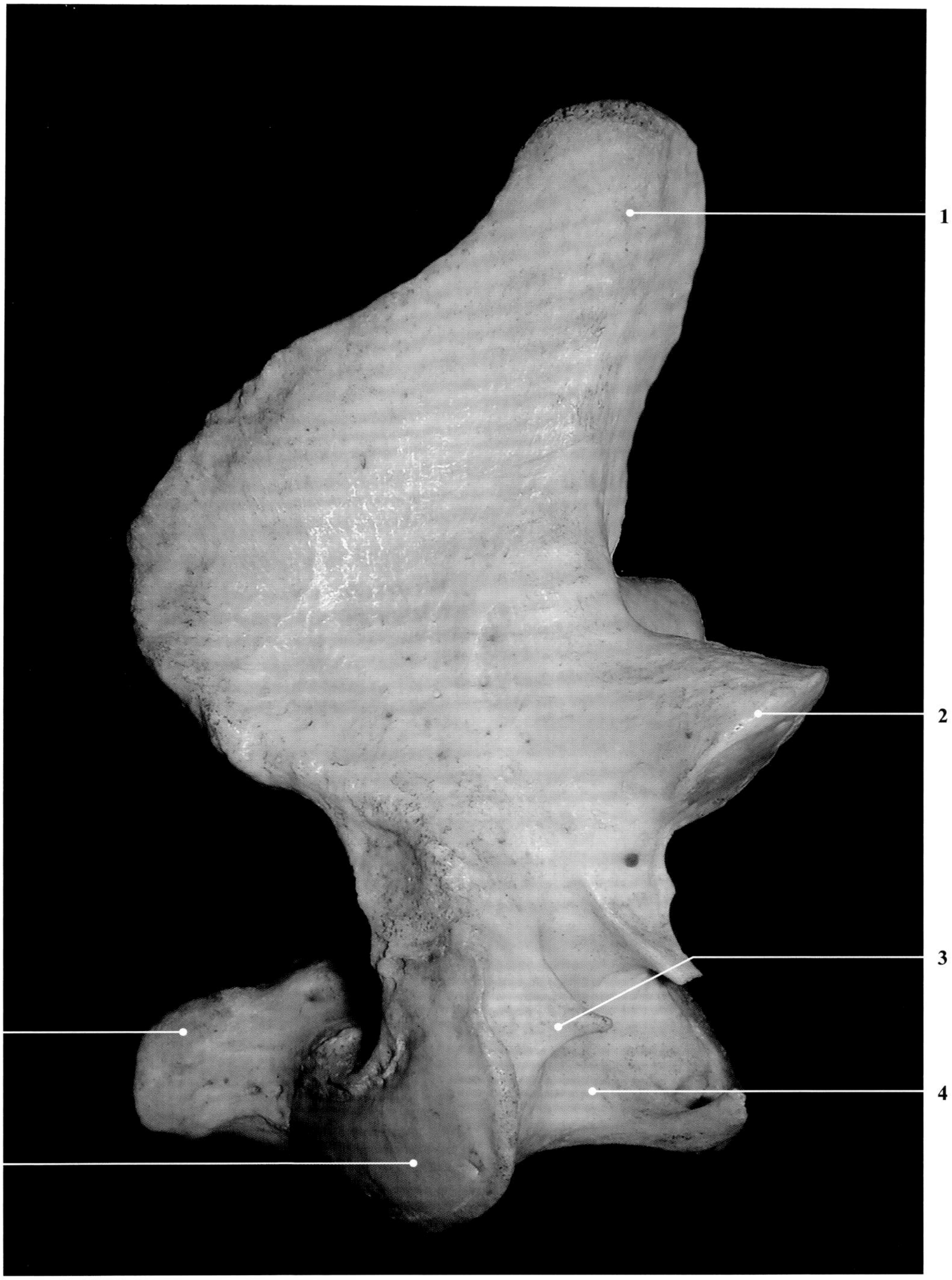

图 2-18　第 2 颈椎（枢椎）左外侧观
Figure 2-18　Left lateral view of the second cervical vertebra (axis).

1- 棘突 spinous process
2- 后关节突 caudal articular process
3- 横突 transverse process
4- 椎体 vertebral body
5- 前关节突 cranial articular process
6- 齿突 dens

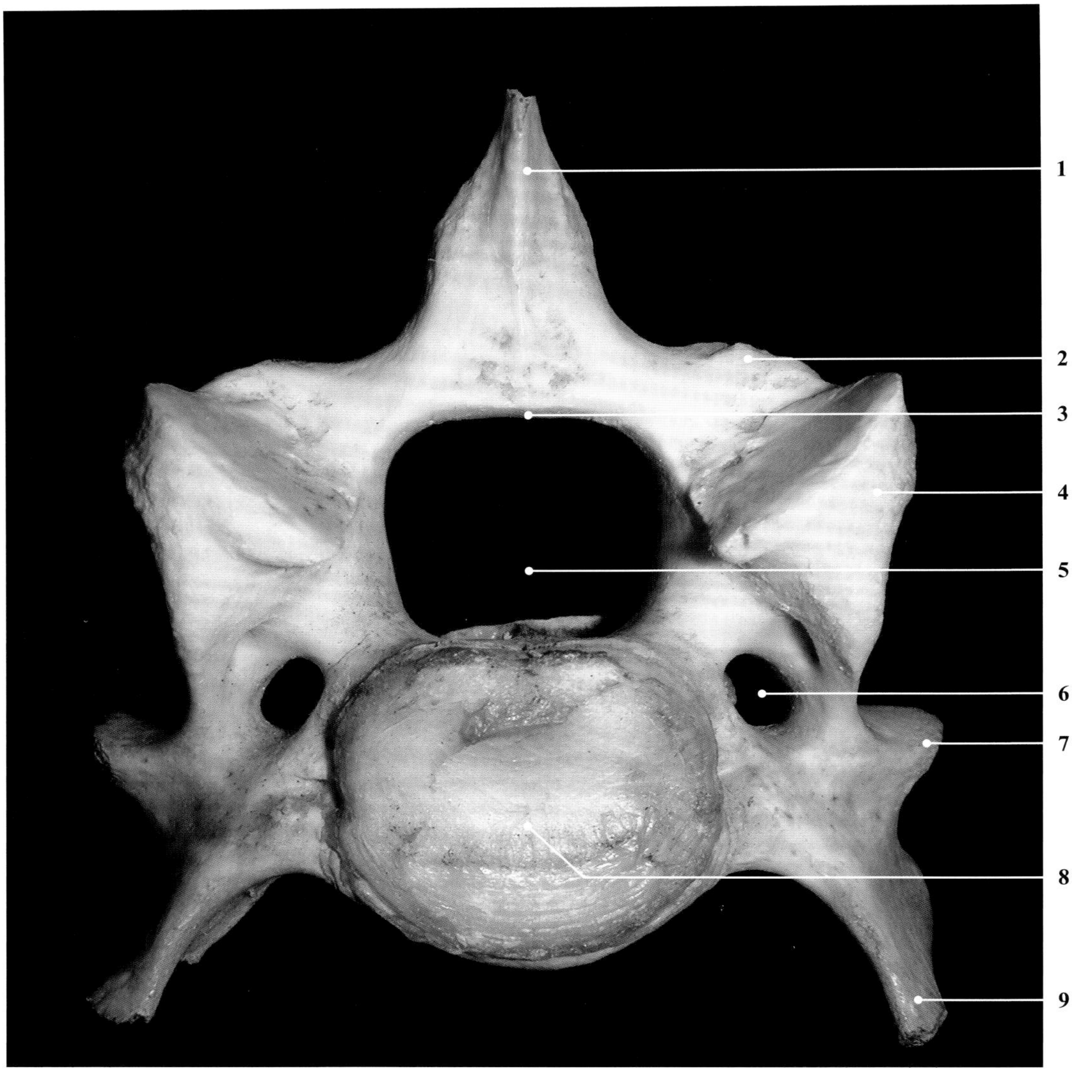

图 2-19　第 4 颈椎前面观
Figure 2-19　Cranial view of the fourth cervical vertebra.

1- 棘突 spinous process
2- 后关节突 caudal articular process
3- 椎弓 vertebral arch
4- 前关节突 cranial articular process
5- 椎孔 vertebral foramen
6- 横突孔 transverse foramen
7- 横突后支 caudal branch of transverse process
8- 椎头 vertebral head
9- 横突前支 cranial branch of transverse process

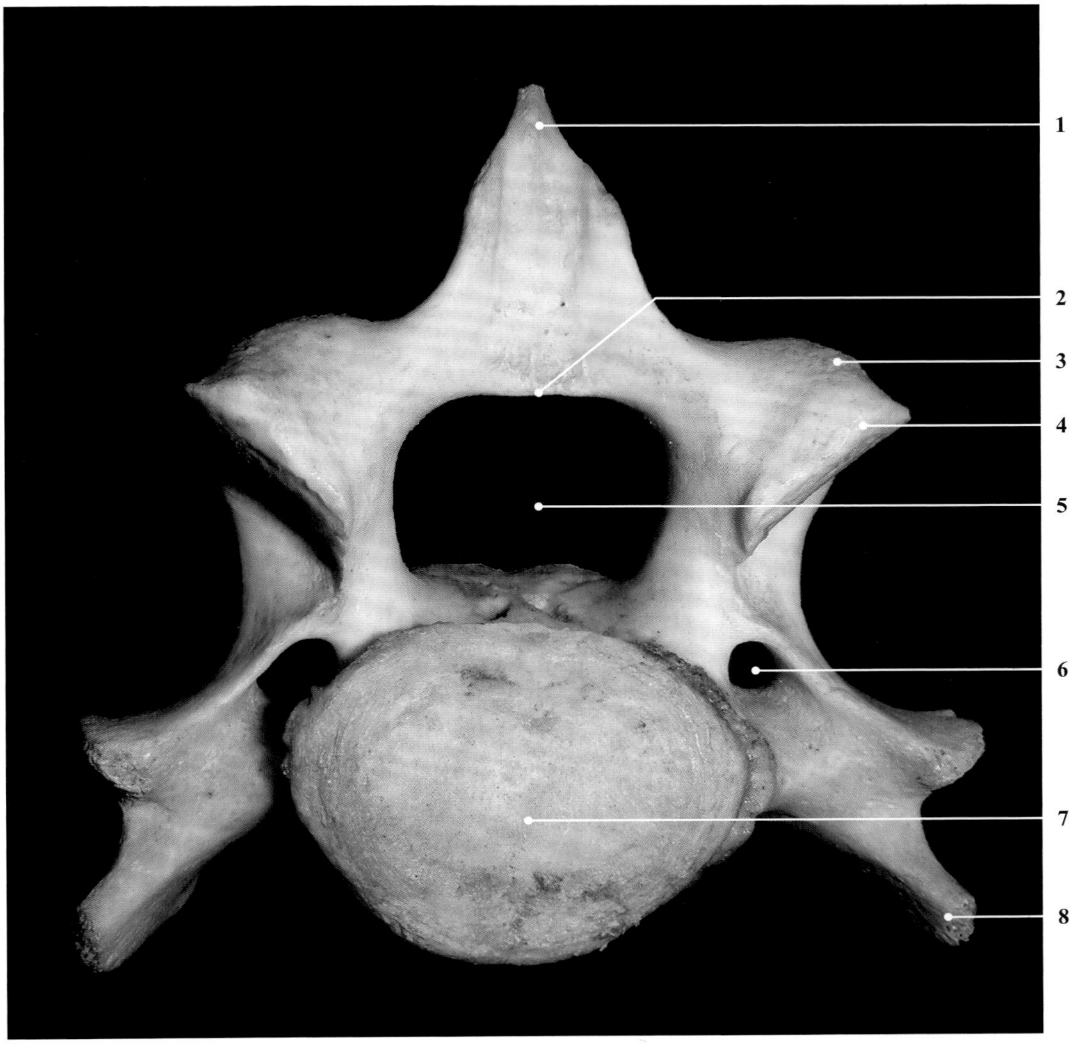

图 2-20　第 4 颈椎后面观
Figure 2-20　Caudal view of the fourth cervical vertebra.

1- 棘突 spinous process
2- 椎弓 vertebral arch
3- 前关节突 cranial articular process
4- 后关节突 caudal articular process
5- 椎孔 vertebral foramen
6- 横突孔 transverse foramen
7- 椎窝 vertebral fossa
8- 横突后支 caudal branch of transverse process

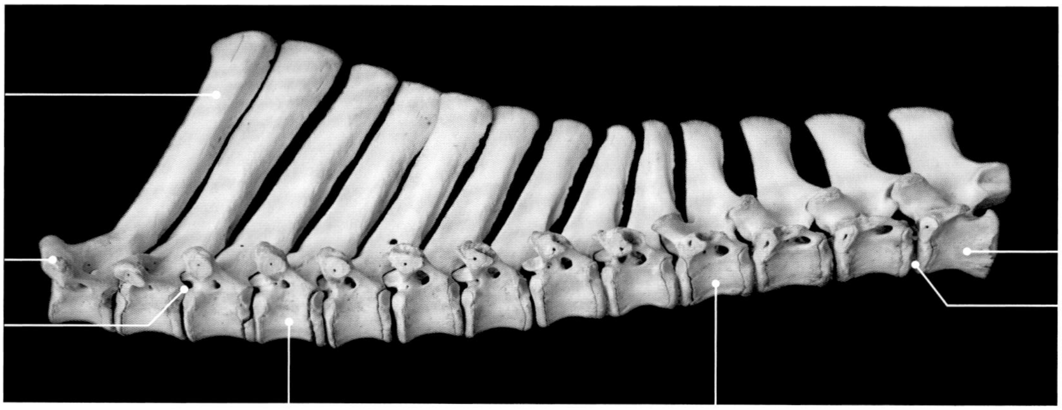

图 2-21　胸椎左外侧观
Figure 2-21　Left lateral view of the thoracic vertebrae.

1- 椎体 vertebral body
2- 椎间盘 intervertebral disc
3- 第 10 胸椎 10th thoracic vertebra
4- 第 4 胸椎 4th thoracic vertebra
5- 椎间孔 intervertebral foramen
6- 前关节突 cranial articular process
7- 胸椎棘突 spinous process of thoracic vertebra

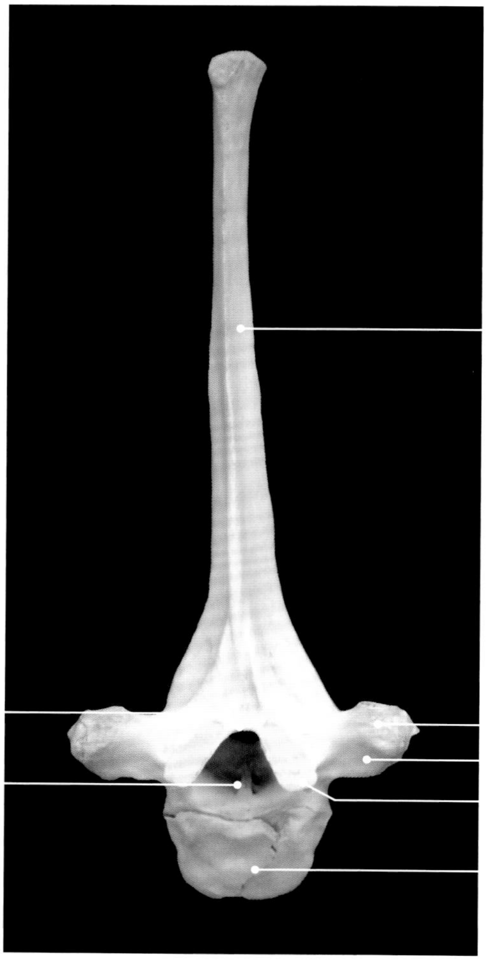

图 2-22　第 3 胸椎前面观
Figure 2-22　Cranial view of the third thoracic vertebra.

1- 棘突 spinous process
2- 横突 transverse process
3- 前肋窝 cranial costal fossa
4- 前关节突 cranial articular process
5- 椎头 vertebral head
6- 椎孔 vertebral foramen
7- 椎弓 vertebral arch

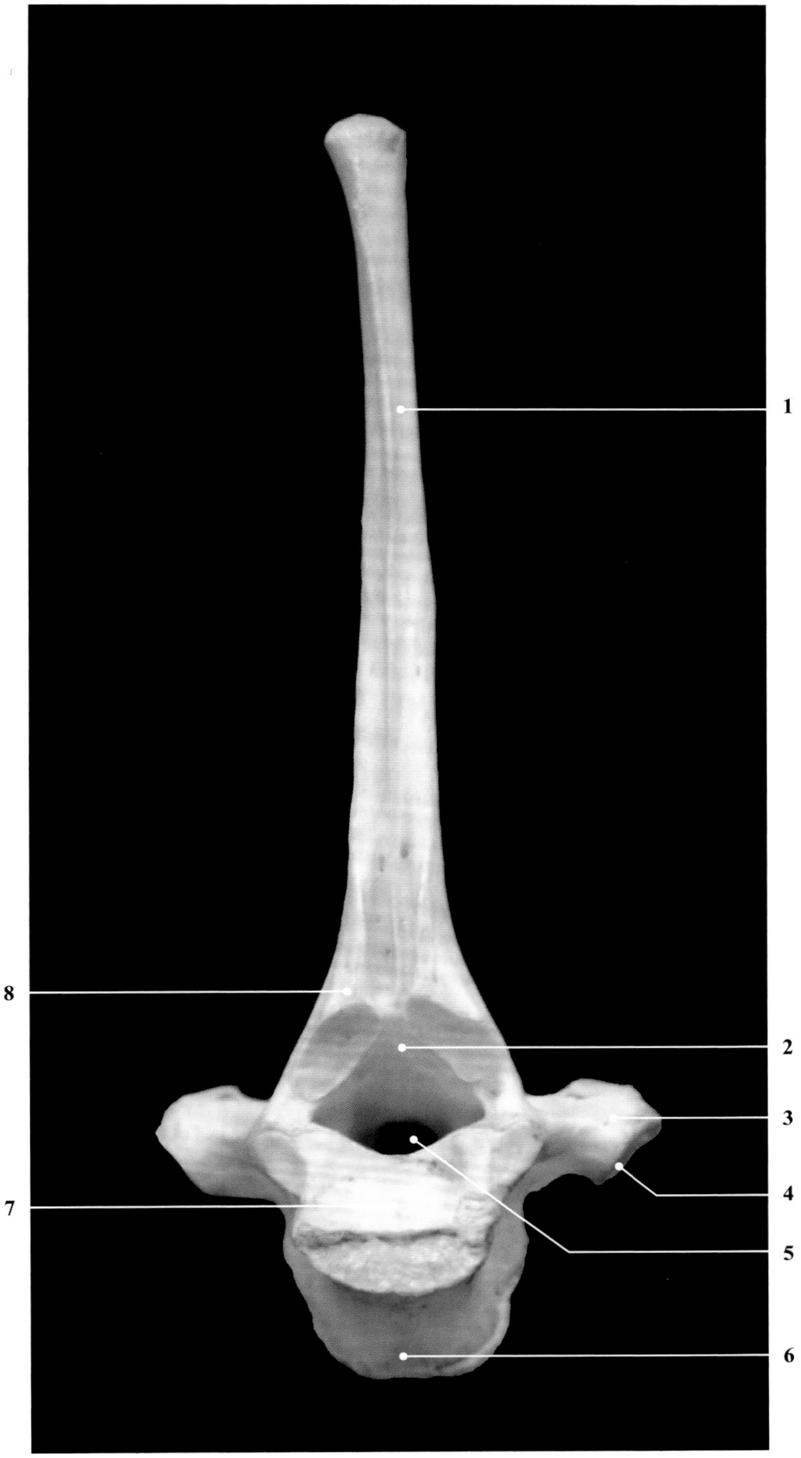

图 2-23 第 3 胸椎后面观
Figure 2-23 Caudal view of the third thoracic vertebra.

1- 棘突 spinous process
2- 椎弓 vertebral arch
3- 横突 transverse process
4- 后肋窝 caudal costal fossa
5- 椎孔 vertebral foramen
6- 椎体 vertebral body
7- 椎窝 vertebral fossa
8- 后关节突 caudal articular process

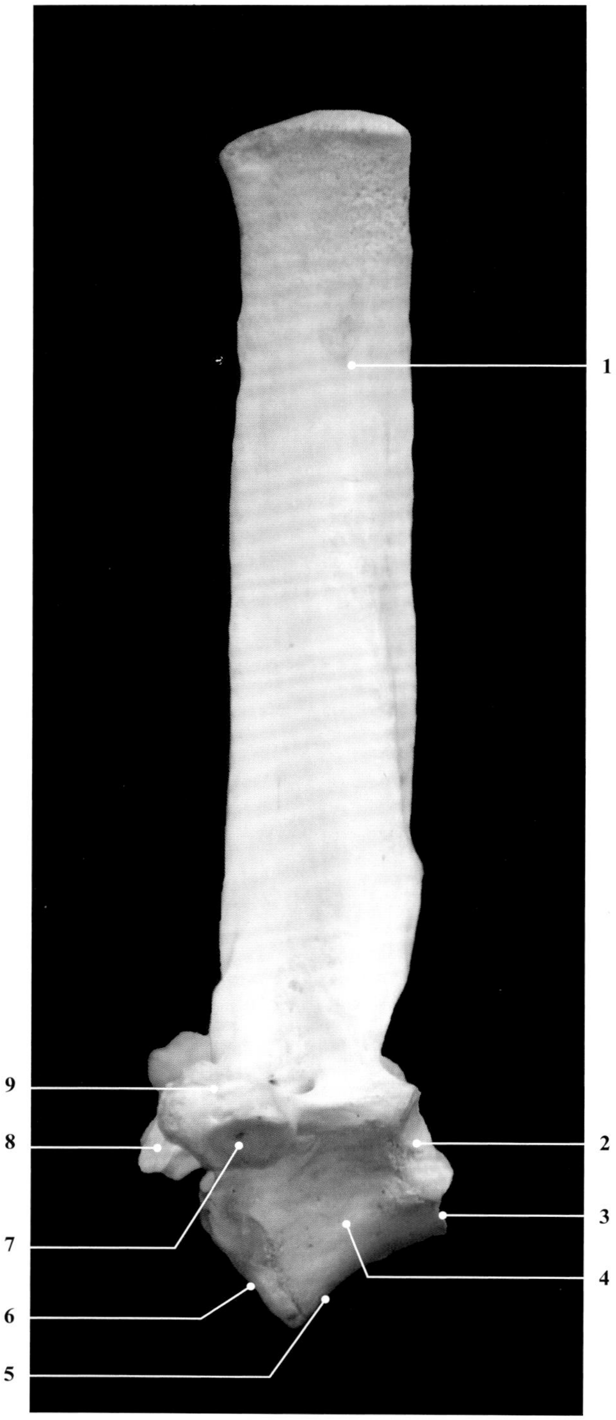

图 2-24　第 3 胸椎左外侧观
Figure 2-24　Left lateral view of the third thoracic vertebra.

1- 棘突 spinous process
2- 后肋窝 caudal costal fossa
3- 椎窝 vertebral fossa
4- 椎体 vertebral body
5- 腹侧嵴 ventral crest
6- 椎头 vertebral head
7- 前肋窝 cranial costal fossa
8- 前关节突 cranial articular process
9- 横突 transverse process

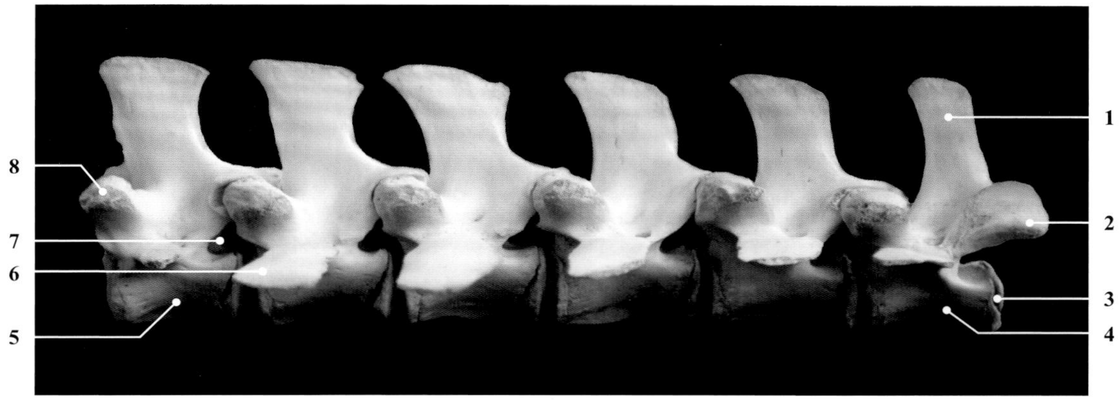

图 2-25 腰椎左外侧观
Figure 2-25 Left lateral view of the lumbar vertebrae.

1- 棘突 spinous process
2- 后关节突 caudal articular process
3- 椎窝 vertebral fossa
4- 椎体 vertebral body
5- 第 1 腰椎 1st lumbar vertebra
6- 横突 transverse process
7- 椎间孔 intervertebral foramen
8- 前关节突 cranial articular process

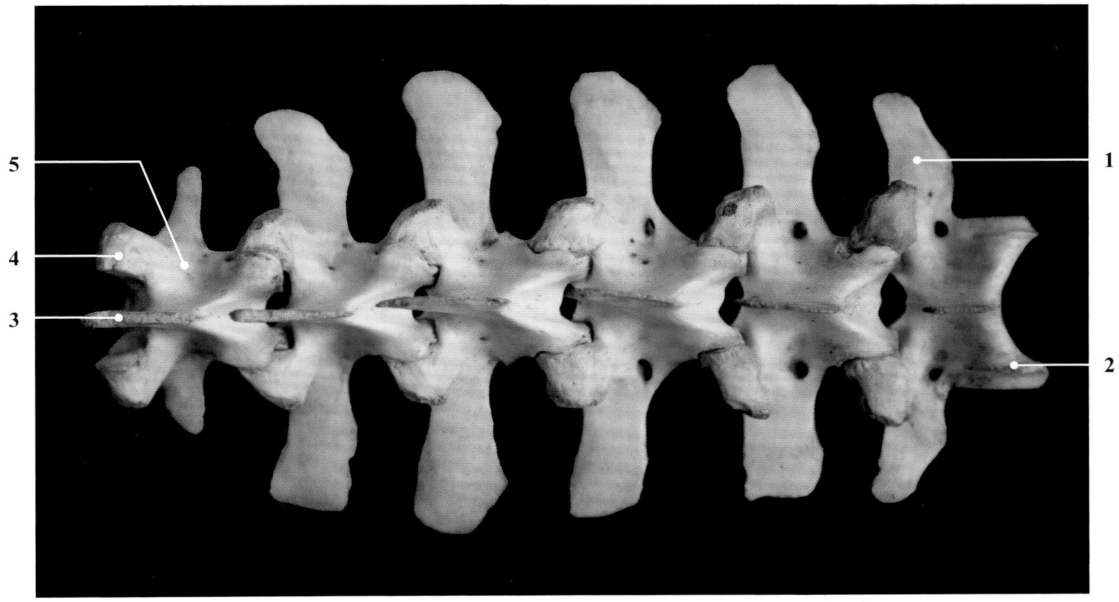

图 2-26 腰椎背侧观
Figure 2-26 Dorsal view of the lumbar vertebrae.

1- 横突 transverse process
2- 后关节突 caudal articular process
3- 棘突 spinous process
4- 前关节突 cranial articular process
5- 第 1 腰椎 1st lumbar vertebra

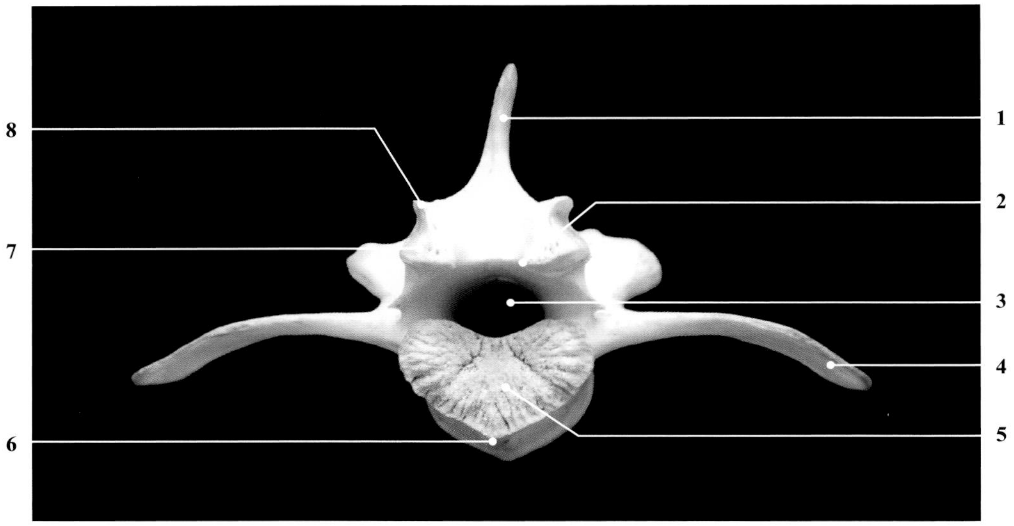

图 2-27　第 3 腰椎前面观
Figure 2-27　Cranial view of the third lumbar vertebra.

1- 棘突 spinous process
2- 椎弓 vertebral arch
3- 椎孔 vertebral foramen
4- 横突 transverse process
5- 椎头 vertebral head
6- 腹侧嵴 ventral crest
7- 前关节突 cranial articular process
8- 乳突 mamilloarticular process

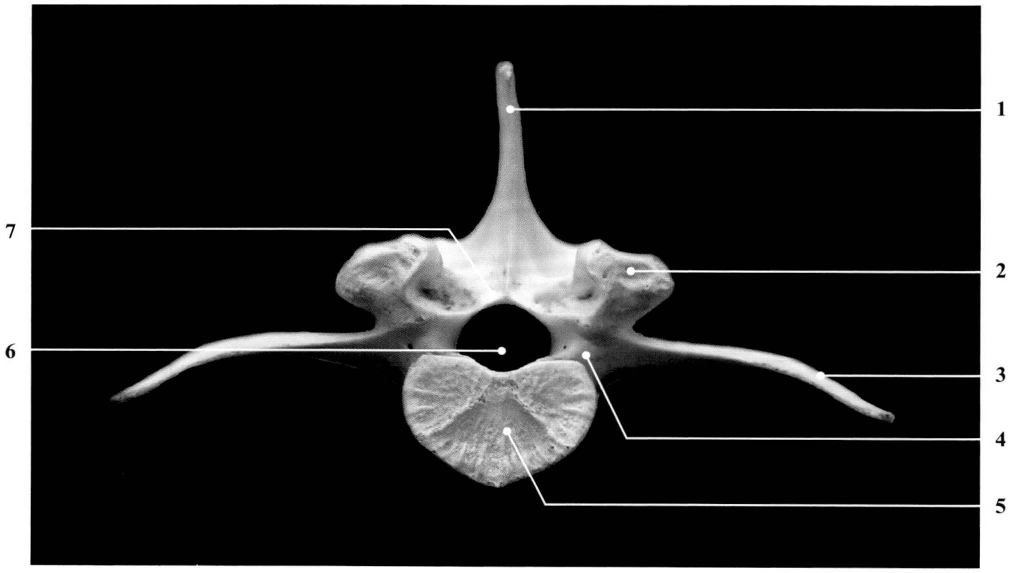

图 2-28　第 3 腰椎后面观
Figure 2-28　Caudal view of the third lumbar vertebra.

1- 棘突 spinous process
2- 后关节突 caudal articular process
3- 横突 transverse process
4- 椎后切迹 caudal vertebral notch
5- 椎窝 vertebral fossa
6- 椎孔 vertebral foramen
7- 椎弓 vertebral arch

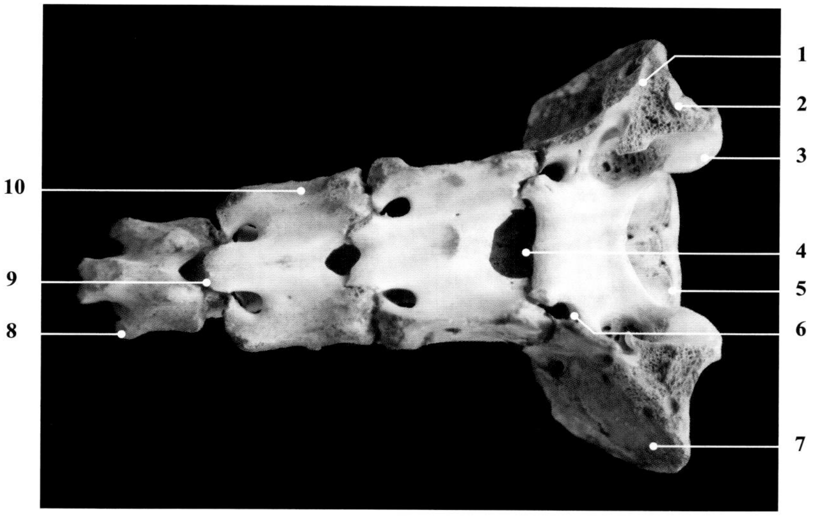

图 2-29 荐骨背侧观
Figure 2-29 Dorsal view of the sacrum.

1- 荐外侧嵴 lateral sacral crest
2- 荐骨翼 wing of sacrum
3- 前关节突 cranial articular process
4- 弓间隙 interarcuate space
5- 第 1 荐骨椎头 vertebral head of 1st sacral vertebra
6- 荐背侧孔 dorsal sacral foramen
7- 耳状面 auricular articular surface
8- 第 1 尾椎 1st coccygeal vertebra
9- 后关节突 caudal articular process
10- 侧部 lateral part

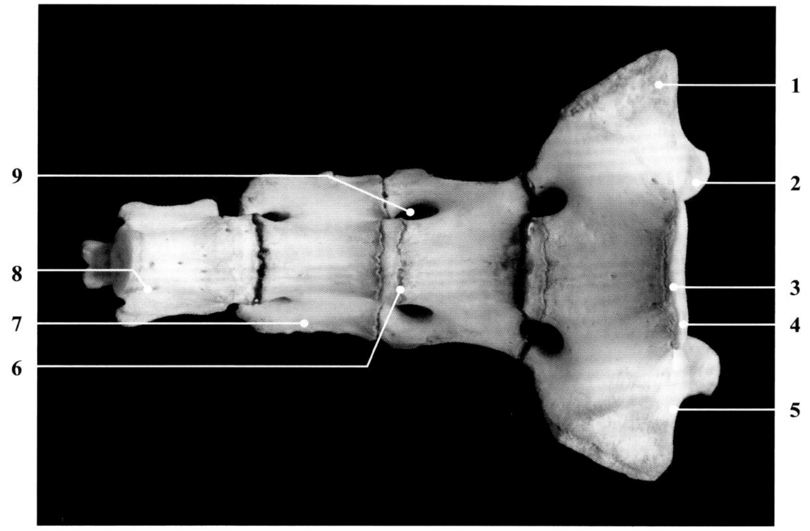

图 2-30 荐骨腹侧观
Figure 2-30 Ventral view of the sacrum.

1- 荐骨翼 wing of sacrum
2- 前关节突 cranial articular process
3- 荐骨岬 promontory of sacrum
4- 第 1 荐骨椎头 vertebral head of 1st sacral vertebra
5- 切迹 notch
6- 横线 transverse line
7- 侧部 lateral part
8- 第 1 尾椎 1st coccygeal vertebra
9- 荐腹侧孔 ventral sacral foramen

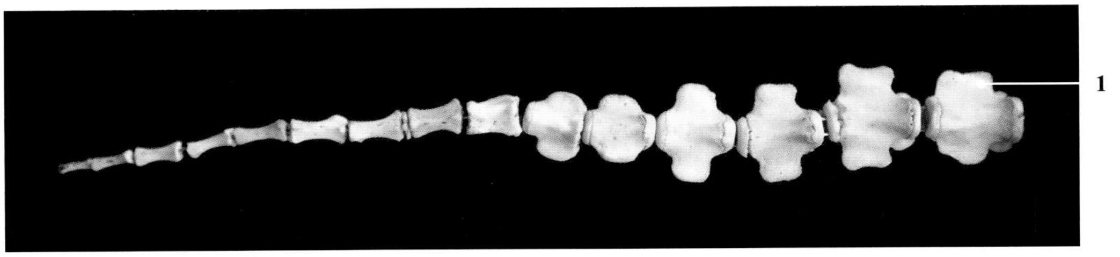

图 2-31　尾椎背侧观
Figure 2-31　Dorsal view of the coccygeal vertebrae.

1- 横突 transverse process

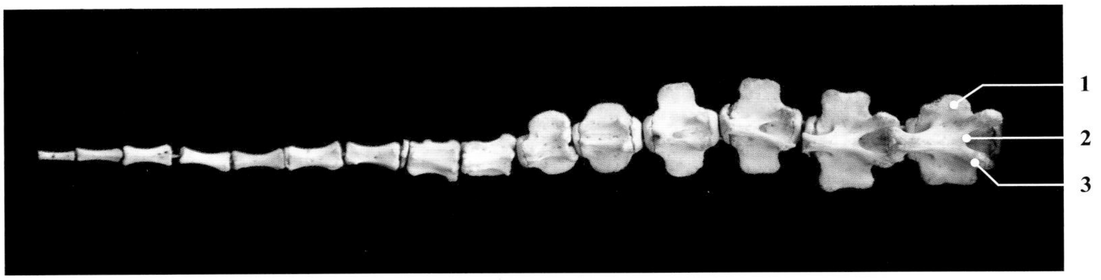

图 2-32　尾椎腹侧观
Figure 2-32　Ventral view of the coccygeal vertebrae.

1- 横突 transverse process
2- 血管弓 hemal arch
3- 血管突 hemal process

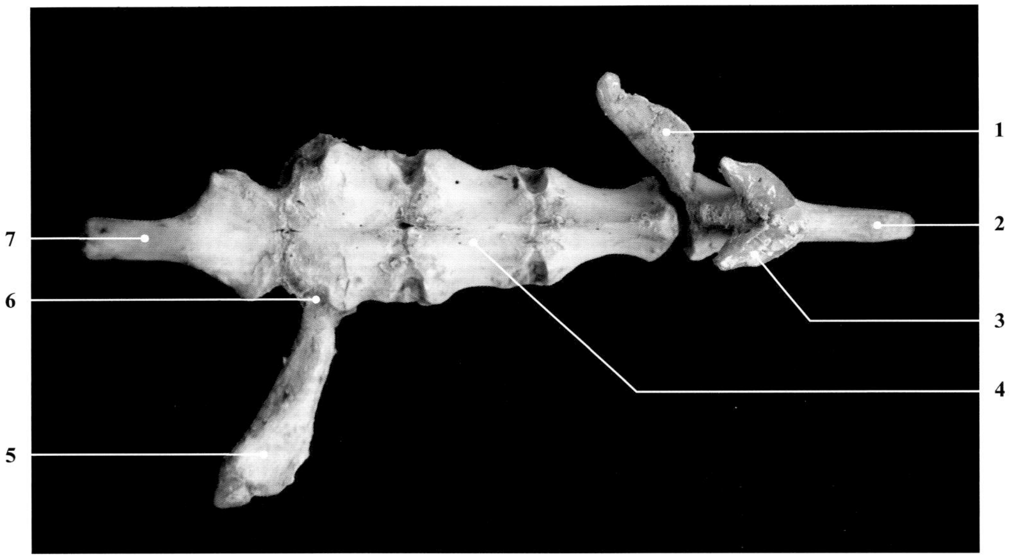

图 2-33　胸骨背侧观
Figure 2-33　Dorsal view of the sternum.

1- 第 2 肋软骨 2nd costal cartilage
2- 胸骨柄软骨 manubrian cartilage
3- 第 1 肋软骨 1st costal cartilage
4- 胸骨体 body of sternum
5- 第 5 肋软骨 5th costal cartilage
6- 胸骨软骨结合 manubriogladiolar junction
7- 剑状软骨 xiphoid cartilage

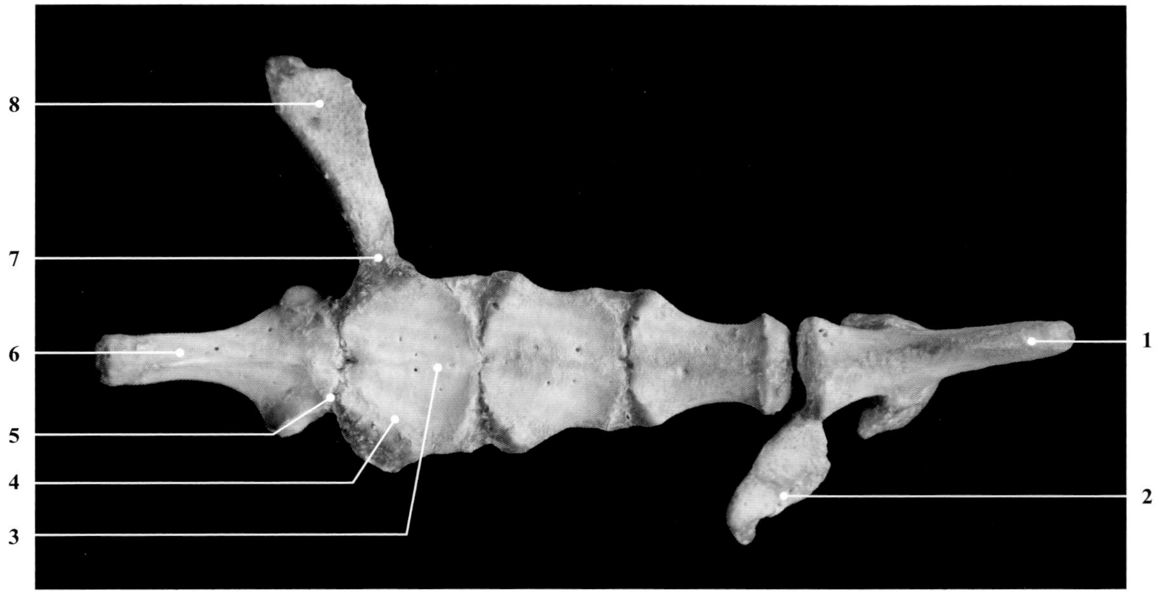

图 2-34　胸骨腹侧观
Figure 2-34　Ventral view of the sternum.

1- 胸骨柄软骨 manubrian cartilage
2- 第 2 肋软骨 2nd costal cartilage
3- 胸骨嵴 sternal crest
4- 胸骨体 body of sternum
5- 剑胸软骨结合 xiphoid sternum synchondrosis
6- 剑状软骨 xiphoid cartilage
7- 胸骨软骨结合 manubriogladiolar junction
8- 第 5 肋软骨 5th costal cartilage

图 2-35　胸骨及肋软骨

Figure 2-35　Sternum and costal cartilages.

1- 肋间隙 intercostal space
2- 第 10 肋 10th rib
3- 肩胛骨 scapula
4- 肱骨 humerus
5- 肋软骨 costal cartilage
6- 鹰嘴 olecranon
7- 桡骨 radius
8- 尺骨 ulna
9- 前臂间隙 space of forearm
10- 剑状软骨 xiphoid cartilage
11- 胸骨 sternum
12- 第 1 肋 1st rib
13- 胸椎 thoracic vertebrae

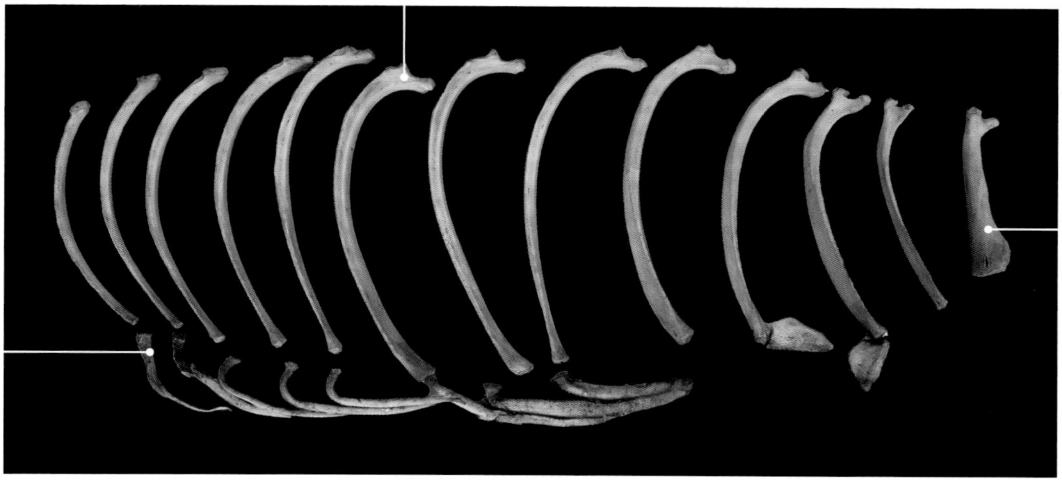

图 2-36 右侧肋
Figure 2-36 Right side of the ribs.

1- 第 1 肋骨 1th costal bone
2- 肋软骨 costal cartilage
3- 第 8 肋骨 8th costal bone

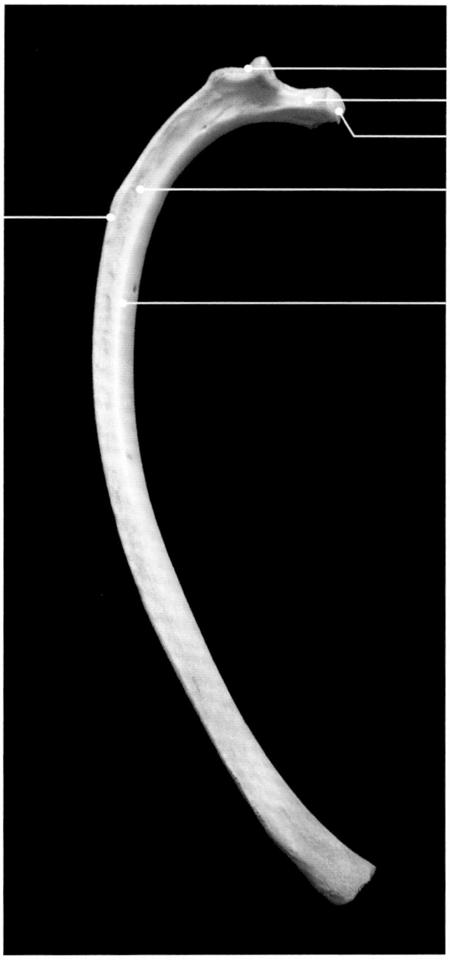

图 2-37 第 8 肋骨后面观
Figure 2-37 Caudal view of the 8th costal bone.

1- 肋（骨）结节 costal tubercle
2- 肋（骨）颈 neck of rib
3- 肋（骨）小头 head of rib
4- 肋（骨）沟 costal groove
5- 肋骨干 shaft of rib
6- 肋（骨）角 angle of rib

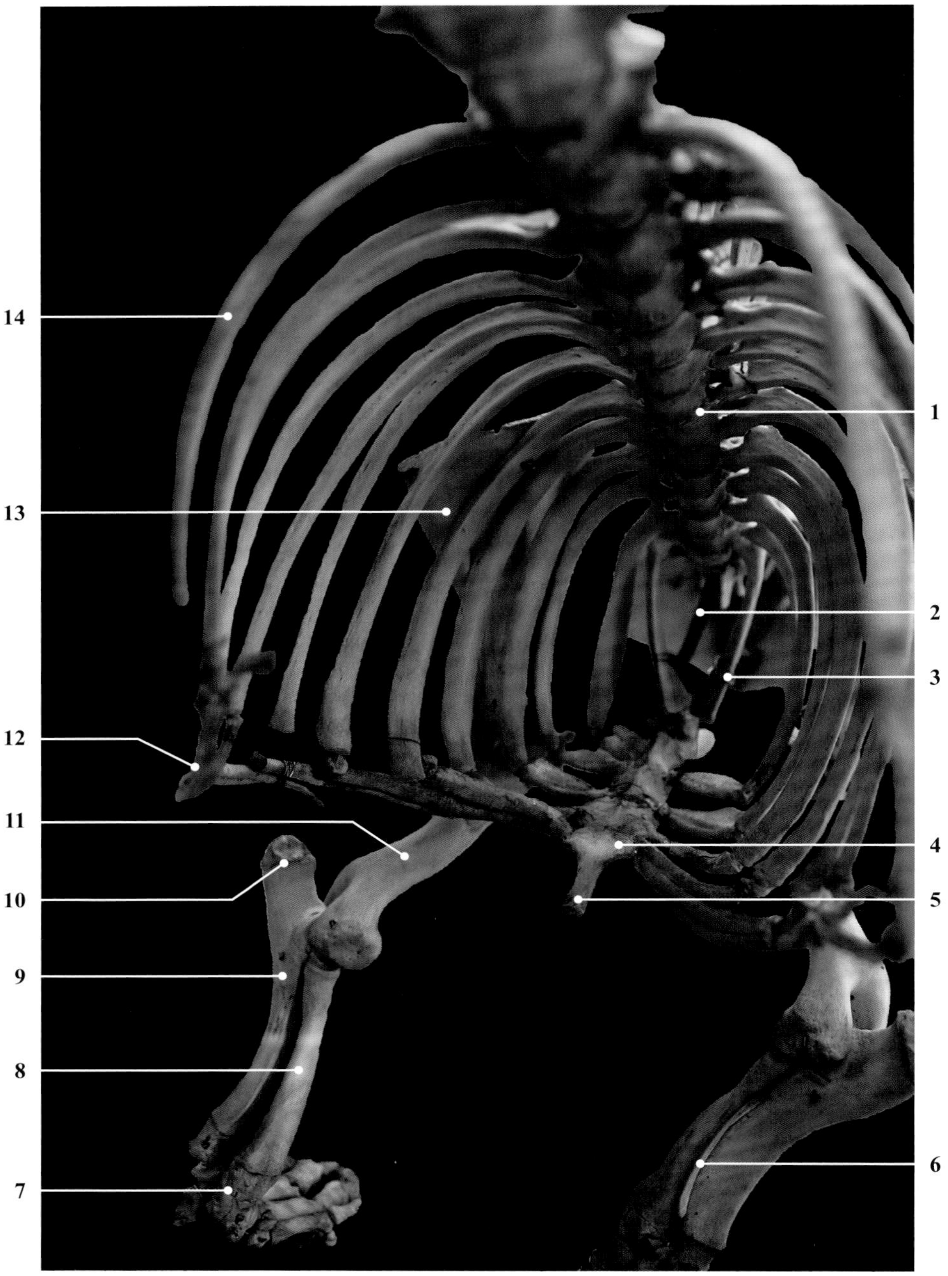

图 2-38　胸廓后口
Figure 2-38　Caudal opening of the thoracic cage.

1- 胸椎 thoracic vertebrae
2- 胸（廓）前口 cranial opening of the thoracic cage
3- 第 1 肋 1st rib
4- 胸骨 sternum
5- 剑状软骨 xiphoid cartilage
6- 前臂间隙 space of forearm
7- 腕骨 carpal bone
8- 桡骨 radius
9- 尺骨 ulna
10- 鹰嘴 olecranon
11- 肱骨 humerus
12- 肋软骨 costal cartilage
13- 肩胛骨 scapula
14- 肋骨 costal bone

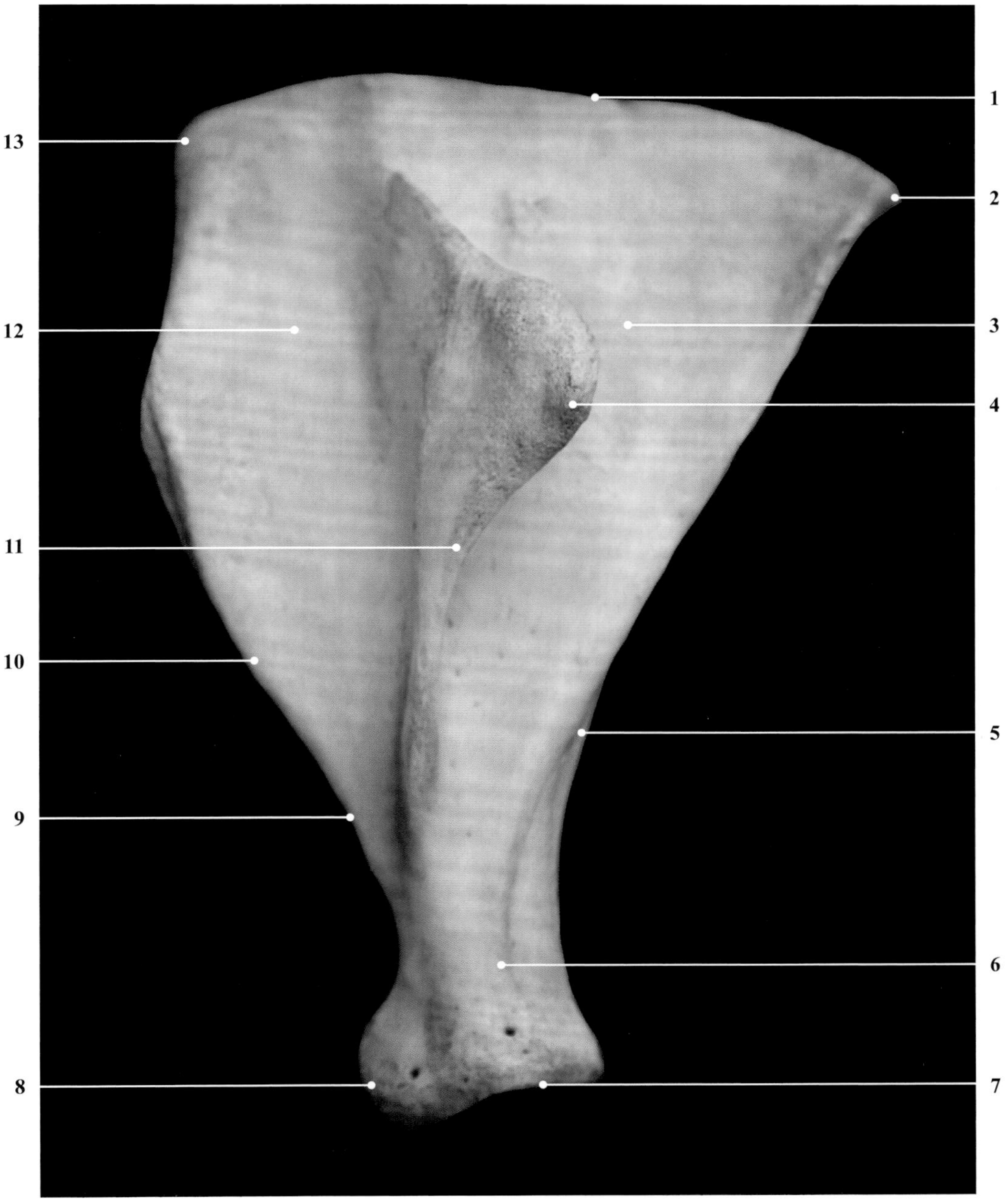

图 2-39　左侧肩胛骨外侧观
Figure 2-39　Lateral view of the left scapula.

1- 背侧缘 dorsal border
2- 后角 caudal angle
3- 冈下窝 infraspinous fossa
4- 肩胛冈结节 tuberosity of scapular spine
5- 后缘 caudal border
6- 肩胛颈 neck of scapula
7- 关节盂（肩臼）glenoid cavity (glenoid fossa)
8- 盂上结节（肩胛结节）supraglenoid tubercle (scapular tuber)
9- 肩胛切迹 scapular notch
10- 前缘 cranial border
11- 肩胛冈 spine of scapula
12- 冈上窝 supraspinous fossa
13- 前角 cranial angle

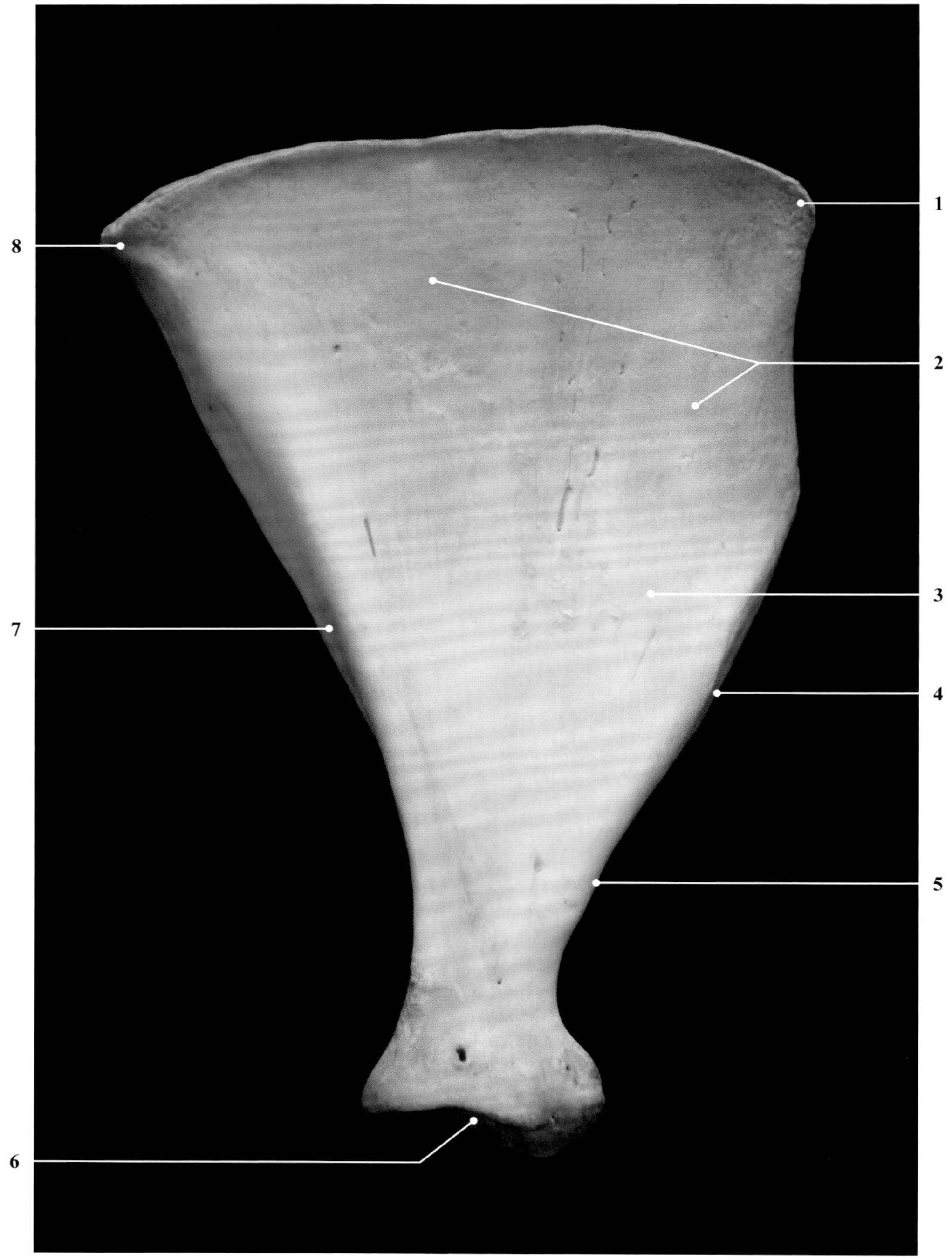

图 2-40　左侧肩胛骨内侧观
Figure 2-40　Medial view of the left scapula.

1- 前角 cranial angle
2- 锯肌面 face for serrate muscle
3- 肩胛下窝 subscapular fossa
4- 前缘 cranial border
5- 肩胛切迹 scapular notch
6- 关节盂（肩臼）glenoid cavity (glenoid fossa)
7- 后缘 caudal border
8- 后角 caudal angle

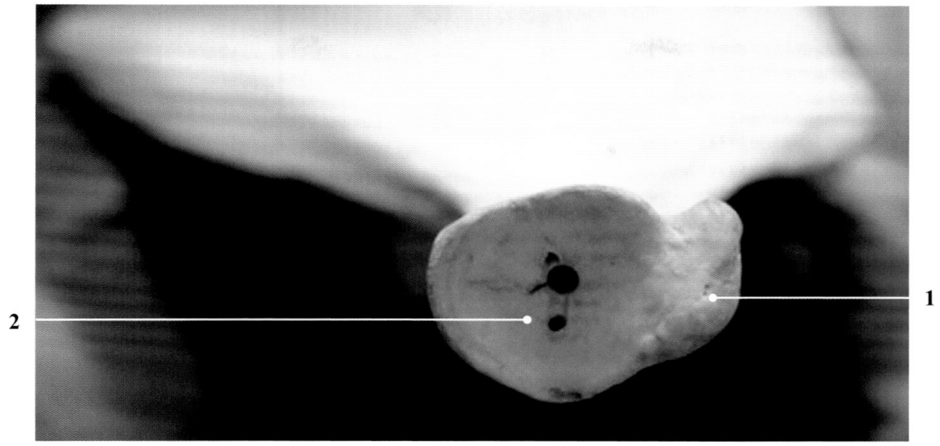

图 2-41 左侧肩胛骨关节盂
Figure 2-41 Glenoid cavity of the left scapula.

1- 盂上结节（肩胛结节）supraglenoid tubercle (scapular tuber)
2- 关节盂（肩臼）glenoid cavity (glenoid fossa)

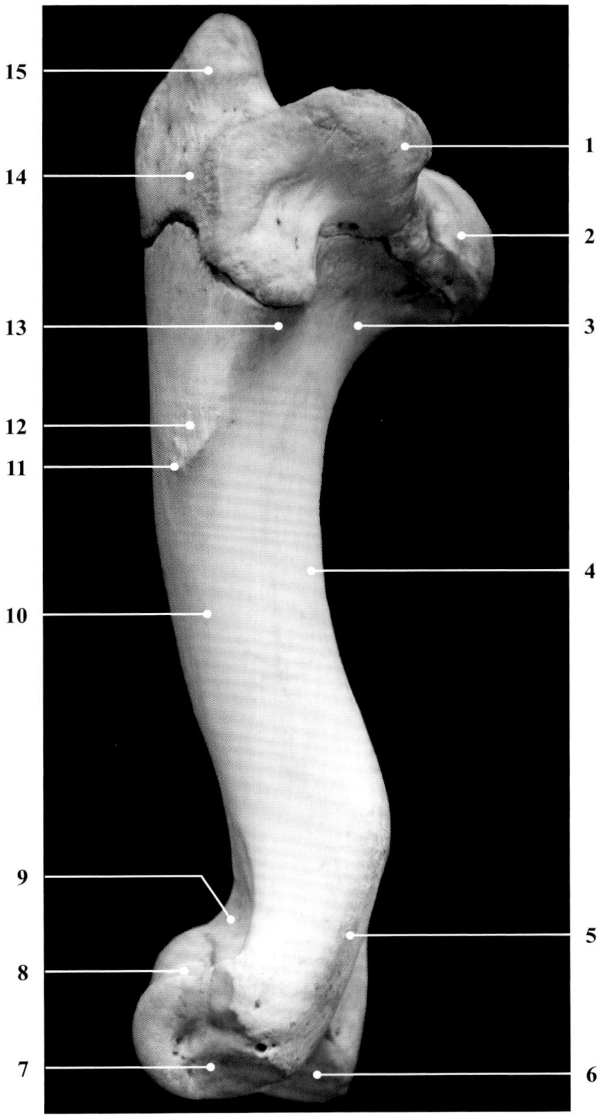

图 2-42 左侧肱骨外侧观
Figure 2-42 Lateral view of the left humerus.

1- 大结节后部 caudal part of greater tubercle
2- 肱骨头 head of humerus
3- 肱骨颈 neck of humerus
4- 肱骨干（体）shaft of humerus
5- 外侧上髁嵴 lateral supracondylar crest
6- 鹰嘴窝（肘窝）olecranon fossa (cubital fossa)
7- 韧带窝 ligament fossa
8- 肱骨髁 humeral condyle
9- 桡骨窝 radial fossa
10- 臂肌沟 bravhial muscle groove
11- 肱骨嵴 humerus crest
12- 三角肌粗隆 deltoid tuberosity
13- 三角肌线 deltoid line
14- 冈下肌止点区 area of insertion for infraspinatus muscle
15- 大结节前部 cranial part of greater tubercle

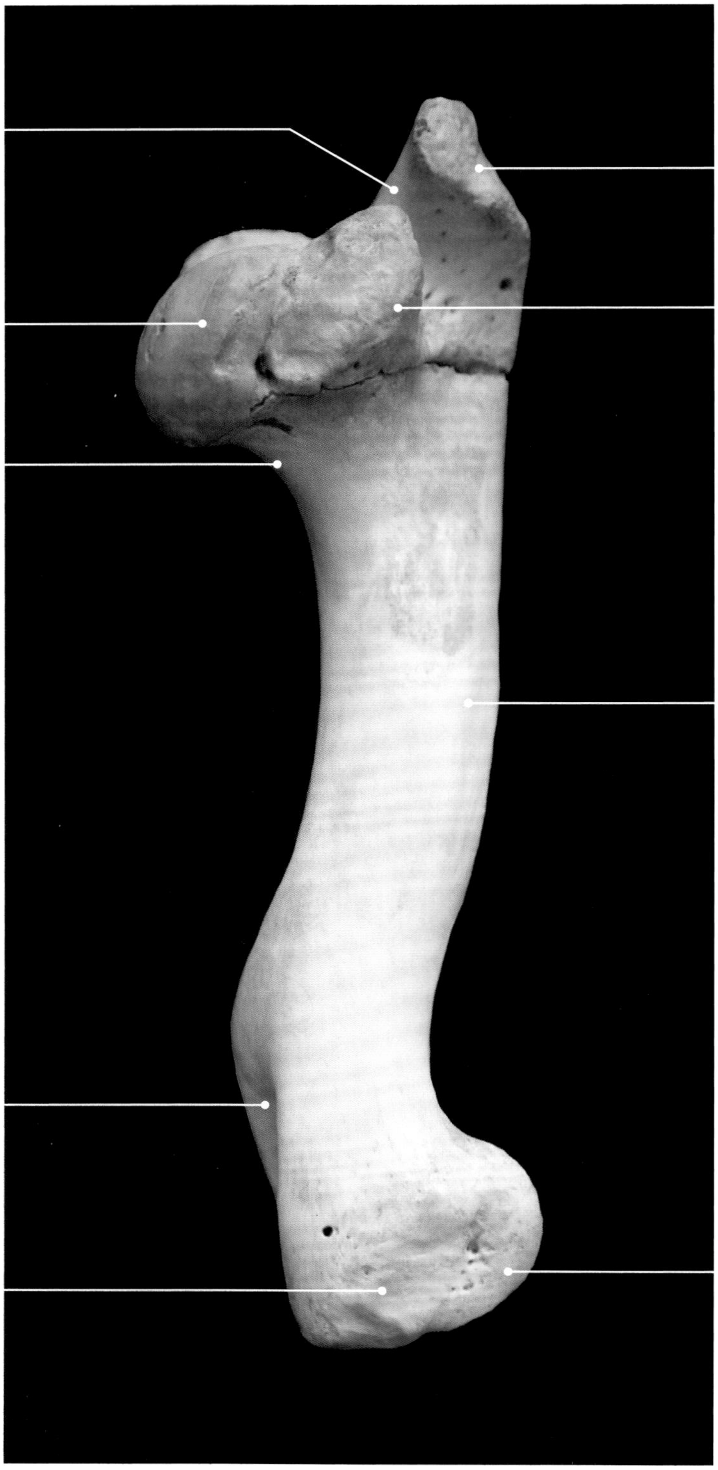

图 2-43　左侧肱骨内侧观
Figure 2-43　Medial view of the left humerus.

1- 大结节 greater tubercle
2- 小结节 less tubercle
3- 肱骨干（体）shaft of humerus
4- 肱骨髁 humeral condyle
5- 韧带窝 ligament fossa
6- 鹰嘴窝（肘窝）olecranon fossa (cubital fossa)
7- 肱骨颈 neck of humerus
8- 肱骨头 head of humerus
9- 二头肌沟（结节间沟）bicipital groove (intertuberc ular sulcus)

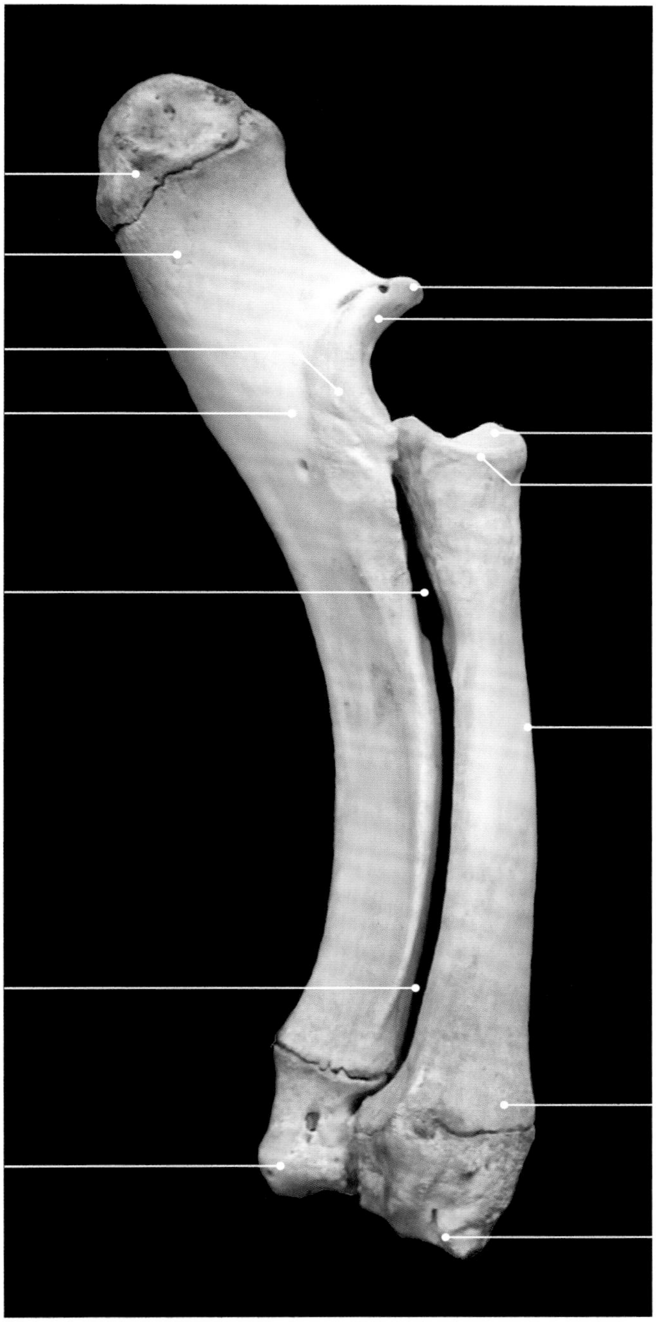

图 2-44 左侧桡骨和尺骨内侧观
Figure 2-44 Medial view of the left radius and ulna.

1- 肘突 anconeal process
2- 滑车切迹 trochlear notch
3- 桡骨凹（关节面）radial foveae (articular surface)
4- 桡骨内侧粗隆 medial eminence of radius
5- 桡骨体（干）body of radius
6- 桡骨远端（横嵴）distal extremity of radius (transverse ridge)
7- 桡骨滑车 radial trochlea
8- 尺骨茎突 styloid process of ulna
9- 远侧前臂间隙 distal space of forearm
10- 近侧前臂间隙 proximal space of forearm
11- 尺骨体（干） body of ulna
12- 内侧冠突 medial coronoid process
13- 鹰嘴 olecranon
14- 鹰嘴结节 olecranal tuber

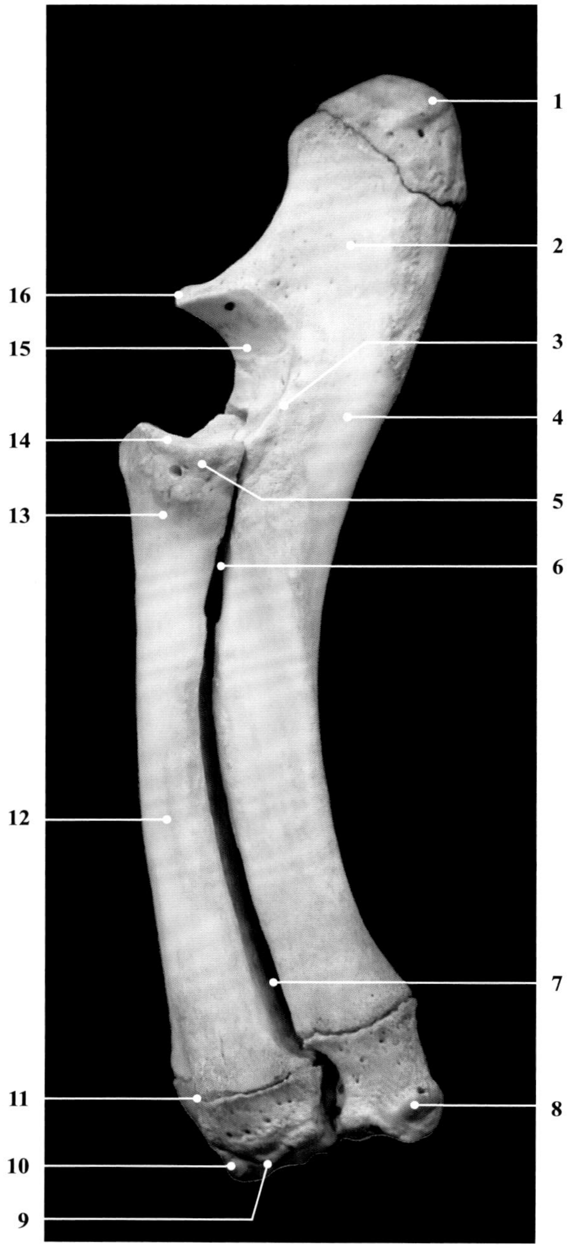

图 2-45　左侧桡骨和尺骨外侧观
Figure 2-45　Lateral view of the left radius and ulna.

1- 鹰嘴结节 olecranon tuber
2- 鹰嘴 olecranon
3- 外侧冠突 lateral coronoid process
4- 尺骨体（干）body of ulna
5- 桡骨外侧粗隆 lateral eminence of the radius
6- 近侧前臂间隙 proximal space of forearm
7- 远侧前臂间隙 distal space of forearm
8- 尺骨茎突 styloid process of ulna
9- 桡骨滑车 radial trochlea
10- 桡骨茎突 styloid process of radius
11- 桡骨远端（横嵴） distal extremity of radius (trans verse ridge)
12- 桡骨体（干） body of radius
13- 桡骨颈 neck of radius
14- 桡骨凹 radial foveae
15- 滑车切迹 trochlear notch
16- 肘突 anconeal process

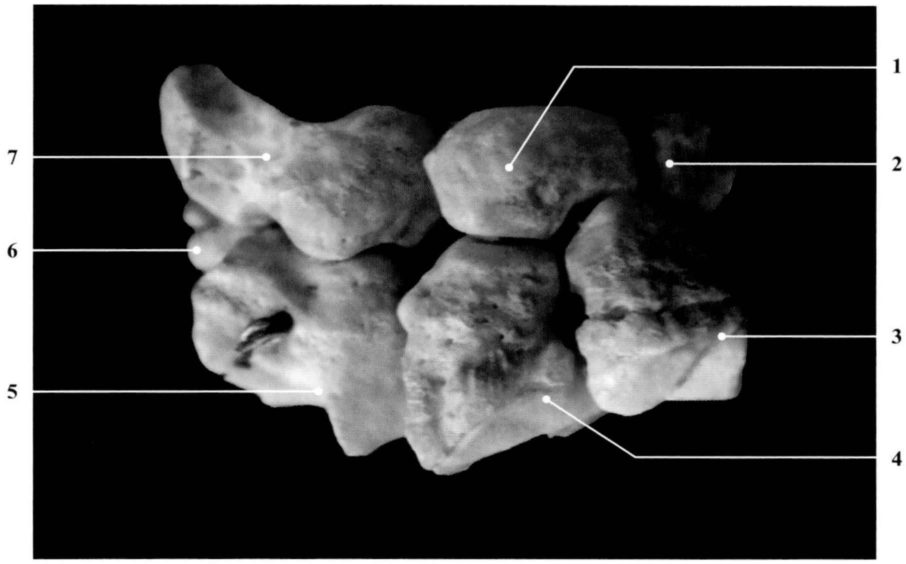

图 2-46 右侧腕骨前面观
Figure 2-46 Cranial view of the right carpal bones.

1- 中间腕骨 intermediate carpal bone
2- 桡腕骨 radial carpal bone
3- 第 2 腕骨 2nd carpal bone
4- 第 3 腕骨 3rd carpal bone
5- 第 4 腕骨 4th carpal bone
6- 副腕骨 accessory carpal bone
7- 尺腕骨 ulnar carpal bone

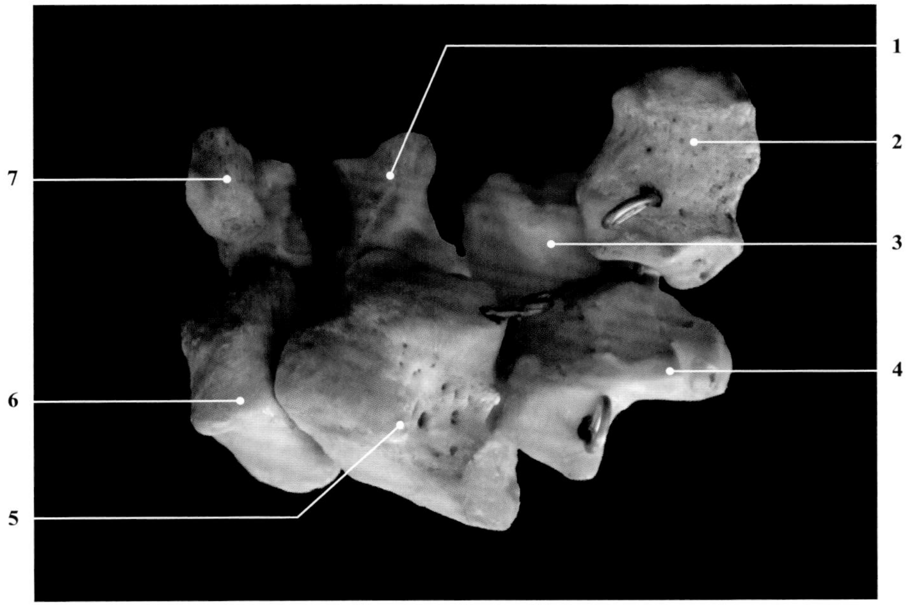

图 2-47 右侧腕骨后面观
Figure 2-47 Caudal view of the right carpal bones.

1- 中间腕骨 intermediate carpal bone
2- 副腕骨 accessory carpal bone
3- 尺腕骨 ulnar carpal bone
4- 第 4 腕骨 4th carpal bone
5- 第 3 腕骨 3rd carpal bone
6- 第 2 腕骨 2nd carpal bone
7- 桡腕骨 radial carpal bone

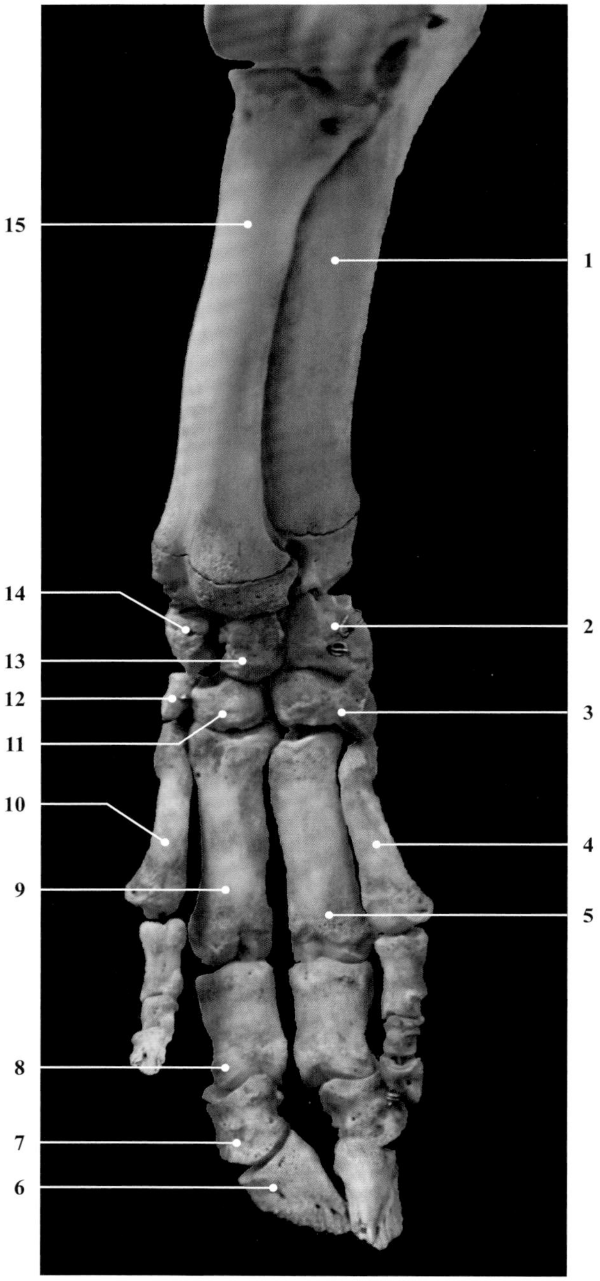

图 2-48　左前脚背侧观
Figure 2-48　Dorsal view of the left front foot.

1- 尺骨 ulna
2- 尺腕骨 ulnar carpal bone
3- 第 4 腕骨 4th carpal bone
4- 第 5 掌骨 5th metacarpal bone
5- 第 4 掌骨 4th metacarpal bone
6- 第 3 指远指节骨（蹄骨）3rd distal phalanx (coffin bone)
7- 第 3 指中指节骨（冠骨）3rd middle phalanx (os coronale)
8- 第 3 指近指节骨（系骨）3rd proximal phalanx (os compedale)
9- 第 3 掌骨 3rd metacarpal bone
10- 第 2 掌骨 2nd metacarpal bone
11- 第 3 腕骨 3rd carpal bone
12- 第 2 腕骨 2nd carpal bone
13- 中间腕骨 intermediate carpal bone
14- 桡腕骨 radial carpal bone
15- 桡骨 radius

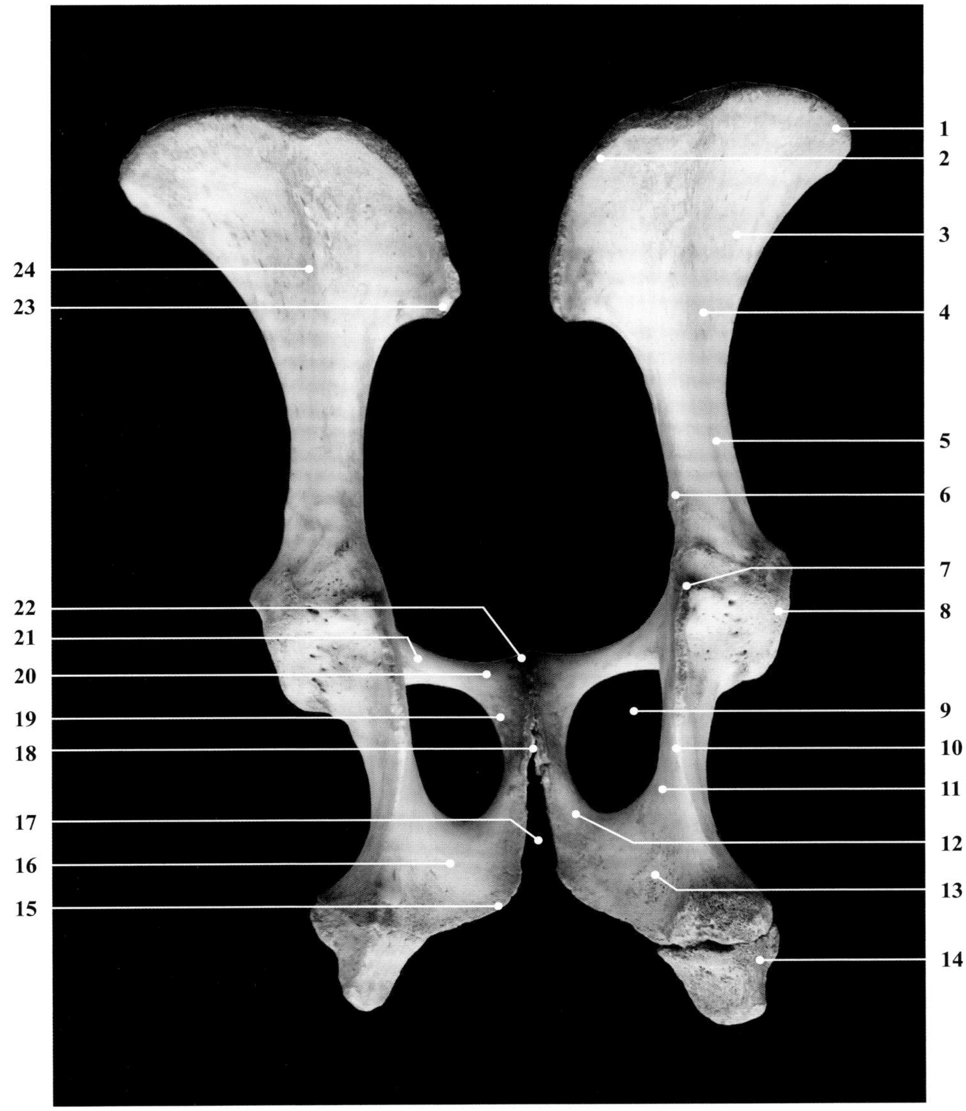

图 2-49 髋骨背侧观
Figure 2-49 Dorsal view of the hip bones.

1- 髋结节 coxal tuberosity
2- 髂嵴 iliac crest
3- 髂骨翼 wing of ilium
4- 臀线 gluteal line
5- 髂骨体 body of ilium
6- 坐骨大切迹 greater sciatic notch
7- 坐骨棘 ischial spine
8- 髋臼 acetabulum
9- 闭孔 obturator foramen
10- 坐骨小切迹 lesser sciatic notch
11- 坐骨体 body of ischium
12- 坐骨支 ramus of ischium
13- 坐骨板 plate of ischium
14- 坐骨结节 ischial tuberosity
15- 坐骨弓 ischial arch
16- 坐骨 ischium
17- 坐骨联合 ischiatic symphysis
18- 耻骨联合 pubic symphysis
19- 耻骨后支 caudal branch of pubis
20- 耻骨前支 cranial branch of pubis
21- 耻骨体 body of pubis
22- 耻骨梳 pecten of pubis
23- 荐结节 sacral tuber
24- 髂骨 ilium

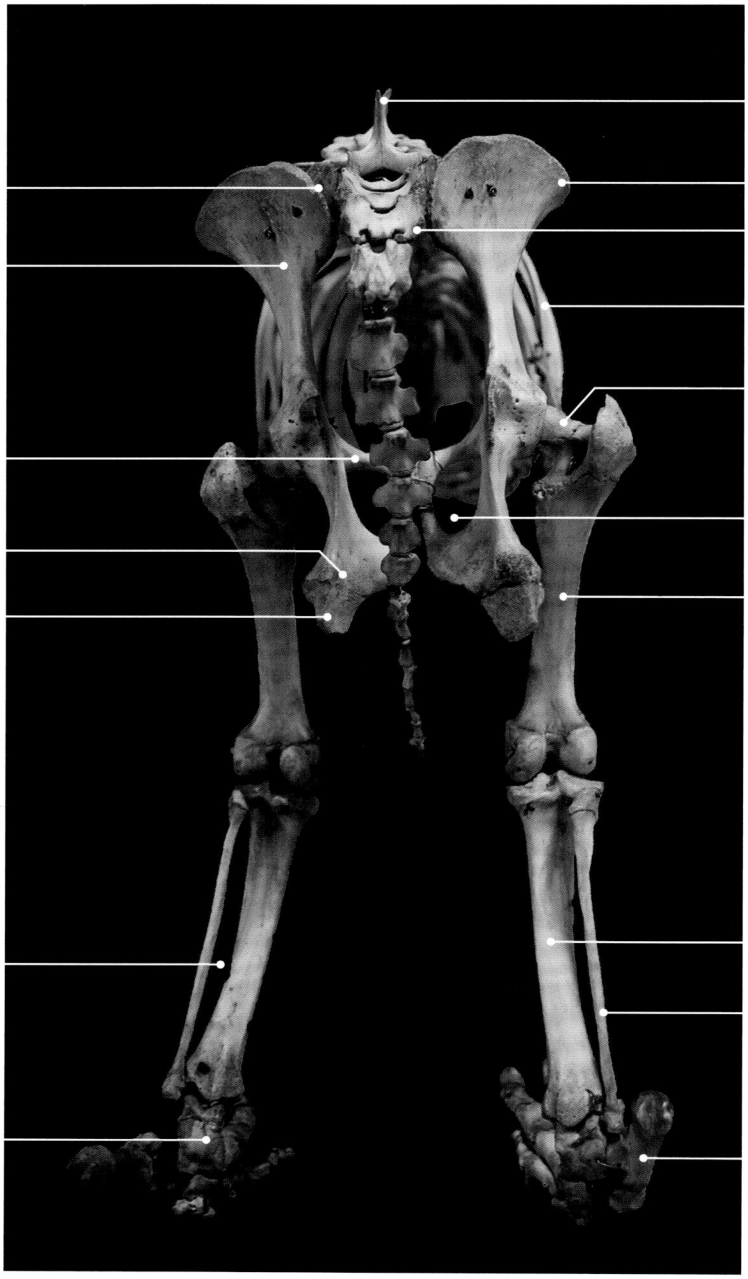

图 2-50　骨盆后口

Figure 2-50　Caudal opening of the pelvis.

1- 荐骨 sacurm

2- 髋结节 coxal tuberosity

3- 尾骨 coccygeal bone

4- 肋 rib

5- 股骨头 head of the femur

6- 闭孔 obturator foramen

7- 股骨 femoral bone, femur

8- 胫骨 tibia

9- 腓骨 fibula

10- 跟骨 calcaneus

11- 跗骨 tarsal bone

12- 骨间隙 interosseous space

13- 坐骨结节 ischial tuberosity

14- 坐骨 ischium

15- 耻骨 pubis

16- 髂骨 ilium

17- 荐结节 sacral tuber

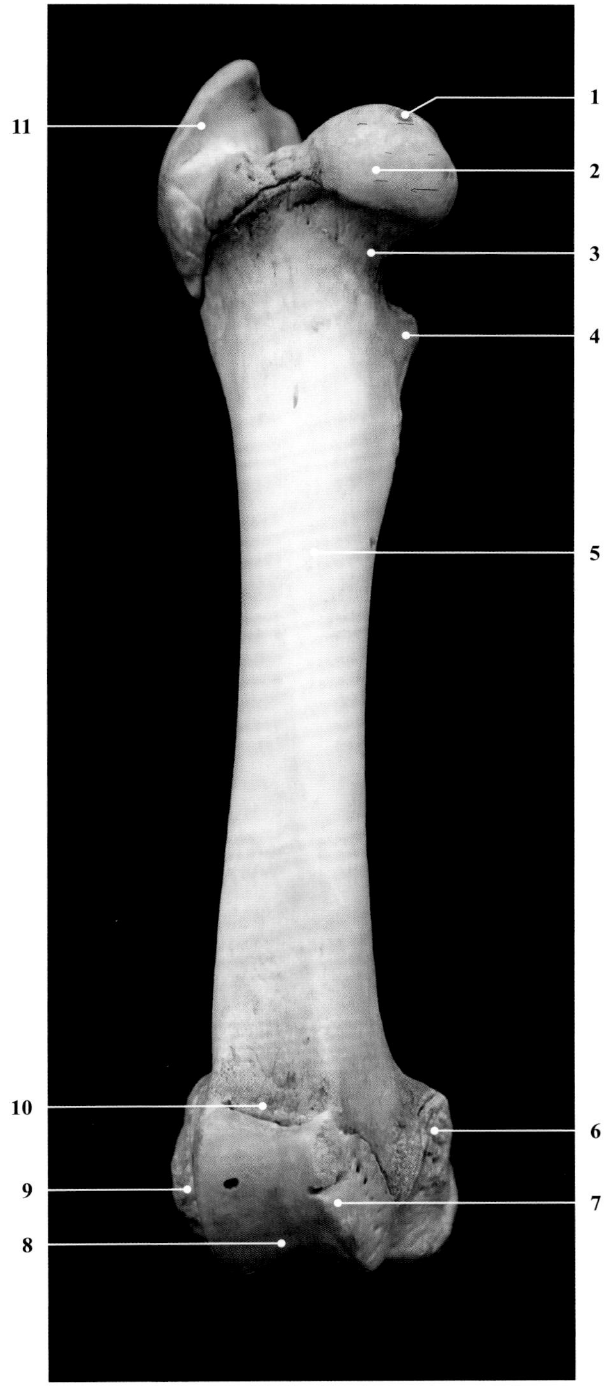

图 2-51 右侧股骨前面观
Figure 2-51 Cranial view of the right femur.

1- 股骨头窝 fovea of femoral head
2- 股骨头 head of femur
3- 股骨颈 neck of femur
4- 小转子 lesser trochanter
5- 股骨体（干）shaft of femur
6- 内侧上髁 medial epicondyle
7- 股骨滑车粗隆 trochlea tuberosity of femur
8- 股骨滑车 trochlea of femur
9- 外侧上髁 lateral epicondyle
10- 膝上窝 suprapatellar fossa
11- 大转子 greater trochanter

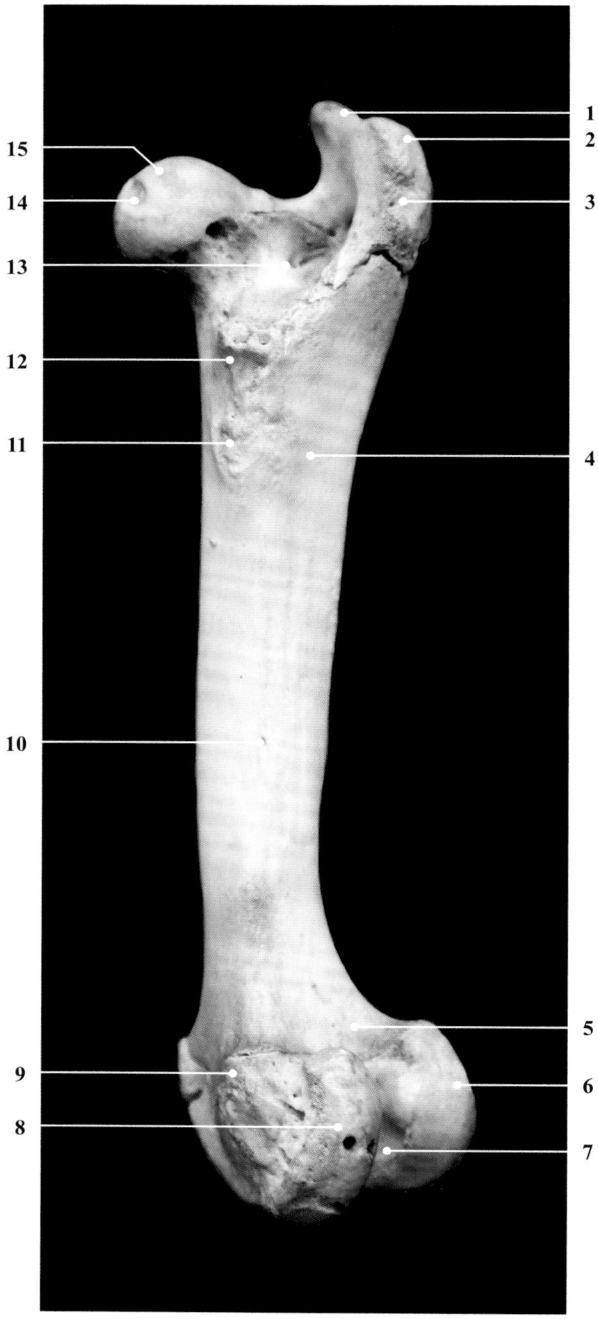

图 2-52　右侧股骨内侧观

Figure 2-52　Medial view of the right femur.

1- 大转子后部 caudal part of greater trochanter
2- 大转子前部 cranial part of greater trochanter
3- 大转子 greater trochanter
4- 股骨体（干）shaft of femur
5- 腘肌面 popliteal surface
6- 外侧髁 lateral condyle
7- 髁间窝 intercondylar fossa
8- 内侧髁 medial condyle
9- 内侧上髁 medial epicondyle
10- 滋养孔 nutrient foramen
11- 二头肌粗隆 bicipital tuberosity
12- 小转子 lesser trochanter
13- 转子窝 trochanteric fossa
14- 股骨头窝 fovea of femoral head
15- 股骨头 head of femur

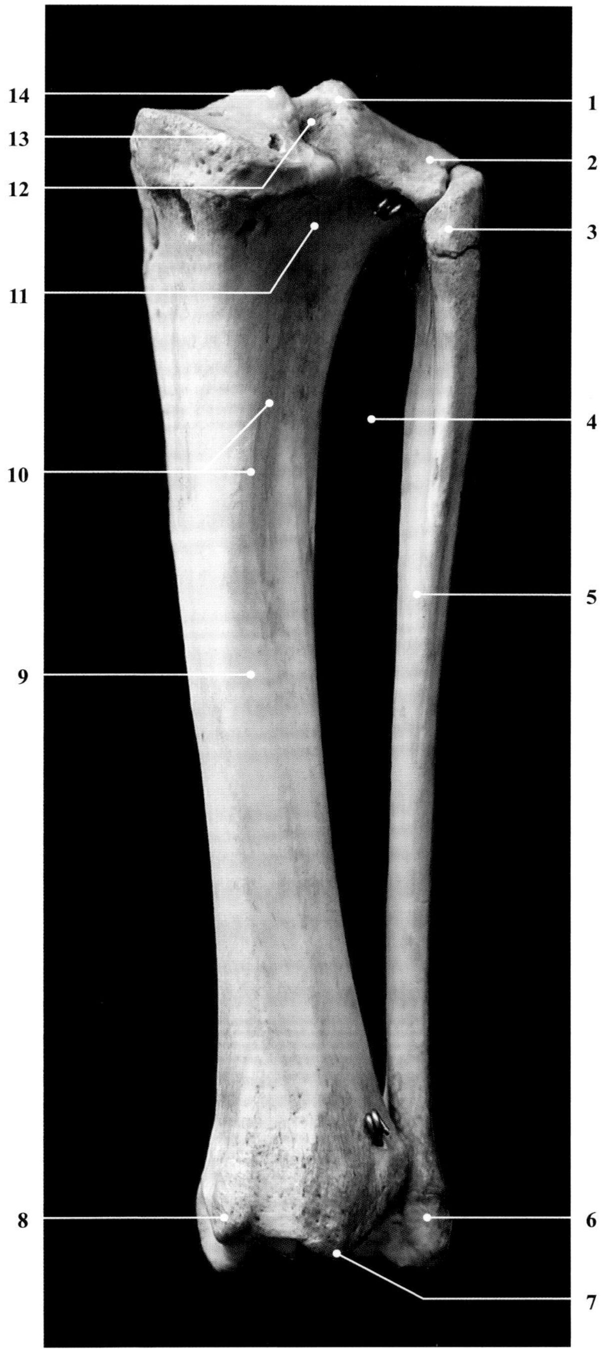

图 2-53 右侧胫骨和腓骨后面观
Figure 2-53 Caudal view of the right tibia and fibula.

1- 外侧髁间结节 lateral intercondyloid tubercle
2- 胫骨外髁 lateral condyle of tibia
3- 腓骨头 head of fibula
4- 骨间隙 interosseous space
5- 腓骨 fibula
6- 外侧踝 lateral malleolus
7- 滑车 trochlea
8- 内侧踝 medial malleolus
9- 胫骨 tibia
10- 肌线 muscular line
11- 腘切迹 popliteal notch
12- 髁间区 intercondyloid area
13- 胫骨内髁 medial condyle of tibia
14- 内侧髁间结节 medial intercondyloid tubercle

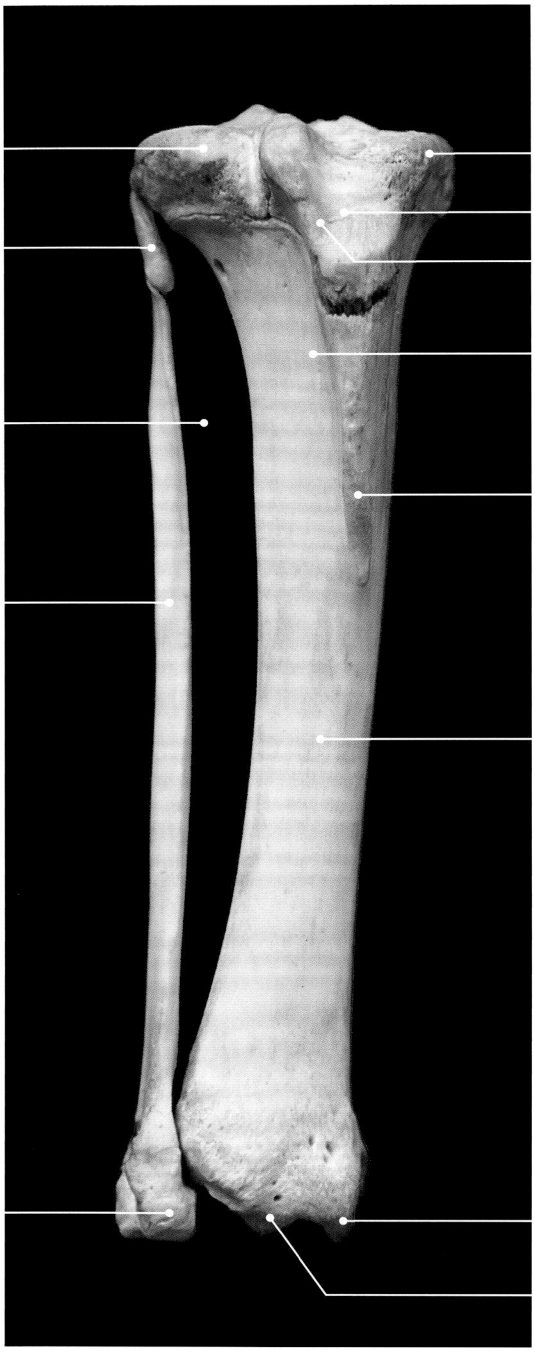

图 2-54　右侧胫骨和腓骨前面观
Figure 2-54　Cranial view of the right tibia and fibula.

1- 胫骨内髁 medial condyle of tibia
2- 粗隆沟 tuberositial groove
3- 胫骨粗隆 tibial tuberosity
4- 伸肌沟 extensor groove
5- 胫骨嵴 tibial crest
6- 胫骨 tibia
7- 内侧踝 medial malleolus
8- 滑车 trochlea
9- 外侧踝 lateral malleolus
10- 腓骨 fibula
11- 骨间隙 interosseous space
12- 腓骨头 head of fibula
13- 胫骨外髁 lateral condyle of tibia

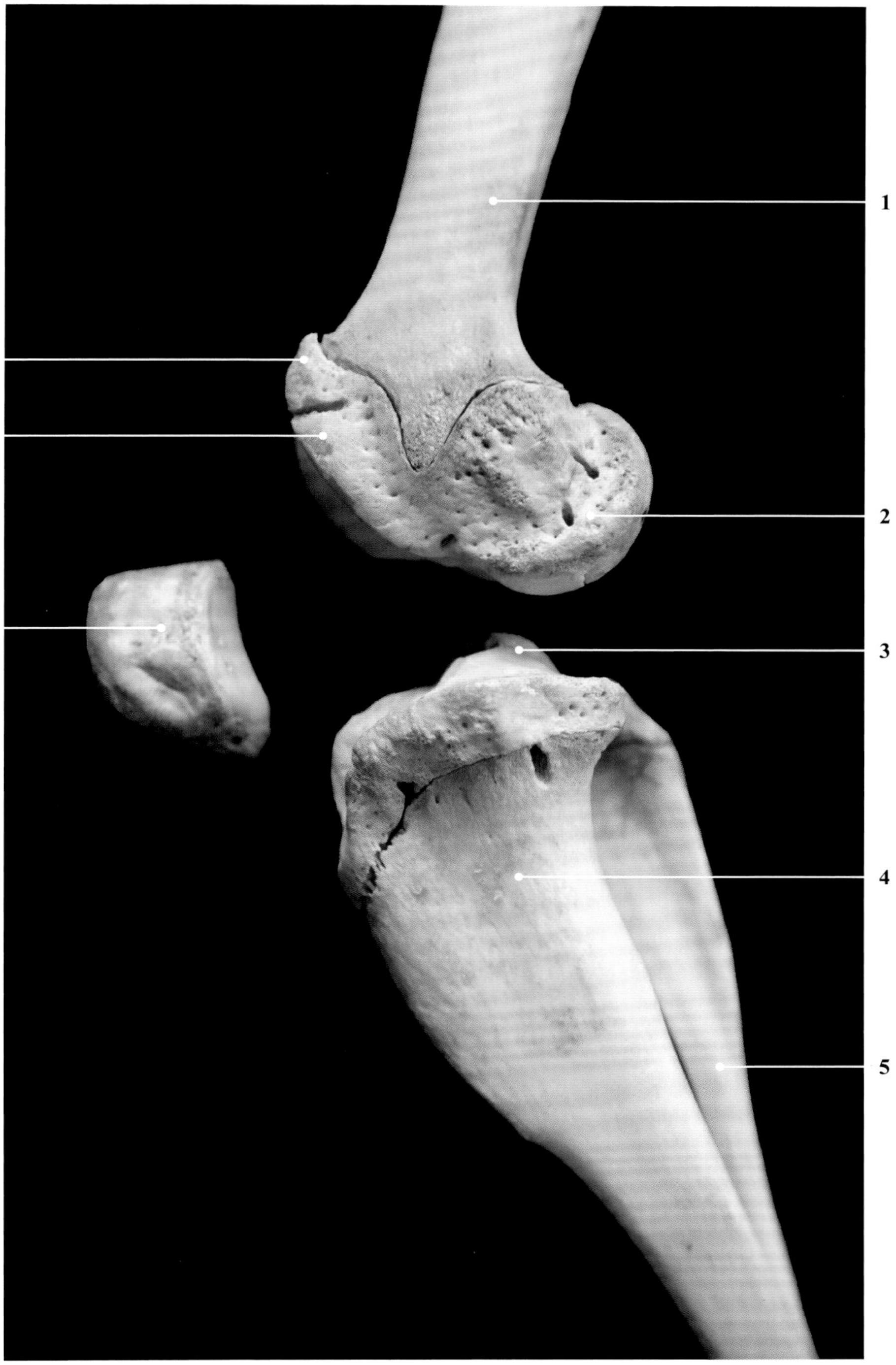

图 2-55 右侧股骨远端、膝盖骨和胫骨近端内侧观

Figure 2-55 Medial view of the distal extremity of the right femur, patella and proximal extremity of the tibia.

1- 股骨 femoral bone, femur
2- 股骨内侧髁 medial condyle of femur
3- 髁间隆起 intercondylar eminence
4- 胫骨 tibia
5- 腓骨 fibula
6- 膝盖骨（髌骨）kneecap (patella)
7- 股骨滑车 trochlea of femur
8- 滑车结节 trochlear tubercle

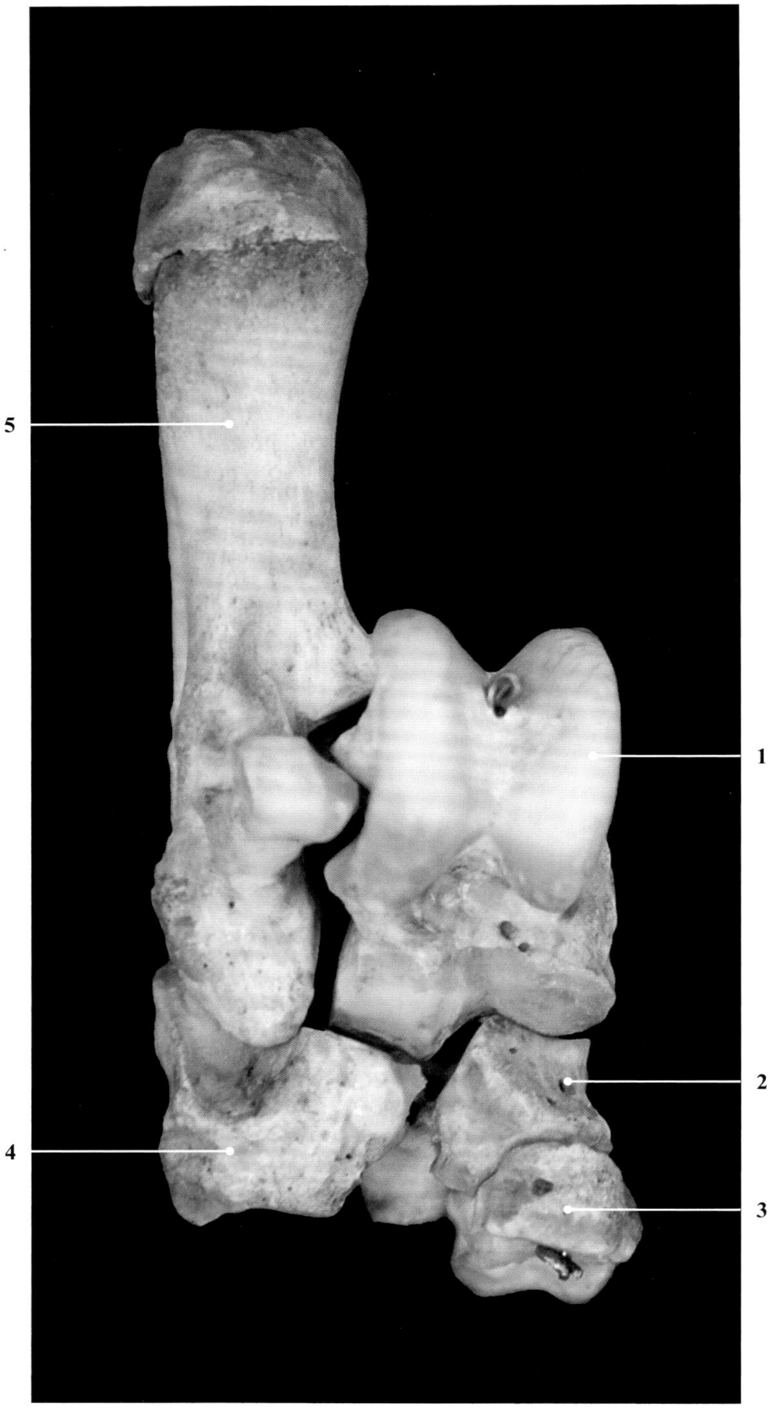

图 2-56　右侧跗骨前面观
Figure 2-56　Cranial view of the right tarsal bone.

1- 距骨 talus
2- 中央跗骨 central tarsal bone
3- 第 3 跗骨 3rd tarsal bone
4- 第 4 跗骨 4th tarsal bone
5- 跟骨 calcaneus

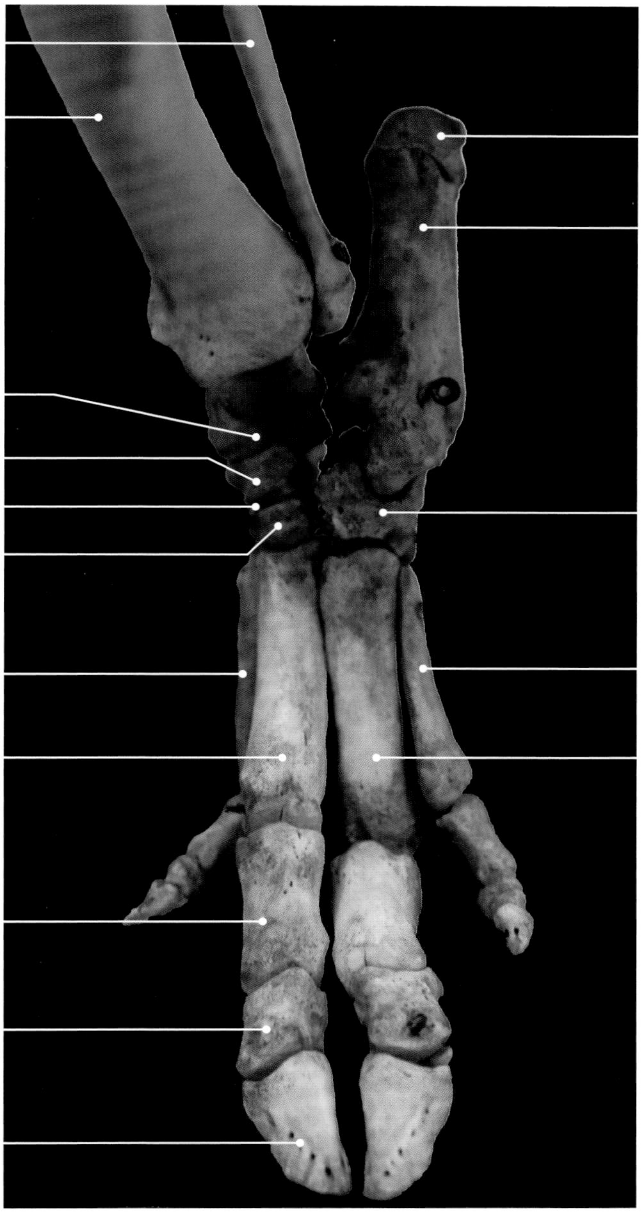

图 2-57　左后脚跖背侧观
Figure 2-57　Dorsometatarsal view of the left posterior limb.

1- 跟结节 calcaneal tuberosity
2- 跟骨 calcaneus
3- 第 4 跗骨 4th tarsal bone
4- 第 5 跖骨 5th metatarsal bone
5- 第 4 跖骨 4th metatarsal bone
6- 第 3 趾远趾节骨（蹄骨）3rd distal phalanx (coffin bone)
7- 第 3 趾中趾节骨（冠骨）3rd middle phalanx (os coronale)
8- 第 3 趾近趾节骨（系骨）3rd proximal phalanx (os compedale)
9- 第 3 跖骨 3rd metatarsal bone
10- 第 2 跖骨 2nd metatarsal bone
11- 第 3 跗骨 3rd tarsal bone
12- 第 2 跗骨 2nd tarsal bone
13- 中央跗骨 central tarsal bone
14- 距骨 talus
15- 胫骨 tibia
16- 腓骨 fibula

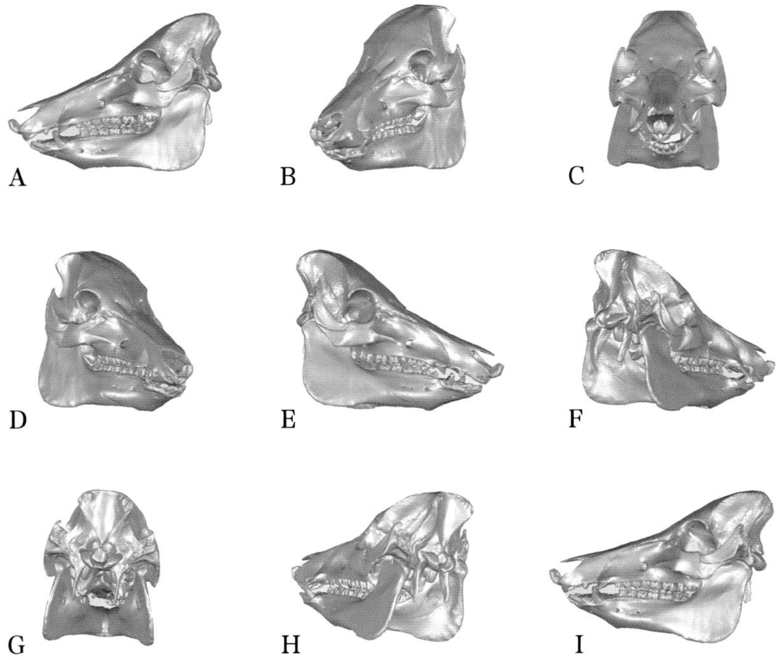

图 2-58　头骨三维水平旋转图像。A-I 为每隔 45° 水平旋转一次的图像
Figure 2-58　Three-dimensional horizontal rotating images of the skull. The images marked as A-I were horizontal rotated at 45° intervals.

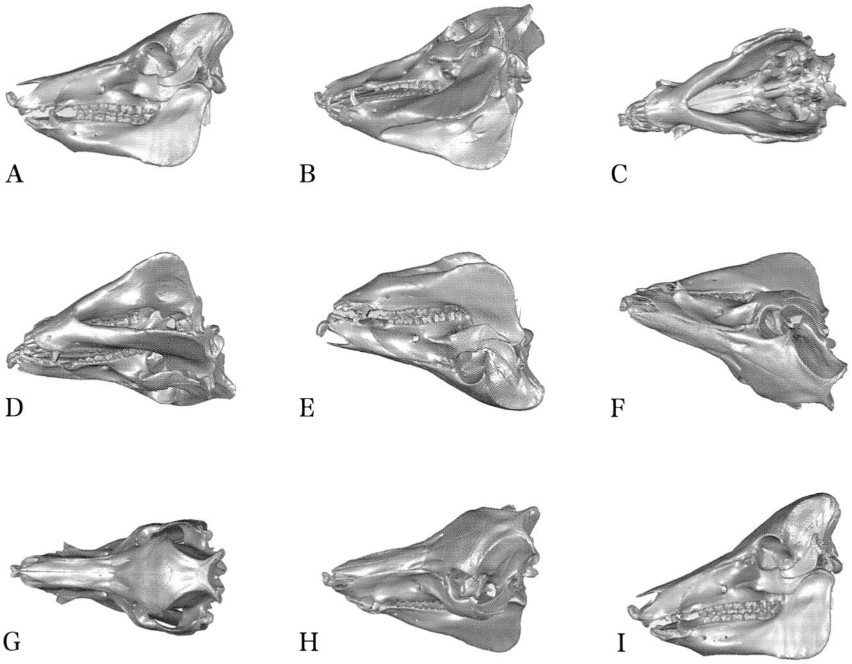

图 2-59　头骨三维垂直旋转图像。A-I 为每隔 45° 垂直旋转一次的图像
Figure 2-59　Three-dimensional vertical rotating images of the skull. The images marked as A-I were vertical rotated at 45° intervals.

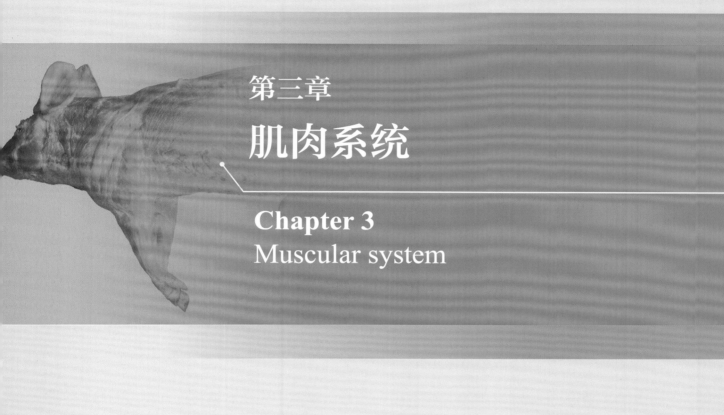

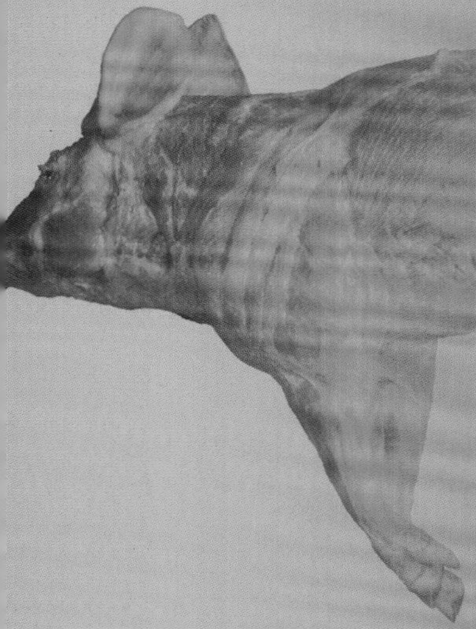

第三章

肌肉系统

Chapter 3
Muscular system

运动系统（locomotor system）中的肌肉由横纹肌构成，附着于骨骼上，又称骨骼肌（skeletal muscle），是运动的动力部分。猪的全身肌肉分为皮肌、头部肌、躯干肌、前肢肌和后肢肌。

一、肌肉的构造

每一块肌肉就是一个肌器官，主要由骨骼肌纤维（肌细胞）构成，还有结缔组织、血管和神经。肌器官可分为能收缩的肌腹和不能收缩的肌腱两部分。

1. 肌腹（muscle belly）：是肌器官的主要部分，位于肌器官的中间，由无数骨骼肌纤维借结缔组织结合而成，具有收缩功能。肌纤维为肌器官的实质部分，在肌肉内部先集合成肌束，肌束再集合成一块肌肉。肌肉的结缔组织形成肌膜，构成肌器官的间质部分。包裹在每一条肌纤维外面的肌膜为肌内膜（endomysium）。若干肌纤维组成肌束，肌束外面包裹有肌束膜（perimysium）。肌外膜（epimysium）包裹在整块肌肉外面。肌膜是肌肉的支持组织，使肌肉具有一定的形状。血管、淋巴管和神经随着肌膜进入肌肉内，对肌肉的代谢和功能调节有重要意义。

骨骼肌属横纹肌，其肌纤维为长圆柱形的多核细胞，胞核位于细胞周围近基膜处，呈扁椭圆形。肌浆内还有许多与细胞长轴平行排列的肌原纤维，肌原纤维由粗、细2种肌丝有规律的平行排列组成，因此纵切的肌纤维呈明暗相间的横纹（cross striation）。横纹由明带（I带）和暗带（A带）组成。在明带中央有一条暗线，为Z线。两条相邻Z线之间的一段肌原纤维为肌节（sarcomere）。每个肌节包括1/2 I带+A带+1/2 I带，它是骨骼肌收缩的基本结构单位。

2. 肌腱（muscle tendon）：位于肌腹的两端或一端，由规则的致密结缔组织构成。在四肢多呈条索状；在躯干多呈薄板状，又称腱膜。腱纤维借肌内膜直接连接肌纤维的两端或贯穿于肌腹中。腱不能收缩，但有很强的韧性和抗张力，不易疲劳。它传导肌腹的收缩力，以提高肌腹的工作效力。其纤维伸入骨膜和骨质中，使肌肉牢固附着于骨上。

3. 肌肉的辅助器官：包括筋膜、黏液囊和腱鞘等，其作用是保护和辅助肌肉的工作。

（1）筋膜（fascia）被覆在肌肉表面的结缔组织膜，分浅筋膜和深筋膜。浅筋膜（superficial fascia）位于皮下（也称皮下组织），由疏松结缔组织构成，覆盖在全身肌的表面。深筋膜（deep fascia）由致密结缔组织构成，位于浅筋膜下方。在某些部位深筋膜形成包围肌群的筋膜鞘；或伸入肌间，附着于骨上，形成肌间隔；或提供肌肉的附着面。

（2）黏液囊（mucous bursa）位于骨的突起与肌肉、腱和皮肤之间，是密闭的结缔组织囊，囊壁内衬有滑膜，腔内有滑液，起减少摩擦的作用。位于关节附近的黏液囊多与关节腔相通，为滑膜囊（synovial bursa）。

（3）腱鞘（tendon sheath）由黏液囊包裹于腱外而成，呈筒状包裹于腱的周围，位于腱通过活动范围较大的关节处，可减少腱活动时的摩擦。

二、皮肌

皮肌（cutaneous muscle）是分布于浅筋膜内的薄板状肌，与皮肤深层紧密相连，有面皮肌（cutaneous muscle of face）、颈皮肌（cutaneous muscle of neck）、肩臂皮肌（cutaneous omobrachial muscle）和躯干皮肌（cutaneous muscle of trunk）。

三、头部肌

头部肌分为面部肌、咀嚼肌和舌骨肌。面部肌位于口和鼻腔周围，主要有鼻唇提肌（nasolabial levator muscle）、上唇提肌（levator muscle of the upper lip）、犬齿肌（canine muscle）、下唇降肌（depressor muscle of the lower lip）、口轮匝肌（orbicular muscle of the mouth）和颊肌（buccinator muscle）。咀嚼肌包括闭口肌〔咬肌（masseter muscle）、翼肌（pterygoideus）和颞肌（temporal muscle）〕和开口肌〔枕颌肌（occipitomandibular muscle）和二腹肌（digastric muscle）〕。舌骨肌主要包括下颌舌骨肌（mylohyoid muscle）和茎舌骨肌（stylohyoid muscle）。

四、躯干肌

躯干肌包括脊柱肌、颈腹侧肌、胸廓肌和腹壁肌。

1. 脊柱肌：支配脊柱活动的肌肉，分为脊柱背侧肌群和脊柱腹侧肌群。脊柱背侧肌群很发达，位于脊柱的背外侧，包括背腰最长肌（dorsal-lumbus longest muscle）、髂肋肌（iliocostal muscle）、夹肌（splenius muscle）、头半棘肌（semispinal muscle of the head）〔又称复肌、颈多裂肌（cervical multifidus muscle）〕。脊柱腹侧肌群不发达，仅位于颈部和腰部脊柱椎体的腹侧，包括颈部的斜角肌（scalene muscle）、头长肌（long muscle of head）和颈长肌（long muscle of neck），腰部的腰大肌（major psoas muscle ）、腰方肌（lumbar quadrate muscle）和腰小肌（minor psoas muscle）。

2. 颈腹侧肌：包括胸头肌（sternocephalic muscle）、胸骨甲状舌骨肌（sterno-thyrohyoid muscle）和肩胛舌骨肌（omohyoid muscle）。胸骨甲状舌骨肌的外侧支止于喉的甲状软骨，称为胸骨甲状肌（sternothyroid muscle）；内侧支止于舌骨，称为胸骨舌骨肌（sternohyoid muscle）。

3. 胸廓肌：位于胸侧壁和胸腔后壁，参与呼吸，分为吸气肌和呼气肌。

（1）吸气肌有肋间外肌（external intercostal muscle）、前背侧锯肌（cranial dorsal serrate muscle）和膈（diaphragm）。膈是一圆拱形凸向胸腔的板状肌，构成胸腔和腹腔间的分界，其周围是由肌纤维构成的肉质缘，中央是强韧的中心腱（central tendon）。膈上有主动脉裂孔（aortic hiatus）、食管裂孔（esophageal hiatus）和后腔静脉裂孔（postcaval vein hiatus），分别供主动脉、食管和后腔静脉通过。吸气肌除膈外，均位于胸侧壁，肌纤维斜向后下方，收缩时肋骨前移，使胸腔横径增大。

（2）呼气肌有后背侧锯肌（caudal dorsal serrate muscle）和肋间内肌（internal intercostal muscle），肌纤维斜向前下方，可向后牵引肋骨，使胸腔横径减小。

4. 腹壁肌：构成腹侧壁和腹底壁，由4层纤维方向不同的板状肌构成，自浅至深分别有腹外斜肌（external oblique abdominal muscle）、腹内斜肌（internal oblique abdominal muscle）、腹直肌（straight abdominal muscle）和腹横肌（transverse abdominal muscle）。腹

壁肌表面覆盖有腹壁筋膜。左右两侧腹壁肌在腹底正中线上，以腱质相连，形成腹白线（abdominal linea alba）。

腹股沟管（inguinal canal）位于腹底壁后部，耻前腱两侧，是腹内斜肌（形成管的前内侧壁）与腹股沟韧带（形成管的后外侧壁）之间的斜行裂隙。公猪的腹股沟管明显，是胎儿时期睾丸从腹腔下降到阴囊的通道，内有精索、总鞘膜、提睾肌、血管和神经通过。母猪的腹股沟管仅供脉管、神经通过。

五、前肢肌

前肢肌按部位分为肩带肌、肩部肌、臂部肌、前臂部肌和前脚部肌。

1. 肩带肌：连接前肢与躯干的肌肉，一般起于躯干，止于肩部和臂部。主要包括斜方肌（trapezius muscle）、菱形肌（rhomboid muscle）、背阔肌（broadest muscle of the back）、臂头肌（brachiocephalic muscle）、肩胛横突肌（omotransverse muscle）、胸肌（pectoral muscle）和腹侧锯肌（ventral serrate muscle）。胸肌位于臂和前臂内侧与胸骨之间，分为胸降肌（胸前浅肌）、胸横肌（胸后浅肌）、锁骨下肌（胸前深肌）和胸升肌（胸后深肌）。

2. 肩部肌：分布于肩胛骨的内侧及外侧面，起自肩胛骨，止于肱骨，跨越肩关节，可伸、屈肩关节和内收、外展前肢。可分为外侧和内侧两组。外侧组有冈上肌（supraspinous muscle）、冈下肌（infraspinous muscle）、三角肌（deltoid muscle）和小圆肌（minor teres muscle）。内侧组有肩胛下肌（subscapular muscle）、大圆肌（major teres muscle）和喙臂肌（coracobrachial muscle）。

3. 臂部肌：分布于肱骨周围，起于肩胛骨和肱骨，跨越肩关节及肘关节，止于肱骨，主要作用在肘关节。可分伸、屈两组。伸肌组位于肱骨后方，有臂三头肌（triceps muscle of the forearm）、前臂筋膜张肌（tensor muscle of the antebrachial fascia）和肘肌（anconeus muscle）。臂三头肌位于肩胛骨和肱骨后方的夹角内，呈三角形，肌腹大，分长头、外侧头和内侧头。屈肌组在肱骨前方，有臂二头肌（biceps muscle of the forearm）和臂肌（brachial muscle）。

4. 前臂部肌及前脚部肌：肌腹分布于前臂骨的背侧、外侧和掌侧面，多为纺锤形。均起自肱骨远端和前臂骨近端，在腕关节上部变为肌腱，作用于腕关节的肌肉的腱较短，止于腕骨及掌骨；作用于指关节的肌肉，其腱较长，跨过腕关节和指关节，止于指骨。除腕尺侧屈肌外，其他各肌的肌腱在经过腕关节时，均包有腱鞘。前臂及前脚部肌可分为背外侧肌群和掌内侧肌群。

（1）背外侧肌群分布于前臂骨的背侧和外侧面，由前向后依次为腕桡侧伸肌（radial extensor muscle of the carpus）、指总伸肌（common digital extensor muscle）和指外侧伸肌（lateral digital extensor muscle），在前臂下部还有腕斜伸肌（extensor carpiobliquus muscle，又称拇长外展肌abductor pollicis longus muscle ）。它们是作用于腕、指关节的伸肌。

（2）掌内侧肌群分布于前臂骨的掌侧面，为腕和指关节的屈肌。肌群的浅层为屈腕的肌肉，包括腕外侧屈肌（lateral flexor muscle of the carpus，又称尺外侧肌）、腕尺侧屈肌（ulnar flexor muscle of the carpus）和腕桡侧屈肌（radial flexor muscle of the carpus）；深层为屈指的肌肉，有指浅屈肌（superficial digital flexor muscle）和指深屈肌（deep digital flexor muscle）。

前肢肌还有骨间肌（interosseus，或称悬韧带suspensory ligament）和屈肌间肌（interflexor muscle），前者位于掌骨的掌侧面，下部接近籽骨，主要由腱质组成；后者分腕上部和腕下部，由指浅屈肌腱走向指深屈肌腱。

六、后肢肌

后肢肌较前肢肌发达，分为臀部肌、股部肌、小腿和后脚部肌。

1．臀部肌：有臀浅肌（superficial gluteal muscle）、臀中肌（middle gluteal muscle）、臀深肌（deep gluteal muscle）和髂肌（iliac muscle），分布于臀部，跨越髋关节，止于股骨。可伸、屈髋关节及外旋大腿。髂肌因与腰大肌的止部紧密结合在一起，故常合称为髂腰肌（iliopsoas）。

2．股部肌：分布于股骨周围，可分为股前、股后和股内侧肌群。

（1）股前肌群位于股骨前面，有阔筋膜张肌（tensor muscle of the fascia lata）和股四头肌（quadriceps muscle of the thigh）。股四头肌有4个肌头，包括股直肌（rectus femoris muscle）、股内侧肌（medial vastus muscle）、股外侧肌（lateral vastus muscle）和股中间肌（intermedial vastus muscle）。

（2）股后肌群位于股后部，有股二头肌（biceps muscle of the thigh）、半腱肌（semitendinous muscle）和半膜肌（semimembranous muscle）。股二头肌长而宽大，位于股后外侧，有两个头，一是椎骨头（长头），起于荐骨和荐结节阔韧带，与臀浅肌融合，形成臀股二头肌（glutaeofemorales biceps muscle），二是坐骨头（短头），起自坐骨结节。

（3）股内侧肌群位于股部内侧，有股薄肌（gracilis muscle）、耻骨肌（pectineal muscle）、内收肌（adductor muscle）和缝匠肌（sartorius muscle）。

股内侧肌群还包括股方肌（quadrate muscle of the thigh），呈长方形，在内收肌外侧的前上方，由坐骨腹侧面至股骨小转子，可内收后肢，并使股骨向外传动。在深层、骨盆底壁和股骨之间还有一些小肌，其作用是旋外股骨，包括闭孔外肌（external obturator muscle）、闭孔内肌（internal obturator muscle）和孖肌（gemellus）。

3．小腿和后脚部肌：多为纺锤形肌，肌腹位于小腿部，在跗关节均变为腱，作用于跗关节和趾关节。可分为背外侧肌群和跖侧肌群。

（1）小腿背外侧肌群有趾长伸肌（long digital extensor muscle）、趾外侧伸肌（lateral digital extensor muscle）、第3腓骨肌（3rd fibular muscle）、胫骨前肌（cranial tibial muscle）和腓骨长肌（long fibular muscle）。

（2）小腿跖侧肌群有腓肠肌（gastrocnemius muscle）、趾浅屈肌（superficial digital flexor muscle）、趾深屈肌（deep digital flexor muscle）、腘肌（popliteus muscle）和比目鱼肌（soleus muscle）等。腓肠肌腱以及附着于跟结节的趾浅屈肌腱、股二头肌腱和半腱肌腱合成一粗而坚硬的腱索，称为跟（总）腱（common calcaneal tendon）。

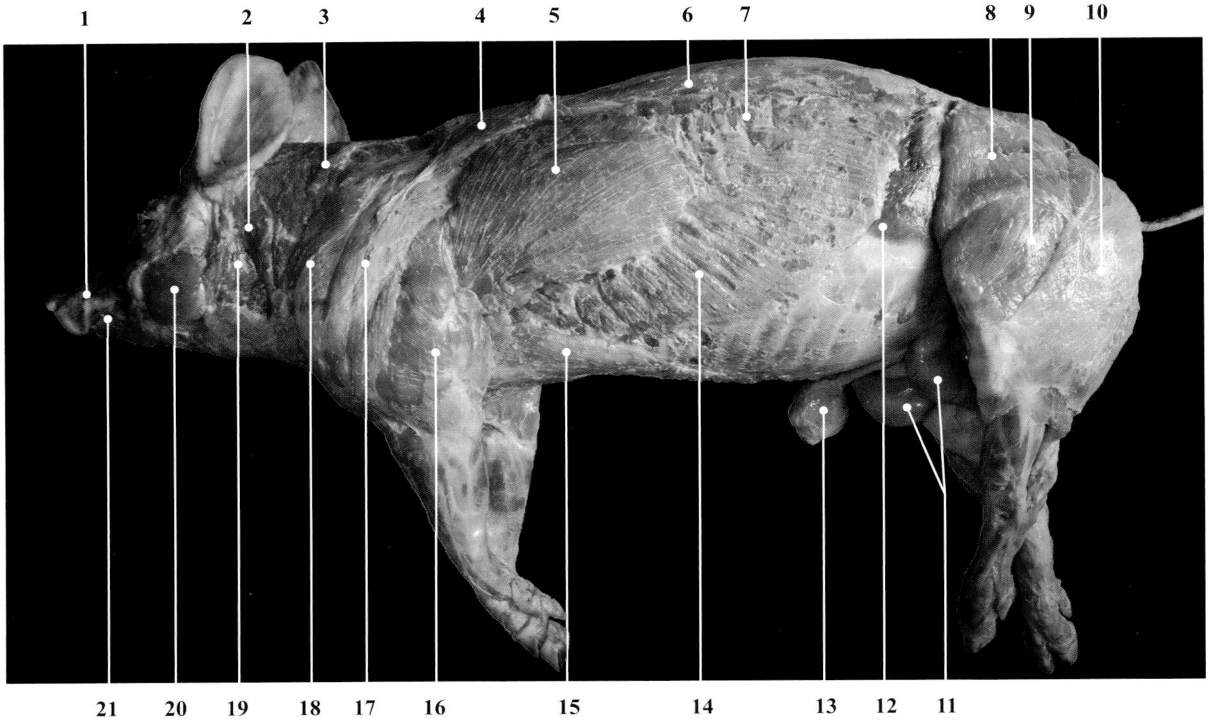

图 3-1 公猪全身浅层肌（左侧观）
Figure 3-1 Superficial muscle groups of the boar (left lateral view).

1- 上唇提肌 levator muscle of the upper lip
2- 臂头肌 brachiocephalic muscle
3- 颈斜方肌 cervical part of trapezius muscle
4- 胸斜方肌 thoracic part of trapezius muscle
5- 背阔肌 broadest muscle of the back
6- 背腰最长肌 dorsal-lumbus longest muscle
7- 后背侧锯肌 caudal dorsal serrate muscle
8- 阔筋膜张肌 tensor muscle of the fascia lata
9- 股四头肌 quadriceps muscle of the thigh
10- 股二头肌 biceps muscle of the thigh
11- 睾丸 testis
12- 腹内斜肌 internal oblique abdominal muscle
13- 包皮憩室（包皮盲囊） preputial diverticulum
14- 腹外斜肌 external oblique abdominal muscle
15- 胸肌 pectoral muscle
16- 臂三头肌 triceps muscle of the forearm
17- 冈下肌 infraspinous muscle
18- 冈上肌 supraspinous muscle
19- 腮腺 parotid gland
20- 咬肌 masseter muscle
21- 颊肌 buccinator muscle

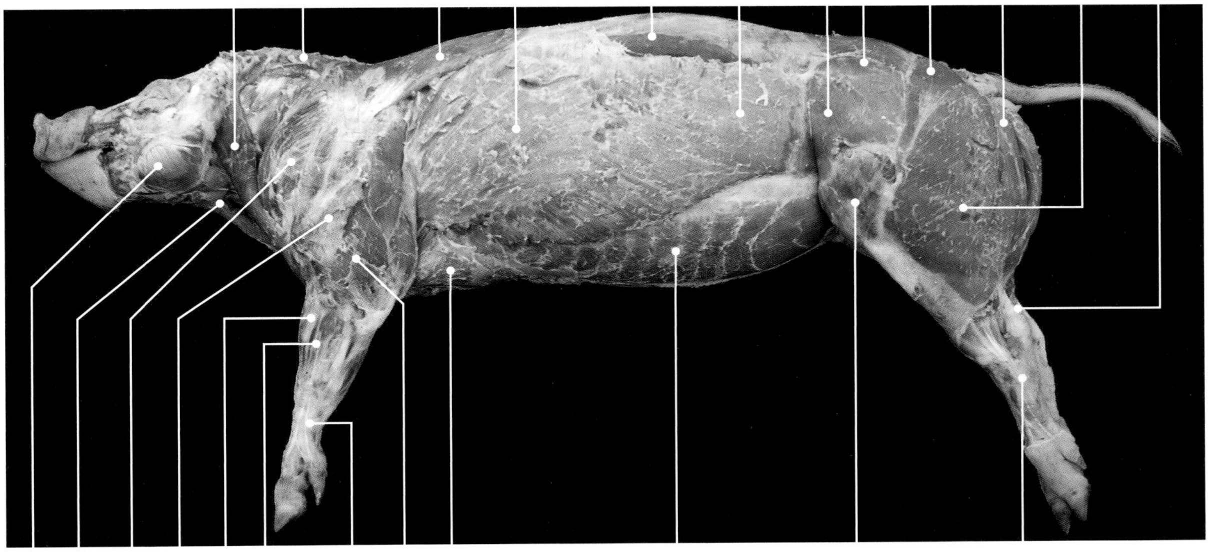

图 3-2 母猪全身浅层肌（左侧观）
Figure 3-2 Superficial muscle groups of the sow (left lateral view).

1- 臂头肌 brachiocephalic muscle
2- 颈斜方肌 cervical part of trapezius muscle
3- 胸斜方肌 thoracic part of trapezius muscle
4- 背阔肌 broadest muscle of the back
5- 背腰最长肌 dorsal-lumbus longest muscle
6- 腹外斜肌 external oblique abdominal muscle
7- 阔筋膜张肌 tensor muscle of the fascia lata
8- 臀中肌 middle gluteal muscle
9- 臀浅肌 superficial gluteal muscle
10- 半腱肌 semitendinous muscle
11- 股二头肌 biceps muscle of the thigh
12- 跟（总）腱 common calcaneal tendon
13- 跗关节 tarsal joint
14- 股四头肌 quadriceps muscle of the thigh
15- 腹直肌 straight abdominal muscle
16- 胸深肌 deep pectoral muscle
17- 臂三头肌 triceps muscle of the forearm
18- 腕关节 carpal joint
19- 指总伸肌 common digital extensor muscle
20- 腕桡侧伸肌 radial extensor muscle of the carpus
21- 三角肌 deltoid muscle
22- 冈上肌 supraspinous muscle
23- 胸头肌 sternocephalic muscle
24- 咬肌 masseter muscle

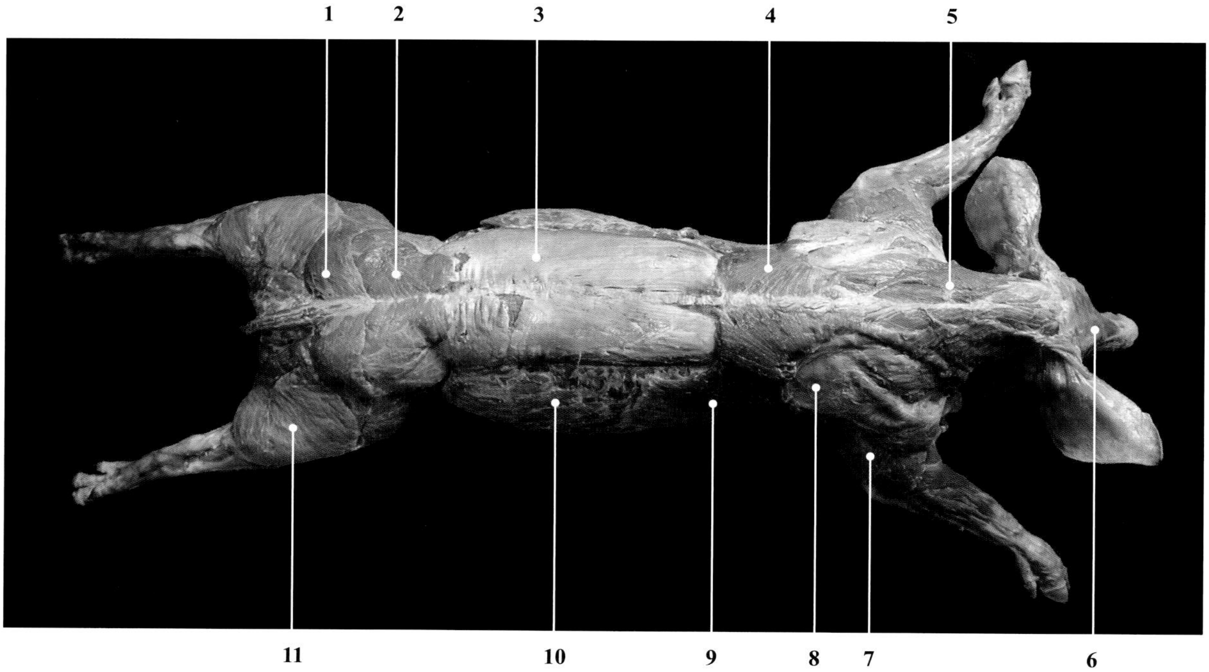

图 3-3　猪全身浅层肌（背侧观）
Figure 3-3　Superficial muscle groups of the pig (dorsal view).

1- 臀浅肌 superficial gluteal muscle
2- 臀中肌 middle gluteal muscle
3- 背腰最长肌 dorsal-lumbus longest muscle
4- 胸斜方肌 thoracic part of trapezius muscle
5- 颈斜方肌 cervical part of trapezius muscle
6- 额皮肌 cutaneous muscle of forehead
7- 臂三头肌 triceps muscle of the forearm
8- 冈下肌 infraspinous muscle
9- 背阔肌 broadest muscle of the back
10- 腹外斜肌 external oblique abdominal muscle
11- 股二头肌 biceps muscle of the thigh

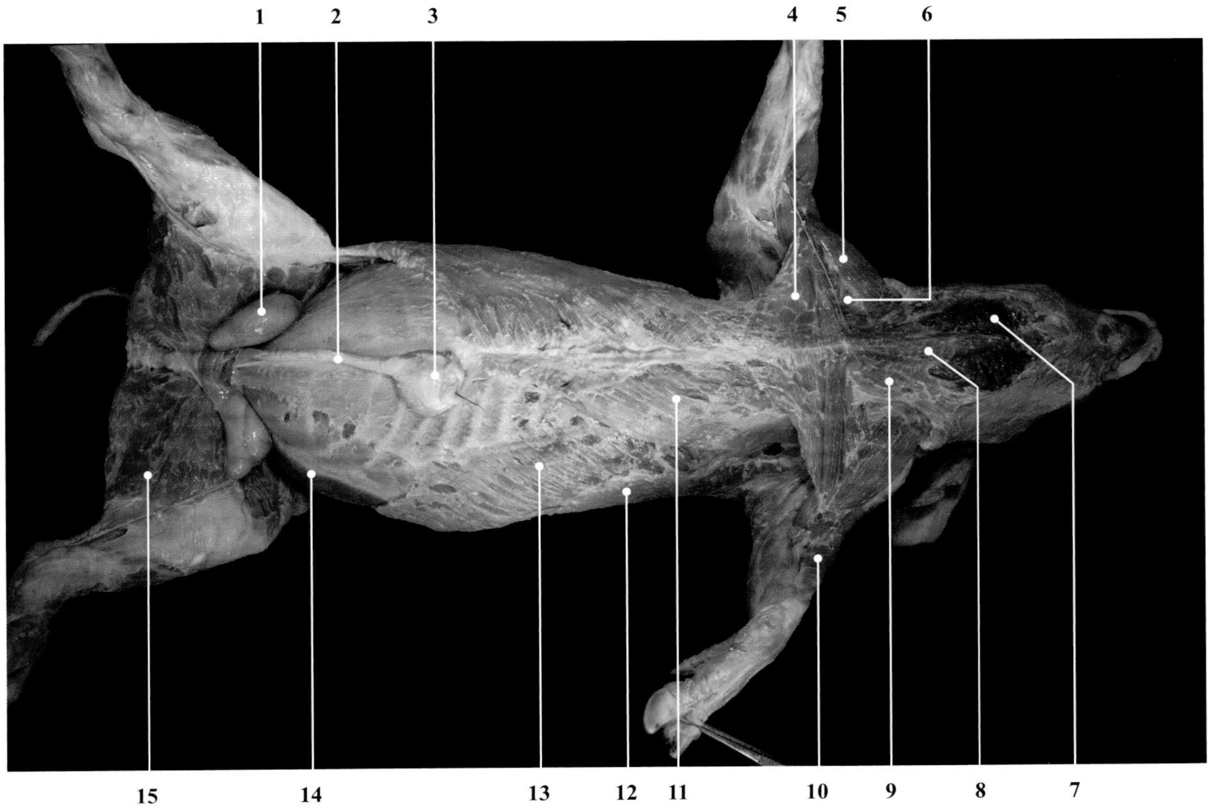

图 3-4 公猪全身浅层肌（腹侧观）
Figure 3-4 Superficial muscle groups of the boar (ventral view).

1- 睾丸 testis
2- 阴茎 penis
3- 包皮憩室（包皮盲囊） preputial diverticulum
4- 胸横肌 transverse pectoral muscle
5- 臂头肌 brachinocephalic muscle
6- 胸降肌 descending pectoral muscle
7- 腮腺 parotid gland
8- 胸骨甲状舌骨肌 sterno-thyrohyoid muscle
9- 胸头肌 sternocephalic muscle
10- 腕桡侧伸肌 radial extensor muscle of the carpus
11- 胸深肌 deep pectoral muscle
12- 背阔肌 broadest muscle of the back
13- 腹外斜肌 external oblique abdominal muscle
14- 腹内斜肌 internal oblique abdominal muscle
15- 股薄肌 gracilis muscle

1- 犬齿肌 canine muscle
2- 上唇提肌 levator muscle of the upper lip
3- 上唇降肌 depressor muscle of the upper lip
4- 腮耳肌 parotido-auricular muscle
5- 臂头肌 brachiocephalic muscle
6- 腮腺 parotid gland
7- 咬肌 masseter muscle
8- 皮肌 cutaneous muscle
9- 颊肌 buccinator muscle

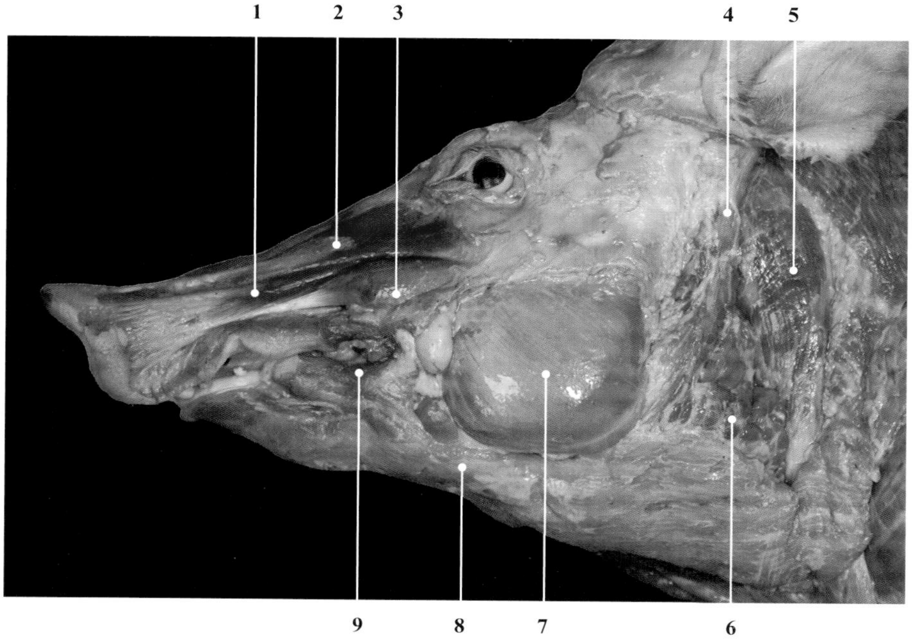

图 3-5 头部浅层肌肉（左外侧观）
Figure 3-5 Superficial muscles of the head (left lateral view).

1- 胸斜方肌 thoracic part of trapezius muscle
2- 背阔肌 broadest muscle of the back
3- 髂肋肌 iliocostal muscle
4- 髂肋肌沟 iliocostal muscle sulcus
5- 背腰最长肌 dorsal-lumbus longest muscle
6- 腹外斜肌 external oblique abdominal muscle
7- 臀中肌 middle gluteal muscle

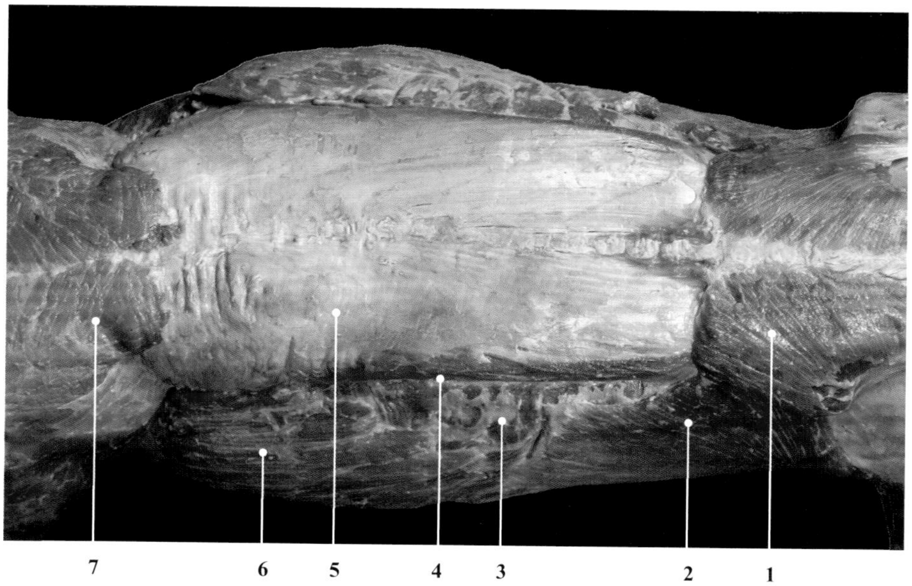

图 3-6 背部浅层肌肉（背侧观）
Figure 3-6 Superficial muscles of the back (dorsal view).

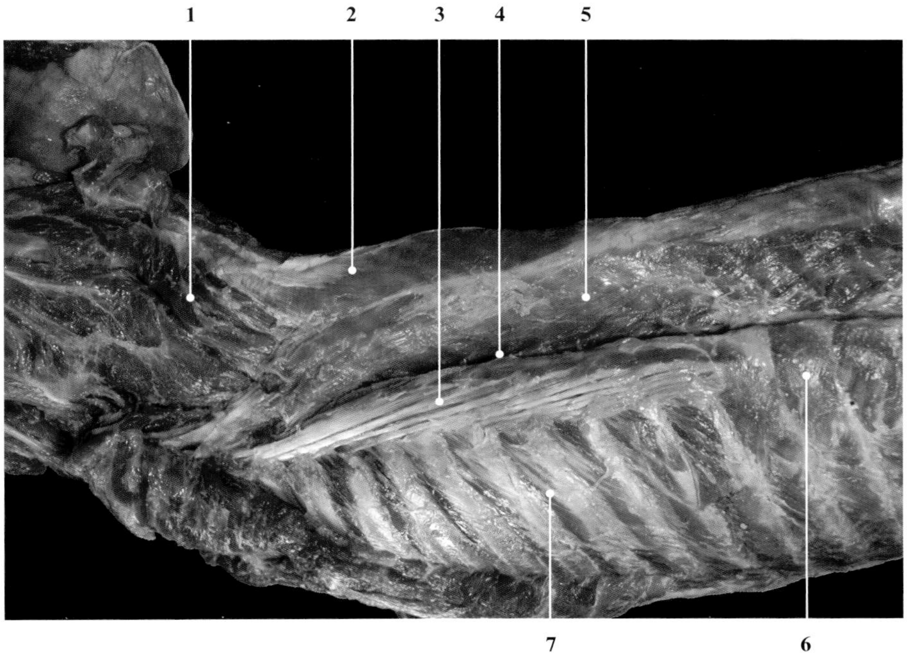

1- 半棘肌 semispinal muscle
2- 夹肌 splenius muscle
3- 髂肋肌 iliocostal muscle
4- 髂肋肌沟 iliocostal muscle sulcus
5- 背腰最长肌 dorsal-lumbus longest muscle
6- 后背侧锯肌 caudal dorsal serrate muscle
7- 肋间外肌 external intercostal muscle

图 3-7 胸背侧肌肉（背外侧观）
Figure 3-7 Muscles of thorax and back (dorsolateral view).

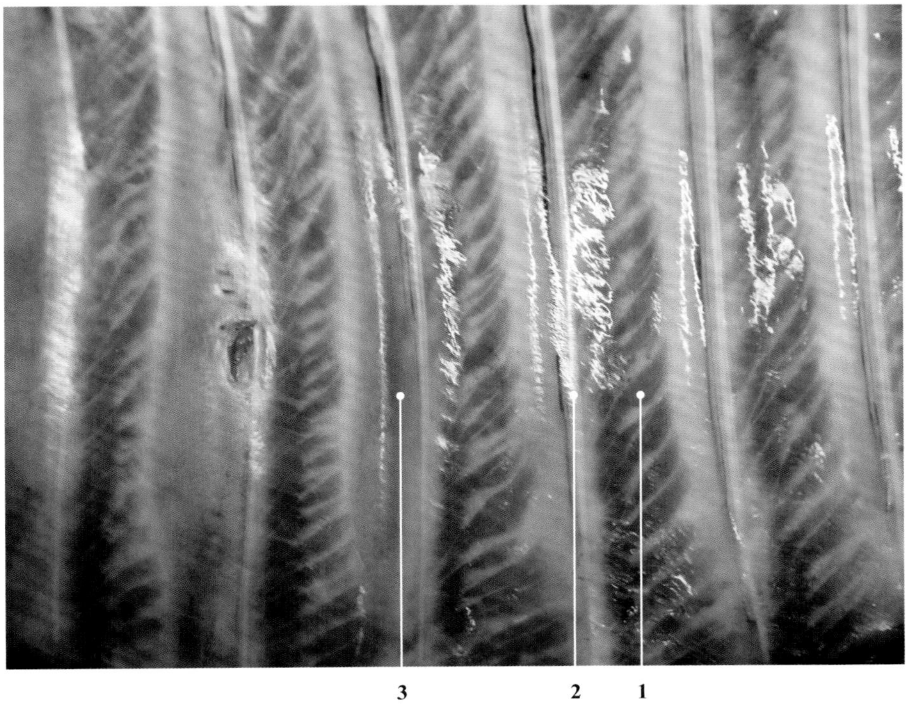

1- 肋间内肌 internal intercostal muscle
2- 肋间动脉、静脉和神经 intercostal artery, vein and nerve
3- 肋骨 costal bone

图 3-8 肋间肌（内侧观）
Figure 3-8 Intercostal muscles (medial view).

1- 肝 liver
2- 肉质缘 pulpa part of diaphragm
3- 中心腱 central tendon
4- 后腔静脉 *caudal vena cava*

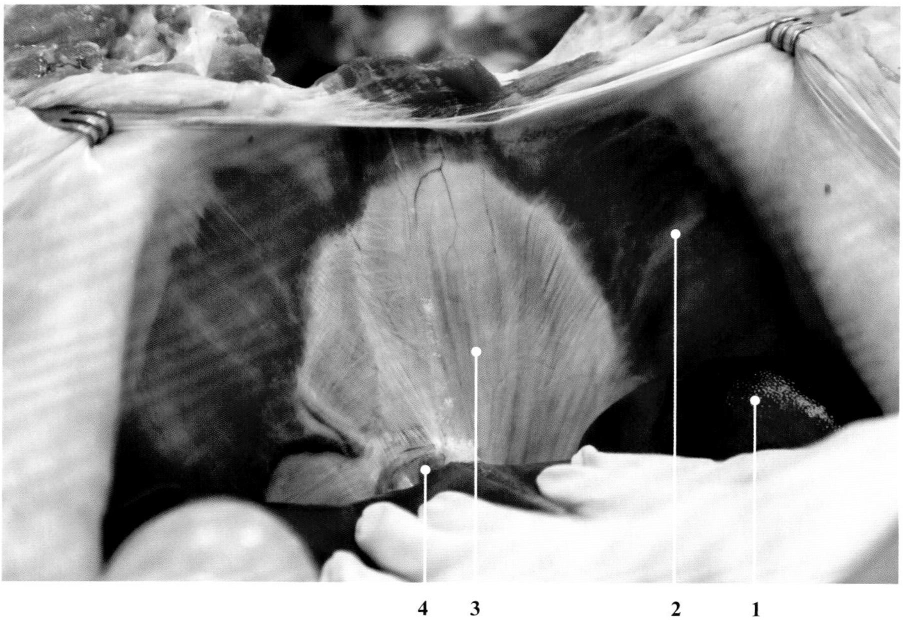

图 3-9　膈（后面观）
Figure 3-9　Diaphragm (caudal view).

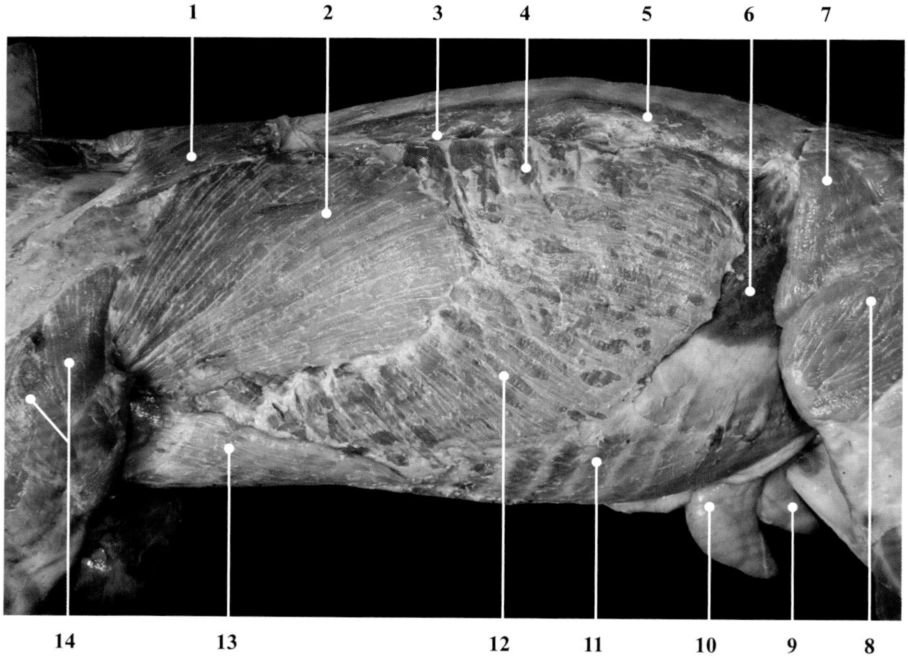

图 3-10　胸壁和腹壁肌（左侧观）
Figure 3-10　Thoracic and abdominal wall muscles (left view).

1- 胸斜方肌 thoracic part of trapezius muscle
2- 背阔肌 broadest muscle of the back
3- 髂肋肌沟 iliocostal muscle sulcus
4- 后背侧锯肌 caudal dorsal serrate muscle
5- 背腰最长肌 dorsal-lumbus longest muscle
6- 腹内斜肌 internal oblique abdominal muscle
7- 阔筋膜张肌 tensor muscle of the fascia lata
8- 股四头肌 quadriceps muscle of the thigh
9- 右侧睾丸 right testis
10- 左侧睾丸 left testis
11- 腹直肌 straight abdominal muscle
12- 腹外斜肌 external oblique abdominal muscle
13- 胸深肌 deep pectoral muscle
14- 臂三头肌 triceps muscle of the forearm

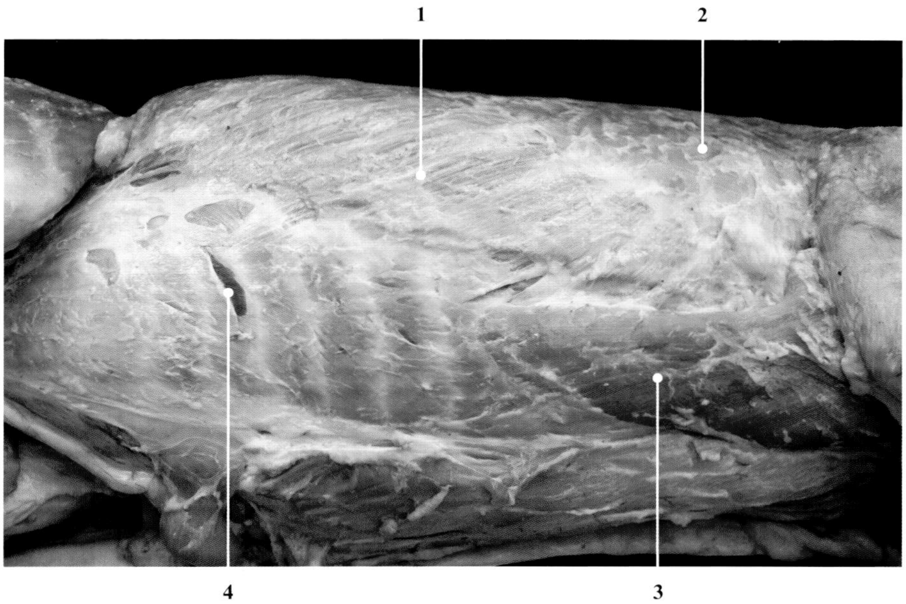

1- 腹外斜肌 external oblique abdominal muscle
2- 背阔肌 broadest muscle of the back
3- 胸深肌 deep pectoral muscle
4- 腹直肌 straight abdominal muscle

图 3-11 腹壁肌肉（腹侧观）
Figure 3-11 Abdominal wall muscles (ventral view).

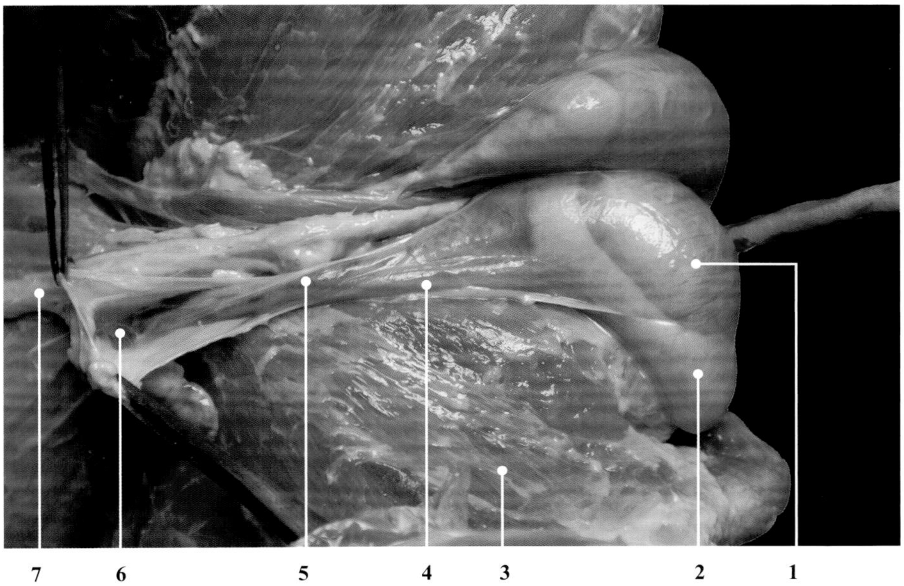

1- 睾丸 testis
2- 附睾 epididymis
3- 股薄肌 gracilis muscle
4- 提睾肌 cremasteric muscle
5- 精索 spermatic cord
6- 腹股沟管皮下环 external ring of inguinal canal
7- 阴茎 penis

图 3-12 腹股沟管
Figure 3-12 Inguinal canal.

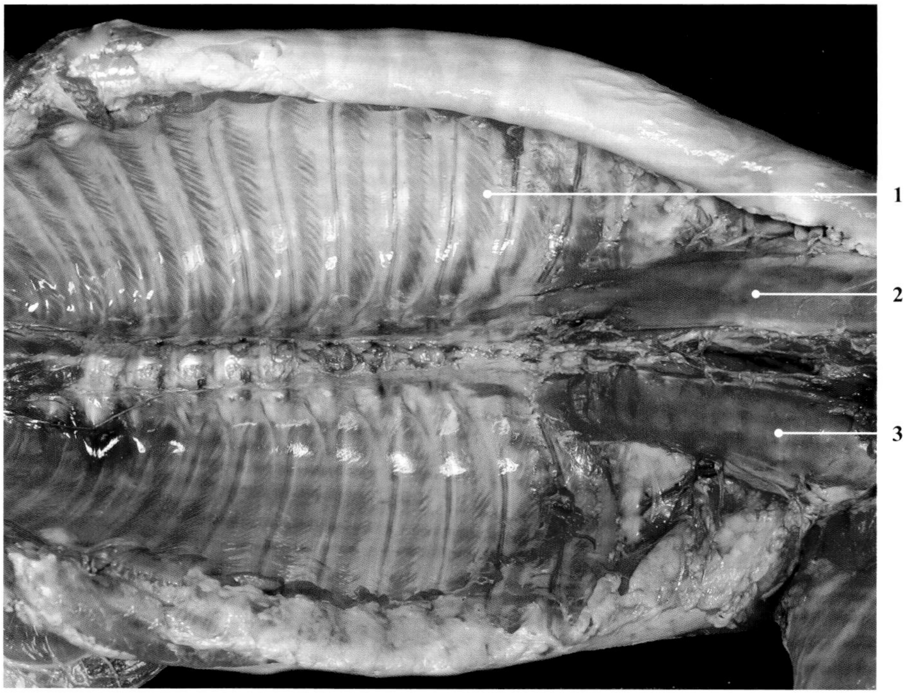

图 3-13　腰大肌（腹侧观）
Figure 3-13　Major psoas muscle (ventral view).

1- 肋间内肌 internal intercostal muscle
2- 左腰大肌 left major psoas muscle
3- 右腰大肌 right major psoas muscle

图 3-14　腰肌（腹侧观）
Figure 3-14　Psoas muscle (ventral view).

1- 左腰大肌 left major psoas muscle
2- 左腰小肌 left minor psoas muscle
3- 右腰小肌 right minor psoas muscle
4- 右腰大肌 right major psoas muscle

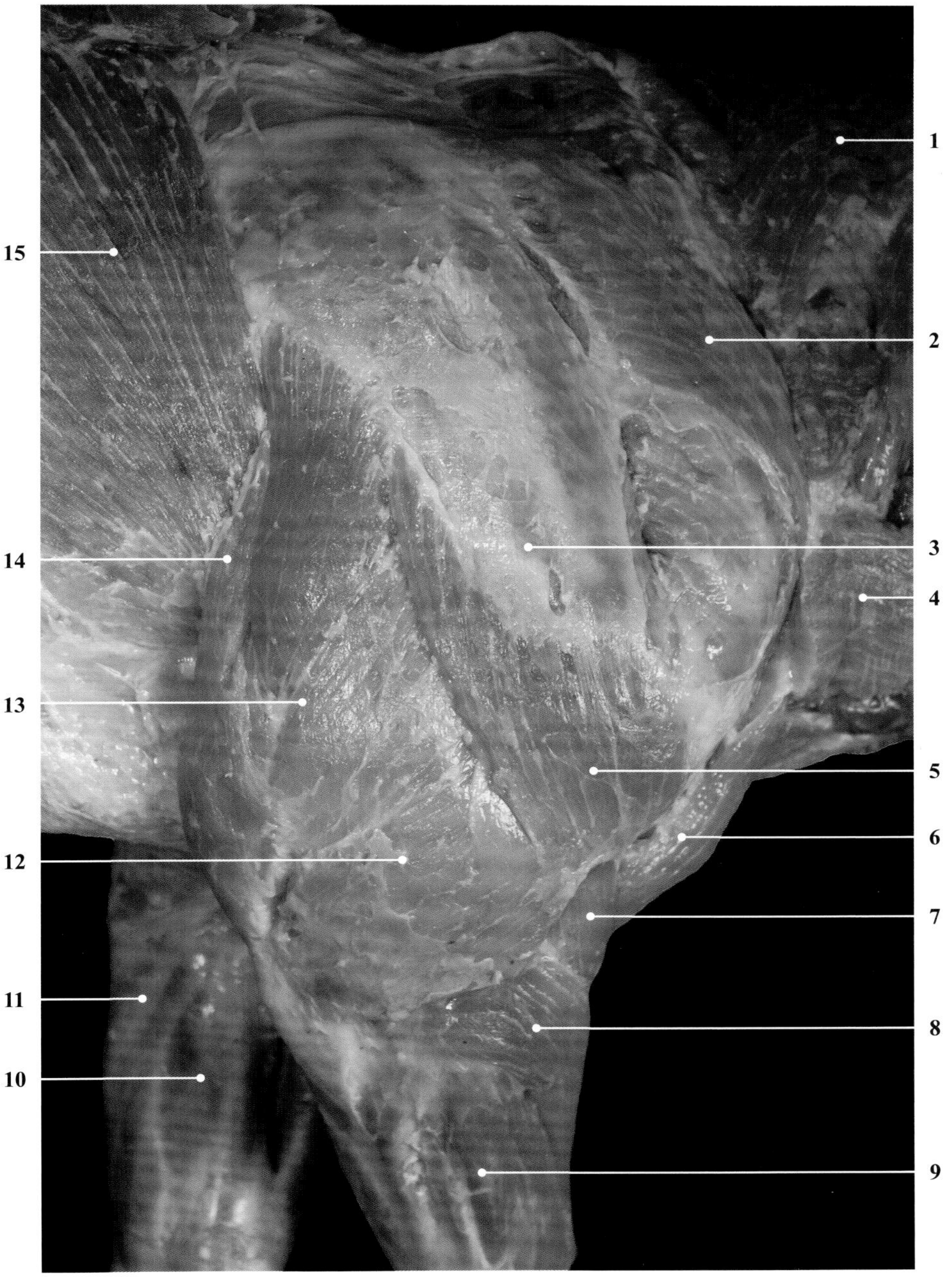

图 3-15 肩带部浅层肌（右外侧观）
Figure 3-15 Superficial muscles of shoulder girdle region (right lateral view).

1- 颈斜方肌 cervical part of trapezius muscle
2- 冈上肌 supraspinous muscle
3- 冈下肌 infraspinous muscle
4- 肩胛横突肌 omotransverse muscle
5- 三角肌 deltoid muscle
6- 臂头肌 brachiocephalic muscle
7- 臂肌 brachial muscle
8- 腕桡侧伸肌 radial extensor muscle of the carpus
9- 指外侧伸肌 lateral digital extensor muscle
10- 腕桡侧屈肌 radial flexor muscle of the carpus
11- 指深屈肌 deep digital flexor muscle
12- 臂三头肌外侧头 lateral head of triceps muscle of the forearm
13- 臂三头肌长头 long head of triceps muscle of the forearm
14- 前臂筋膜张肌 tensor muscle of the antebrachial fascia
15- 背阔肌 broadest muscle of the back

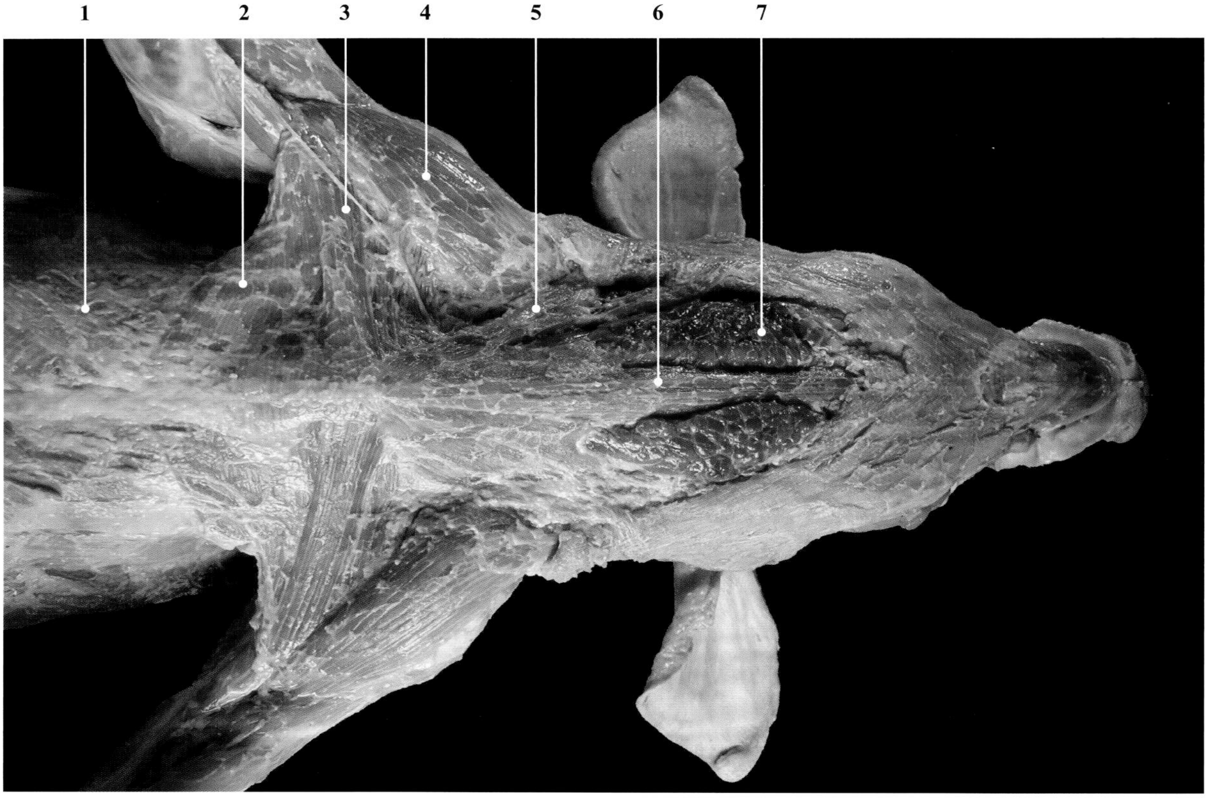

图 3-16 肩带部肌（腹侧观）
Figure 3-16 Shoulder girdle muscles (ventral view).

1- 胸深肌 deep pectoral muscle
2- 胸横肌 transverse pectoral muscle
3- 胸降肌 descending pectoral muscle
4- 臂头肌 brachiocephalic muscle
5- 胸头肌 sternocephalic muscle
6- 胸骨甲状舌骨肌 sterno-thyrohyoid muscle
7- 腮腺 parotid gland

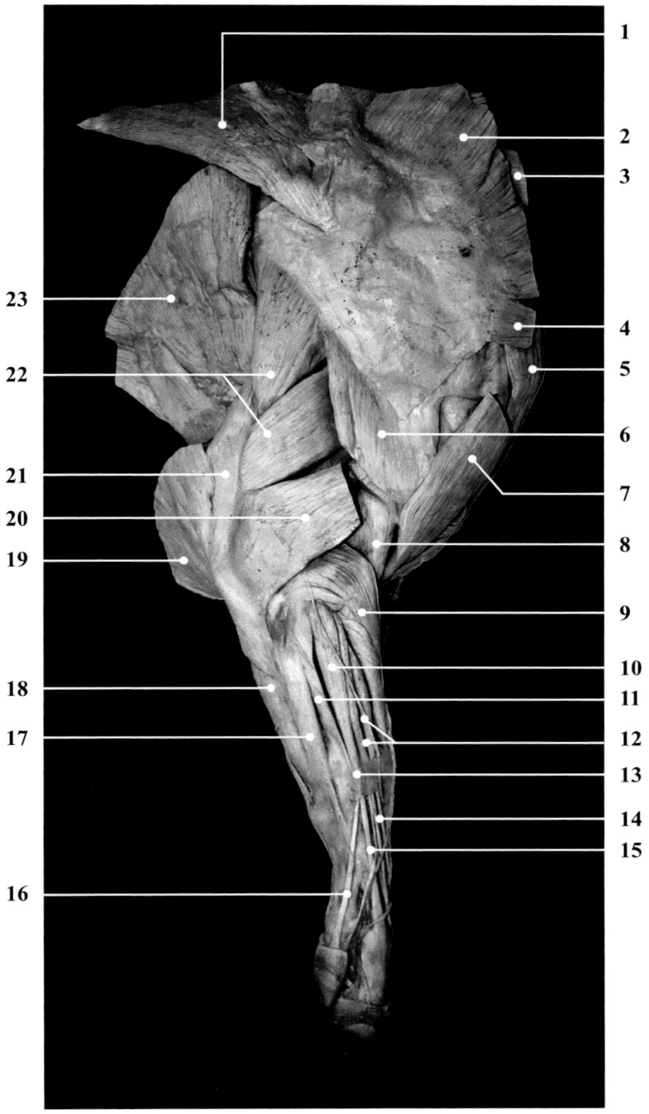

图 3-17 右前肢肌（外侧观，塑化标本）
Figure 3-17 Muscles of right forelimb (lateral view, plasticized preparation).

1- 胸斜方肌 thoracic part of trapezius muscle
2- 颈斜方肌 cervical part of trapezius muscle
3- 颈菱形肌 cervical part of rhomboid muscle
4- 肩胛横突肌 omotransverse muscle
5- 冈上肌 supraspinous muscle
6- 三角肌 deltoid muscle
7- 臂头肌 brachiocephalic muscle
8- 臂肌 brachial muscle
9- 腕桡侧伸肌 radial extensor muscle of the carpus
10- 指外侧伸肌（第 4 指伸肌）lateral digital extensor muscle to 4th digitorum
11- 指外侧伸肌（第 5 指伸肌）lateral digital extensor muscle to 5th digitorum
12- 指总伸肌 common digital extensor muscle
13- 腕关节 carpal joint
14- 指总伸肌腱 common digital extensor tendon
15- 第 4 指伸肌腱 4th digital extensor tendon
16- 第 5 指伸肌腱 5th digital extensor tendon
17- 腕外侧屈肌 lateral flexor muscle of the carpus
18- 指浅屈肌 superficial digital flexor muscle
19- 胸深肌 deep pectoral muscle
20- 臂三头肌外侧头 lateral head of triceps muscle of the forearm
21- 前臂筋膜张肌 tensor muscle of the antebrachial fascia
22- 臂三头肌长头 long head of triceps muscle of the forearm
23- 背阔肌 broadest muscle of the back

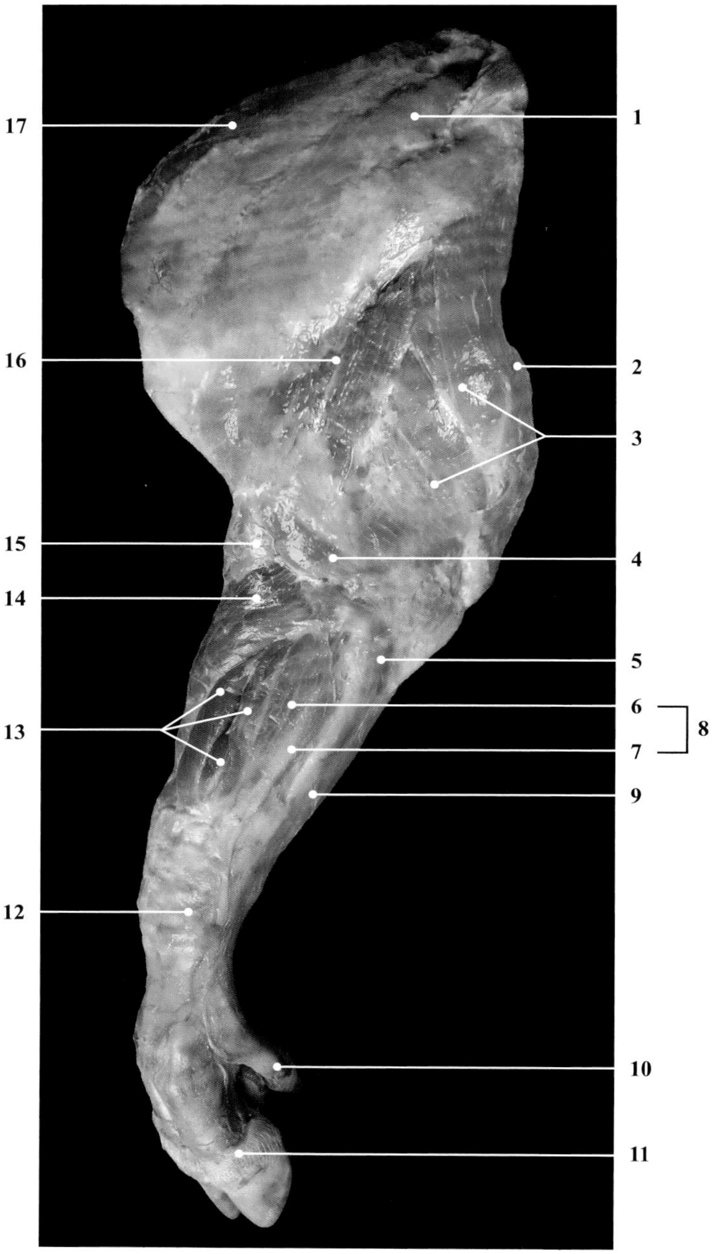

图 3-18　左前肢肌肉（外侧观）
Figure 3-18　Muscles of left forelimb (lateral view).

1- 冈下肌 infraspinous muscle
2- 前臂筋膜张肌 tensor muscle of the antebrachial fascia
3- 臂三头肌长头 long head of triceps muscle
4- 臂三头肌外侧头 lateral head of triceps muscle of the forearm
5- 腕外侧屈肌 lateral flexor muscle of the carpus
6- 第 4 指伸肌 4th digital extensor muscle
7- 第 5 指伸肌 5th digital extensor muscle
8- 指外侧伸肌 lateral digital extensor muscle
9- 指浅屈肌 superficial digital flexor muscle
10- 第 5 指 5th finger (dewclaw)
11- 第 4 指 4th finger (hoof)
12- 腕关节 carpal joint
13- 指总伸肌 common digital extensor muscle
14- 腕桡侧伸肌 radial extensor muscle of the carpus
15- 臂肌 brachial muscle
16- 三角肌 deltoid muscle
17- 冈上肌 supraspinous muscle

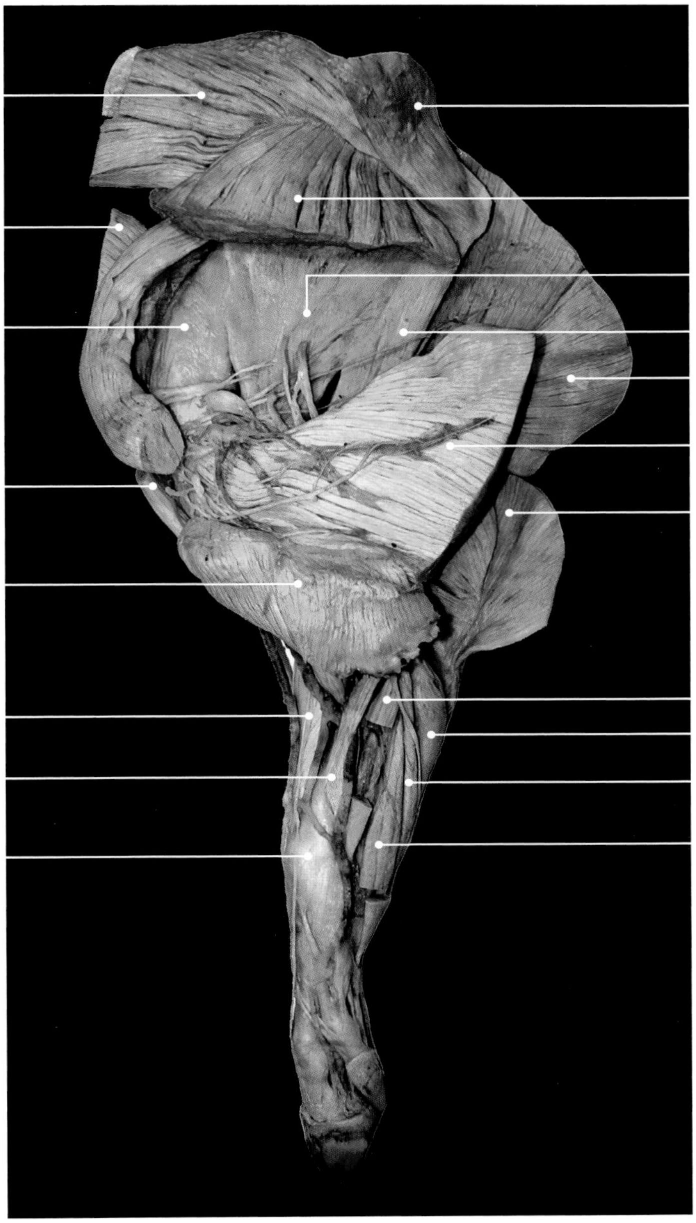

图 3-19 右前肢肌（内侧观，塑化标本）
Figure 3-19 Muscles of right forelimb (medial view, plasticized preparation).

1- 胸菱形肌 thoracic part of the trapezius muscle
2- 腹侧锯肌 ventral serrate muscle
3- 肩胛下肌 subscapular muscle
4- 大圆肌 major teres muscle
5- 背阔肌 broadest muscle of the back
6- 胸深肌 deep pectoral muscle
7- 前臂筋膜张肌 tensor muscle of the antebrachial fascia
8- 腕桡侧屈肌 radial flexor muscle of the carpus
9- 指深屈肌肱骨头 humeral head of deep digital flexor muscle
10- 腕尺侧屈肌 ulnar flexor muscle of the carpus
11- 指浅屈肌 superficial digital flexor muscle
12- 腕关节 carpal joint
13- 桡骨 radius
14- 腕桡侧伸肌 radial extensor muscle of the carpus
15- 胸浅肌 superficial pectoral muscle
16- 臂头肌 brachiocephalic muscle
17- 冈上肌 supraspinous muscle
18- 肩胛横突肌 omotransvere muscle
19- 颈菱形肌 cervical part of trapezius muscle

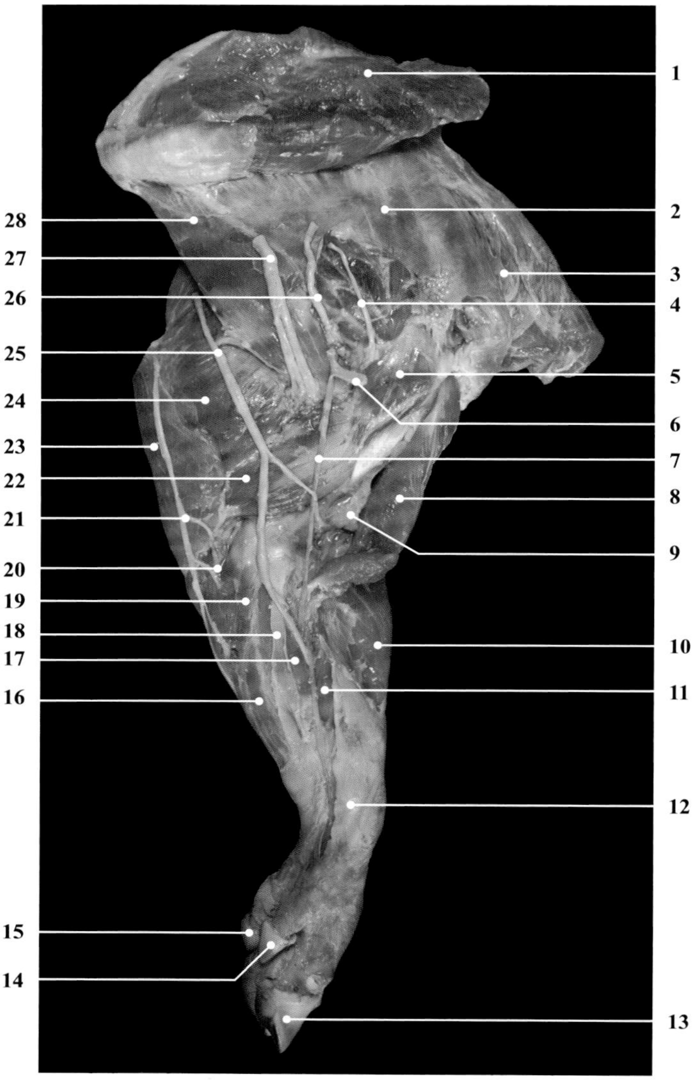

图 3-20　左前肢肌（内侧观）
Figure 3-20　Muscles of left forelimb (medial view).

1- 腹侧锯肌 ventral serrate muscle
2- 肩胛下肌 subscapular muscle
3- 冈上肌 supraspinous muscle
4- 肩胛下神经 subscapular nerve
5- 喙臂肌 coracobrachial muscle
6- 腋动脉 axillary artery
7- 臂动脉 brachial artery
8- 臂二头肌 biceps muscle of the forearm
9- 臂肌 brachial muscle
10- 腕桡侧伸肌 radial extensor muscle of the carpus
11- 指深屈肌桡骨头 radial head of deep digital flexor muscle
12- 腕关节 carpal joint
13- 第 3 指 3rd finger (hoof)
14- 第 2 指 2nd finger (dewclaw)
15- 第 5 指 5th finger (dewclaw)
16- 指浅屈肌 superficial digital flexor muscle
17- 腕桡侧屈肌 radial flexor muscle of the carpus
18- 腕尺侧屈肌 ulnar flexor muscle of the carpus
19- 指深屈肌肱骨头 humeral head of deep digital flexor muscle
20- 指深屈肌尺骨头 ulnar head of deep digital flexor muscle
21- 尺神经 ulnar nerve
22- 臂三头肌内侧头 medial head of triceps muscle of the forearm
23- 前臂筋膜张肌 tensor muscle of the antebrachial fascia
24- 臂三头肌长头 long head of triceps of the forearm muscle
25- 正中神经 median nerve
26- 腋神经 axillary nerve
27- 桡神经 radial nerve
28- 大圆肌 major teres muscle

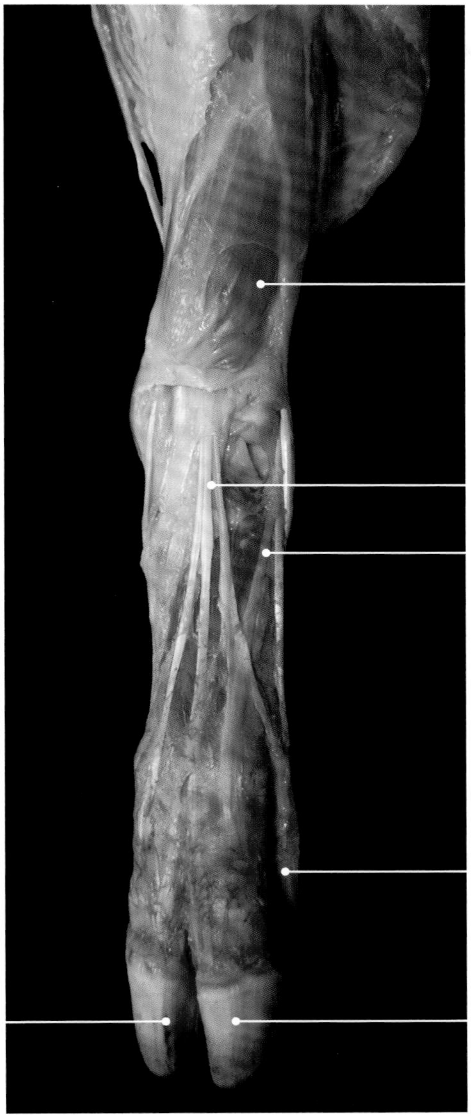

图 3-21 左前肢肌（前面观）
Figure 3-21 Muscles of left forelimb (cranial view).

1- 腕桡侧伸肌 radial extensor muscle of the carpus
2- 指总伸肌腱 common digital extensor tendon
3- 指外侧伸肌腱 lateral digital extensor tendon
4- 第 5 指 5th finger (dewclaw)
5- 第 4 指 4th finger (hoof)
6- 第 3 指 3rd finger (hoof)

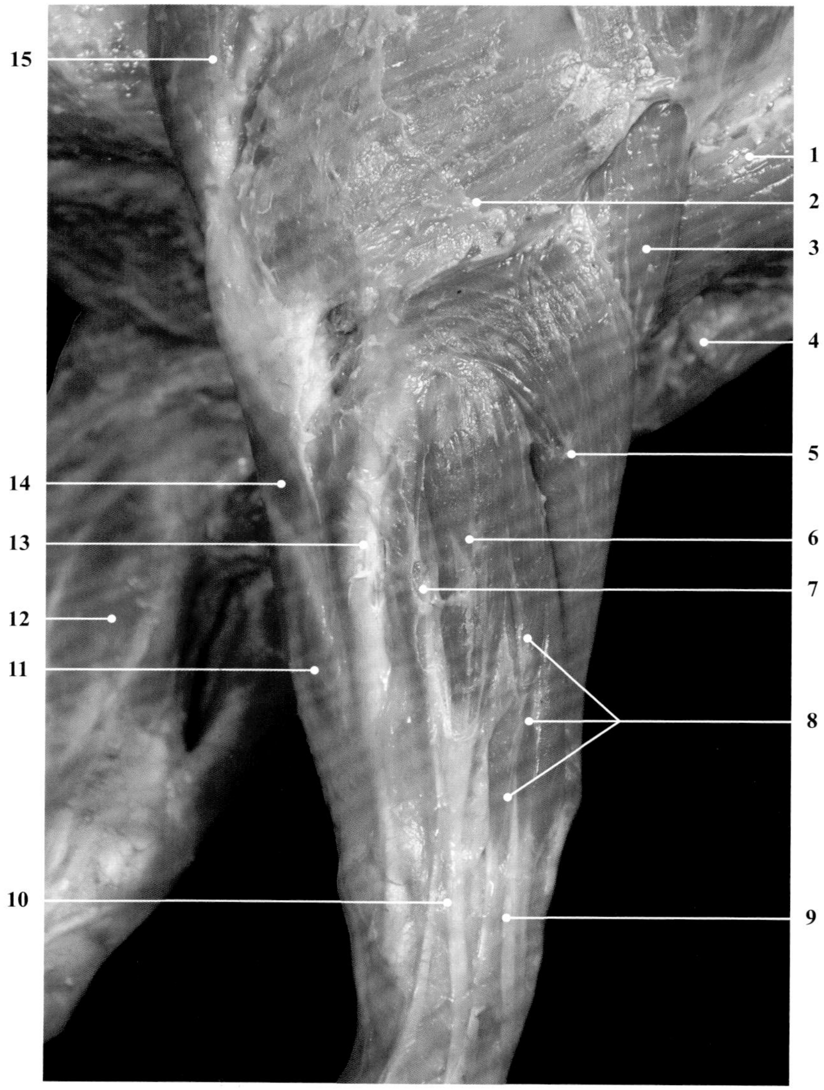

图 3-22　右前臂部肌（外侧观）
Figure 3-22　Muscles of right antebrachium (lateral view).

1- 臂头肌 brachiocephalic muscle
2- 臂三头肌外侧头 lateral head of triceps muscle of the forearm
3- 臂肌 brachial muscle
4- 胸头肌 sternocephalic muscle
5- 腕桡侧伸肌 radial extensor muscle of the carpus
6- 指外侧伸肌（第 4 指伸肌）lateral digital extensor muscle to 4th digitorum
7- 指外侧伸肌（第 5 指伸肌）lateral digital extensor muscle to 5th digitorum
8- 指总伸肌 common digital extensor muscle
9- 指总伸肌腱 common digital extensor tendon
10- 指外侧伸肌腱 lateral digital extensor tendon
11- 指浅屈肌 superficial digital flexor muscle
12- 腕桡侧屈肌 radial flexor muscle of the carpus
13- 腕外侧屈肌 lateral flexor muscle of the carpus
14- 指深屈肌尺骨头 ulnar head of deep digital flexor muscle
15- 臂三头肌长头 long head of triceps muscle of the forearm

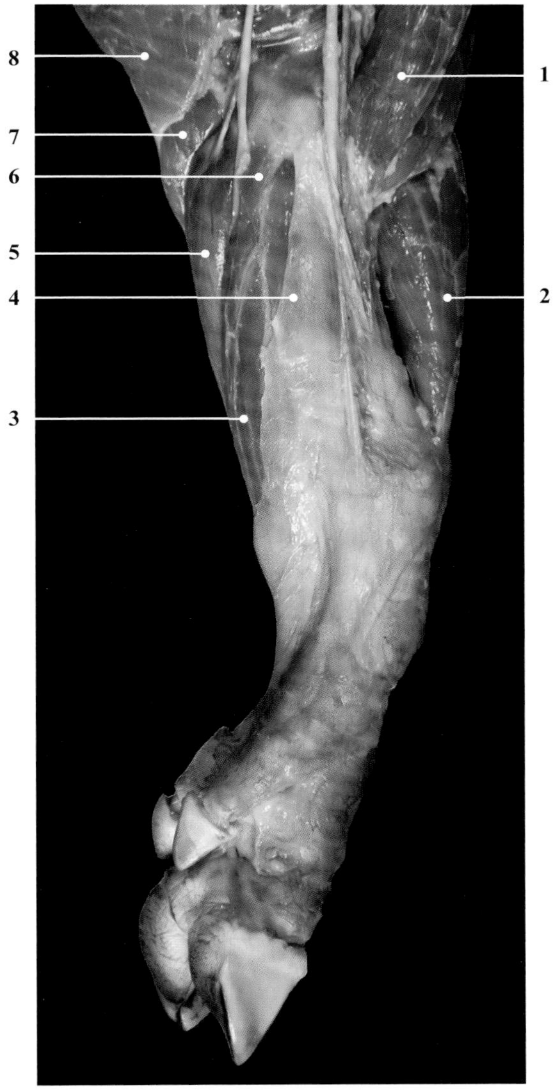

图 3-23 左前臂部肌（内侧观）
Figure 3-23 Muscles of left antebrachium (medial view).

1- 臂二头肌 biceps muscle of the forearm
2- 腕桡侧伸肌 radial extensor muscle of the carpus
3- 指浅屈肌 superficial digital flexor muscle
4- 腕桡侧屈肌 radial flexor muscle of the carpus
5- 指深屈肌肱骨头 humeral head of deep digital flexor muscle
6- 腕尺侧屈肌 ulnar flexor muscle of the carpus
7- 指深屈肌尺骨头 ulnar head of deep digital flexor muscle
8- 臂三头肌 triceps muscle of the forearm

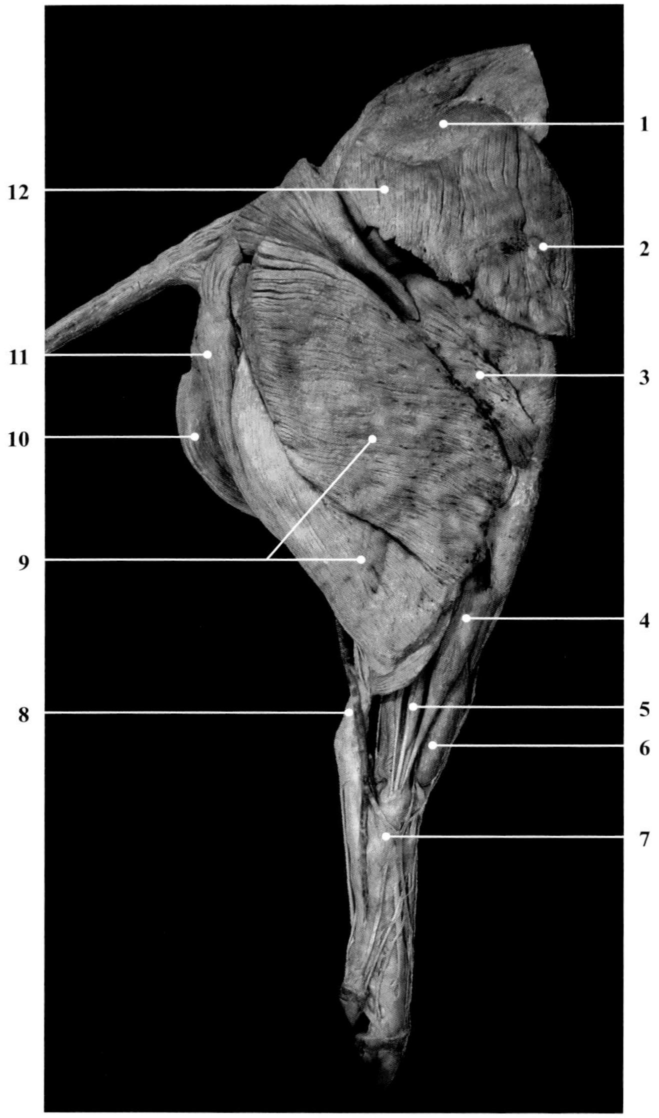

图 3-24　右后肢肌（外侧观，塑化标本）
Figure 3-24　Muscles of right hindlimb (lateral view, plasticized preparation).

1- 臀中肌 middle gluteal muscle
2- 阔筋膜张肌 tensor muscle of the fascia lata
3- 股四头肌 quadriceps muscle of the thigh
4- 腓骨长肌 long fibular muscle
5- 趾外侧伸肌 lateral digital extensor muscle
6- 第 3 腓骨肌 3rd fibular muscle
7- 跗关节 tarsal joint
8- 跟（总）腱 common calcaneal tendon
9- 股二头肌 biceps muscle of the thigh
10- 半膜肌 semimembranous muscle
11- 半腱肌 semitendinous muscle
12- 臀浅肌 superficial gluteal muscle

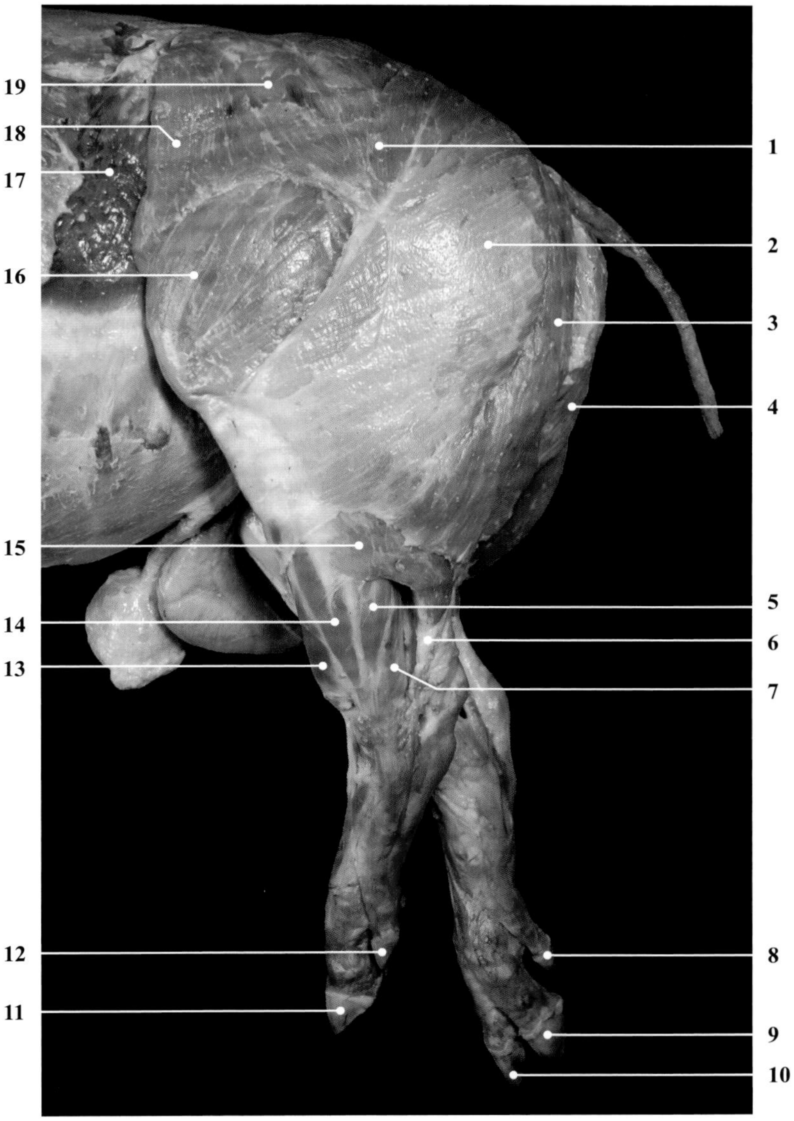

图 3-25　左后肢肌肉（外侧观）
Figure 3-25　Muscles of left hindlimb (lateral view).

1- 臀浅肌 superficial gluteal muscle
2- 股二头肌 biceps muscle of the thigh
3- 半腱肌 semitendinous muscle
4- 半膜肌 semimembranous muscle
5- 趾外侧伸肌 lateral digital extensor muscle
6- 跟（总）腱 common calcaneal tendon
7- 趾深屈肌 deep digital flexor muscle
8- 右第 2 趾 2nd toe (dewclaw), right
9- 右第 3 趾 3rd toe (hoof), right
10- 右第 4 趾 4th toe (hoof), right
11- 左第 4 趾 4th toe (hoof), left
12- 左第 5 趾 5th toe (dewclaw), left
13- 第 3 腓骨肌 3rd fibular muscle
14- 腓骨长肌 long fibular muscle
15- 腓肠肌 gastrocnemius muscle
16- 股外侧肌 lateral vastus muscle
17- 腹内斜肌 internal oblique abdominal muscle
18- 股阔筋膜张肌 tensor muscle of the fascia lata
19- 臀中肌 middle gluteal muscle

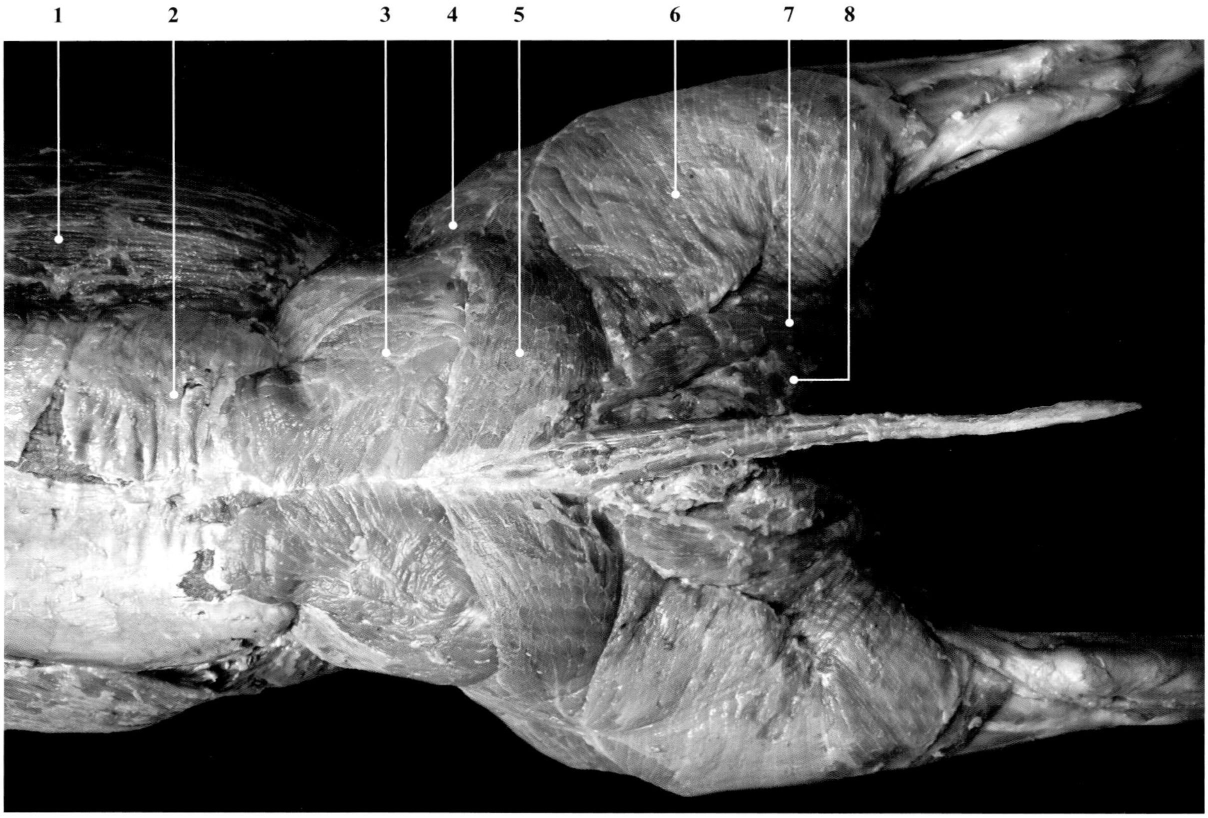

图 3-26 臀部肌（背侧观）
Figure 3-26 Rump muscles (dorsal view).

1- 腹外斜肌 external oblique abdominal muscle
2- 背腰最长肌 dorsal-lumbus longest muscle
3- 臀中肌 middle gluteal muscle
4- 股阔筋膜张肌 tensor muscle of the fascia lata
5- 臀浅肌 superficial gluteal muscle
6- 股二头肌 biceps muscle of the thigh
7- 半腱肌 semitendinous muscle
8- 半膜肌 semimembranous muscle

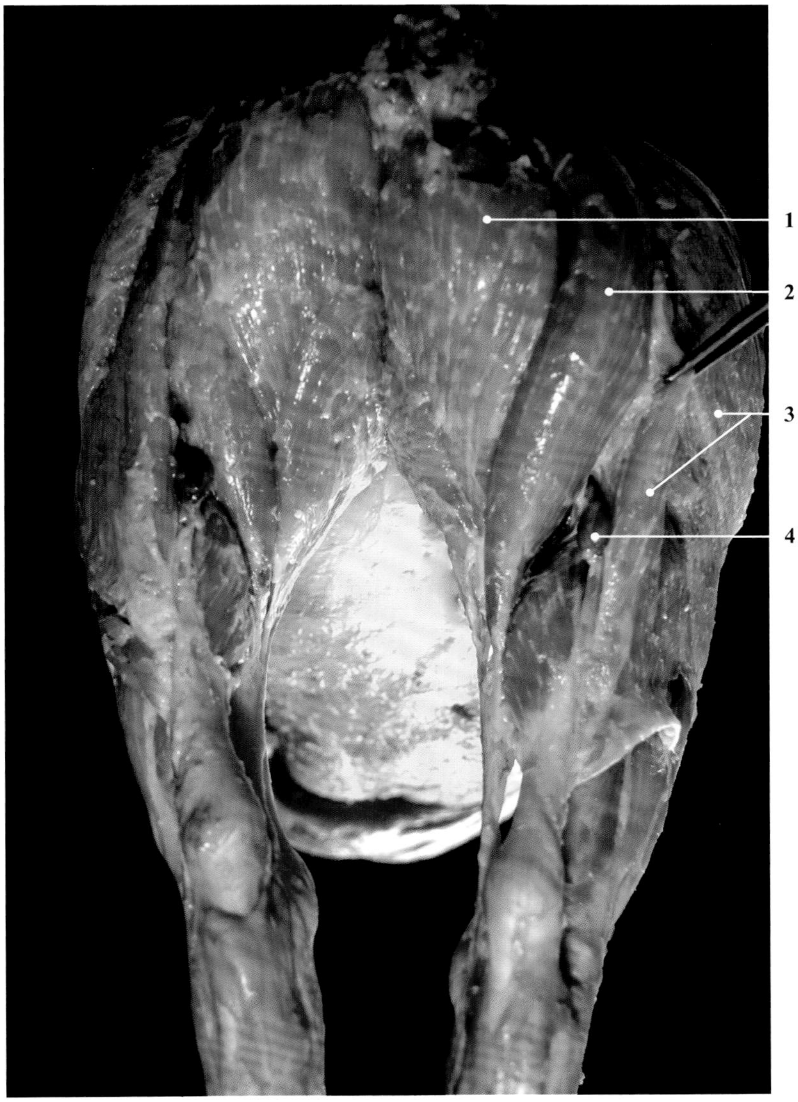

图 3-27 后肢浅层肌（后面观）
Figure 3-27 Superficial muscles of hindlimb (caudal view).

1- 半膜肌 semimembranous muscle
2- 半腱肌 semitendinous muscle
3- 股二头肌 biceps muscle of the thigh
4- 腘淋巴结 popliteal lymph node

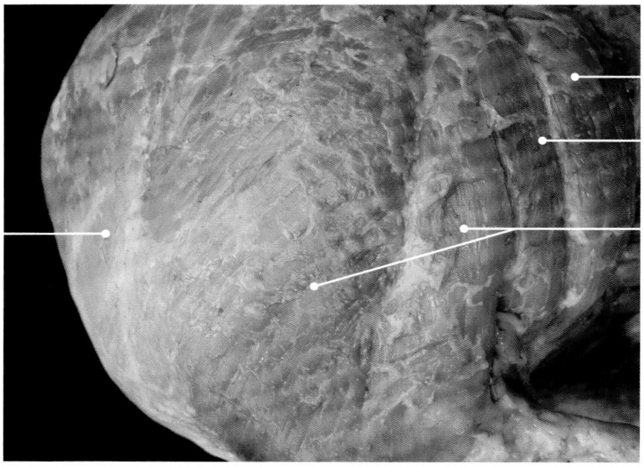

图 3-28　股后肌群（左外侧观）
Figure 3-28　Posterior thigh muscle groups (left lateral view).

1- 半膜肌 semimembranous muscle
2- 半腱肌 semitendinous muscle
3- 股二头肌 biceps muscle of the thigh
4- 股外侧肌 lateral vastus muscle

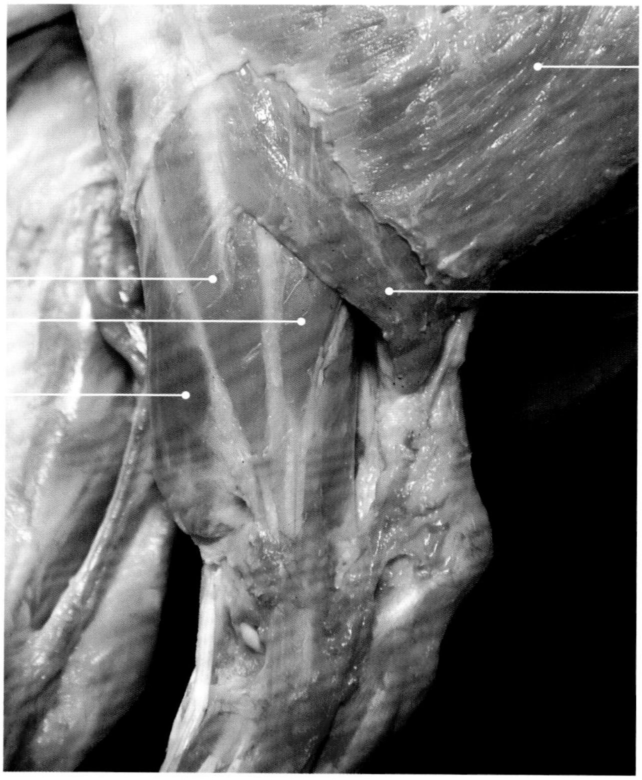

图 3-29　左小腿部肌（外侧观）
Figure 3-29　Muscles of left crus (lateral view).

1- 股二头肌 biceps muscle of the thigh
2- 腓肠肌 gastrocnemius muscle
3- 第 3 腓骨肌 3rd fibular muscle
4- 趾外侧伸肌 lateral digital extensor muscle
5- 腓骨长肌 long fibular muscle

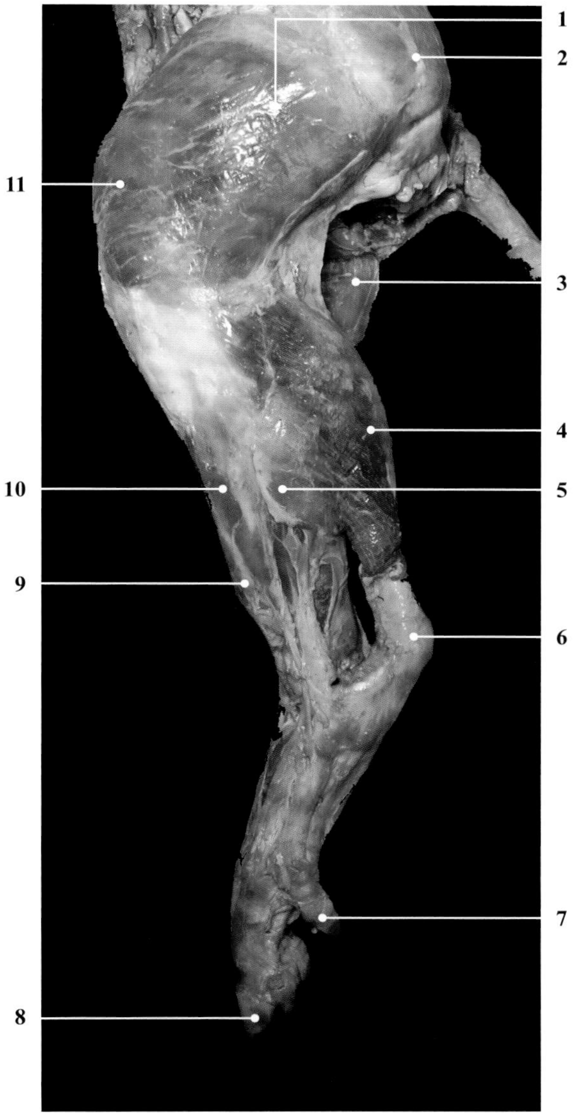

图 3-30 左后肢深层肌（外侧观）
Figure 3-30 Deep muscles of left hindlimb (lateral view).

1- 股外侧肌 lateral vastus muscle
2- 臀中肌 middle gluteal muscle
3- 半膜肌 semimembranous muscle
4- 腓肠肌 gastrocnemius muscle
5- 比目鱼肌 soleus muscle
6- 跟结节 calcaneal tuberosity
7- 第 5 趾 5th toe (dewclaw)
8- 第 4 趾 4th toe (hoof)
9- 第 3 腓骨肌 3rd fibular muscle
10- 腓骨长肌 long fibular muscle
11- 股直肌 rectus femoris muscle

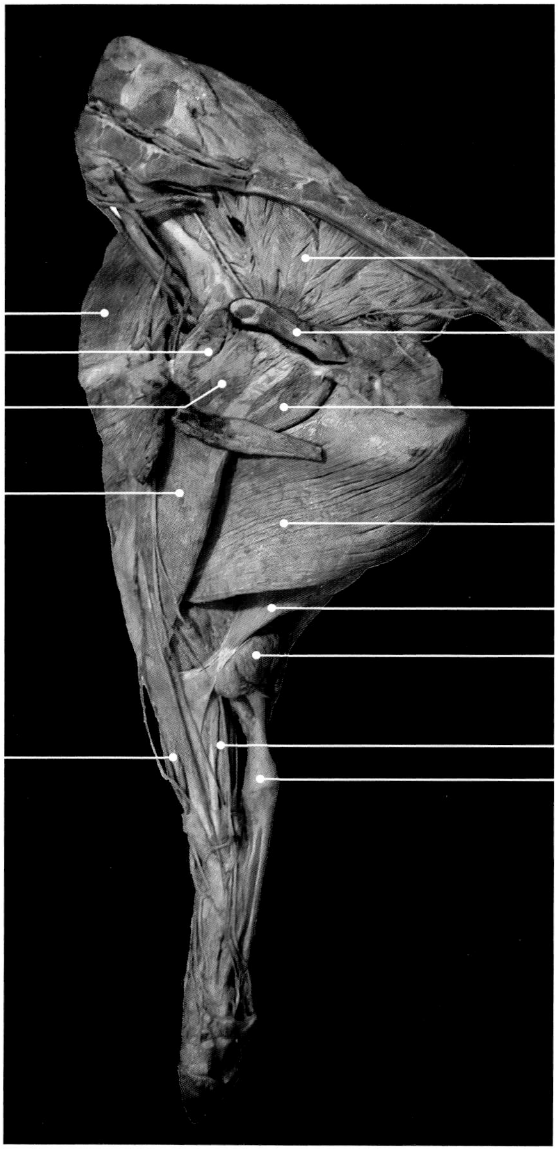

图 3-31　右后肢肌（内侧观，塑化标本）
Figure 3-31　Muscles of right hindlimb (medial view, plasticzed preparation).

1- 闭孔内肌 internal obturator muscle
2- 骨盆联合 pelvic symphysis
3- 内收肌 adductor muscle
4- 半膜肌 semimembranous muscle
5- 半腱肌 semitendinous muscle
6- 腓肠肌 gastrocnemius muscle
7- 趾深屈肌 deep digital flexor muscle
8- 跟结节 calcaneal tuberosity
9- 第 3 腓骨肌 3rd fibular muscle
10- 股内侧肌 medial vastus muscle
11- 耻骨肌 pectineal muscle
12- 缝匠肌 sartorius muscle
13- 腹内斜肌 internal oblique abdominal muscle

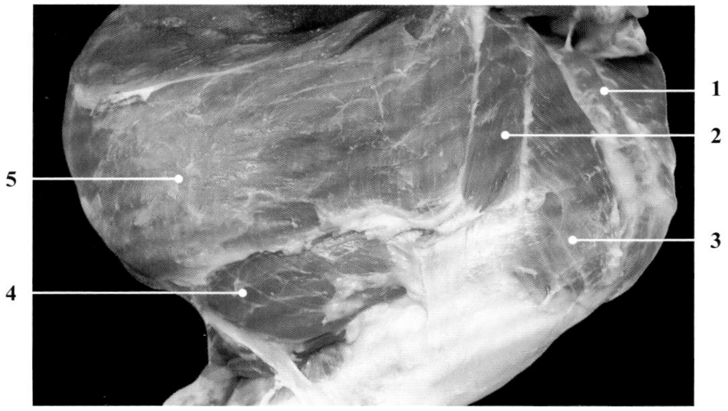

图 3-32　股内侧肌群（内侧观）
Figure 3-32　Interfemus muscle groups (medial view).

1- 股直肌 rectus femoris muscle
2- 缝匠肌 sartorius muscle
3- 股内侧肌 medial vastus muscle
4- 腓肠肌 gastrocnemius muscle
5- 股薄肌 gracilis muscle

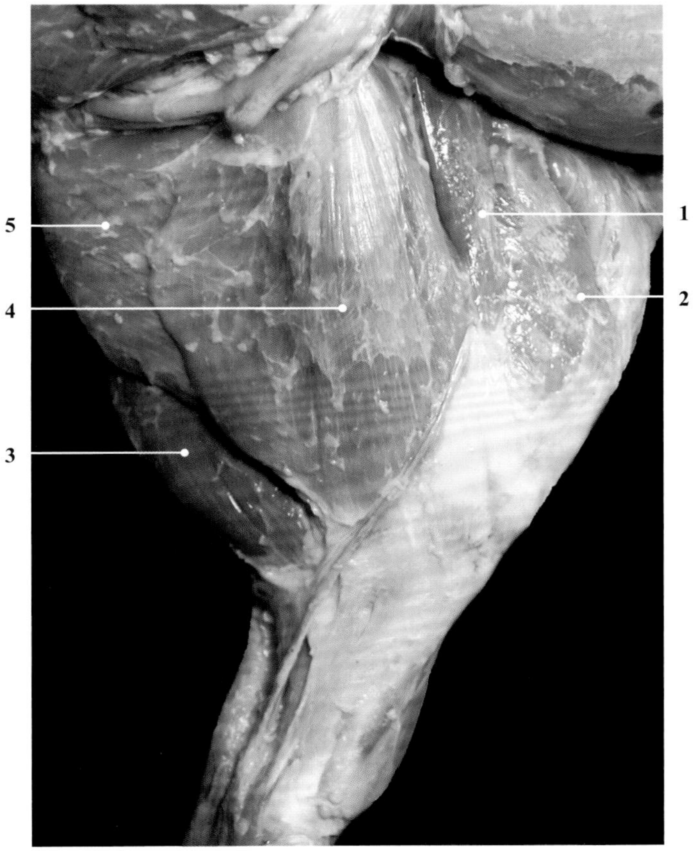

图 3-33　左后肢肌（切除股薄肌，内侧观）
Figure 3-33　Muscles of left hindlimb (Removal of the gracilis muscle, medial view).

1- 缝匠肌 sartorius muscle
2- 股内侧肌 medial vastus muscle
3- 腓肠肌 gastrocnemius muscle
4- 内收肌 adductor muscle
5- 半膜肌 semimembranous muscle

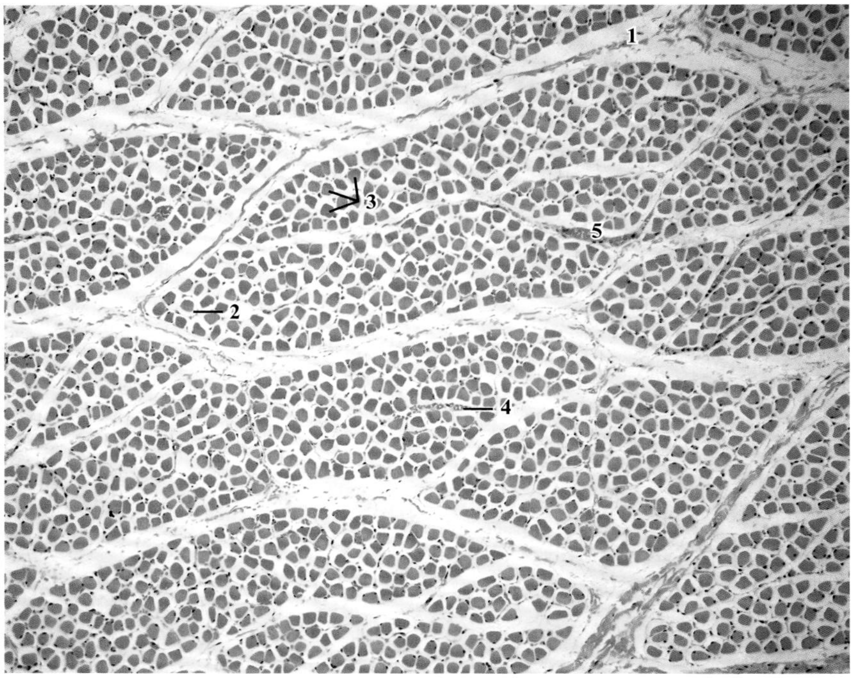

1- 肌束膜 perimysium
2- 肌内膜 endomysium
3- 肌纤维 muscle fiber
4- 毛细血管 capillary
5- 小静脉 small vein

图 3-34　咬肌组织切片（HE 染色，低倍镜）
Figure 3-34　Histological section of the masseter muscle (HE staining, lower power).

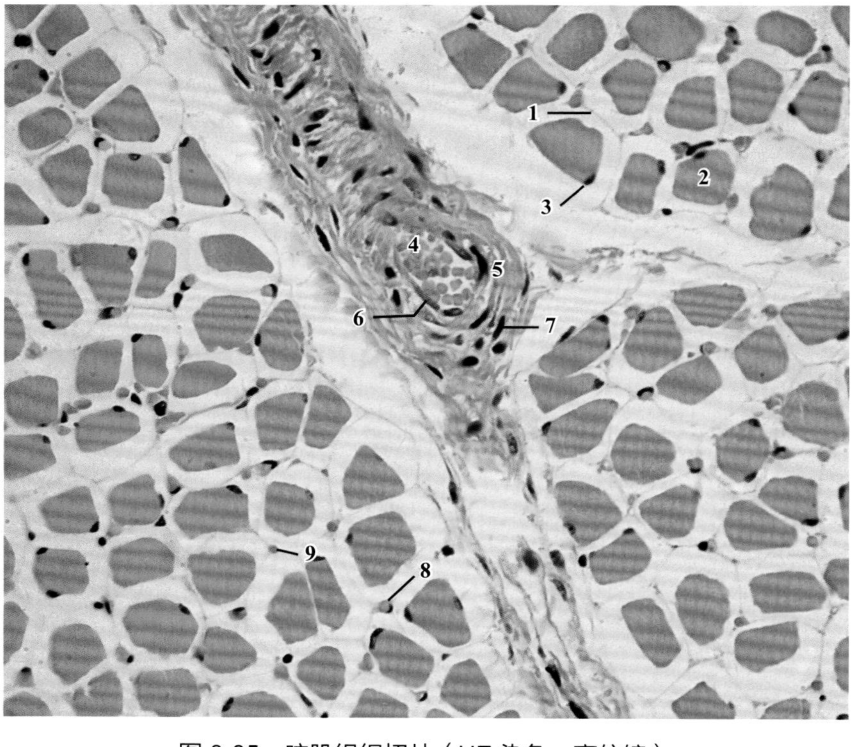

1- 肌内膜 endomysium
2- 肌纤维 muscle fiber
3- 肌细胞核 nucleus of the muscle cell
4- 微动脉 arteriole
5- 微动脉中膜 tunia media of the arteriole
6- 内皮细胞 endotheliocyte
7- 平滑肌细胞核 nucleus of the smooth muscle cell
8- 毛细血管 capillary
9- 红细胞 erythrocyte

图 3-35　咬肌组织切片（HE 染色，高倍镜）
Figure 3-35　Histological section of the masseter muscle (HE staining, higher power).

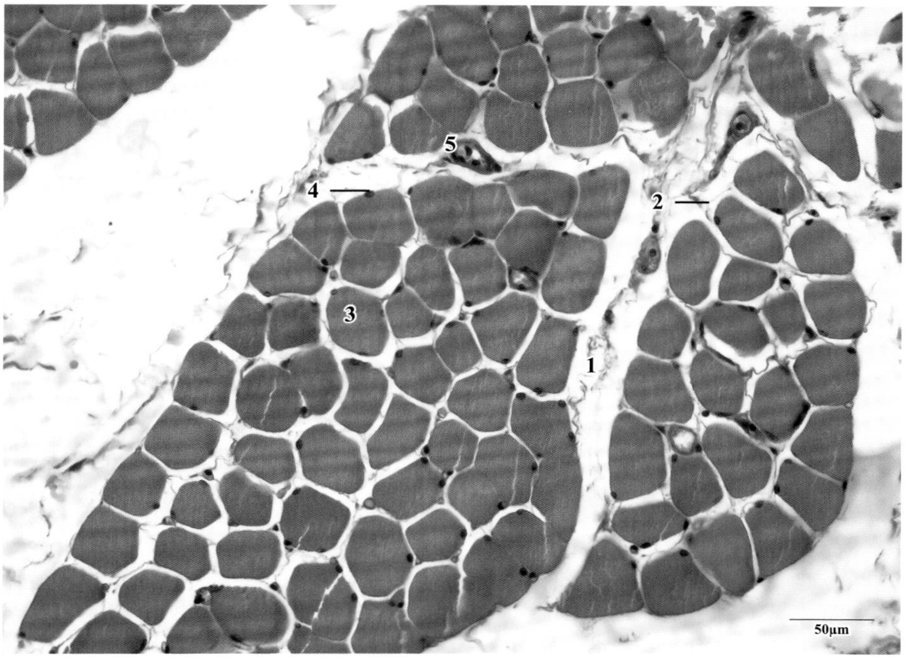

1- 肌束膜 perimysium
2- 肌内膜 endomysium
3- 肌纤维 muscle fiber
4- 肌细胞核 nucleus of the muscle cell
5- 毛细血管 capillary

图 3-36 腓肠肌组织切片（HE 染色，高倍镜）
Figure 3-36 Histological section of the gastrocnemius muscle (HE staining, higher power).

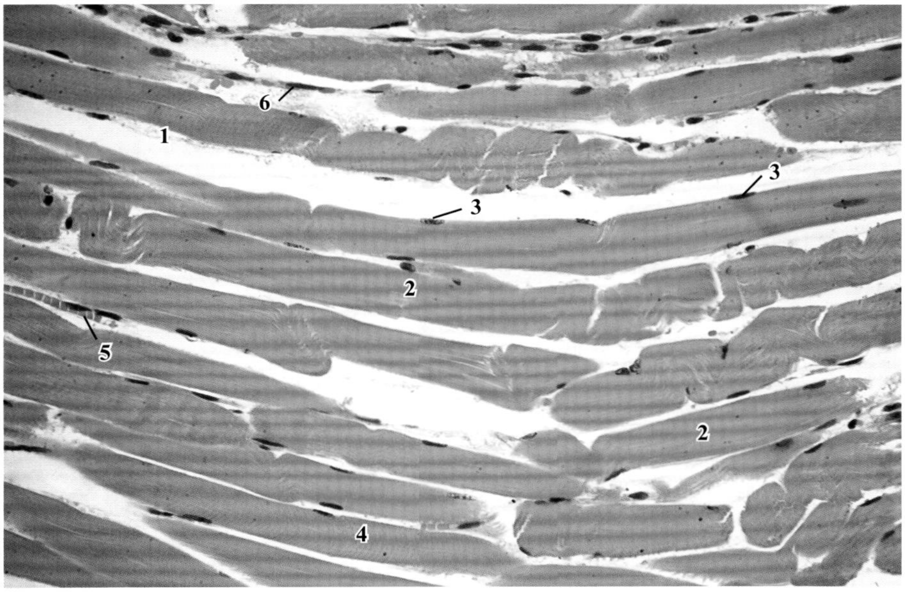

1- 肌内膜 endomysium
2- 肌纤维 muscle fiber
3- 肌细胞核 nucleus of the muscle cell
4- 横纹 cross striation
5- 毛细血管 capillary
6- 内皮细胞 endotheliocyte

图 3-37 三角肌组织切片（HE 染色，高倍镜）
Figure 3-37 Histological section of the deltoid muscle (HE staining, higher power).

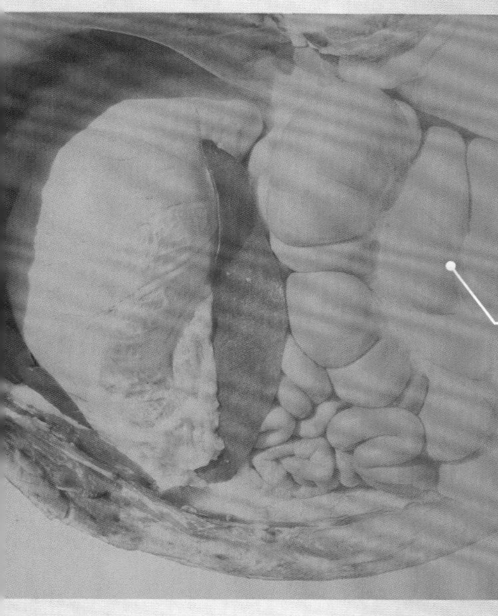

第四章

消化系统

Chapter 4
Digestive system

消化系统（digestive system）包括消化管（digestive tract）和消化腺（digestive gland）。消化管包括口腔、咽、食管、胃、肠和肛门。消化腺包括壁内腺和壁外腺。壁内腺存在于消化管壁中，如食管腺、胃腺、肠腺等。壁外腺包括唾液腺（腮腺、颌下腺、舌下腺等）以及肝、胰。消化系统的主要功能是摄取和消化食物，吸收食物中的营养并排出残渣。

一、口腔

口腔（oral cavity）是消化管的起始部，前壁为唇，侧壁为颊，顶壁为硬腭，底壁的大部分被舌所占据。前端经口裂与外界相通，向后与咽相通。

1. 唇（lip）：分为上唇和下唇，上、下唇之间围成口裂（oral fissure）。口裂大，上唇与鼻连为一体构成吻突（rostral disc, snout）或吻镜（rostral plate），内有一吻骨。下唇短而尖，不灵活。唇以口轮匝肌、唇腺以及结缔组织为基础，内覆黏膜，外被皮肤而成，但唇腺不发达。

2. 颊（cheek）：构成口腔的侧壁，主要由颊肌内衬黏膜、外覆皮肤构成。颊黏膜光滑，颊腺排列于臼齿上下两排。

3. 硬腭（hard palate）：构成口腔的顶壁，向后与软腭相延续。硬腭长而窄，黏膜表面覆以角质化的复层扁平上皮。黏膜面正中有一条纵向的沟，称为腭缝（palatine raphe）。腭缝的两侧为22条横行隆起的腭褶（palatine fold），前端有一隆起，为切齿乳头。切齿乳头的两侧有切齿管，通软腭。软腭（soft palate）是硬腭向后的延伸，为一含有肌组织和腺体的黏膜褶，短而厚，前缘附着于腭骨，后缘游离，在游离缘正中形成明显的腭垂（悬雍垂，uvula）。软腭腹侧口腔面有一浅矢状沟，沟的两侧是发达的腭帆扁桃体（tonsil of palatine velum），但猪没有腭扁桃体。

4. 口腔底：大部分被舌占据。舌尖后部与口腔底之间有2条舌系带相连。舌下肉阜小或缺，颌下腺管开口于舌系带。

5. 舌（tongue）：为一肌性器官。舌长而狭，分为舌尖、舌体和舌根3部分。舌尖薄，舌根的后部连接会厌，此处的黏膜形成纵褶，即舌会厌褶，旁有舌扁桃体（lingual tonsil）。舌的背侧面中后部稍隆起，称为舌背（dorsum of tongue）。舌由舌肌和黏膜构成。舌肌为横纹肌。黏膜表面覆有复层扁平上皮，黏膜角质化形成舌乳头（lingual papilla）。黏膜固有层的结缔组织向上皮形成次级乳头。舌背黏膜有丝状乳头（filiform papilla）、轮廓乳头（vallate papilla）、菌状乳头（fungiform papilla）、叶状乳头（foliate papilla）和锥状乳头（conical papillae）。丝状乳头和锥状乳头的表层细胞角质化明显，可进行机械作用；而菌状乳头、轮廓乳头和叶状乳头含有味蕾（taste bud），具有味觉功能。

6. 齿（tooth）：嵌合于上、下颌骨及切齿骨的齿槽内，是体内最坚硬的器官。上、下列齿排列形成弓状，称为上、下齿弓。根据齿的功能、位置和形态结构，可将齿分为切齿（incisor）、犬齿（canine）和颊齿（cheek tooth）。颊齿又分为前臼齿（premolar）和臼齿（molar）。上、下切齿每侧各有3枚，上切齿较小，下切齿较大，中间齿最大，边齿最小。犬齿在公猪发育完好，上犬齿长5～18cm，下犬齿长6～10cm，都向后外方弯曲，相互摩擦，愈磨愈尖。颊齿由前向后，递次增大。第1前臼齿较小，也称狼齿，无乳狼齿。

猪乳齿齿式 $2\left[\frac{3(I)\quad 1(C)\quad 3(P)\quad 0(M)}{3(I)\quad 1(C)\quad 3(P)\quad 0(M)}\right]=28$

猪恒齿齿式 $2\left[\frac{3(I)\quad 1(C)\quad 4(P)\quad 3(M)}{3(I)\quad 1(C)\quad 4(P)\quad 3(M)}\right]=44$

7. 唾液腺（salivary gland）：分布于口腔周围，分为小唾液腺和大唾液腺两类。前者如颊腺、唇腺、舌腺等壁内腺，直接位于黏膜下；后者有腮腺、颌下腺和舌下腺，以较长的腺管开口于口腔。

（1）腮腺（parotid gland）较发达，呈倒三角形，淡红色，位于耳根下方。腮腺管经下颌骨腹侧缘转至咬肌前缘，开口于第4上前臼齿相对的颊黏膜上。大部分属于浆液腺，但常见小的黏液性细胞群。

（2）颌下腺（mandibular gland）较小，呈卵圆形，色淡红，位于腮腺的深面。腺管开口于舌系带附着处。属于混合腺，既有黏液腺泡，又有浆液性腺泡，但黏液性腺泡占多数。

（3）舌下腺（sublingual gland）位于舌体和下颌骨之间的黏膜下，分前部的多管舌下腺和后部的单管舌下腺，腺管开口于口腔底黏膜上。舌下腺是混合腺，但黏液性腺泡占多数。

二、咽

咽（pharynx）是消化管与呼吸道共同的通道。由于软腭的伸入，咽腔被分为鼻咽部、口咽部和喉咽部3部分。咽通过鼻咽部与一对鼻后孔和一对咽鼓管相通，通过口咽部与口腔相通，后端经喉口和食管口通喉腔和食管。

咽狭长，鼻咽部正中有矢状的咽中隔，其侧壁上有咽漏斗；喉咽部的喉口及会厌向前突出，在其侧面形成凹陷的梨状隐窝（piriform recess）或咽后隐窝（pharyngeal recess）。咽黏膜内淋巴组织发达，形成扁桃体（tonsil），包括成对的舌扁桃体（lingual tonsil）、腭帆扁桃体（tonsil of palatine velum）、咽鼓管扁桃体（tubal tonsil）和不成对的咽扁桃体（pharyngeal tonsil）。

三、食管

食管（esophagus）短而直，始终位于气管的背侧，前端以漏斗状与口咽相连，末端与胃的贲门相连。管壁可分黏膜、黏膜下层、肌层、外膜或浆膜4层。黏膜表面为复层扁平上皮，轻度角质化；固有层由致密结缔组织构成；黏膜肌层为散在的纵行平滑肌束，可作为黏膜层与黏膜下层之间的分界线，食管前段缺乏黏膜肌层，后段变得较发达。黏膜下层为疏松结缔组织，十分发达，内有食管腺（esophageal gland）。食管腺多集中在食管的前半部，属黏液细胞为主的混合腺。肌层几乎全部为横纹肌，仅腹腔段有平滑肌分布。内环行肌层较厚，外环行肌层较薄。颈段食管肌层外面为外膜，胸腹段则为浆膜。

四、胃

猪胃（stomach）为单室混合型胃，大部分位于左季肋部，小部分位于右季肋部。呈左右横向弯曲的囊状。胃小弯短而凹，胃大弯向下突出，饱食时可接触到腹腔底。胃左端后上方突出一圆锥状的胃憩室（gastric diverticulum）。贲门位于胃小弯左侧面。幽门位于右季肋部，其内腔有一鞍形黏膜隆起，为幽门圆枕（torus pyloricus），具有关闭幽门的作用。胃黏膜分为有腺部和无腺部。无腺部小，黏膜内无腺体，仅分布于贲门周围，黏膜苍白，向上可延伸至胃憩室。有腺部的面积大，分为贲门腺区、胃底腺区和幽门腺区。贲门腺区最大，约占胃黏膜的1/3，呈淡灰色；胃底腺区较小，沿胃大弯分布，呈棕红色；幽门腺区最小，位于幽门部，呈灰白色至黄色。

胃壁的结构具有消化道典型的4层结构：黏膜层（包括上皮、固有层和黏膜肌层）、黏膜下层、肌层（包括内斜行、中环行和外纵行3层平滑肌）和浆膜。无腺部的黏膜上皮为复层扁平上皮，而有腺部则为单层柱状上皮。在有腺部，黏膜表面有许多由上皮凹陷形成的胃小凹（gastric pit），是胃腺的开口。柱状上皮主要是表面黏液细胞，是胃黏膜屏障的主要成分。固有层由富含网状纤维的结缔组织构成，有大量浸润的白细胞和淋巴小结。黏膜下层为疏松结缔组织，分布着较多的血管、淋巴管、黏膜下神经丛和孤立淋巴小结，尤以胃盲囊的憩室内最多。

胃腺（gastric gland）分布于黏膜固有层，分贲门腺、胃底腺和幽门腺。胃底腺（fundic gland）数量最多，分布在胃底部，属分枝单管状腺，分颈、体、底3部分。由壁细胞、主细胞、颈黏液细胞、干细胞和内分泌细胞组成。壁细胞（parietal cell）又称泌酸细胞（oxyntic cell），细胞较大，呈锥形，多位于腺的颈和体部，胞质嗜酸性，分泌盐酸和内因子。主细胞（chief cell）又称胃酶原细胞（zymogenic cell），数量多，呈柱状，多位于腺的体部和底部，胞质呈酸碱性，分泌胃蛋白酶原。猪的颈黏液细胞（mucus neck cell）分布于腺体各部，以腺底部最多，呈矮柱状，分泌酸性黏液。贲门腺（cardiac gland）为分枝单管状腺，腺细胞呈立方形或柱状，分泌黏液。幽门腺（pyloric gland）属分枝泡状腺，腺细胞呈柱状，胃小凹深，分泌黏液。

五、小肠

小肠（small intestine）是食物消化和吸收的主要场所，起于幽门，后端止于盲肠，长15～20m，可分为十二指肠（duodenum）、空肠（jejunum）和回肠（ileum）3部分，主要位于腹腔的右侧。

小肠是典型的管状器官，管壁分黏膜、黏膜下层、肌层和浆膜。黏膜的结构特点是有环行皱襞、肠绒毛、微绒毛和小肠腺。黏膜和部分黏膜下层向肠腔内隆起形成环行皱襞。在皱襞的黏膜表面布满了由上皮和固有层突向肠腔内形成的指状突起——肠绒毛（intestinal villus）。绒毛表面为单层柱状上皮，由吸收细胞（absorptive cell）、杯状细胞（goblet cell）和少量的内分泌细胞组成。吸收细胞朝向肠腔的游离缘有明显的纹状缘（striated border），是由密集排列的微绒毛构成。绒毛根部的肠上皮下陷至固有层内形成单管状腺的小肠腺（small intestinal gland），又称肠隐窝（intestinal crypt）或李氏隐窝（crypt of Lieberkühn）。这些结构极大地扩大了小肠的分泌表面和吸收表面。固有层分布于肠腺之间并构成绒毛的中轴，为富含网状纤维的疏松结缔组织，内有血管、淋巴管、中央乳糜管（central lacteal）、神经、淋巴细胞和巨噬细胞等。黏膜下层为疏松结缔组织，有黏膜下腺（submucosal gland），为黏液腺，主要分布在十二指肠和部分空肠，又称十二指肠腺（duodenal gland，或布伦纳腺，Brunner's gland）。回肠的黏膜下层和固有层有大量的集合淋巴小结（派尔结，Peyer's patch）分布；空肠也有，但较少。

六、大肠

大肠（large intestine）主要位于腹腔的左侧，包括盲肠（caecum）、结肠（colon）和直肠（rectum），末端以肛门（anus）开口于外界。盲肠短而直，呈圆筒状。结肠分为升结肠（ascending colon）、横结肠（transverse colon）和降结肠（decending colon）。升结肠分为结肠旋襻和结肠终襻；旋襻呈螺旋状盘曲形成结肠圆锥，由向心回和离心回组成，向心回较粗。直肠有直肠壶腹（rectal ampulla）。肛门不向外突出。

大肠壁的组织学结构与小肠的相似，不同点主要表现在黏膜。大肠黏膜没有绒毛，但杯状细胞多，肠腺发达，固有层和黏膜下层有大量的孤立淋巴小结。

七、肝

肝（liver）较大且发达，占体重的1.5%～2.1%。大部分位于右季肋区，小部分位于左季肋区。中央厚而周边薄锐，壁面向前凸，与膈相贴；脏面凹，朝向后方。肝以3个深的切迹分为4叶，由左向右分别是左外叶、左内叶、右内叶和右外叶。其中右内叶的脏面上有嵌入胆囊窝内的胆囊（gallbladder）。胆囊以及胆囊管与食管压迹之间围成狭长的中叶，又以肝门分为尾叶和方叶。尾叶的尾状突较小，不与肾接触。胆囊管在肝门处与肝管汇合形成胆总管，开口于十二指肠乳头。

肝脏为实质性器官，表面覆盖有一层富含弹性纤维的结缔组织被膜。被膜表面大部分有浆膜，其结缔组织在肝门处与门静脉、肝动脉、肝管的分支和淋巴管等伸入肝实质，将其分成许多肝小叶。猪的肝小叶发达，界线明显。肝小叶（hepatic lobule）是肝的基本结构和功能单位，呈多面棱柱体，横断面为不规则多边形。小叶的中央有一条纵贯长轴的中央静脉（central vein）。外周是呈放射状相间排列的肝板（hepatic plate）和血窦（sinusoid）。肝板由单行肝细胞排列而成。肝细胞（hepatocyte）呈多角形，有1～2个胞核。肝血窦为肝板之间相互吻合的网状血管，窦壁由一层内皮细胞构成，窦内有一种巨噬细胞，称枯否细胞（Kuffer's cell），具有吞噬功能。血窦内皮细胞与肝细胞之间的微小裂隙称为窦周间隙，其内充满由血窦渗出的血浆成分，是肝细胞与血液间进行物质交换的场所。间隙内有散在的网状纤维和贮脂细胞。贮脂细胞有储存脂肪和维生素A的作用。相邻几个肝小叶之间的结缔组织组成门管区（portal area），内有小叶间动脉、小叶间静脉和小叶间胆管伴行。

八、胰

胰（pancreas）位于胃小弯后上方，十二指肠左侧，略呈三角形，分为胰体和左右两叶。胰体（胰头）位于中间，左叶和右叶伸向侧方。门静脉穿过胰头前部，形成门静脉环（annulus portae），或称胰环（pancreatic ring）。胰管（pancreatic duct）自右叶走出，开口于十二指肠黏膜。

胰由外分泌部和内分泌部构成。外分泌部是重要的消化腺，分泌胰液，参与消化；内分泌部则分泌激素，参与糖代谢的调节。胰为实质性器官，表面被覆一薄层结缔组织被膜，后者深入胰的实质，将胰分成许多小叶。外分泌部构成胰小叶的绝大部分，为管泡状腺，腺泡为单层锥形的浆液性细胞，分泌胰液；导管为输送胰液至十二指肠的管道，包括闰管、小叶内导管、小叶间导管和胰管。内分泌部为分散于外分泌部腺泡之间的细胞团，称胰岛或朗格汉斯岛（pancreatic islet, islet of Langerhans），由A细胞（α细胞，分泌胰高血糖素）、B细胞（β细胞，数量最多，分泌胰岛素）、D细胞（δ细胞，分泌生长抑素）和PP细胞（F细胞，数量最少，分泌胰多肽）等组成。

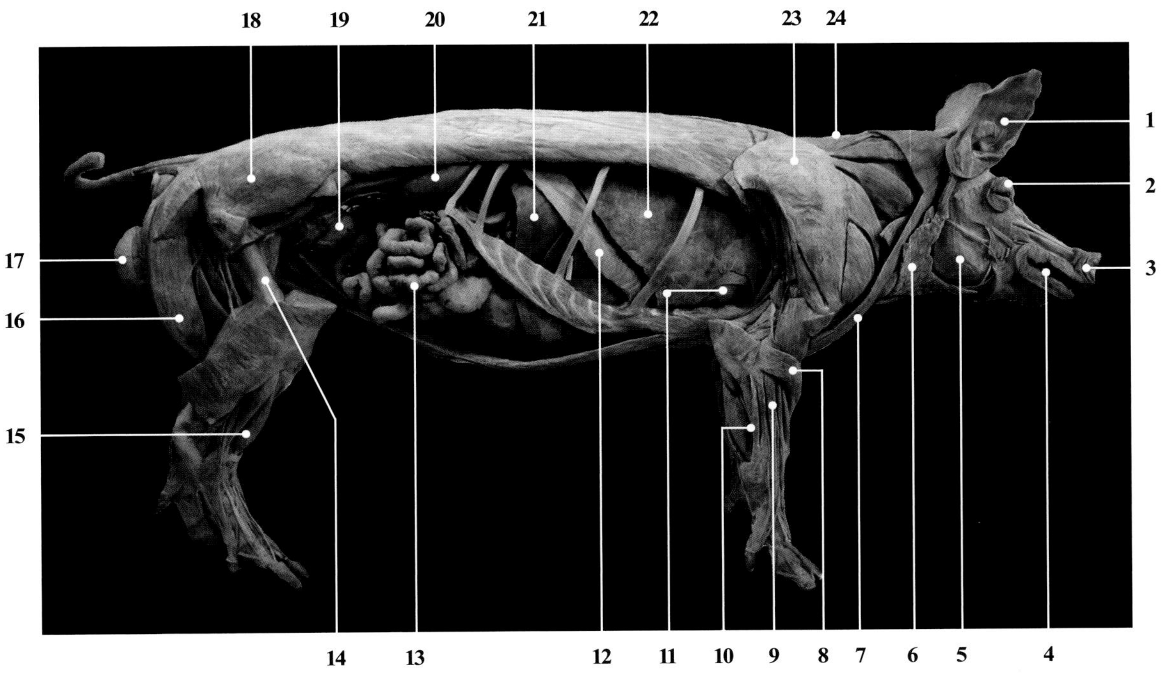

图 4-1 内脏器官（右侧观）
Figure 4-1 Visceral organs (right lateral view).

1- 耳 ear
2- 眼 eye
3- 吻突 rostral disc, snout
4- 口腔 oral cavity
5- 咬肌 masseter muscle
6- 腮腺 parotid gland
7- 臂头肌 brachiocephalic muscle
8- 腕桡侧伸肌 radial extensor muscle of the carpus
9- 指总伸肌 common digital extensor muscle
10- 腕外侧屈肌 lateral flexor muscle of the carpus
11- 心脏 heart
12- 膈 diaphragm
13- 小肠 small intestine
14- 股骨 femoral bone, femur
15- 第 3 腓骨肌 3rd fibular muscle
16- 半腱肌 semitendinous muscle
17- 睾丸 testis
18- 臀浅肌 superficial gluteal muscle
19- 膀胱 urinary bladder
20- 肾 kidney
21- 肝 liver
22- 肺 lung
23- 肩胛骨 scapula
24- 斜方肌 trapezius muscle

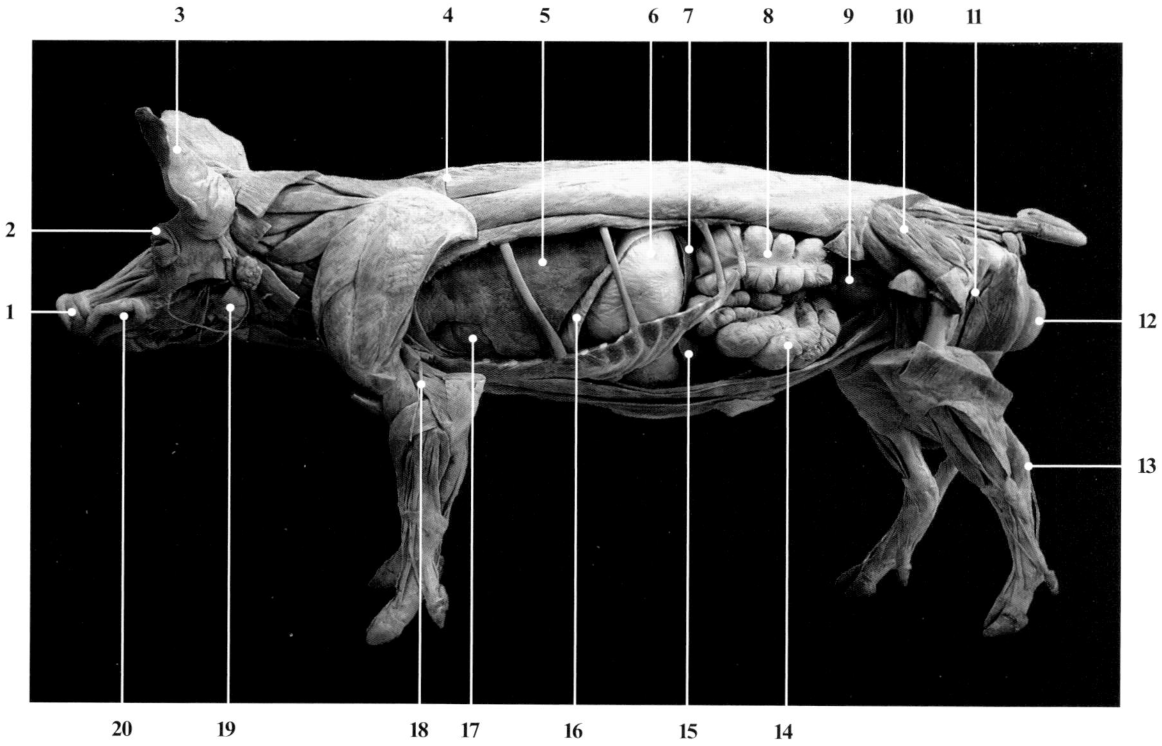

图 4-2　内脏器官（左侧观）
Figure 4-2　Visceral organs (left lateral view).

1- 吻突 rostral disc, snout
2- 眼 eye
3- 耳 ear
4- 斜方肌 trapezius muscle
5- 肺 lung
6- 胃 stomach
7- 脾 spleen
8- 盲肠 caecum
9- 膀胱 urinary bladder
10- 臀中肌 middle gluteal muscle
11- 坐骨神经 sciatic nerve
12- 睾丸 testis
13- 跟结节 calcaneal tuberosity
14- 结肠 colon
15- 肝 liver
16- 膈 diaphragm
17- 心脏 heart
18- 桡神经 radial nerve
19- 颌下腺 mandibular gland
20- 口腔 oral cavity

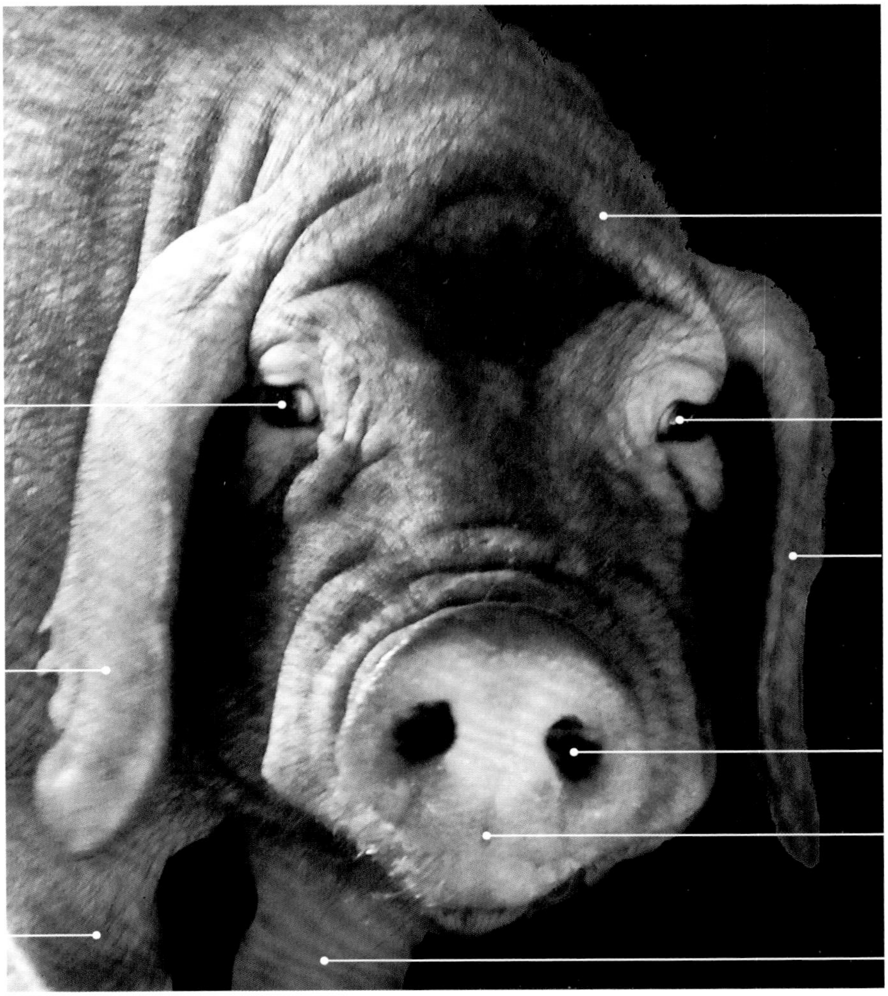

图 4-3 吻突
Figure 4-3 Rostral disc.

1- 被毛和皮肤 clothing hair and skin
2- 左眼 left eye
3- 左耳 left ear
4- 鼻孔 nostril
5- 吻突 rostral disc, snout
6- 左前肢 left forelimb
7- 右前肢 right forelimb
8- 右耳 right ear
9- 右眼 right eye

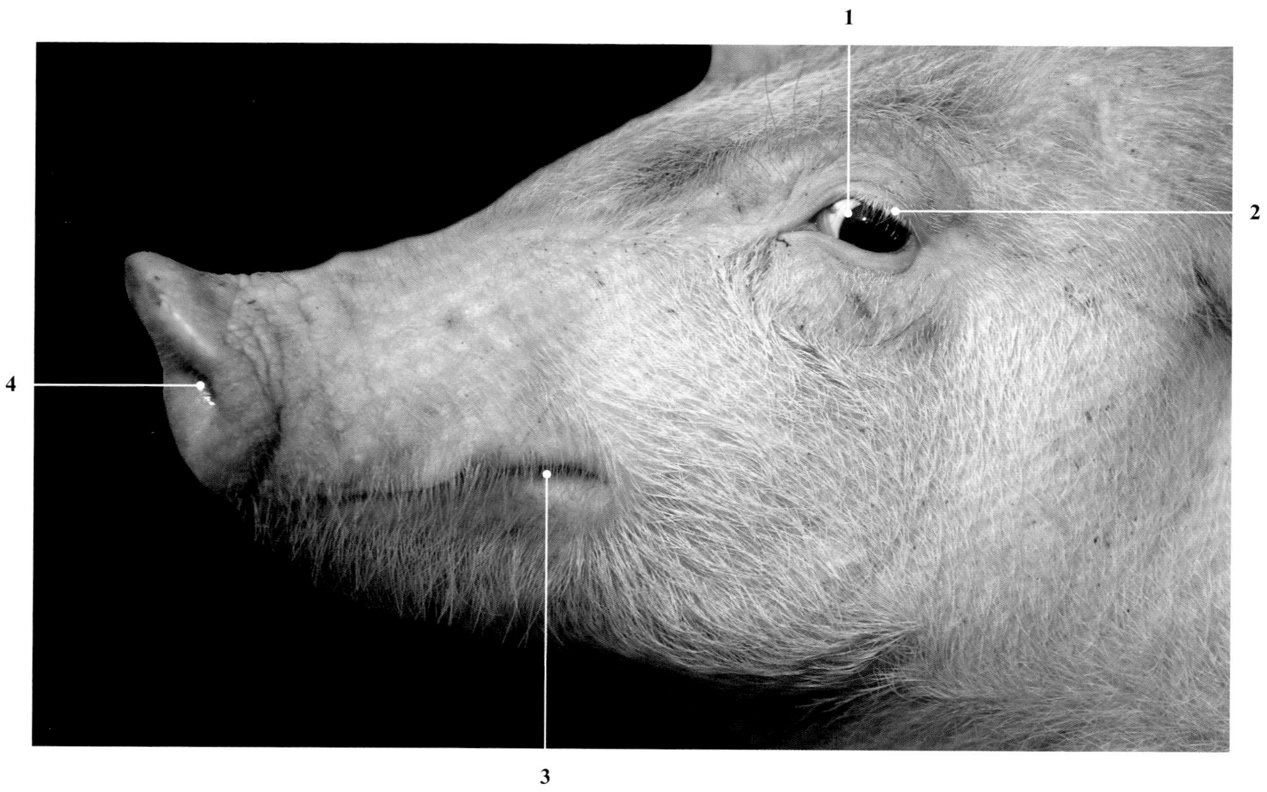

图 4-4　口裂
Figure 4-4　Oral fissure.

1- 眼 eye
2- 眼睫毛 eye lash
3- 口裂 oral fissure
4- 鼻孔 nostril

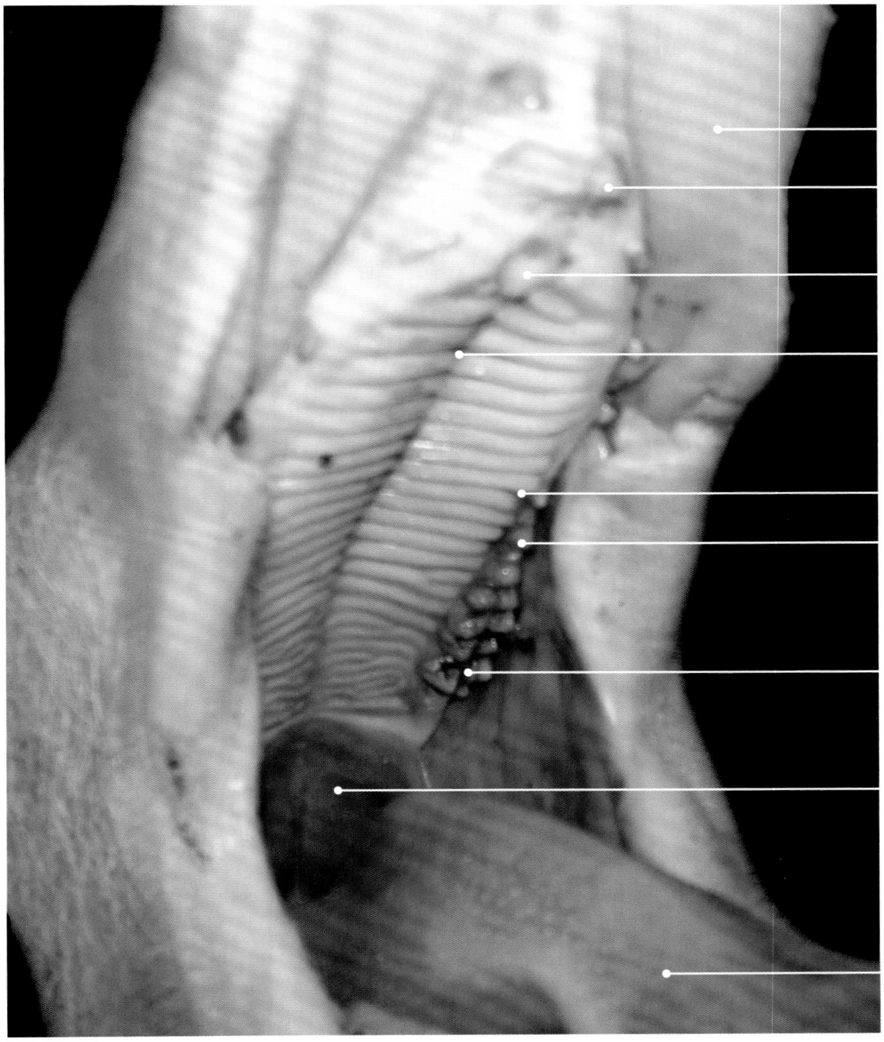

图 4-5 口腔
Figure 4-5 Oral cavity.

1- 上唇 upper lip
2- 切齿 incisor
3- 切齿乳头 incisive papilla
4- 腭缝 palatine raphe
5- 腭褶 palatine fold
6- 前臼齿 premolar
7- 臼齿 molar
8- 咽 pharynx
9- 舌 tongue

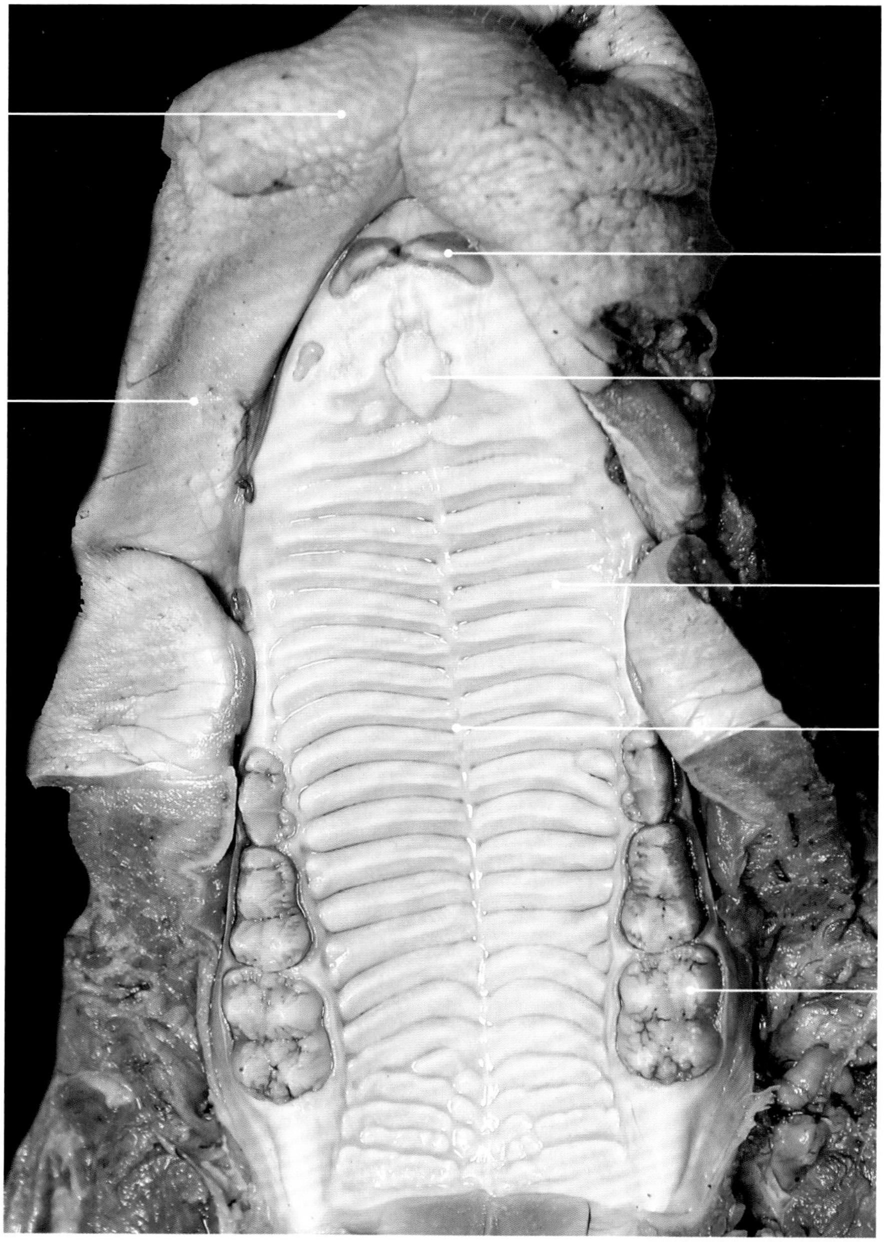

图 4-6 硬腭
Figure 4-6 Hard palate.

1- 切齿 incisor
2- 切齿乳头 incisive papilla
3- 腭褶 palatine fold
4- 腭缝 palatine raphe
5- 臼齿 molar
6- 上唇 upper lip
7- 吻突 rostral disc, snout

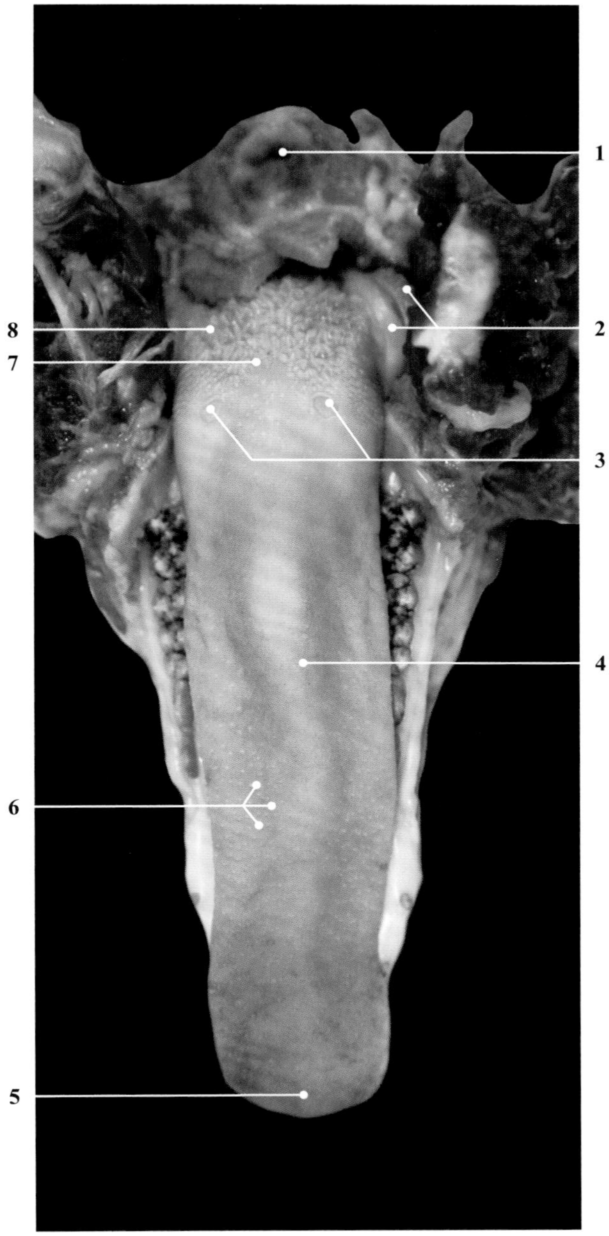

图 4-7　母猪舌（背侧面）
Figure 4-7　Tongue of a sow (dorsal aspect).

1- 喉口 aperture of larynx
2- 腭帆扁桃体 tonsil of palatine velum
3- 轮廓乳头 vallate papilla
4- 舌体 body of tongue (lingual body)
5- 舌尖 apex of tongue (lingual apex)
6- 菌状乳头 fungiform papilla
7- 舌根 root of tongue (lingual root)
8- 锥状乳头 conical papillae

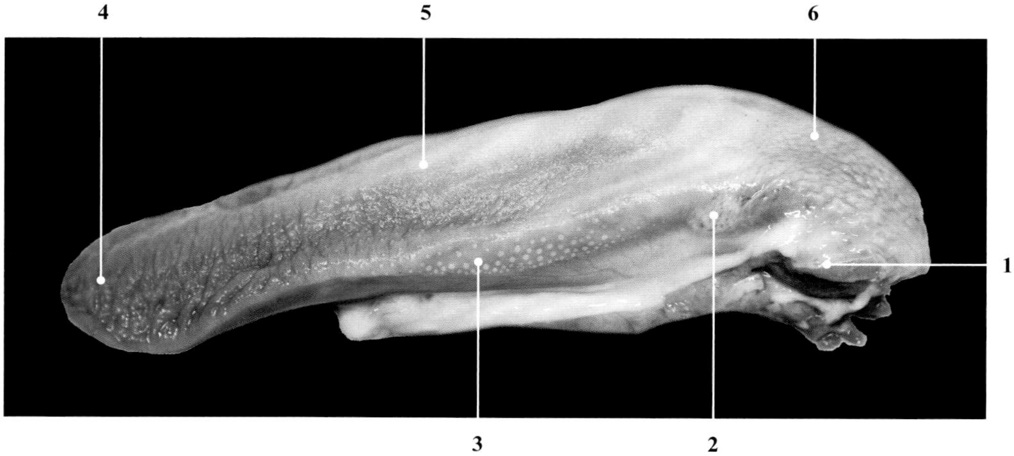

图 4-8　母猪舌（外侧面）
Figure 4-8　Tongue of a sow (lateral aspect).

1- 腭帆扁桃体 tonsil of palatine velum
2- 叶状乳头 foliate papilla
3- 菌状乳头 fungiform papilla
4- 舌尖 apex of tongue
5- 舌体 body of tongue
6- 舌根 root of tongue

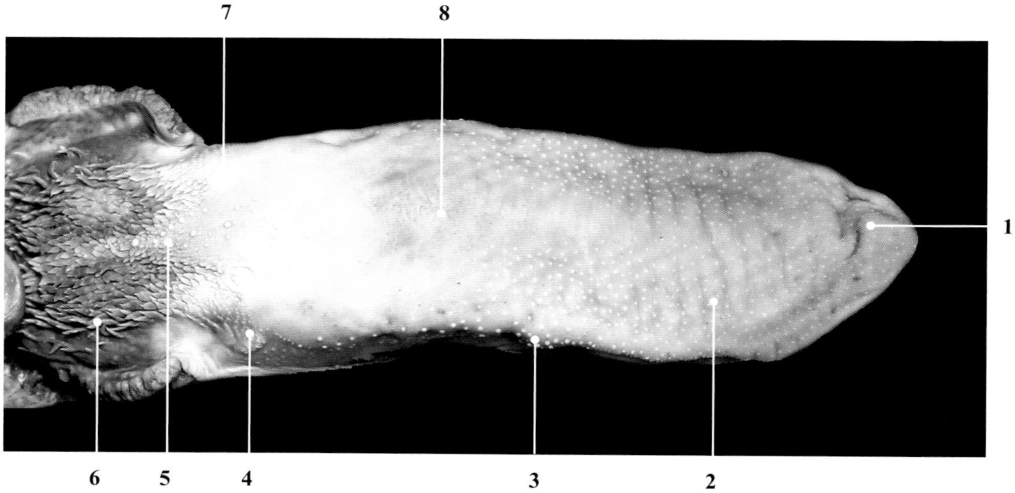

图 4-9　母猪舌（背外侧面）
Figure 4-9　Tongue of a sow (dorsolateral aspect).

1- 舌尖 apex of tongue
2- 丝状乳头 filiform papilla
3- 菌状乳头 fungiform papilla
4- 叶状乳头 foliate papilla
5- 舌根 root of tongue
6- 锥状乳头 conical papillae
7- 轮廓乳头 vallate papilla
8- 舌体 body of tongue

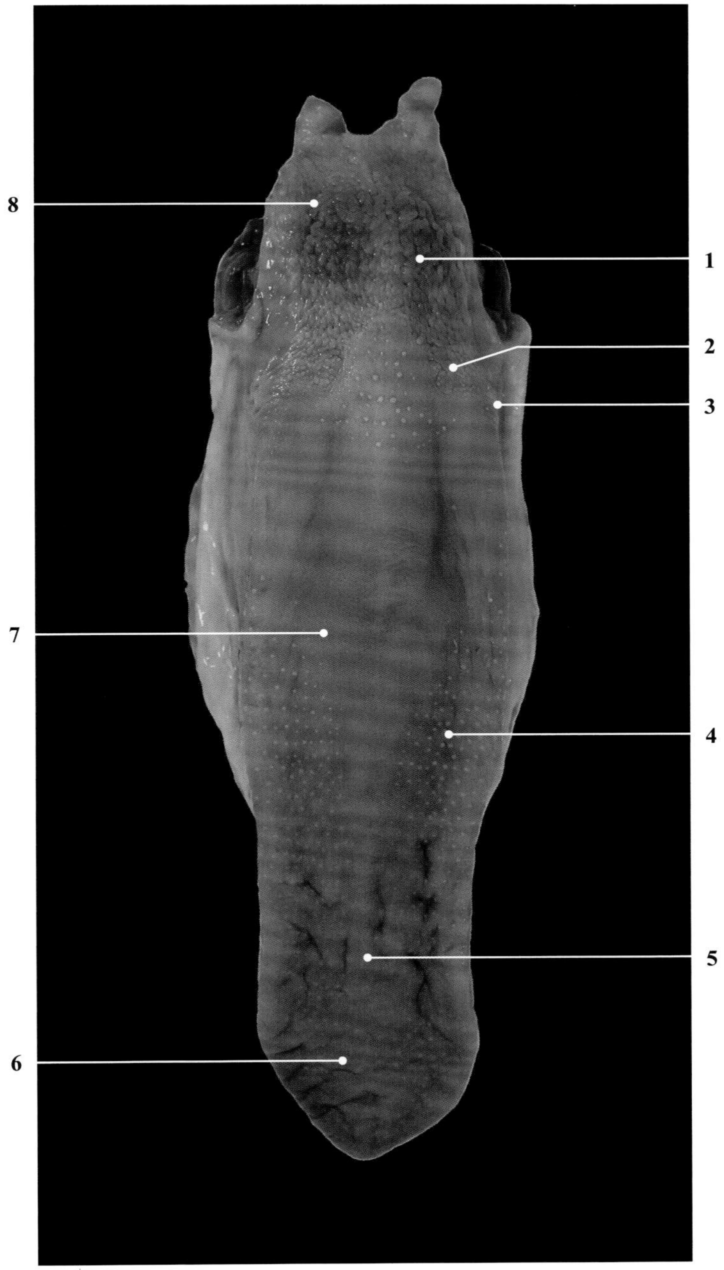

图 4-10 公猪舌（背侧面）
Figure 4-10 Tongue of a boar (dorsal aspect).

1- 锥状乳头 conical papilla
2- 轮廓乳头 vallate papilla
3- 叶状乳头 foliate papilla
4- 菌状乳头 fungiform papilla
5- 丝状乳头 filiform papilla
6- 舌尖 apex of tongue
7- 舌体 body of tongue
8- 舌根 root of tongue

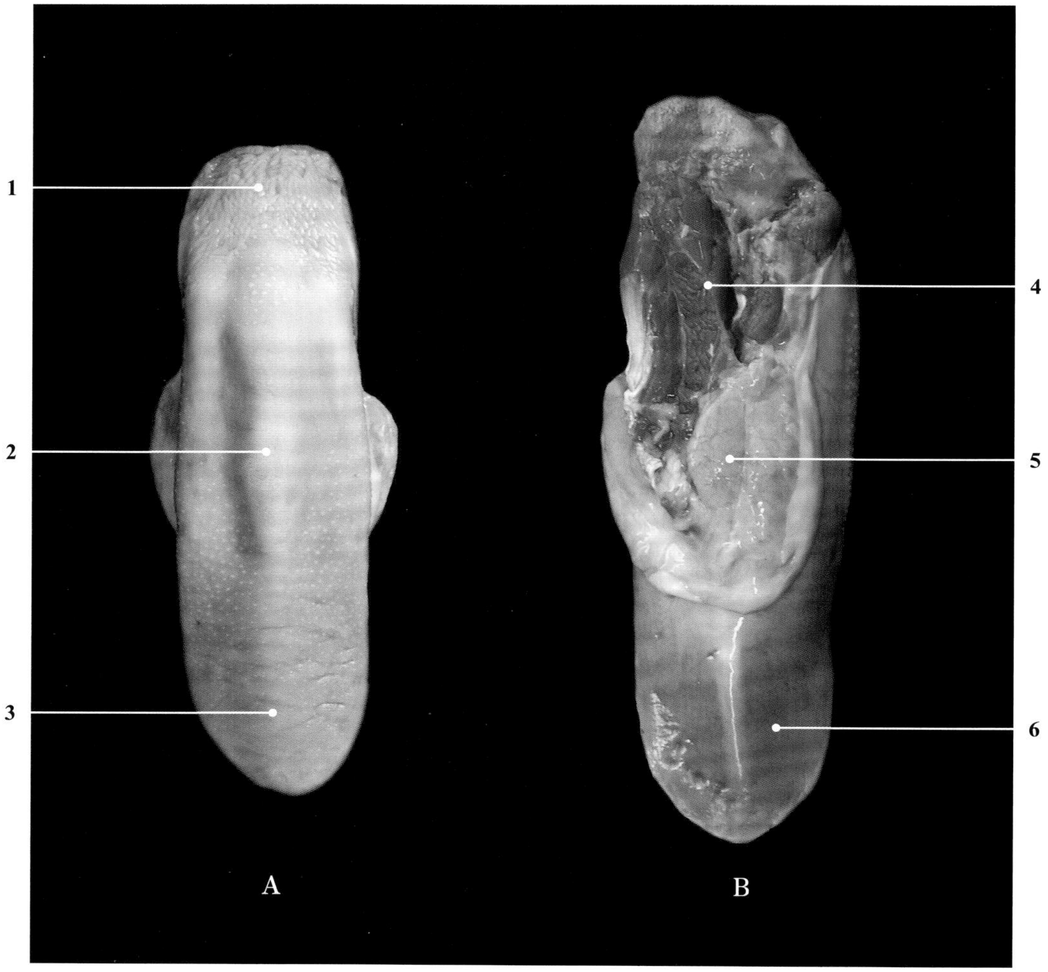

图 4-11 舌的形态
Figure 4-11 Morphology of the tongue.

A－腹侧面 (ventral aspect)
1- 舌根 root of tongue
2- 舌体 body of tongue
3- 舌尖 apex of tongue

B－背侧面 (dorsal aspect)
4- 舌肌 muscles of tongue
5- 舌下腺 sublingual gland
6- 舌黏膜 mucous membrane of the tongue

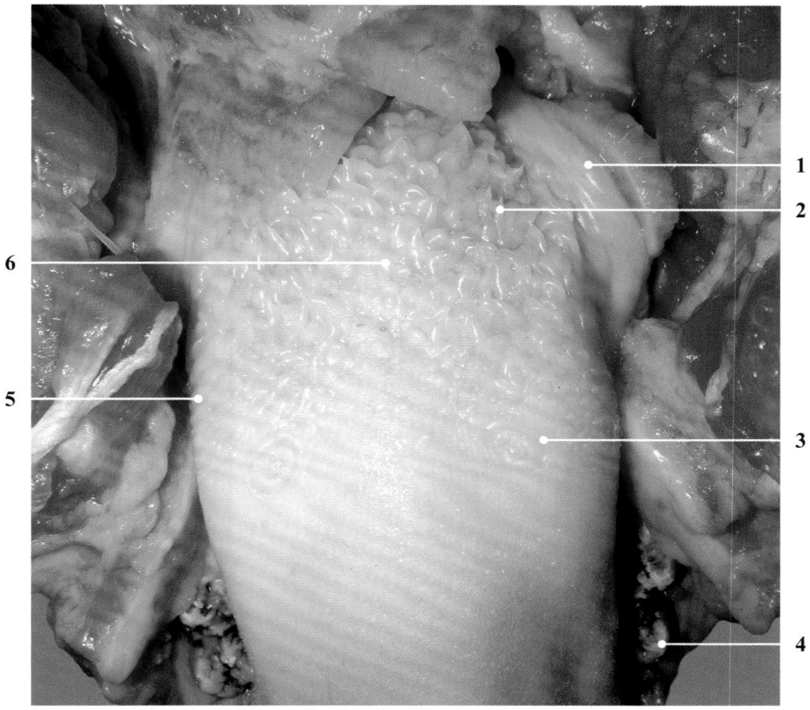

图 4-12　母猪舌根
Figure 4-12　Root of the tongue of a sow.

1- 腭帆扁桃体 tonsil of palatine velum
2- 锥状乳头 conical papilla
3- 轮廓乳头 vallate papilla
4- 臼齿 molar
5- 叶状乳头 foliate papilla
6- 舌根 root of tongue

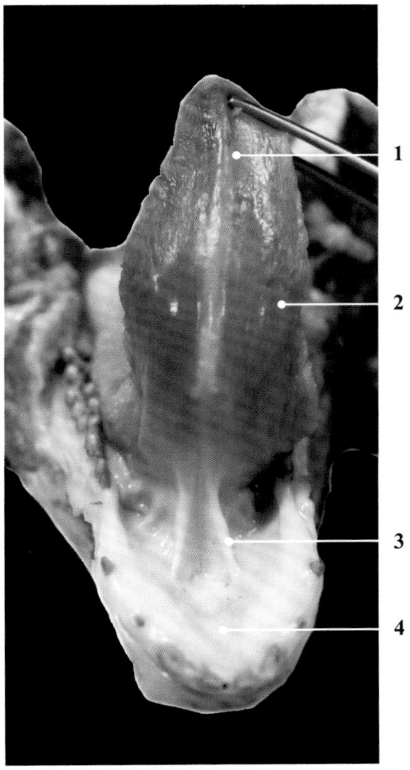

图 4-13　母猪舌系带
Figure 4-13　Lingual frenum of a sow.

1- 舌尖 apex of tongue
2- 舌体 body of tongue
3- 舌系带 lingual frenum
4- 口腔底 basis cavum oris

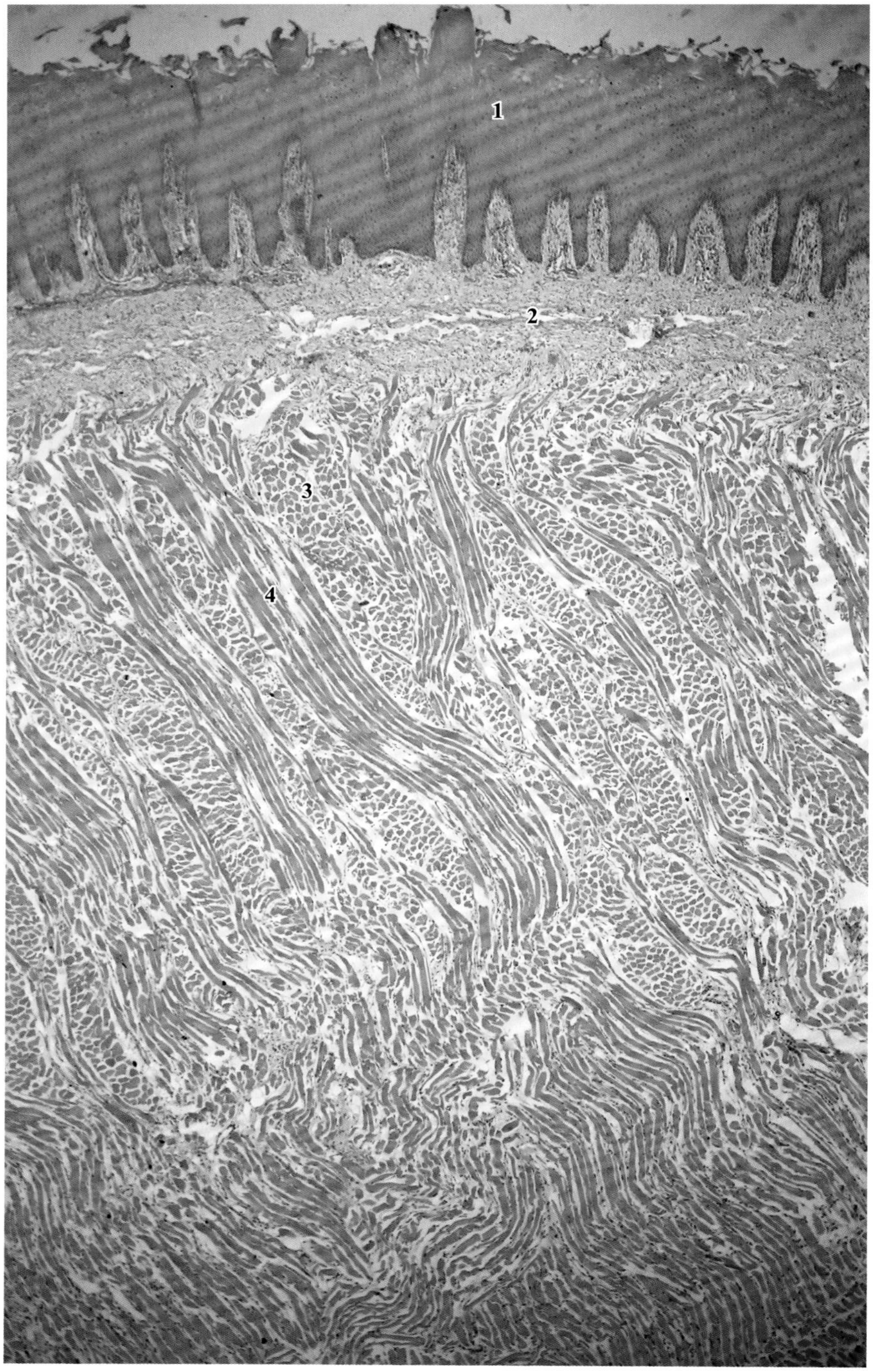

图 4-14　舌体组织切片（HE 染色，低倍镜）
Figure 4-14　Histological section of body of tongue (HE staining, lower power).

1- 角化的复层扁平上皮 keratinized stratified squamous epithelium
2- 固有层 lamina propria
3- 纵行肌 longitudinal muscle
4- 垂直肌 vertical muscle

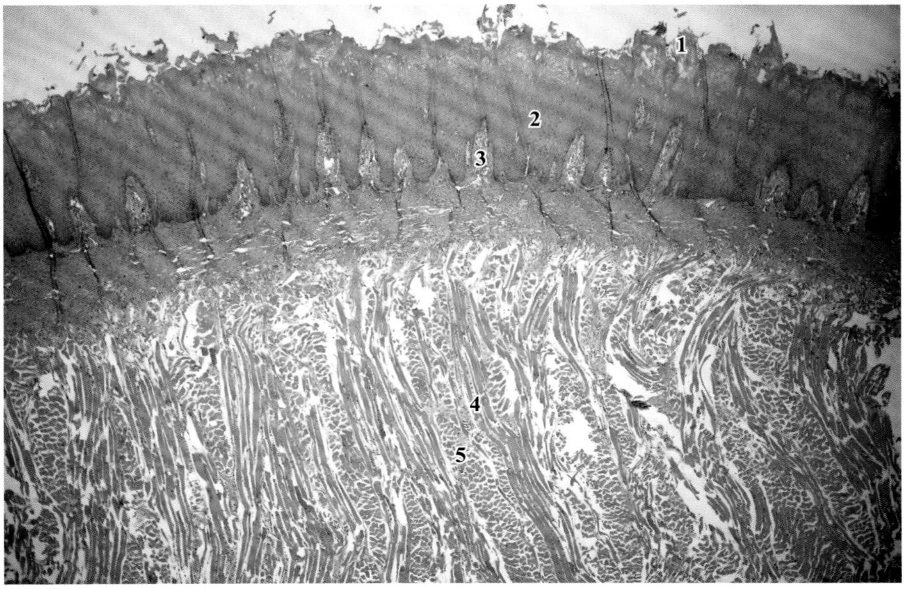

1- 舌乳头 lingual papilla
2- 复层扁平上皮 stratified squamous epithelium
3- 固有层 lamina propria
4- 垂直肌 vertical muscle
5- 纵行肌 longitudinal muscle

图 4-15 舌体组织切片（HE 染色，低倍镜）
Figure 4-15 Histological section of body of tongue (HE staining, lower power).

1- 味蕾 taste bud
2- 复层扁平上皮 stratified squamous epithelium
3- 固有层 lamina propria
4- 固有层初级乳头 primary papilla of lamina propria
5- 固有层次级乳头 secondary papilla of lamina propria
6- 肌层 muscular layer
7- 舌乳头 lingual papilla

图 4-16 舌体组织切片，示黏膜（HE 染色，高倍镜）
Figure 4-16 Histological section of body of tongue showing the mucosa (HE staining, higher power).

1- 复层扁平上皮 stratified squamous epithelium
2- 固有层 lamina propri
3- 微静脉 venule
4- 毛细血管 blood capillary

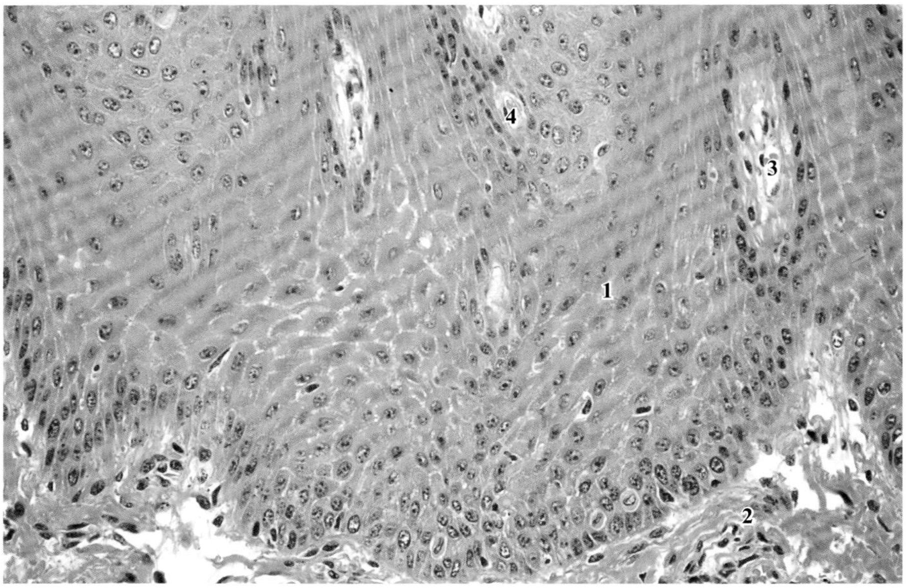

图 4-17　舌体组织切片，示黏膜（HE 染色，高倍镜）
Figure 4-17　Histological section of body of tongue showing the mucosa (HE staining, higher power).

1- 脂肪组织 adipose tissue
2- 骨骼肌 skeletal muscle
3- 黏液腺泡 mucous acinus

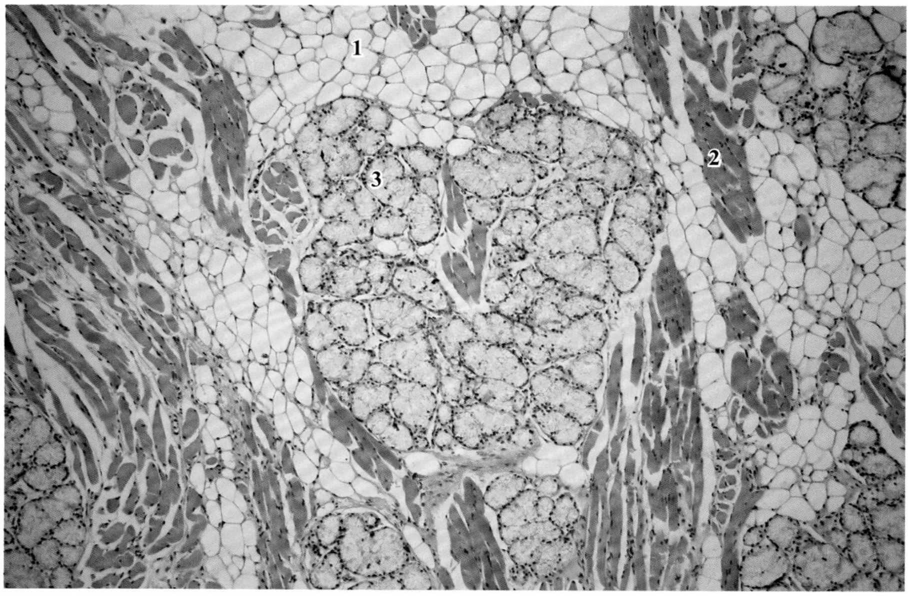

图 4-18　舌根组织切片，示舌腺（HE 染色，高倍镜）
Figure 4-18　Histological section of root of tongue showing the lingual gland (HE staining, higher power).

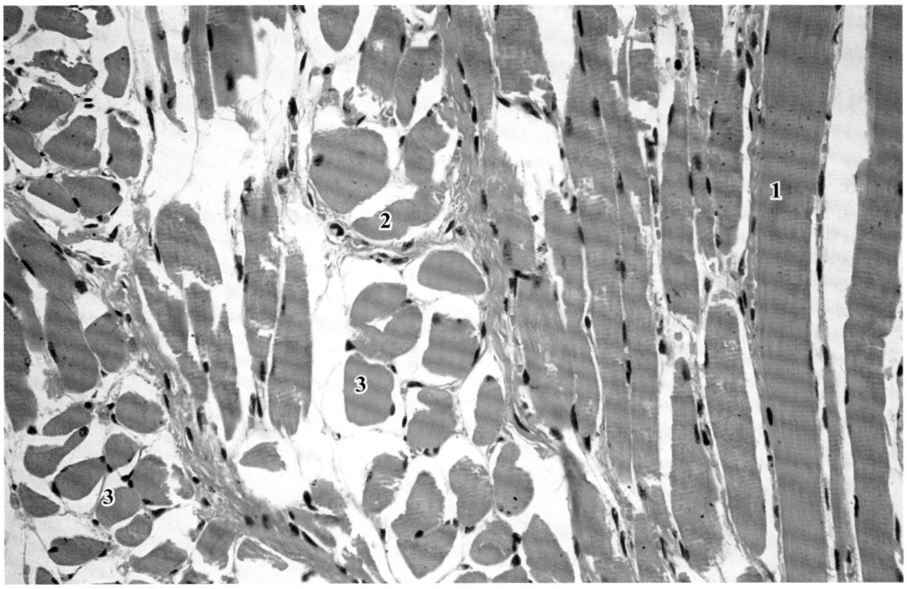

1- 垂直肌 vertical muscle
2- 横行肌 transversal muscle
3- 纵行肌 longitudinal muscle

图 4-19 舌体组织切片，示横纹肌（HE 染色，高倍镜）
Figure 4-19 Histological section of body of tongue showing the striated muscle (HE staining, higher power).

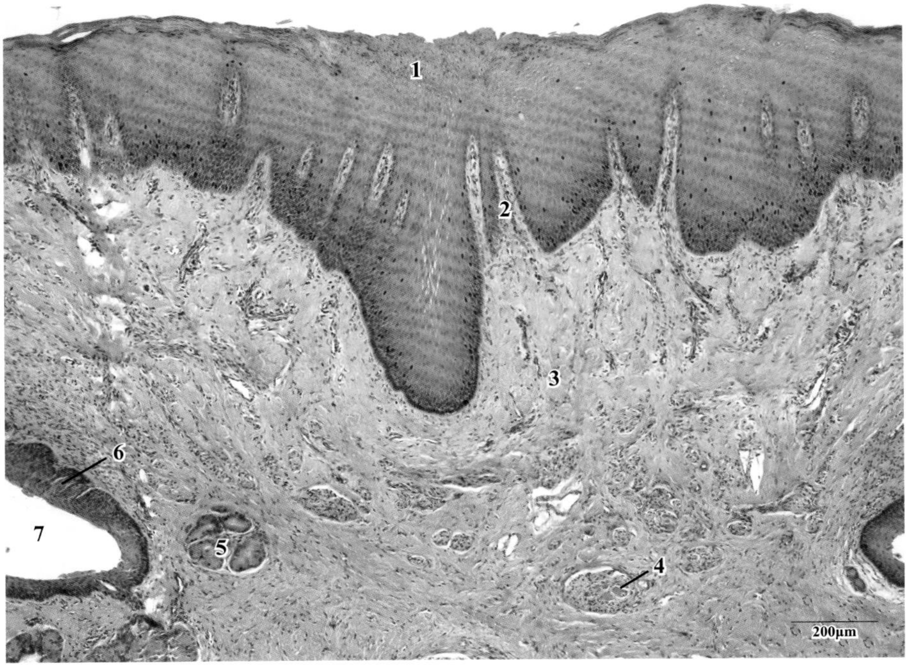

1- 复层扁平上皮 stratified squamous epithelium
2- 固有层乳头 papilla of lamina propria
3- 轮廓乳头 vallate papilla
4- 神经元 neuron
5- 舌腺 lingual gland
6- 味蕾 taste bud
7- 味沟 gustatory furrow

图 4-20 舌轮廓乳头组织切片（HE 染色，高倍镜）
Figure 4-20 Histological section of a vallate papilla of tongue (HE staining, higher power).

1- 复层扁平上皮 stratified squamous epithelium
2- 固有层 lamina propria
3- 固有层乳头 papilla of lamina propria

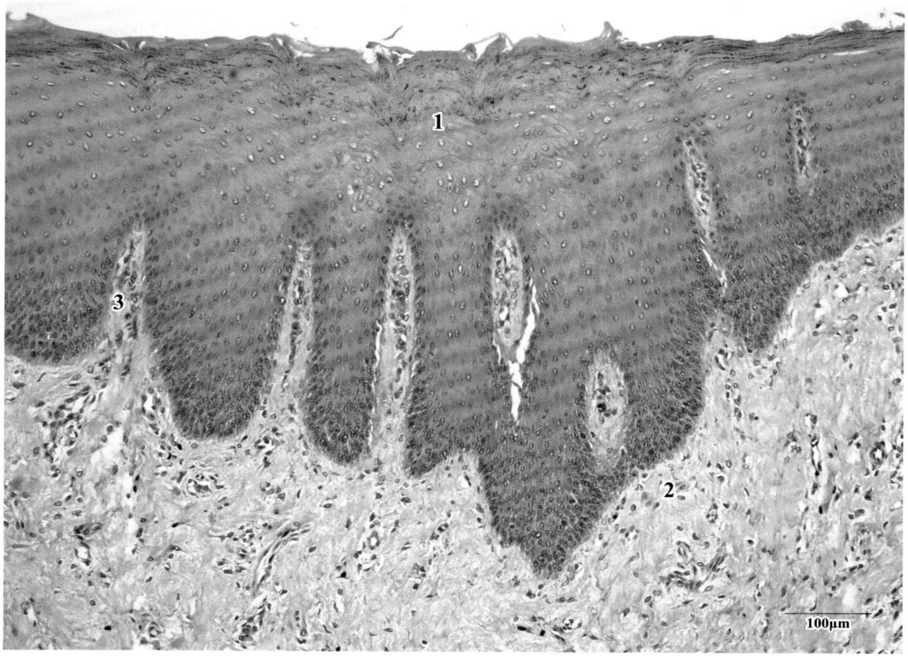

图 4-21　舌轮廓乳头组织切片，示黏膜上皮（HE 染色，高倍镜）
Figure 4-21　Histological section of a vallate papilla of tongue showing the mucosal epithelium (HE staining, higher power).

1- 固有层 lamina propria
2- 神经元 neuron
3- 血管 blood vessel
4- 味腺（冯·埃布纳腺）taste gland (von Ebner's gland)

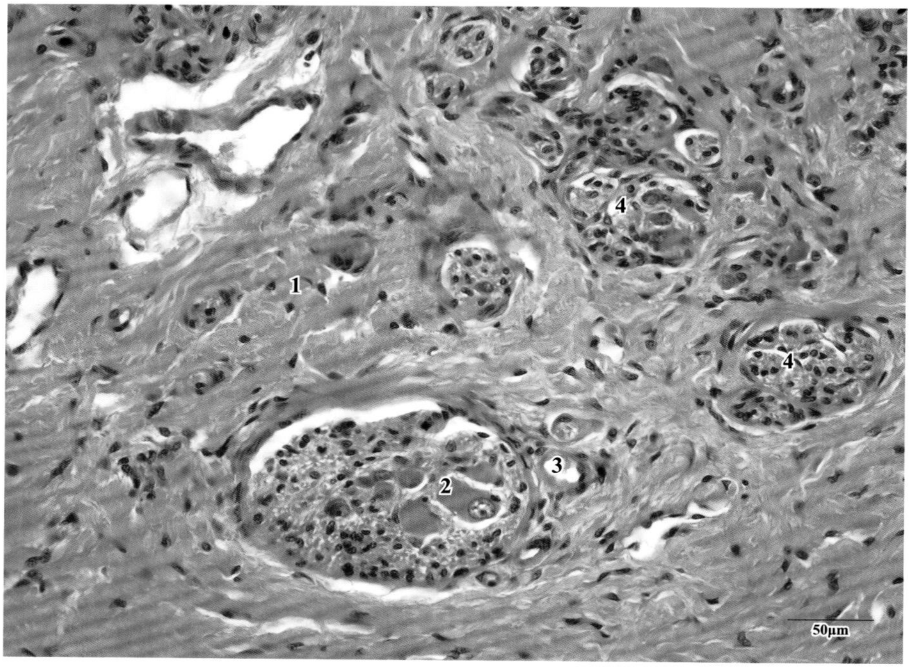

图 4-22　舌轮廓乳头组织切片，示固有层（HE 染色，高倍镜）
Figure 4-22　Histological section of a vallate papilla of tongue showing the lamina propria (HE staining, higher power).

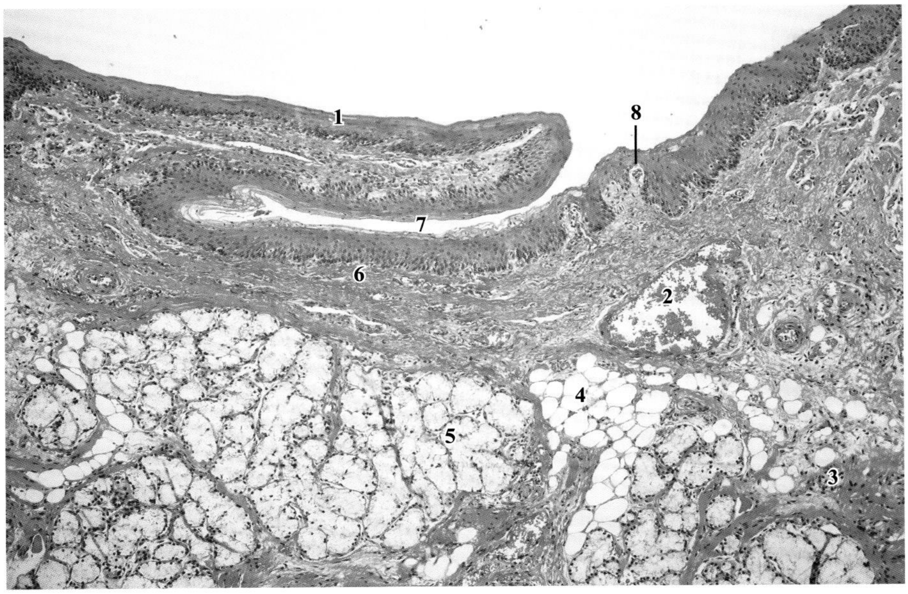

1- 黏膜上皮 mucosal epithelium
2- 血管 blood vessel
3- 浆液腺泡 serous acinus
4- 脂肪组织 adipose tissue
5- 黏液性腺泡 mucous acinus
6- 固有层 lamina propria
7- 味沟 gustatory furrow
8- 味蕾 taste bud

图 4-23 舌叶状乳头组织切片（HE 染色，低倍镜）
Figure 4-23 Histological section of a foliate papilla of the tongue (HE staining, lower power).

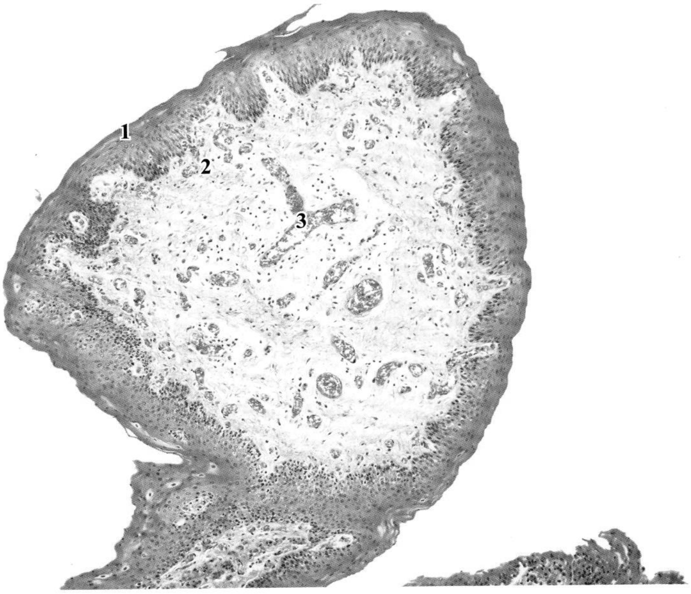

1- 复层扁平上皮 stratified squamous epithelium
2- 固有层 lamina propria
3- 血管 blood vessel

图 4-24 舌乳头组织切片（HE 染色，低倍镜）
Figure 4-24 Histological section of the lingual papilla (HE staining, lower power).

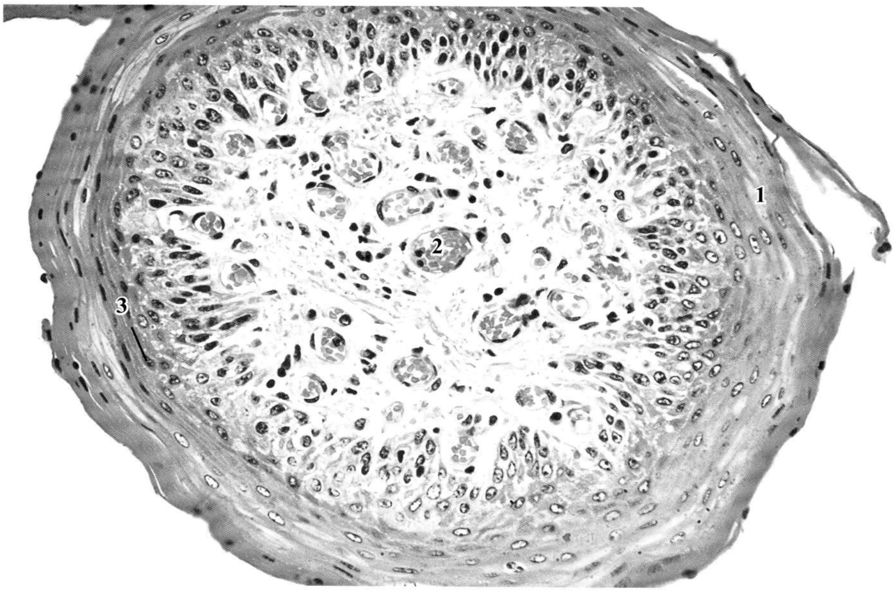

1- 复层扁平上皮 stratified squamous epithelium
2- 血管 blood vessel
3- 细胞核 nucleus

图 4-25 舌乳头组织切片，横截面（HE 染色，高倍镜）
Figure 4-25 Transverse cross section of the lingual papilla (HE staining, higher power).

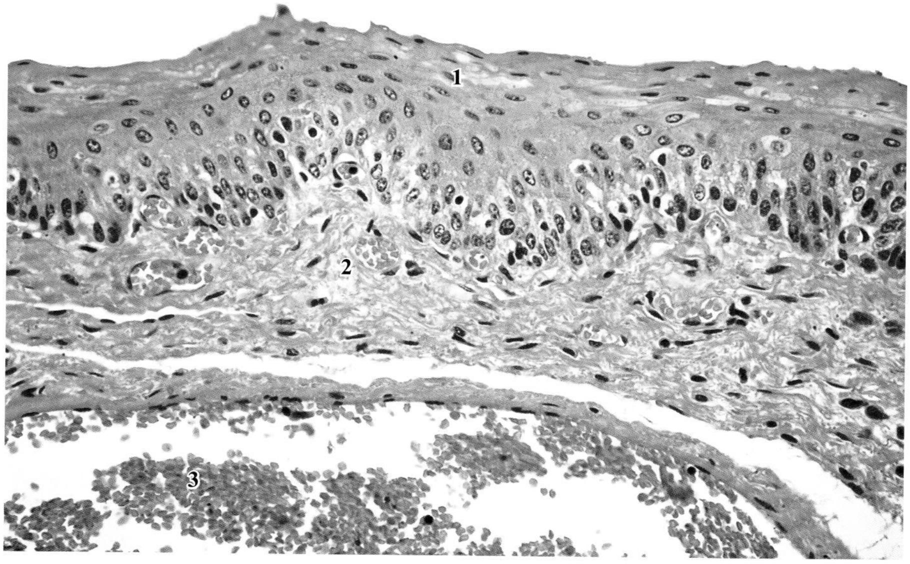

1- 复层扁平上皮 stratified squamous epithelium
2- 固有层 lamina propria
3- 血管 blood vessel

图 4-26 舌黏膜组织切片（HE 染色，高倍镜）
Figure 4-26 Histological section of the mucosa of tongue (HE staining, higher power).

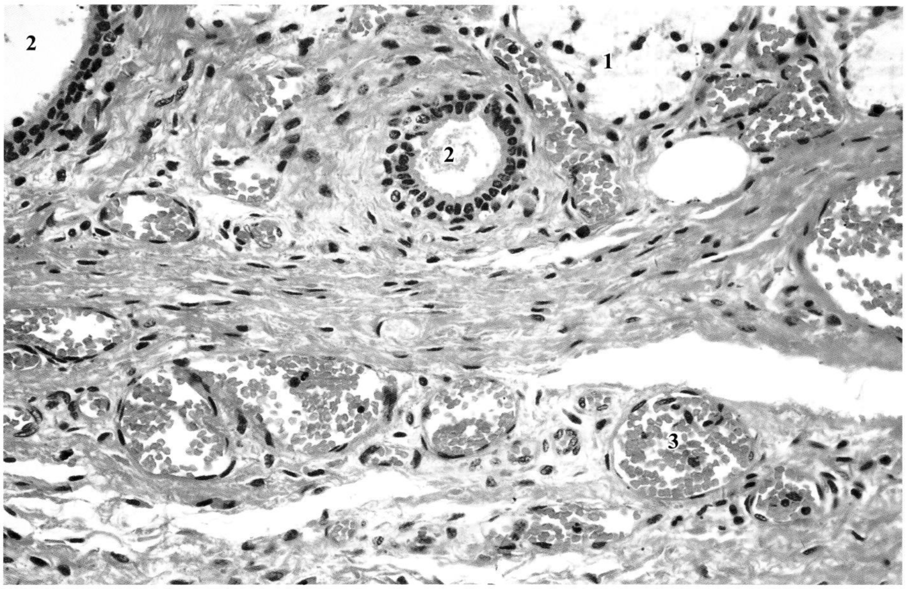

1- 黏液腺泡 mucous acinus
2- 导管 duct
3- 血管 blood vessel

图 4-27 舌下腺管组织切片（HE 染色，高倍镜）
Figure 4-27 Histological section of the sublingual duct (HE staining, higher power).

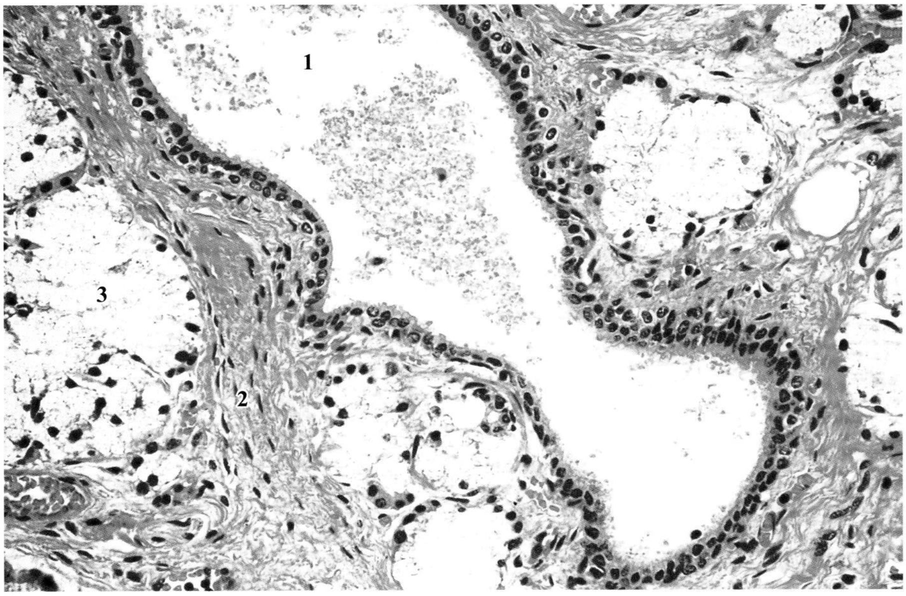

1- 导管 duct
2- 小叶间结缔组织 interlobular connective tissue
3- 黏液腺泡 mucous acinus

图 4-28 舌下腺管组织切片（HE 染色，高倍镜）
Figure 4-28 Histological section of the sublingual duct (HE staining, higher power).

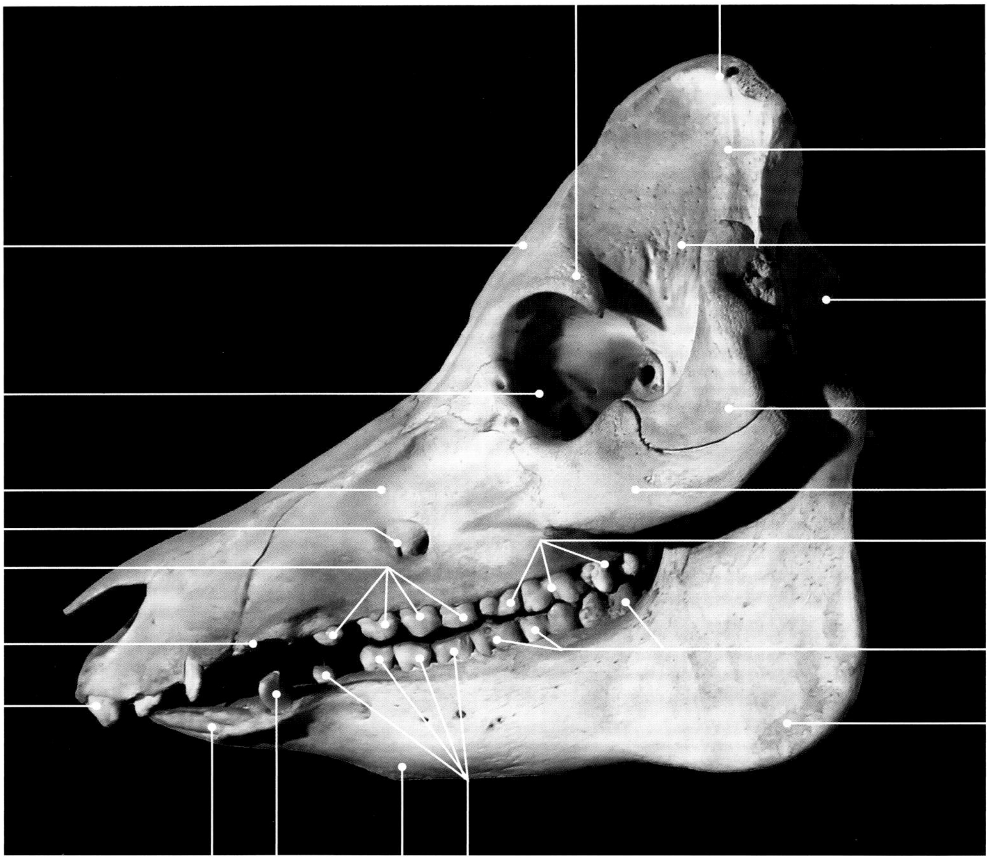

图 4-29　恒齿 -1（左外侧观）
Figure 4-29　The permanent tooth-1 (left lateral view).

1- 顶骨 parietal bone
2- 颞窝 temporal fossa
3- 枕骨 occipital bone
4- 颞骨颧突 zygomatic process of temporal bone
5- 颧骨 zygomatic bone
6- 上臼齿 upper molar
7- 下臼齿 lower molar
8- 下颌骨角 angle of mandible
9- 下前臼齿 lower premolar
10- 颏结节 mental tubercle
11- 下犬齿 lower canine
12- 下切齿 lower incisors
13- 上切齿 upper incisors
14- 上犬齿窝 upper canine fossa
15- 上前臼齿 upper premolar
16- 眶下孔 infraorbital foramen
17- 上颌骨 maxilla
18- 眼窝 orbit
19- 额骨 frontal bone
20- 额骨颧突 zygomatic process of frontal bone
21- 顶嵴 parietal crest

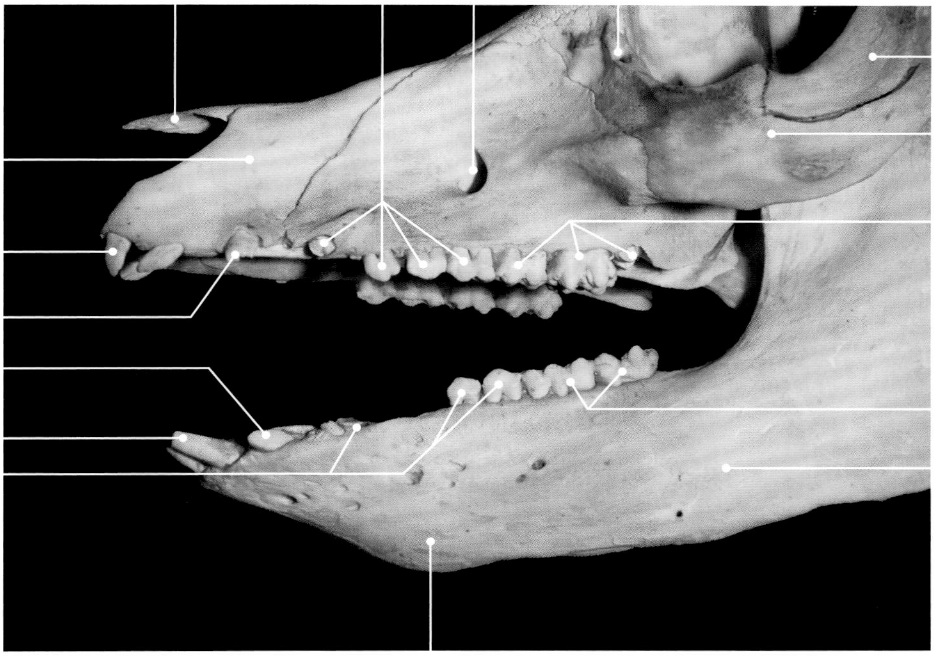

图 4-30 恒齿 -2（左外侧观）
Figure 4-30 The permanent tooth-2 (left lateral view).

1- 颞骨 temporal bone
2- 颧骨 zygomatic bone
3- 上臼齿 upper molar
4- 下臼齿 lower molar
5- 下颌骨 mandible
6- 颏结节 mental tubercle
7- 下前臼齿 lower premolar
8- 下切齿 lower incisors
9- 下犬齿 lower canine
10- 上犬齿 upper canine
11- 上切齿 upper incisors
12- 切齿骨 incisive bone
13- 鼻骨 nasal bone
14- 上前臼齿 upper premolar
15- 眶下孔 infraorbital foramen
16- 泪孔 lacrimal foramen

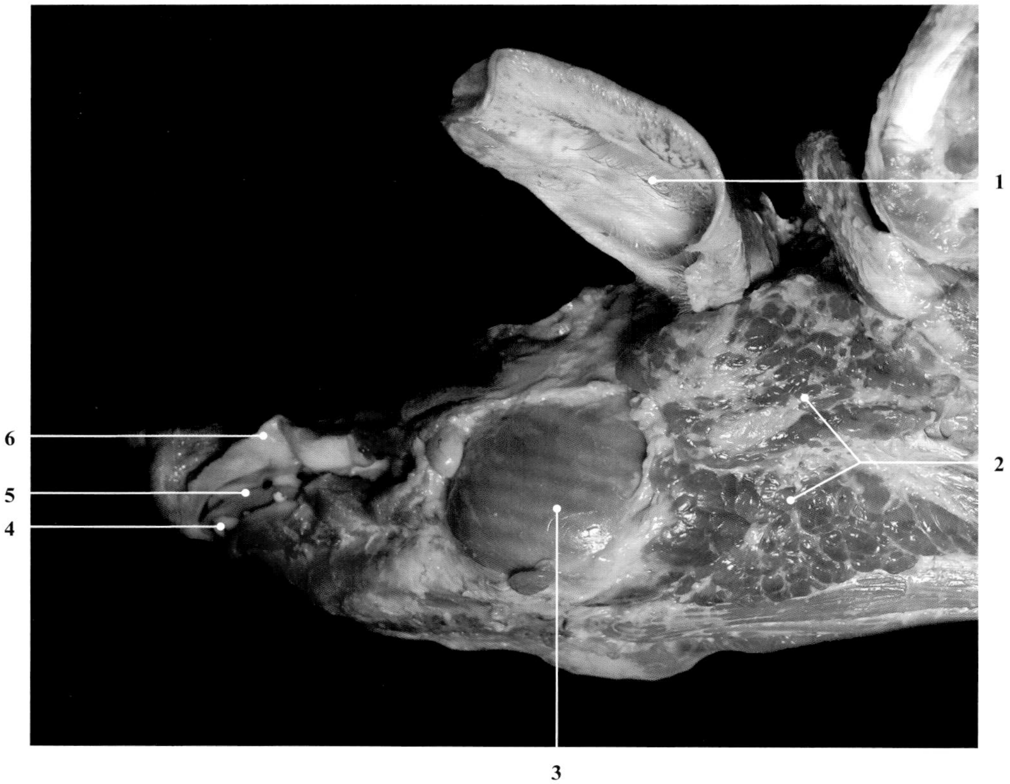

图 4-31 腮腺的位置
Figure 4-31 Anatomical position of the parotid gland.

1- 耳 ear
2- 腮腺 parotid gland
3- 咬肌 masseter muscle
4- 切齿 incisors
5- 舌 tongue
6- 颊 cheek

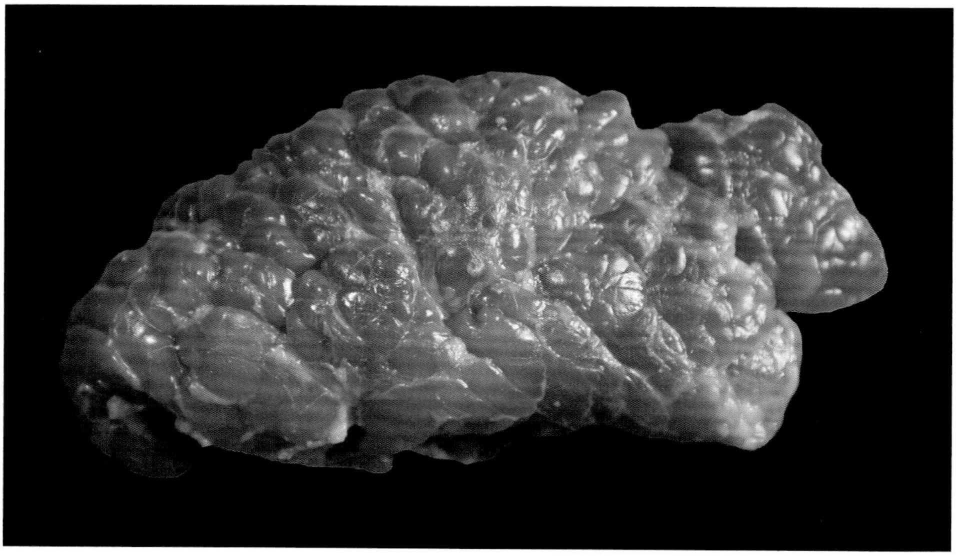

图 4-32 腮腺的形态
Figure 4-32 Morphology of the parotid gland.

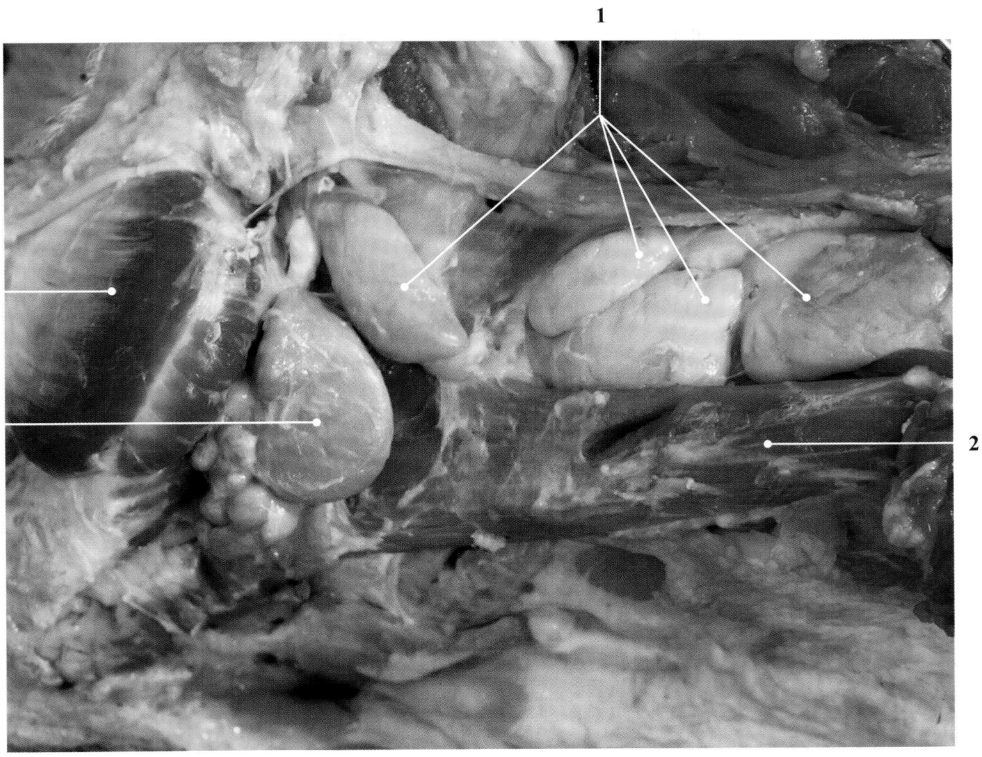

图 4-33　颌下腺（左外侧观）
Figure 4-33　Mandibular gland (left lateral view).

1- 胸腺 thymus
2- 胸骨甲状舌骨肌 sternothyrohyoid muscle
3- 颌下腺 mandibular gland
4- 咬肌 masseter muscle

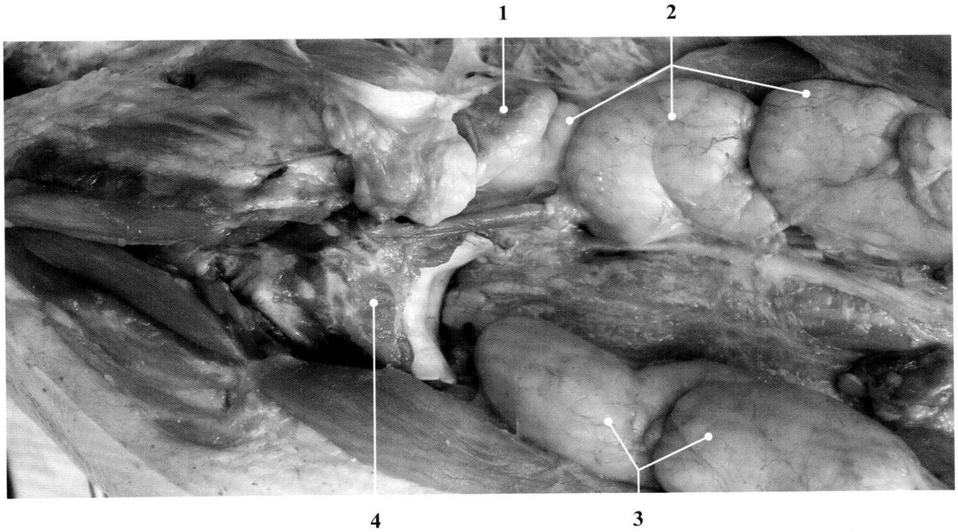

图 4-34　颌下腺（腹侧观）
Figure 4-34　Mandibular gland (ventral view).

1- 左侧颌下腺 left mandibular gland
2- 左侧胸腺 left thymus
3- 右侧胸腺 right thymus
4- 喉 larynx

1- 下颌淋巴结 mandibular lymph node
2- 颌下腺 mandibular gland

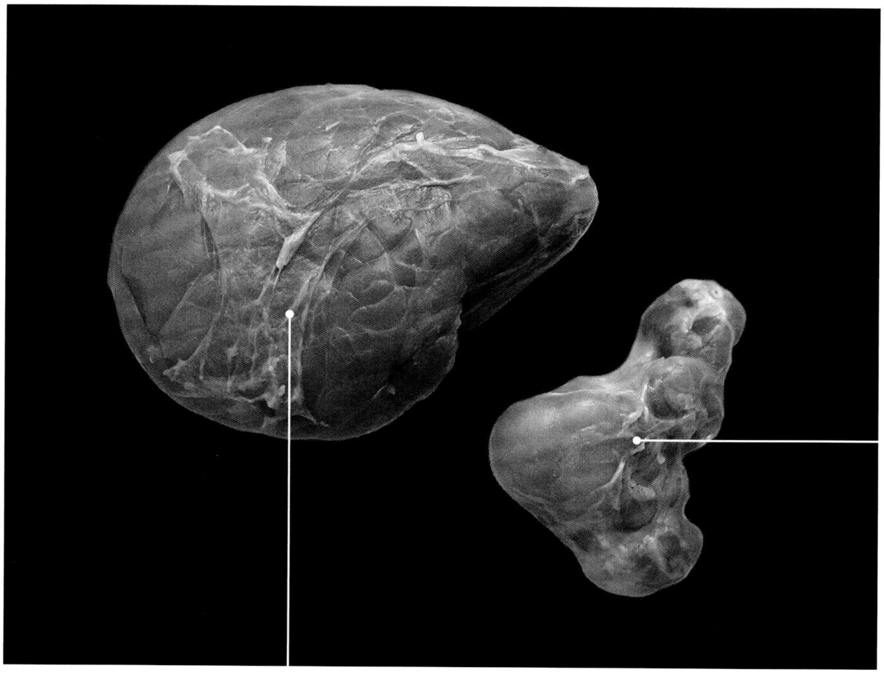

图 4-35 颌下腺和下颌淋巴结的形态
Figure 4-35 Morphology of the mandibular gland and lymph node.

1- 纹状管 striated duct
2- 浆液腺泡 serous acinus
3- 闰管 intercalated duct
4- 基底纹 basal striation

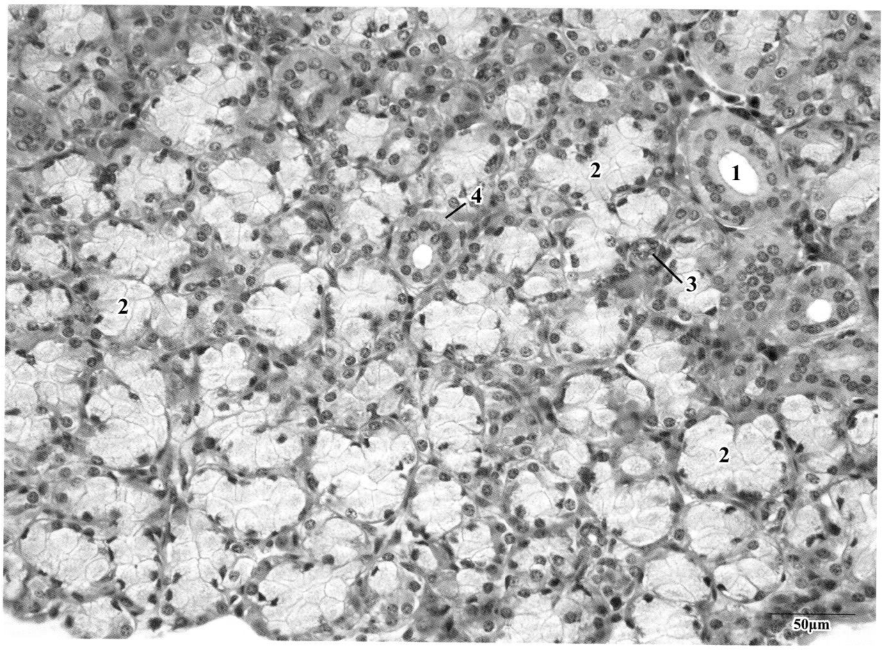

图 4-36 腮腺组织切片（HE 染色，高倍镜）
Figure 4-36 Histological section of the parotid gland (HE staining, higher power).

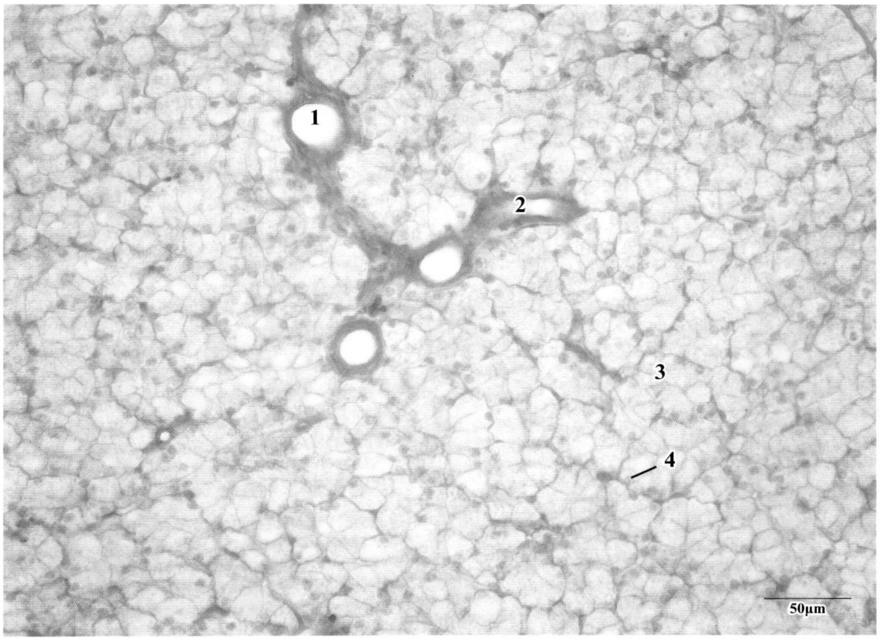

1- 纹状管 striated duct
2- 闰管 intercalated duct
3- 浆液腺泡 serous acinus
4- 细胞核 nucleus

图 4-37 腮腺组织切片（Mallory 三色染色，高倍镜）
Figure 4-37 Histological section of the parotid gland (Mallory staining, higher power).

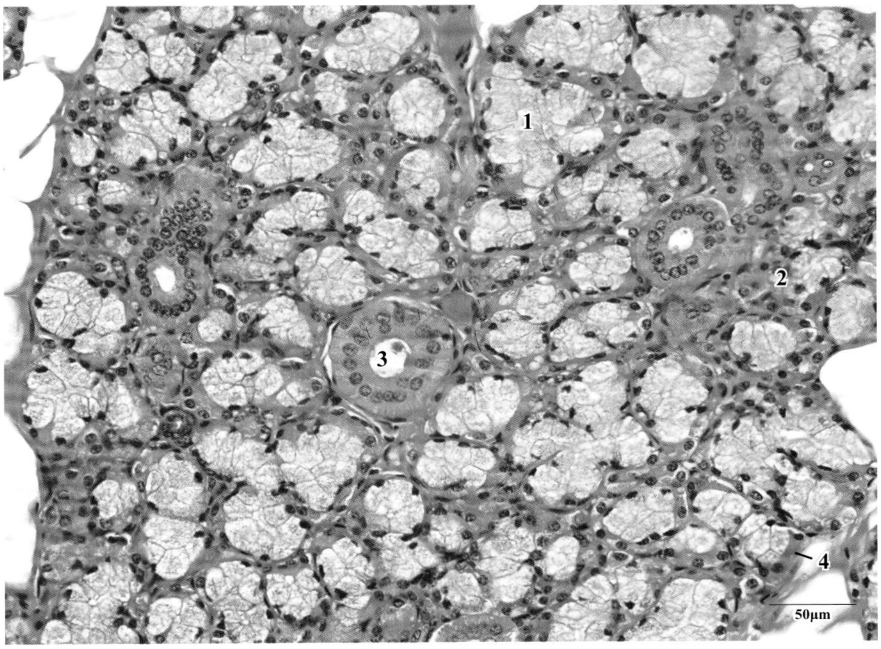

1- 黏液腺泡 mucous acinus
2- 混合腺泡 mixed acinus
3- 导管 duct
4- 浆半月 serous demilune

图 4-38 颌下腺组织切片（HE 染色，高倍镜）
Figure 4-38 Histological section of the mandibular gland (HE staining, higher power).

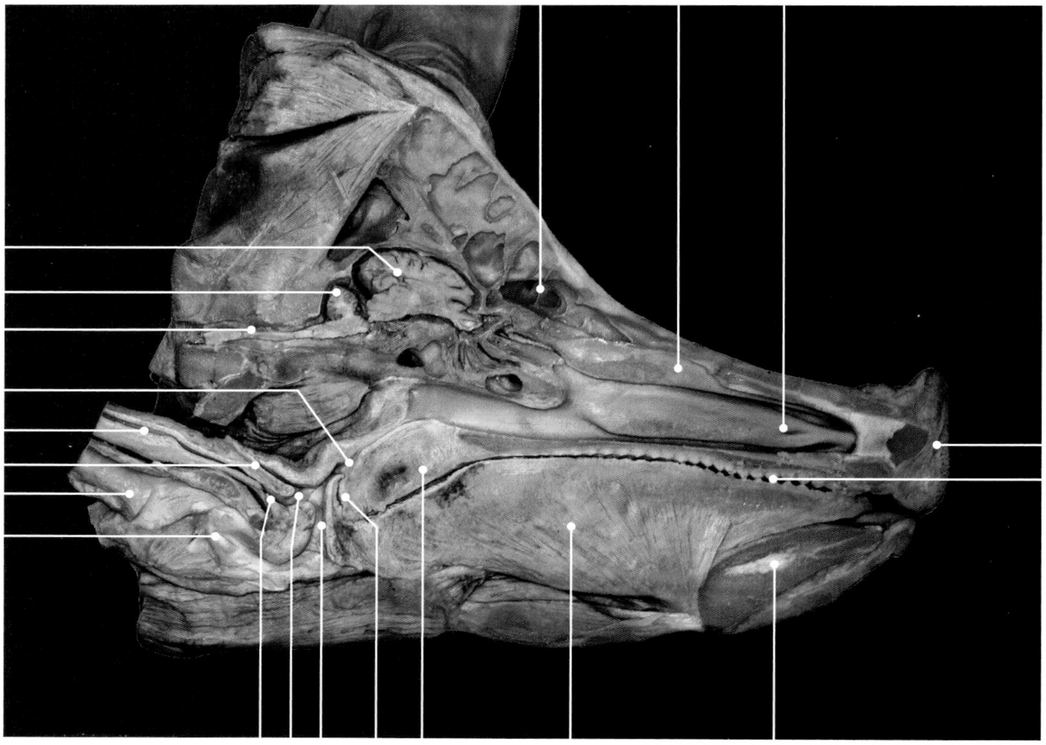

图 4-39　咽
Figure 4-39　The pharynx.

1- 吻突 rostral disc, snout
2- 硬腭 hard palate
3- 下颌骨 mandible
4- 舌 tongue
5- 软腭 soft palate
6- 咽峡 isthmus of the fauces
7- 喉口 aperture of larynx
8- 咽 pharynx
9- 食管口 pharyngeal opening of the esophagus
10- 喉 larynx
11- 气管 trachea
12- 咽后隐窝 pharyngeal recess
13- 食管 esophagus
14- 鼻后孔 posterior nasal apertures
15- 脊髓 spinal cord
16- 小脑 cerebellum
17- 大脑 cerebrum
18- 额窦 frontal sinus
19- 上鼻甲 dorsal nasal concha
20- 下鼻甲 ventral nasal concha

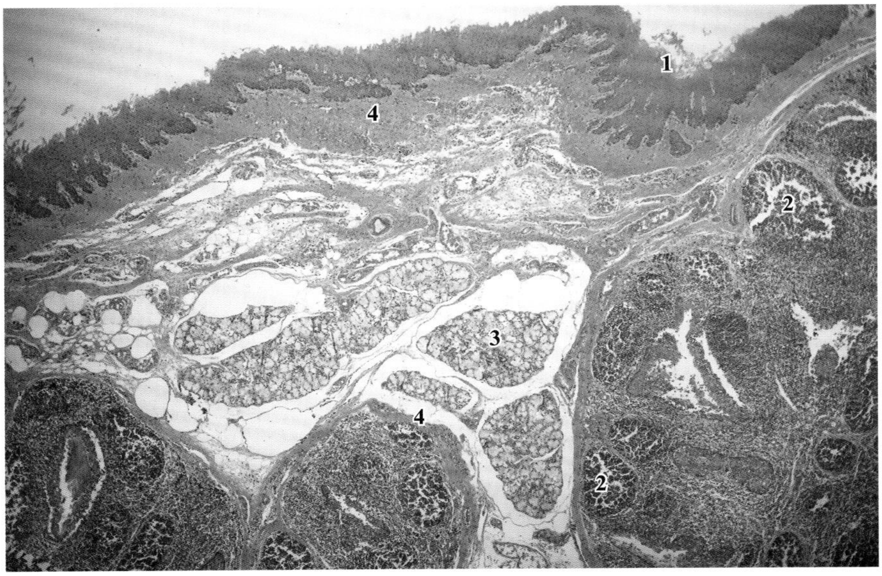

1- 复层扁平上皮 stratified squamous epithelium
2- 淋巴小结 lymphatic nodule
3- 混合腺 mixed gland
4- 固有层 lamina propria

图 4-40　咽组织切片（HE 染色，低倍镜）
Figure 4-40　Histological section of the pharynx (HE staining, lower power).

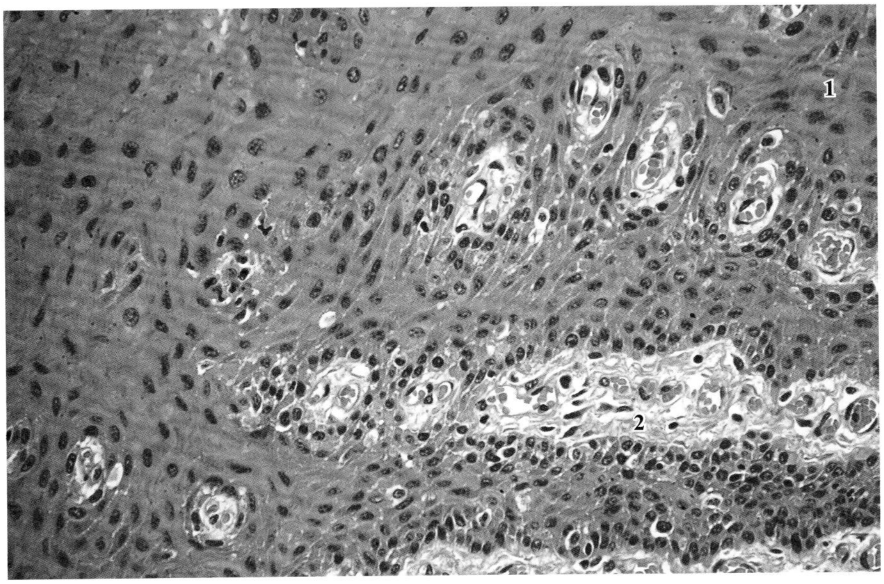

1- 复层扁平上皮 stratified squamous epithelium
2- 固有层 lamina propria

图 4-41　咽组织切片，示黏膜（HE 染色，高倍镜）
Figure 4-41　Histological section of the pharynx showing the mucosa (HE staining, higher power).

1- 固有层 lamina propria
2- 黏膜上皮 mucosa epithelium
3- 毛细血管 blood capillary

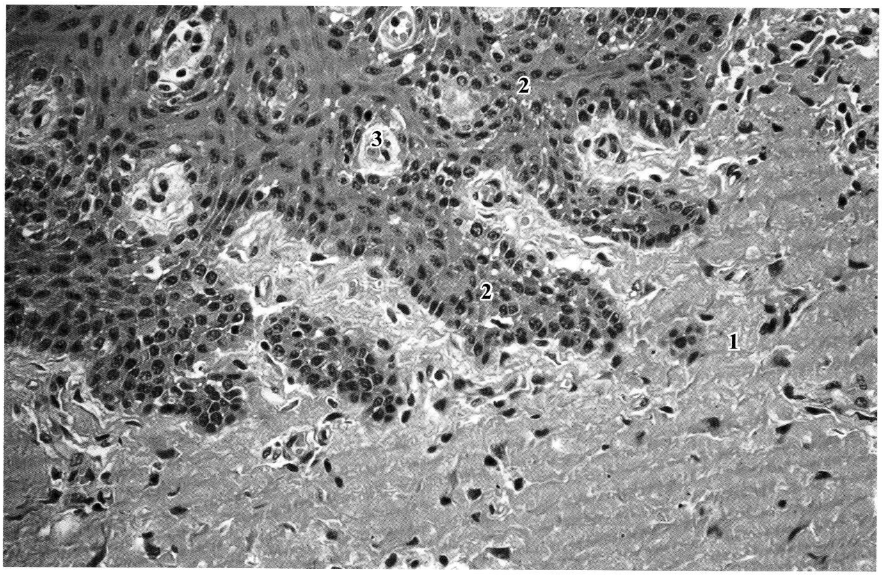

图 4-42　咽组织切片，示黏膜（HE 染色，高倍镜）
Figure 4-42　Histological section of the pharynx showing the mucosa (HE staining, higher power).

1- 淋巴细胞 lymphocyte
2- 血管 blood vessel
3- 弥散淋巴组织 diffuse lymphoid tissue
4- 固有层 lamina propria

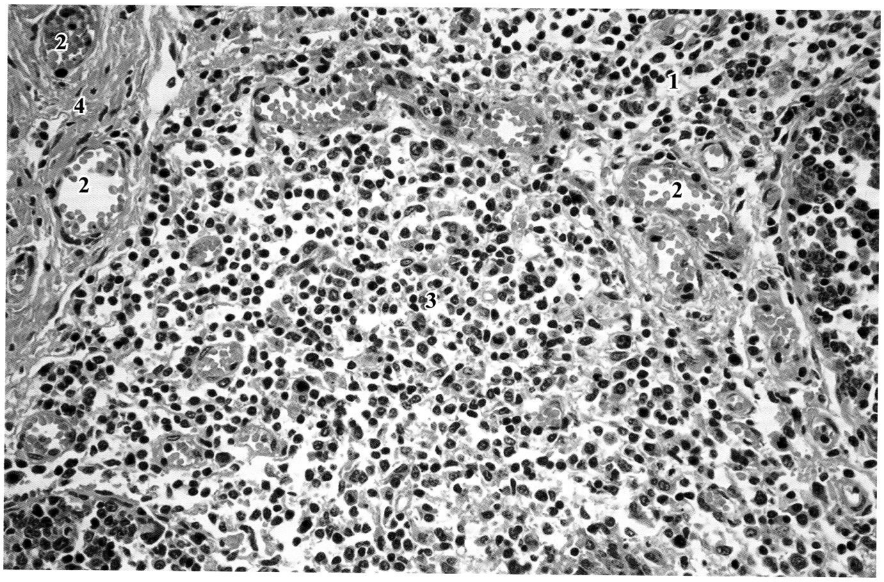

图 4-43　咽组织切片，示淋巴组织（HE 染色，高倍镜）
Figure 4-43　Histological section of the pharynx showing the lymphatic tissue (HE staining, higher power).

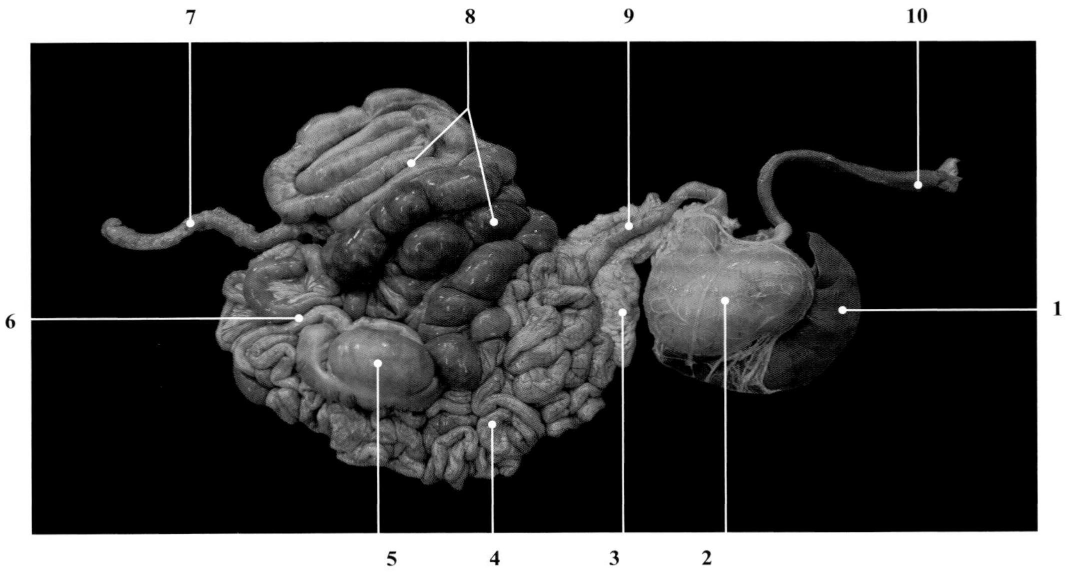

图 4-44　消化管
Figure 4-44　The digestive tract.

1- 肝 liver
2- 胃 stomach
3- 胰 pancreas
4- 空肠 jejunum
5- 盲肠 caecum
6- 回肠 ileum
7- 直肠 rectum
8- 结肠 colon
9- 十二指肠 duodenum
10- 食管 esophagus

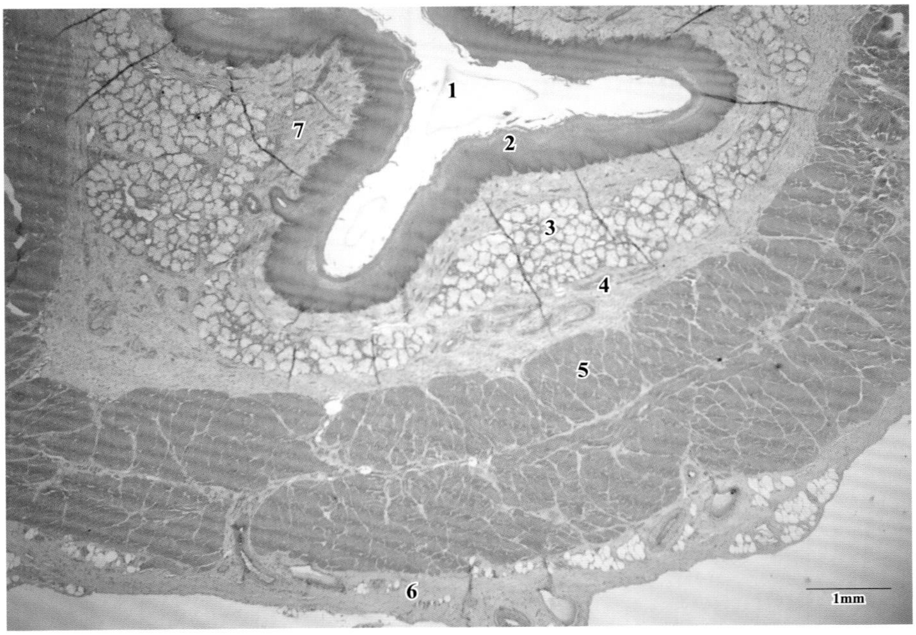

1- 食管腔 esophageal lumen
2- 黏膜上皮 mucosal epithelium
3- 食管腺 esophageal gland
4- 黏膜下层 submucosa
5- 肌层 muscular layer
6- 外膜 adventitia
7- 固有层 lamina propria

图 4-45　食管组织切片（HE 染色，低倍镜）
Figure 4-45　Histological section of the esophagus (HE staining, lower power).

1- 复层扁平上皮 stratified squamous epithelium
2- 食管腺 esophageal gland
3- 固有层 lamina propria
4- 导管 duct
5- 小叶间结缔组织 interlobular connective tissue
6- 食管腔 esophageal lumen

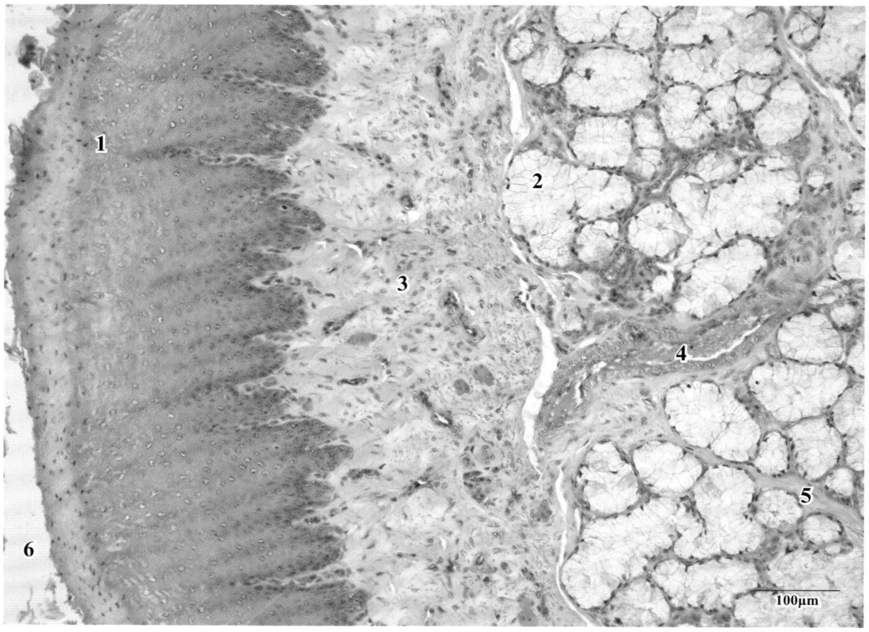

图 4-46　食管组织切片（HE 染色，高倍镜）
Figure 4-46　Histological section of the esophagus (HE staining, higher power).

1- 导管 duct
2- 黏液腺 mucous gland
3- 小叶间结缔组织 interlobular connective tissue

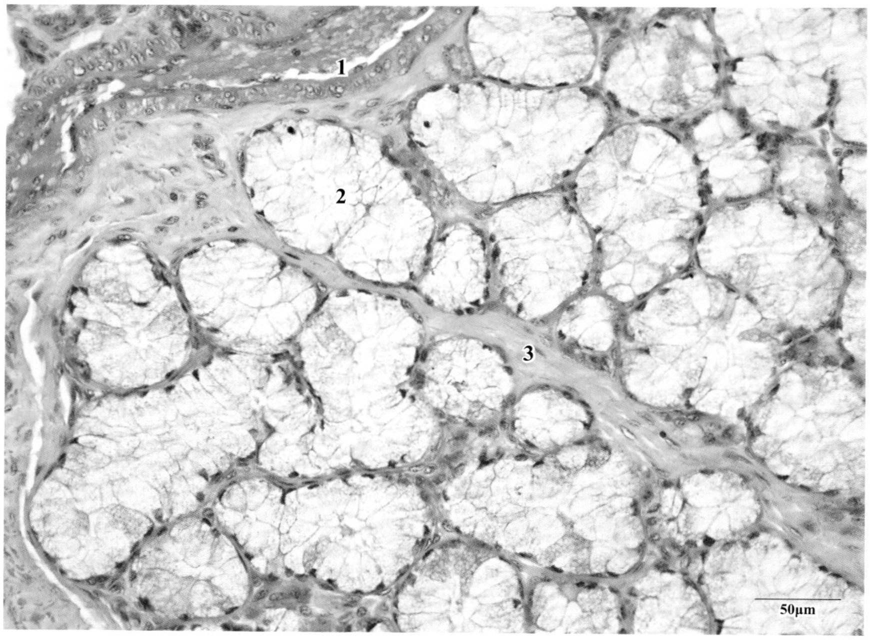

图 4-47　食管腺组织切片（HE 染色，高倍镜）
Figure 4-47　Histological section of the esophageal gland (HE staining, higher power).

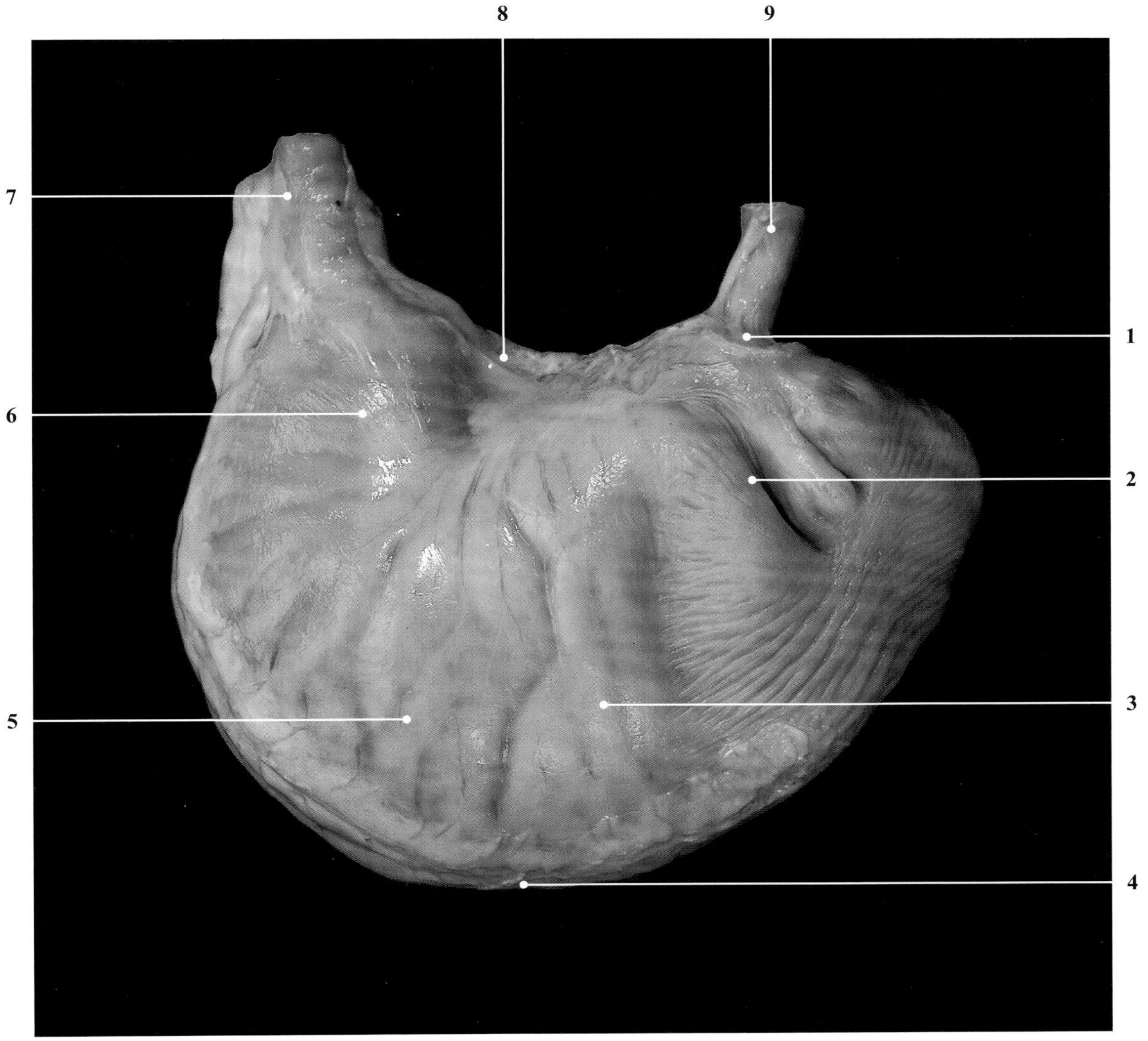

图 4-48 母猪胃（膈面观）
Figure 4-48 The stomach of a sow (diaphragmatic view).

1- 贲门 cardia
2- 贲门部 cardiac part
3- 胃体 body of stomach
4- 胃大弯 greater curvature of stomach
5- 胃底部 fundus of stomach
6- 幽门部 pyloric part
7- 幽门 pylorus
8- 胃小弯 lesser curvature of stomach
9- 食管 esophagus

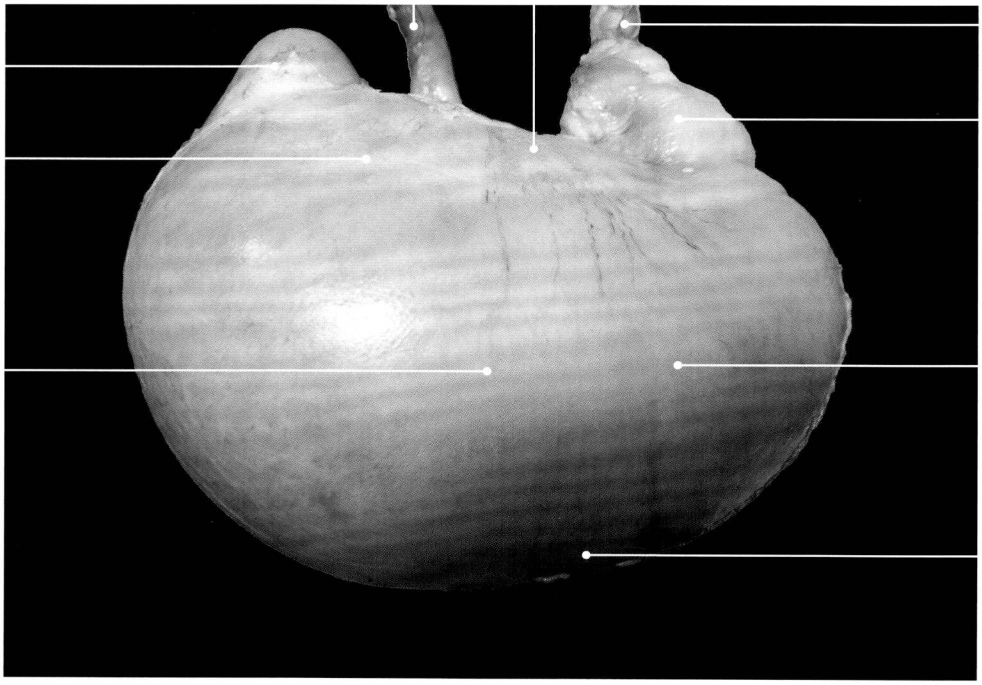

图 4-49　母猪胃（脏面观）
Figure 4-49　The stomach of a sow (visceral view).

1- 十二指肠 duodenum
2- 幽门部 pyloric part
3- 胃底部 fundus of stomach
4- 胃大弯 greater curvature of stomach
5- 胃体 body of stomach
6- 贲门部 cardiac part
7- 胃憩室 gastric diverticulum
8- 食管 esophagus
9- 胃小弯 lesser curvature of stomach

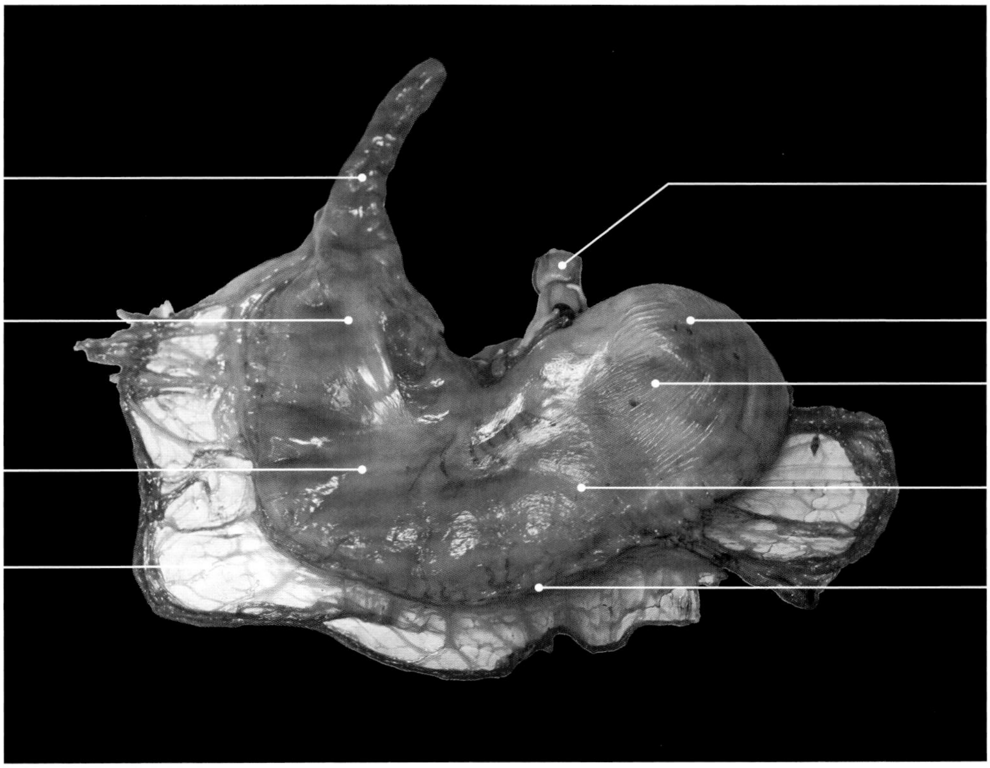

图 4-50 公猪胃（膈面观）
Figure 4-50 The stomach of a boar (diaphragmatic view).

1- 食管 esophagus
2- 胃憩室 gastric diverticulum
3- 贲门部 cardiac part
4- 胃体 body of stomach
5- 胃大弯 greater curvature of stomach
6- 大网膜 greater omentum
7- 胃底部 fundus of stomach
8- 幽门部 pyloric part
9- 十二指肠 duodenum

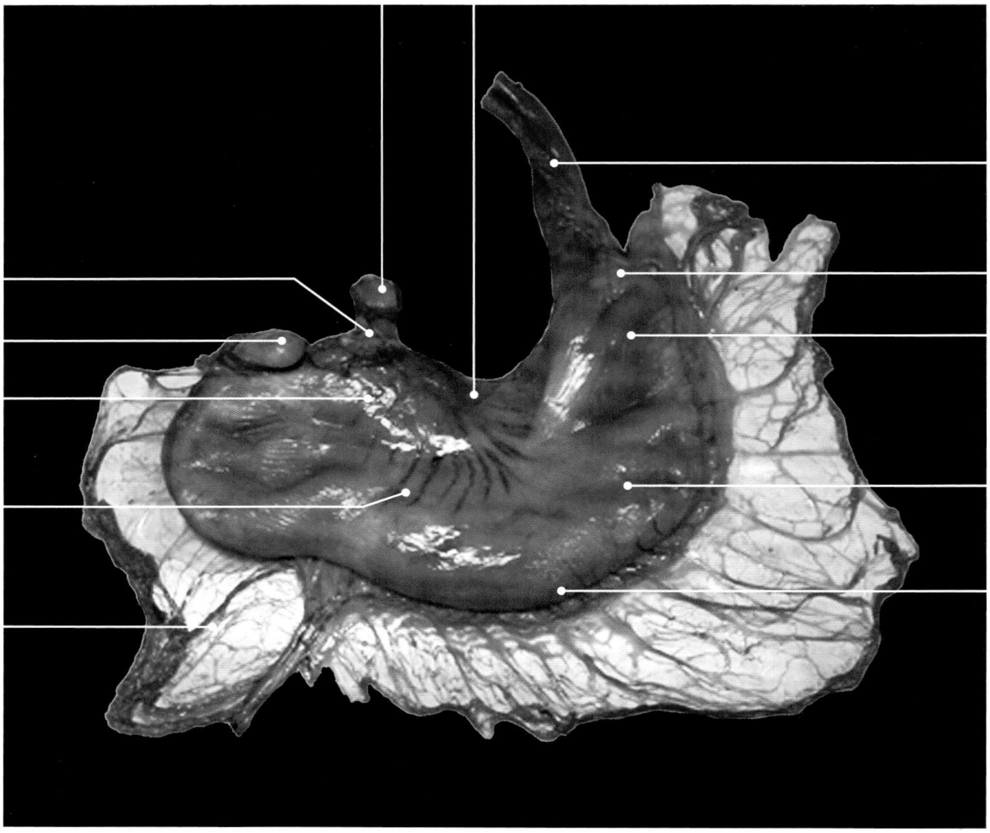

图 4-51　公猪胃（脏面观）
Figure 4-51　The stomach of a boar (visceral view).

1- 十二指肠 duodenum
2- 幽门 pylorus
3- 幽门部 pyloric part
4- 胃底部 fundus of stomach
5- 胃大弯 greater curvature of stomach
6- 大网膜 greater omentum
7- 胃体 body of stomach
8- 贲门部 cardiac part
9- 胃憩室 gastric diverticulum
10- 贲门 cardia
11- 食管 esophagus
12- 胃小弯 lesser curvature of stomach

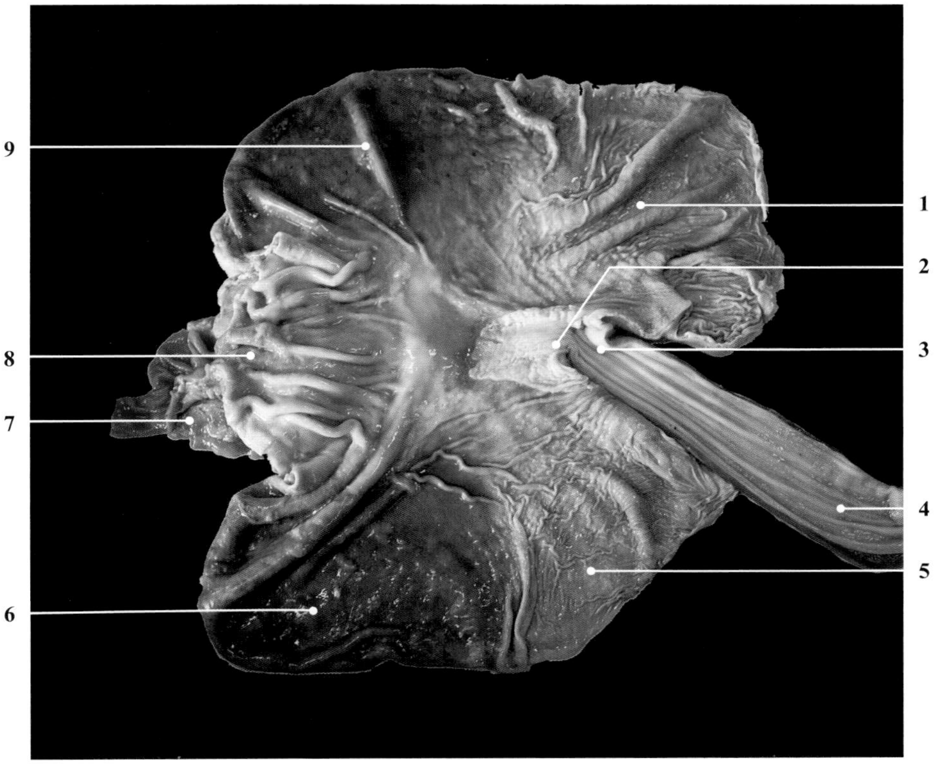

图 4-52 母猪胃黏膜面
Figure 4-52 The mucosal surface of stomach of a sow.

1- 贲门腺区 region of cardiac gland
2- 无腺部 non-glandular part
3- 贲门 cardia
4- 食管 esophagus
5- 贲门腺区 region of cardiac gland
6- 胃底腺区 region of fundic gland
7- 十二指肠 duodenum
8- 幽门腺区 region of pyloric gland
9- 胃底腺区 region of fundic gland

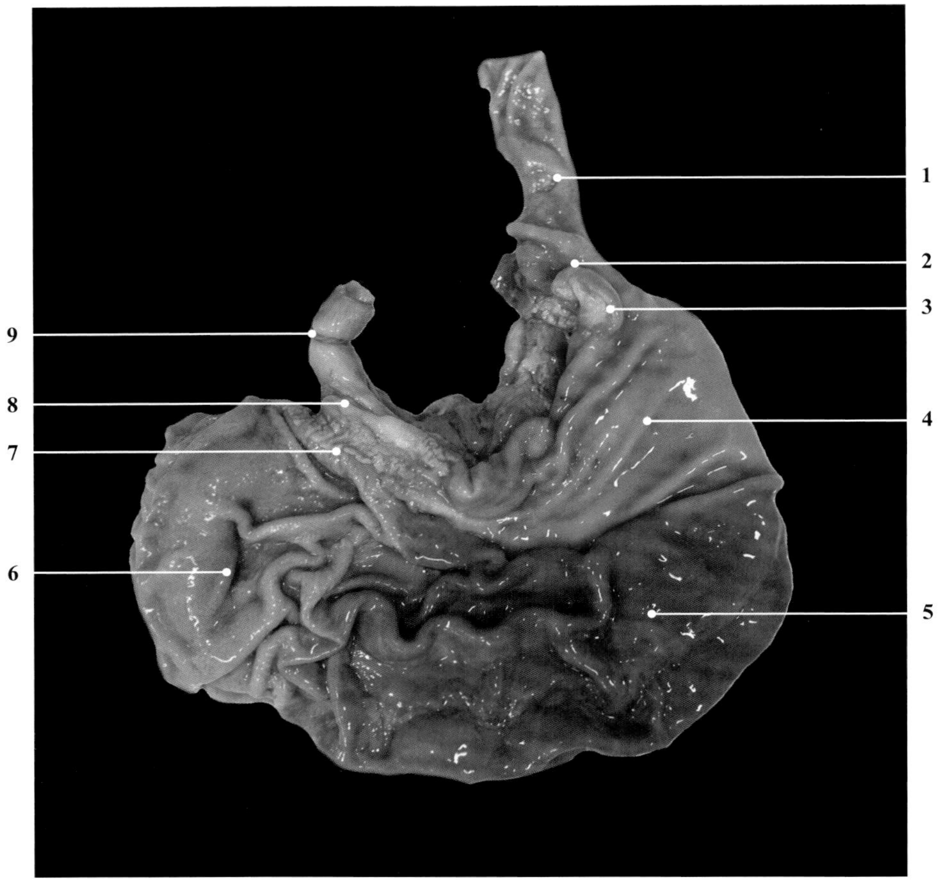

图 4-53　公猪胃黏膜面
Figure 4-53　The mucosal surface of stomach of a boar.

1- 十二指肠 duodenum
2- 幽门 pylorus
3- 幽门圆枕 torus pyloricus
4- 幽门腺区 region of pyloric gland
5- 胃底腺区 region of fundic gland
6- 贲门腺区 region of cardiac gland
7- 无腺部 non-glandular part
8- 贲门 cardia
9- 食管 esophagus

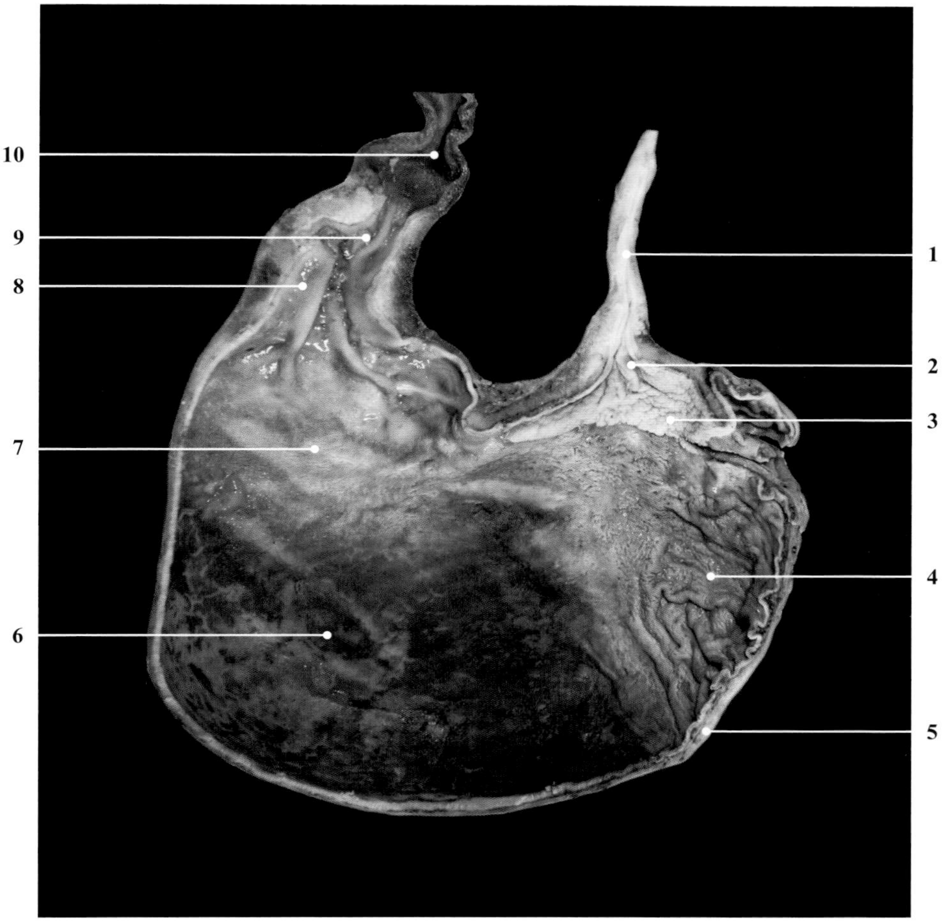

图 4-54 胃黏膜面（固定标本）
Figure 4-54 The mucosal surface of stomach (fixed preparation).

1- 食管 esophagus
2- 贲门 cardia
3- 无腺部 non-glandular part
4- 贲门腺区 region of cardiac gland
5- 胃壁 gastric wall
6- 胃底腺区 region of fundic gland
7- 幽门腺区 region of pyloric gland
8- 幽门圆枕 torus pyloricus
9- 幽门 pylorus
10- 十二指肠 duodenum

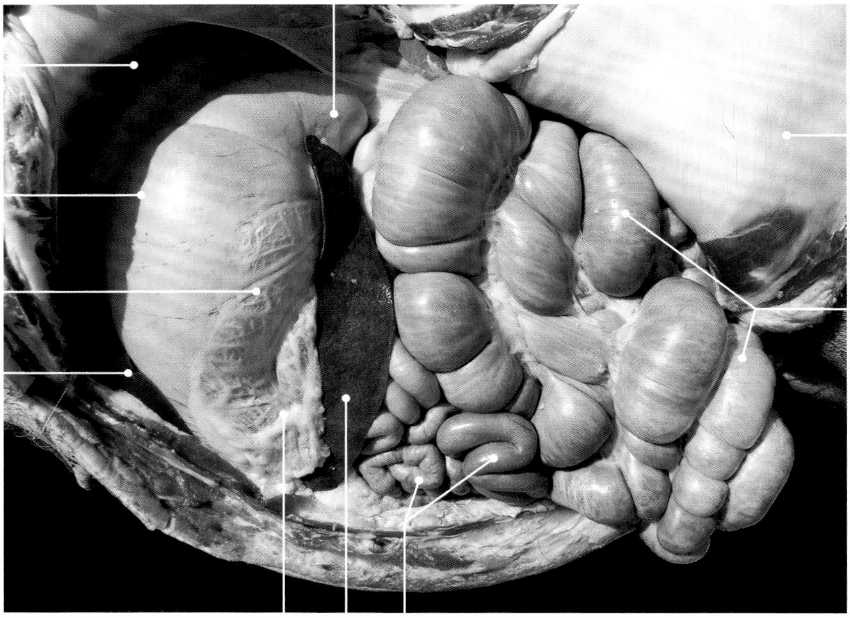

图 4-55　胃的位置（左侧观）
Figure 4-55　Anatomical position of the stomach (left view).

1- 腹壁 abdominal wall
2- 结肠 colon
3- 空肠 jejunum
4- 脾 spleen
5- 大网膜 greater omentum
6- 肝 liver
7- 胃大弯 greater curvature of stomach
8- 胃壁面 diaphragmatic surface of stomach
9- 膈 diaphragm
10- 胃憩室 gastric diverticulum

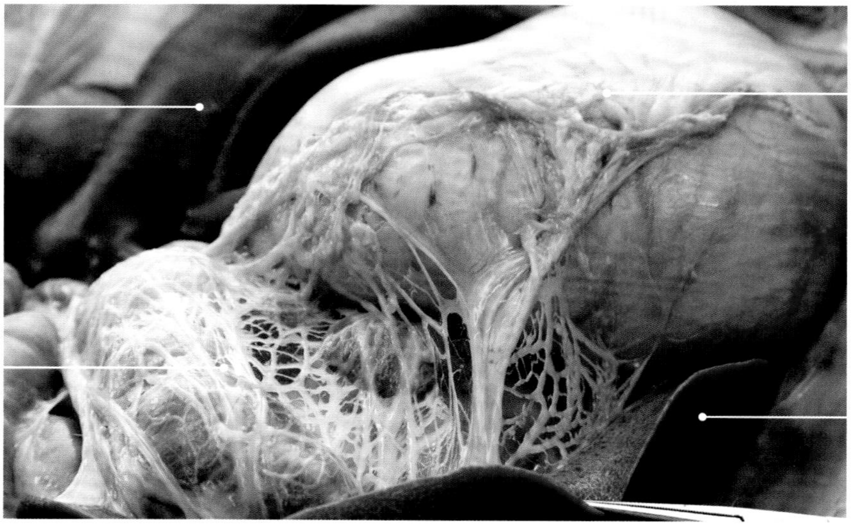

图 4-56　大网膜 -1
Figure 4-56　The greater omentum-1.

1- 胃大弯 greater curvature of stomach
2- 脾 spleen
3- 大网膜 greater omentum
4- 肝 liver

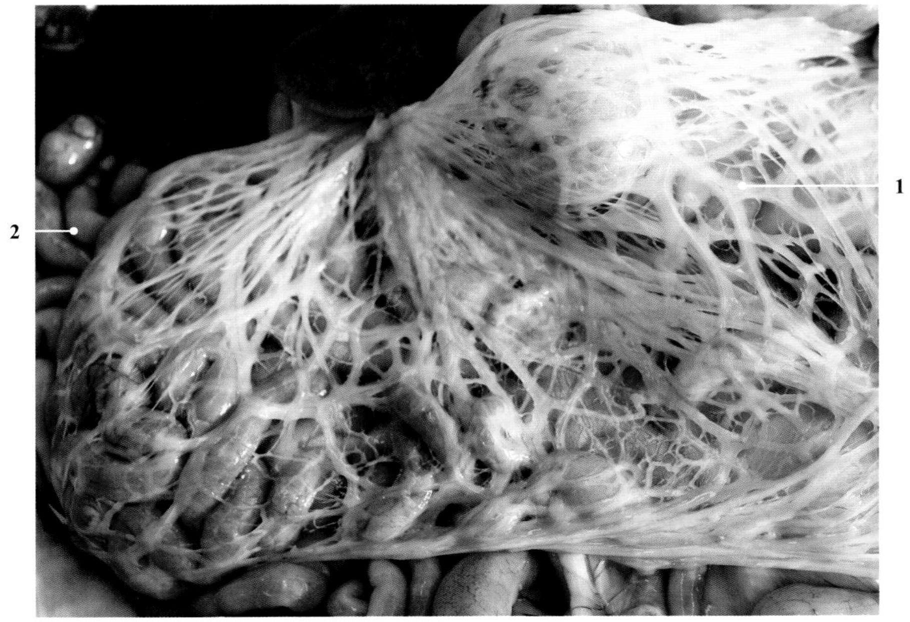

图 4-57 大网膜 -2
Figure 4-57 The greater omentum-2.

1- 大网膜 greater omentum
2- 小肠 small intestine

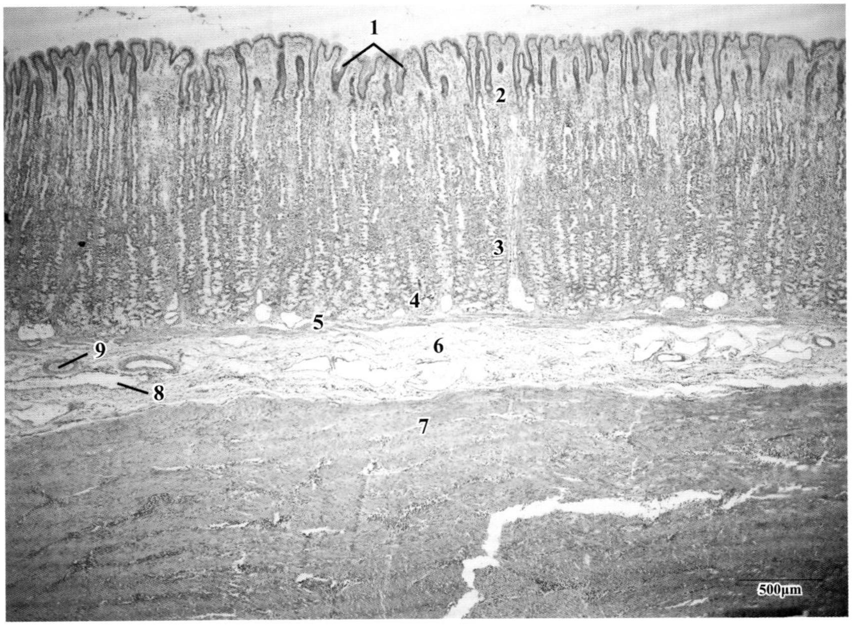

图 4-58 胃底部组织切片（HE 染色，低倍镜）
Figure 4-58 Histological section of the fundus of stomach (HE staining, lower power).

1- 胃小凹 gastric pit
2- 固有层 lamina propria
3- 胃底腺 fundic gland
4- 胃底腺基部 base of fundic gland
5- 黏膜肌层 lamina muscularis mucosae
6- 黏膜下层 submucosa
7- 肌层 muscular layer
8- 小静脉 small vein
9- 小动脉 small artery

1- 胃底腺 fundic gland
2- 胃小凹 gastric pit
3- 固有层 lamina propria
4- 胃底腺基部 base of fundic gland
5- 黏膜肌层 lamina muscularis mucosae
6- 黏膜下层 submucosa

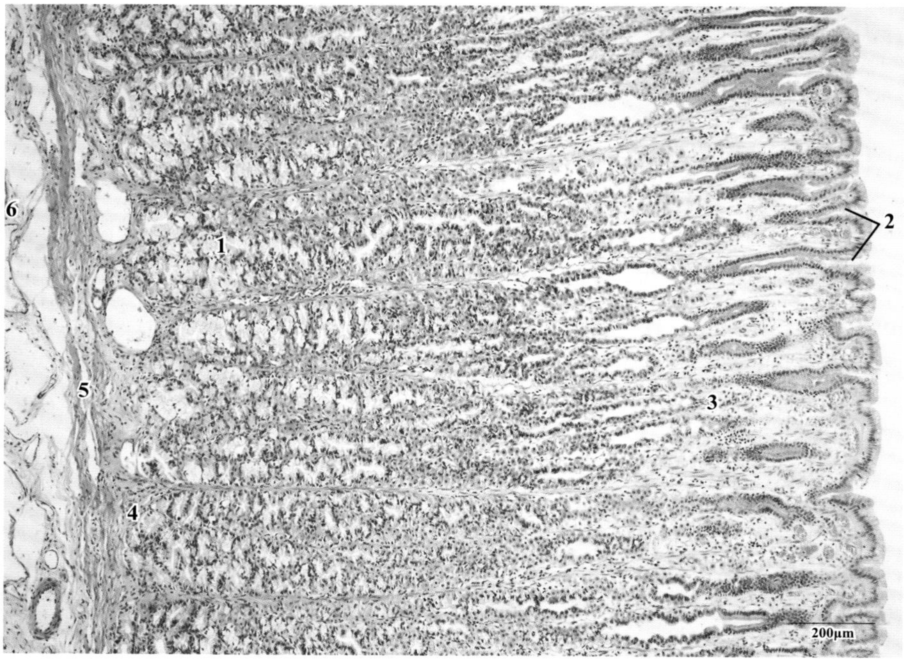

图 4-59　胃底部组织切片，示黏膜（HE 染色，低倍镜）
Figure 4-59　Histological section of the fundus of stomach showing the mucosa (HE staining, lower power).

1- 单层柱状上皮 simple columnar epithelium
2- 胃小凹 gastric pit
3- 壁细胞 parietal cell
4- 固有层的血管 blood vessel within lamina propria

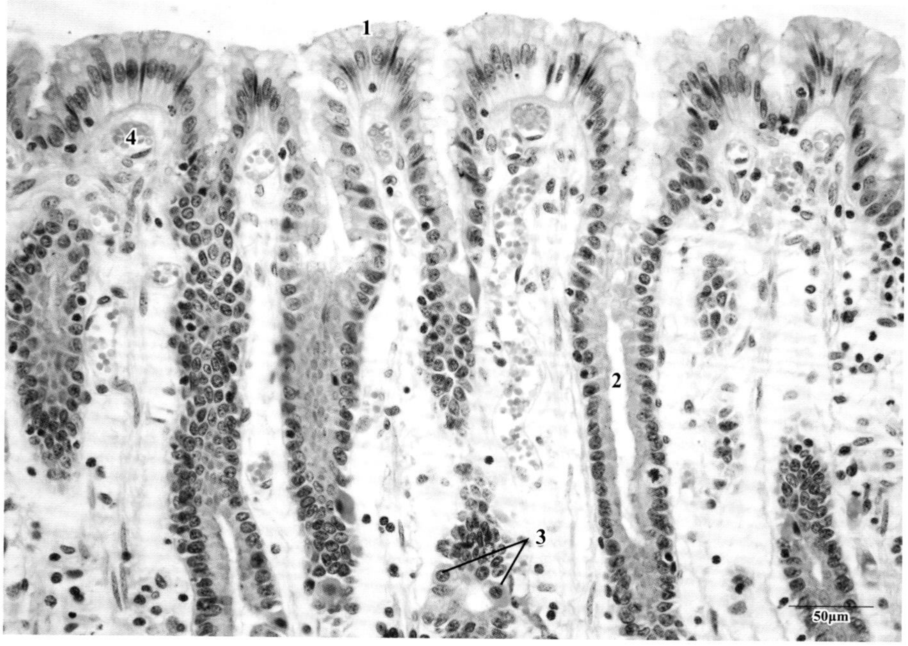

图 4-60　胃底部组织切片，示黏膜上皮（HE 染色，高倍镜）
Figure 4-60　Histological section of the fundus of stomach showing the mucosal epithelium (HE staining, higher power).

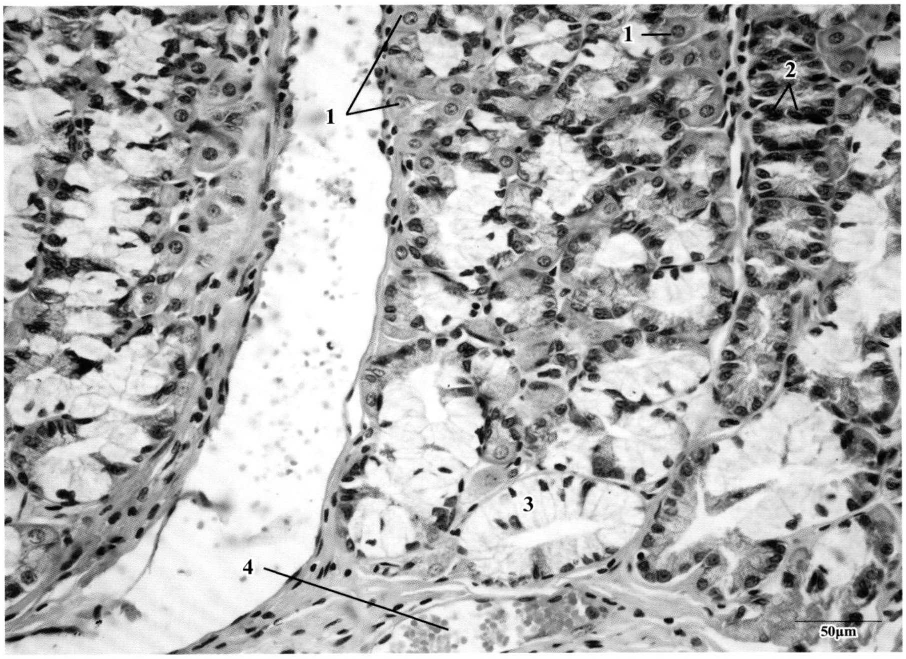

1- 壁细胞 parietal cell
2- 主细胞 chief cell
3- 颈黏液腺细胞 mucous neck cell
4- 血管 blood vessel

图 4-61 胃底腺组织切片（HE 染色，高倍镜）
Figure 4-61 Histological section of the fundic gland (HE staining, higher power).

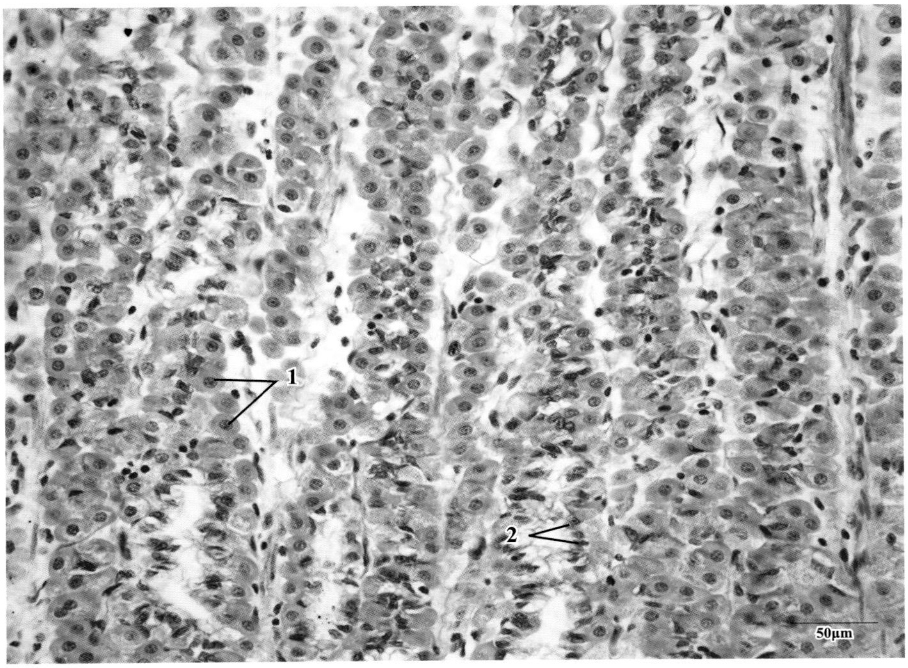

1- 壁细胞 parietal cell
2- 主细胞 chief cell

图 4-62 胃底腺组织切片（HE 染色，高倍镜）
Figure 4-62 Histological section of the fundic gland (HE staining, higher power).

1- 壁细胞 parietal cell
2- 主细胞 chief cell
3- 颈黏液腺细胞 mucous neck cell
4- 黏膜肌层 lamina muscularis mucosae

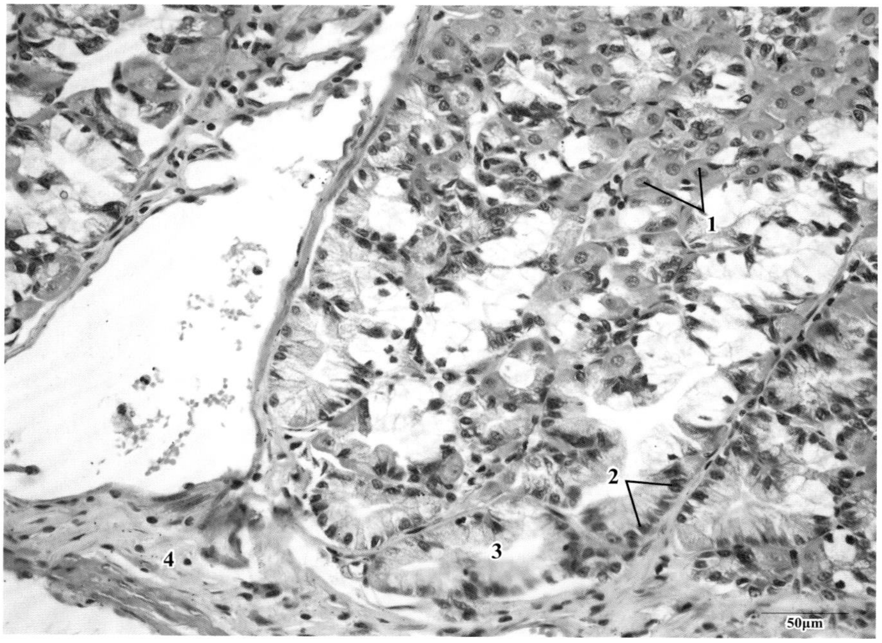

图 4-63　胃底腺组织切片（HE 染色，高倍镜）
Figure 4-63　Histological section of the fundic gland (HE staining, higher power).

1- 浆液腺泡 serous acinus
2- 黏液腺泡 mucous acinus
3- 固有层 lamina propria

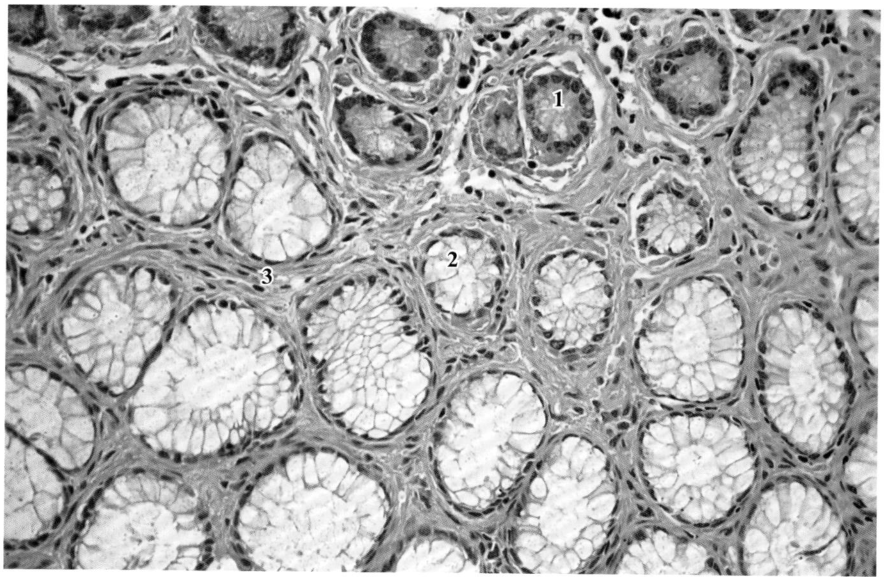

图 4-64　贲门腺组织切片（HE 染色，高倍镜）
Figure 4-64　Histological section of the cardiac gland (HE staining, higher power).

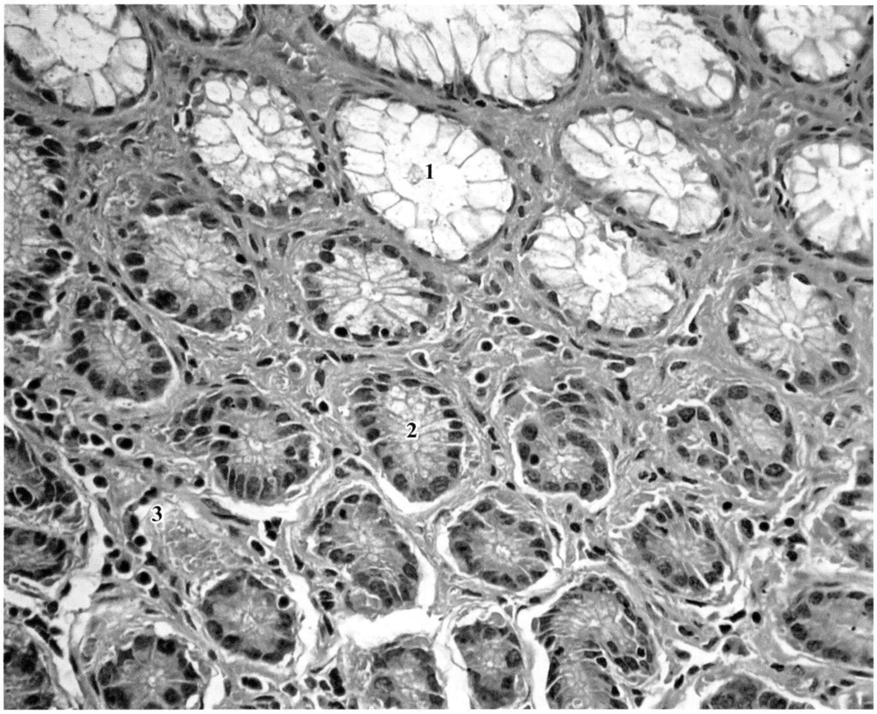

1- 黏液腺泡 mucous acinus
2- 浆液腺泡 serous acinus
3- 固有层 lamina propria

图 4-65 贲门腺组织切片（HE 染色，高倍镜）
Figure 4-65 Histological section of the cardiac gland (HE staining, higher power).

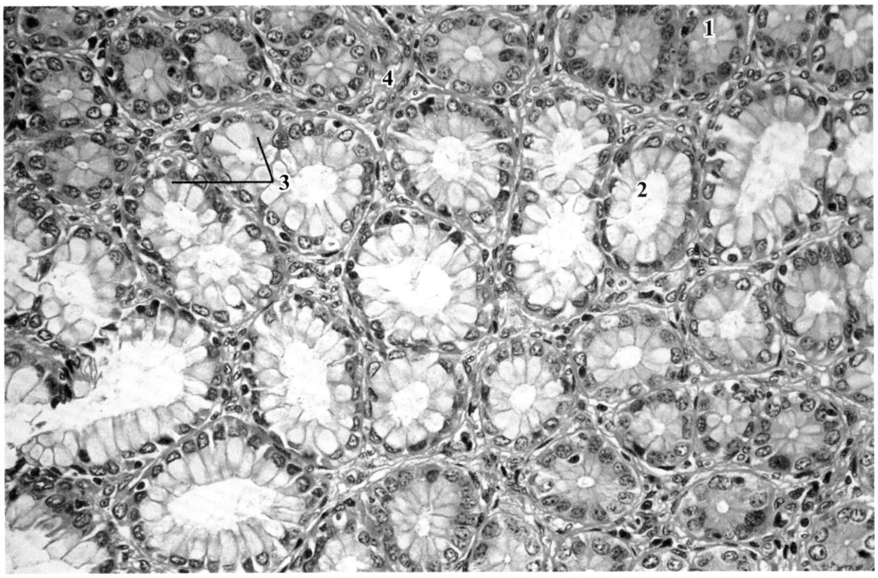

1- 浆液腺泡 serous acinus
2- 黏液腺泡 mucous acinus
3- 支持细胞 sustentacular cell
4- 固有层 lamina propria

图 4-66 幽门腺组织切片（HE 染色，高倍镜）
Figure 4-66 Histological section of the pyloric gland (HE staining, higher power).

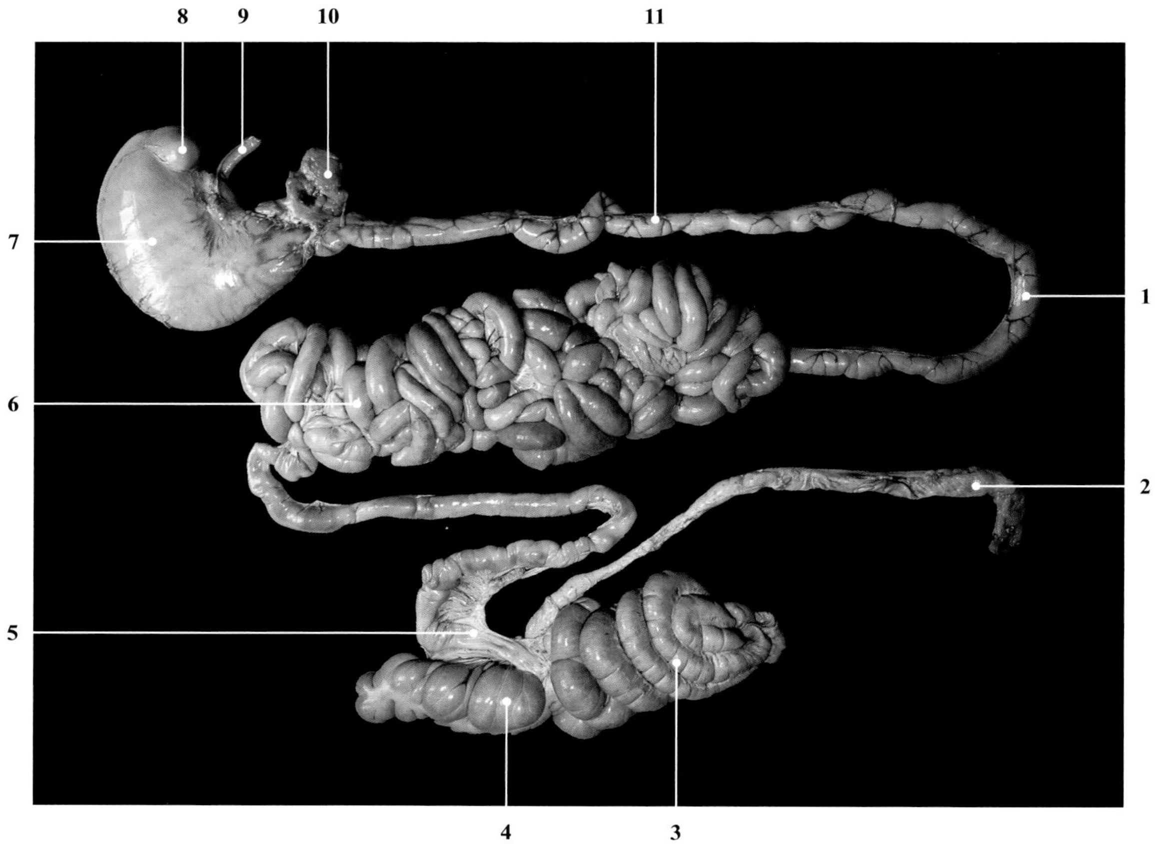

图 4-67　胃肠道
Figure 4-67　The gastrointestinal tract.

1- 十二指肠后曲 caudal duodenal flexure
2- 直肠 rectum
3- 结肠 colon
4- 盲肠 caecum
5- 回肠 ileum
6- 空肠 jejunum
7- 胃 stomach
8- 胃憩室 gastric diverticulum
9- 食管 esophagus
10- 胰 pancreas
11- 十二指肠 duodenum

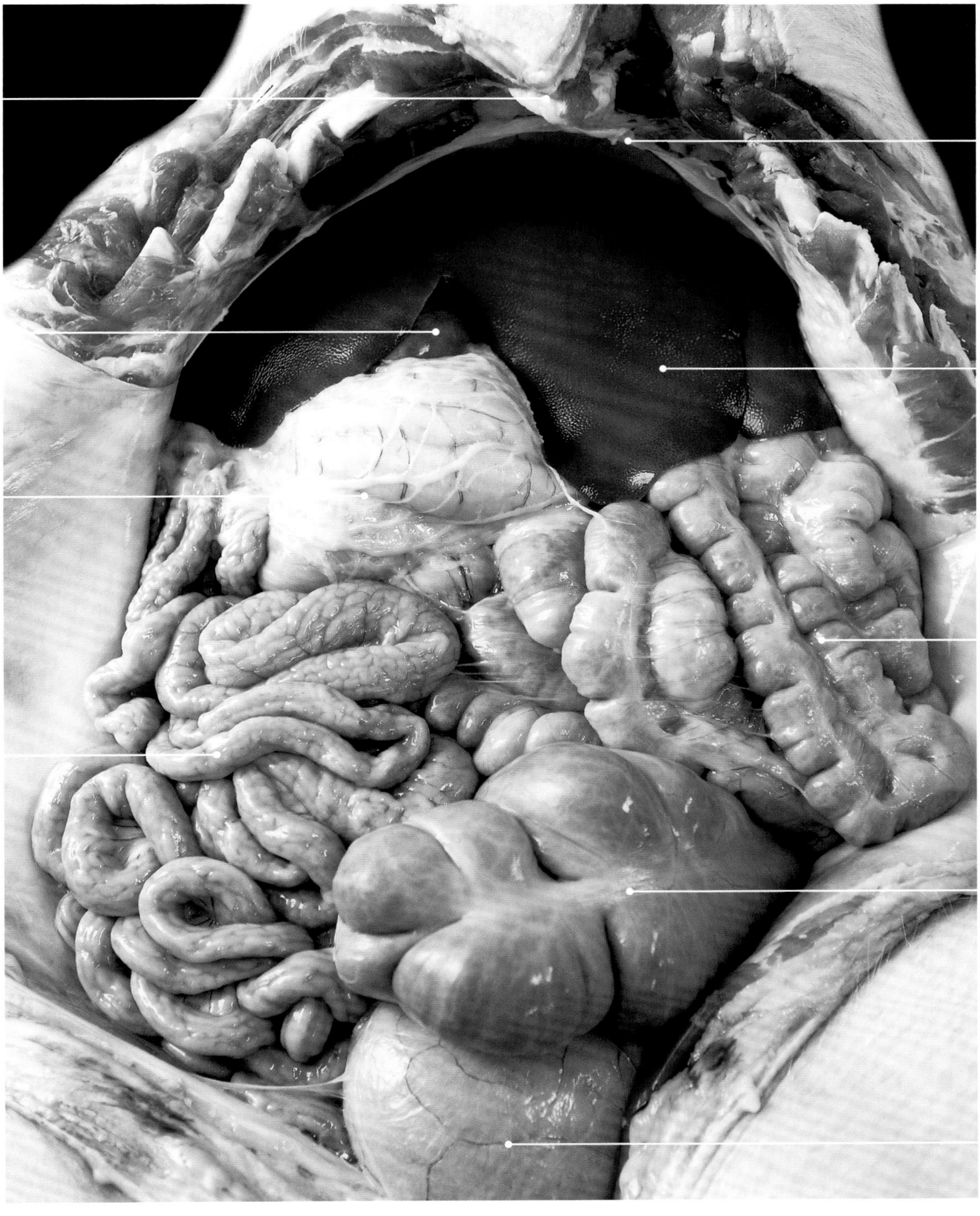

图 4-68　公猪腹腔内脏器（腹侧观）
Figure 4-68　Abdominal visceral organs of the boar (ventral view).

1- 膈 diaphragm
2- 肝 liver
3- 结肠 colon
4- 盲肠 caecum
5- 膀胱 urinary bladder
6- 小肠 small intestine
7- 胃 stomach
8- 胆囊 gallbladder
9- 剑状软骨 xiphoid cartilage

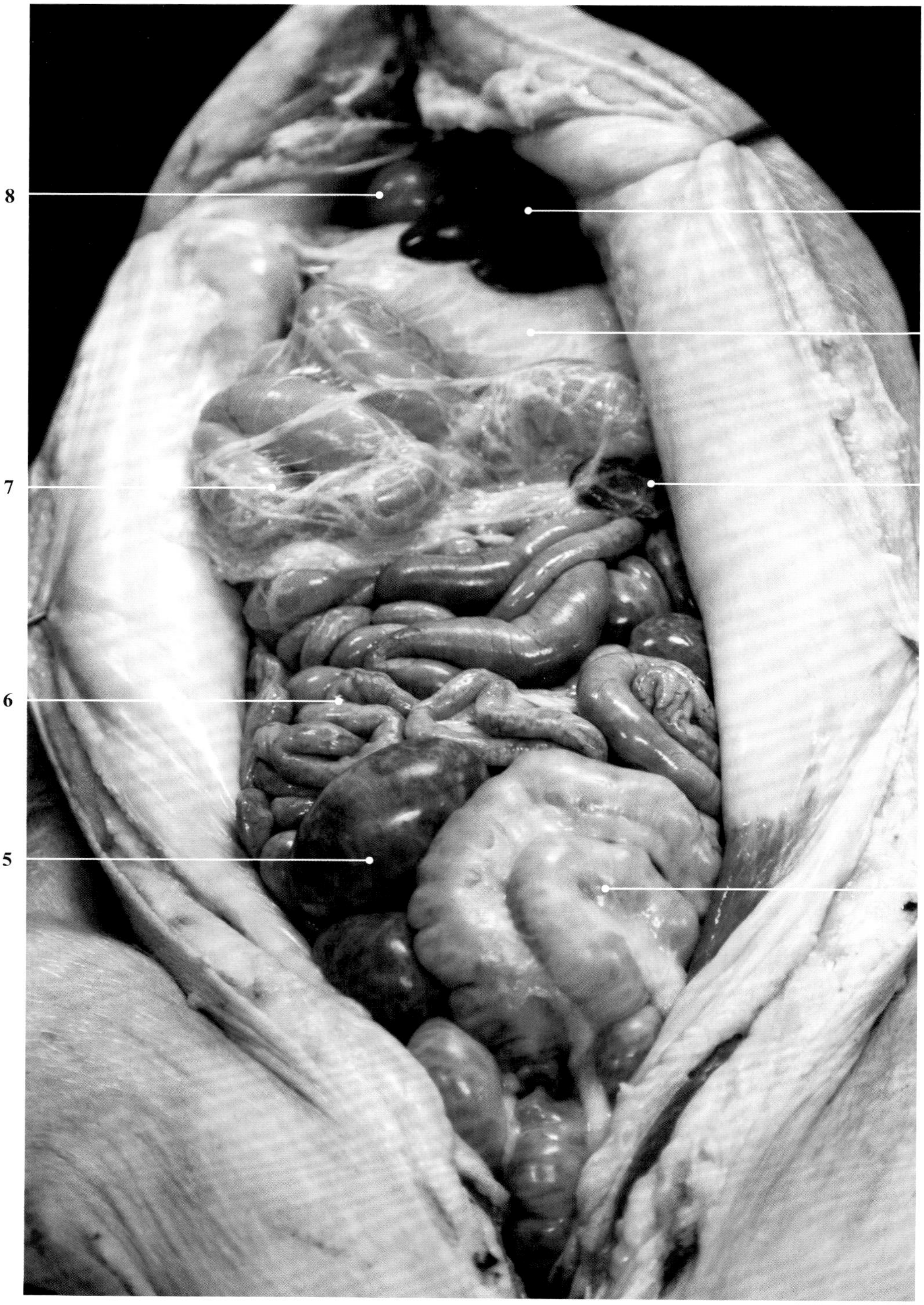

图 4-69　母猪腹腔内脏器（腹侧观）
Figure 4-69　Abdominal visceral organs of the sow (ventral view).

1- 肝 liver
2- 胃 stomach
3- 脾 spleen
4- 结肠 colon
5- 盲肠 caecum
6- 小肠 small intestine
7- 大网膜 greater omentum
8- 胆囊 gallbladder

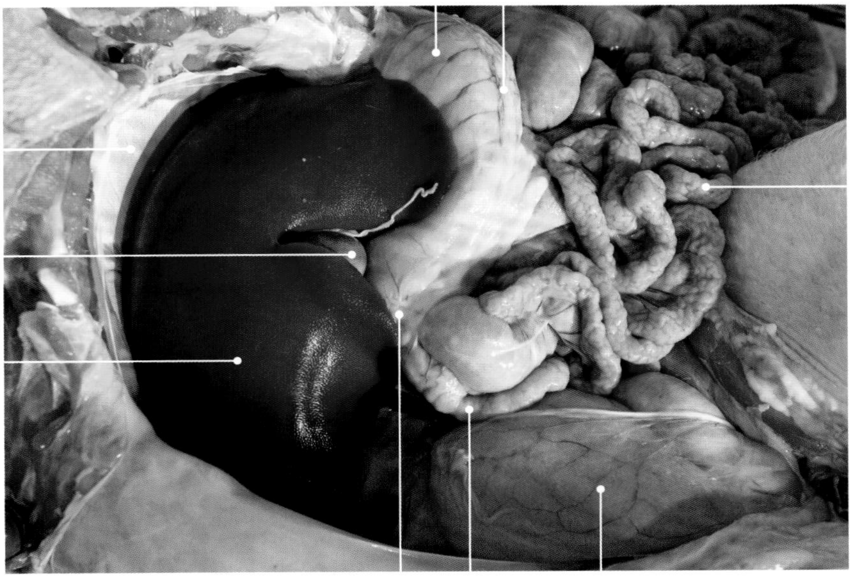

图 4-70 肝、胃和小肠（腹侧观）
Figure 4-70 The liver, stomach and small intestine (ventral view).

1- 空肠 jejunum
2- 膀胱 urinary bladder
3- 十二指肠 duodenum
4- 幽门 pylorus
5- 肝 liver
6- 胆囊 gallbladder
7- 膈 diaphragm
8- 胃 stomach
9- 胃大弯 greater curvature of stomach

图 4-71 公猪小肠（腹侧观）
Figure 4-71 The small intestine of a boar (ventral view).

1- 大肠 large intestine
2- 空肠 jejunum
3- 胃 stomach

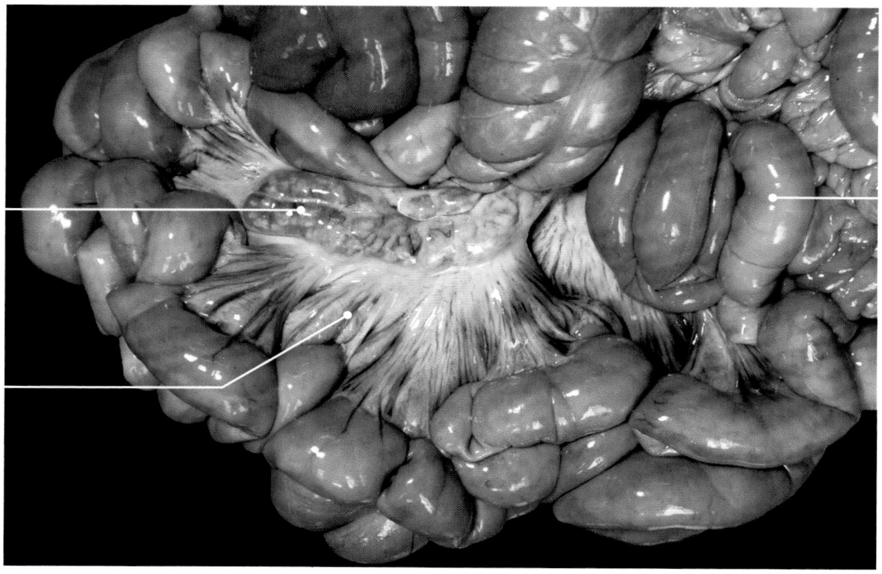

图 4-72　肠系膜
Figure 4-72　The mesentery.

1- 空肠 jejunum
2- 空肠系膜 mesojejunum
3- 肠系膜淋巴结 mesenteric lymph node

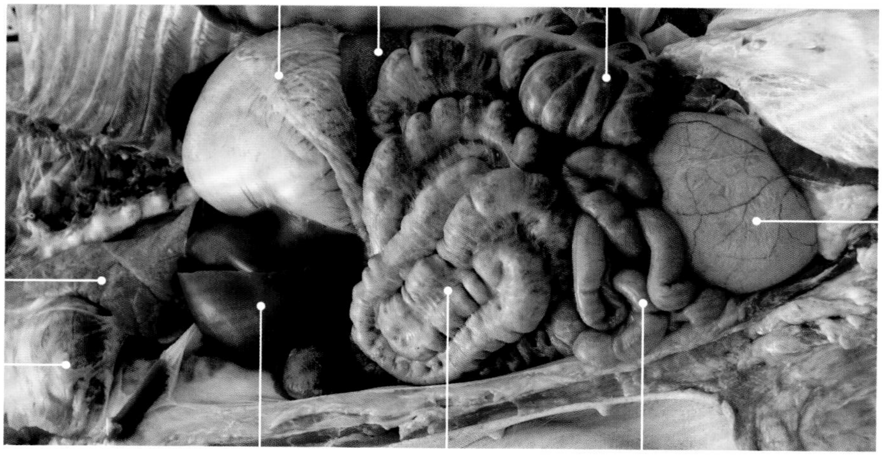

图 4-73　盲肠和结肠襻
Figure 4-73　The caecum and colon loop.

1- 膀胱 urinary bladder
2- 空肠 jejunum
3- 结肠襻 colon loop
4- 肝 liver
5- 心脏 heart
6- 肺 lung
7- 胃 stomach
8- 脾 spleen
9- 盲肠 caecum

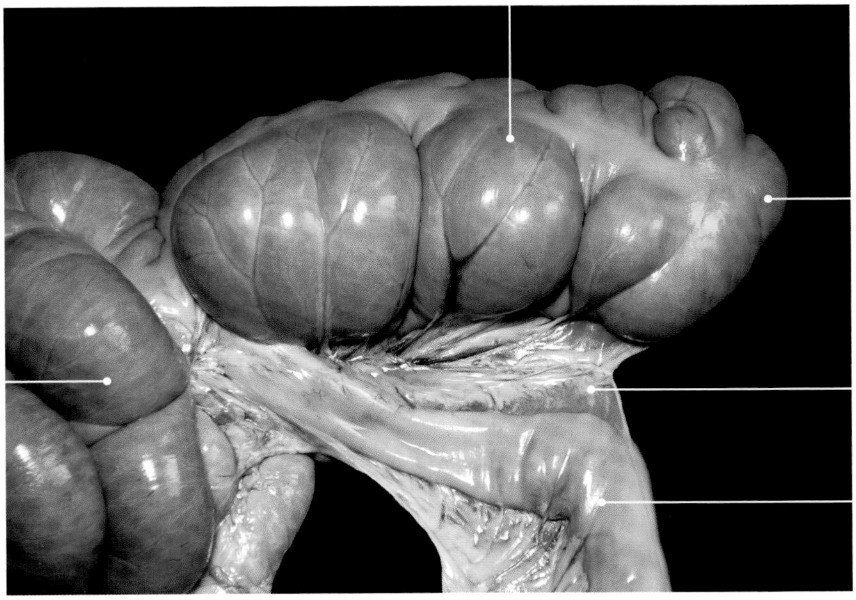

图 4-74　回盲韧带
Figure 4-74　The ileocecal ligament.

1- 盲肠 caecum
2- 盲肠尖 apex of caecum
3- 回盲韧带 ileoacecal ligament
4- 回肠 ileum
5- 结肠 colon

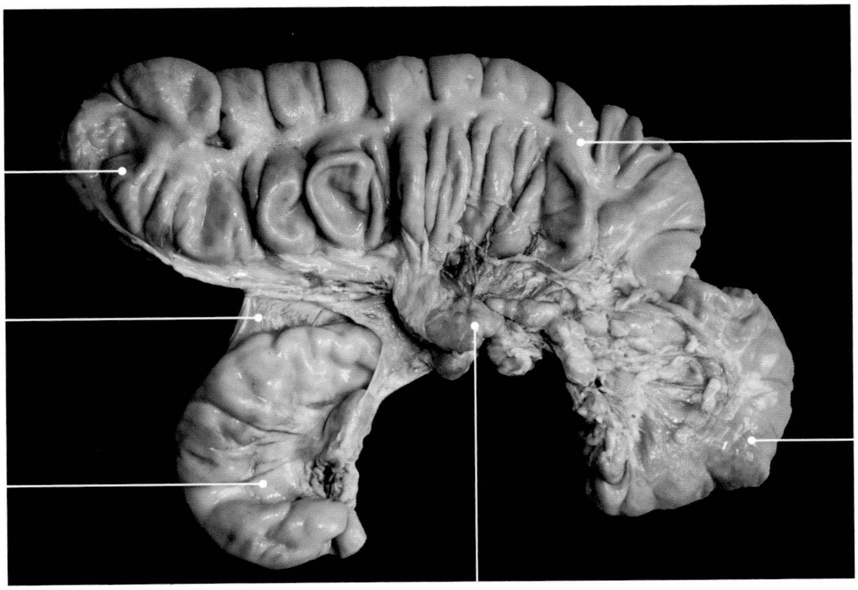

图 4-75　盲肠
Figure 4-75　The caecum.

1- 盲肠 caecum
2- 结肠 colon
3- 盲肠淋巴结 caecal lymph nodes
4- 回肠 ileum
5- 回盲韧带 ileocaecal ligament
6- 盲肠尖 apex of caecum

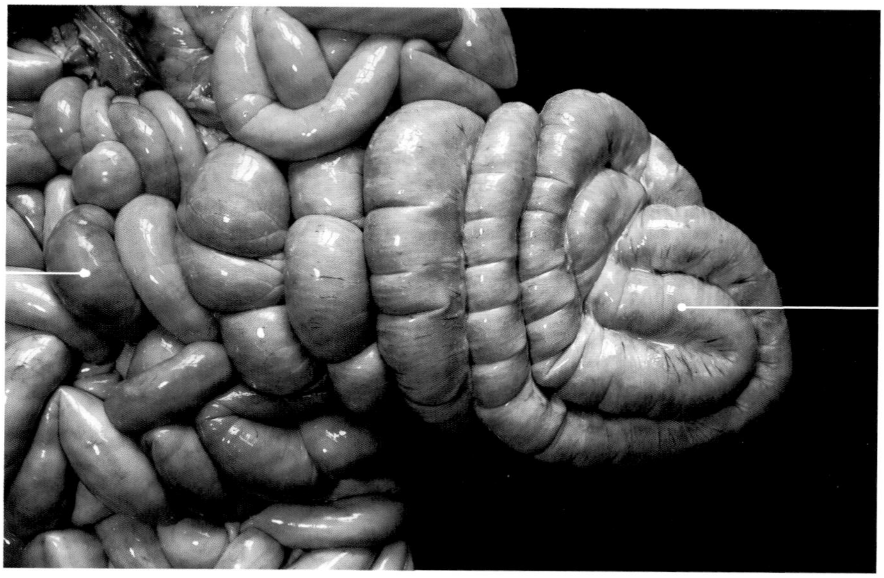

图 4-76　结肠襻（壁面观）
Figure 4-76　The colon loop (parietal view).

1- 结肠襻 colon loop　　**2-** 空肠 jejunum

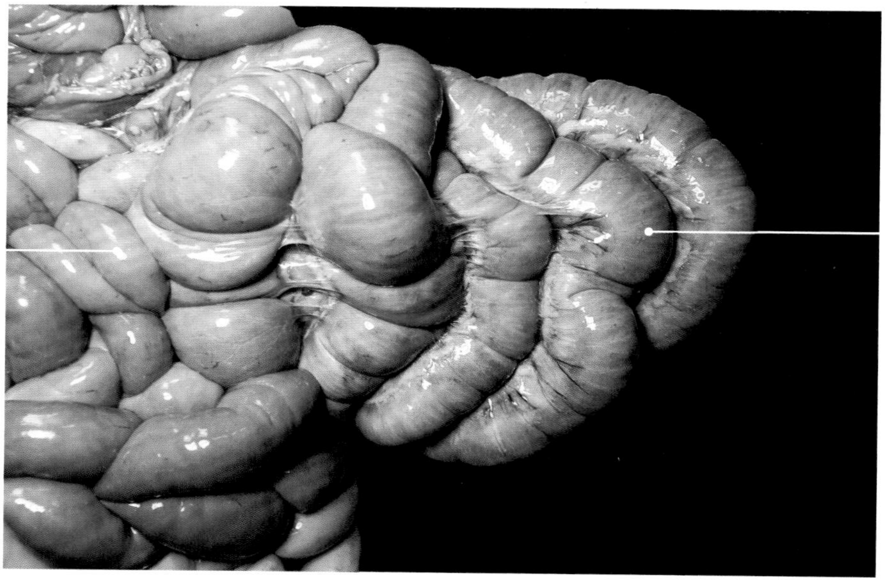

图 4-77　结肠襻（脏面观）
Figure 4-77　The colon loop (visceral view).

1- 结肠襻 colon loop　　**2-** 空肠 jejunum

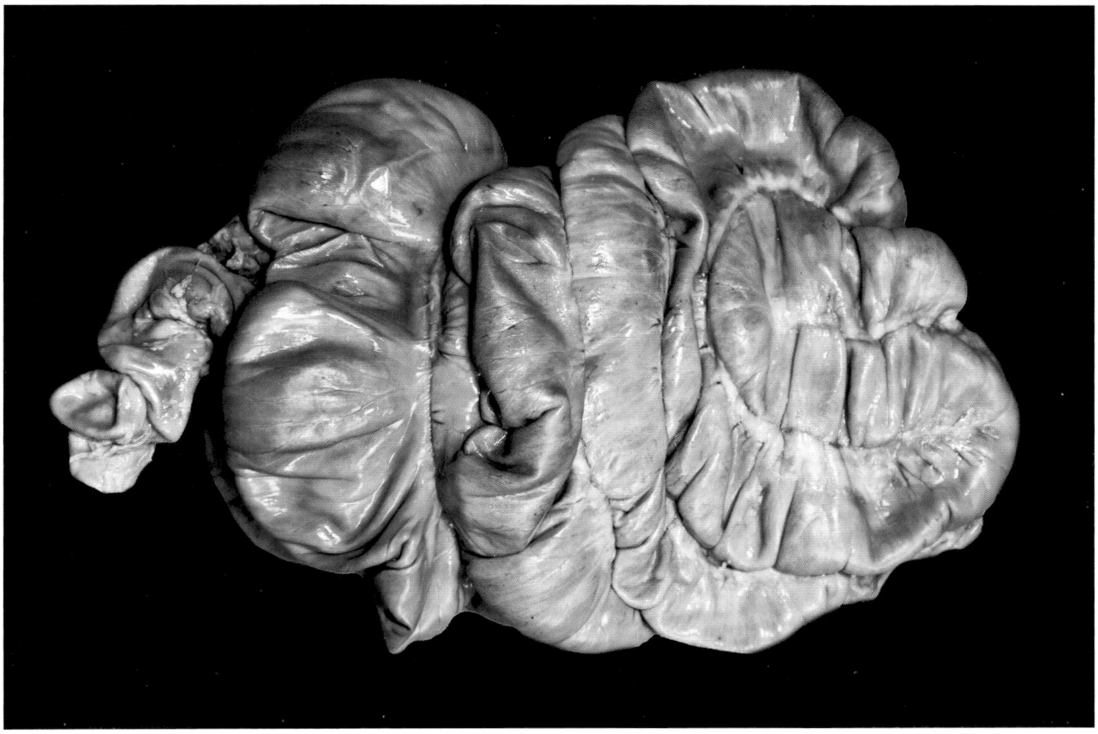

图 4-78 结肠襻（固定标本，壁面观）
Figure 4-78 The colon loop (fixed preparation, parietal view).

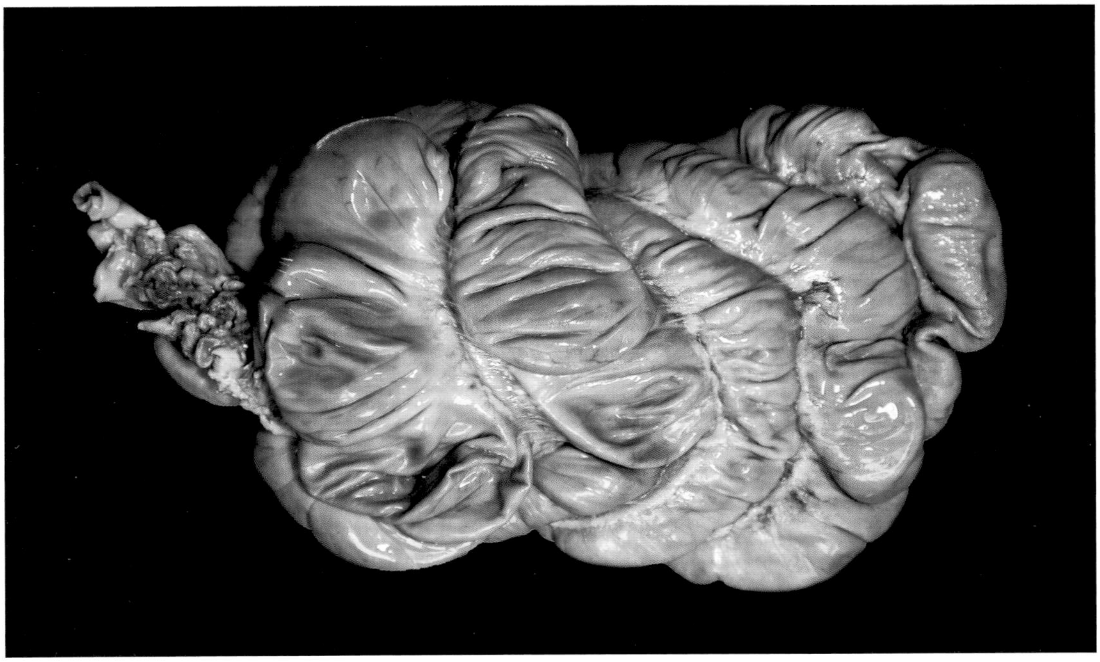

图 4-79 结肠襻（固定标本，脏面观）
Figure 4-79 The colon loop (fixed preparation, visceral view).

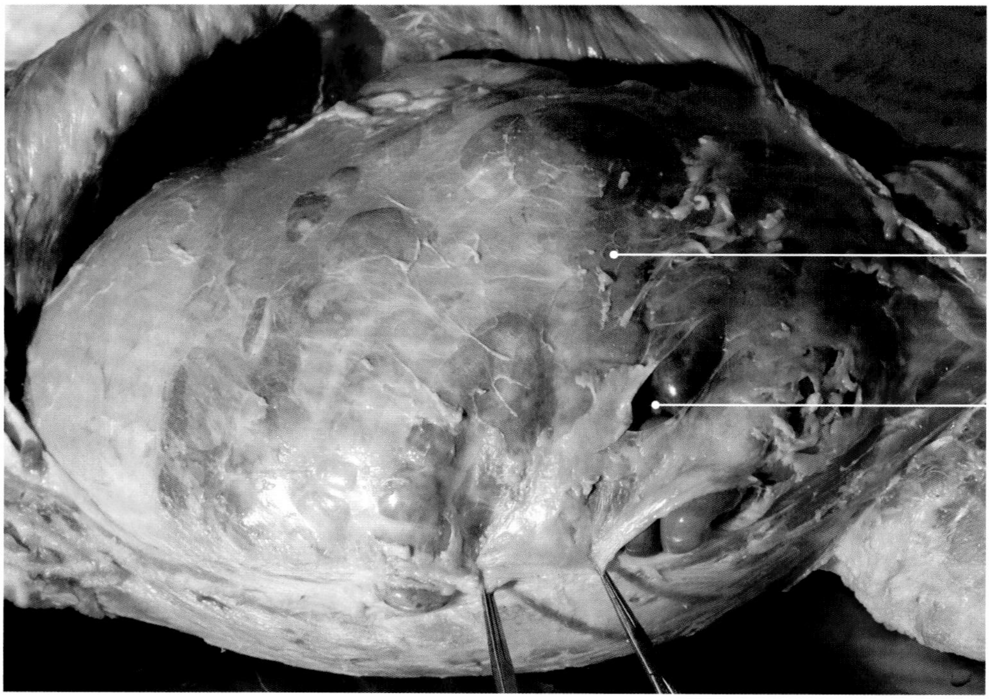

图 4-80 腹膜
Figure 4-80 The peritoneum.

1- 腹膜壁层 parietal peritoneum
2- 腹膜腔 peritoneal cavity

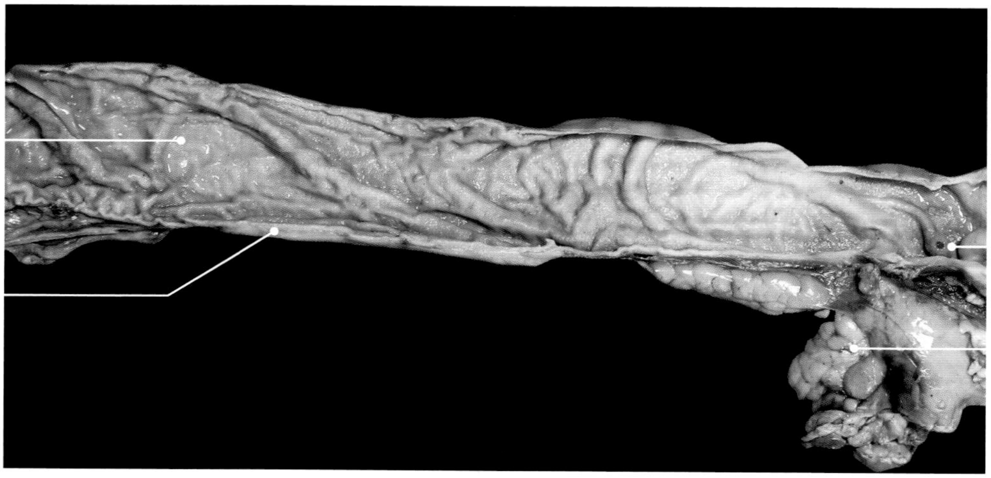

图 4-81 十二指肠黏膜面
Figure 4-81 The mucosal surface of the duodenum.

1- 十二指肠乳头 duodenal papilla
2- 胰 pancreas
3- 十二指肠壁 duodenal wall
4- 十二指肠黏膜 duodenal mucosa

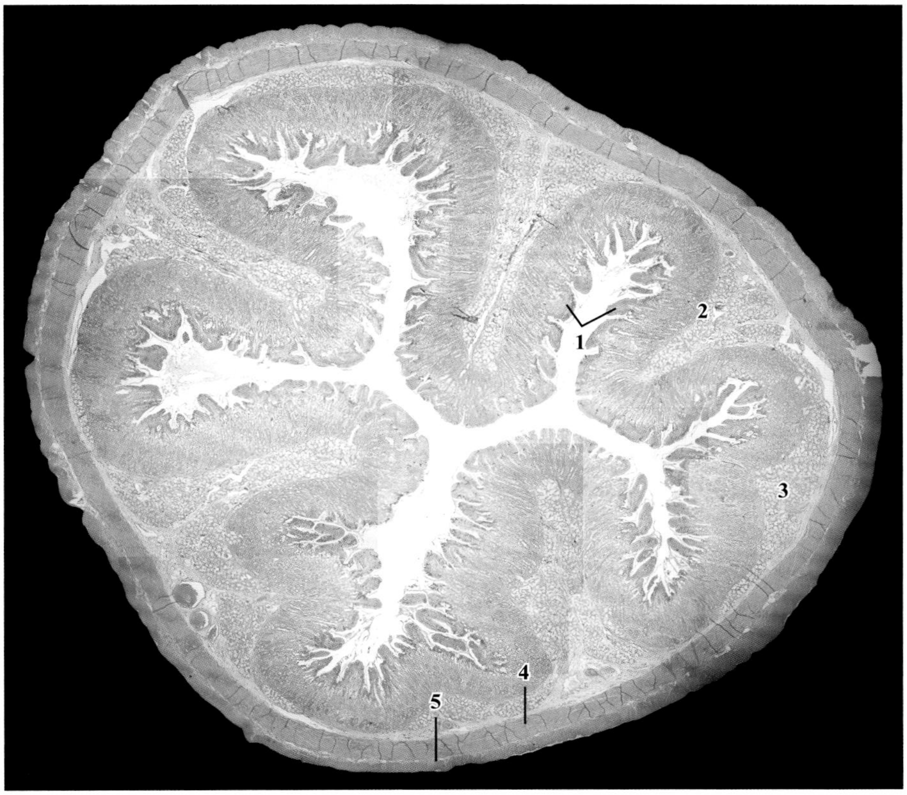

1- 绒毛 villus
2- 黏膜肌层 lamina muscularis mucosae
3- 黏膜下层 submucosa
4- 内环肌层 inner circular stratum
5- 外纵肌层 outer longitudinal stratum

图 4-82 十二指肠组织切片（HE 染色，低倍镜）
Figure 4-82 Histological section of the duodenum (HE staining, lower power).

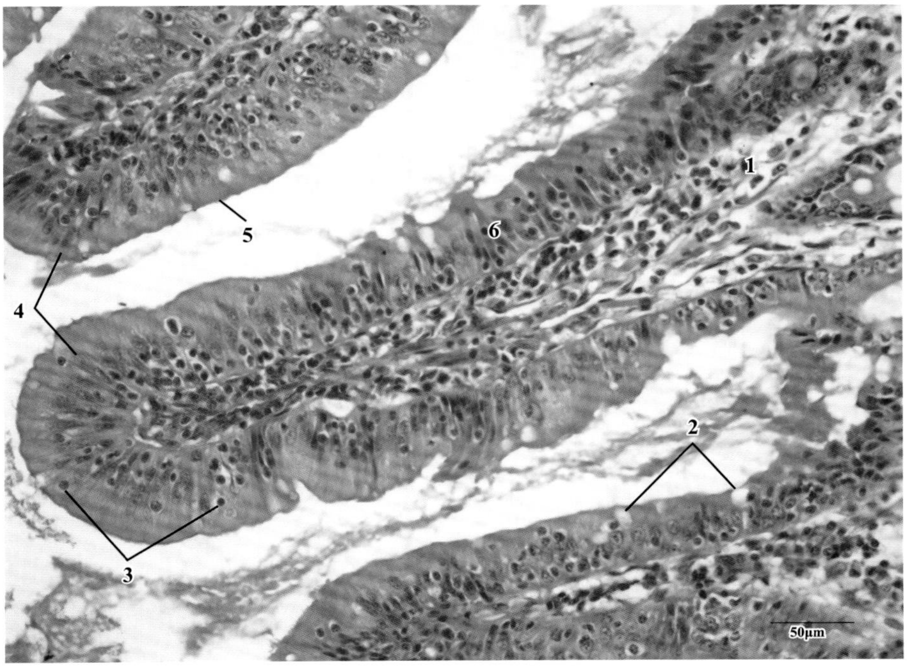

1- 固有层 lamina propria
2- 杯状细胞 goblet cell
3- 上皮内淋巴细胞 intraepithelial lymphocyte
4- 绒毛 villus
5- 纹状缘 striated border
6- 单层柱状上皮 simple columnar epithelium

图 4-83 十二指肠组织切片，示肠绒毛（HE 染色，高倍镜）
Figure 4-83 Histological section of the duodenum showing the intestinal villus (HE staining, higher power).

1- 分泌单位 secretory unit
2- 十二指肠腺 / 布伦纳腺 duodenal gland / Brunner's gland
3- 黏液腺泡 mucous acinus
4- 黏膜肌层 lamina muscularis mucosae
5- 小肠腺 small intestinal gland
6- 浆液腺泡 serous acinus
7- 固有层 lamina propria

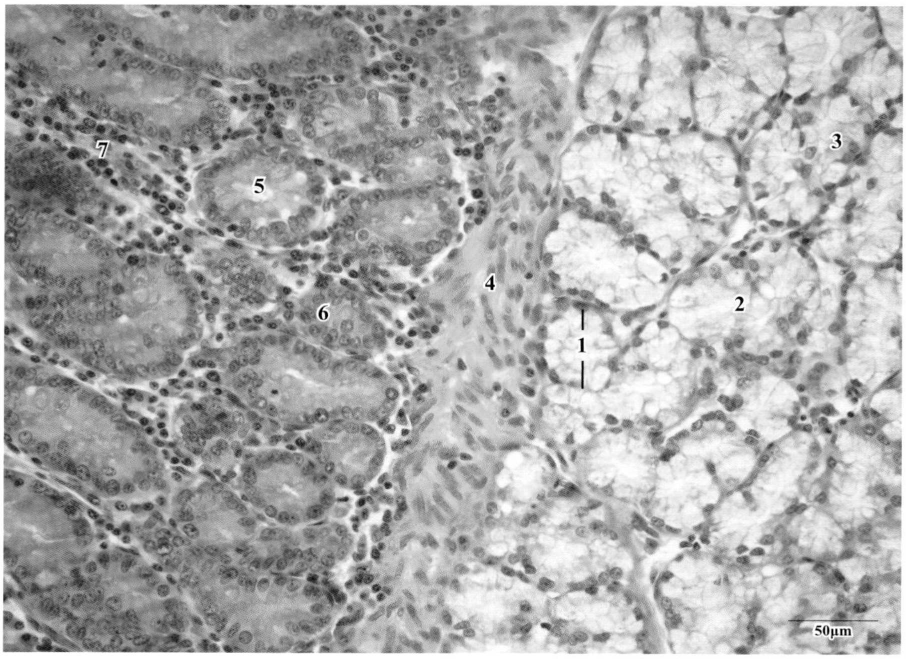

图 4-84　十二指肠组织切片，示十二指肠腺（布伦纳腺）（HE 染色，高倍镜）
Figure 4-84　Histological section of the duodenum showing the duodenal gland (Brunner's gland) (HE staining, higher power).

1- 导管 duct
2- 黏液腺泡 mucous acinus
3- 小叶间结缔组织 interlobular connective tissue

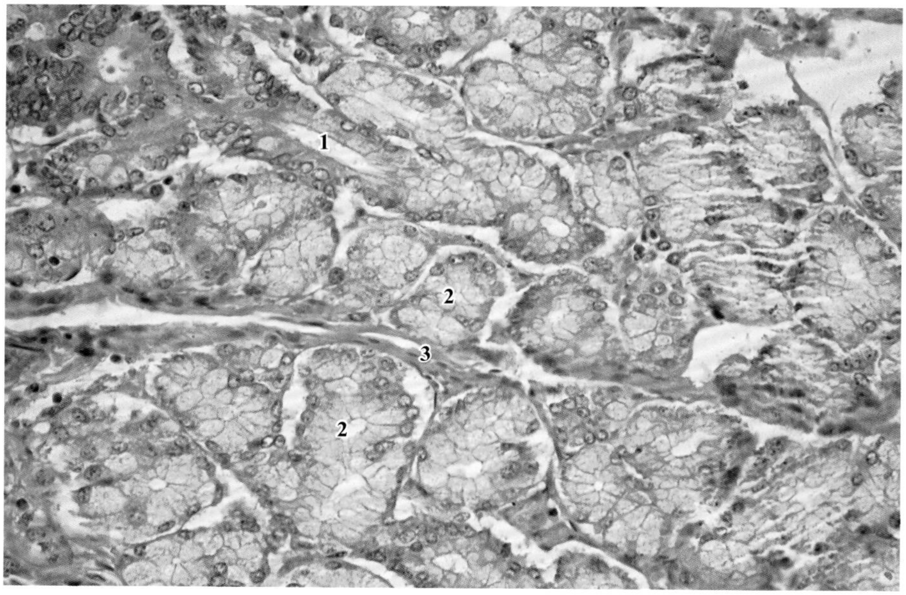

图 4-85　十二指肠腺（布伦纳腺）组织切片（HE 染色，高倍镜）
Figure 4-85　Histological section of the duodenal gland (Brunner's gland) (HE staining, higher power).

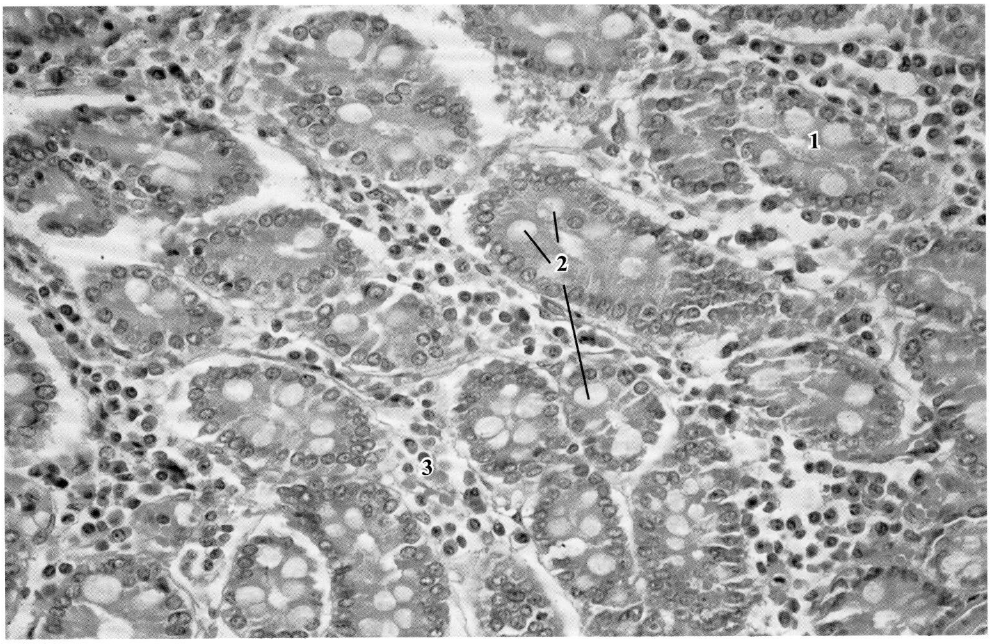

图 4-86　十二指肠的小肠腺组织切片（HE 染色，高倍镜）
Figure 4-86　Histological section of the small intestinal gland in duodenum (HE staining, higher power).

1- 小肠腺 small intestinal gland
2- 杯状细胞 goblet cell
3- 固有层 lamina propria

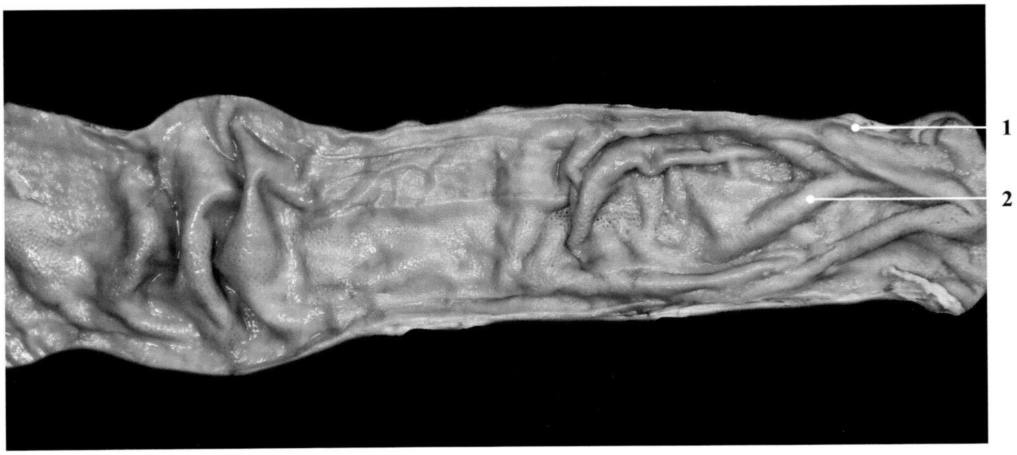

图 4-87　空肠黏膜面
Figure 4-87　The mucosal surface of jejunum.

1- 空肠壁 jejunal wall
2- 空肠黏膜 jejunal mucosa

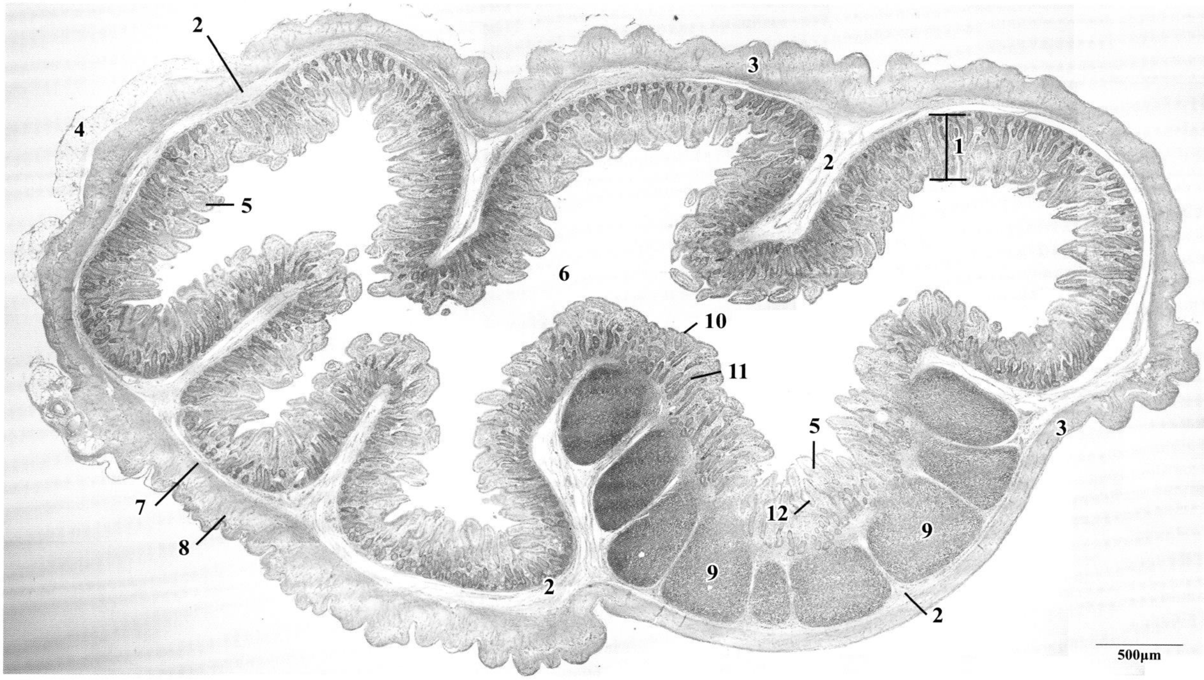

图 4-88　空肠组织切片（HE 染色，低倍镜）
Figure 4-88　Histological section of the jejunum (HE staining, lower power).

1- 黏膜 mucosa
2- 黏膜下层 submucosa
3- 肌层 muscular layer
4- 浆膜 serosa
5- 肠绒毛 intestinal villi
6- 肠腔 intestinal lumen
7- 内环肌层 inner circular stratum
8- 外纵肌层 outer longitudinal stratum
9- 淋巴小结 lymphatic nodule
10- 黏膜上皮 mucosal epithelium
11- 小肠腺 small intestinal gland
12- 肠隐窝 / 李氏隐窝 intestine crypt (crypt of Lieberkühn)

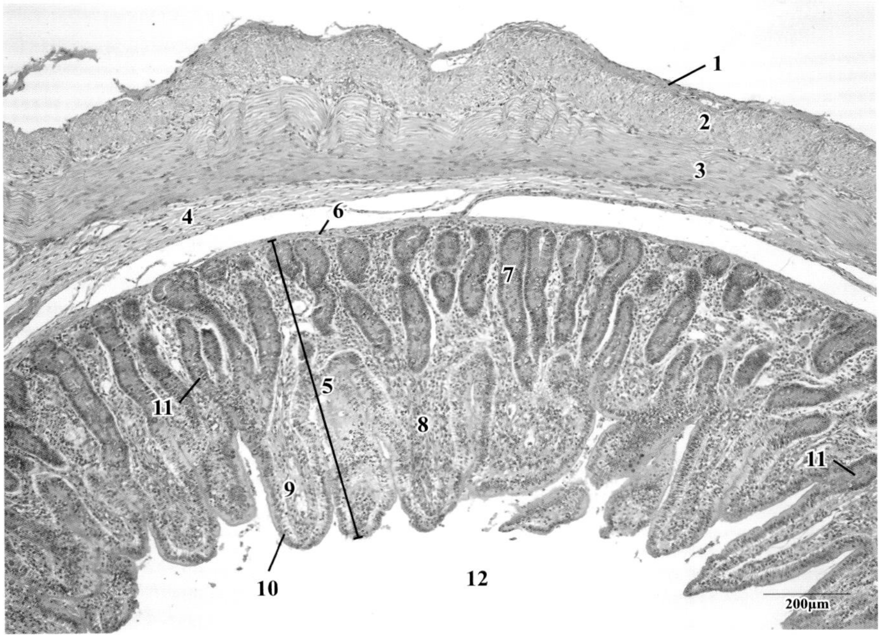

1- 浆膜 serosa
2- 外纵肌层 outer longitudinal stratum
3- 内环肌层 inner circular stratum
4- 黏膜下层 submucosa
5- 黏膜 mucosa
6- 黏膜肌层 lamina muscularis mucosae
7- 小肠腺 small intestinal gland
8- 固有层 lamina propria
9- 肠绒毛 intestinal villus
10- 黏膜上皮 mucosal epithelium
11- 肠隐窝 / 李氏隐窝 intestine crypt (crypt of Lieberkühn)
12- 肠腔 intestinal lumen

图 4-89　空肠组织切片（HE 染色，低倍镜）
Figure 4-89　Histological section of the jejunum (HE staining, lower power).

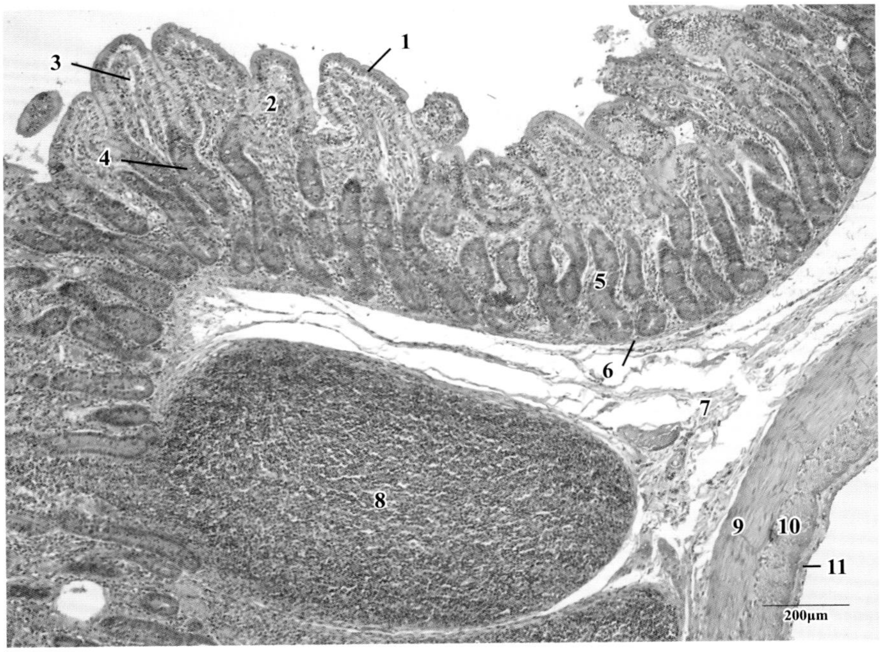

1- 黏膜上皮 mucosal epithelium
2- 固有层 lamina propria
3- 中央乳糜管 central lacteal
4- 肠隐窝 / 李氏隐窝 intestine crypt (crypt of Lieberkühn)
5- 小肠腺 small intestinal gland
6- 黏膜肌层 lamina muscularis mucosae
7- 黏膜下层 submucosa
8- 淋巴小结 lymphatic nodule
9- 内环肌层 inner circular stratum
10- 外纵肌层 outer longitudinal stratum
11- 浆膜 serosa

图 4-90　空肠组织切片（HE 染色，低倍镜）
Figure 4-90　Histological section of the jejunum (HE staining, lower power).

1- 肠隐窝 / 李氏隐窝 intestine crypt（crypt of Lieberkühn）
2- 中央乳糜管 central lacteal
3- 绒毛 villus
4- 上皮内淋巴细胞 intraepithelial lymphocyte
5- 黏膜肌层 lamina muscularis mucosae
6- 固有层 lamina propria

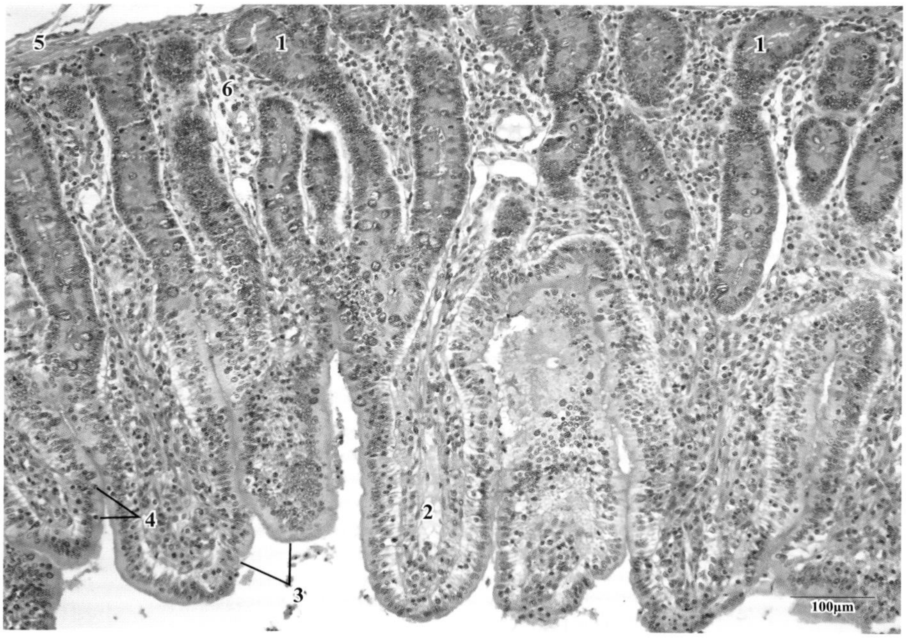

图 4-91　空肠组织切片（HE 染色，高倍镜）
Figure 4-91　Histological section of the jejunum (HE staining, higher power).

1- 吸收细胞 absorptive cell
2- 上皮内淋巴细胞 intraepithelial lymphocyte
3- 中央乳糜管 central lacteal
4- 纹状缘 striated border
5- 固有层 lamina propria
6- 基膜 basal lamina
7- 杯状细胞 goblet cell
8- 肠隐窝 / 李氏隐窝 intestine crypt (crypt of Lieberkühn)

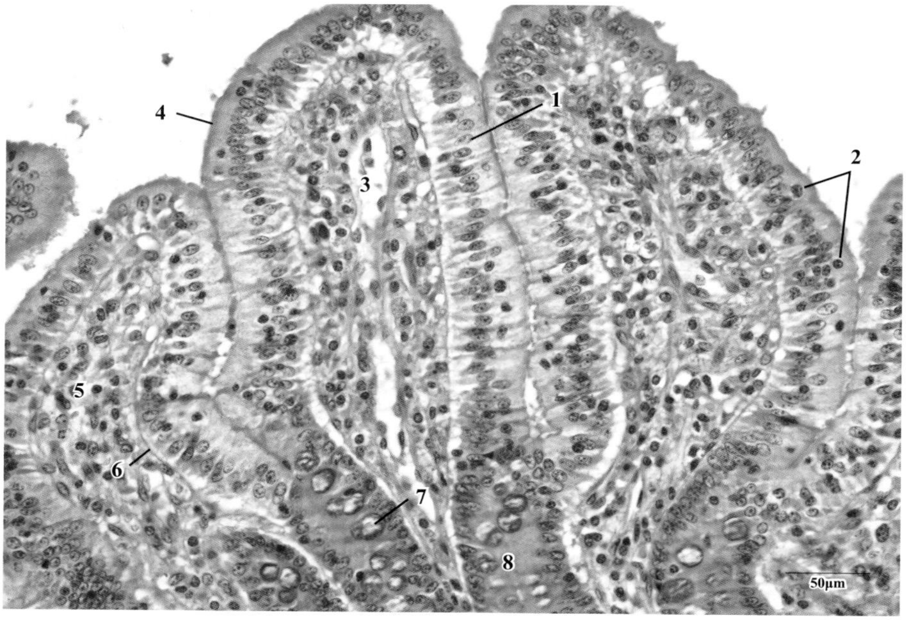

图 4-92　空肠绒毛组织切片（HE 染色，高倍镜）
Figure 4-92　Histological section of the villus of jejunum (HE staining, higher power).

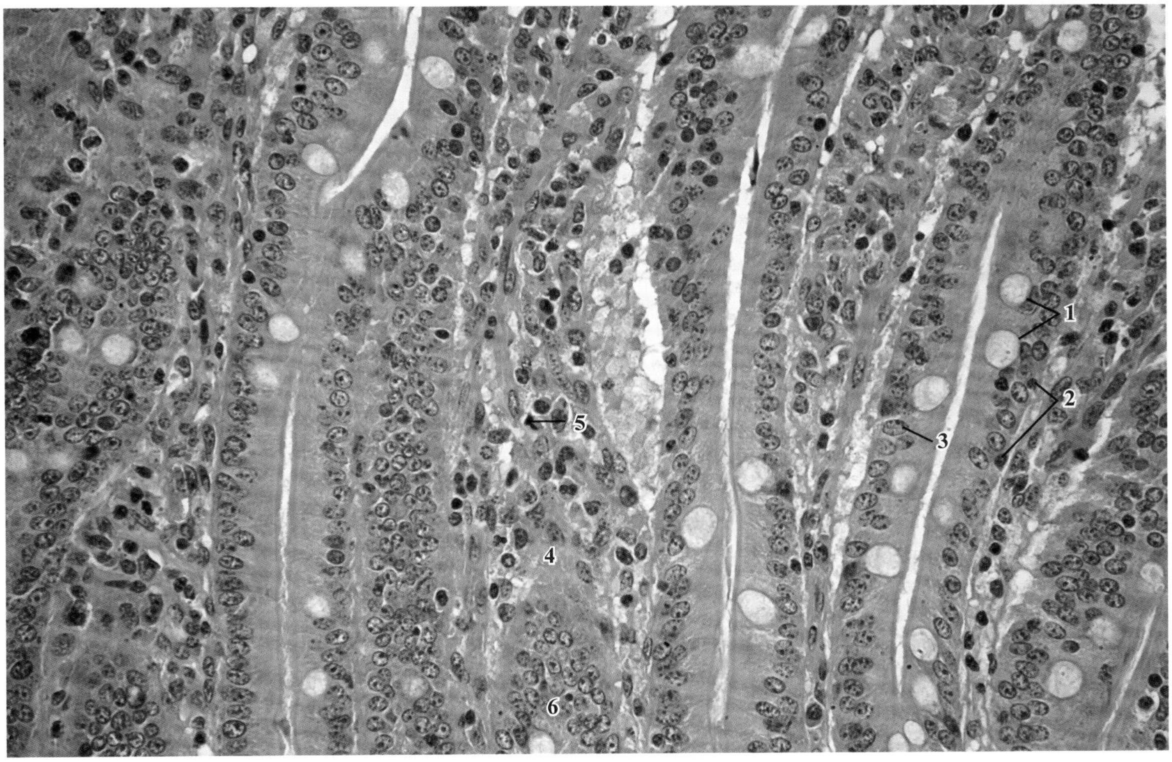

图 4-93　空肠绒毛组织切片（HE 染色，高倍镜）
Figure 4-93　Histological section of the villus of jejunum (HE staining, higher power).

1- 杯状细胞 goblet cell
2- 上皮内淋巴细胞 intraepithelial lymphocyte
3- 吸收细胞核 nucleus of the absorptive cell
4- 固有层 lamina propria
5- 嗜伊红细胞 eosinophil
6- 肠隐窝 / 李氏隐窝 intestine crypt (crypt of Lieberkühn)

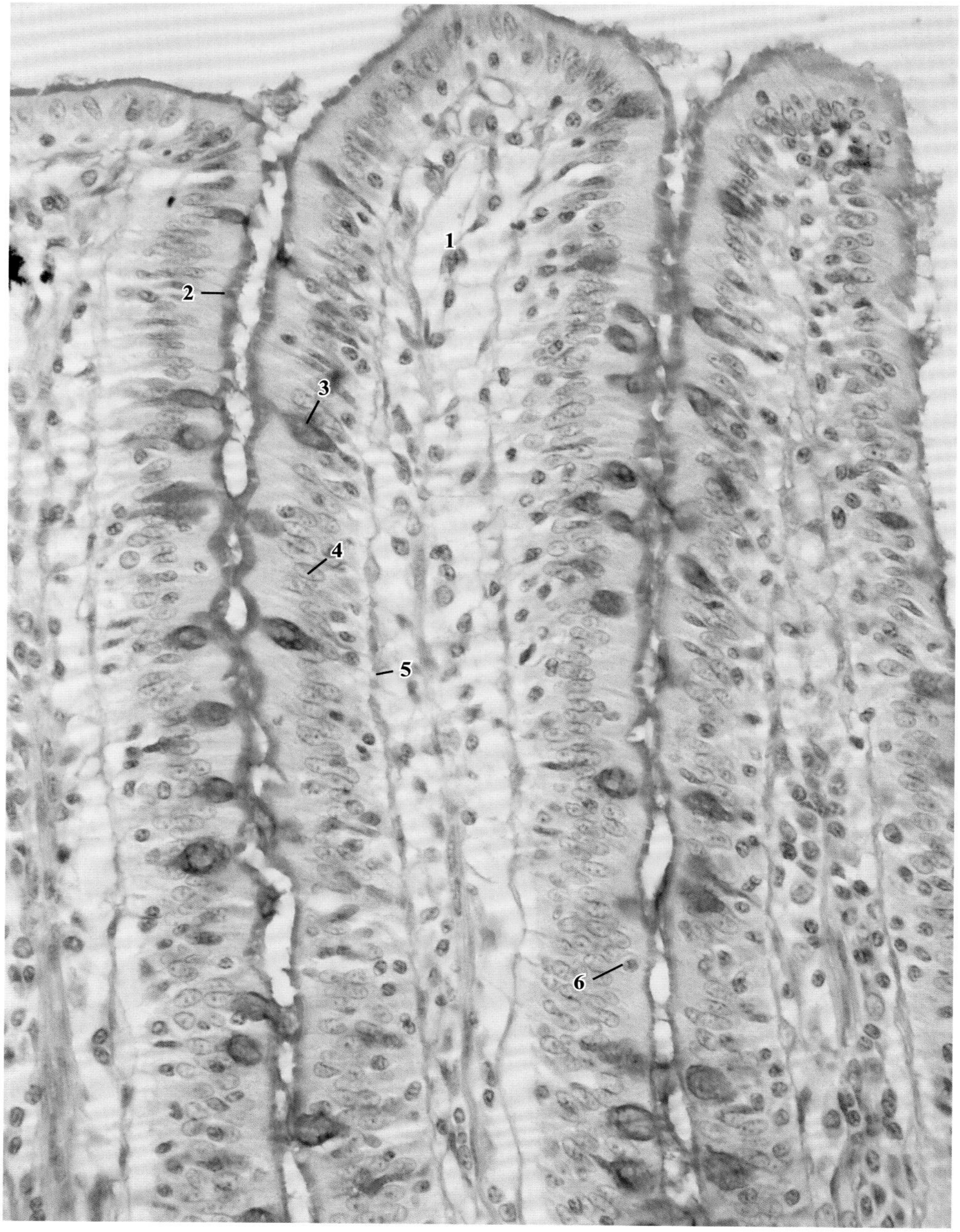

图 4-94　空肠绒毛组织切片（PAS 染色，高倍镜）
Figure 4-94　Histological section of the villus of jejunum (PAS staining, higher power).

1- 中央乳糜管 central lacteal
2- 纹状缘 striated border
3- 杯状细胞 goblet cell
4- 吸收细胞核 nucleus of the absorptive cell
5- 基膜 basal lamina
6- 上皮内淋巴细胞 intraepithelial lymphocyte

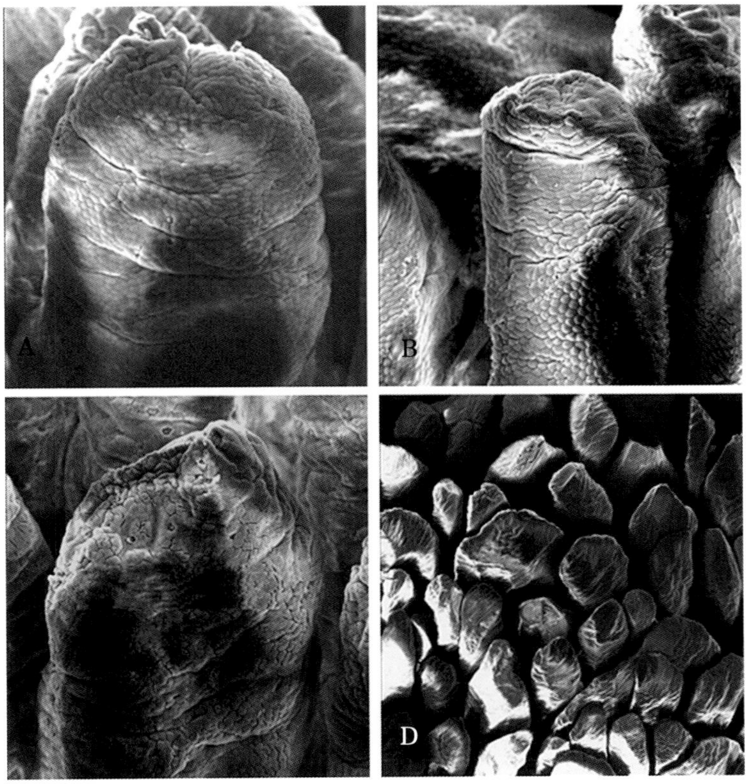

图 4-95 空肠绒毛扫描电镜
Figure 4-95 Scanning electron microscopy (SEM) of the villus of jejunum.

A ~ C：高倍镜 high power **D：低倍镜 lower power**

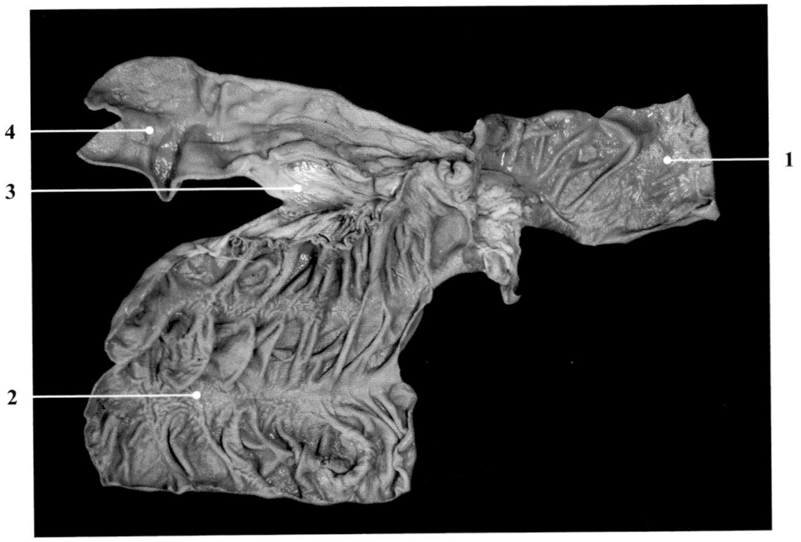

图 4-96 回肠、盲肠和结肠黏膜面
Figure 4-96 Mucosal surface of the ileum, caecum and colon.

1- 结肠黏膜 colonic mucosa
2- 盲肠黏膜 caecal mucosa
3- 回盲韧带 ileocaecal ligament
4- 回肠黏膜 ileal mucosa

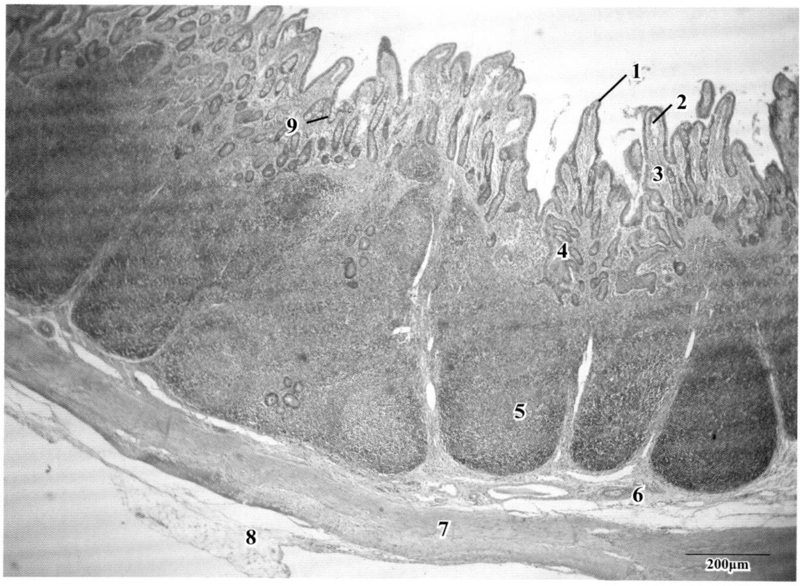

图 4-97　回肠组织切片（HE 染色，低倍镜）
Figure 4-97　Histological section of the ileum (HE staining, lower power).

1- 肠绒毛 intestinal villus
2- 中央乳糜管 central lacteal
3- 固有层 lamina propria
4- 小肠腺 small intestinal gland
5- 淋巴小结 lymphatic nodule
6- 黏膜下层 submucosa
7- 肌层 muscular layer
8- 浆膜 serosa
9- 肠隐窝 intestine crypt

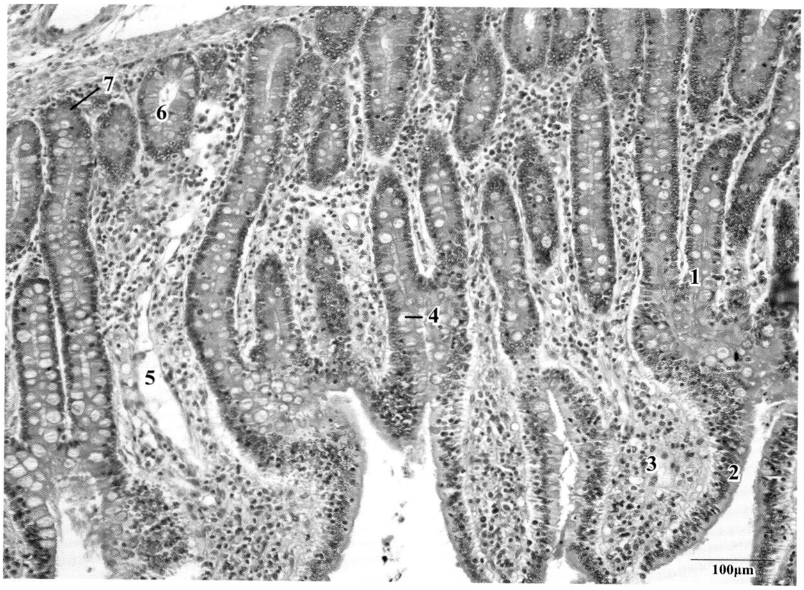

图 4-98　回肠组织切片（HE 染色，高倍镜）
Figure 4-98　Histological section of the ileum (HE staining, higher power).

1- 肠隐窝 intestinal crypt
2- 单层柱状上皮 simple columnar epithelium
3- 固有层 lamina propria
4- 杯状细胞 goblet cell
5- 中央乳糜管 central lacteal
6- 小肠腺 small intestinal gland
7- 未分化细胞 undifferentiated cell

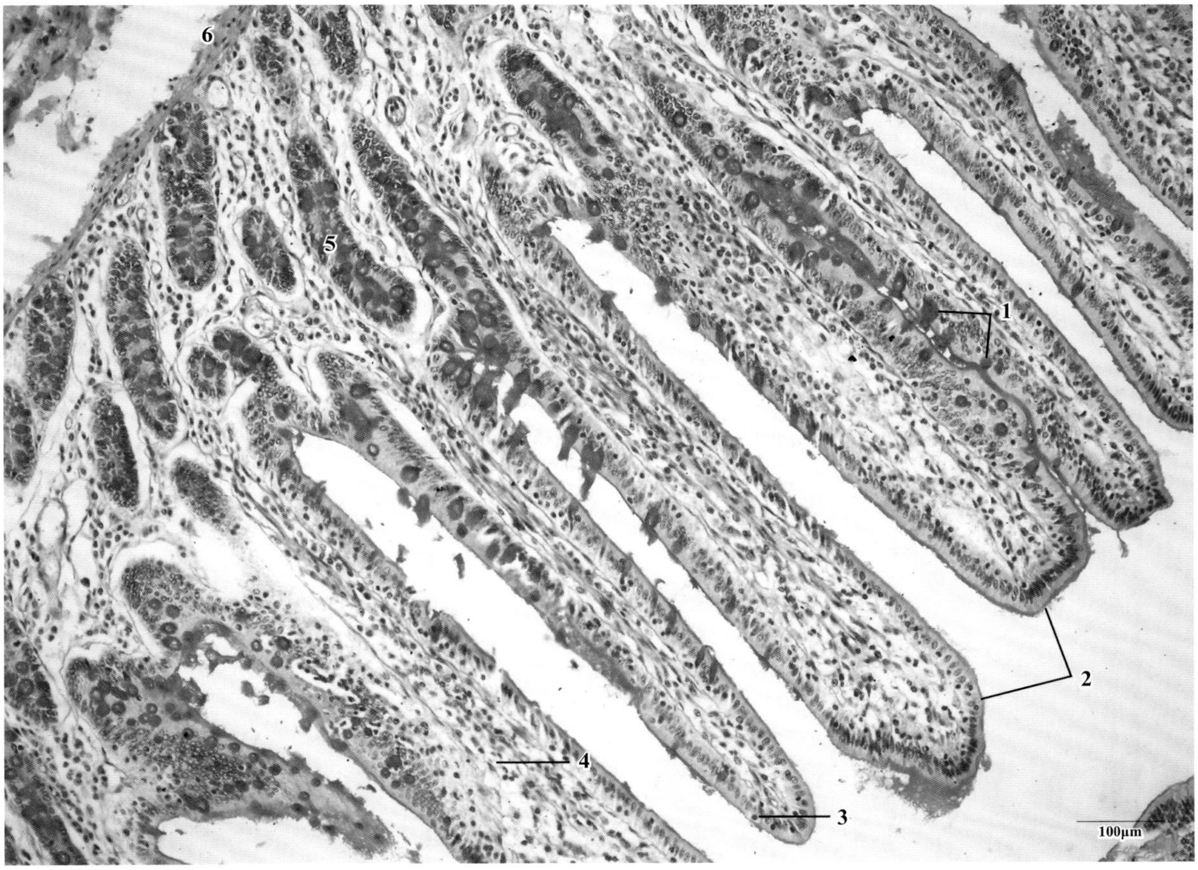

图 4-99 回肠组织切片（PAS 染色，高倍镜）
Figure 4-99 Histological section of the ileum (PAS staining, higher power).

1- 杯状细胞 goblet cell
2- 绒毛和纹状缘 villus and striated border
3- 淋巴细胞 lymphocyte
4- 固有层 lamina propria
5- 小肠腺 small intestinal gland
6- 黏膜肌层 lamina muscularis mucosae

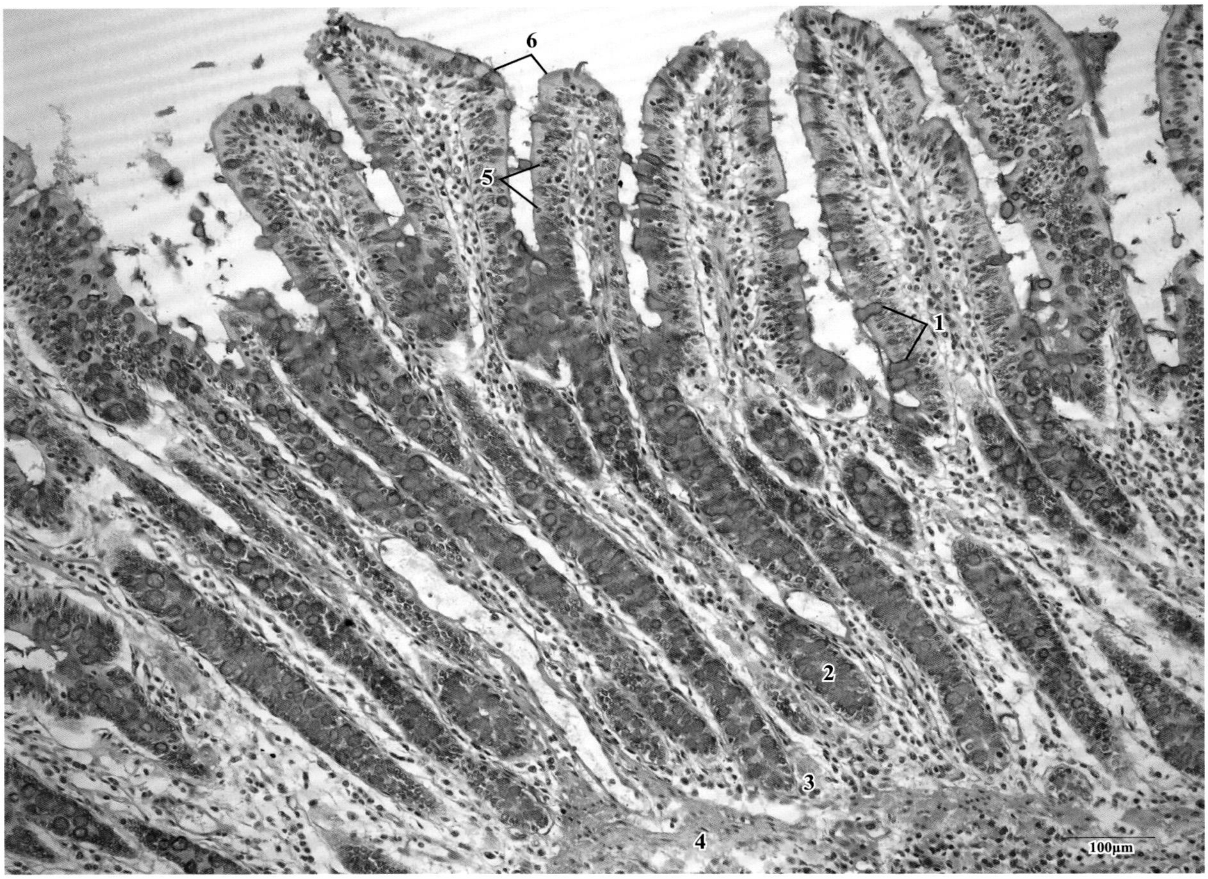

图 4-100　回肠组织切片（PAS 染色，高倍镜）
Figure 4-100　Histological section of the ileum (PAS staining, higher power).

1- 杯状细胞 goblet cell
2- 小肠腺 small intestinal gland
3- 固有层 lamina propria
4- 黏膜肌层 lamina muscularis mucosae
5- 上皮内淋巴细胞 intraepithelial lymphocyte
6- 纹状缘 striated border

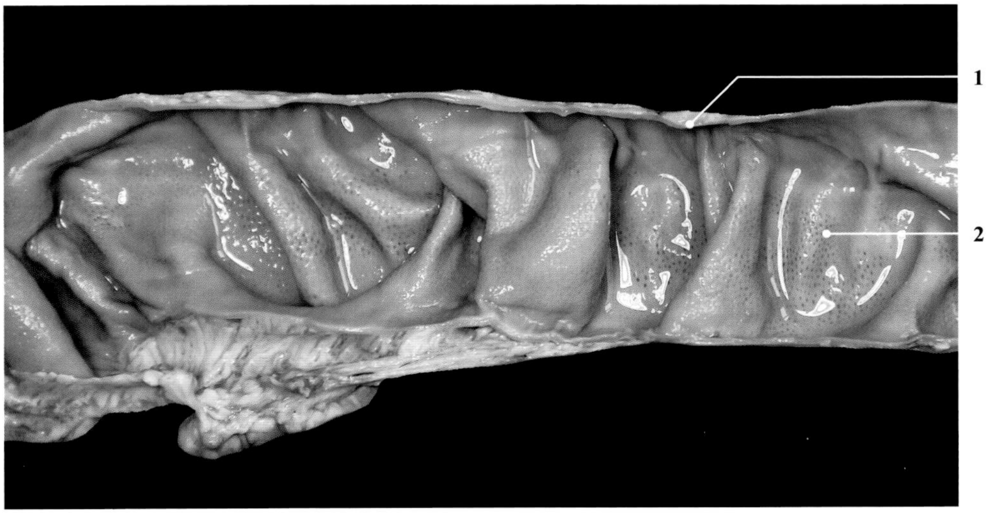

图 4-101　结肠黏膜面
Figure 4-101　The mucosal surface of the colon.

1- 结肠壁 colonic wall

2- 结肠黏膜 colonic mucosa

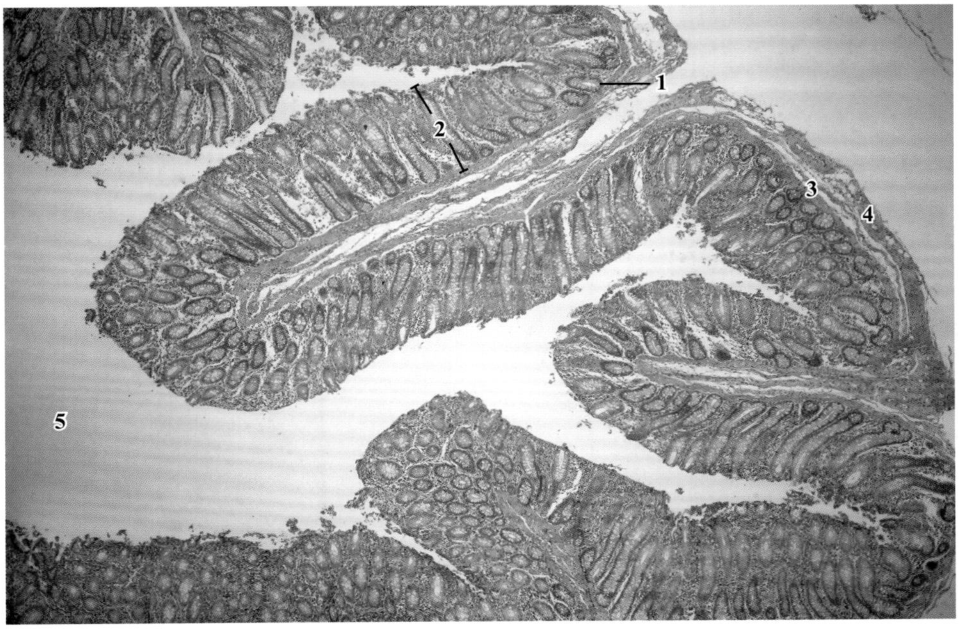

图 4-102　结肠组织切片（HE 染色，低倍镜）
Figure 4-102　Histological section of the colon (HE staining, lower power).

1- 肠腺 intestinal gland

2- 黏膜 mucosa

3- 黏膜肌层 lamina muscularis mucosae

4- 黏膜下层 submucosa

5- 肠腔 intestinal lumen

1- 黏膜 mucosa
2- 肠腺 intestinal gland
3- 黏膜肌层 lamina muscularis mucosae
4- 杯状细胞 goblet cell
5- 黏膜下层 submucosa

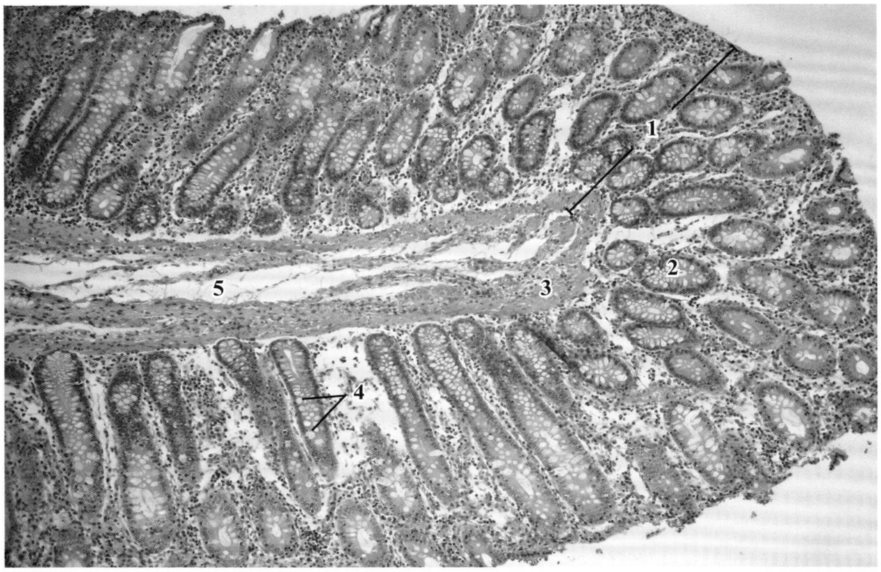

图 4-103　结肠绒毛组织切片（HE 染色，高倍镜）
Figure 4-103　Histological section of the villus of colon (HE staining, higher power).

1- 杯状细胞 goblet cell
2- 肠腺 intestinal gland
3- 固有层 lamina propria
4- 嗜伊红细胞 eosinophil
5- 黏膜肌层 lamina muscularis mucosae
6- 黏膜下层 submucosa

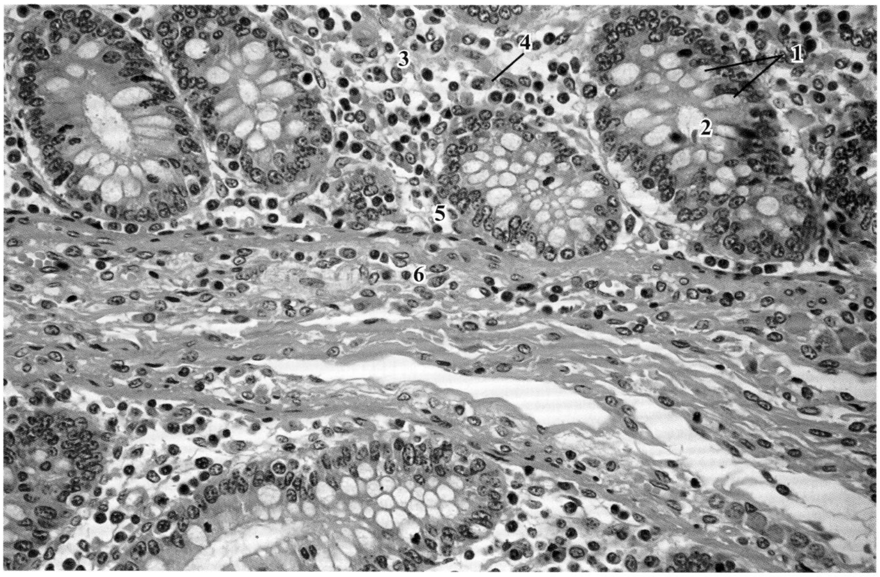

图 4-104　结肠组织切片，示肠腺（HE 染色，高倍镜）
Figure 4-104　Histological section of the colon showing the intestinal gland (HE staining, higher power).

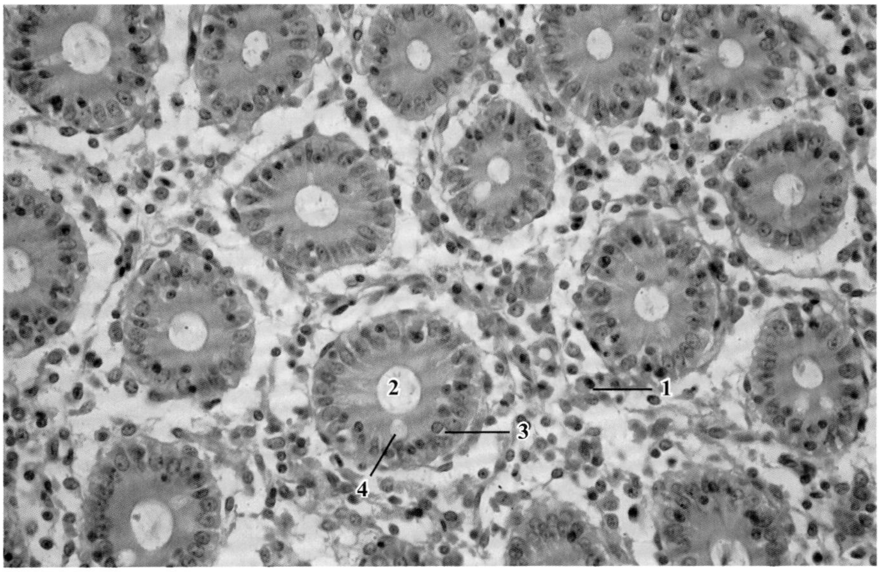

1- 嗜伊红细胞 eosinophil
2- 肠腺 intestinal gland
3- 细胞核 nucleus
4- 杯状细胞 goblet cell

图 4-105 结肠腺组织切片（HE 染色，高倍镜）
Figure 4-105 Histological section of the colonic gland (HE staining, higher power).

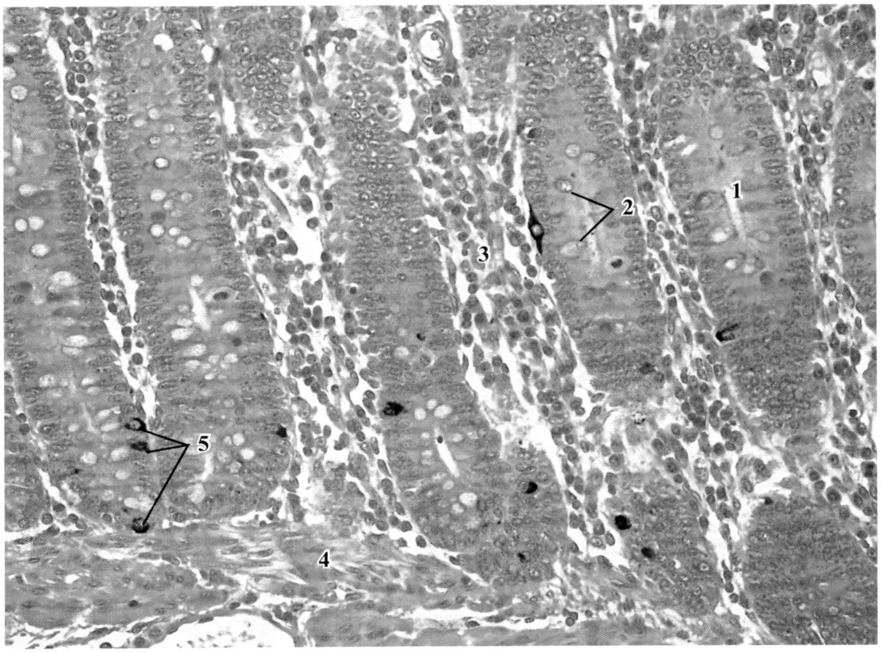

1- 肠腺 intestinal gland
2- 杯状细胞 goblet cell
3- 固有层 lamina propria
4- 黏膜肌层 lamina muscularis mucosae
5- 嗜银细胞 argyrophilic cell

图 4-106 直肠组织切片（嗜银染色，高倍镜）
Figure 4-106 Histological section of the rectum (silver staining, higher power).

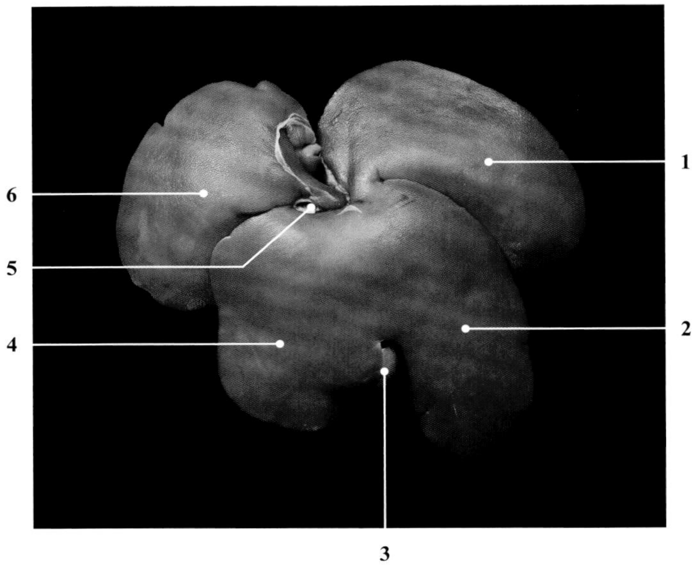

图 4-107　公猪肝（膈面观）
Figure 4-107　The liver of a boar (diaphragmatic view).

1- 肝左外叶 left lateral hepatic lobe
2- 肝左内叶 left medial hepatic lobe
3- 胆囊 gallbladder
4- 肝右内叶 right medial hepatic lobe
5- 肝静脉 hepatic vein
6- 肝右外叶 right lateral hepatic lobe

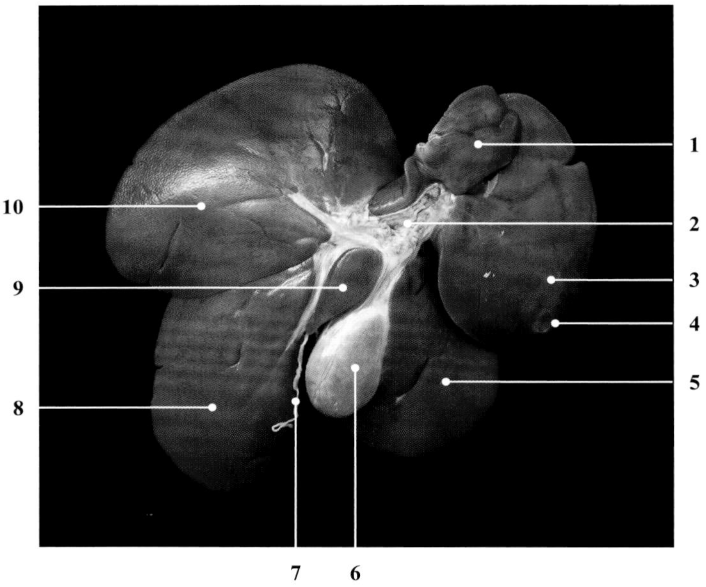

图 4-108　公猪肝（脏面观）
Figure 4-108　The liver of a boar (visceral view).

1- 尾叶 caudate hepatic lobe
2- 肝门 hepatic porta
3- 肝右外叶 right lateral hepatic lobe
4- 囊肿 cyst sac
5- 肝右内叶 right medial hepatic lobe
6- 胆囊 gallbladder
7- 肝圆韧带 round ligament of liver
8- 肝左内叶 left medial hepatic lobe
9- 方叶 quadrate hepatic lobe
10- 肝左外叶 left lateral hepatic lobe

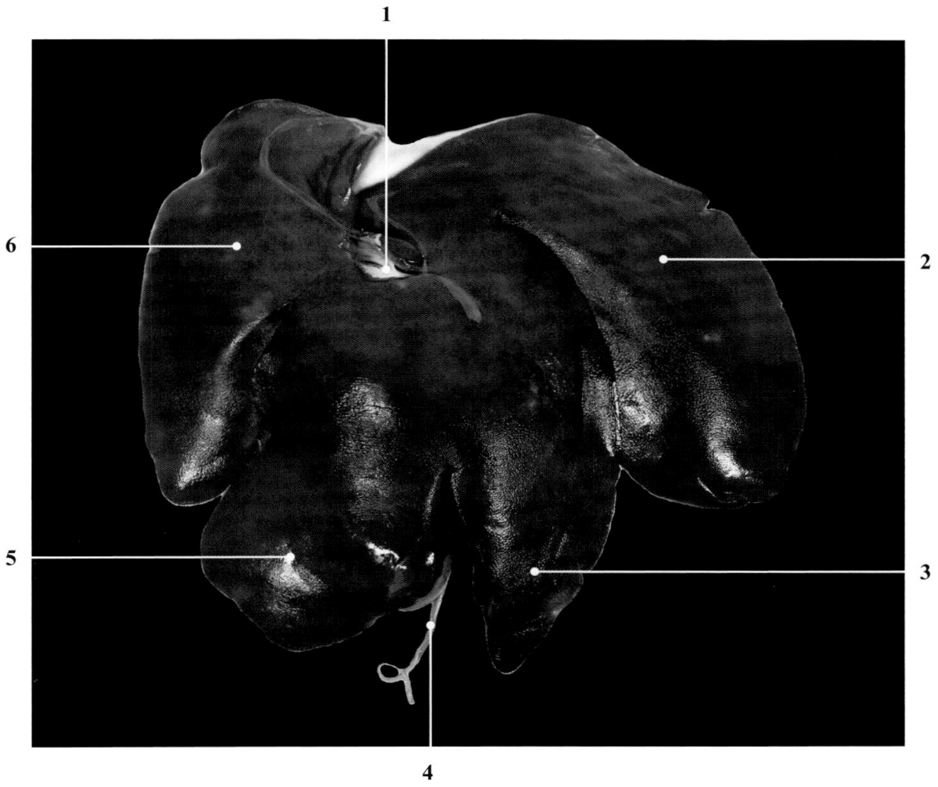

图 4-109 母猪肝（膈面观）
Figure 4-109 The liver of a sow (diaphragmatic view).

1- 肝静脉 hepatic vein
2- 肝左外叶 left lateral hepatic lobe
3- 肝左内叶 left medial hepatic lobe
4- 肝圆韧带 round ligament of liver
5- 肝右内叶 right medial hepatic lobe
6- 肝右外叶 right lateral hepatic lobe

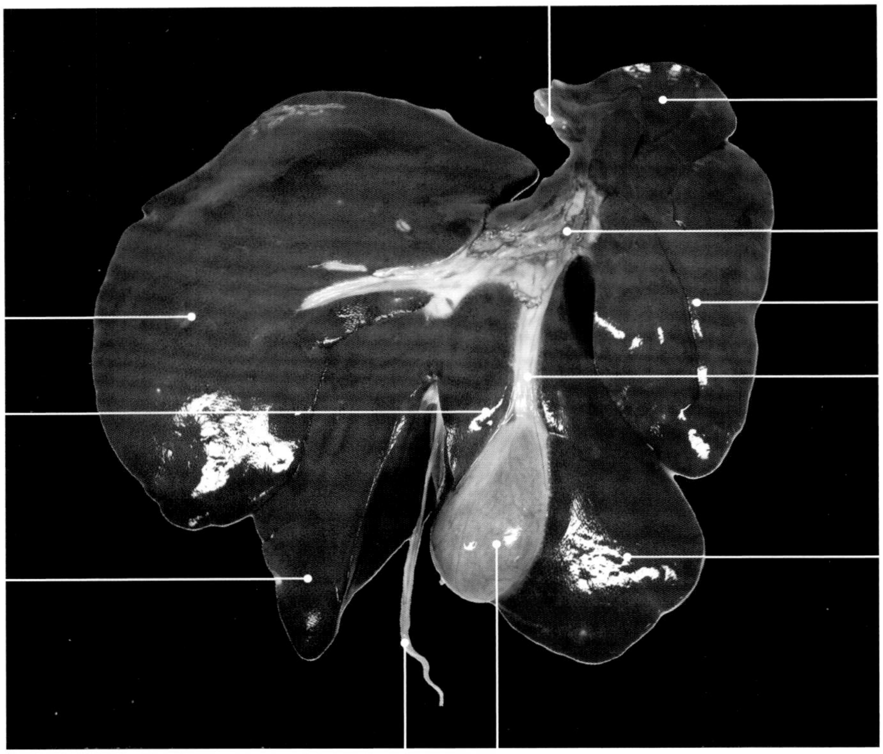

图 4-110　母猪肝（脏面观）
Figure 4-110　The liver of a sow (visceral view).

1- 肝尾叶 caudate hepatic lobe
2- 肝门 hepatic porta
3- 肝右外叶 right lateral hepatic lobe
4- 胆囊管 cystic duct
5- 肝右内叶 right medial hepatic lobe
6- 胆囊 gallbladder
7- 肝圆韧带 round ligament of liver
8- 肝左内叶 left medial hepatic lobe
9- 方叶 quadrate hepatic lobe
10- 肝左外叶 left lateral hepatic lobe
11- 右三角韧带 right triangular ligament

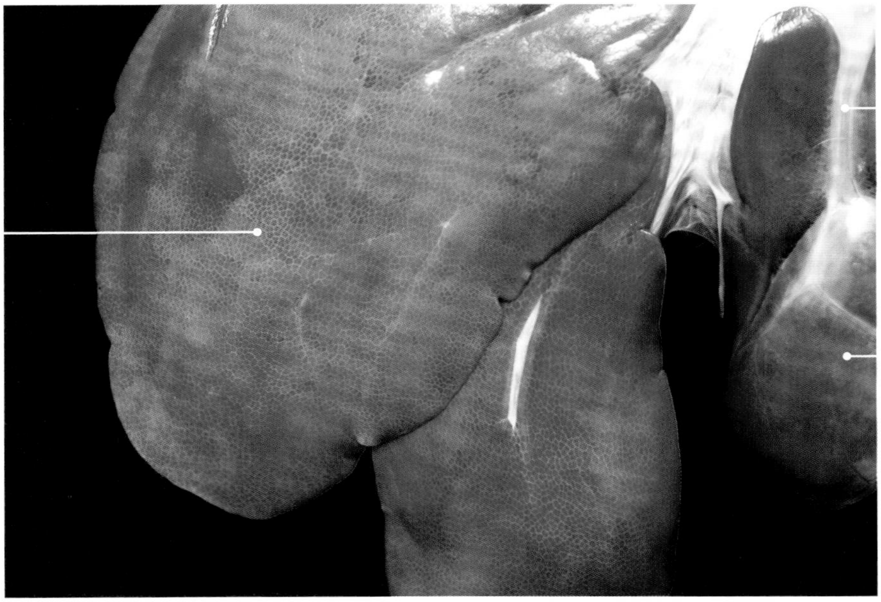

图 4-111 肝小叶
Figure 4-111 The hepatic lobule.

1- 胆囊管 cystic duct
2- 胆囊 gallbladder
3- 肝左外叶 left lateral hepatic lobe

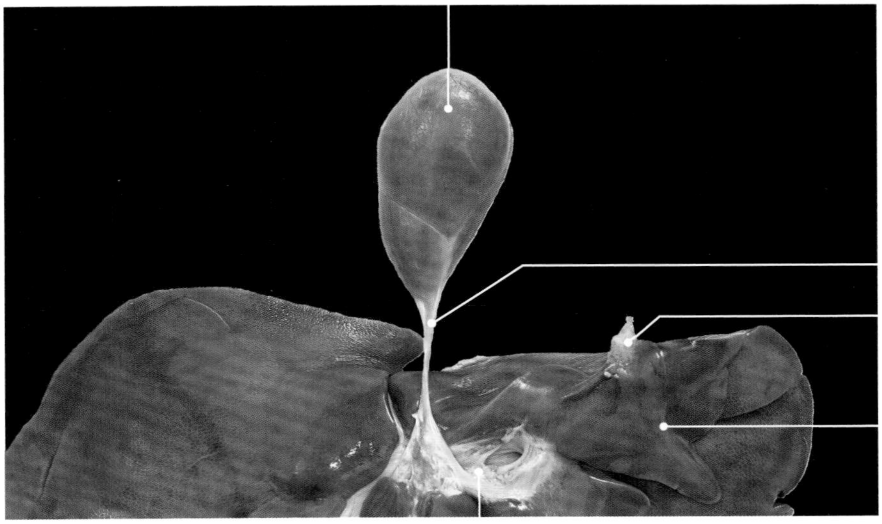

图 4-112 胆囊
Figure 4-112 The gallbladder.

1- 胆囊 gallbladder
2- 胆囊管 cystic duct
3- 右三角韧带 right triangular ligament
4- 肝尾叶 caudate hepatic lobe
5- 肝门 hepatic porta

9 10

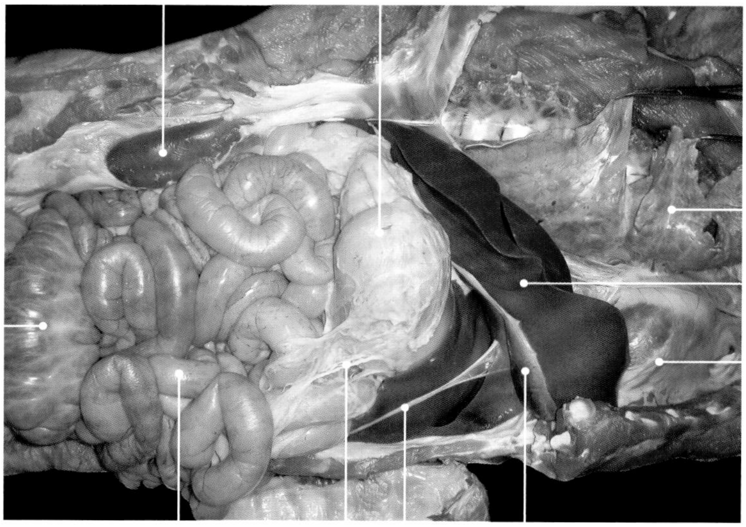

1
2
3
8

7 6 5 4

图 4-113　母猪肝圆韧带
Figure 4-113　The round ligament of liver of a sow.

1- 肺 lung
2- 肝 liver
3- 心脏 heart
4- 胆囊 gallbladder
5- 肝圆韧带 round ligament of liver
6- 大网膜 greater omentum
7- 空肠 jejunum
8- 大肠 large intestine
9- 右肾 right kidney
10- 胃 stomach

10
9
8
7

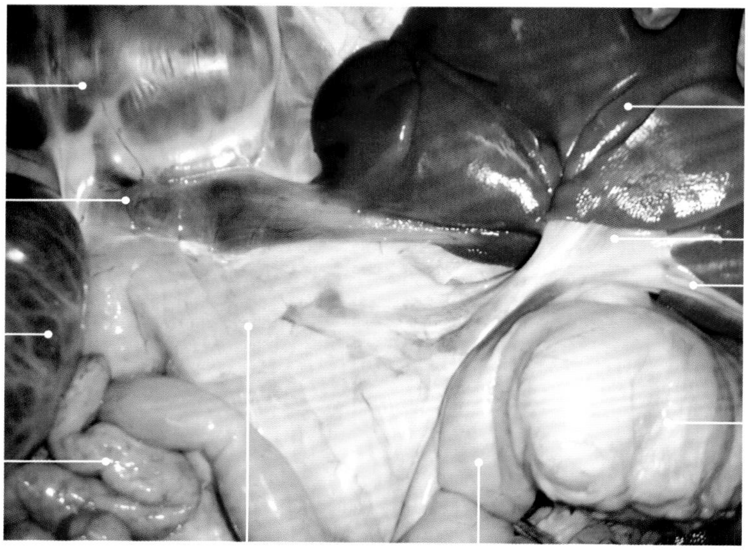

1
2
3
4

6 5

图 4-114　公猪胆囊管
Figure 4-114　The cystic duct of a boar.

1- 肝 liver
2- 肝门静脉 hepatic portal vein
3- 胆囊管 cystic duct
4- 胃 stomach
5- 十二指肠 duodenum
6- 胰 pancreas
7- 空肠 jejunum
8- 大肠 large intestine
9- 后腔静脉 *caudal vena cava*
10- 肾 kidney

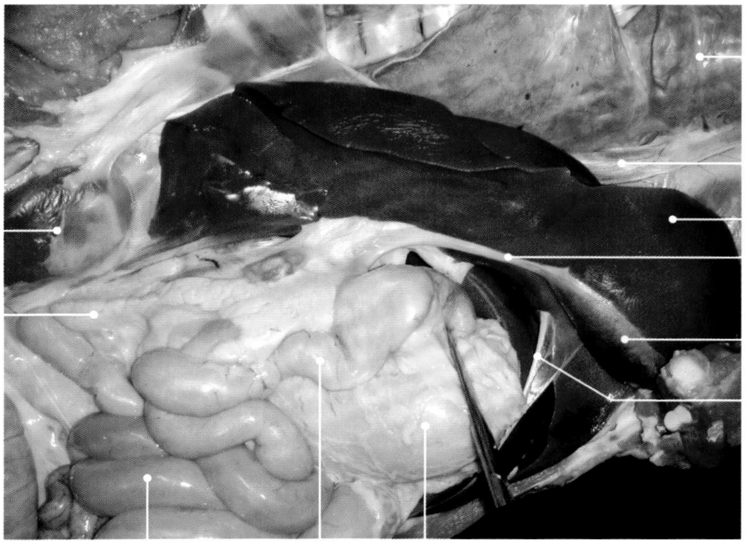

图 4-115 母猪胆囊管
Figure 4-115 The cystic duct of a sow.

1- 肺 lung
2- 后腔静脉 *caudal vena cava*
3- 肝 liver
4- 胆囊管 cystic duct
5- 胆囊 gallbladder
6- 肝圆韧带 round ligament of liver
7- 胃 stomach
8- 十二指肠 duodenum
9- 空肠 jejunum
10- 胰 pancreas
11- 肾 kidney

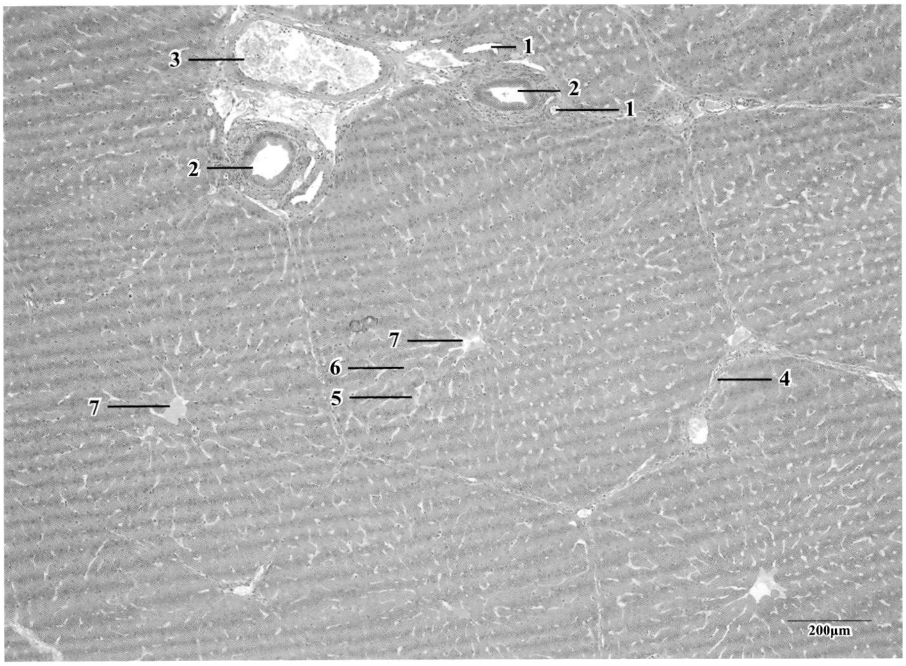

1- 小叶间动脉 interlobular artery
2- 小叶间胆管 interlobular bile duct
3- 小叶间静脉 interlobular vein
4- 小叶间结缔组织 interlobular connective tissue
5- 血窦 sinusoid
6- 肝板 hepatic plate
7- 中央静脉 central vein

图 4-116 肝组织切片（HE 染色，低倍镜）
Figure 4-116 Histological section of the liver (HE staining, lower power).

1- 中央静脉 central vein
2- 血窦 sinusoid
3- 肝板 hepatic plate
4- 淋巴管 lymphatic vessel
5- 小叶间静脉 interlobular vein
6- 小叶间胆管 interlobular bile duct
7- 小叶间结缔组织 interlobular connective tissue

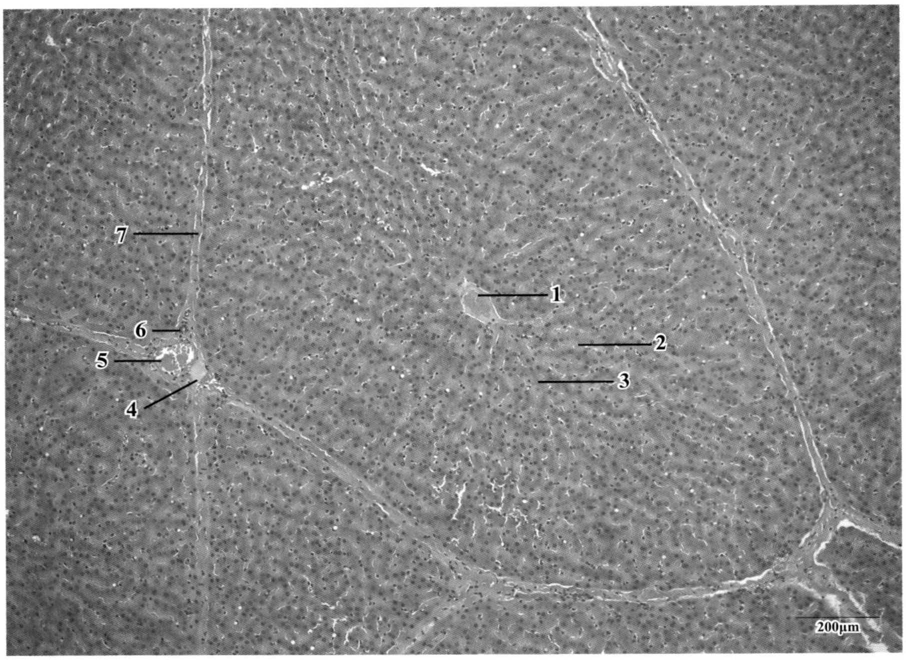

图 4-117 肝组织切片，示肝小叶（HE 染色，低倍镜）
Figure 4-117 Histological section of the liver showing the hepatic lobule (HE staining, lower power).

1- 中央静脉 central vein
2- 内皮细胞 endothelial cell
3- 肝板 hepatic plate
4- 单核肝细胞 hepatocyte with single nucleus
5- 双核肝细胞 binucleate hepatocyte
6- 血窦 sinusoid
7- 枯否细胞 Kuffer's cell
8- 血细胞 blood cell, hemocyte

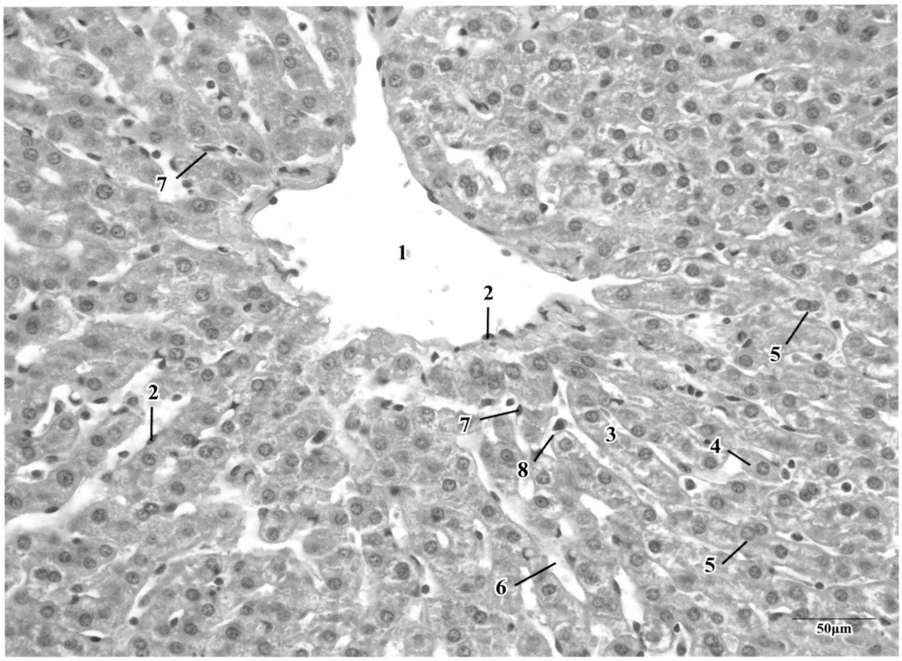

图 4-118 肝小叶组织切片，示中央静脉（HE 染色，高倍镜）
Figure 4-118 Histological section of the hepatic lobule showing central vein (HE staining, higher power).

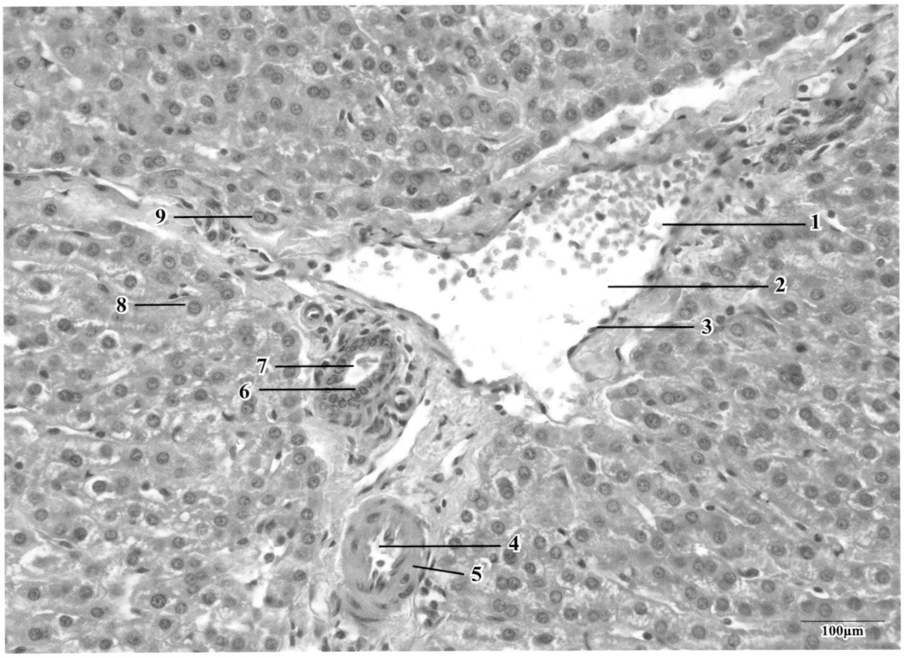

1- 红细胞 erythrocyte
2- 小叶间静脉 interlobular vein
3- 内皮细胞 endothelial cell
4- 小叶间动脉 interlobular artery
5- 平滑肌 smooth muscle
6- 单层立方上皮 simple cuboidal epithelium
7- 小叶间胆管 interlobular bile duct
8- 单核肝细胞 hepatocyte with single nucleus
9- 双核肝细胞 binucleate hepatocyte

图 4-119 肝小叶组织切片，示门管区（HE 染色，高倍镜）
Figure 4-119 Histological section of the hepatic lobule showing portal area (HE staining, higher power).

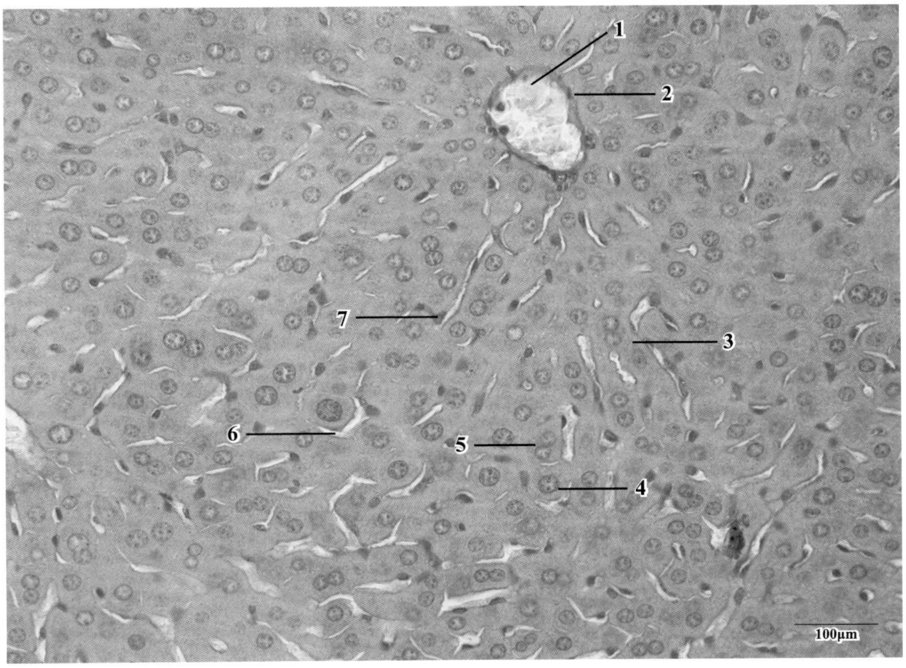

1- 中央静脉 central vein
2- 内皮细胞 endothelial cell
3- 肝板 hepatic plate
4- 单核肝细胞 hepatocyte with single nucleus
5- 双核肝细胞 binucleate hepatocyte
6- 血窦 sinusoid
7- 枯否细胞 Kupffer's cell

图 4-120 肝小叶组织切片，示中央静脉（PAS 染色，高倍镜）
Figure 4-120 Histological section of the hepatic lobule showing central vein (PAS staining, higher power).

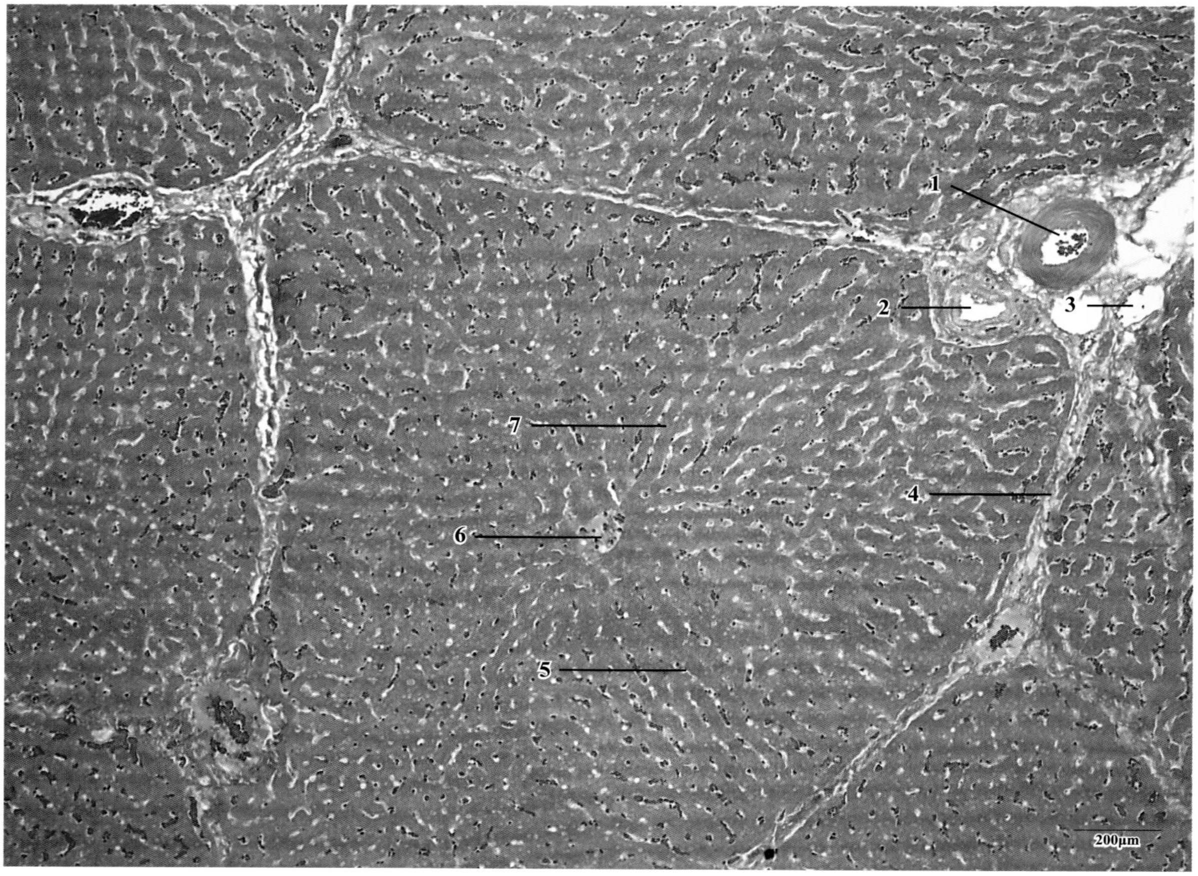

图 4-121　肝小叶组织切片（Mallory 三色染色，高倍镜）
Figure 4-121　Histological section of the hepatic lobule (Mallory staining, higher power).

1- 小叶间动脉 interlobular artery
2- 小叶间胆管 interlobular bile duct
3- 小叶间静脉 interlobular vein
4- 小叶间结缔组织 interlobular connective tissue
5- 血窦 sinusoid
6- 中央静脉 central vein
7- 肝板 hepatic plate

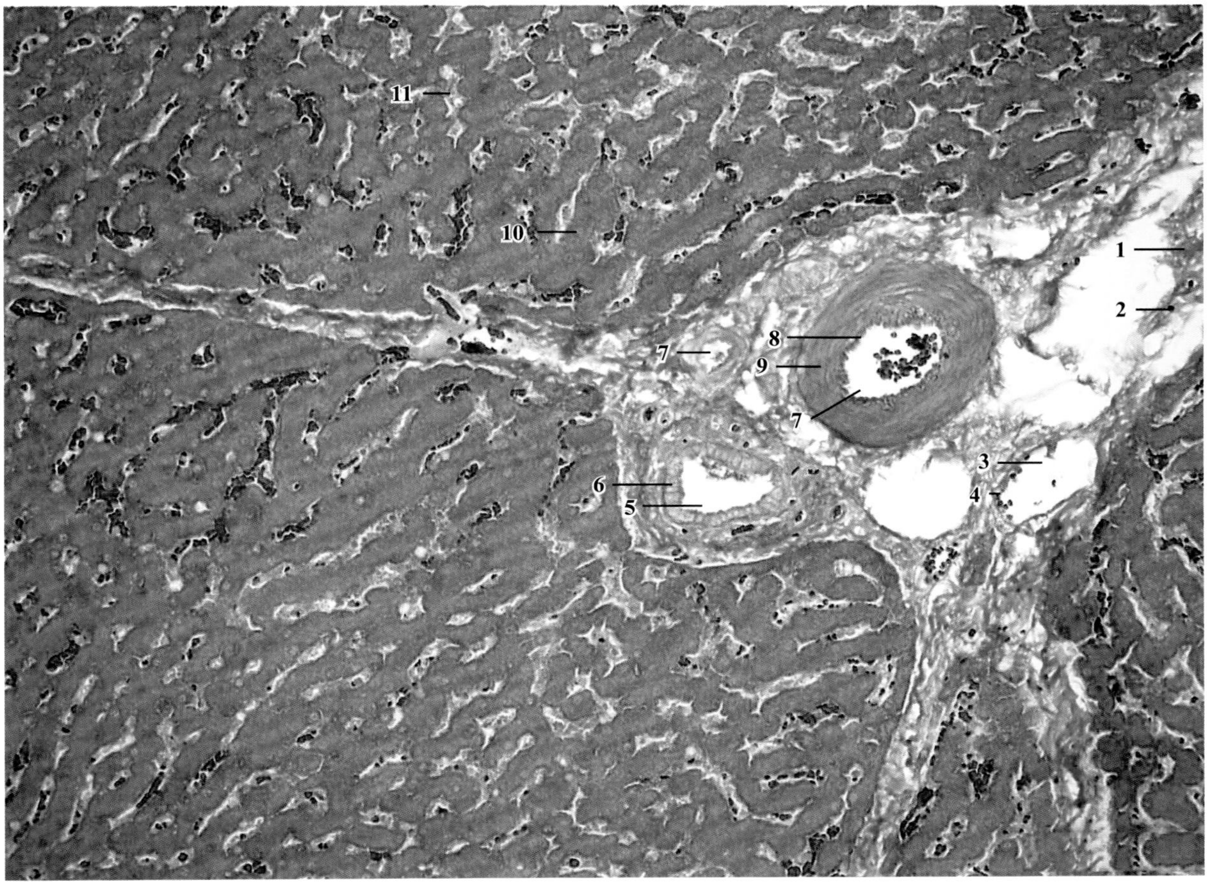

图 4-122 肝小叶组织切片，示门管区（Mallory 三色染色，高倍镜）
Figure 4-122 Histological section of the hepatic lobule showing portal area (Mallory staining, higher power).

1- 胶原纤维 collagenous fiber
2- 成纤维细胞 fibroblast
3- 小叶间静脉 interlobular vein
4- 红细胞 erythrocyte
5- 小叶间胆管 interlobular bile duct
6- 单层立方上皮 simple cuboidal epithelium
7- 小叶间动脉 interlobular artery
8- 内弹性膜 internal elastic membrane
9- 平滑肌 smooth muscle
10- 肝细胞 hepatocyte
11- 血窦 sinusoid

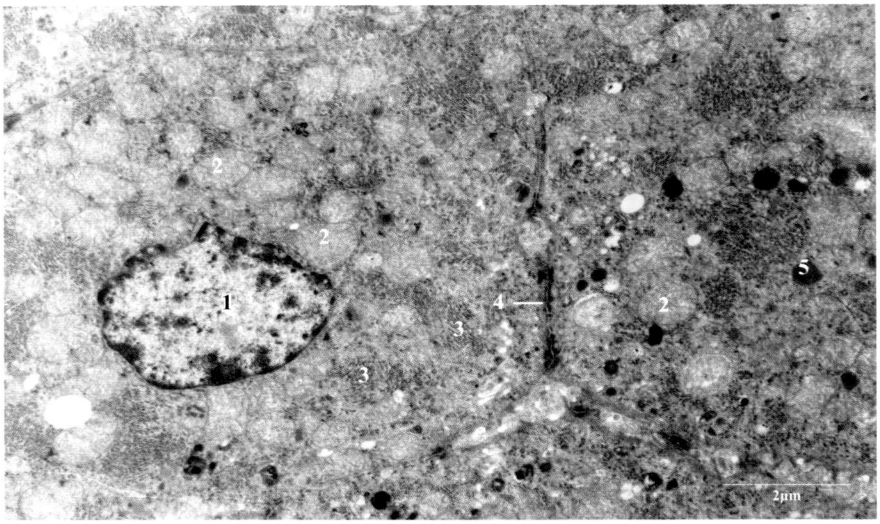

图 4-123　肝脏透射电镜，示肝细胞

Figure 4-123　Transmission electron microscope (TEM) of the liver showing hepatocyte.

1- 肝细胞核 nucleus of hepatocyte
2- 线粒体 mitochondrion
3- 粗面内质网 rough endoplasmic reticulum
4- 细胞连接 cell junction
5- 溶酶体 lysosome

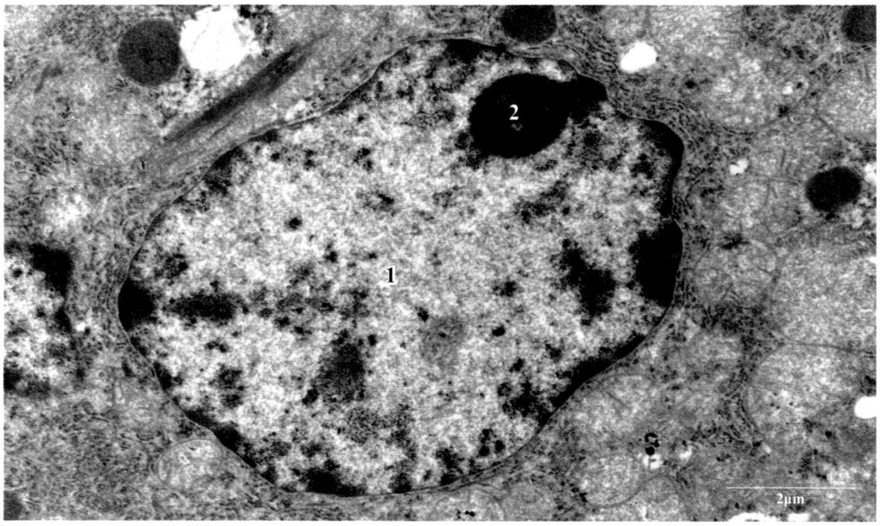

图 4-124　肝脏透射电镜，示肝细胞核

Figure 4-124　Transmission electron microscope (TEM) of the liver showing nucleus of hepatocyte.

1- 肝细胞核 nucleus of hepatocyte
2- 核仁 nucleolus

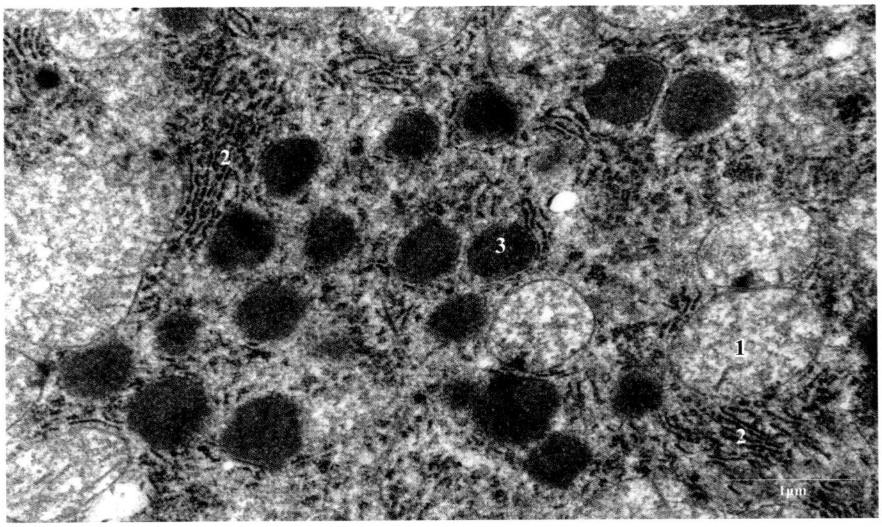

图 4-125　肝脏透射电镜，示肝细胞细胞器
Figure 4-125　Transmission electron microscope (TEM) of the liver showing organelle of hepatocyte.

1- 线粒体 mitochondrion
2- 粗面内质网 rough endoplasmic reticulum
3- 溶酶体 lysosome

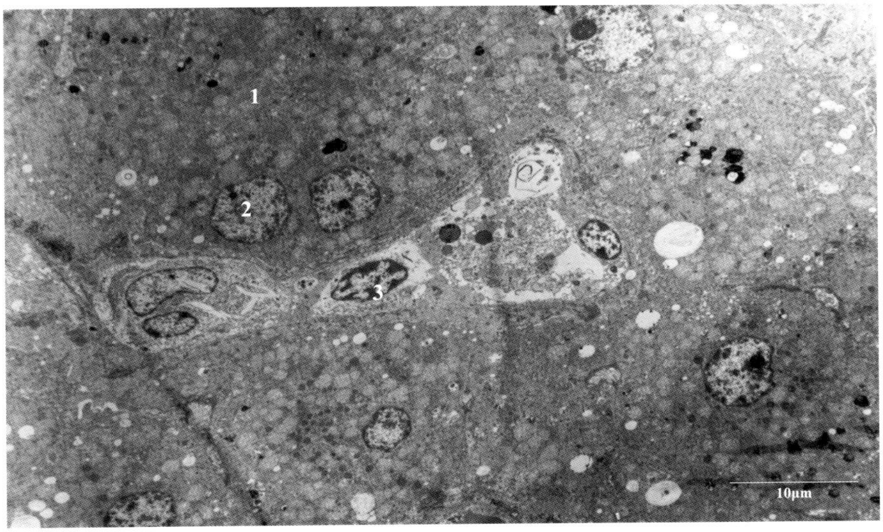

图 4-126　肝脏透射电镜，示枯否细胞
Figure 4-126　Transmission electron microscope (TEM) of the liver showing Kupffer's cell.

1- 肝细胞 hepatocyte
2- 细胞核 nucleus
3- 枯否细胞 Kupffer's cell

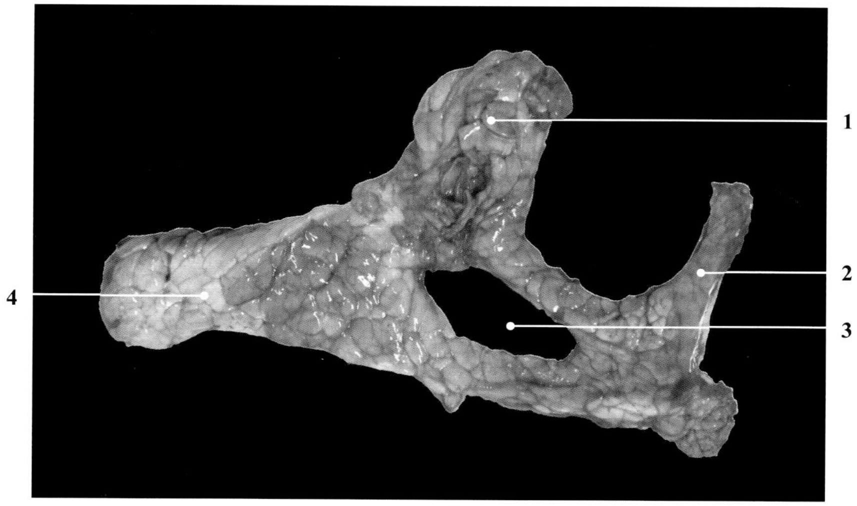

图 4-127　母猪胰（腹侧观）
Figure 4-127　The pancreas of a sow (ventral view).

1- 胰体 body of the pancreas
2- 胰右叶 right pancreatic lobe
3- 胰环 pancreatic ring
4- 胰左叶 left pancreatic lobe

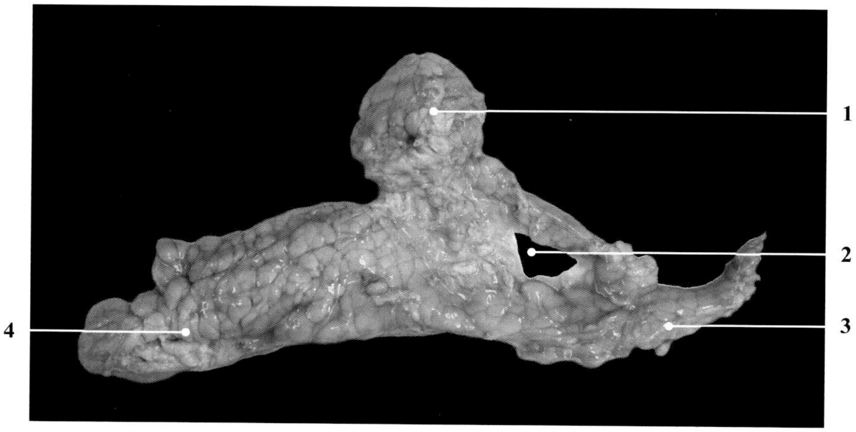

图 4-128　公猪胰（腹侧观）
Figure 4-128　The pancreas of a boar (ventral view).

1- 胰体 body of the panereas
2- 胰环 pancreatic ring
3- 胰右叶 right pancreatic lobe
4- 胰左叶 left pancreatic lobe

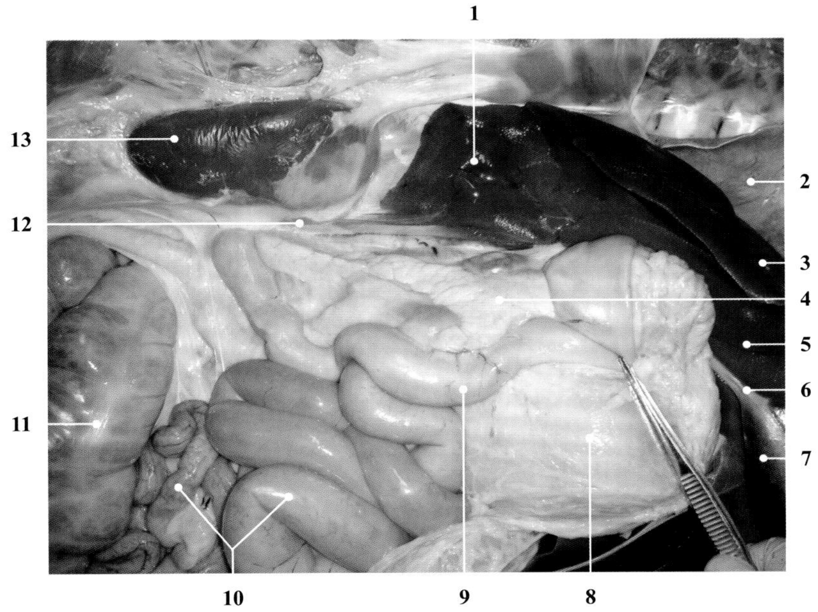

图 4-129 母猪胰的位置
Figure 4-129 Anatomical position of the pancreas of a sow.

1- 肝尾叶 caudate hepatic lobe
2- 肺 lung
3- 肝右外叶 right lateral hepatic lobe
4- 胰 pancreas
5- 肝右内叶 right medial hepatic lobe
6- 胆囊管 cystic duct
7- 胆囊 gallbladder
8- 胃 stomach
9- 十二指肠 duodenum
10- 空肠 jejunum
11- 大肠 large intestine
12- 后腔静脉 *caudal vena cava*
13- 肾 kidney

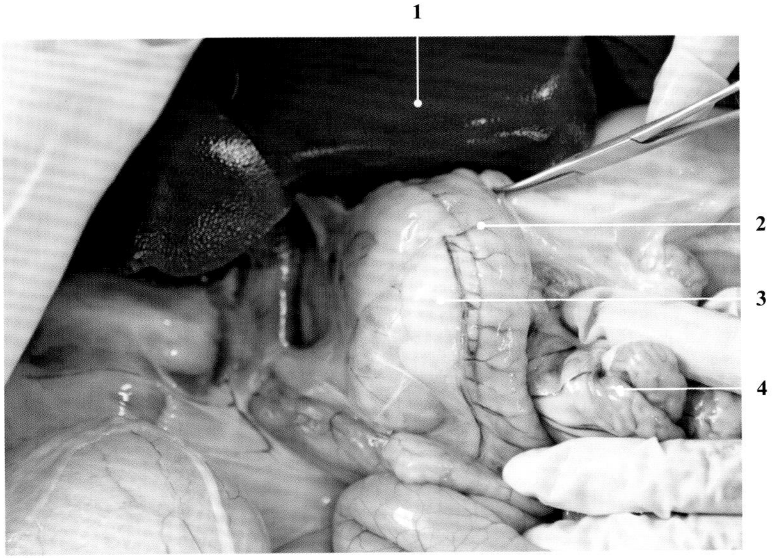

图 4-130 公猪胰的位置
Figure 4-130 Anatomical position of the pancreas of a boar.

1- 肝 liver
2- 十二指肠 duodenum
3- 胰 pancreas
4- 空肠 jejunum

1- 胰腺细胞 acinar cell of the pancreas
2- 泡心细胞 centroacinar cell
3- 闰管 intercalated duct
4- 小动脉 small antery
5- 小静脉 small vein
6- 小叶间导管 interlobular duct

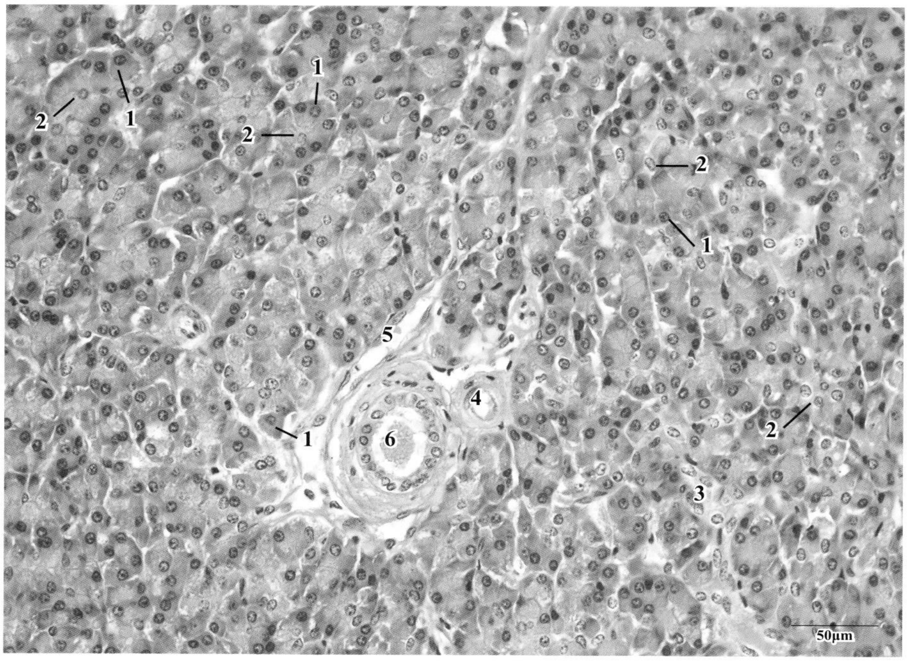

图 4-131　胰组织切片（HE 染色，高倍镜）
Figure 4-131　Histological section of the pancreas (HE staining, higher power).

1- 胰腺细胞 acinar cell of the pancreas
2- 泡心细胞 centroacinar cell
3- 闰管 intercalated duct
4- 小叶间导管 interlobular duct
5- 小叶间结缔组织 interlobular connective tissue

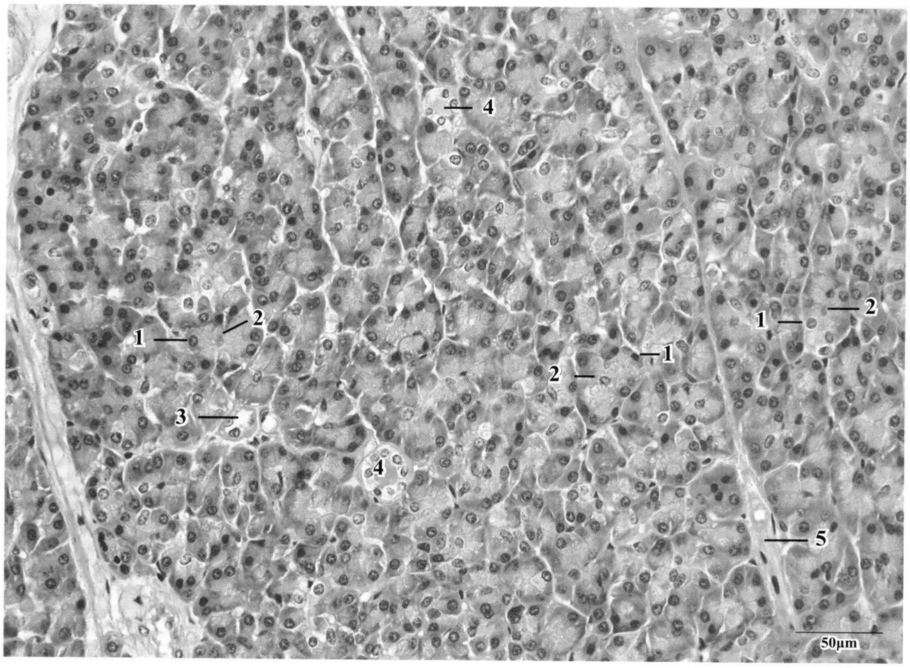

图 4-132　胰组织切片（HE 染色，高倍镜）
Figure 4-132　Histological section of the pancreas (HE staining, higher power).

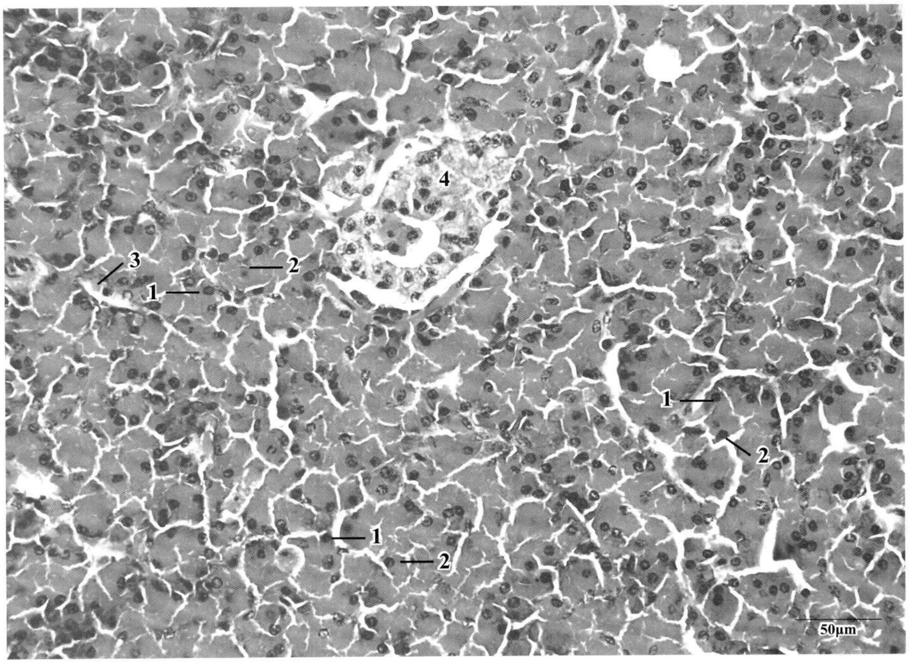

1- 胰腺细胞 acinar cell of the pancreas
2- 泡心细胞 centroacinar cell
3- 闰管 intercalated duct
4- 胰岛 pancreatic islet (islet of Langerhans)

图 4-133　胰组织切片，示胰岛（HE 染色，高倍镜）
Figure 4-133　Histological section of the pancreas showing pancreatic islet (HE staining, higher power).

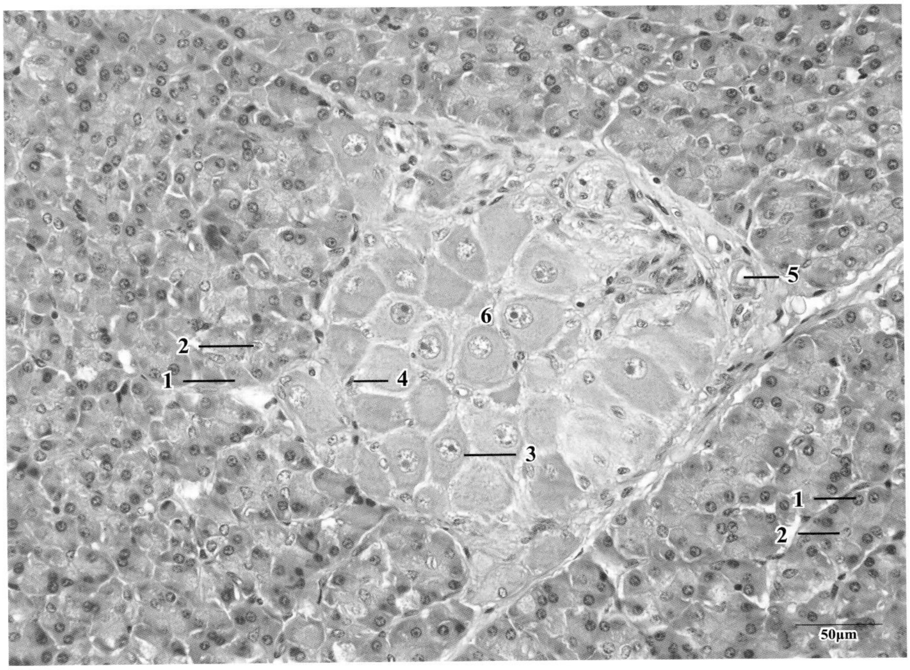

1- 胰腺细胞 acinar cell of the pancreas
2- 泡心细胞 centroacinar cell
3- 神经元胞体 soma of the neuron
4- 卫星细胞 satellite cell
5- 血管 blood vessel
6- 神经节 ganglion

图 4-134　胰组织切片，示神经节（HE 染色，高倍镜）
Figure 4-134　Histological section of the pancreas showing ganglion (HE staining, higher power).

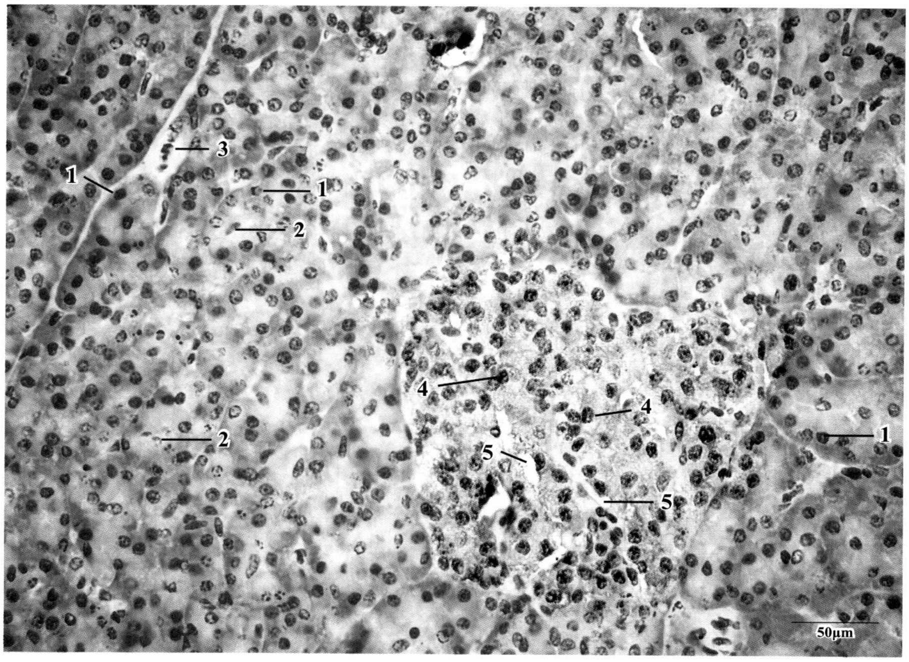

图 4-135 胰组织切片，示胰岛素细胞（免疫组织化学染色，高倍镜）
Figure 4-135 Histological section of the pancreas showing insulin-cell (Immunohistostaining staining, higher power).

1- 胰腺细胞 acinar cell of the pancreas
2- 泡心细胞 centroacinar cell
3- 闰管 intercalated duct
4- 胰岛 B 细胞（胰岛素细胞） B cell (insulin cell)
5- 毛细血管 capillary

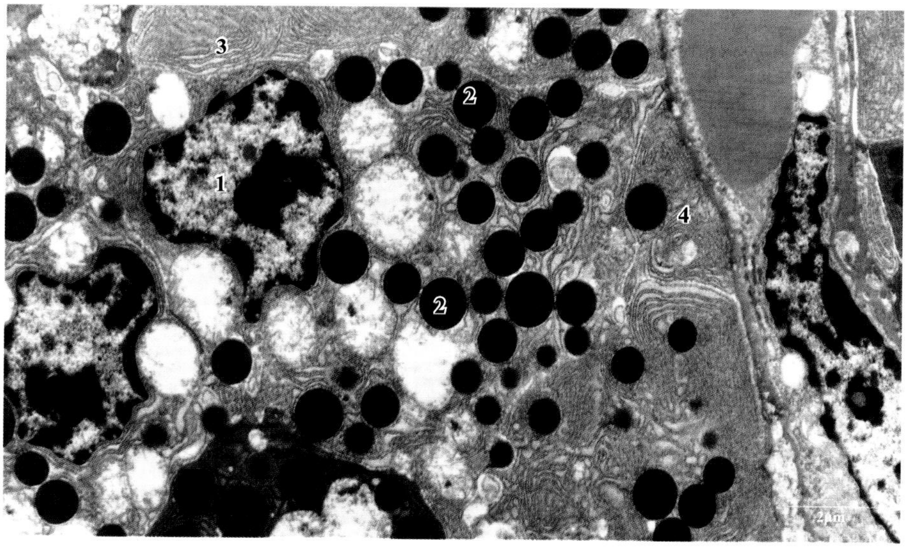

图 4-136 胰透射电镜，示胰腺细胞
Figure 4-136 Transmission electron microscope (TEM) of the pancreas showing acinar cell.

1- 胰腺细胞核 nucleus of the acinar cell
2- 酶原颗粒 zymogen granule
3- 粗面内质网 rough endoplasmic reticulum
4- 高尔基复合体 Golgi complex

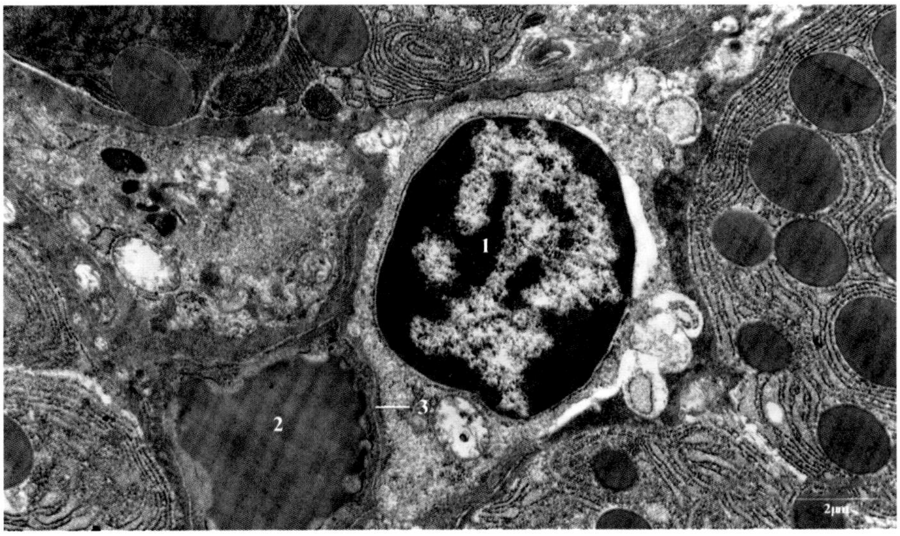

图 4-137 胰透射电镜，示泡心细胞
Figure 4-137 Transmission electron microscope (TEM) of the pancreas showing centroacinar cell.

1- 泡心细胞 centroacinar cell
2- 腺腔 lumen of gland
3- 细胞连接 cell junction

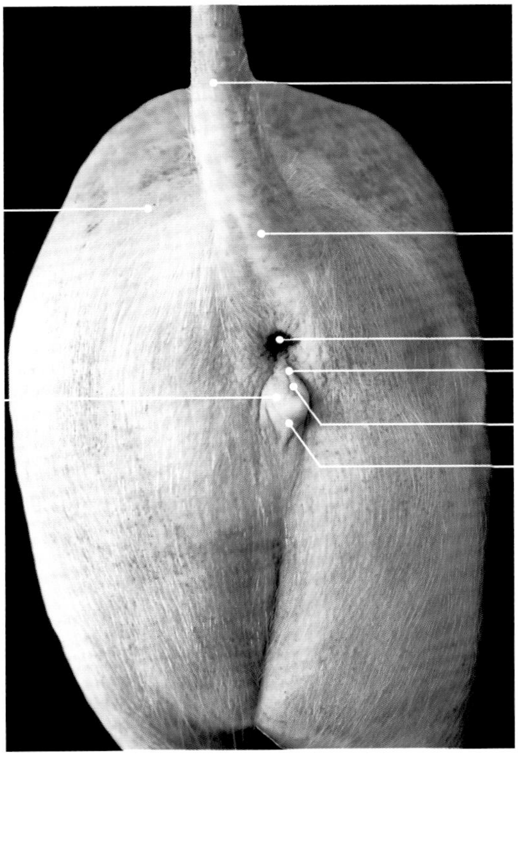

图 4-138 母猪肛门
Figure 4-138 The anus of a sow.

1- 尾 tail
2- 尾根 root of the tail
3- 肛门 anus
4- 阴唇背侧联合 dorsal commissure of labium
5- 阴门 vulva
6- 阴唇腹侧联合 ventral commissure of labium
7- 阴唇 labium of vulva
8- 被毛 clothing hair

第五章

呼吸系统

Chapter 5
Respiratory system

猪的呼吸系统（respiratory system）包括呼吸道和肺。呼吸道是气体进出肺的通道，包括鼻、鼻咽部、喉、气管和支气管。临床上，将鼻和鼻咽部称上呼吸道，喉、气管和肺称下呼吸道。呼吸道的特征是由骨或软骨构成支架，围成管腔，内表面衬以黏膜，以保证气体出入通畅。肺是进行气体交换的场所，包括支气管树和肺泡，总面积很大，有利于气体交换。

一、鼻

鼻（nose）包括外鼻（external nose）、鼻腔（nasal cavity）和鼻旁窦（paranasal sinus）。

外鼻由骨、软骨构成支架，外覆软组织和皮肤，呈圆盘状突出，多动，称吻突（rostral disc）或吻镜（rostral plate）。吻突内有吻骨支撑，表面皮肤特化，表面具触毛，深处有吻腺，确保吻镜湿润。在吻突平面中部有2个椭圆形的鼻孔（nostril, nasal opening）。猪的鼻孔以鼻软骨为支架，故不能充分开张。鼻软骨（nasal cartilage）包括鼻背外侧软骨和鼻腹外侧软骨，两者相互连接，决定鼻孔开口形状。

鼻腔从鼻孔一直延伸至筛骨筛板，被鼻中隔分成左右两部分。每侧鼻腔的内部有鼻甲（nasal conchae）突入。鼻甲是表面覆有鼻黏膜的卷曲状鼻甲骨，把鼻腔分成3个鼻道。上鼻道（dorsal nasal meatus）位于鼻腔顶壁和上鼻甲之间，与鼻腔底部相通，是空气与嗅觉黏膜接触的直接通道。中鼻道（middle nasal meatus）位于上、下鼻甲骨之间，通鼻旁窦。下鼻道（ventral nasal meatus）是空气进入咽的主要通道，位于下鼻甲和鼻腔底部之间。鼻腔内表面和鼻甲表面有鼻黏膜覆盖，黏膜下分布由多重吻合静脉构成的血管丛和黏液腺，可以分泌黏液，使吸入鼻腔内的空气变得温暖、湿润，减少对肺的刺激。鼻黏膜分为呼吸区和嗅区两部分。呼吸区为鼻腔前部的黏膜，呈微红色，含有色素；假复层柱状纤毛上皮，杯状细胞较多，基膜较厚，固有层内有混合性的鼻腺和静脉丛。嗅区较小，位于呼吸区之后，呈褐色；上皮含有嗅细胞（上皮内的感觉神经元）、支持细胞和基细胞，固有层内有浆液性嗅腺。

鼻旁窦位于鼻腔周围颅骨（额骨、蝶骨、上颌骨、筛骨）内的含气空腔，均有窦口与鼻腔相通。左右对称，共4对，分别称为额窦（frontal sinus）、上颌窦（maxillary sinus）、蝶窦（sphenoidal sinus）和筛窦（ethmoid sinus），对发音起共鸣作用。

二、喉

喉（larynx）是一个双侧对称的管状肌性、软骨器官，前面连接咽，后面通气管，有通气和发声的功能。喉壁由喉软骨及与其相连的肌肉和韧带组成，围成喉腔（laryngeal cavity），喉腔依靠声带（vocal cords）进行收缩。

喉软骨（laryngeal cartilage）构成喉的支架，包括单一的甲状软骨（thyroid cartilage）、环状软骨（cricoid cartilage）、会厌软骨（epiglottic cartilage）和成对的杓状软骨（arytenoid cartilage）。喉软骨彼此借关节、韧带和纤维相连，构成喉的支架。喉肌属横纹肌，包括环甲肌、环杓后肌、环杓侧肌 、甲杓肌和杓肌，运动喉的软骨和关节，使声带紧张或松弛，开大或缩小声门裂，从而调节喉口的大小。

喉黏膜为咽部的黏膜经喉口向喉腔内的延续。会厌表面及喉前部表面的黏膜上皮为复层扁平上皮，喉后部表面上皮为假复层纤毛柱状上皮。固有层为富含弹性纤维的疏松结缔组织，有混合腺和淋巴组织分布。在会厌黏膜、喉侧室和会厌底处有孤立淋巴小结，即会厌旁扁桃体。

声带（vocal cords）包括游离缘较薄的膜部和基部的软骨部。膜部黏膜表面覆盖复层扁平上皮，固有层较厚，包括浅部的疏松结缔组织和深部富含弹性纤维的致密结缔组织。固有层无腺体分布，有少量血管分布，深层骼肌纤维构成声带肌。软骨部黏膜表面覆盖假复层纤毛柱状上皮，黏膜下层的疏松结缔组织内有混合腺分布，外膜有软骨和骨骼肌分布。

三、气管和支气管

气管（trachea）是一系列由一纵向纤维带连接的C形透明软骨构成的管状结构。猪的C形软骨环有29～36个，气管环开口于气管背侧，游离的两端由横向的气管肌和结缔组织相连。气管起于喉的环状软骨，沿颈椎腹侧后伸入胸腔，于第5肋间隙水平处的心基背侧分支成支气管（bronchus），经肺门进入左右肺。在气管叉前端分出一支独立的右尖叶支气管，通向右肺副叶。

气管和支气管的管壁可分为黏膜、黏膜下层和外膜。黏膜的上皮为假复层纤毛柱状上皮，内有纤毛细胞（ciliated cell）、杯状细胞（goblet cell）、刷细胞（brush cell）、小颗粒细胞（small granule cell）和基细胞分布。固有层的结缔组织内有弥散的淋巴组织和弹性纤维。黏膜下层由疏松结缔组织组成，其中含有气管腺、血管及神经等。外膜最厚，由透明软骨和疏松结缔组织组成，在软骨环缺口处可见平滑肌纤维束。胸段气管的外层为浆膜。

四、肺

肺（lung）位于胸腔内纵隔两侧，占据胸腔大部。肺的表面有浆膜（脏胸膜），光滑、湿润。左、右肺大体相似，两者在气管叉处相连。肺的颜色取决于肺中血液的含量，放血后呈苍白色或橘红色，血液充盈时呈深红色。肺有三个面，分别为肋面，隆凸，与胸侧壁相对；纵隔面，与胸腔纵隔相对；膈面，与膈相对。纵隔面，有心脏、食管和大血管等纵隔器官的压迹，心压迹最大，其上方有食管压迹和主动脉压迹；右肺纵隔面尚有后腔静脉沟，供同名静脉通过。在心压迹的后上方有肺门（hilum of the lung），为主支气管、肺动脉、肺静脉、淋巴管和神经等出入肺的地方；这些出入肺门的结构，被结缔组织包裹在一起，称肺根。纵隔面和肋面在背侧以厚而圆的背侧缘（钝缘）相连，在腹侧以薄的腹侧缘（锐缘）相连。背侧缘位于肋骨和椎骨之间的凹槽内。腹侧缘于心脏处向内凹陷，形成心切迹。左肺心切迹较大，体表投影位于第4～6肋骨/肋间隙，是心脏听诊的适宜部位；右肺心切迹小，体表投影位于3～4肋骨/肋间隙。左、右肺上有叶间裂，把每个肺分成若干个叶。猪的左肺分前叶、中叶和后叶，右肺分尖叶、心叶、膈叶和副叶。

肺由被膜和实质构成。被膜为肺表面的一层浆膜，称肺胸膜（pulmonary pleura），其深部为结缔组织，内含血管、神经、淋巴管、弹性纤维和平滑肌纤维。实质由导管部和呼吸部组成。导管部为支气管经肺门入肺后反复分支，依次为肺叶支气管（lobar bronchus）、肺段支气管（segmental bronchus）、细支气管（bronchiole）、终末细支气管（terminal bronchiole），统称为支气管树（bronchial tree），是气体在肺内流通的管道。叶支气管至小支气管的结构与支气管相似，上皮为假复层纤毛柱状上皮，固有层内富含弹性纤维、淋巴组织。平滑肌增多，为不成环的环形平滑肌束。外膜的软骨环变成软骨片，且逐渐减少。细支气管的黏膜上皮为假复层或单层纤毛柱状上皮；较少或完全没有杯状细胞、软骨片、腺体分布；环行平滑肌明显增多。终末细支气管的上皮为单层柱状上皮，以克拉拉细胞（Clara cell）为主，杯状细胞、软骨片和腺体消失；有完整的环行平滑肌。克拉拉细胞为柱状细胞，有分泌颗粒，分泌糖蛋白，形成终末细支气管壁的保护膜。

呼吸部由终末细支气管的逐级分支组成，包括呼吸性细支气管（respiratory bronchiole）、肺泡管（alveolar duct）、肺泡囊（alveolar sac）和肺泡（pulmonary alveolus），是肺内气体交换的功能部位。呼吸性细支气管的结构特点为管壁出现少量的肺泡，上皮为单层立方上皮，肌层为薄层平滑肌。肺泡管由较多肺泡囊和肺泡的开口共同形成的管道结构，管壁上肺泡隔末端膨大，由单层立方上皮和平滑肌束构成。肺泡囊是由2个以上肺泡共同形成的囊状结构。肺泡为半球形小囊，由单层肺泡上皮构成。肺泡上皮包含Ⅰ型肺泡细胞和Ⅱ型肺泡细胞。Ⅰ型肺泡细胞数量较少，呈扁平状，细胞的无胞核处较薄。细胞器少，有较多吞饮小泡，无分裂增殖能力，主要参与气体交换。Ⅱ型肺泡细胞数量较多，呈立方形，含板层小体，可分泌表面活性物质（磷脂），降低肺泡表面张力，稳定肺泡大小；可增殖分化为Ⅰ型肺泡细胞。相邻肺泡之间的结缔组织为肺泡隔，含丰富的连续毛细血管、弹性纤维、肺巨噬细胞等。肺巨噬细胞（pulmonary macrophage）属单核吞噬细胞系统，位于肺泡隔和肺泡腔内，吞噬大量尘埃颗粒后称为尘细胞（dust cell）。肺泡与血液之间进行气体交换所通过的结构称为气–血屏障（blood–air barrier），包括4层结构：① 肺泡表面液体层；② Ⅰ型肺泡细胞与基膜；③ 薄层结缔组织；④ 连续毛细血管基膜与内皮。

结缔组织伸入肺的实质内，将实质分为许多肺小叶（pulmonary lobule）。每一细支气管连同其分支，包括细支气管、终末细支气管、呼吸性细支气管、肺泡管、肺泡囊和肺泡组成一个肺小叶（pulmonary lobules）。肺小叶呈多边锥体形，底朝向肺的表面，顶朝向肺门。猪的小叶间结缔组织发达。临床上的小叶性肺炎即为肺小叶的局部性炎症。

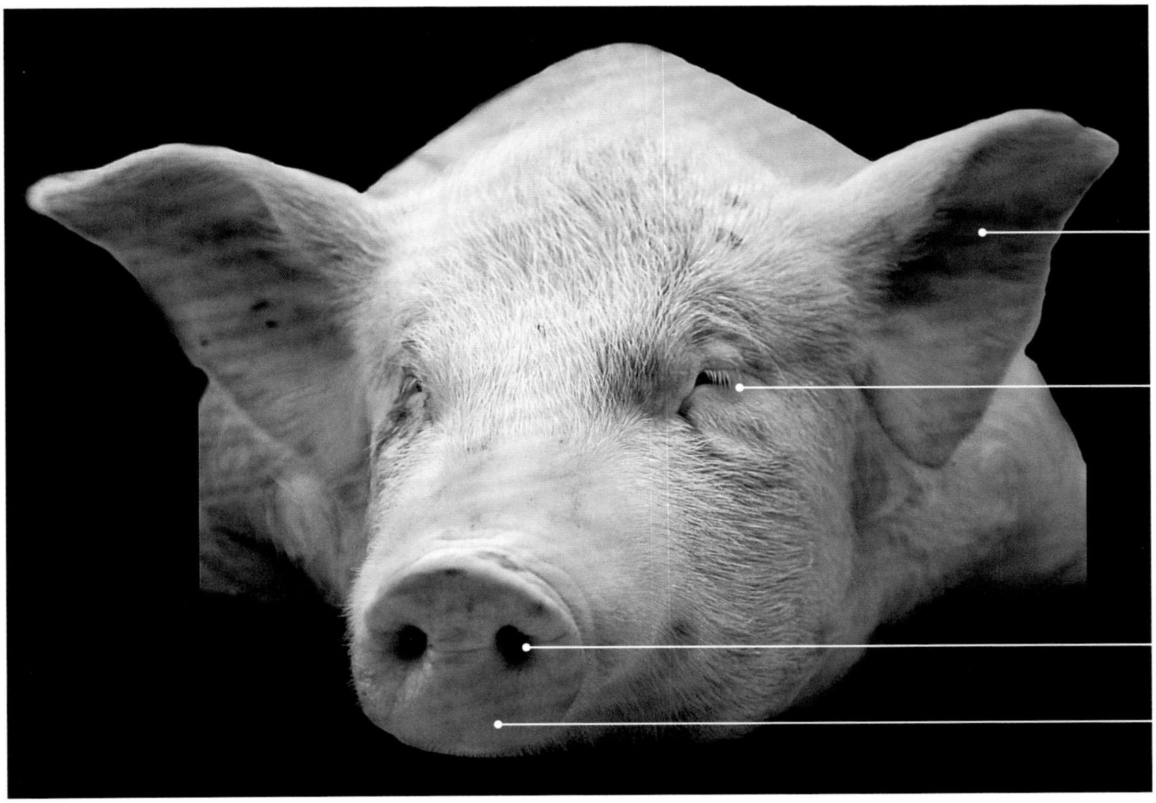

图 5-1 鼻孔与吻突（正面观）
Figure 5-1 Nostril and rostral disc (anterior view).

1- 耳 ear
2- 眼 eye
3- 鼻孔 nostril, nasal opening
4- 吻突 rostral disc

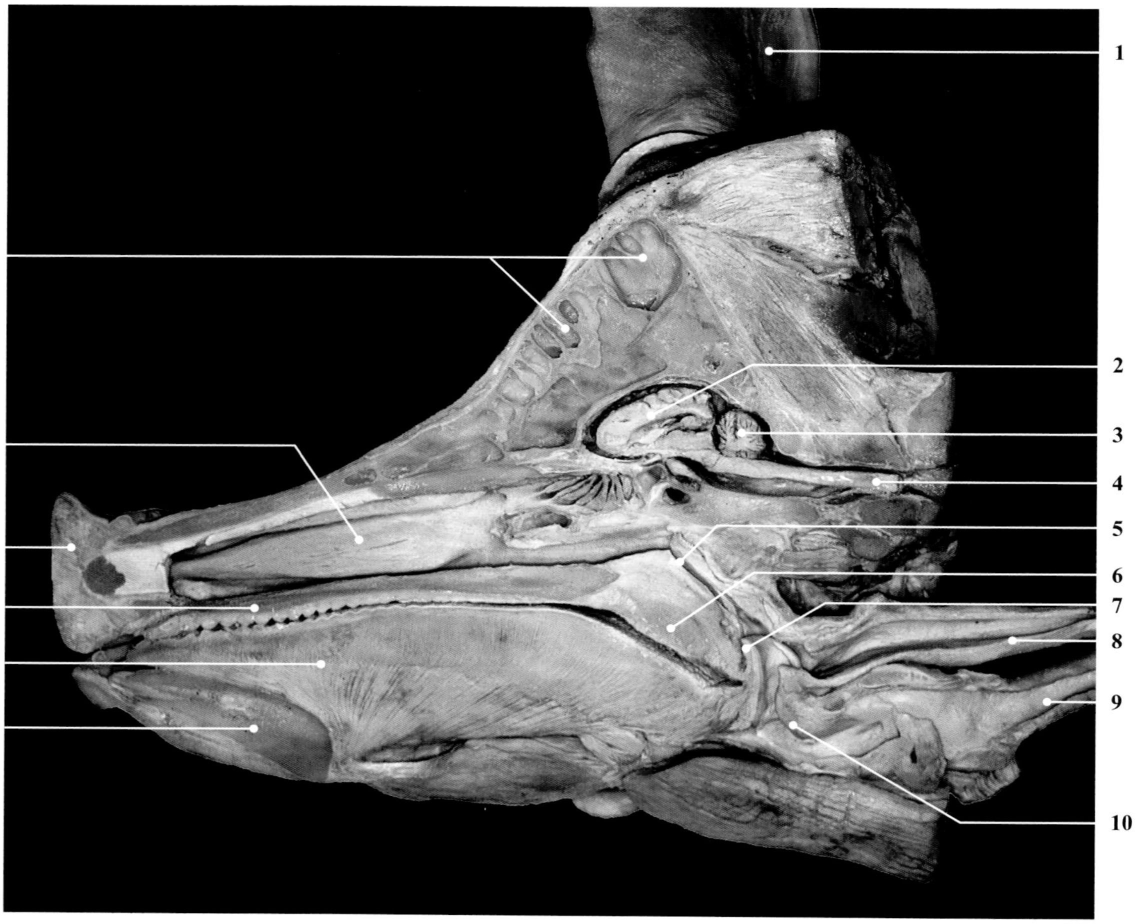

图 5-2 头部矢状面，示鼻腔
Figure 5-2 Sagittal plane of the head showing nasal cavity.

1- 耳 ear
2- 大脑 cerebrum
3- 小脑 cerebellum
4- 脊髓 spinal cord
5- 鼻后孔 posterior nasal apertures
6- 软腭 soft palate
7- 咽 pharynx
8- 食管 esophagus
9- 气管 trachea
10- 喉 larynx
11- 下颌骨 mandible
12- 舌 tongue
13- 硬腭 hard palate
14- 吻突 rostral disc
15- 鼻甲 nasal conchae
16- 鼻旁窦 paranasal sinus

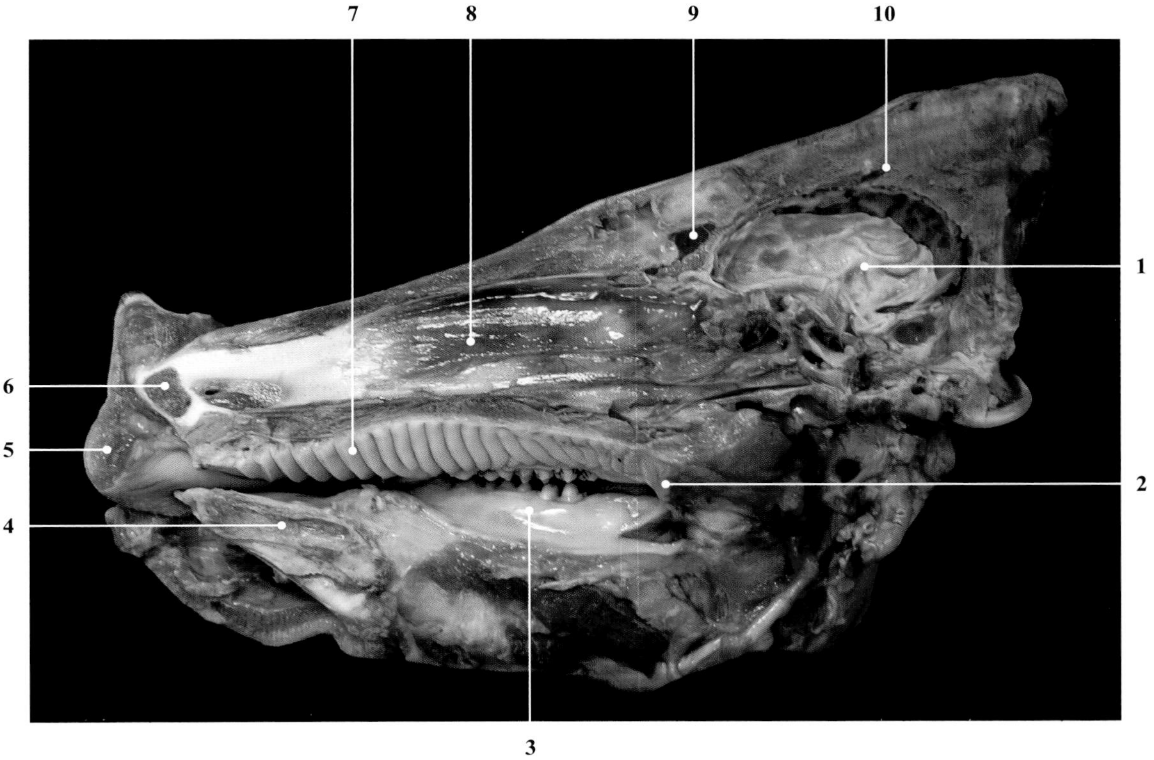

图 5-3　头部矢状面 - 示鼻中隔
Figure 5-3　Sagittal plane of the head showing nasal septum.

1- 颅腔 cranial cavity
2- 软腭 soft palate
3- 口腔 oral cavity
4- 下颌骨 mandible
5- 吻突 rostral disc
6- 吻骨 rostral bone
7- 硬腭 hard palate
8- 鼻中隔 nasal septum
9- 额窦 frontal sinus
10- 颅骨 cranial bone

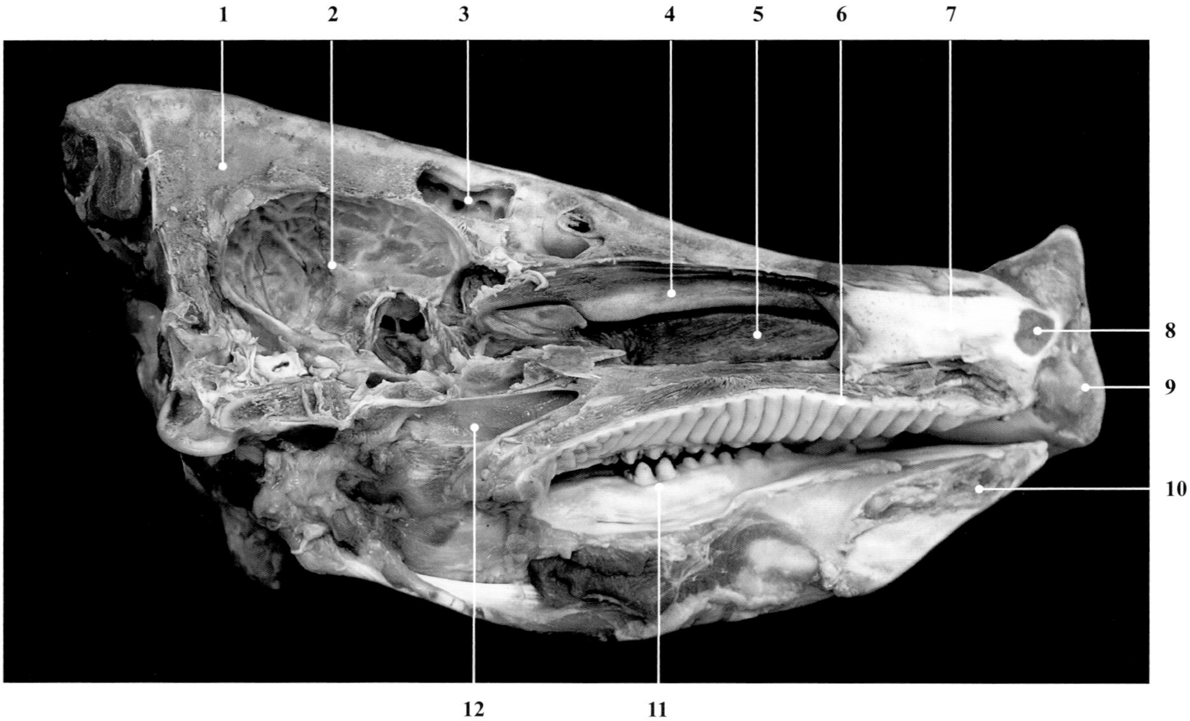

图 5-4 头部矢状面 - 示鼻甲
Figure 5-4 Sagittal plane of the head showing nasal concha.

1- 颅骨 cranial bone
2- 颅腔 cranial cavity
3- 额窦 frontal sinus
4- 上鼻甲 dorsal nasal concha
5- 下鼻甲 ventral nasal concha
6- 硬腭 hard palate
7- 鼻中隔 nasal septum
8- 吻骨 rostral bone
9- 吻突 rostral disc
10- 下颌骨 mandible
11- 口腔 oral cavity
12- 鼻后孔 posterior nasal apertures

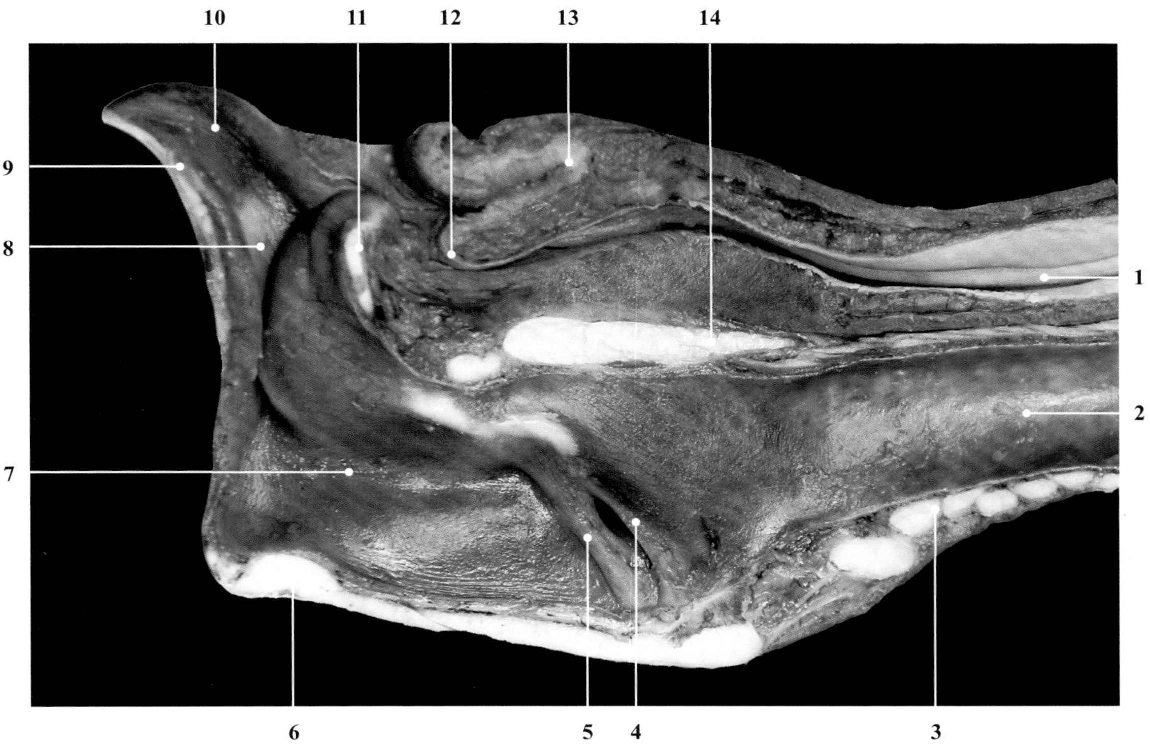

图 5-5 咽、喉和气管纵剖面
Figure 5-5 Longitudinal section of the pharynx, larynx and trachea.

1- 食管 esophagus
2- 气管 trachea
3- 气管软骨 tracheal cartilage
4- 声带 vocal cords
5- 前庭襞 vestibular fold
6- 甲状软骨 thyroid cartilage
7- 喉腔 larynx carvity
8- 喉口 aperture of larynx
9- 会厌软骨 epiglottic cartilage
10- 会厌 epiglottis
11- 杓状软骨 arytenoid cartilage
12- 食管咽口 pharynx entrance of the esophagus
13- 咽后隐窝 pharyngeal recess
14- 环状软骨 cricoid cartilage

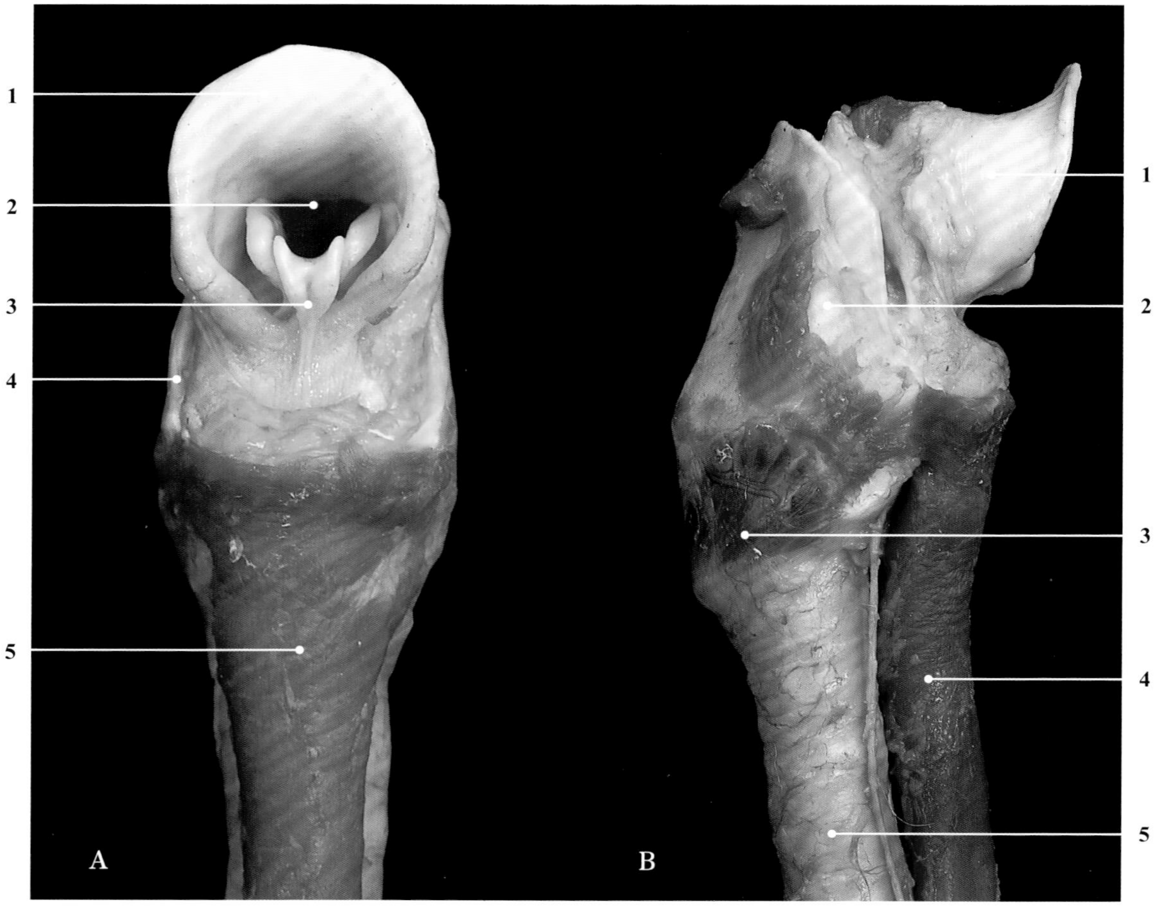

图 5-6 喉
Figure 5-6 Larynx.

A- 背侧观 Dorsal view
1- 会厌软骨 epiglottic cartilage
2- 喉口 aperture of larynx
3- 杓状软骨 arytenoid cartilage
4- 甲状软骨 thyroid cartilage
5- 食管 esophagus

B- 左侧观 Left view
1- 会厌软骨 epiglottic cartilage
2- 甲状软骨 thyroid cartilage
3- 环甲肌 cricothyreoid muscle
4- 食管 esophagus
5- 气管 trachea

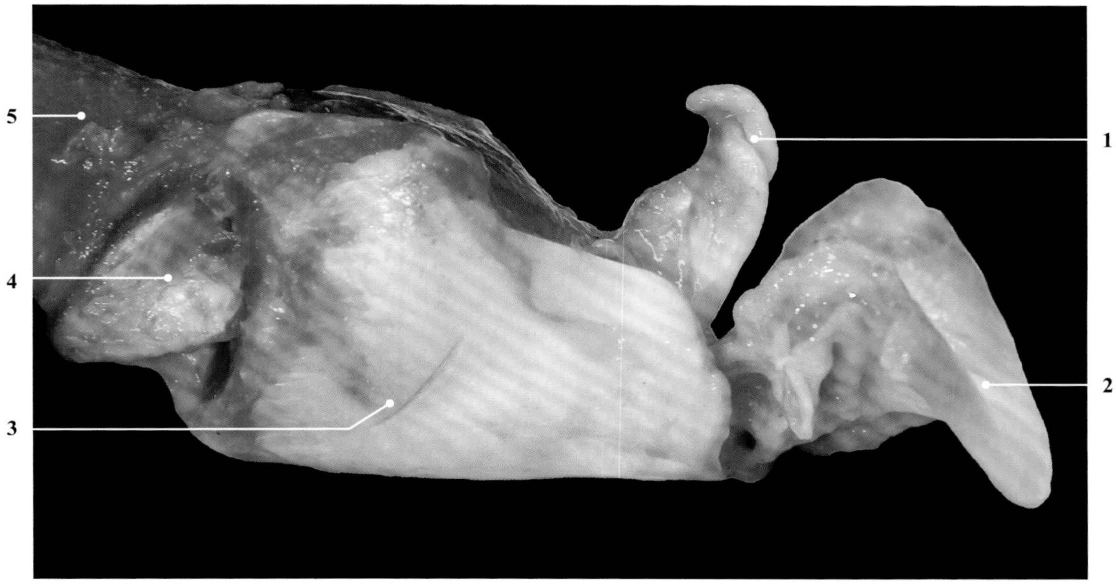

图 5-7 喉软骨
Figure 5-7 Laryngeal cartilage.

1- 杓状软骨 arytenoid cartilage
2- 会厌软骨 epiglottic cartilage
3- 甲状软骨 thyroid cartilage
4- 环状软骨 cricoid cartilage
5- 气管 trachea

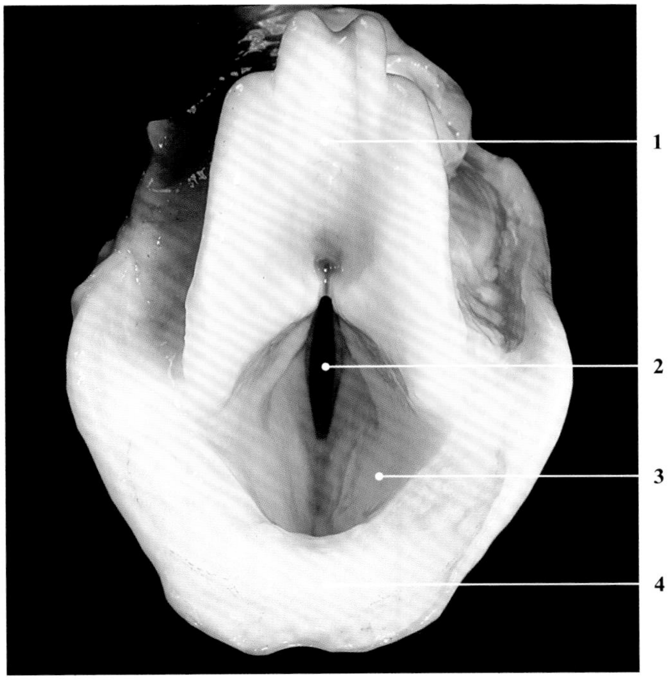

图 5-8 喉口
Figure 5-8 laryngeal aperture.

1- 杓状软骨 arytenoid cartilage
2- 声门裂 rima glottidis
3- 喉口 aperture of larynx
4- 会厌软骨 epiglottic cartilage

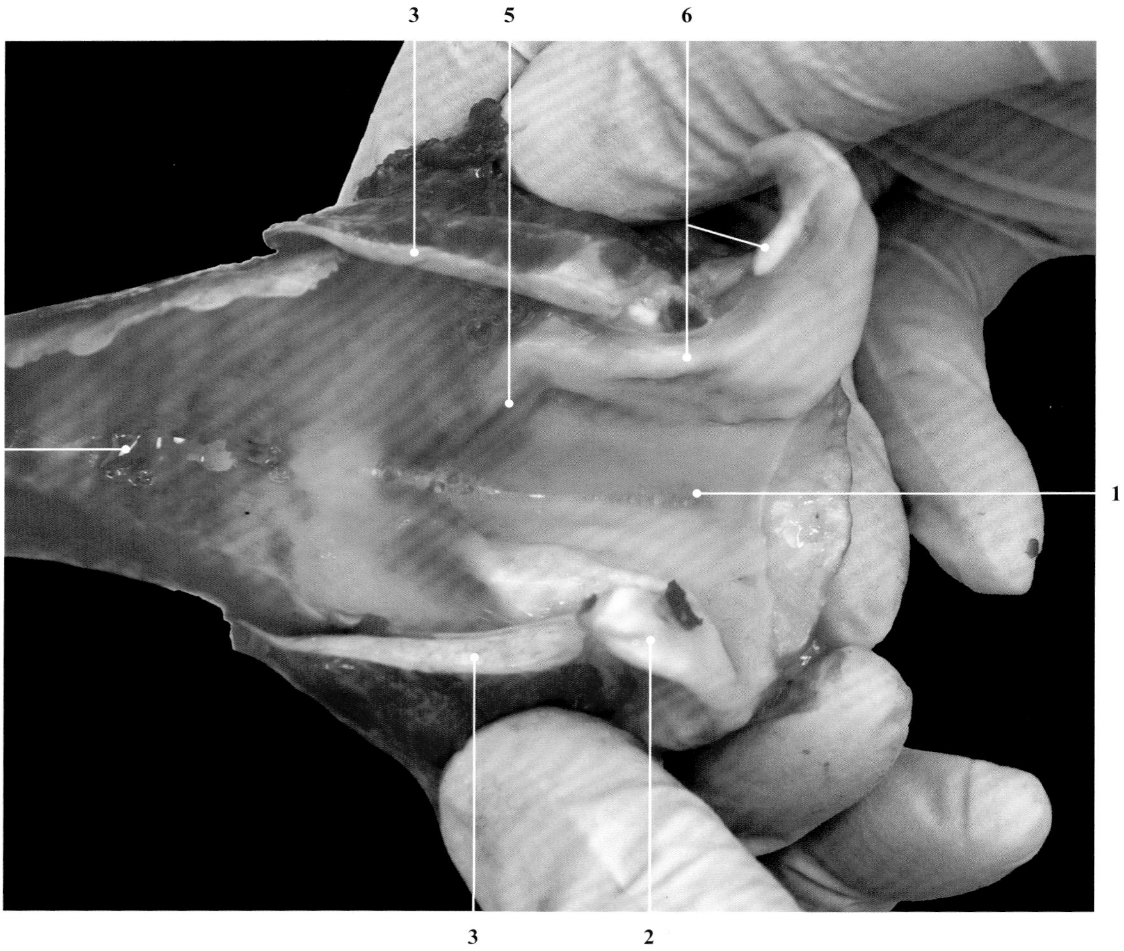

图 5-9　喉腔黏膜面（背侧观）
Figure 5-9　Mucous membrane surface of the laryngeal cavity (dorsal view).

1- 喉口 aperture of larynx
2- 杓状软骨 arytenoid cartilage
3- 环状软骨 cricoid cartilage
4- 气管 trachea
5- 声带 vocal cords
6- 杓状软骨 arytenoid cartilage

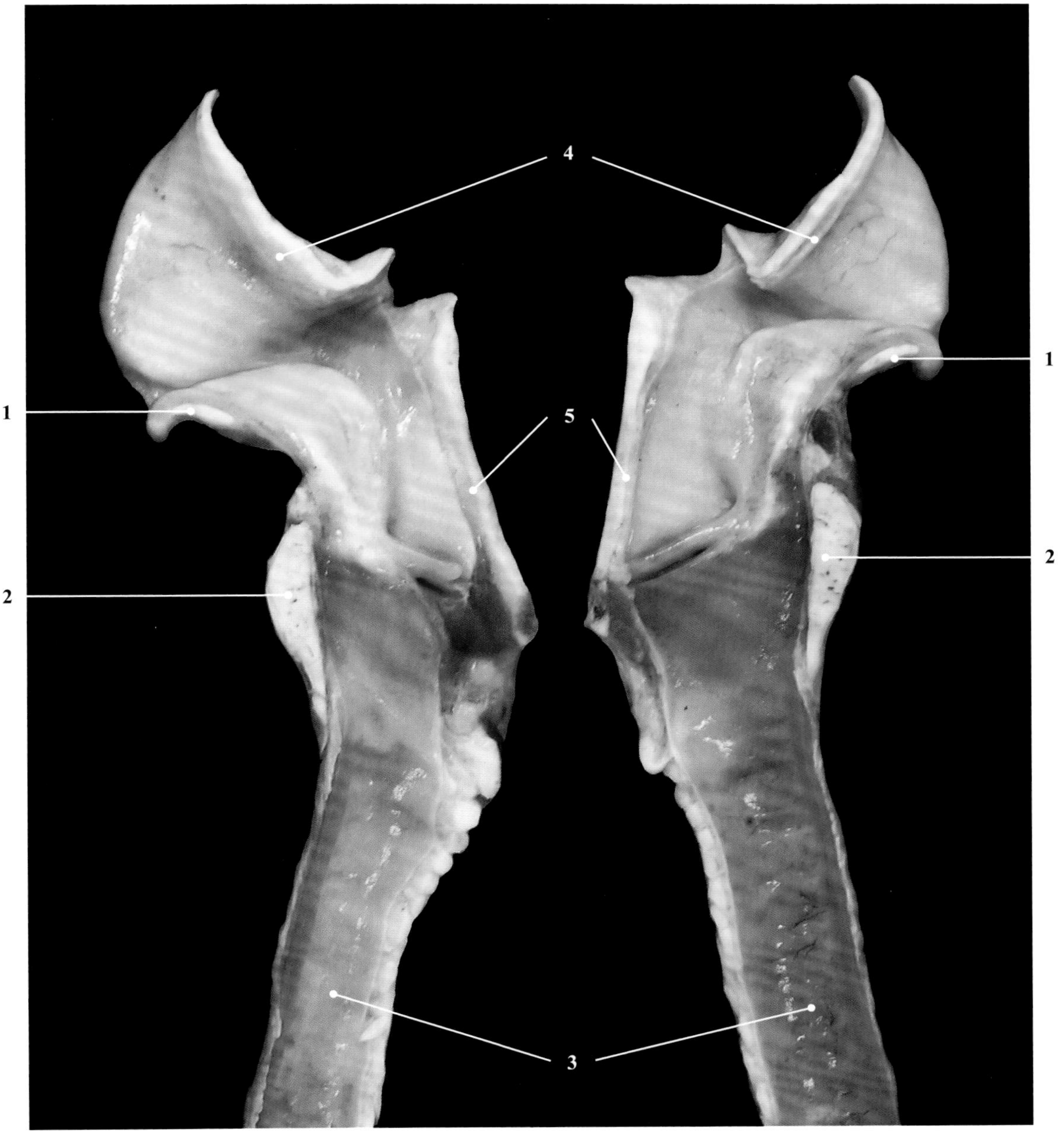

图 5-10 喉纵剖面
Figure 5-10 Longitudinal section of the larynx.

1- 杓状软骨 arytenoid cartilage
2- 环状软骨 cricoid cartilage
3- 气管 trachea
4- 会厌软骨 epiglottic cartilage
5- 甲状软骨 thyroid cartilage

1- 复层扁平上皮 stratified squamous epithelium
2- 固有层 lamina propria
3- 黏膜下层 submucosa
4- 静脉 vein
5- 浆液腺泡 serous acinus
6- 黏液腺泡 mucous acinus
7- 混合腺泡 mixed acinus
8- 分泌导管 excretory duct
9- 胶原纤维束 collagenous fiber bundle

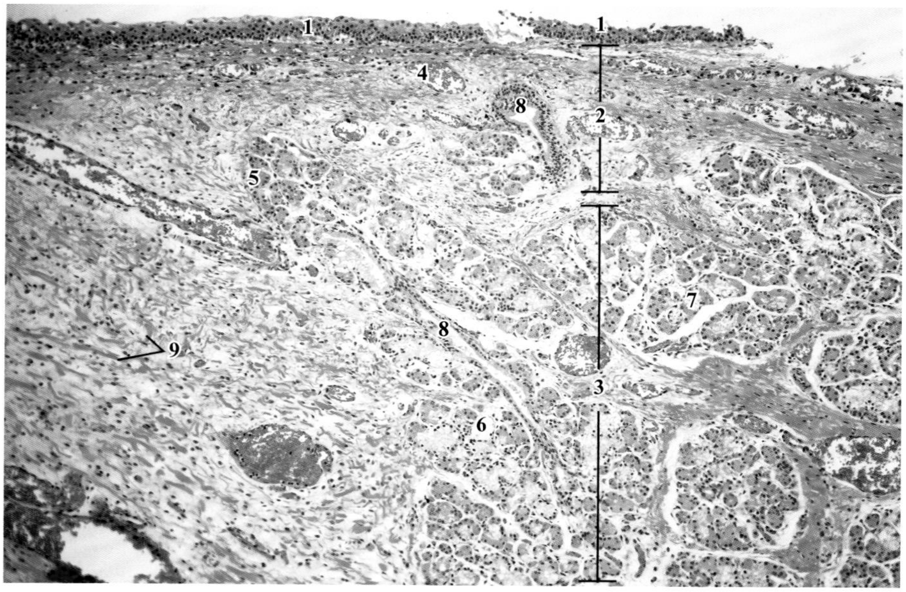

图 5-11　声带组织切片（HE 染色，低倍镜）
Figure 5-11　Histological section of the vocal cords (HE staining, lower power).

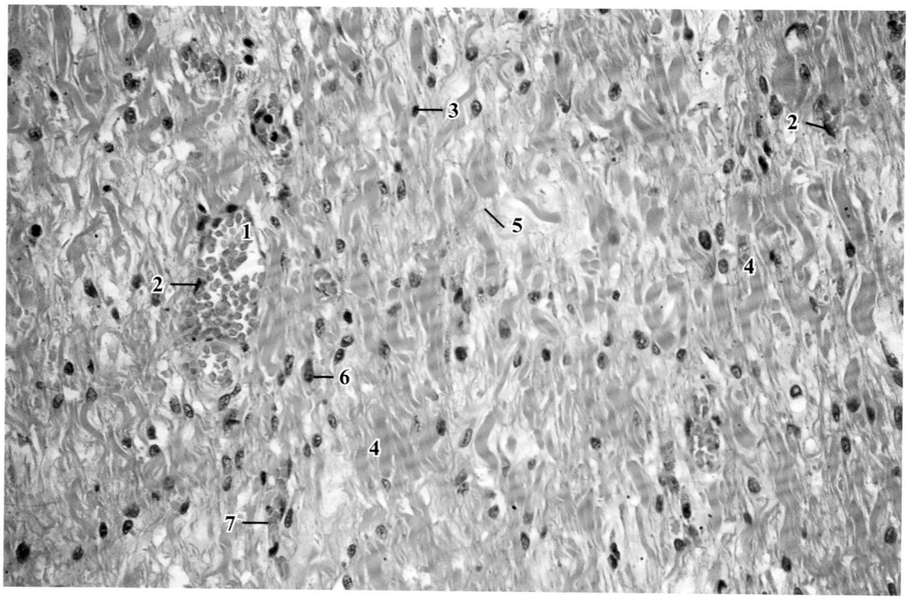

1- 静脉 vein
2- 内皮细胞 endothelium
3- 成纤维细胞 fibroblast
4- 胶原纤维束 collagen fiber bundle
5- 弹性纤维 elastic fiber
6- 巨噬细胞 macrophage
7- 毛细血管 capillary

图 5-12　声带组织切片（HE 染色，高倍镜）
Figure 5-12　Histological section of the vocal cords (HE staining, higher power).

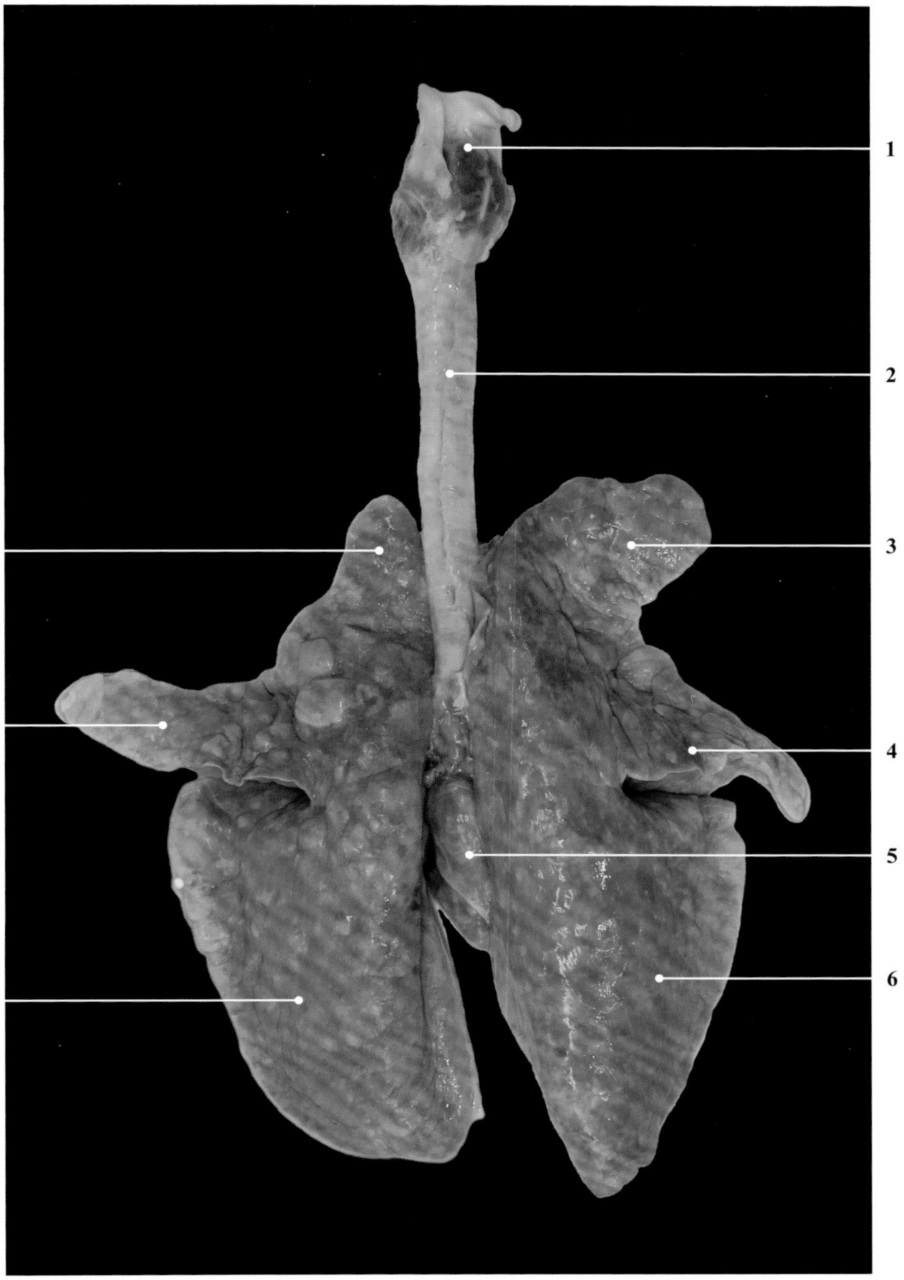

图 5-13 喉、气管和肺背侧观
Figure 5-13 Dorsal view of the larynx, trachea and lung.

1- 喉 larynx
2- 气管 trachea
3- 右肺前叶（尖叶） right cranial lobe (apical lobe) of the lung
4- 右肺中叶（心叶） right middle lobe (cardiac lobe) of the lung
5- 副叶 accessory lobe
6- 右肺后叶（膈叶） right caudal lobe (diaphragmatic lobe) of the lung
7- 左肺后叶（膈叶） left caudal lobe (diaphragmatic lobe) of the lung
8- 左肺中叶（心叶） left middle lobe (cardiac lobe) of the lung
9- 左肺前叶（尖叶） left cranial lobe (apical lobe) of the lung

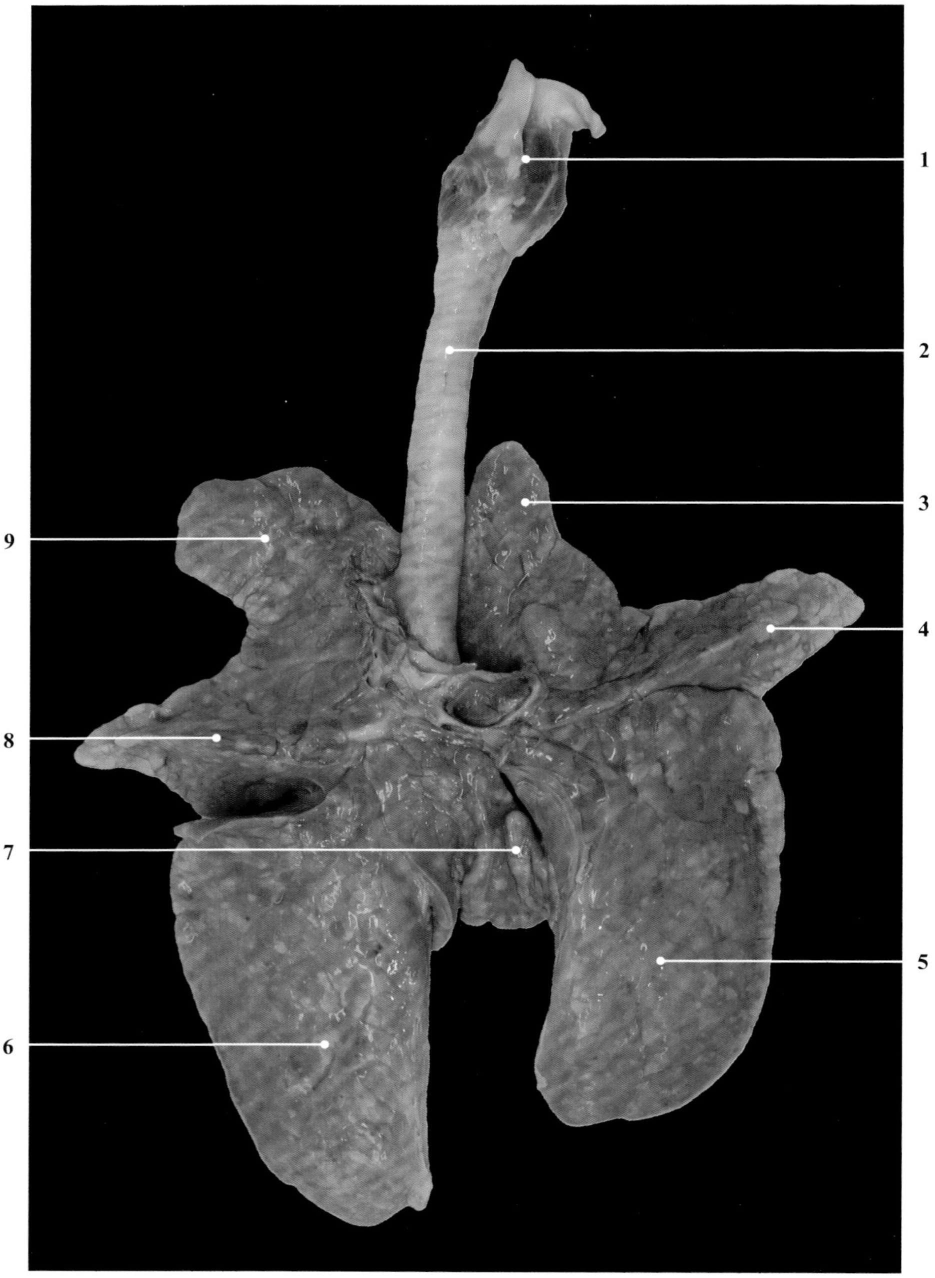

图 5-14　喉、气管和肺腹侧观
Figure 5-14　Ventral view of the larynx, trachea and lung.

1- 喉 larynx
2- 气管 trachea
3- 左肺前叶（尖叶） left cranial lobe (apical lobe) of the lung
4- 左肺中叶（心叶） left middle lobe (cardiac lobe) of the lung
5- 左肺后叶（膈叶） left caudal lobe (diaphragmatic lobe) of the lung
6- 右肺后叶（膈叶） right caudal lobe (diaphragmatic lobe) of the lung
7- 副叶 accessory lobe
8- 右肺中叶（心叶） right middle lobe (cardiac lobe) of the lung
9- 右肺前叶（尖叶） right cranial lobe (apical lobe) of the lung

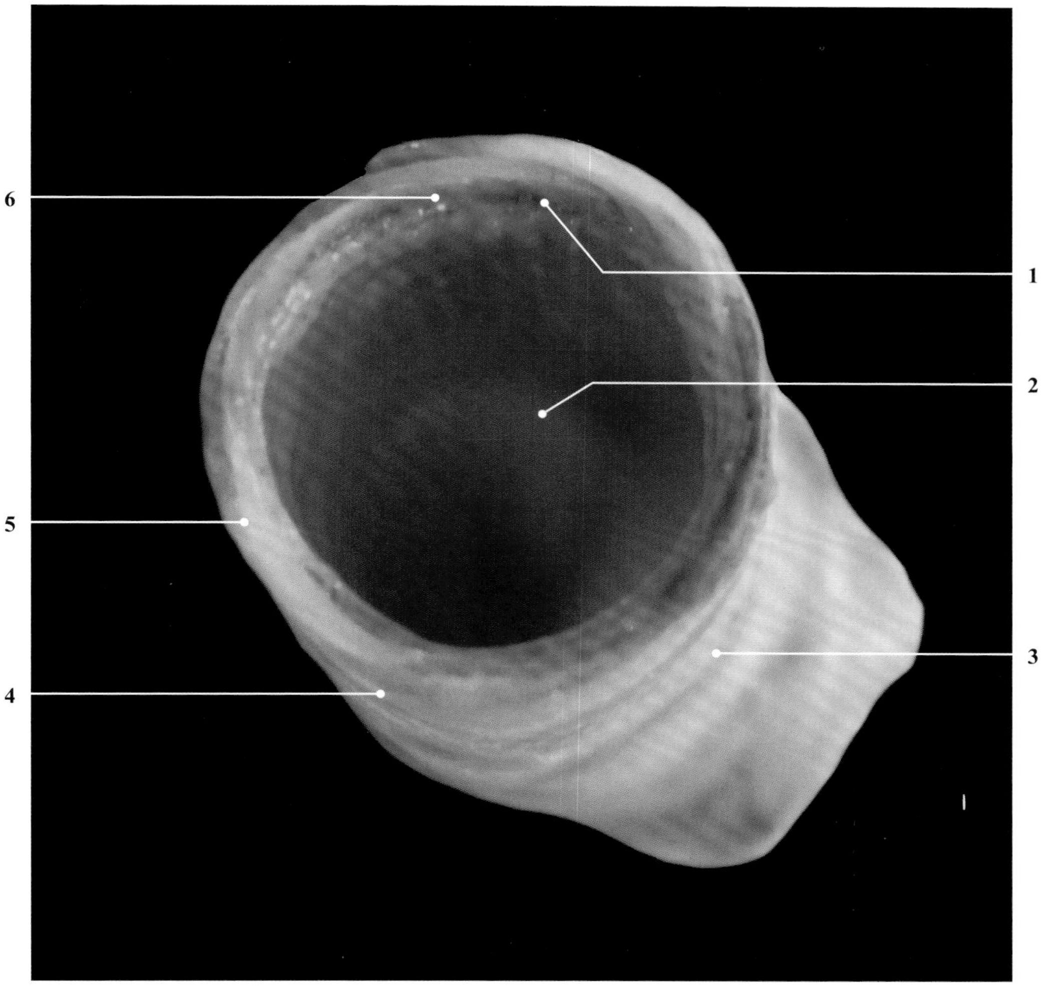

图 5-15　气管横断面
Figure 5-15　Cross section of the trachea.

1- 气管肌 tracheal muscle
2- 气管腔 tracheal cavity
3- 气管 trachea
4- 外膜 adventitia
5- 气管软骨 tracheal cartilage
6- 黏膜 mucosa

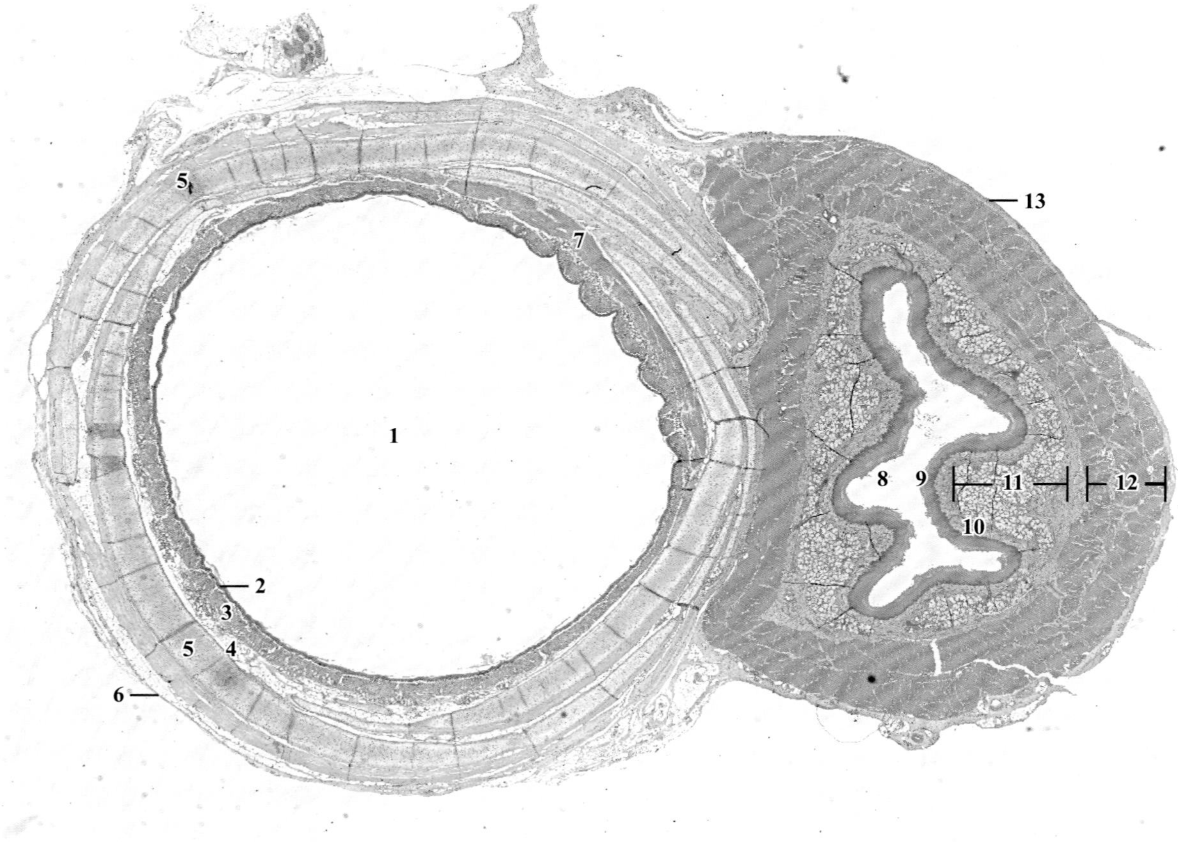

图 5-16　气管与食管组织切片（HE 染色，低倍镜）
Figure 5-16　Histological section of the trachea and esophagus (HE staining, lower power).

1- 气管腔 tracheal lumen
2- 气管黏膜上皮 mucosal epithelium of the trachea
3- 气管黏膜固有层 mucosal lamina propria of the trachea
4- 气管黏膜下层 submucous layer of the trachea
5- 透明软骨 hyaline cartilage
6- 气管外膜 adventitia of trachea
7- 气管肌 tracheal muscle
8- 食管腔 esophagus lumen
9- 食管黏膜上皮 mucosal epithelium of the esophagus
10- 食管黏膜固有层 mucosal lamina propria of the esophagus
11- 食管黏膜下层 submucous layer of the esophagus
12- 食管肌层 muscular layer of the esophagus
13- 食管外膜 adventitia of the esophagus

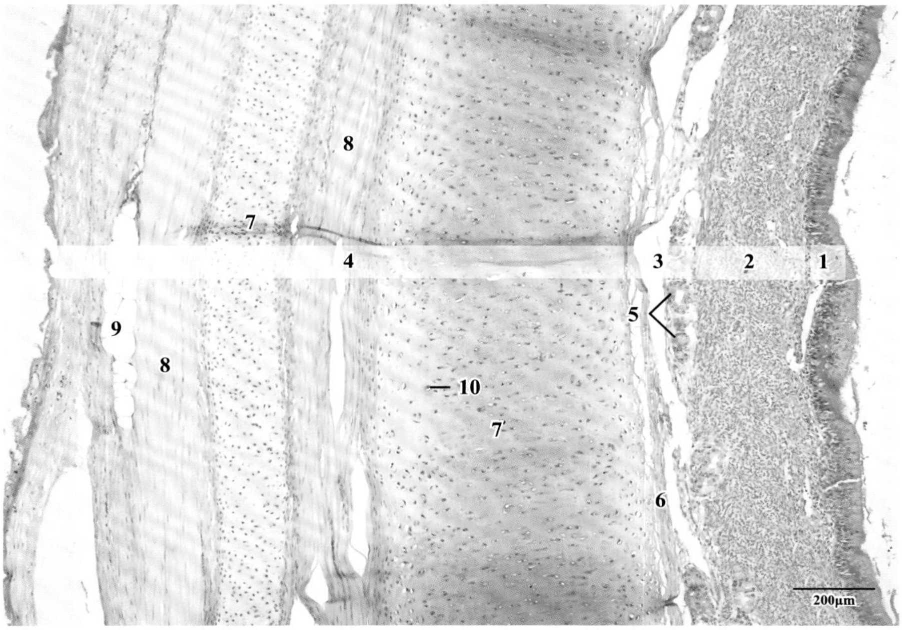

图 5-17 气管组织切片（HE 染色，低倍镜）
Figure 5-17 Histological section of the trachea (HE staining, lower power).

1- 黏膜上皮，假复层纤毛柱状上皮 mucosal epithelium, pseudostratified ciliated columnar epithelium
2- 黏膜固有层 mucosal lamina propria
3- 黏膜下层 submucous layer
4- 外膜 adventitia
5- 气管腺 tracheal gland
6- 软骨膜 perichondrium
7- 透明软骨 hyaline cartilage
8- 平滑肌 smooth muscle
9- 脂肪细胞 adipocyte
10- 软骨细胞 chondrocyte

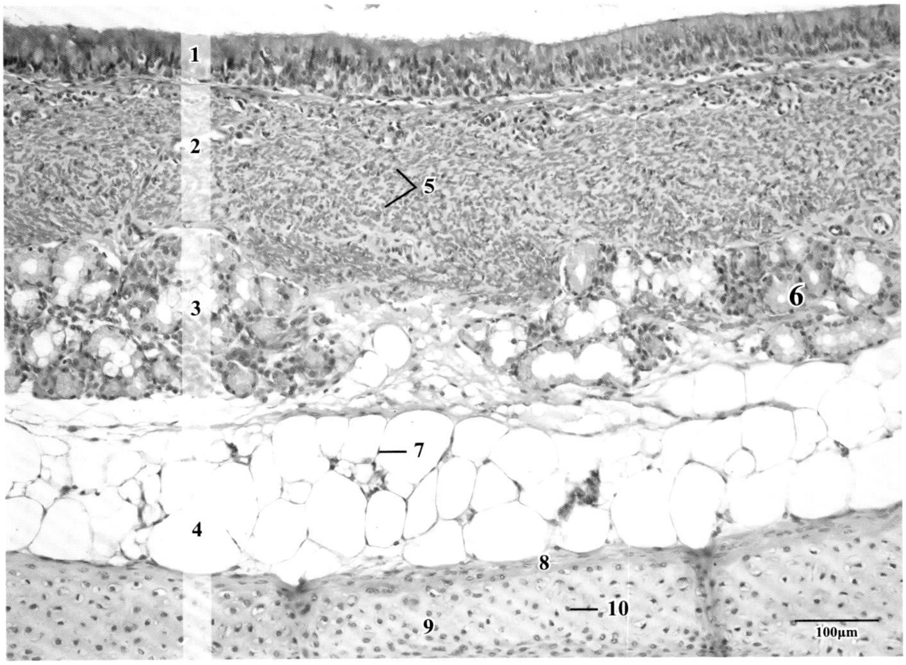

图 5-18 气管组织切片（HE 染色，高倍镜）
Figure 5-18 Histological section of the trachea (HE staining, higher power).

1- 黏膜上皮，假复层纤毛柱状上皮 mucosal epithelium, pseudostratified ciliated columnar epithelium
2- 黏膜固有层 mucosal lamina propria
3- 黏膜下层 submucous layer
4- 外膜 adventitia
5- 弹性纤维 elastic fiber
6- 气管腺 tracheal gland
7- 脂肪细胞核 nucleus, adipocyte
8- 软骨膜 perichondrium
9- 透明软骨 hyaline cartilage
10- 软骨细胞 chondrocyte

1- 假复层纤毛柱状上皮 pseudostratified ciliated columnar epithelium
2- 假杯状细胞 goblet cell
3- 基膜 basement membrane
4- 基底细胞 basal cell
5- 动脉 artery
6- 弹性纤维 elastic fiber
7- 静脉 vein
8- 软骨膜 perichondrium
9- 透明软骨 hyaline cartilage
10- 软骨细胞 chondrocyte
11- 软骨陷窝 cartilage lacuna
12- 同源细胞群 isogenous group

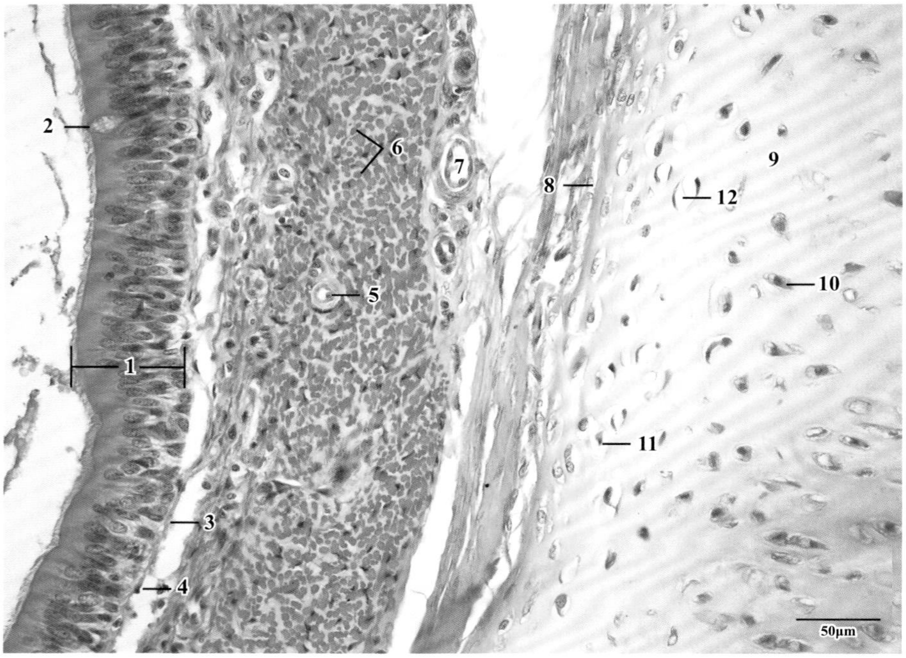

图 5-19 气管组织切片（HE 染色，高倍镜）
Figure 5-19 Histological section of the trachea (HE staining, higher power).

1- 假复层纤毛柱状上皮 pseudostratified ciliated columnar epithelium
2- 杯状细胞 goblet cell
3- 基膜 basement membrane
4- 基底细胞 basal cell
5- 固有层 lamina propria
6- 弹性纤维 elastic fiber
7- 气管腺 tracheal gland

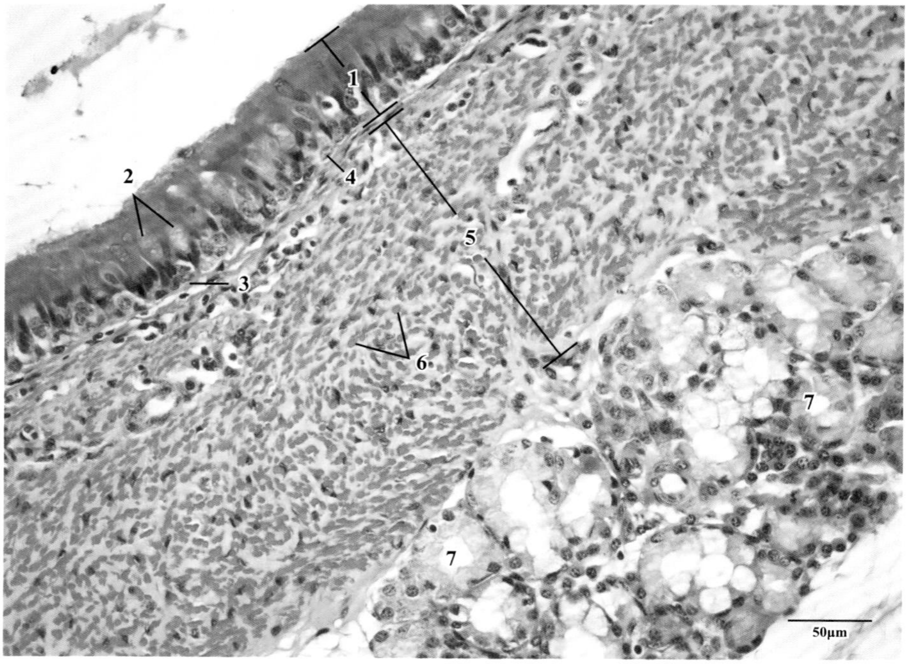

图 5-20 气管组织切片（HE 染色，高倍镜）
Figure 5-20 Histological section of the trachea (HE staining, higher power).

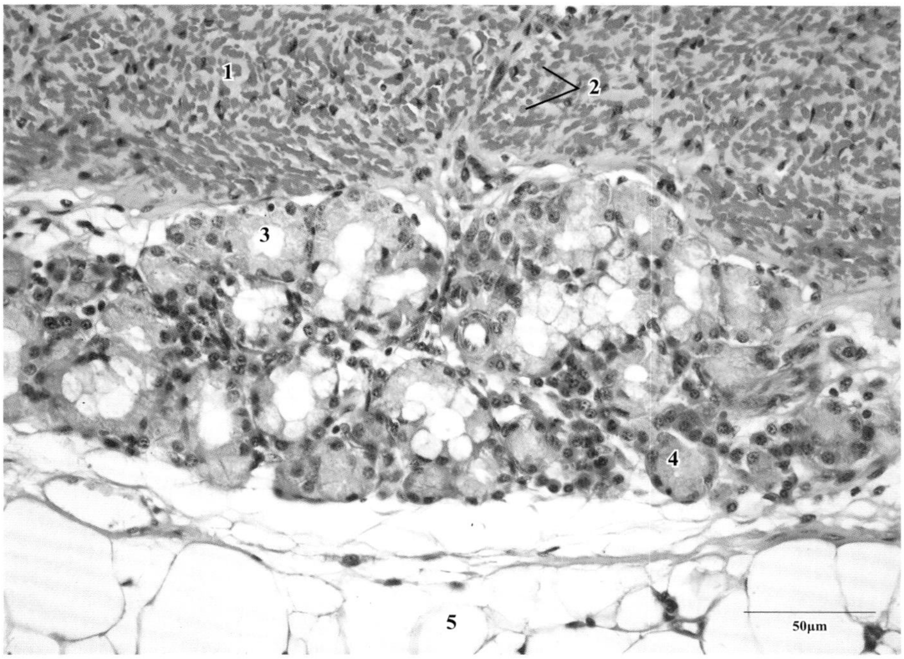

1- 固有层 lamina propria
2- 弹性纤维 elastic fiber
3- 黏液腺泡 mucous acinus
4- 浆液腺泡 serous acinus
5- 脂肪细胞 adipocyte

图 5-21　气管腺组织切片（HE 染色，高倍镜）
Figure 5-21　Histological section of the tracheal gland (HE staining, higher power).

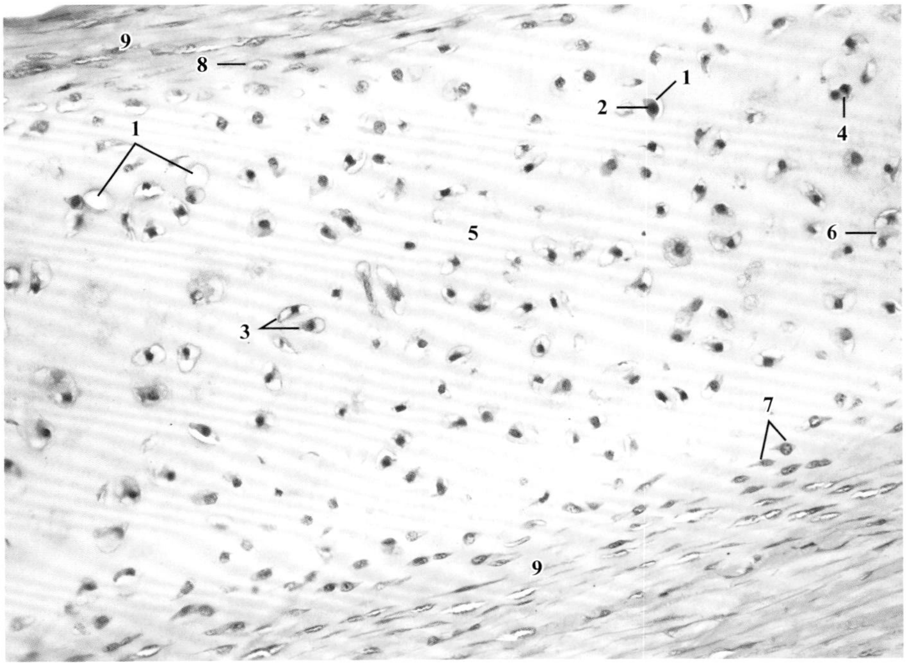

1- 软骨陷窝 cartilage lacuna
2- 软骨细胞 chondrocyte
3- 软骨囊 cartilage capsule
4- 细胞分裂 cell division
5- 软骨基质 cartilage matrix
6- 同源细胞群 isogenous group
7- 成纤维细胞 fibroblast
8- 幼稚的软骨细胞 naive chondrocyte
9- 软骨膜 perichondrium

图 5-22　气管软骨（透明软骨）组织切片（HE 染色，高倍镜）
Figure 5-22　Histological section of the tracheal cartilage (hyaline cartilage) (HE staining, higher power).

1- 神经元胞体 cell body, neuron
2- 神经元突起 process of the neuron
3- 神经细胞核 nucleus of the neuron
4- 毛细血管 capillary
5- 被囊细胞 / 卫星细胞 capsular cell /satellite cell
6- 神经 nerve
7- 成纤维细胞 fibroblast
8- 施万细胞 / 神经膜细胞 Schwann cell/ neurolemmal cell
9- 神经束膜 perineurium

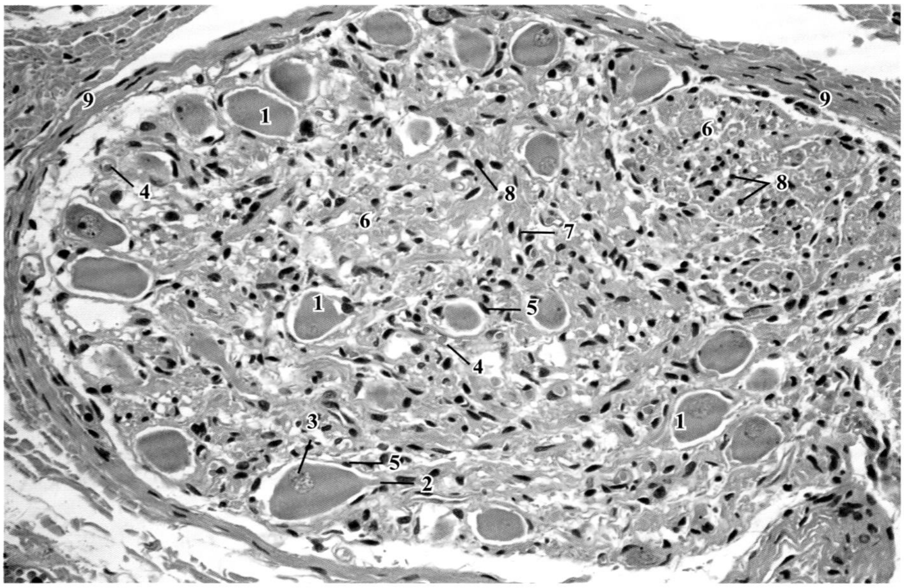

图 5-23　气管神经节组织切片（HE 染色，高倍镜）
Figure 5-23　Histological section of the tracheal ganglion (HE staining, higher power).

1- 假复层纤毛柱状上皮 pseadostratified ciliated columnar epithelium
2- 固有层 lamina propria
3- 黏膜下层 submucous layer
4- 外膜 adventitia
5- 平滑肌 smooth muscle
6- 混合腺 mixed gland
7- 软骨膜 perichondrium
8- 透明软骨片 hyaline cartilage plate
9- 静脉 vein
10- 脂肪细胞 adipocyte

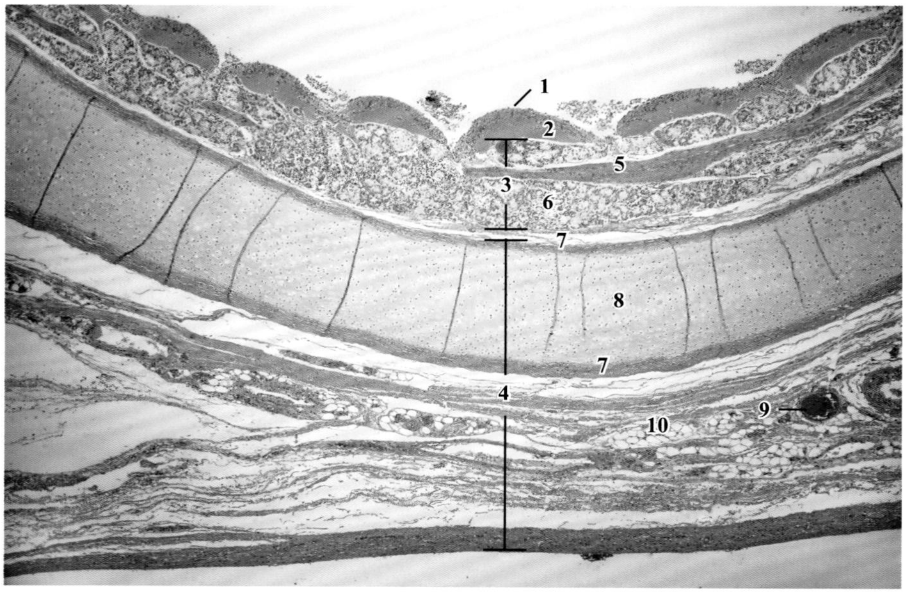

图 5-24　支气管组织切片（HE 染色，高倍镜）
Figure 5-24　Histological section of the branchus (HE staining, higher power).

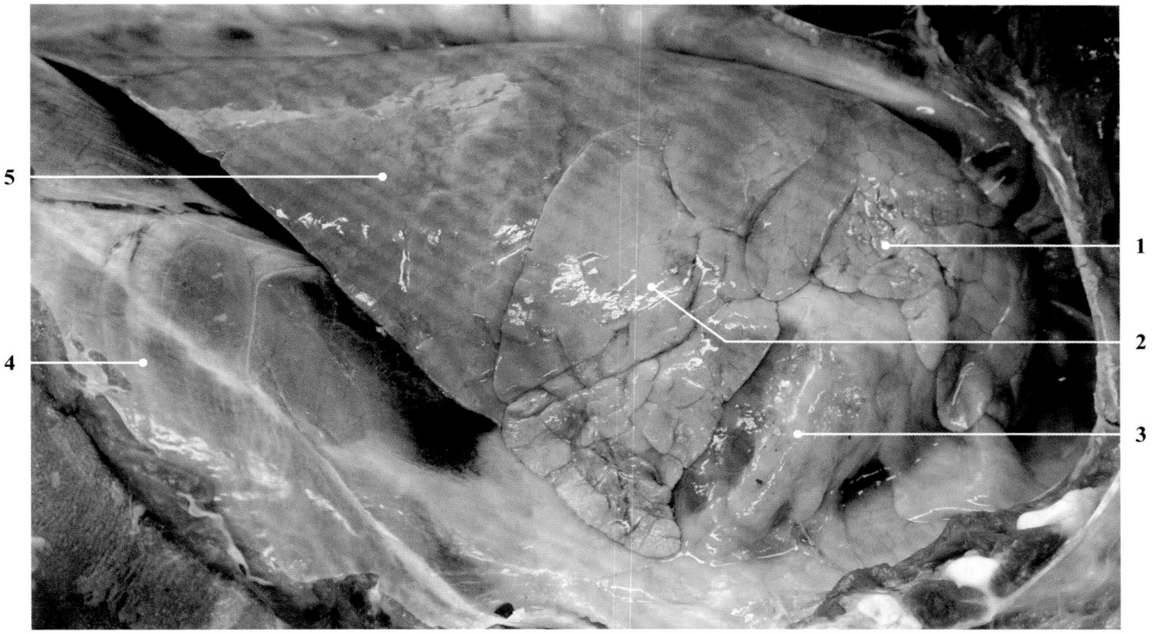

图 5-25 胸腔右侧观 - 示肺
Figure 5-25 Right view of the troracic cavity showing the lung.

1- 右肺前叶（尖叶） right cranial lobe (apical lobe) of the lung
2- 右肺中叶（心叶） right middle lobe (cardiac lobe) of the lung
3- 心脏和心包 heart and pericardium
4- 膈 diaphragm
5- 右肺后叶（膈叶） right caudal lobe (diaphragmatic lobe) of the lung

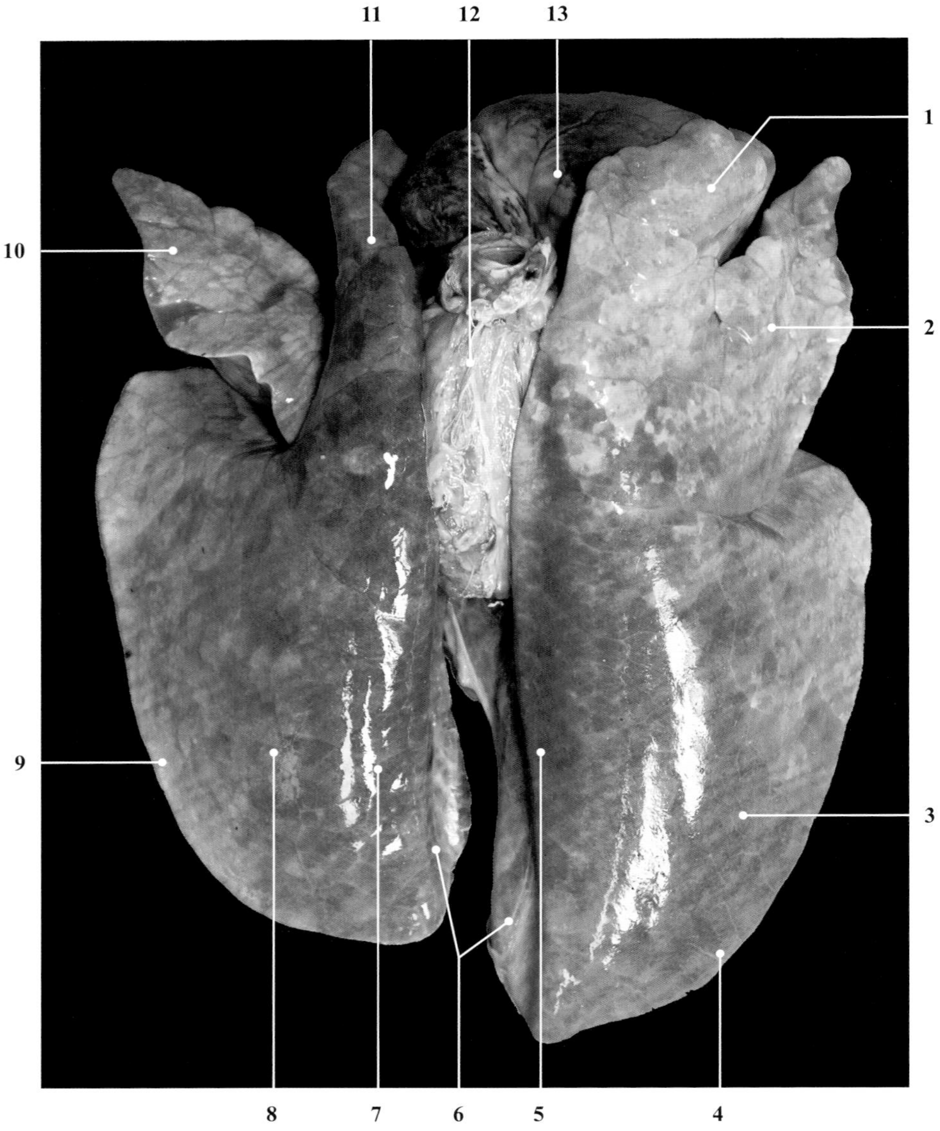

图 5-26　心肺壁面观
Figure 5-26　Parietal view of the heart and lung.

1- 右肺前叶（尖叶） right cranial lobe (apical lobe) of the lung
2- 右肺中叶（心叶） right middle lobe (cardiac lobe) of the lung
3- 右肺后叶（膈叶） right caudal lobe (diaphragmatic lobe) of the lung
4- 右肺锐缘（腹侧缘） right sharp edge (ventral margin) of the lung
5- 右肺钝缘（背侧缘） right blunt edge (dorsal margin) of the lung
6- 纵隔面 mediastinal surface
7- 左肺钝缘（背侧缘） left blunt edge (dorsal margin) of the lung
8- 左肺后叶（膈叶） left caudal lobe (diaphragmatic lobe) of the lung
9- 左肺锐缘（腹侧缘） left sharp edge (ventral margin) of the lung
10- 左肺中叶（心叶） left middle lobe (cardiac lobe) of the lung
11- 左肺前叶（尖叶）left cranial lobe (apical lobe) of the lung
12- 纵隔 mediastinum
13- 心脏 heart

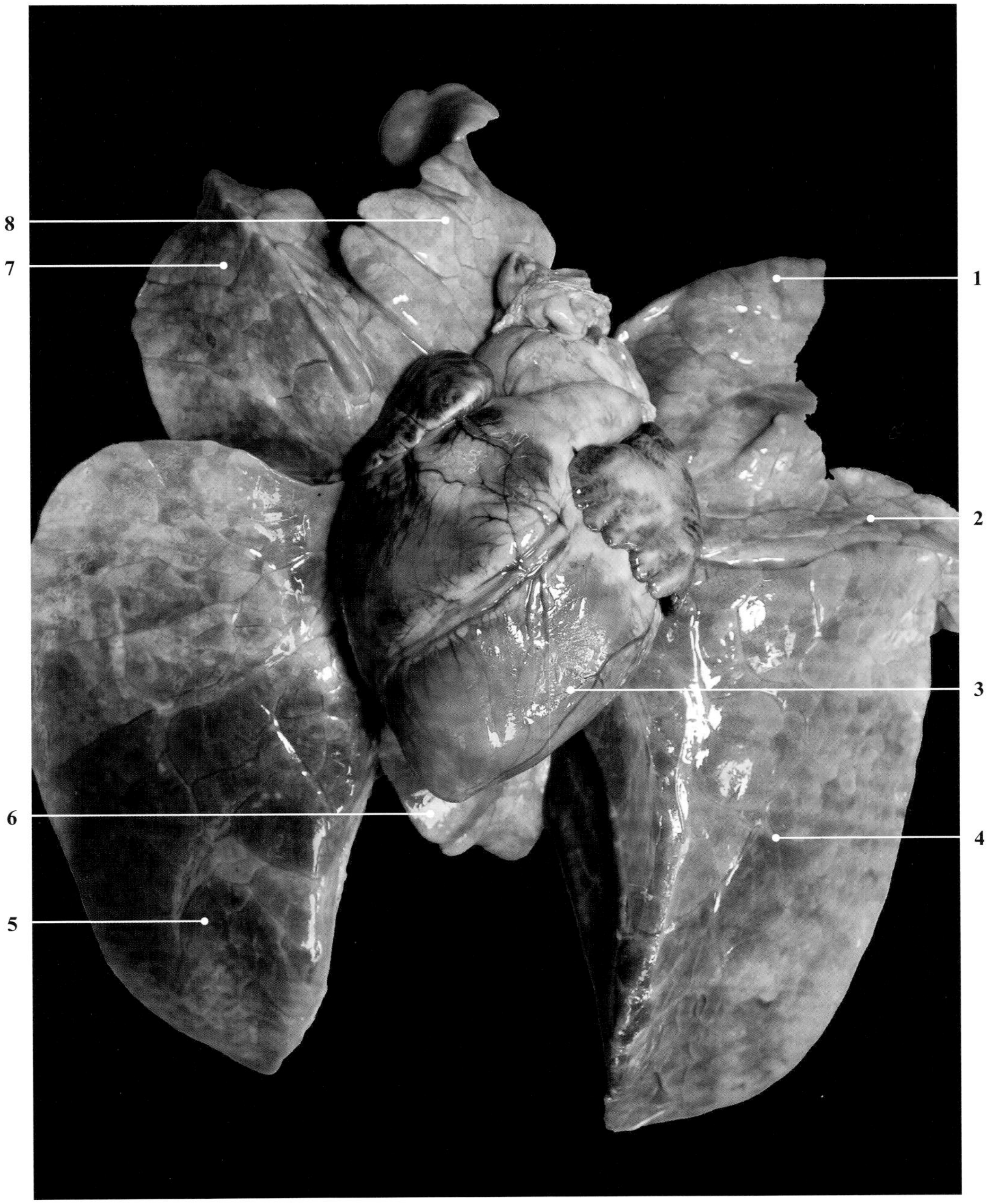

图 5-27 心肺脏面观
Figure 5-27 Visceral view of the heart and lung.

1- 左肺前叶（尖叶） left cranial lobe (apical lobe) of the lung
2- 左肺中叶（心叶） left middle lobe (cardiac lobe) of the lung
3- 心脏 heart
4- 左肺后叶（膈叶） left caudal lobe (diaphragmatic lobe) of the lung
5- 右肺后叶（膈叶） right caudal lobe (diaphragmatic lobe) of the lung
6- 副叶 accessory lobe
7- 右肺中叶（心叶） right middle lobe (cardiac lobe) of the lung
8- 右肺前叶（尖叶） right cranial lobe (apical lobe) of the lung

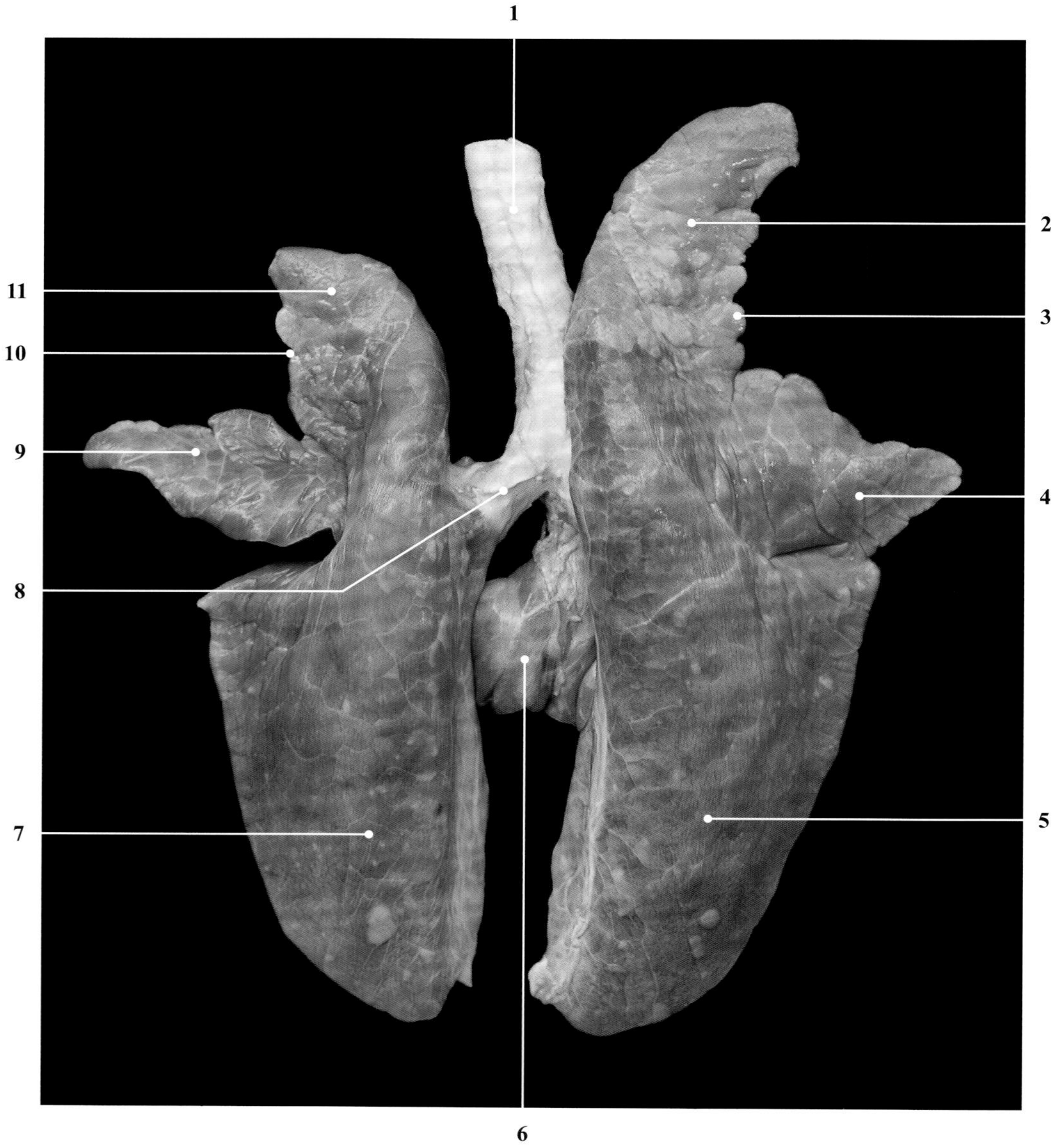

图 5-28 肺壁面观
Figure 5-28 Parietal view of the lung.

1- 气管 trachea
2- 右肺前叶（尖叶） right cranial lobe (apical lobe) of the lung
3- 右心切迹 right cardiac notch
4- 右肺中叶（心叶） right middle lobe (cardiac lobe) of the lung
5- 右肺后叶（膈叶） right caudal lobe (diaphragmatic lobe) of the lung
6- 副叶 accessory lobe
7- 左肺后叶（膈叶） left caudal lobe (diaphragmatic lobe) of the lung
8- 左支气管 left bronchus
9- 左肺中叶（心叶） left middle lobe (cardiac lobe) of the lung
10- 左心切迹 left cardiac notch
11- 左肺前叶（尖叶） left cranial lobe (apical lobe) of the lung

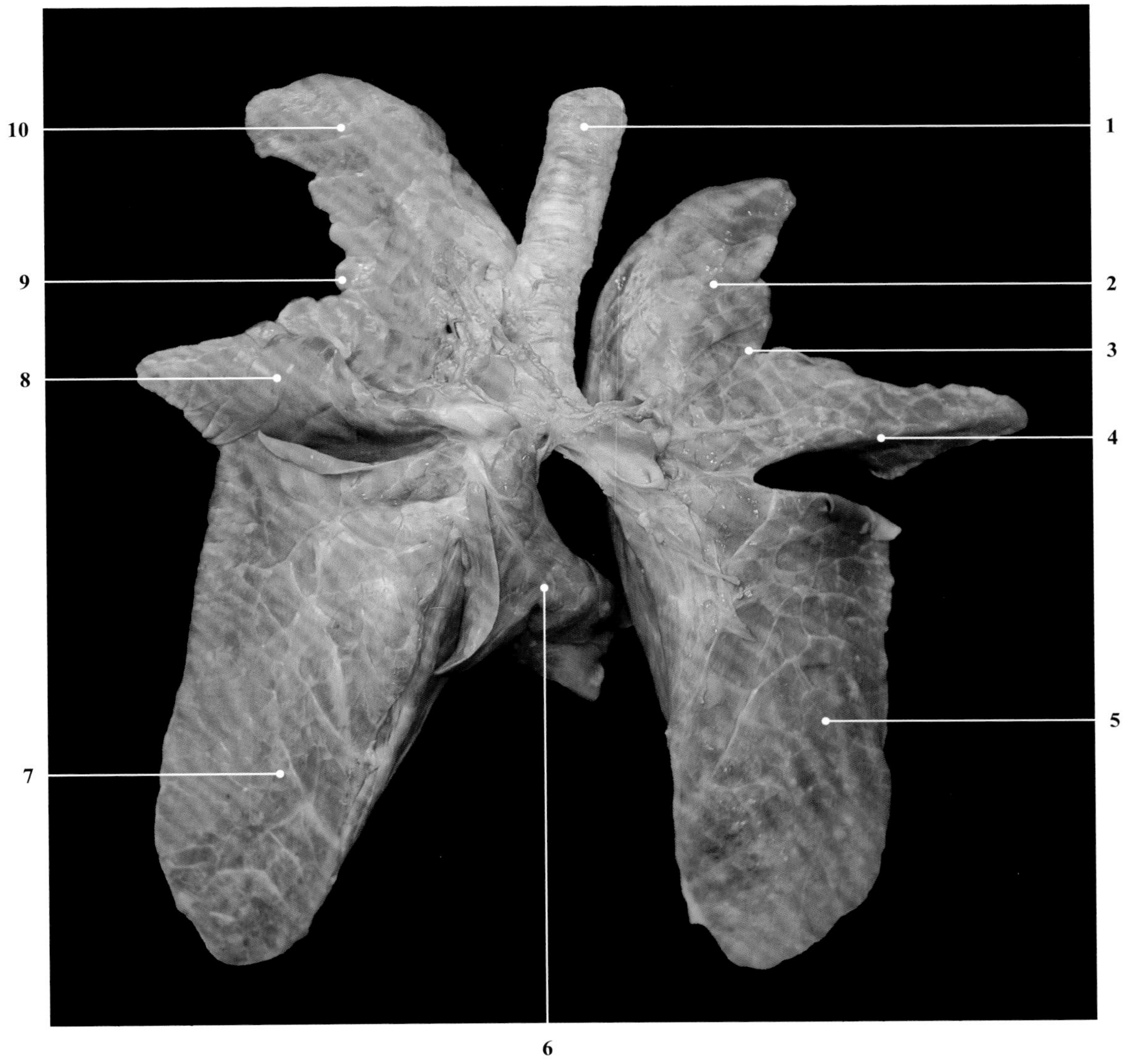

图 5-29 肺脏面观
Figure 5-29 Visceral view the lung.

1- 气管 trachea
2- 左肺前叶（尖叶） left cranial lobe (apical lobe) of the lung
3- 左心切迹 left cardiac notch
4- 左肺中叶（心叶） left middle lobe (cardiac lobe) of the lung
5- 左肺后叶（膈叶） left caudal lobe (diaphragmatic lobe) of the lung
6- 副叶 accessory lobe
7- 右肺后叶（膈叶） right caudal lobe (diaphragmatic lobe) of the lung
8- 右肺中叶（心叶） right middle lobe (cardiac lobe) of the lung
9- 右心切迹 right cardiac notch
10- 右肺前叶（尖叶） right cranial lobe (apical lobe) of the lung

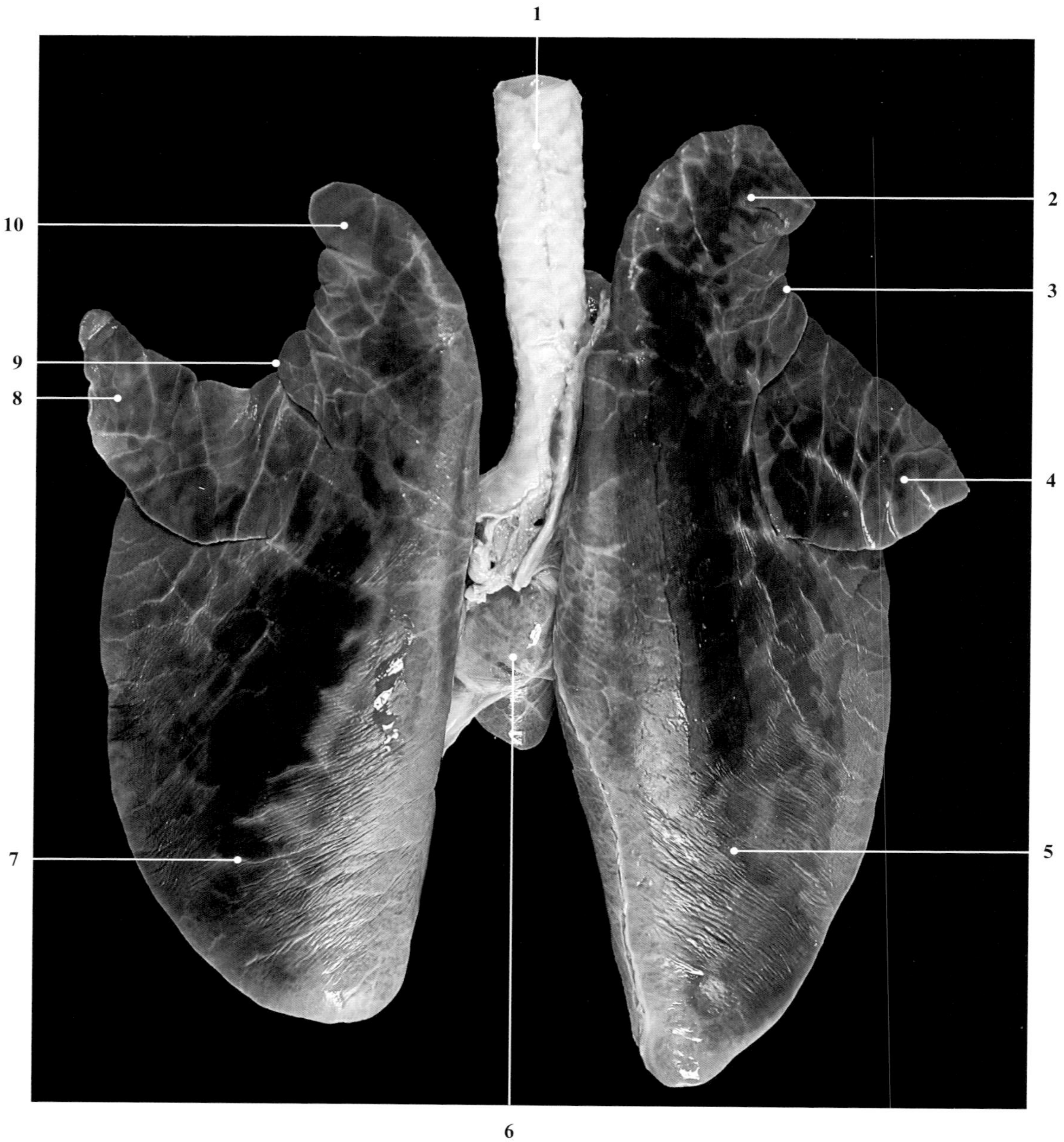

图 5-30　肺壁面观（固定标本）
Figure 5-30　Parietal view of the lung (fixed preparation).

1- 气管 trachea
2- 右肺前叶（尖叶） right cranial lobe (apical lobe) of the lung
3- 右心切迹 right cardiac notch
4- 右肺中叶（心叶） right middle lobe (cardiac lobe) of the lung
5- 右肺后叶（膈叶） right caudal lobe (diaphragmatic lobe) of the lung
6- 副叶 accessory lobe
7- 左肺后叶（膈叶） left caudal lobe (diaphragmatic lobe) of the lung
8- 左肺中叶（心叶） left middle lobe (cardiac lobe) of the lung
9- 左心切迹 left cardiac notch
10- 左肺前叶（尖叶） left cranial lobe (apical lobe) of the lung

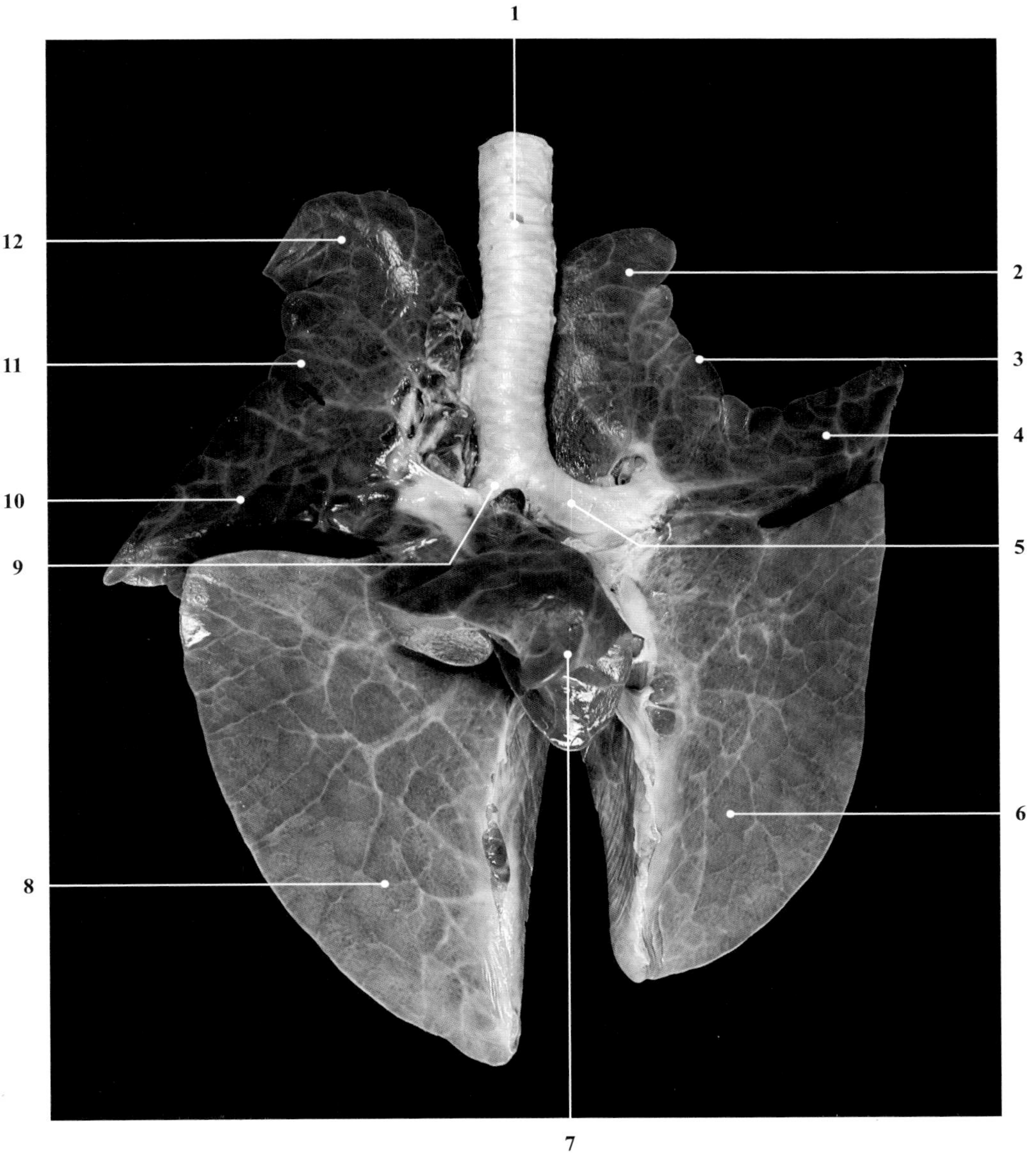

图 5-31 肺脏面观（固定标本）
Figure 5-31 Visceral view the lung (fixed preparation).

1- 气管 trachea
2- 左肺前叶（尖叶） left cranial lobe (apical lobe) of the lung
3- 左心切迹 left cardiac notch
4- 左肺中叶（心叶） left middle lobe (cardiac lobe) of the lung
5- 左支气管 left bronchus
6- 左肺后叶（膈叶） left caudal lobe (diaphragmatic lobe) of the lung
7- 副叶 accessory lobe
8- 右肺后叶（膈叶） right caudal lobe (diaphragmatic lobe) of the lung
9- 右支气管 light bronchus
10- 右肺中叶（心叶） right middle lobe (cardiac lobe) of the lung
11- 右心切迹 right cardiac notch
12- 右肺前叶（尖叶） right cranial lobe (apical lobe) of the lung

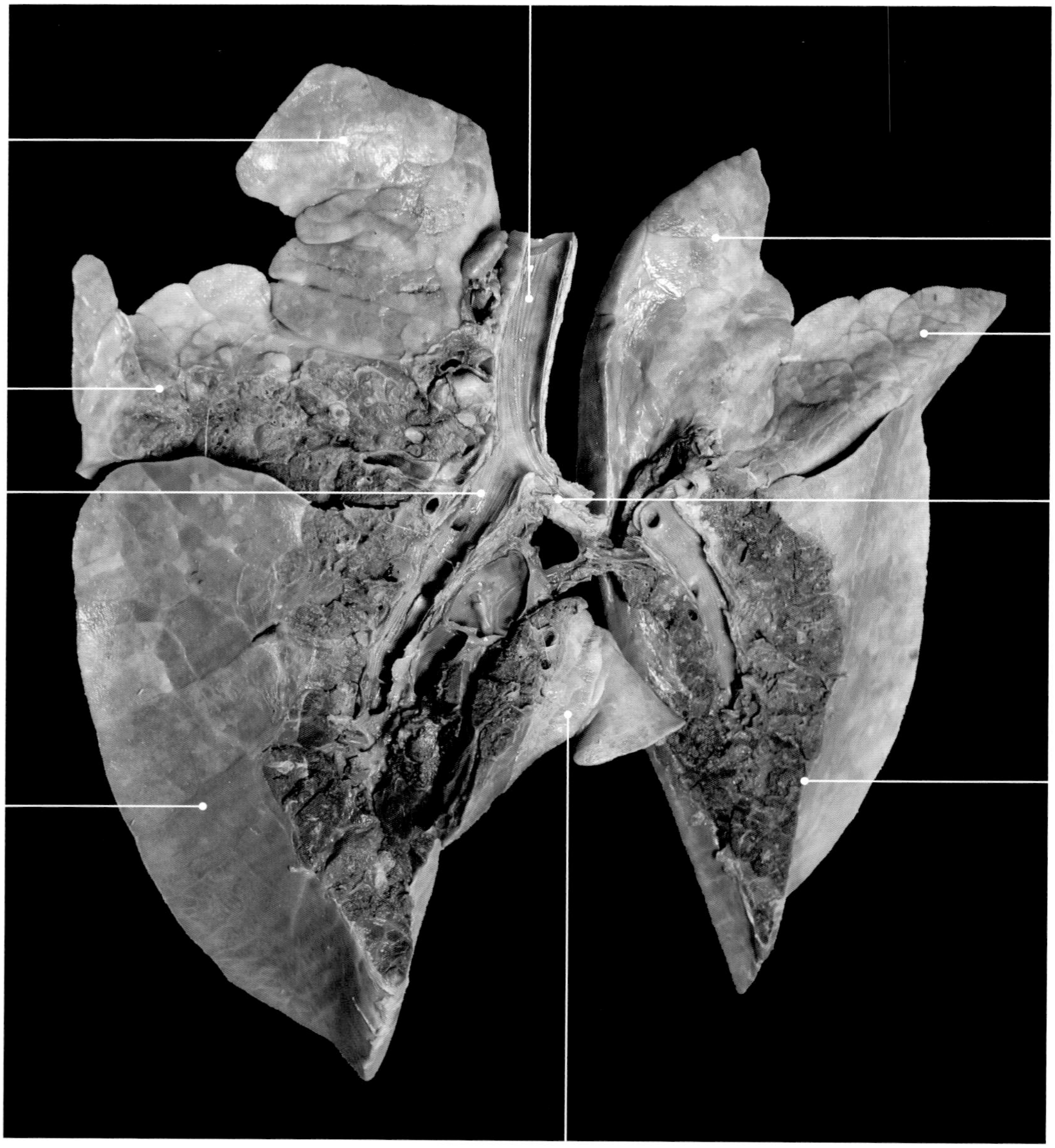

图 5-32　肺脏面切开（固定肺）
Figure 5-32　Visceral incision of the lung (fixed preparation).

1- 气管 trachea
2- 左肺前叶（尖叶） left cranial lobe (apical lobe) of the lung
3- 左肺中叶（心叶） left middle lobe (cardiac lobe) of the lung
4- 左支气管 left bronchus
5- 左肺后叶（膈叶） left caudal lobe (diaphragmatic lobe) of the lung
6- 副叶 accessory lobe
7- 右肺后叶（膈叶） right caudal lobe (diaphragmatic lobe) of the lung
8- 右支气管 light bronchus
9- 右肺中叶（心叶） right middle lobe (cardiac lobe) of the lung
10- 右肺前叶（尖叶） right cranial lobe (apical lobe) of the lung

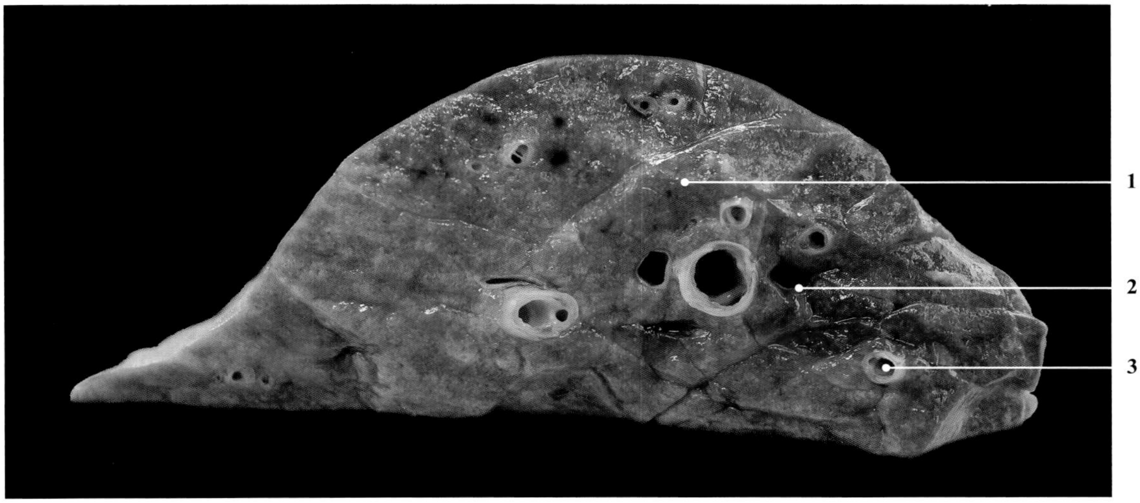

图 5-33 肺的切面
Figure 5-33 Cross section of the lung.

1- 肺实质 pulmonary parenchyma
2- 静脉 vein
3- 小支气管 small bronchiole

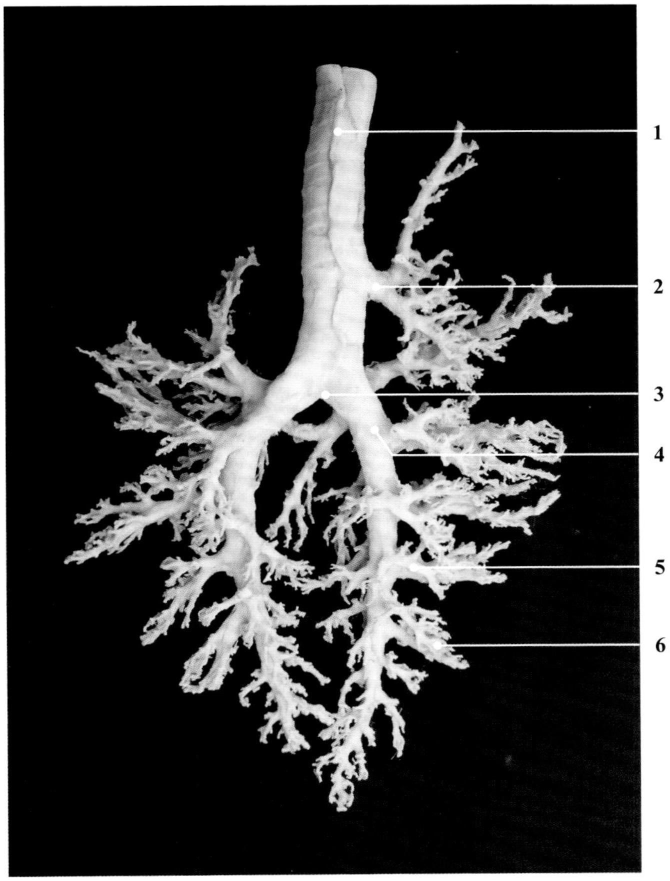

图 5-34 支气管树
Figure 5-34 Bronchial tree.

1- 气管 trachea
2- 右尖叶支气管 right apical lobar bronchus
3- 气管分叉处 divaricate site of trachea
4- 主支气管 primary bronchus
5- 肺叶支气管 lobar bronchus, secondary buonchus
6- 节段性支气管 segmental bronchus, tertiary bronchus

1- 支气管 bronchus
2- 终末细支气管 terminal bronchiole
3- 呼吸性细支气管 respiratory bronchiole
4- 肺泡管 alveolar duct
5- 肺泡囊 alveolar sac
6- 肺泡 pulmonary alveoli
7- 肺泡隔 alveolar septum
8- 软骨片 cartilage plates
9- 平滑肌 smooth muscle

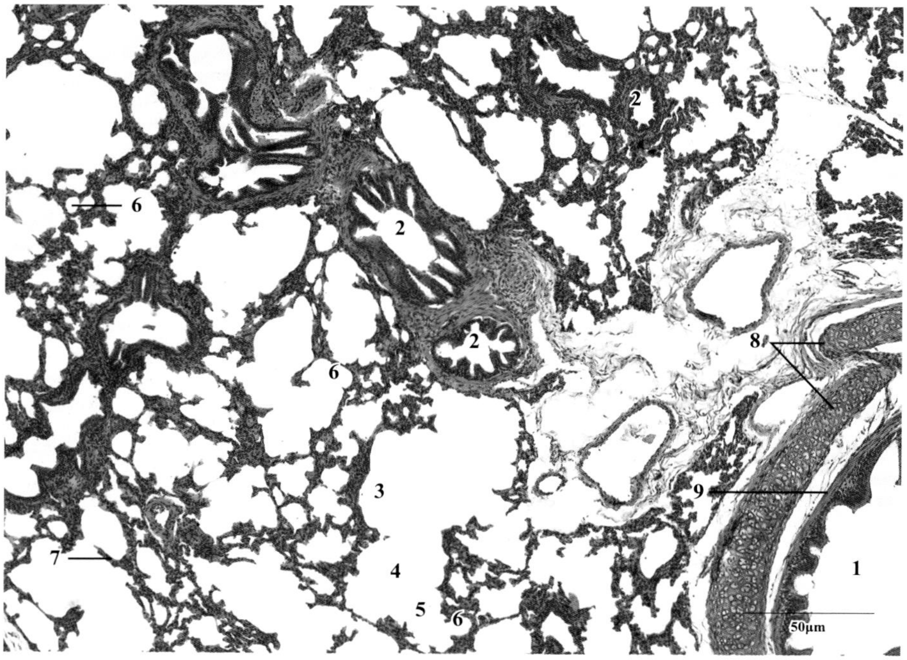

图 5-35 肺组织切片（HE 染色，低倍镜）
Figure 5-35 Histological section of the lung (HE staining, lower power).

1- 支气管 bronchus
2- 假复层纤毛柱状上皮 pseudostratified ciliated columnar epithelium
3- 固有层 lamina propria
4- 平滑肌 smooth muscle
5- 软骨片 cartilage plate
6- 细支气管 bronchiole
7- 肺泡 pulmonary alveoli
8- 静脉 vein

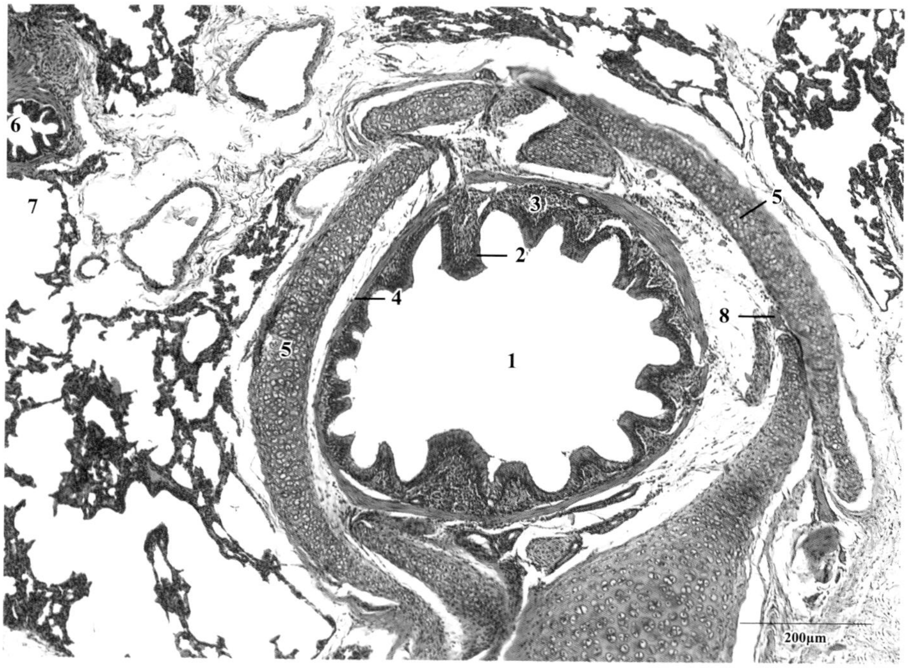

图 5-36 肺支气管组织切片（HE 染色，低倍镜）
Figure 5-36 Histological section of a bronchus in the lung (HE staining, lower power).

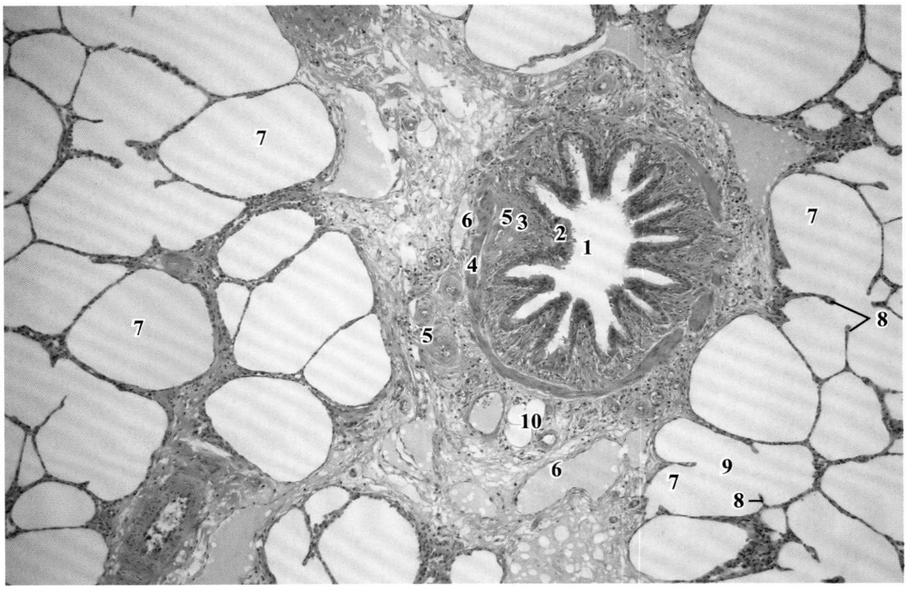

1- 细支气管 bronchiole
2- 假复层纤毛柱状上皮 pseudostratified ciliated columnar epithelium
3- 固有层 lamina propria
4- 平滑肌 smooth muscle
5- 动脉 artery
6- 静脉 vein
7- 肺泡 pulmonary alveoli
8- 肺泡隔 alveolar septum
9- 肺泡囊 alveolar sac
10- 脂肪细胞 adipocyte

图 5-37 细支气管组织切片（HE 染色，低倍镜）
Figure 5-37 Histological section of the bronchiole (HE staining, lower power).

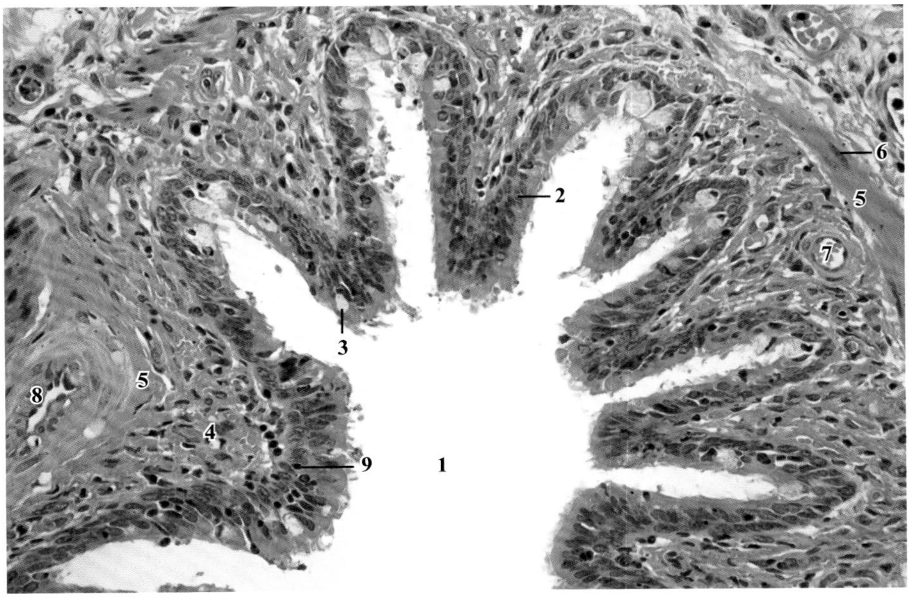

1- 细支气管管腔 lumen of the bronchiole
2- 假复层纤毛柱状上皮 pseudostratified ciliated columnar epithelium
3- 杯状细胞 goblet cell
4- 固有层 lamina propria
5- 平滑肌 smooth muscle
6- 平滑肌细胞核 nucleus of the smooth muscle cell
7- 静脉 vein
8- 动脉 artery
9- 淋巴细胞 lymphocyte

图 5-38 细支气管组织切片（HE 染色，高倍镜）
Figure 5-38 Histological section of the bronchiole (HE staining, higher power).

1- 肺泡 pulmonary alveoli
2- Ⅰ型肺泡细胞 type Ⅰ pneumocyte
3- Ⅱ型肺泡细胞 type Ⅱ pneumocyte
4- 肺泡巨噬细胞 / 尘细胞 pulmonary alveolar macrophage/dust cell
5- 红细胞 erythrocyte

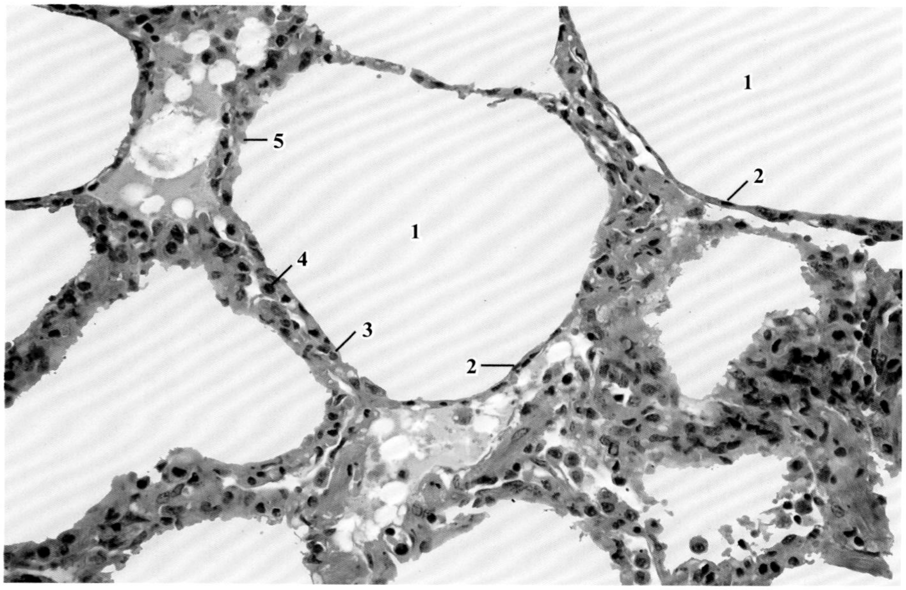

图 5-39　肺泡组织切片（HE 染色，高倍镜）
Figure 5-39　Histological section of the pulmonary alveoli (HE staining, higher power).

1- 肺泡 pulmonary alveoli
2- Ⅰ型肺泡细胞 type Ⅰ pneumocyte
3- Ⅱ型肺泡细胞 type Ⅱ pneumocyte
4- 肺泡巨噬细胞 / 尘细胞 pulmonary alveolar macrophage/dust cell
5- 胶原纤维 collagenous fiber
6- 弹性纤维 elastic fiber
7- 血管 blood vessel
8- 淋巴细胞 lymphocyte

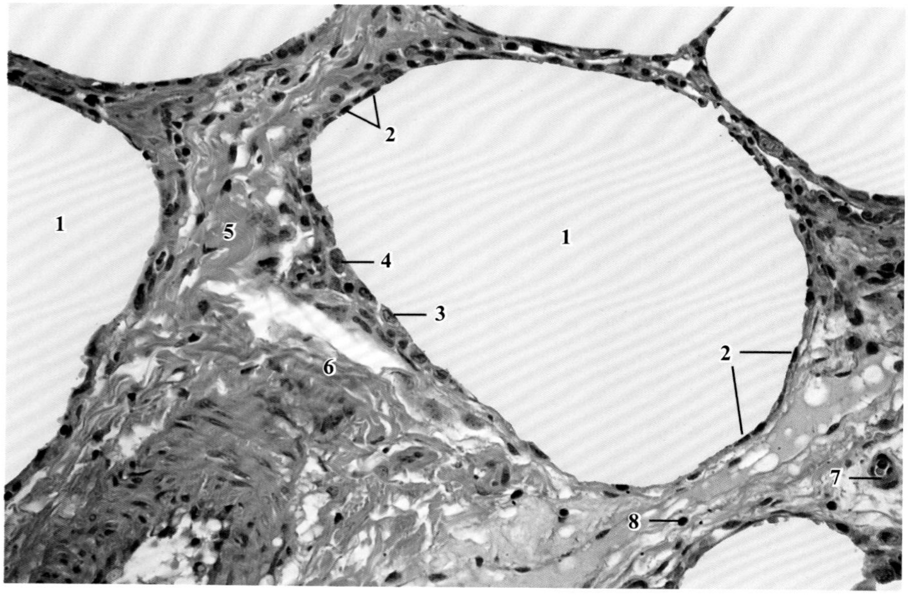

图 5-40　肺泡组织切片（HE 染色，高倍镜）
Figure 5-40　Histological section of the pulmonary alveoli (HE staining, higher power).

第六章

泌尿系统

Chapter 6

Urinary system

猪的泌尿系统（urinary system）包括肾、输尿管、膀胱和尿道，主要功能为排泄，维持机体内环境的相对稳定。

一、肾

肾（kidney）是生成尿的器官。猪体在新陈代谢过程中，产生的代谢产物（如尿素、尿酸等）和多余的水分，由血液带到肾，在肾内形成尿液，经排尿管道排出体外。肾除了排泄功能外，在维持机体水盐代谢、渗透压和酸碱平衡方面也起着重要作用。此外还具有内分泌功能，能产生多种生物活性物质如肾素、前列腺素等，对机体的某些生理功能起调节作用。

（一）肾的形态与位置

肾是成对的实质性器官，左右各一，呈长扁豆形，位于最后胸椎和前3个腰椎的腹侧，腹主动脉和后腔静脉的两侧。营养状况良好时，肾周围有脂肪包裹，叫肾脂囊（*renal adipose capsule*）。肾的内侧缘中部凹陷形成肾门（renal hilum），是输尿管、血管（肾动脉和肾静脉）、淋巴管和神经出入肾的地方。肾门深入肾内形成肾窦（renal sinus），是由肾实质围成的腔隙，以容纳肾盏和肾盂等。肾的表面包有一层薄而坚韧的纤维膜，称为纤维囊（fibrous capsule），亦称被膜。健康猪肾的纤维囊容易剥离。

肾的实质分为浅部的皮质和深部的髓质。肾皮质（renal cortex）因富含血管，故新鲜标本呈红褐色，切面上有许多细小颗粒状小体，叫肾小体（renal corpuscle）。肾髓质（renal medulla）颜色较浅，切面上可见许多纵向条纹，它是由许多肾小管构成。猪肾属于平滑多乳头肾。肾的皮质部完全合并，但髓质则是分开，肾乳头（renal papillae）单独存在。每个肾乳头与一个肾小盏相对，肾小盏汇入两个肾大盏，后者汇成肾盂，延接输尿管。

（二）肾的结构

肾实质主要是由肾单位和集合小管组成，其间有少量结缔组织和血管、神经等构成的肾间质。肾间质细胞呈星形，有突起，胞质内有较多脂滴，能合成髓脂Ⅰ，分泌后在肝中转化为髓脂Ⅱ，是一种血管舒张剂，可降低血压。肾小管周围的血管内皮细胞能产生红细胞生成素，刺激骨髓中红细胞生成。

1. 肾单位（nephron）：是肾的结构和功能单位，由肾小体和肾小管组成。肾小体为肾单位起始端，为膨大的小球，每个肾小体与一条长而粗细不等的肾小管相连。肾小管又依次分近端小管、细段和远端小管，其末端与集合小管相接。

根据肾小体在皮质中深浅位置不同，可将肾单位分为浅表肾单位和髓旁肾单位。浅表肾单位（superfacial nephron）位于皮质浅层，又称皮质肾单位（cortical nephron），体积较小，髓襻和细段较短，数量多（占肾单位总数的80%），在尿液的形成过程中起重要作用。髓旁肾单位（juxtamedullary nephron）位于皮质深部，体积较大，髓襻和细段较长，数量少（占肾单位总数的20%），在尿液的浓缩过程中起重要作用。

（1）肾小体（renal corpuscle）：由血管球和肾小囊组成，形似球形，直径约200 μm。肾小体具有两极，微动脉出入的一端称血管极，另一端位于血管极的对侧，称尿极，其与近端小管相连。

血管球（glomus），旧称肾小球（renal glomerulus），为盘曲成球状的一团动脉性毛细血管网，入球微动脉发出分支形成襻状毛细血管后汇聚成出球微动脉。毛细血管为孔径50～100 nm的有孔型毛细血管，入球微动脉粗，出球微动脉细，毛细血管之间有血管系膜（mesangium）支撑。血管系膜（球内系膜 intraglomerular）由球内系膜细胞和系膜基质构成。球内系膜细胞（intraglomerular mesangial cell）呈星型、多突起，突起可伸至内皮细胞与基膜之间，胞核小，染色深，其功能有合成基膜和系膜基质的成分；吞噬、降解沉积在基膜上的免疫复合物，维持基膜通透性和参与基膜的更新和修复；收缩功能，调节毛细血管管径，影响血管球内的血流量；分泌肾素和酶等生物活性物质。血管球基膜（glomerular basement membrane）位于足细胞突起与毛细血管内皮细胞之间，或足细胞突起与血管系膜之间，较厚，为均质状，呈PAS反应阳性，内含Ⅳ型胶原蛋白、蛋白多糖、层粘连蛋白等，构成以蛋白质为骨架的分子筛，参与组成滤过屏障。

肾小囊（renal capsule）又称鲍曼囊（Bowman's capsule），是肾小管起始部膨大并凹陷而成的双层杯状囊。外层（壁层）由单层扁平上皮组成，与近端小管相连。内层（脏层）由足细胞组成，包在血管球毛细血管外。两层之间为肾小囊腔。足细胞（podocyte）的胞体大，核染色深，胞体伸出初级突起及次级突起，包裹在毛细血管外。相邻细胞突起相互交叉呈指状镶嵌形成栅栏状，紧贴在毛细血管基膜外突起之间有直径约25 nm的裂隙称裂孔（slit pore），孔上覆盖一层厚4～6 nm的裂孔膜（slit membrane）。

肾小体以滤过方式形成滤液。血管球毛细血管内的血浆成分滤入肾小囊必须经过滤过屏障。滤过屏障（filtration barrier）又称滤过膜（filtration membrance），由毛细血管有孔内皮、血管球基膜和足细胞裂孔膜组成。其功能为滤过血液形成原尿。原尿成分与血浆类似，即除去血细胞及其他大分子物质。滤过动力源自入球微动脉粗于出球微动脉。相对分子质量70000以下的物质可通过滤过膜，而相对分子质量69000的白蛋白可少量通过，相对分子质量在150000～200000的免疫球蛋白则停滞在基膜内不能通过。但是，当滤过屏障受损，一些大分子蛋白和血细胞也可能漏出，导致蛋白尿或血尿。

（2）肾小管（renal tubule）：为单层上皮细胞所围成的管道，分为近端小管、细部和远端小管三段。

1）近端小管（proximal tubule）：较粗较长，分为近端小管曲部和近端小管直部。近端小管曲部（proximal convoluted tubule）又称近曲小管，位于皮质，起始于肾小体尿极，蟠曲行走在所属的肾小体附近而后进入髓放线。腔小、不规则，管径为50～60 μm。细胞为单层立方或锥体形，细胞体积较大，界限不清，胞核大而圆，染色较浅，位于基底部，细胞质嗜酸性，细胞游离面有刷状缘，基底面有基底纵纹。近端小管直部直行于髓放线和肾锥体内，结构与曲部类似，上皮高度较低。微绒毛、指状交叉侧突及质膜内褶不发达，细胞器少。

2）细段（thin segment）：位于髓放线和肾锥体内。管径细，直径10～15 μm，由单层扁平上皮细胞构成，胞核凸向腔内，胞质染色浅，无刷状缘。

3）远端小管（distal tubule）：光镜下上皮细胞为单层立方形，细胞界限较清楚，无刷状缘，基底纵纹较明显，胞质弱嗜酸性，染色浅，胞核圆形，在小管横截面上常见较多的胞核。

肾小管的功能包括选择性重吸收作用、分泌和排泄作用。近端小管主要吸收100%的葡萄糖、氨基酸和蛋白质，85%的钠、水，分泌氢、氨、肌酐及马尿酸等。排泄钾及外源物（青霉素、酚红等）。细段利于水和离子的通透。远端小管吸收水、钠；排泄钾、氢及氨等。

2. 集合小管（collecting tubule）：分弓形集合小管、直集合小管和乳头管三段。细胞染色浅，界限清晰，胞核居中，上皮细胞形状依次为立方形、柱状和高柱状。其功能为在肾上腺醛固酮和垂体抗利尿激素调节下重吸收水、分泌H^+和HCO_3^-。

3. 球旁复合体（juxtaglomerular complex）：位于肾小体血管极处，由球旁细胞、致密斑和球外系膜细胞组成。

（1）球旁细胞（juxtaglomerular cell）：为入球微动脉近血管极处中膜平滑肌细胞转变为上皮样并有内分泌功能的细胞。细胞大，呈立方形；胞核大而圆；胞质弱嗜碱性，富含PAS^+分泌颗粒。其作用为分泌肾素（蛋白水解酶），使血液中的血管紧张素原转变成血管紧张素Ⅰ和Ⅱ，一方面使血管收缩，血压增高；另一方面使肾上腺分泌醛固酮增加，肾小管排K^+保Na^+功能增强，肾小管钠水重吸收增加，血压增高。

（2）致密斑（macula densa）：为远端小管靠近肾小体侧的上皮细胞增高变窄形成的椭圆形斑状隆起。细胞高柱状，胞质色浅，胞核椭圆，位于细胞顶部。基膜不完整，基部小突起可与邻近细胞突起镶嵌，而成为传递“信息”的场所。致密斑为离子感受器，能够感受小管内的钠离子浓度，调节球旁细胞分泌肾素。

（3）球外系膜细胞（extraglomerular mesangial cell）：位于血管极三角区内的一群形态不规则的细胞，与球内系膜延续，与球旁细胞、球内系膜细胞之间有缝隙连接，在复合体中起信息传递作用。

二、输尿管

输尿管（ureter）是把肾脏生成的尿液输送到膀胱的细长管道，左右各一条，起于肾盂，出肾门后，沿腹腔顶壁向后伸延。左侧输尿管在腹主动脉的外侧，右侧输尿管在后腔静脉的外侧，横过髂内动脉的腹侧面进入骨盆腔，向后伸达膀胱颈的背侧，斜向穿入膀胱壁。

输尿管管壁分黏膜、肌层和外膜。黏膜形成许多皱襞，上皮为变移上皮，固有层为疏松结缔组织。肌层包括内、外两层纵行的平滑肌和中层为环形的平滑肌。

三、膀胱

膀胱（urinary bladder）可分为膀胱顶（膀胱尖 apex of bladder）、膀胱体（body of bladder）和膀胱颈（neck of bladder）三部分。输尿管在膀胱壁内斜向延伸一段距离，在靠近膀胱颈的部位开口于膀胱背侧壁。这种结构特点可防止尿液逆流。膀胱颈延接尿道。在膀胱两侧与盆腔侧壁之间有膀胱侧韧带。在膀胱侧韧带的游离缘有一圆索状物，称为膀胱圆韧带（ligamentum teres vesicae），是胎儿时期脐动脉的遗迹。

猪的膀胱比较大，但随着贮存尿液量的不同，膀胱的形状、大小和位置均有变化。膀胱空虚时，呈梨状，位于骨盆腔内。充满尿液时，大部分突入腹腔内。

膀胱是一个囊形储尿器官，管壁分黏膜、肌层和外膜。黏膜形成许多不规则皱襞，上皮为变移上皮，其厚度随膀胱充盈度而异。固有层由富含弹性纤维的疏松结缔组织组成。黏膜肌层很薄。肌层发达，但分层不规则，一般包括内、外两层纵行的平滑肌和中层为环形的平滑肌，以中层最厚，在膀胱与尿道的交界处有括约肌，可以控制尿液的排出。

四、尿道

尿道（urethra）前端以尿道内口连接膀胱，后端以尿道外口连于体外，将尿排出体外。公猪尿道也可将精液导出，因此也是生殖器官之一，称为尿生殖道，分为尿生殖道骨盆部和尿生殖道阴茎部。尿生殖道在阴囊前方形成乙状弯曲。母猪尿道较短，前端以尿道内口连于膀胱，后端开口于阴道前庭起始部腹侧壁，为尿道外口。

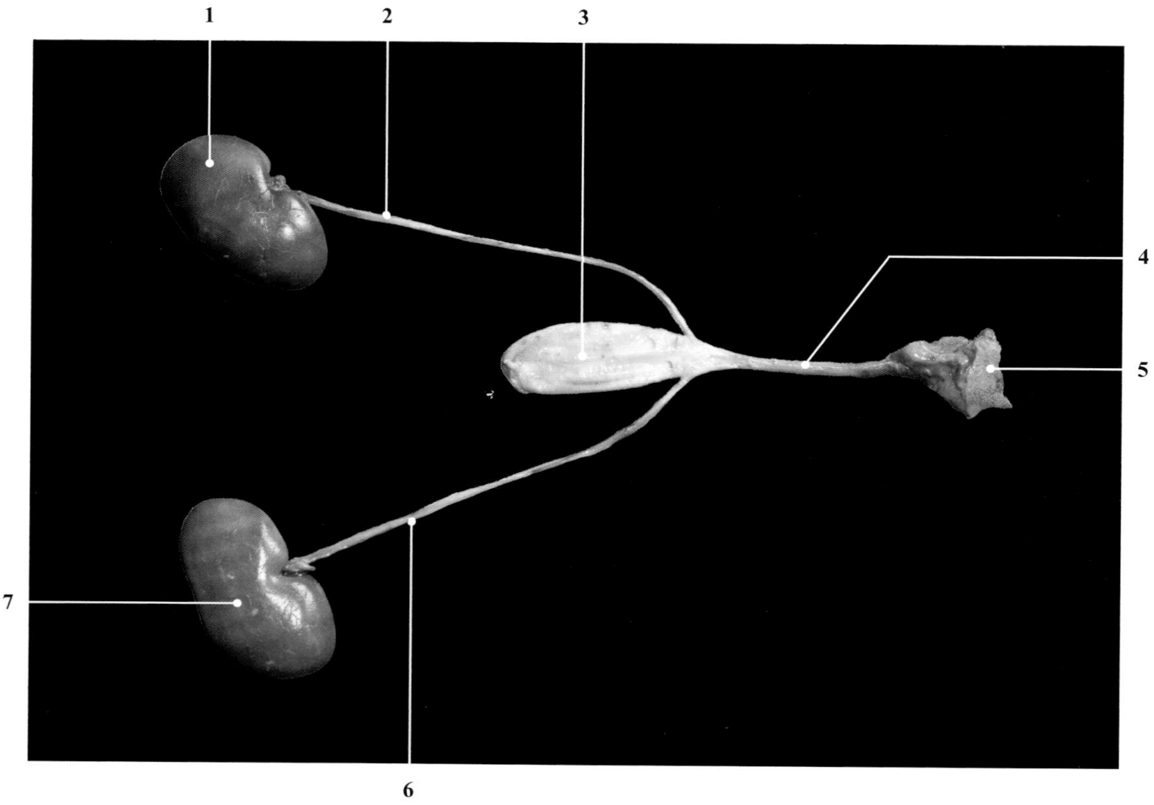

图 6-1 母猪泌尿系统
Figure 6-1 Urinary system of the sow.

1- 右肾 right kidney
2- 右输尿管 right ureter
3- 膀胱 urinary bladder
4- 尿道 urethra
5- 阴门 vulva
6- 左输尿管 left ureter
7- 左肾 left kidney

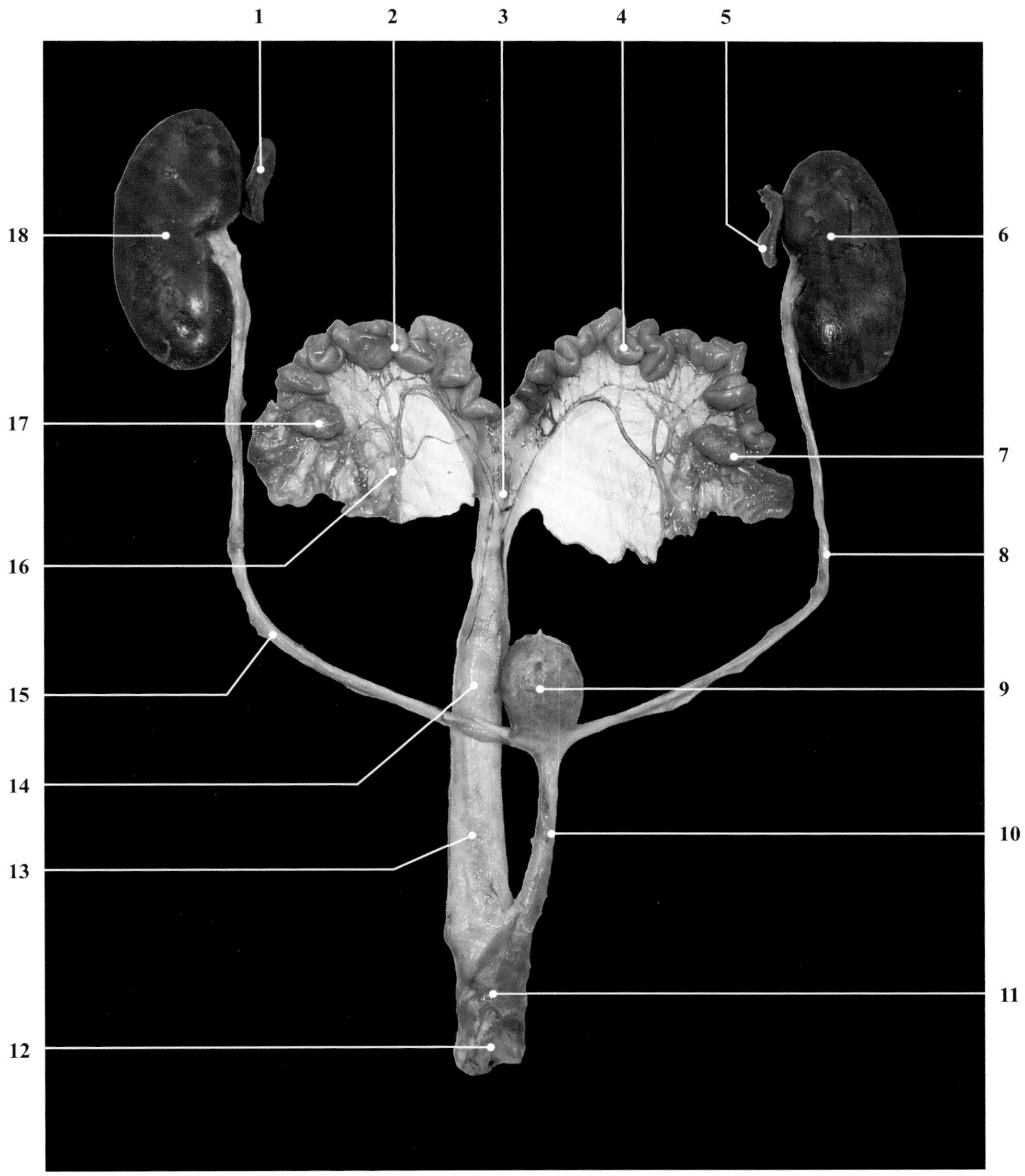

图 6-2 母猪泌尿生殖器官
Figure 6-2 Urogenital organs of the sow.

1- 右侧肾上腺 right adrenal gland
2- 右侧子宫角 right uterine horn
3- 子宫体 uterine body
4- 左侧子宫角 left uterine horn
5- 左侧肾上腺 left adrenal gland
6- 左肾 left kidney
7- 左侧卵巢 left ovary
8- 左侧输尿管 left ureter
9- 膀胱 urinary bladder
10- 尿道 urethra
11- 阴道前庭 vaginal vestibulum
12- 阴门 vulva
13- 阴道 vagina
14- 子宫颈 uterine cervix
15- 右侧输尿管 right ureter
16- 子宫阔韧带 broad ligament of uterus
17- 右侧卵巢 right ovary
18- 右肾 right kidney

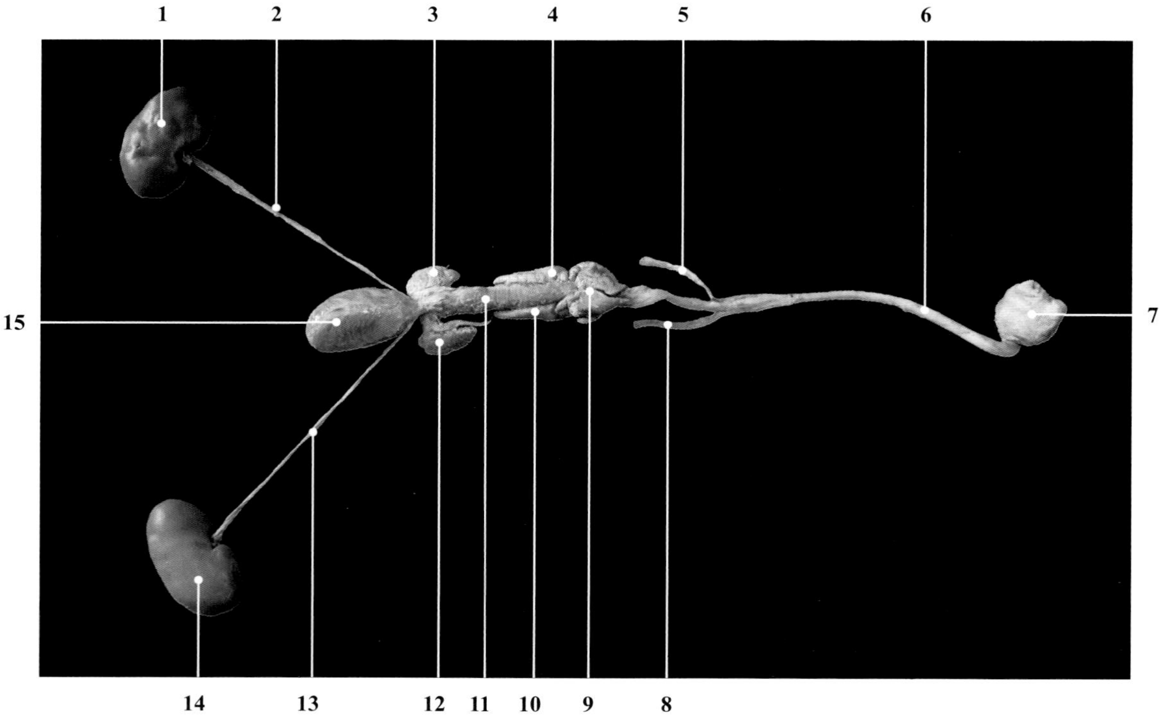

图 6-3 公猪泌尿生殖器官（固定标本）
Figure 6-3 Urogenital organs of the boar (fixed preparation).

1- 右肾 right kidney
2- 右侧输尿管 right ureter
3- 右侧精囊腺 right vesicular gland
4- 右侧尿道球腺 right bulbourethral gland
5- 右侧阴茎缩肌 right retractor penis muscle
6- 阴茎 penis
7- 包皮憩室（包皮盲囊）preputial diverticulum
8- 左侧阴茎缩肌 left retractor penis muscle
9- 阴茎脚 crus of penis
10- 左侧尿道球腺 left bulbourethral gland
11- 尿道骨盆部 pelvis part of urethra
12- 左侧精囊腺 left vesicular gland
13- 左侧输尿管 left ureter
14- 左肾 left kidney
15- 膀胱 urinary bladder

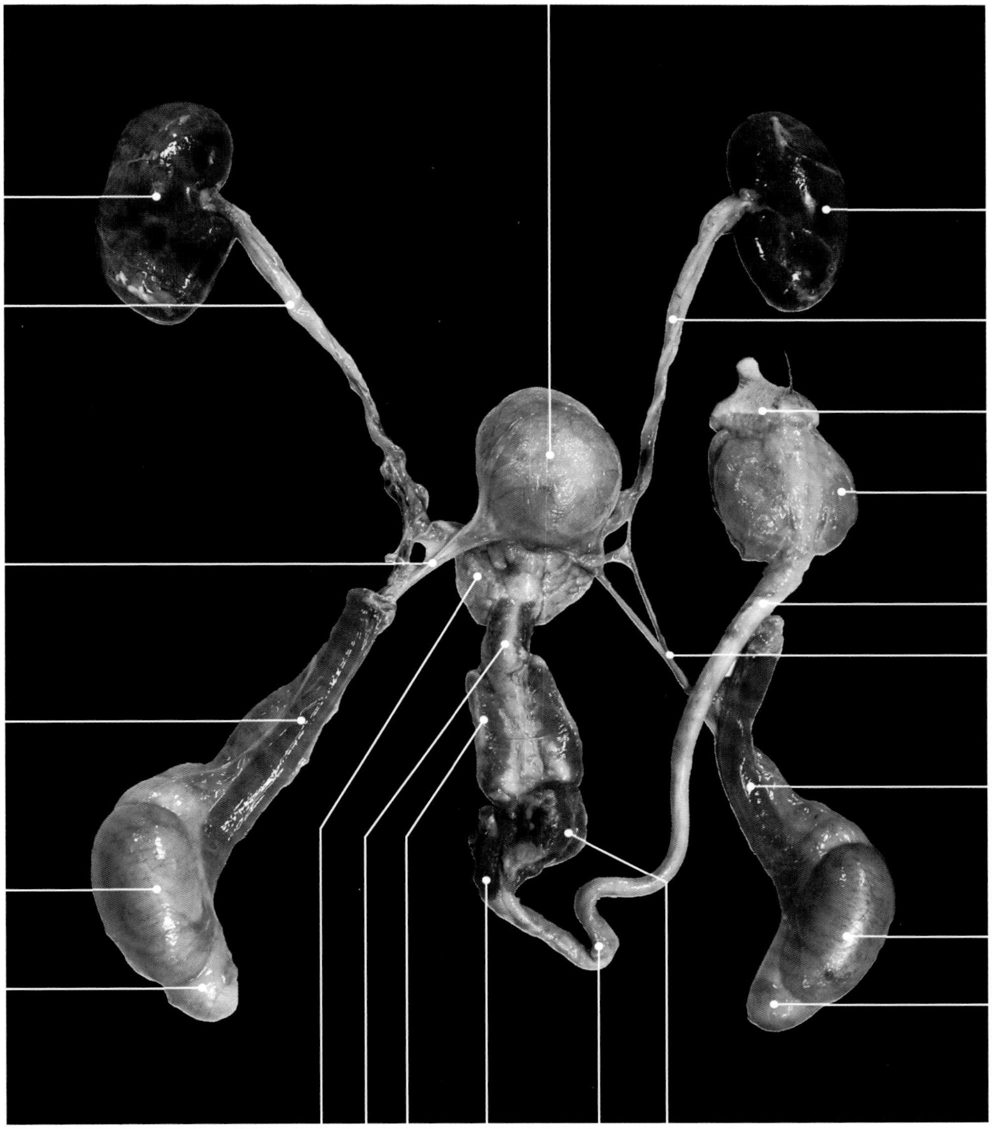

图 6-4 公猪泌尿生殖器官
Figure 6-4 Urogenital organs of the boar.

1- 膀胱 urinary bladder
2- 右肾 right kidney
3- 右侧输尿管 right ureter
4- 包皮 prepuce
5- 包皮憩室（包皮盲囊）preputial diverticulum
6- 阴茎 penis
7- 右侧输精管 right deferent duct
8- 右侧精索 right spermatic cord
9- 右侧睾丸 right testis
10- 右侧附睾 right epididymis
11- 阴茎球 penis bulb
12- 阴茎乙状弯曲 sigmoid flexure of the penis
13- 阴茎脚 crus of penis
14- 尿道球腺 bulbourethral gland
15- 尿道骨盆部 pelvis part of urethra
16- 精囊腺 vesicular gland
17- 左侧附睾 left epididymis
18- 左侧睾丸 left testis
19- 左侧精索 left spermatic cord
20- 左侧输精管 left deferent duct
21- 左侧输尿管 left ureter
22- 左肾 left kidney

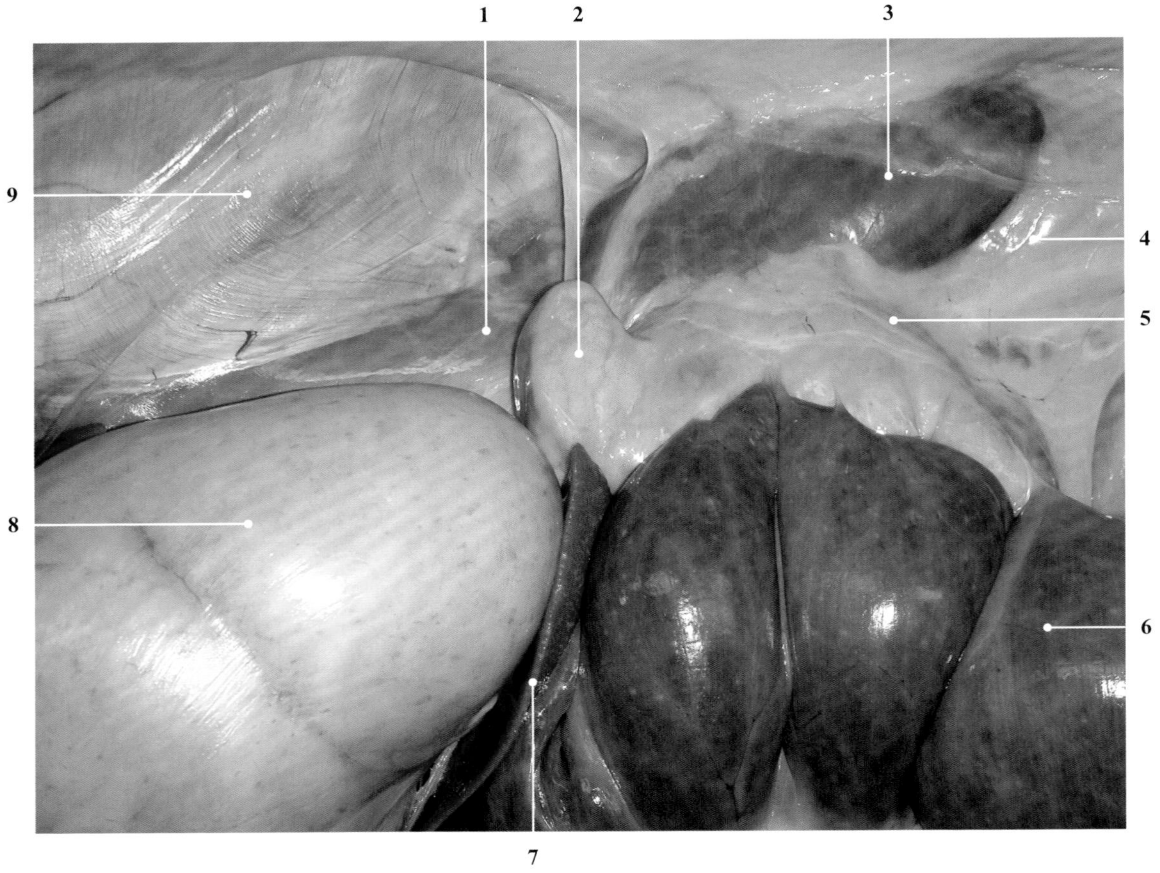

图 6-5　肾脂囊
Figure 6-5　Renal adipose capsule.

1- 膈脚 crus of diaphragm
2- 胰 pancreas
3- 肾 kidney
4- 肾脂囊 *renal adipose capsule*
5- 输尿管 ureter
6- 大肠 large intestine
7- 脾 spleen
8- 胃 stomach
9- 膈中心腱 central tendon of diaphragm

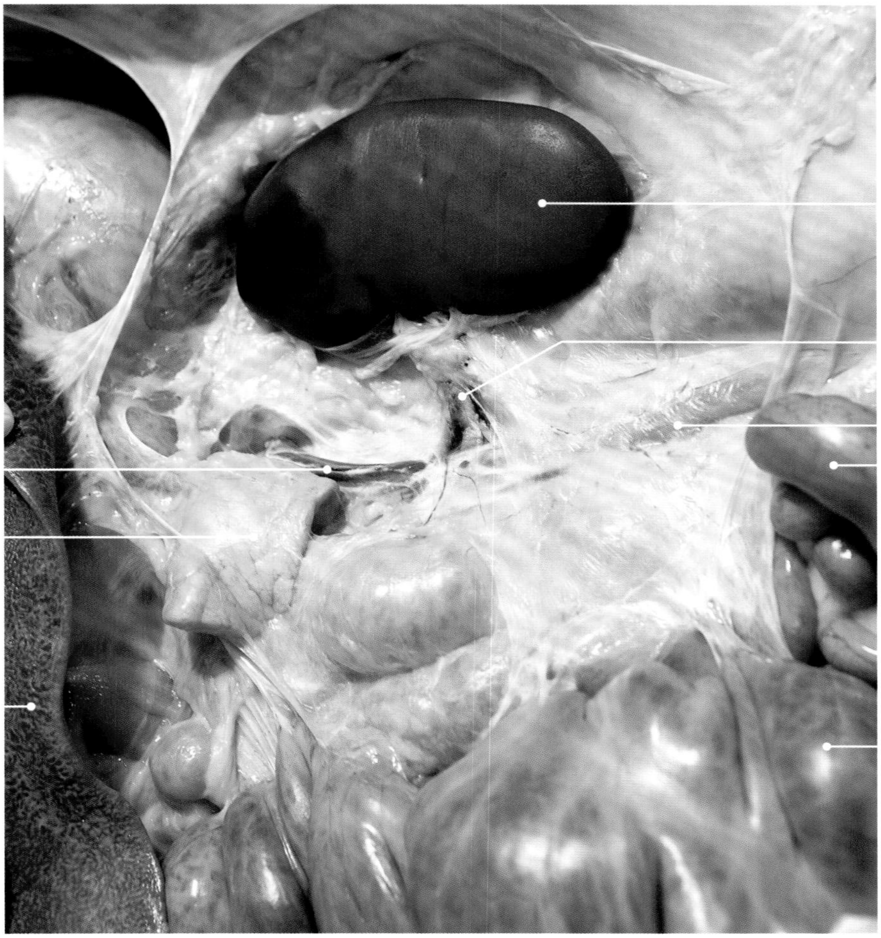

图 6-6 肾的位置
Figure 6-6 Location of the kidney.

1- 肾 kidney
2- 肾动、静脉 renal artery and vein
3- 腹主动脉 abdominal aorta
4- 小肠 small intestine
5- 大肠 large intestine
6- 脾 spleen
7- 胰 pancreas
8- 肾上腺 adrenal gland

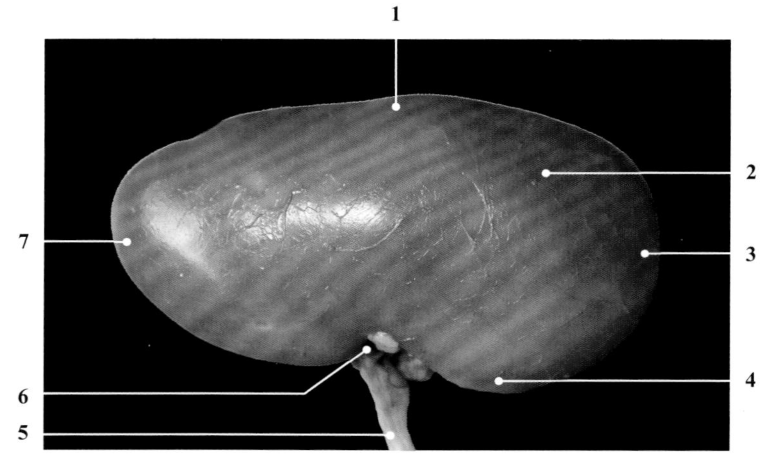

图 6-7 肾被膜
Figure 6-7 Kidney capsule.

1- 外侧缘 lateral border
2- 被膜 capsule
3- 前端 cranial extremity
4- 内侧缘 medial border
5- 输尿管 ureter
6- 肾门 renal hilum
7- 后端 caudal extremity

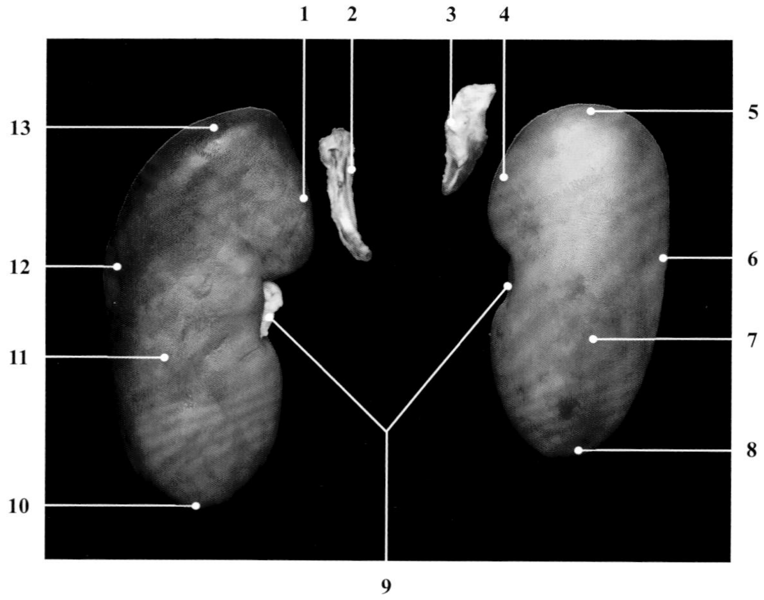

图 6-8 肾的外形
Figure 6-8 Shape of the kidney.

1- 左肾内侧缘 medial border of left kidney
2- 左肾上腺 left adrenal gland
3- 右肾上腺 right adrenal gland
4- 右肾内侧缘 medial border of right kidney
5- 右肾前端 cranial extremity of right kidney
6- 右肾外侧缘 lateral border of right kidney
7- 右肾实质 right renal parenchyma
8- 右肾后端 caudal extremity of right kidney
9- 肾门 renal hilum
10- 左肾后端 caudal extremity of left kidney
11- 左肾实质 left renal parenchyma
12- 左肾外侧缘 lateral border of left kidney
13- 左肾前端 cranial extremity of left kidney

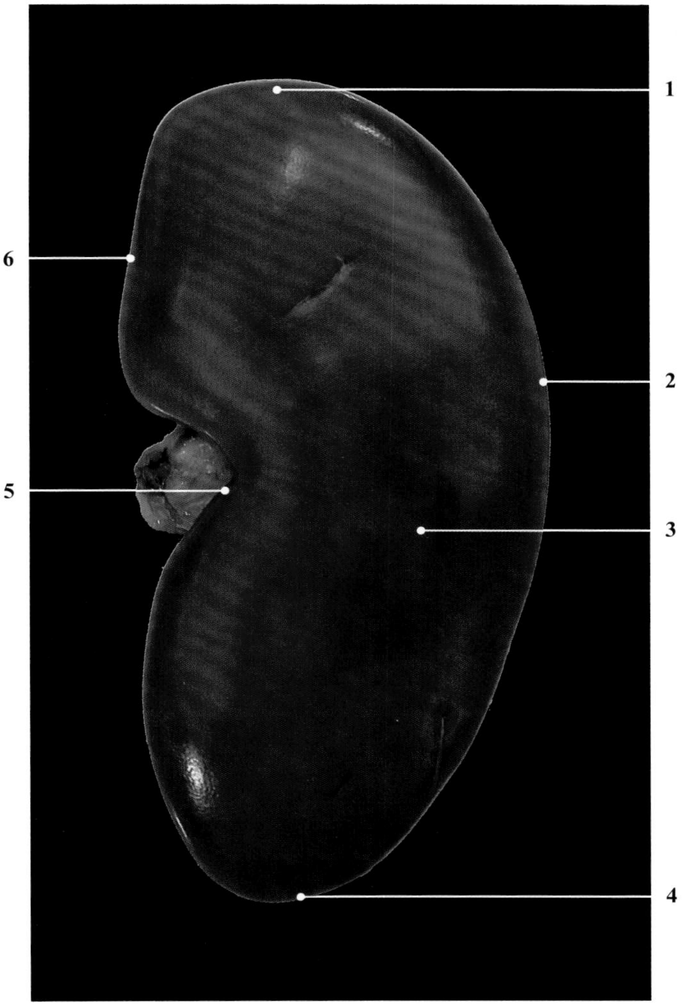

图 6-9　肾表面结构（除去被膜）
Figure 6-9　Surface structure of the kidney, renal capsule removed.

1- 前端 cranial extremity
2- 外侧缘 lateral border
3- 实质 parenchyma
4- 后端 caudal extremity
5- 肾门 renal hilum
6- 内侧缘 medial border

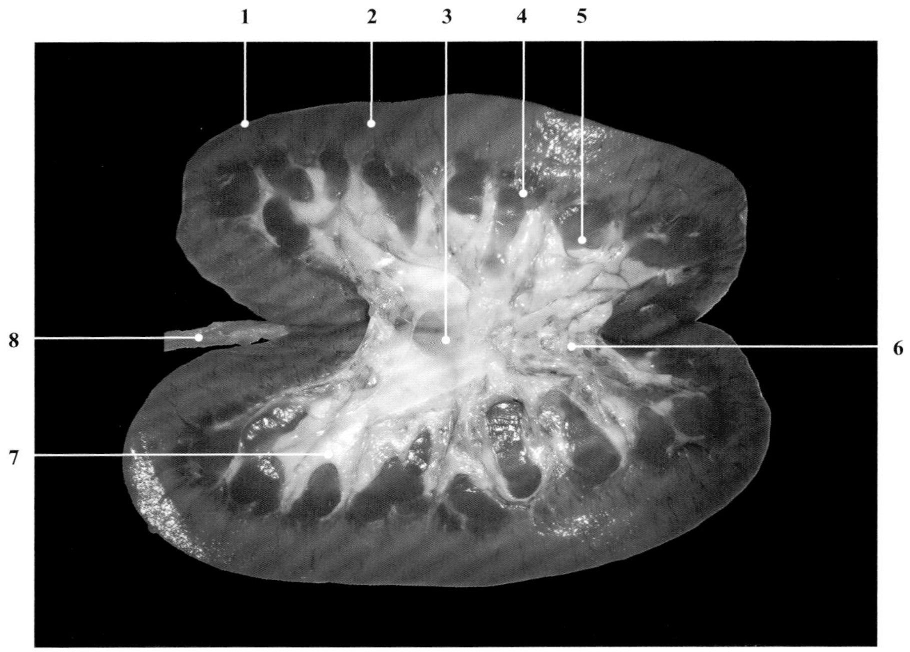

图 6-10　母猪肾剖面观
Figure 6-10　Cross section of the sow kidney.

1- 被膜 capsule
2- 肾皮质 renal cortex
3- 肾盂 renal pelvis
4- 肾髓质 renal medulla
5- 肾乳头 renal papillae
6- 肾窦 renal sinus
7- 肾盏 renal calices
8- 输尿管 ureter

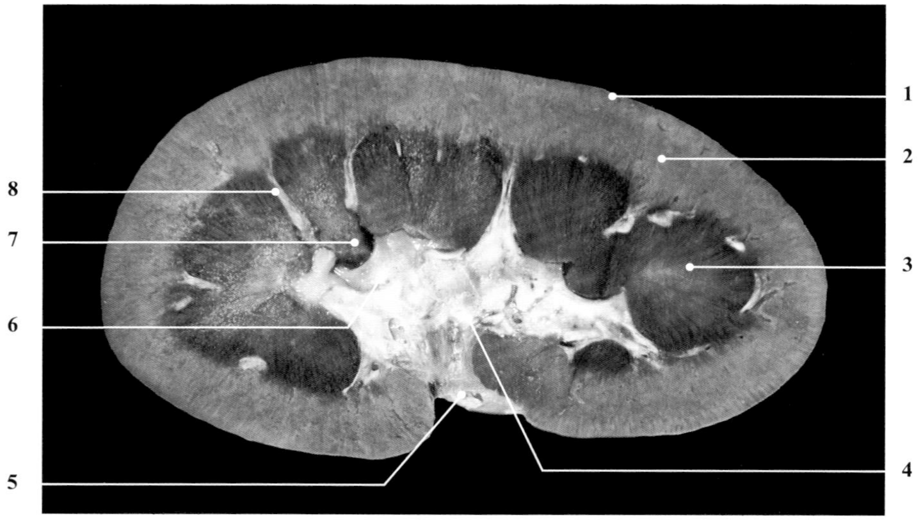

图 6-11　公猪肾剖面观
Figure 6-11　Cross section of the boar kidney.

1- 被膜 capsule
2- 肾皮质 renal cortex
3- 肾髓质 renal medulla
4- 肾窦 renal sinus
5- 肾门 renal hilum
6- 肾盏 renal calices
7- 肾乳头 renal papillae
8- 肾柱 renal column

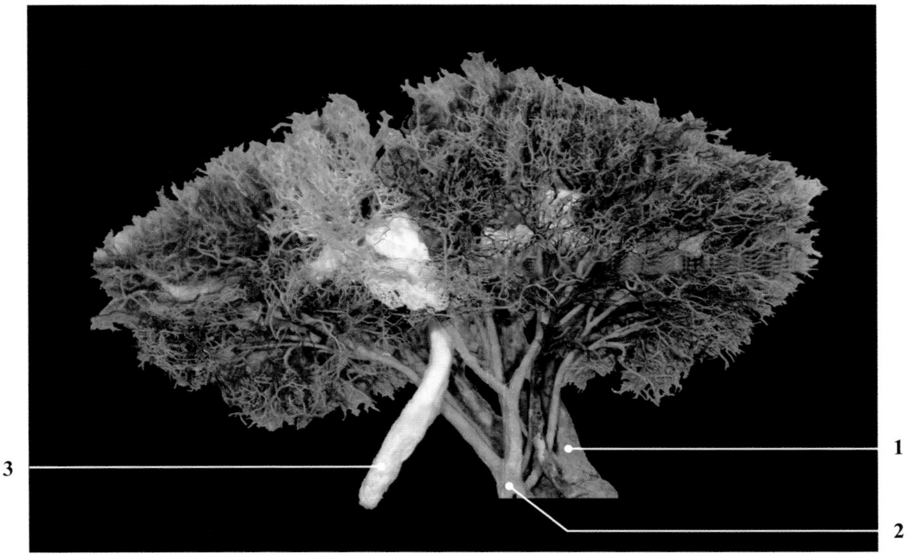

图 6-12 肾铸型
Figure 6-12 Kidney cast.

1- 肾静脉 renal vein
2- 肾动脉 renal artery
3- 输尿管 ureter

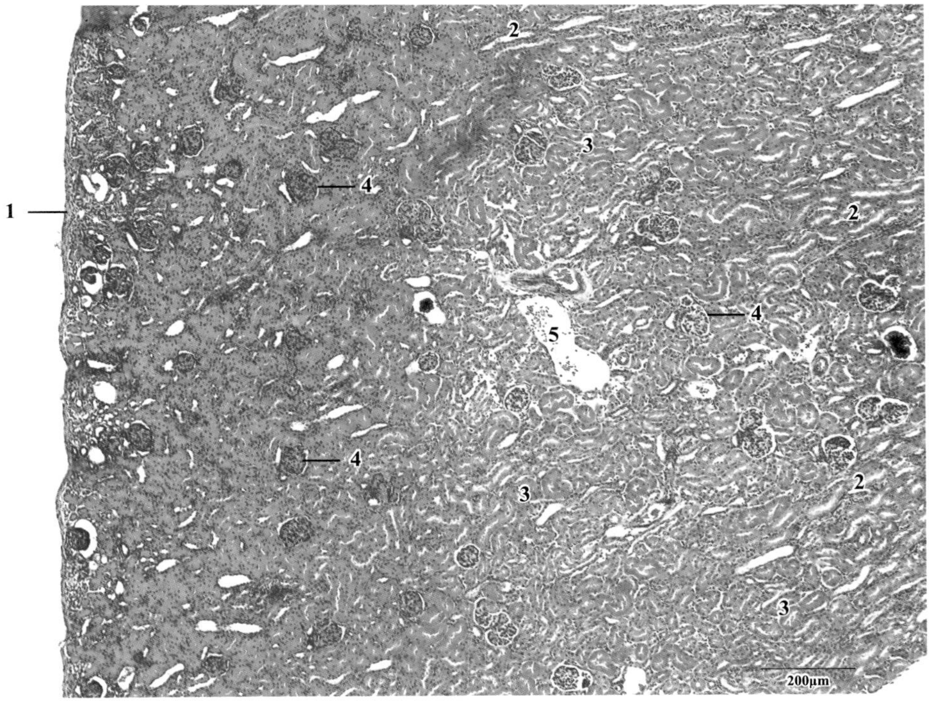

图 6-13 肾皮质组织切片（HE 染色，低倍镜）
Figure 6-13 Histological section of the renal cortex (HE staining, lower power).

1- 被膜 capsule
2- 髓放线 medullary ray
3- 皮质迷路 cortical labyrinth
4- 肾小体 renal corpuscle
5- 静脉 vein

1- 血管球 glomerulus
2- 肾小囊腔（鲍曼腔）capsular space (Bowman's space)
3- 肾小囊壁层 parietal layer of renal capsule
4- 血管极 vascular pole
5- 致密斑 macula densa
6- 近端小管 proximal tubule
7- 远端小管 distal tubule
8- 细段 thin segment
9- 动脉 artery

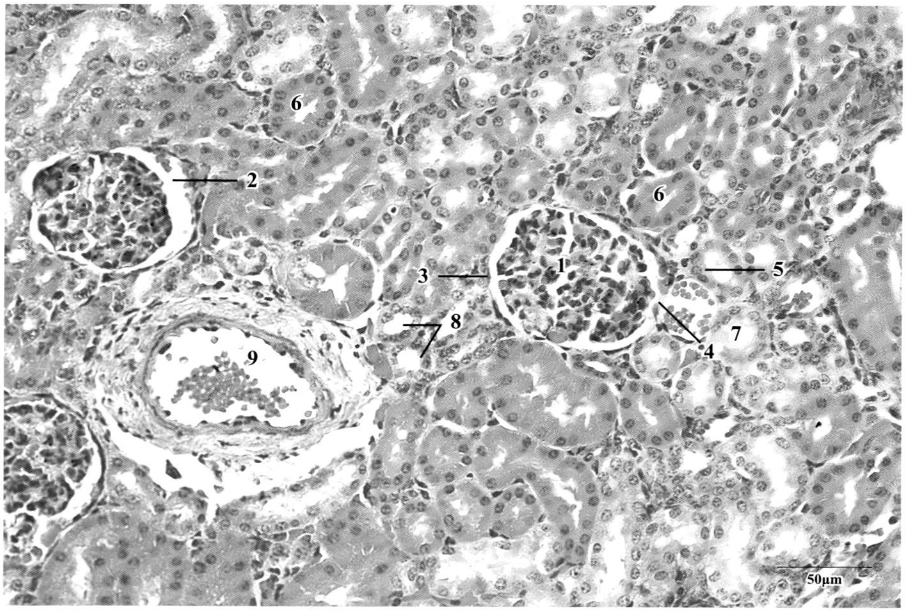

图 6-14　肾皮质组织切片（HE 染色，高倍镜）
Figure 6-14　Histological section of the renal cortex (HE staining, higher power).

1- 血管球 glomerulus
2- 肾小囊腔（鲍曼腔）capsular space (Bowman's space)
3- 肾小囊壁层 parietal layer of renal capsule
4- 足细胞 podocyte
5- 血管极 vascular pole
6- 致密斑 macula densa
7- 球外系膜细胞 extroglomerular mesangial cell
8- 小动脉 small artery

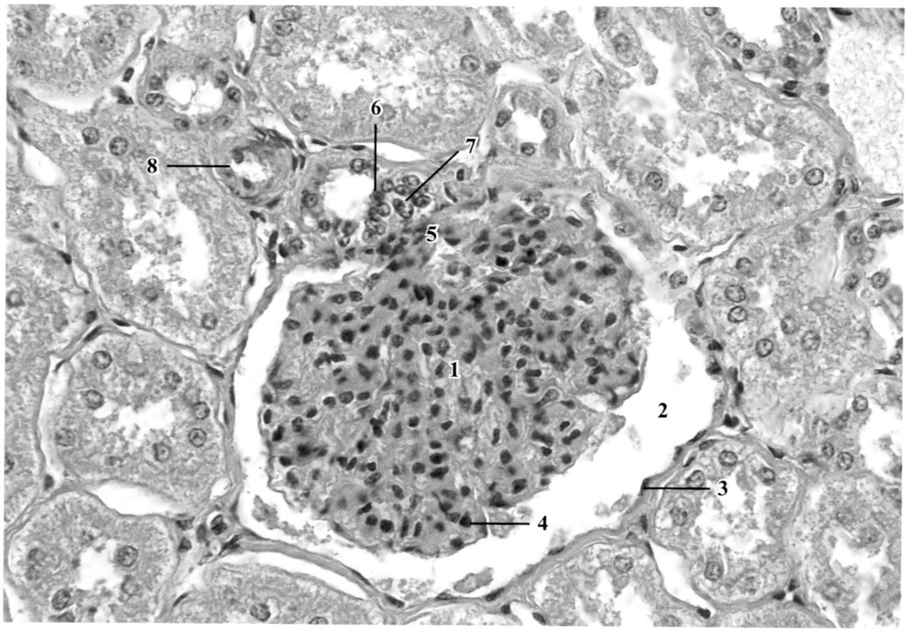

图 6-15　肾皮质组织切片，示肾小体（HE 染色，高倍镜）
Figure 6-15　Histological section of the renal cortex showing the renal corpuscle (HE staining, higher power).

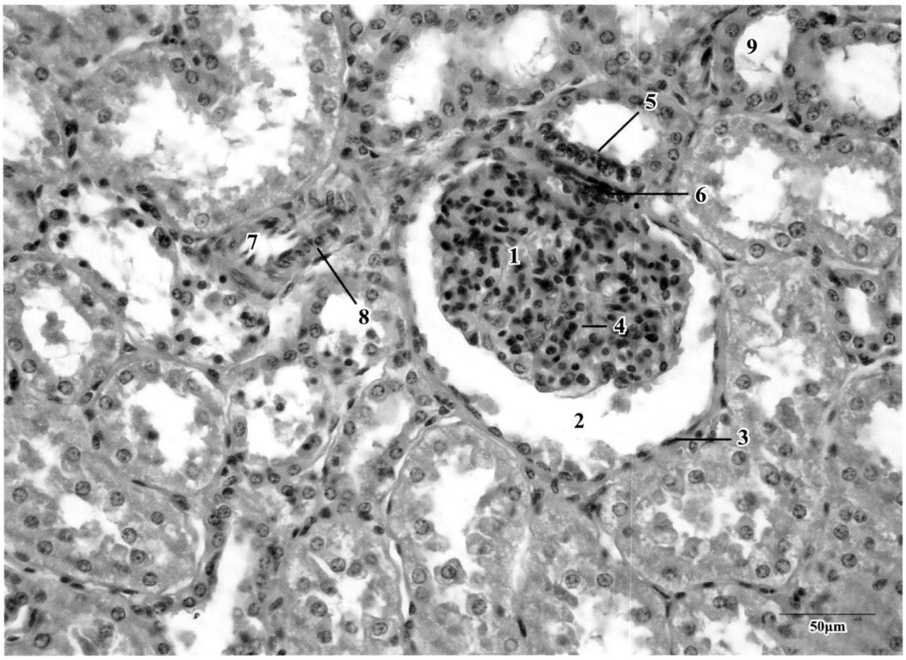

1- 血管球 glomerulus
2- 肾小囊腔（鲍曼腔）capsular space (Bowman's space)
3- 肾小囊壁层 parietal layer of renal capsule
4- 足细胞 podocyte
5- 致密斑 macula densa
6- 血管极 vascular pole
7- 入球小动脉 afferent arteriole
8- 球旁细胞 juxtaglomerular cell
9- 远端小管 distal tubule

图 6-16 肾小体组织切片（HE 染色，高倍镜）
Figure 6-16 Histological section of the renal corpuscle (HE staining, higher power).

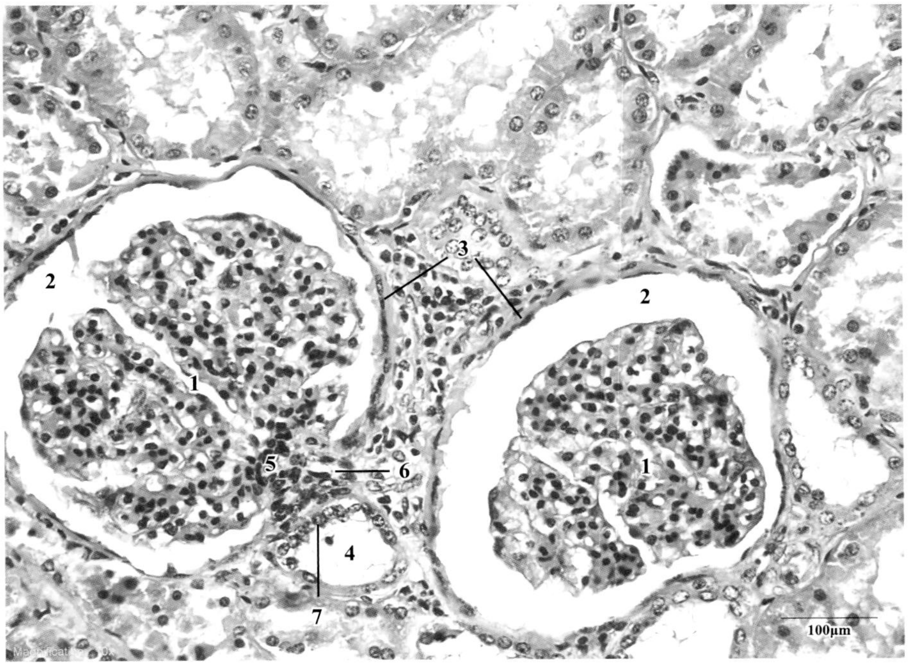

1- 血管球 glomerulus
2- 肾小囊腔（鲍曼腔）capsular space (Bowman's space)
3- 肾小囊壁层 parietal layer of renal capsule
4- 远端小管 distal tubule
5- 血管极 vascular pole
6- 出球小动脉 efferent arteriole
7- 致密斑 macula densa

图 6-17 肾小体组织切片（HE 染色，高倍镜）
Figure 6-17 Histological section of the renal corpuscle (HE staining, higher power).

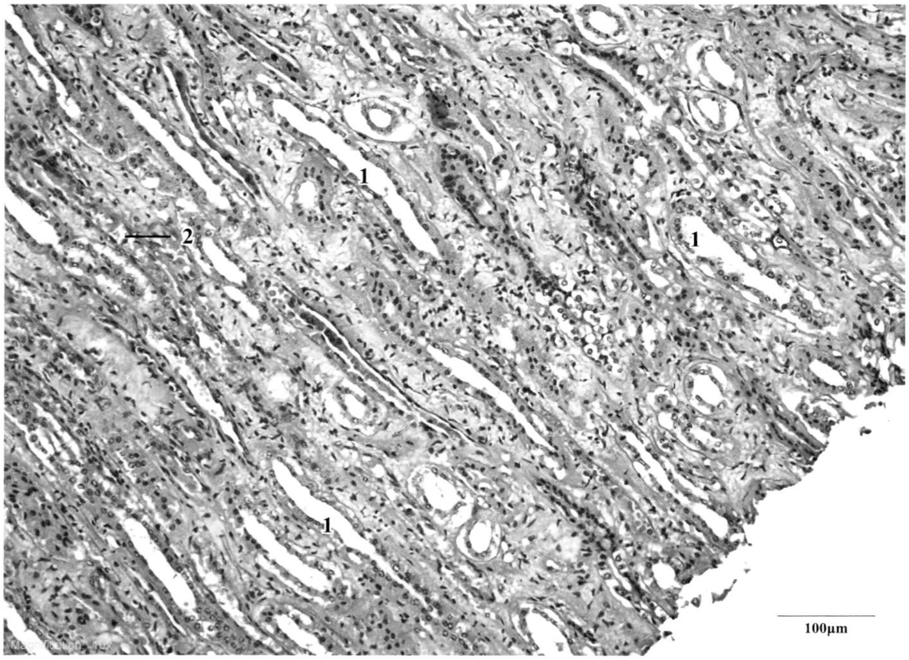

1- 细段 thin segment
2- 直小静脉 straight venule

图 6-18 肾髓质组织切片（HE 染色，低倍镜）
Figure 6-18 Histological section of the renal medulla (HE staining, lower power).

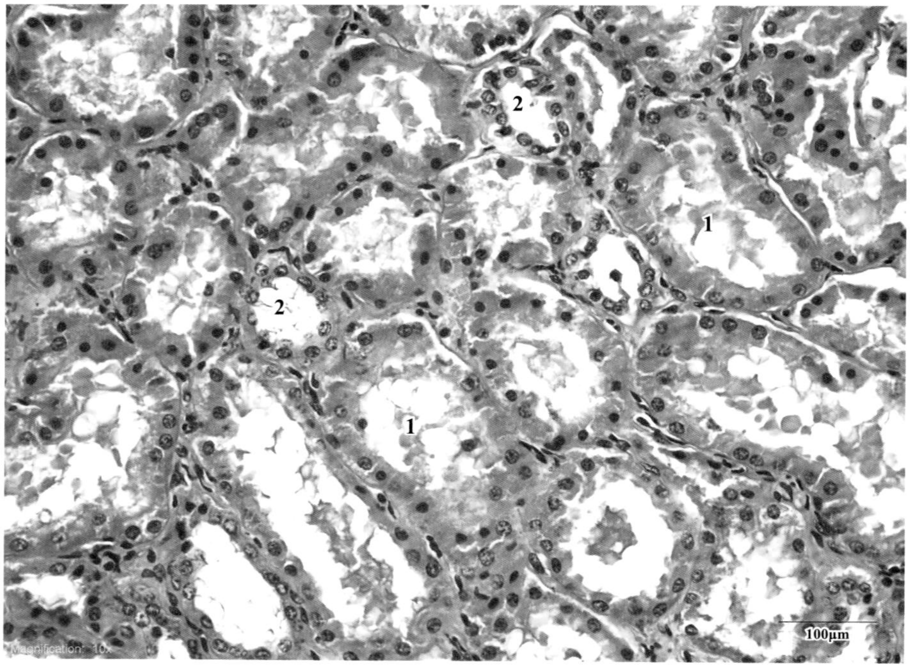

1- 远直小管 distal straight tubule
2- 细段 thin segment

图 6-19 肾小管组织切片（HE 染色，高倍镜）
Figure 6-19 Histological section of the renal tubule (HE staining, higher power).

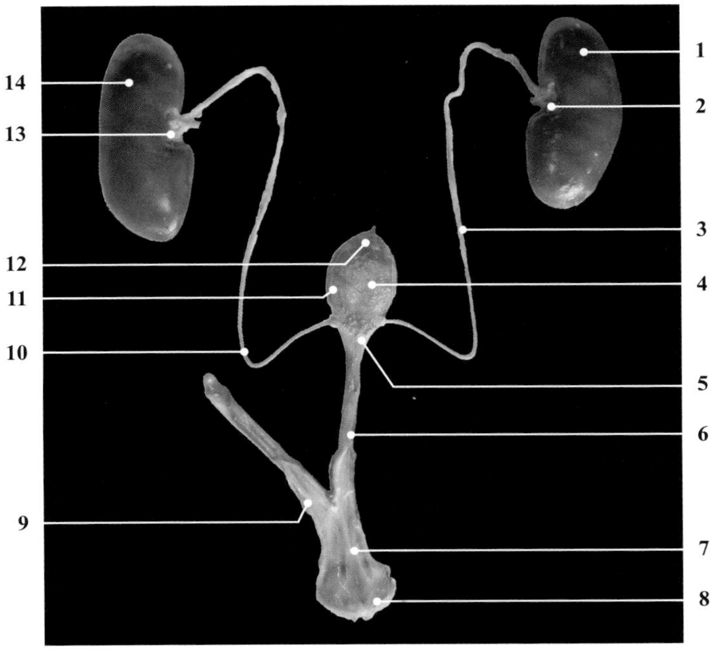

图 6-20 母猪输尿管

Figure 6-20 Ureter of the sow.

1- 右肾 right kidney
2- 右肾门 right renal hilum
3- 右侧输尿管 right ureter
4- 膀胱 urinary bladde
5- 膀胱颈 bladder neck
6- 尿道 urethra
7- 阴道前庭 vaginal vestibulum
8- 阴唇 labia
9- 阴道 vagina
10- 左侧输尿管 left ureter
11- 膀胱体 body of bladder
12- 膀胱顶 apex of bladder
13- 左肾门 left renal hilum
14- 左肾 left kidney

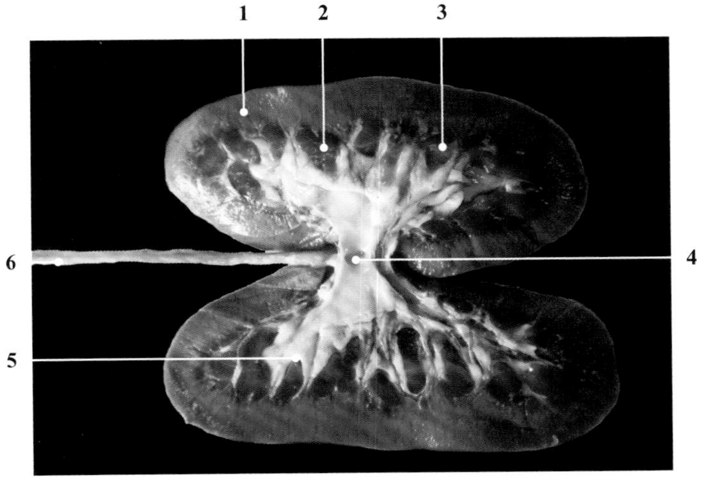

图 6-21 肾盂和肾盏

Figure 6-21 Renal pelvis and calices.

1- 肾皮质 renal cortex
2- 肾髓质 renal medulla
3- 肾乳头 renal papillae
4- 肾盂 renal pelvis
5- 肾盏 renal calices
6- 输尿管 ureter

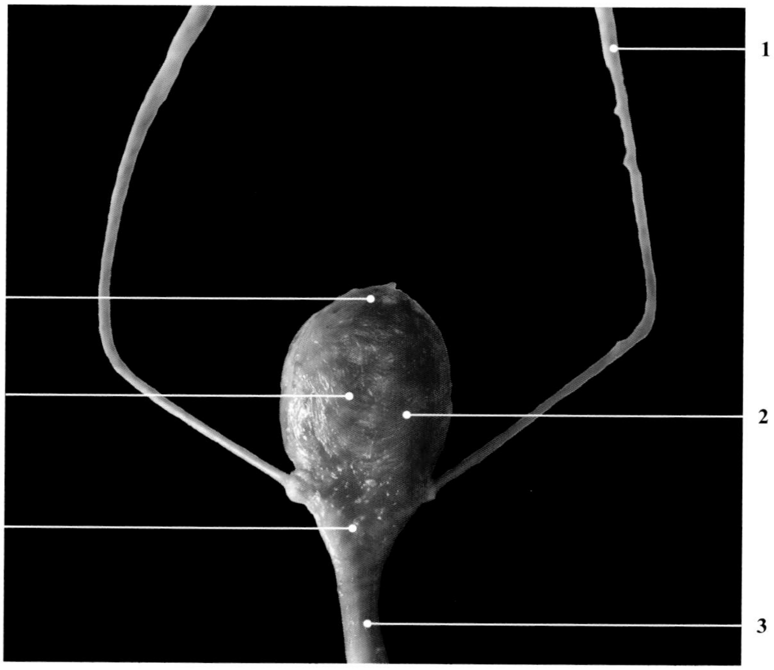

1- 输尿管 ureter
2- 膀胱 urinary bladder
3- 尿道 urethra
4- 膀胱颈 neck of bladder
5- 膀胱体 body of bladder
6- 膀胱顶 apex of bladder

图 6-22 输尿管末段
Figure 6-22 Final segment of the ureter.

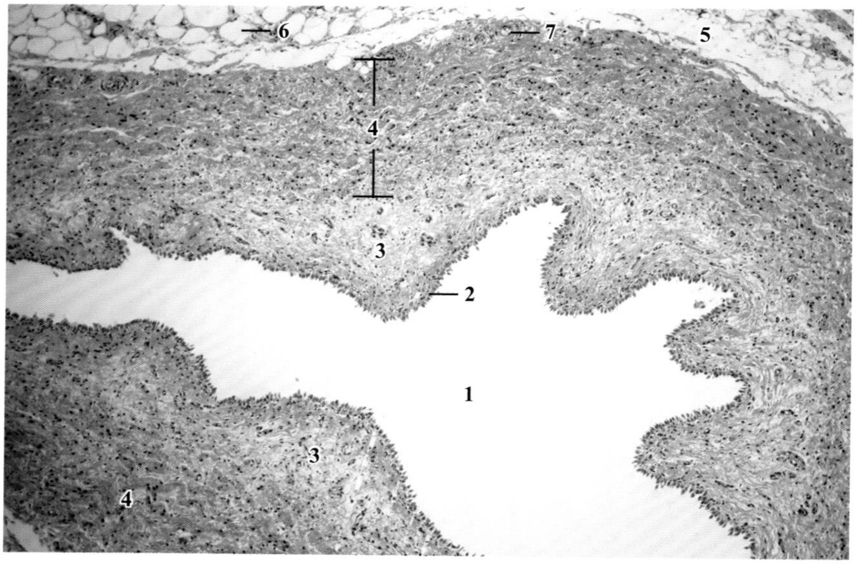

1- 输尿管管腔 ureteral lumen
2- 变移上皮 transitional epithelium
3- 固有层 lamina propria
4- 肌层 muscular layer
5- 外膜 adventitia
6- 脂肪细胞 adipocyte
7- 动脉 artery

图 6-23 输尿管组织切片（HE 染色，低倍镜）
Figure 6-23 Histological section of the ureter (HE staining, lower power).

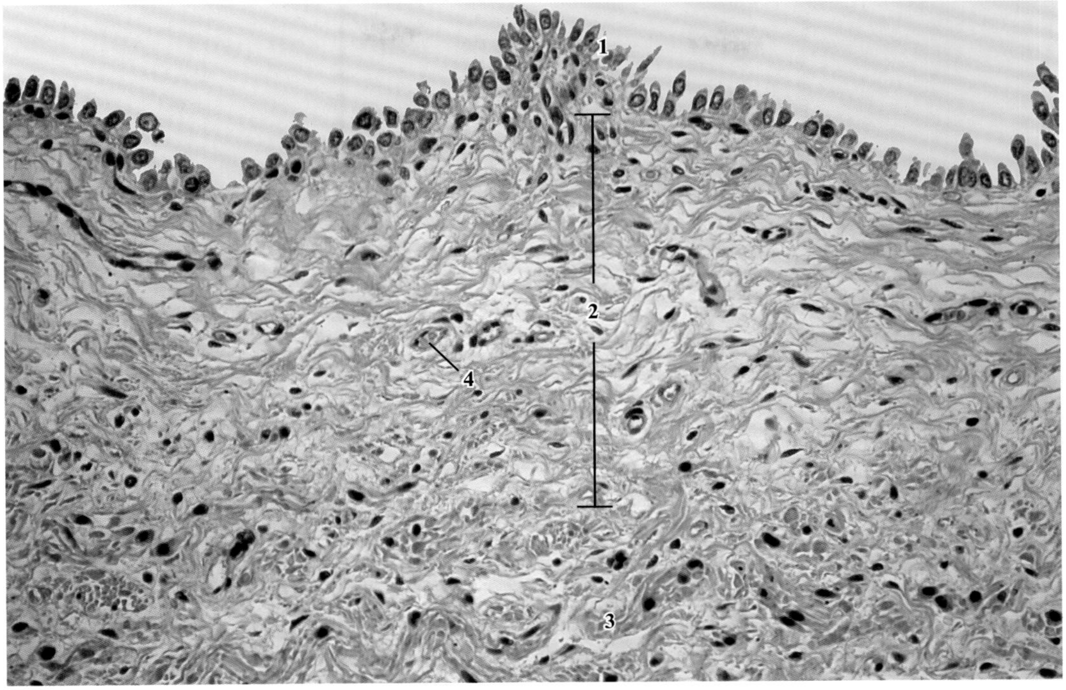

图 6-24 输尿管组织切片（HE 染色，高倍镜）
Figure 6-24 Histological section of the ureter (HE staining, higher power).

1- 变移上皮 transitional epithelium
2- 固有层 lamina propria
3- 肌层 muscular layer
4- 血管 blood vessel

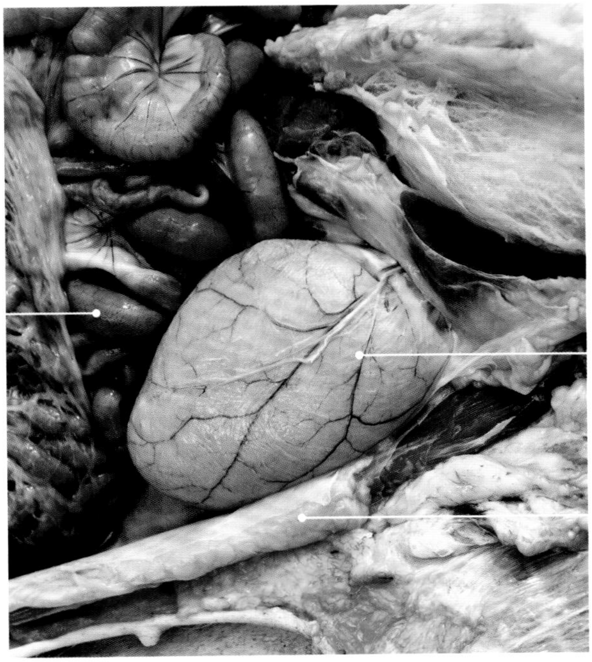

图 6-25 膀胱
Figure 6-25 Urinary bladder.

1- 膀胱 urinary bladder
2- 腹底壁 bottom of the abdomen wall
3- 小肠 small intestine

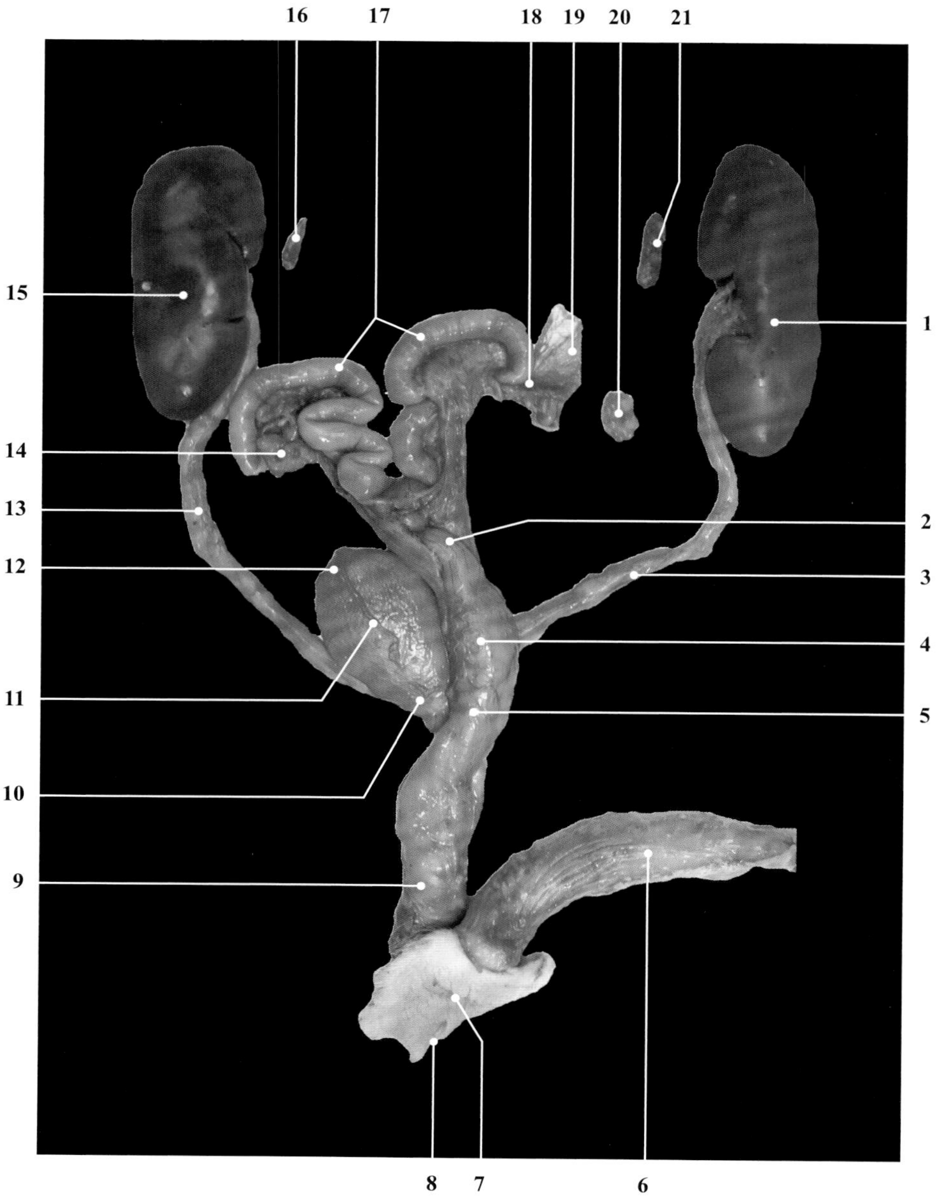

图 6-26 膀胱外形
Figure 6-26 Shape of the urinary bladder.

1- 右肾 right kidney
2- 子宫体 uterine body
3- 右输尿管 right ureter
4- 子宫颈 uterine cervix
5- 阴道 vagina
6- 直肠 rectum
7- 肛门 anus
8- 阴门 vulva
9- 阴道前庭 vaginal vestibulum
10- 膀胱颈 neck of bladder
11- 膀胱体 body of bladder
12- 膀胱顶 apex of bladder
13- 左输尿管 left ureter
14- 左卵巢 left ovary
15- 左肾 left kidney
16- 左肾上腺 left adrenal gland
17- 子宫角 uterine horn
18- 右输卵管 right oviduct
19- 输卵管伞 fimbriae of uterine tube
20- 右侧卵巢 right ovary
21- 右肾上腺 right adrenal gland

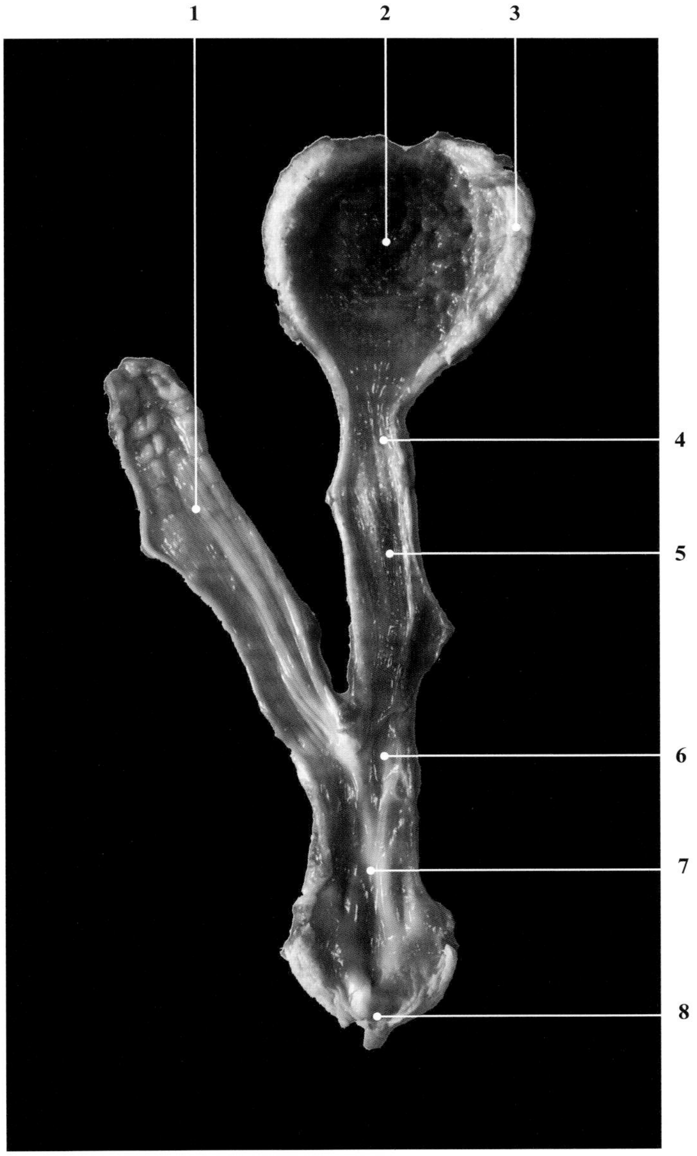

图 6-27 膀胱黏膜面
Figure 6-27 Mucosal surface of the urinary bladder.

1- 阴道 vagina
2- 膀胱黏膜 mucosa of bladder
3- 膀胱壁 bladder wall
4- 膀胱颈 neck of bladder
5- 尿道 urethra
6- 尿道外口 external urethral orifice
7- 阴道前庭 vaginal vestibulum
8- 阴蒂 clitoris

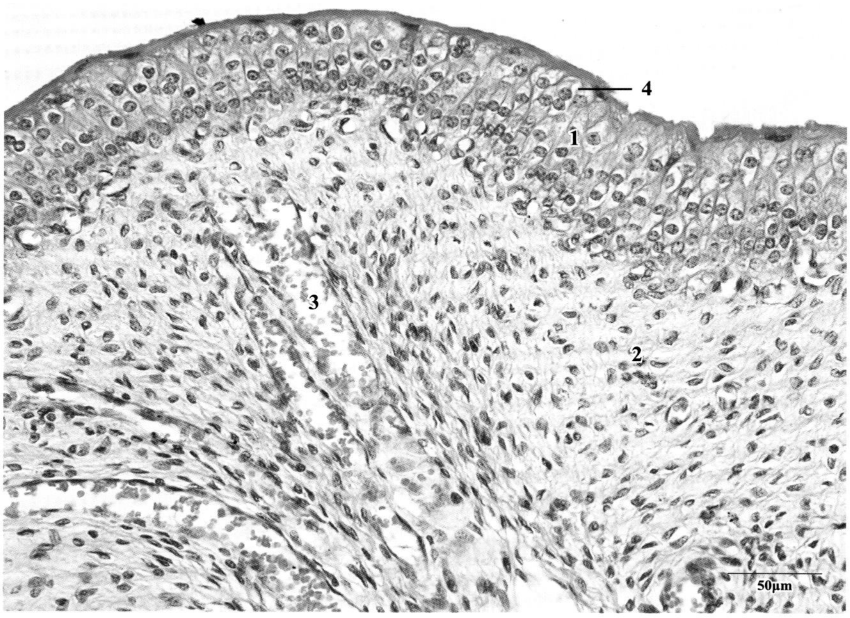

图 6-28　空虚的膀胱上皮组织切片（HE 染色，高倍镜）
Figure 6-28　Histological section of the empty bladder epithelium (HE staining, higher power).

1- 变移上皮 transitional epithelium
2- 固有层 lamina propria
3- 血管 blood vessel
4- 双核细胞 binucleate cell

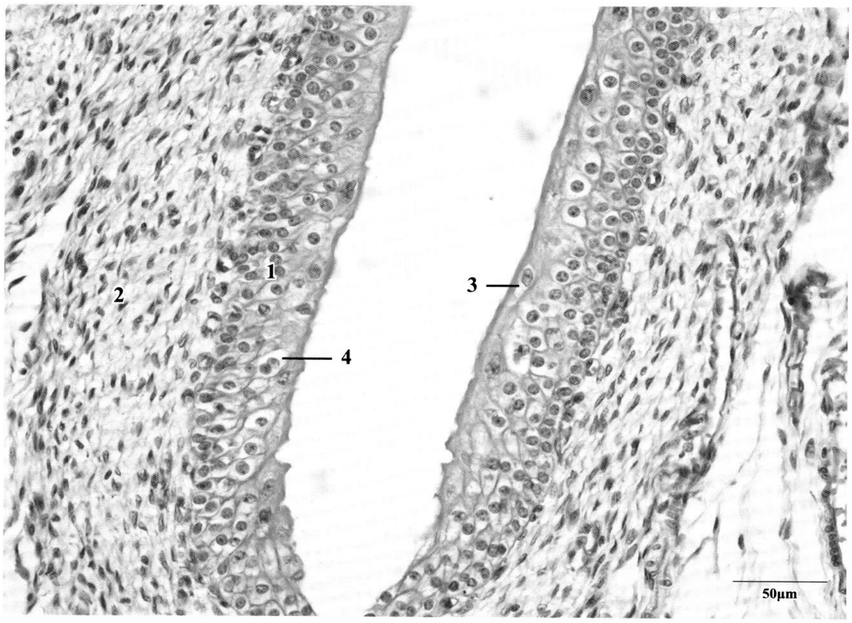

图 6-29　空虚的膀胱上皮组织切片（HE 染色，高倍镜）
Figure 6-29　Histological section of the empty bladder epithelium (HE staining, higher power).

1- 变移上皮 transitional epithelium
2- 固有层 lamina propria
3- 盖细胞 tectorial cell
4- 双核细胞 binucleate cell

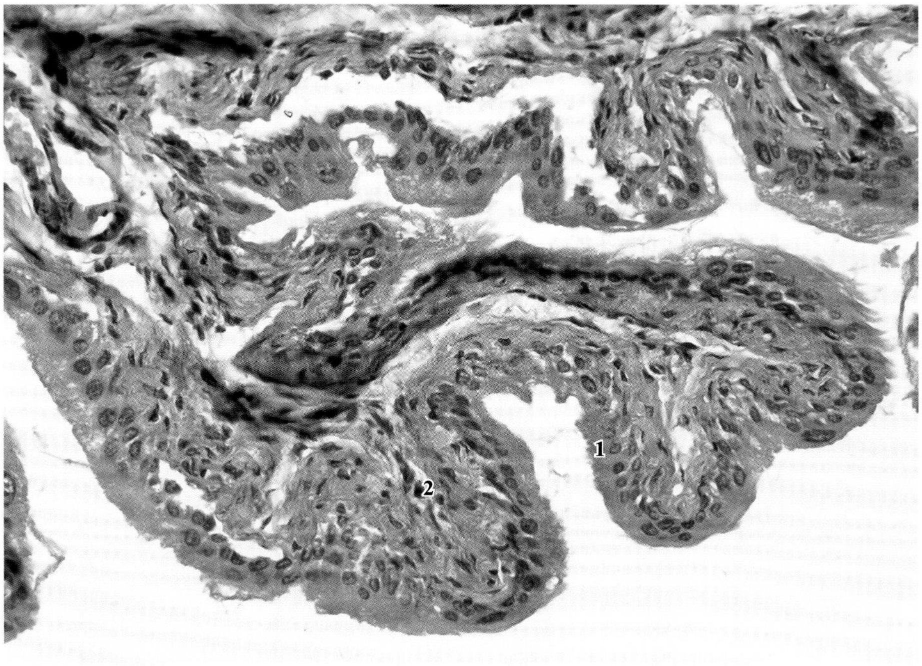

图 6-30 充盈的膀胱上皮组织切片（HE 染色，高倍镜）
Figure 6-30 Histological section of the distended bladder epithelium (HE staining, higher power).

1- 变移上皮 transitional epithelium **2-** 固有层 lamina propria

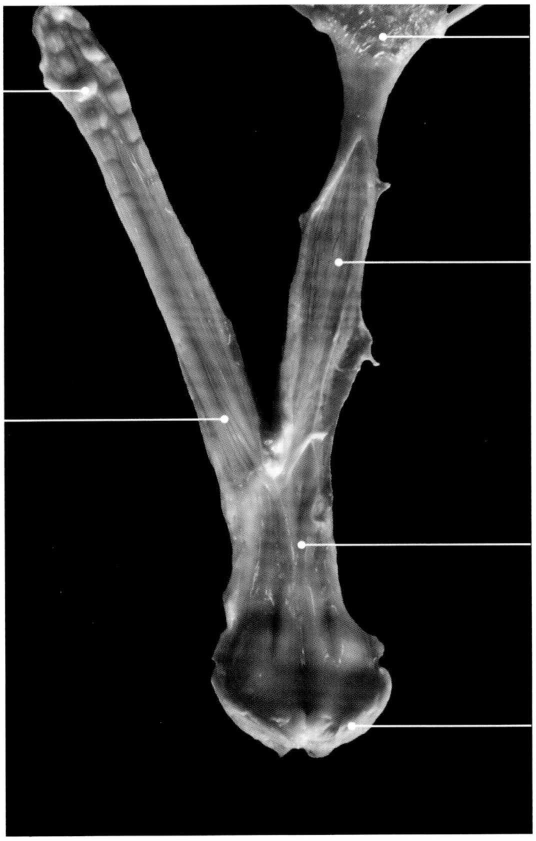

图 6-31 母猪尿道
Figure 6-31 Urethra of the sow.

1- 膀胱 urinary bladder
2- 尿道 urethra
3- 阴道前庭 vaginal vestibulum
4- 阴唇 labia
5- 阴道 vagina
6- 子宫颈 uterine cervix

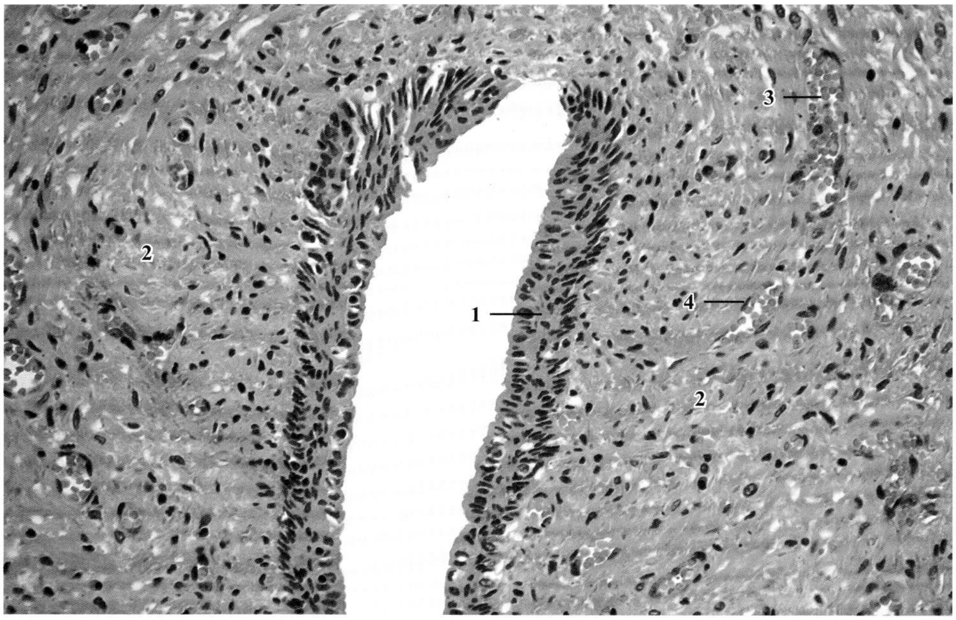

图 6-32 母猪尿道组织切片（HE 染色，高倍镜）
Figure 6-32 Histological section of the urethra of the sow (HE staining, higher power).

1- 复层柱状上皮 stratified columnar epithelium
2- 固有层 lamina propria
3- 血管 blood vessel
4- 内皮细胞 endothelial cell

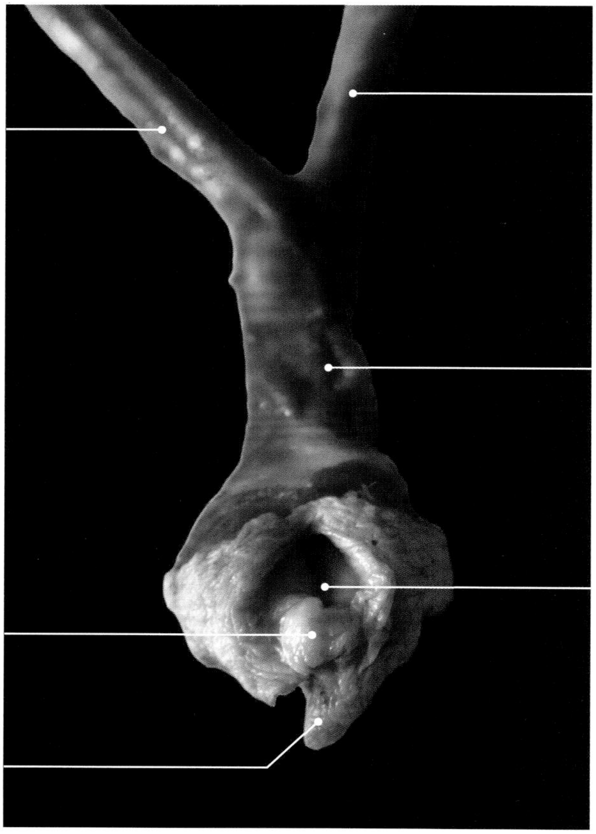

图 6-33 母猪尿道外口
Figure 6-33 External urethral orifice of the sow.

1- 尿道 urethra
2- 阴道前庭 vaginal vestibulum
3- 阴门 vulva
4- 阴唇腹侧联合 ventral commissure of labium
5- 阴蒂 clitoris
6- 阴道 vagina

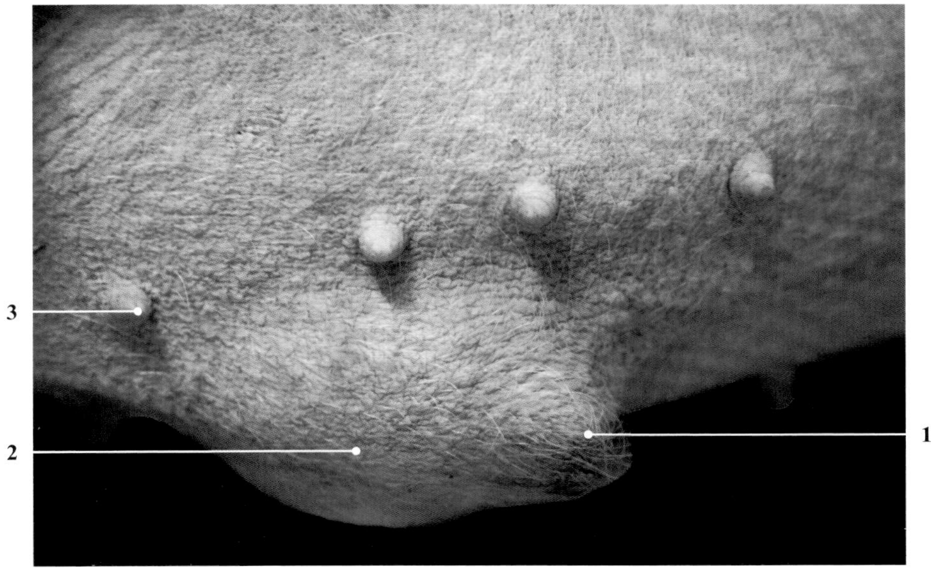

图 6-34 公猪尿道外口和包皮
Figure 6-34 External urethral orifice and prepuce of the boar.

1- 尿道外口 external urethral orifice
2- 包皮 prepuce
3- 乳头 teat

第七章 雄性生殖系统

Chapter 7 Male reproductive system

公猪生殖器官包括睾丸、附睾、输精管、精索、阴囊、尿生殖道、副性腺、阴茎和包皮。

一、睾丸

睾丸（testis）是产生精子和分泌雄性激素的器官。相比较其他动物，公猪睾丸相对较大，位于阴囊内，在肛门腹侧（会阴部），左右各一。呈椭圆形，长轴斜向后上方，睾丸头朝向前下方。一侧有附睾附着，称为附睾缘，另一侧为游离缘。血管和神经进入的一端为睾丸头，接附睾头，另一端为睾丸尾，以睾丸固有韧带（proper ligament of testis）与附睾尾相连。

睾丸表面覆有光滑的浆膜，称固有鞘膜（proper vagina tunica）。鞘膜深面为厚而坚韧的结缔组织白膜（tunica albuginea）。白膜沿睾丸纵轴集中形成网状的睾丸纵隔（mediastinum testis），分出许多发达的呈放射状排列的睾丸小隔（septula testis），将睾丸实质分割成许多锥形的睾丸小叶。每一睾丸小叶内含有数条迂曲的生精小管（seminiferous tubule），或称曲精小管（contorted seminiferous tubule），小管壁的生殖上皮可产生精子；小管之间为睾丸间质，含有间质细胞。在靠近纵隔处，曲精小管变成直而短的直精小管（straight seminiferous tubule），它们进入睾丸纵隔后，相互吻合形成睾丸网（testicular network）。从睾丸网发出10余条睾丸输出小管，穿出睾丸头，形成附睾头。在曲精小管上皮内可见不同发育阶段的生精细胞（spermatogenic cell），从曲精小管的基底部至管腔面依次是精原细胞（spermatogonium）、初级精母细胞（primary spermatocyte）、次级精母细胞（secondary spermatocyte）、精子细胞（spermatid）和精子（spermatozoon）。从精原细胞到形成精子的过程，称精子发生（spermatogenesis），历经精原细胞的增殖、精母细胞的生长与减数分裂以及精子形成等阶段，这一过程约45天。曲精小管之间的疏松结缔组织，由血管、淋巴管、神经和特殊的睾丸间质细胞（interstitial cell, Leydig cell）组成。相比较其他动物，猪的间质细胞数量较多，主要分泌睾酮等雄性激素，有促进精子发生、促进雄性生殖管道的发育与分化、维持第二性征与性功能等作用。

二、附睾

附睾（epididymis）是贮存精子和精子进一步成熟的地方。附睾附着于睾丸的附睾缘，分为附睾头（head of epididymis）、附睾体（body of epididymis）和附睾尾（tail of epididymis）三部分。附睾头膨大，由睾丸输出小管构成。输出小管汇合成一条较粗而长的附睾管，盘曲而成附睾体和附睾尾，在附睾尾处管径增大，最后延续为输精管。附睾尾也膨大，借睾丸固有韧带与睾丸尾相连，借附睾尾韧带（或称阴囊韧带scrotal ligament）与阴囊相连。在附睾的表面也被覆有固有鞘膜和薄的白膜。

三、输精管和精索

输精管（deferent duct）为运送精子的管道，起始于附睾尾的附睾管，沿附睾体走至附睾头，进入精索后缘内侧的输精管褶中，经腹股沟管上行进入腹腔，随即向后进入骨盆腔，末端与精囊腺导管一同开口于尿生殖道起始部背侧壁的精阜上。输精管是附睾管的延续，管腔小而管壁厚。管壁由黏膜、肌层和浆膜组成。黏膜上皮为假复层柱状上皮，肌层厚而发达，为内环行、中斜行和外纵行的平滑肌。

输精管经腹股沟管时，与进入睾丸的脉管、神经和提睾肌等组成精索（spermatic cord），外面包以固有鞘膜，为一扁平的圆锥形索状结构，基部较宽，附着于睾丸和附睾上，向上逐渐变细，顶端达腹股沟管内口（腹环）。

四、阴囊

阴囊（scrotum）斜位于肛门腹侧（会阴部），与周围界限不明显，为袋状的腹壁囊，借助腹股沟管与腹腔相通，内有睾丸、附睾和部分精索。阴囊壁的结构与腹壁相似，由外向内为皮肤、肉膜、阴囊筋膜及提睾肌和总鞘膜。睾丸和附睾通过阴囊韧带（scrotal ligament）与阴囊壁相连。当进行阉割术时，在切开阴囊壁之后，必须剪断附睾尾韧带、睾丸系膜和精索，才能摘除睾丸和附睾。

五、尿生殖道

尿生殖道（urogenital tract）为尿液和精液排出的共同通道，起于膀胱颈，沿骨盆腔底壁向后伸延，绕过坐骨弓，再沿阴茎腹侧的尿道沟向前伸延，以尿道外口开口于外界。尿生殖道分骨盆部和阴茎部两部分，两部间以坐骨弓为界。在交界处，尿生殖道内腔变细，称为尿道峡（urethral isthmus）。尿道峡是临床上尿道结石或尿道阻塞的常发病部位。尿生殖道骨盆部长，位于骨盆腔底壁与直肠之间。在其起始处背侧壁的黏膜上，有一圆形隆起，称为精阜（seminal hillock），是输精管及精囊腺导管的共同开口。在尿生殖道骨盆部黏膜的表面，还有前列腺和尿道球腺的开口。尿生殖道阴茎部是尿道经坐骨弓至阴茎腹侧的一段，末端开口在阴茎头，开口处称尿道外口（external urethral orifice）。在尿道峡后方尿生殖道壁上的海绵体层稍变厚，形成尿道球（urethral bulb），又称阴茎球（penis bulb）。

尿生殖道管壁从内向外由黏膜、海绵体层、肌层和外膜构成。黏膜常集拢成许多皱褶，含有一些小腺体。海绵体层主要是由毛细血管膨大而形成的海绵体腔。肌层由深层的平滑肌和浅层的横纹肌组成。横纹肌在骨盆部的称为尿道肌（urethral muscle），在阴茎部的称为球海绵体肌（bulbocavernous muscle）。横纹肌的收缩在交配时对射精起重要作用，还可帮助排出余尿。

六、副性腺

副性腺包括精囊腺、前列腺和尿道球腺，分泌物构成精液，可稀释精子，营养精子，有利于精子的生存和运动，改善阴道内环境等作用，公猪去势后的副性腺均发育不良。

1. 精囊腺（vesicular gland）：有一对，特别发达，位于膀胱颈背侧的尿生殖褶中，呈棱形三面体，淡红色，由许多腺小叶组成。每侧精囊腺导管单独或与同侧输精管共同开口于精阜。

2. 前列腺（prostate gland）：分腺体部和扩散部。腺体部较小，位于尿生殖道起始部背侧，被精囊腺所遮盖。扩散部很发达，占据尿生殖道骨盆部黏膜与尿道肌之间的海绵层内，切面上呈黄色。两部分均有许多导管，腺体部开口于精阜外侧，扩散部直接开口于尿生殖道骨盆部背侧黏膜。

3．尿道球腺（bulbourethral gland）：有一对，很发达，位于尿生殖道骨盆部后2/3部的两侧，呈长圆柱形，硬而致密。每侧腺体各有一条导管，在坐骨弓处开口于尿生殖道盆部背侧半月形黏膜褶所围成的盲囊内。

七、阴茎

阴茎（penis）为公猪的排尿、排精和交配器官，位于腹底壁皮下，起自坐骨弓，经两股之间，沿中线向前伸达脐区。阴茎分为阴茎根（root of penis）、阴茎体（body of penis）和阴茎头（glans of penis）三部分。阴茎体在阴囊的前方形成阴茎乙状弯曲（sigmoid flexure of the penis）。阴茎头尖细呈螺旋状扭转。尿道外口呈裂隙状，位于阴茎头前端的腹外侧。

阴茎主要由阴茎海绵体、尿生殖道阴茎部和阴茎肌构成，外面包有皮肤。阴茎海绵体（cavernous body of penis）实际上是毛细血管膨大形成的，衬以内皮，与阴茎血管直接相通。当充血时，海绵体膨胀，阴茎变粗变硬而勃起，故海绵体又称勃起组织（erectile tissue）。阴茎肌包括球海绵体肌、坐骨海绵体肌和阴茎缩肌。猪的阴茎缩肌（retractor penis muscle）为两条平行的带状平滑肌，起于前两个尾椎腹侧，向前伸延，止于阴茎乙状弯曲的第二曲处。收缩时可使阴茎退缩，将阴茎头隐藏于包皮内。

八、包皮

包皮（prepuce）为下垂于腹底壁的皮肤折转而形成的管状鞘，有容纳和保护阴茎的作用。包皮的游离缘围成狭窄的包皮口（preputial opening）。包皮口位于脐后稍后方，开口朝前，周围有长毛。包皮外层为腹壁皮肤，在包皮口向包皮腔折转，形成包皮内层。包皮腔很长，前宽后窄，前部背侧壁上有一圆孔，通向椭圆形的包皮憩室（preputial diverticulum）或包皮盲囊。憩室内常聚集有腐败的余尿和脱落的上皮细胞，具有特殊的腥臭味，在猪屠宰后，应将其切除，以免污染肉品。

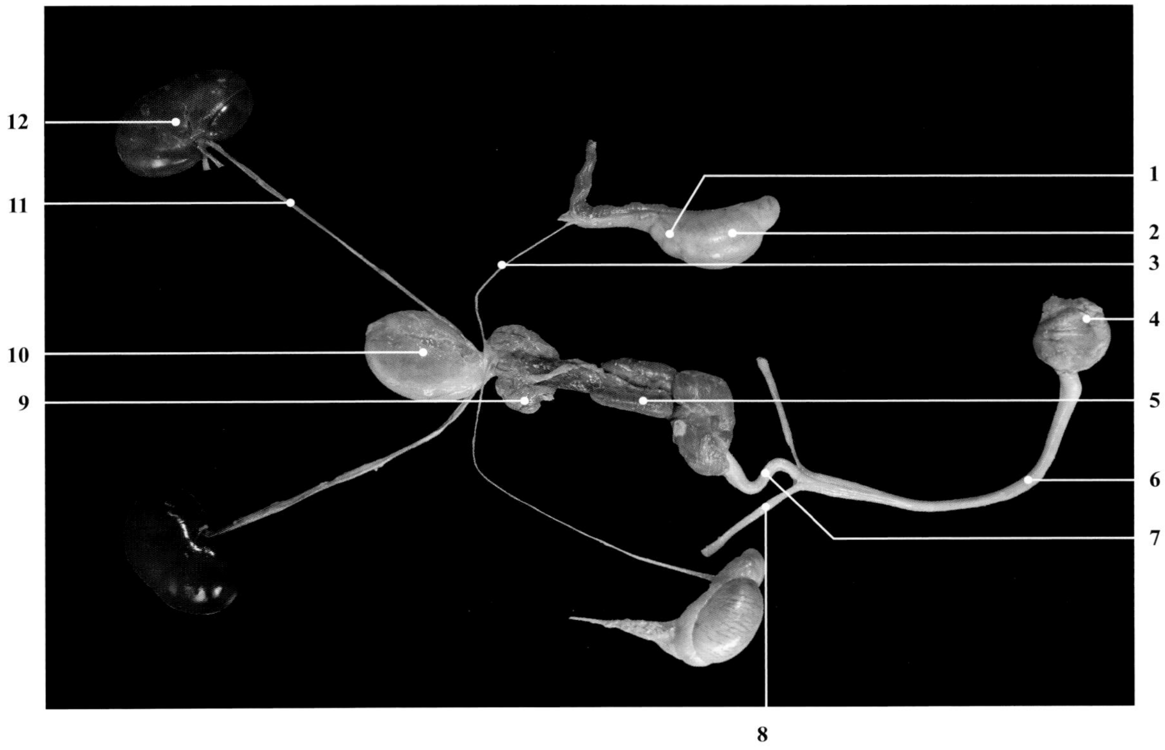

图 7-1　公猪泌尿生殖系统
Figure 7-1　Urogenital system of a boar.

1- 附睾 epididymis
2- 睾丸 testis
3- 输精管 deferent duct
4- 包皮憩室（包皮盲囊） preputial diverticulum
5- 尿道球腺 bulbourethral gland
6- 阴茎 penis
7- 乙状弯曲 sigmoid flexure
8- 阴茎缩肌 retractor penis muscle
9- 精囊腺 vesicular gland
10- 膀胱 urinary bladder
11- 输尿管 ureter
12- 肾 kidney

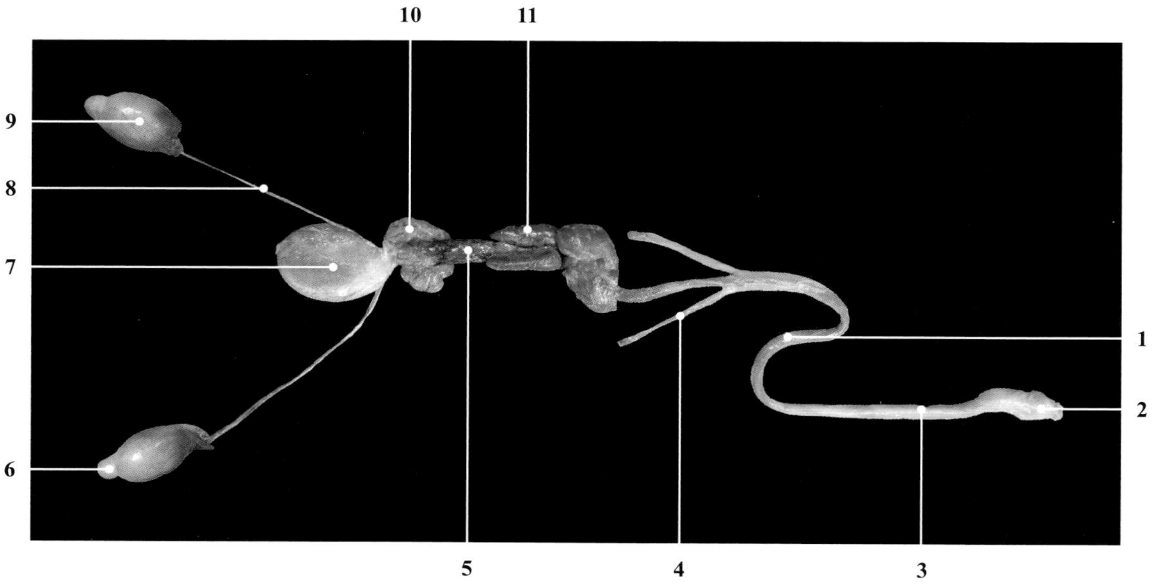

图 7-2 公猪生殖系统
Figure 7-2 Reproductive system of a boar.

1- 乙状弯曲 sigmoid flexure
2- 包皮憩室（包皮盲囊） preputial diverticulum
3- 阴茎 penis
4- 阴茎缩肌 retractor penis muscle
5- 尿生殖骨盆部 pelvis part of urogenital tract
6- 附睾 epididymis
7- 膀胱 urinary bladder
8- 输精管 deferent duct
9- 睾丸 testis
10- 精囊腺 vesicular gland
11- 尿道球腺 bulbourethral gland

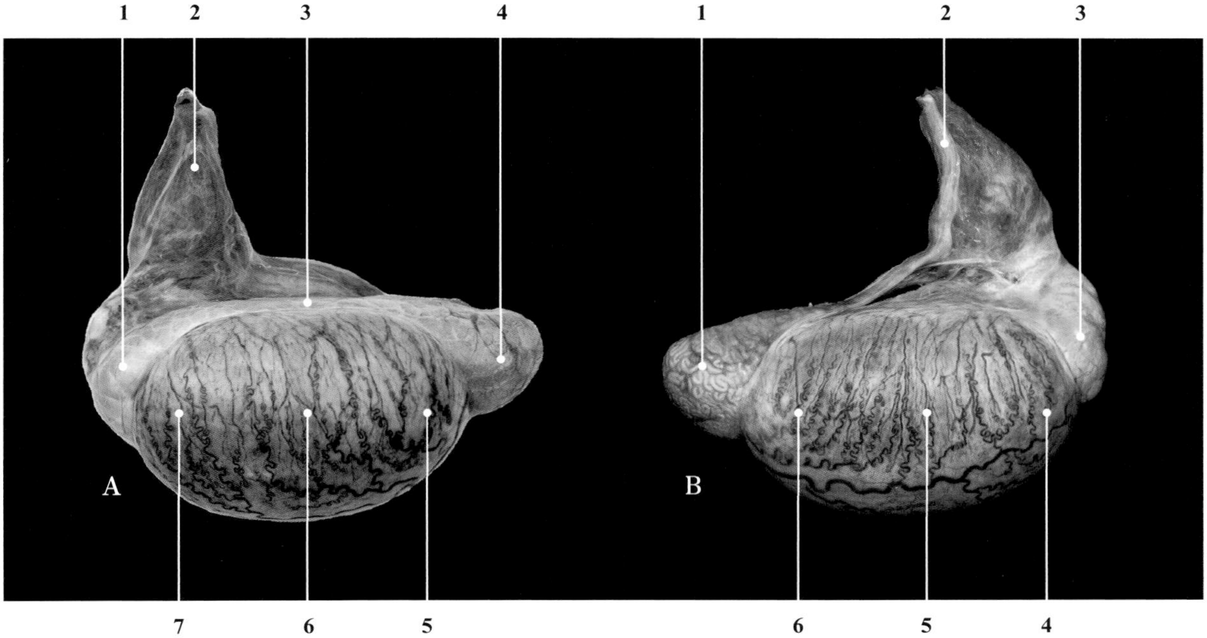

图 7-3　左侧睾丸
Figure 7-3　The left testis.

A：外侧观 lateral view
1- 附睾头 head of epididymis
2- 精索 spermatic cord
3- 附睾体 body of epididymis
4- 附睾尾 tail of epididymis
5- 睾丸尾 tail of testis
6- 睾丸体 body of testis
7- 睾丸头 head of testis

B：内侧观 medial view
1- 附睾尾 tail of epididymis
2- 输精管 deferent duct
3- 附睾头 head of epididymis
4- 睾丸头 head of testis
5- 睾丸体 body of testis
6- 睾丸尾 tail of testis

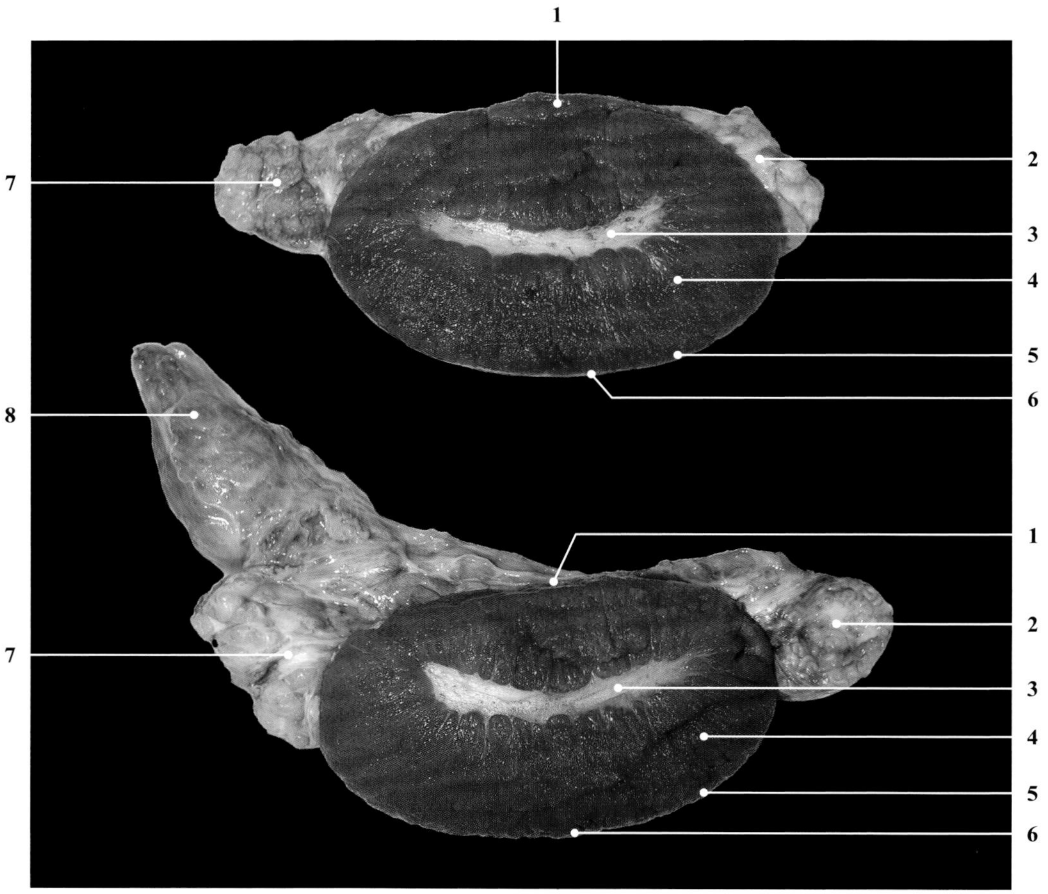

图 7-4　左侧睾丸纵剖面
Figure 7-4　Longitudinal section of the left testis.

1- 睾丸附睾缘 margo epididymidis of testis
2- 附睾尾 tail of epididymis
3- 睾丸纵隔 mediastinum testis
4- 睾丸实质 parenchyma of testis
5- 白膜 tunica albuginea
6- 睾丸游离缘 margo liber of testis
7- 附睾头 head of epididymis
8- 精索 spermatic cord

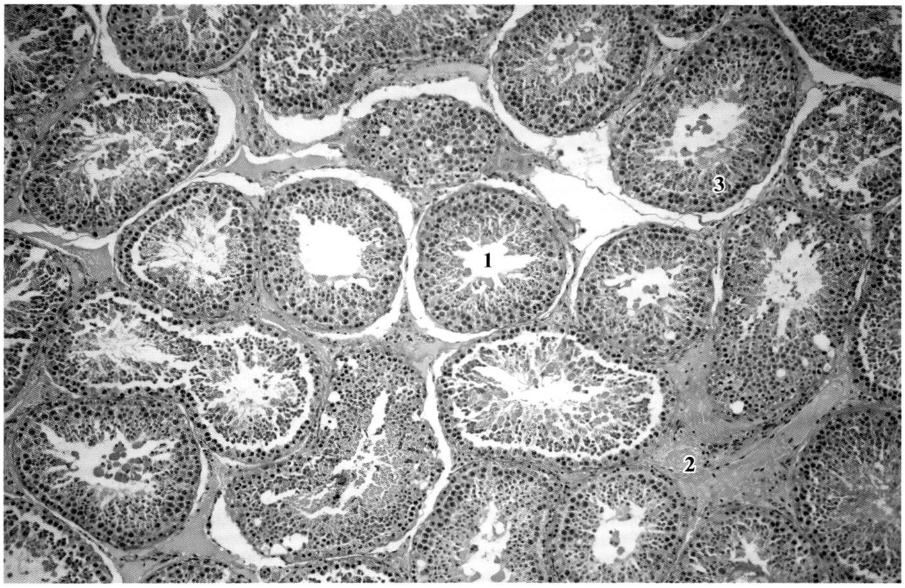

1- 生精小管 seminiferous tubule
2- 间质组织 interstitial tissue
3- 生精上皮 spermatogenic epithelium

图 7-5 睾丸组织切片（HE 染色，低倍镜）
Figure 7-5 Histological section of the testis (HE staining, lower power).

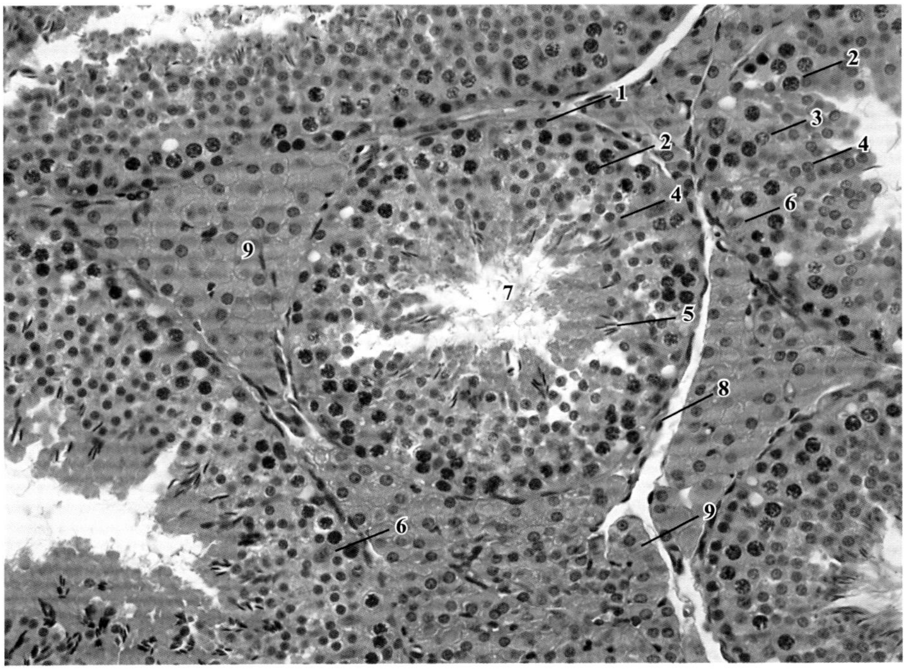

1- 精原细胞 spermatogonium
2- 初级精母细胞 primary spermatocyte
3- 次级精母细胞 secondary spermatocyte
4- 精子细胞 spermatid
5- 精子 spermatozoon
6- 塞尔托利细胞（支持细胞） Sertoli cell (sustentacular cell)
7- 管腔 lumens
8- 肌样细胞的胞核 nucleus of myoid cell
9- 间质细胞 Leydig cell

图 7-6 睾丸组织切片，示生精小管（HE 染色，高倍镜）
Figure 7-6 Histological section of a testis showing the seminiferous tubule (HE staining, higher power).

1- 精原细胞 spermatogonium
2- 初级精母细胞 primary spermatocyte
3- 精子细胞 spermatid
4- 精子 spermatozoon
5- 肌样细胞的细胞核 nucleus of myoid cell
6- 塞尔托利细胞（支持细胞） Sertoli cell (sustentacular cell)
7- 间质细胞 Leydig cell
8- 管腔 lumens

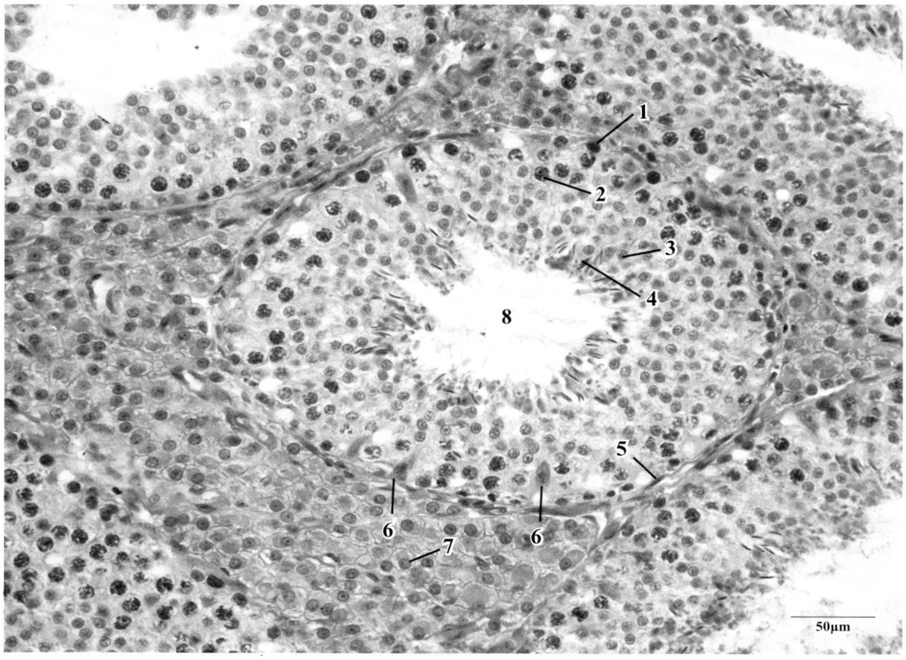

图 7-7 睾丸组织切片，示生精小管（HE 染色，高倍镜）
Figure 7-7 Histological section of a testis showing the seminiferous tubule (HE staining, higher power).

1- 精原细胞 spermatogonium
2- 初级精母细胞 primary spermatocyte
3- 精子细胞 spermatid
4- 塞尔托利细胞 / 支持细胞（核） sertoli cell / sustentacular cell (nucleus)
5- 肌样细胞的胞核 nucleus of myoid cell

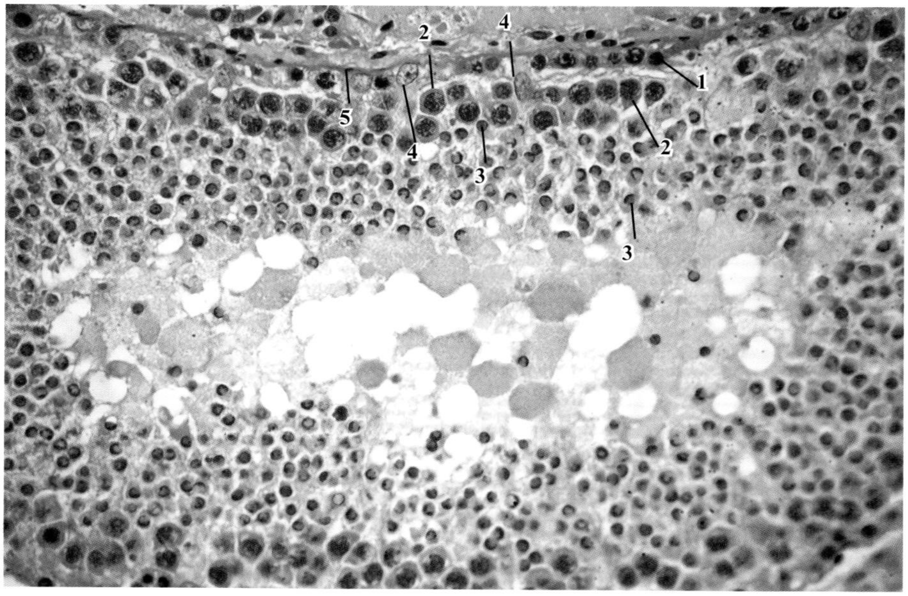

图 7-8 生精小管组织切片，示生精细胞（HE 染色，高倍镜）
Figure 7-8 Histological section of the wall of a seminiferous tubule showing the spermatogenic cell (HE staining, higher power).

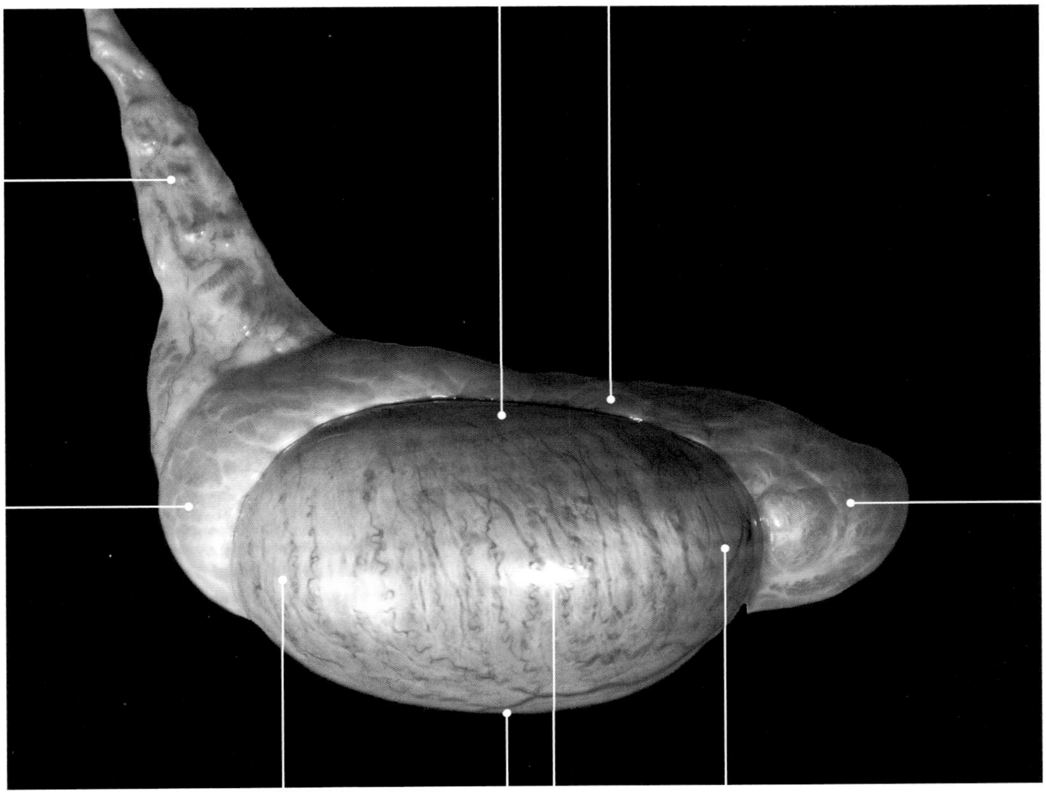

图 7-9 左侧附睾
Figure 7-9 The left epididymis.

1- 睾丸附睾缘 margo epididymis of testis
2- 附睾体 body of epididymis
3- 附睾尾 tail of epididymis
4- 睾丸尾 tail of testis
5- 睾丸体 body of testis
6- 睾丸游离缘 margo liber of testis
7- 睾丸头 head of testis
8- 附睾头 head of epididymis
9- 精索 spermatic cord

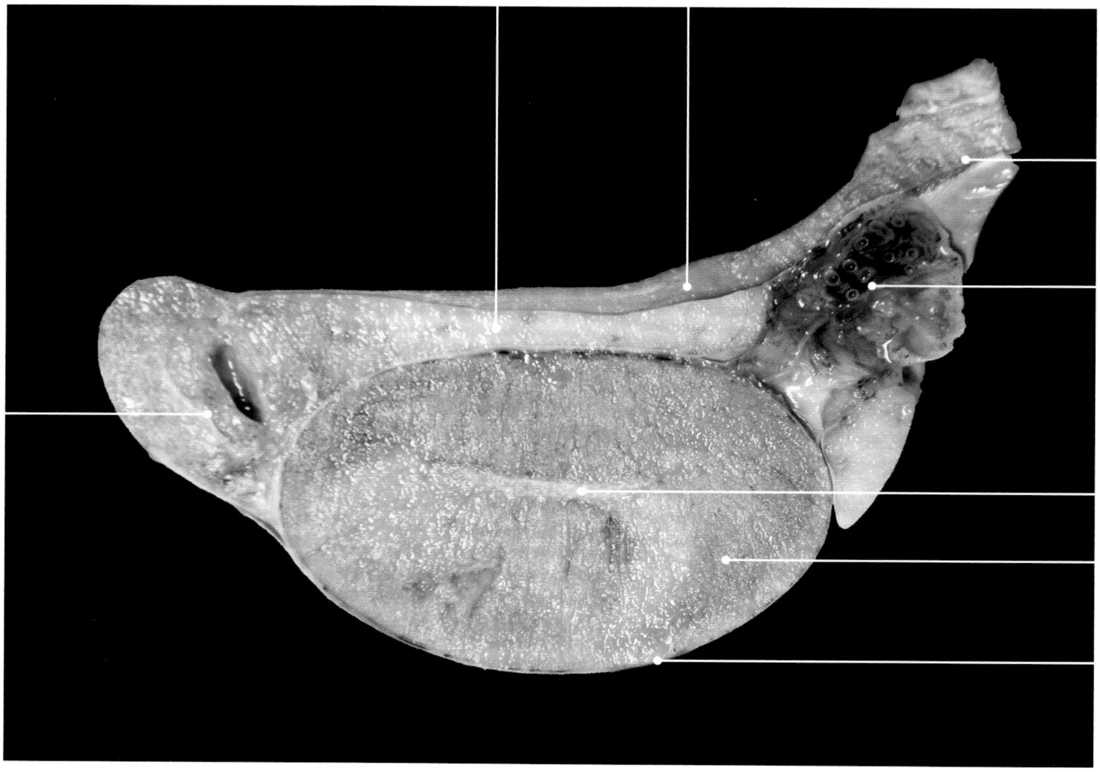

图 7-10　附睾纵剖面
Figure 7-10　Longitudinal section of epididymis.

1- 精索 spermatic cord
2- 睾丸动脉和蔓状丛 testicular artery and pampiniform plexus
3- 睾丸纵隔 mediastinum testis
4- 睾丸实质 parenchyma of testis
5- 白膜 tunica albuginea
6- 附睾尾 tail of epididymis
7- 附睾体 body of epididymis
8- 输精管 deferent duct

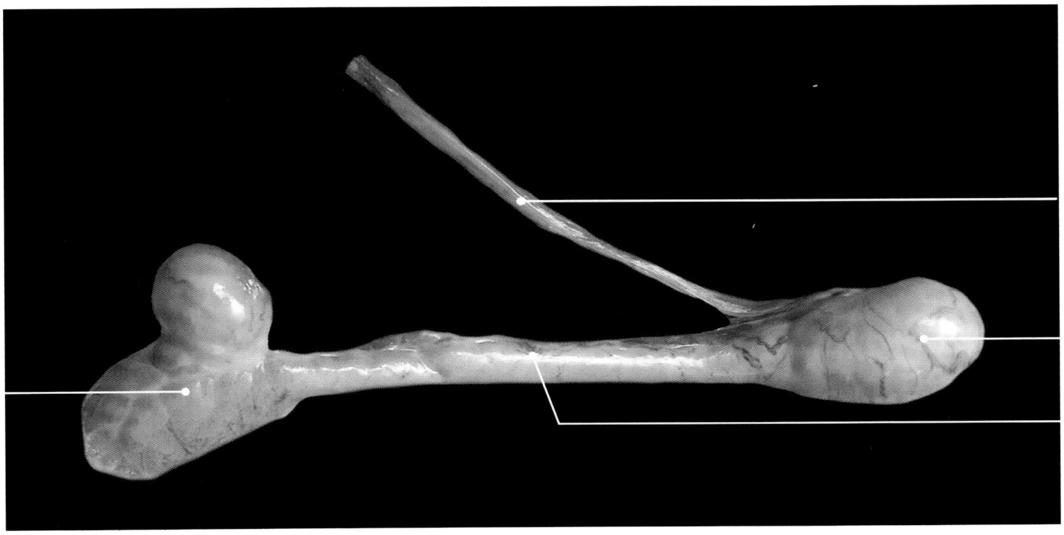

图 7-11 附睾的形态
Figure 7-11 Shape of epididymis.

1- 输精管 deferent duct
2- 附睾尾 tail of epididymis
3- 附睾体 body of epididymis
4- 附睾头 head of epididymis

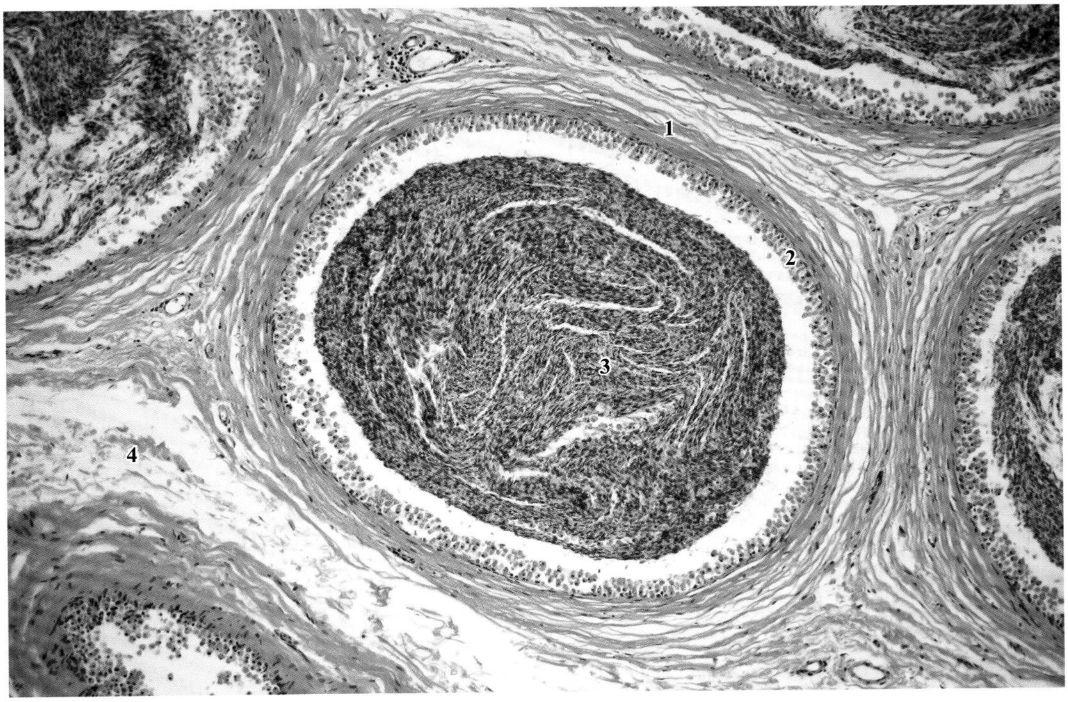

图 7-12 附睾头组织切片，示睾丸输出小管（HE 染色，低倍镜）
Figure 7-12 Histological section of the head of epididymis showing the efferent duct (HE staining, lower power).

1- 平滑肌 smooth muscle
2- 假复层上皮 pseudostratified epithelium
3- 精子 spermatozoa
4- 疏松结缔组织 loose connective tissue

1- 假复层上皮 pseudostratified epithelium
2- 微绒毛 microvillus
3- 基底细胞 basal cell
4- 平滑肌 smooth muscle
5- 疏松结缔组织 loose connective tissue

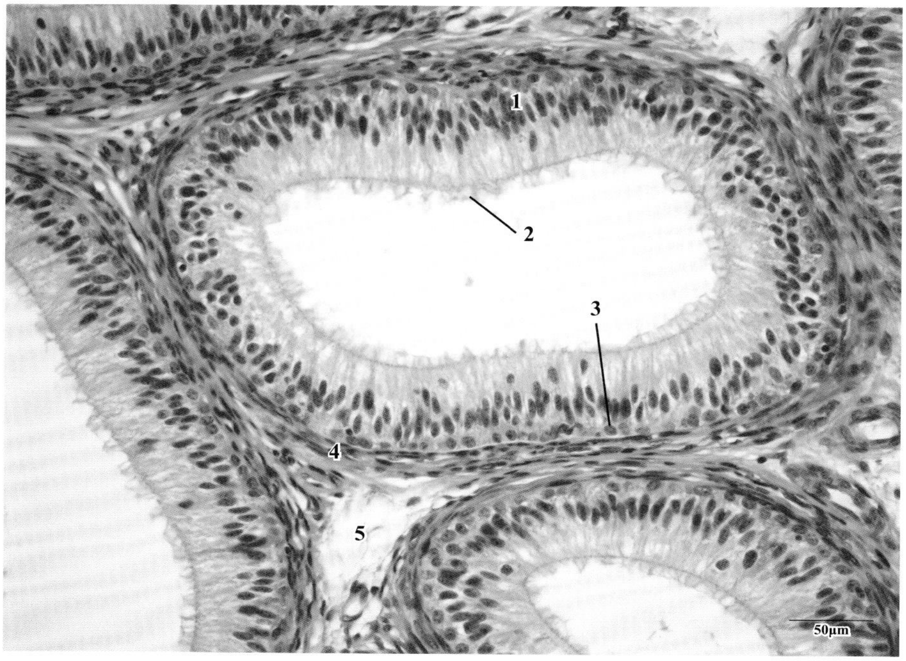

图 7-13　附睾头组织切片，示睾丸输出小管（HE 染色，高倍镜）
Figure 7-13　Histological section of the head of epididymis showing the efferent duct (HE staining, higher power).

1- 平滑肌 smooth muscle
2- 基底细胞 basal cell
3- 假复层上皮 pseudostratified epithelium
4- 微绒毛 microvillus

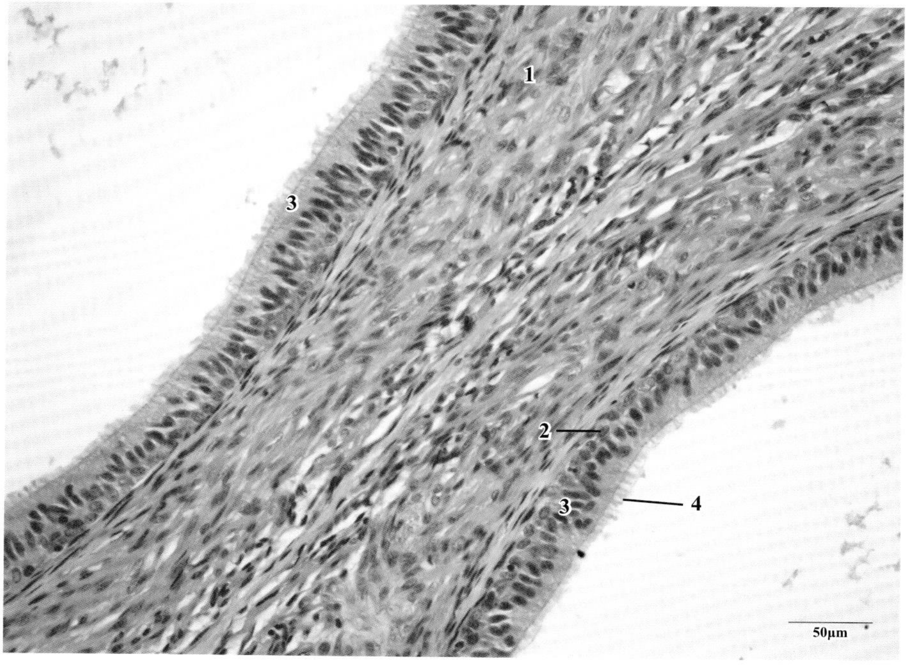

图 7-14　睾丸输出小管组织切片（HE 染色，高倍镜）
Figure 7-14　Histological section of the wall of efferent ductules of testis (HE staining, higher power).

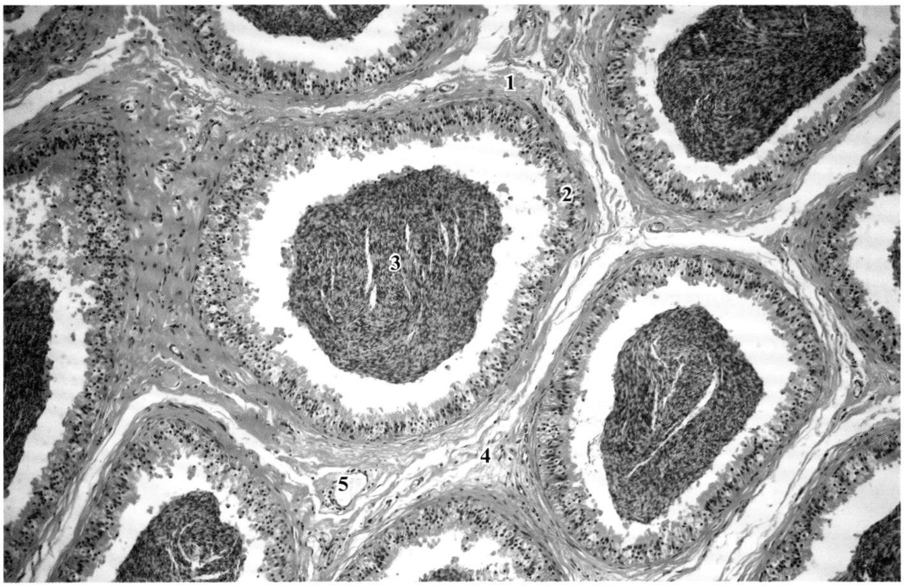

1- 平滑肌 smooth muscle
2- 假复层上皮 pseudostratified epithelium
3- 精子 spermatozoa
4- 疏松结缔组织 loose connective tissue
5- 血管 blood vessel

图 7-15 附睾尾组织切片，示附睾管（HE 染色，低倍镜）
Figure 7-15 Histological section of the tail of epididymis showing the epididymal duct (HE staining, lower power).

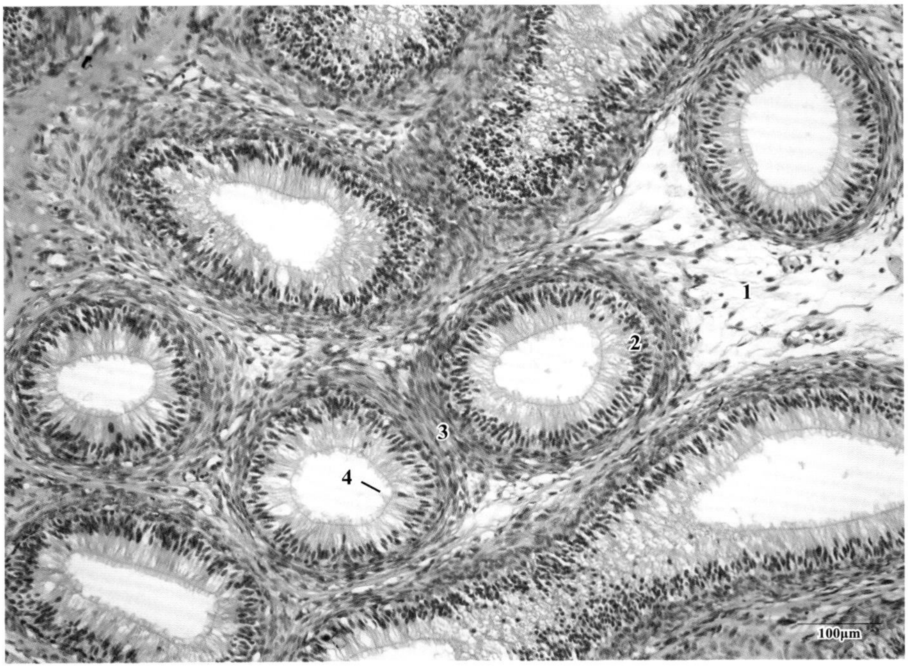

1- 疏松结缔组织 loose connective tissue
2- 假复层上皮 pseudostratified epithelium
3- 平滑肌 smooth muscle
4- 微绒毛 microvillus

图 7-16 附睾尾组织切片，示附睾管（HE 染色，低倍镜）
Figure 7-16 Histological section of the tail of epididymis showing the epididymal duct (HE staining, lower power).

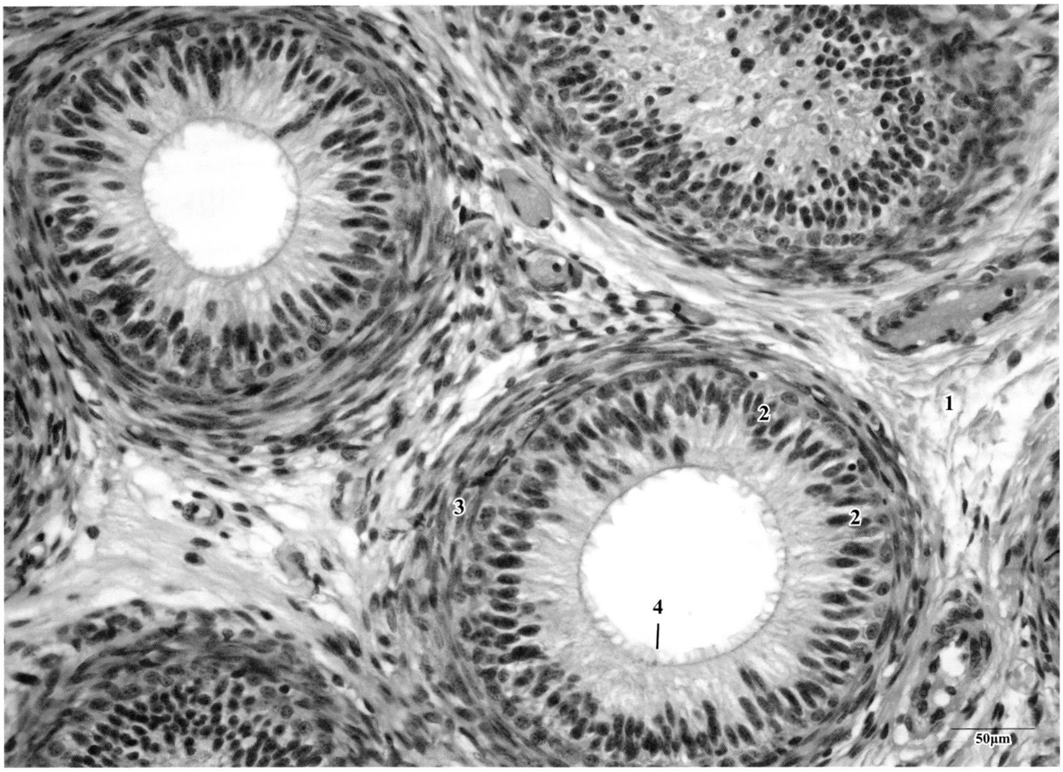

图 7-17　附睾尾组织切片，示附睾管（HE 染色，高倍镜）
Figure 7-17　Histological section of the tail of epididymis showing the epididymal duct (HE staining, higher power).

1- 疏松结缔组织 loose connective tissue
2- 假复层上皮 pseudostratified epithelium
3- 平滑肌 smooth muscle
4- 微绒毛 microvillus

1- 尾根 root of the tail
2- 肛门 anus
3- 阴囊 scrotum
4- 阴囊中缝 scrotal raphe
5- 阴囊内右侧睾丸 right testis within scrotum
6- 阴囊内左侧睾丸 left testis within scrotum

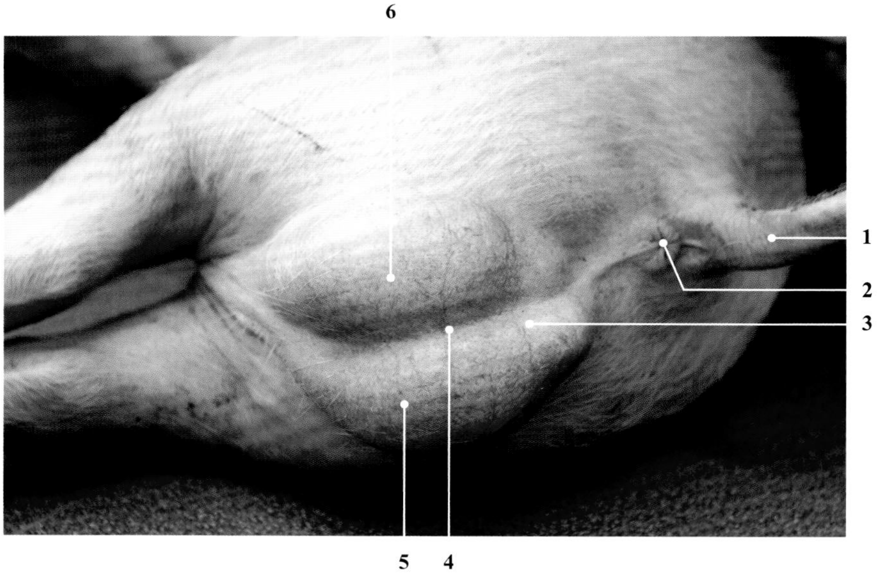

图 7-18　阴囊（后面观，水平位）
Figure 7-18　Scrotum (caudal view, horizontal position).

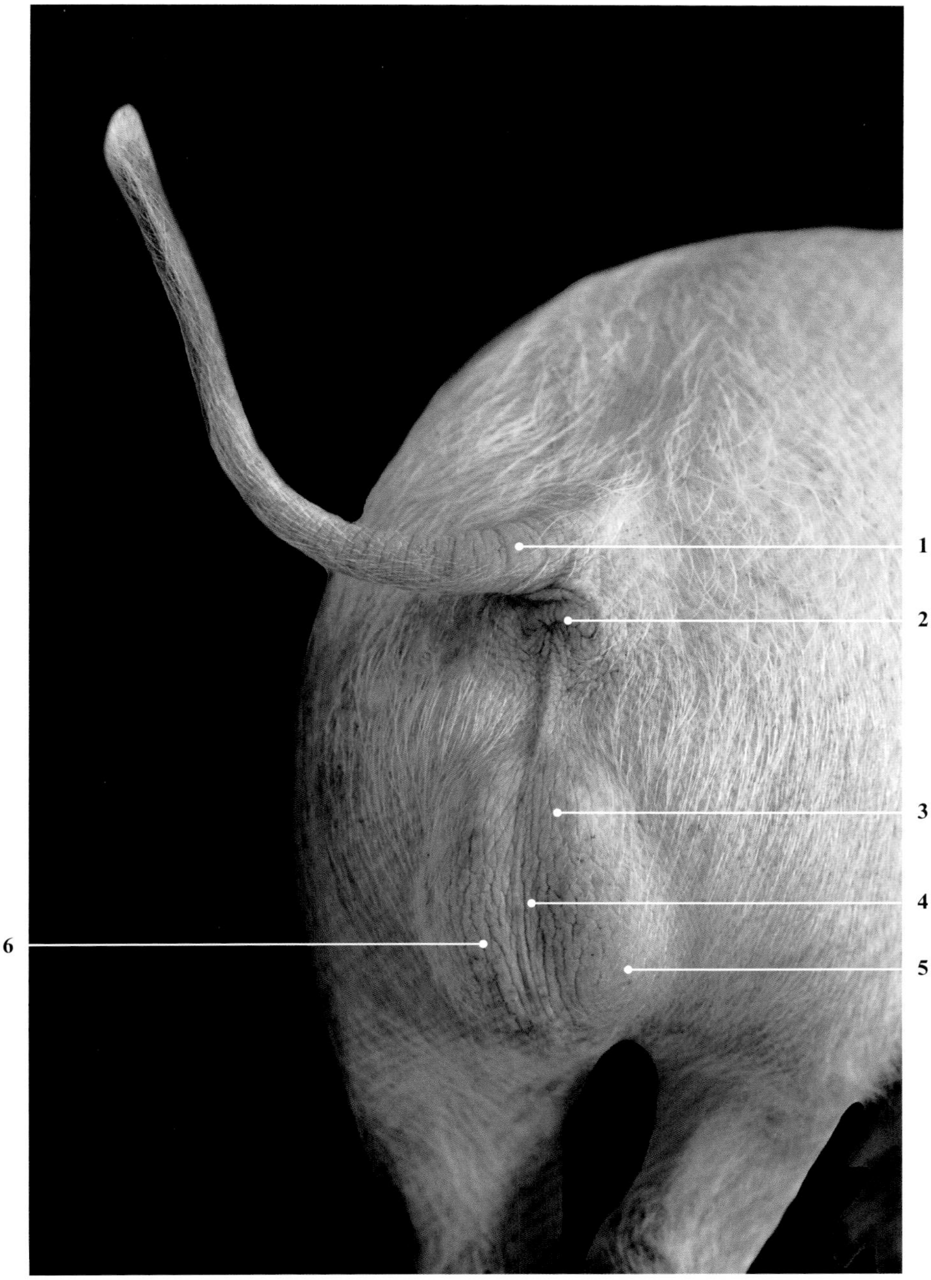

图 7-19　阴囊（后面观，垂直位）
Figure 7-19　Scrotum (caudal view, vertical position).

1- 尾根 root of the tail
2- 肛门 anus
3- 阴囊 scrotum
4- 阴囊中缝 scrotal raphe
5- 阴囊内右侧睾丸 right testis within scrotum
6- 阴囊内左侧睾丸 left testis within scrotum

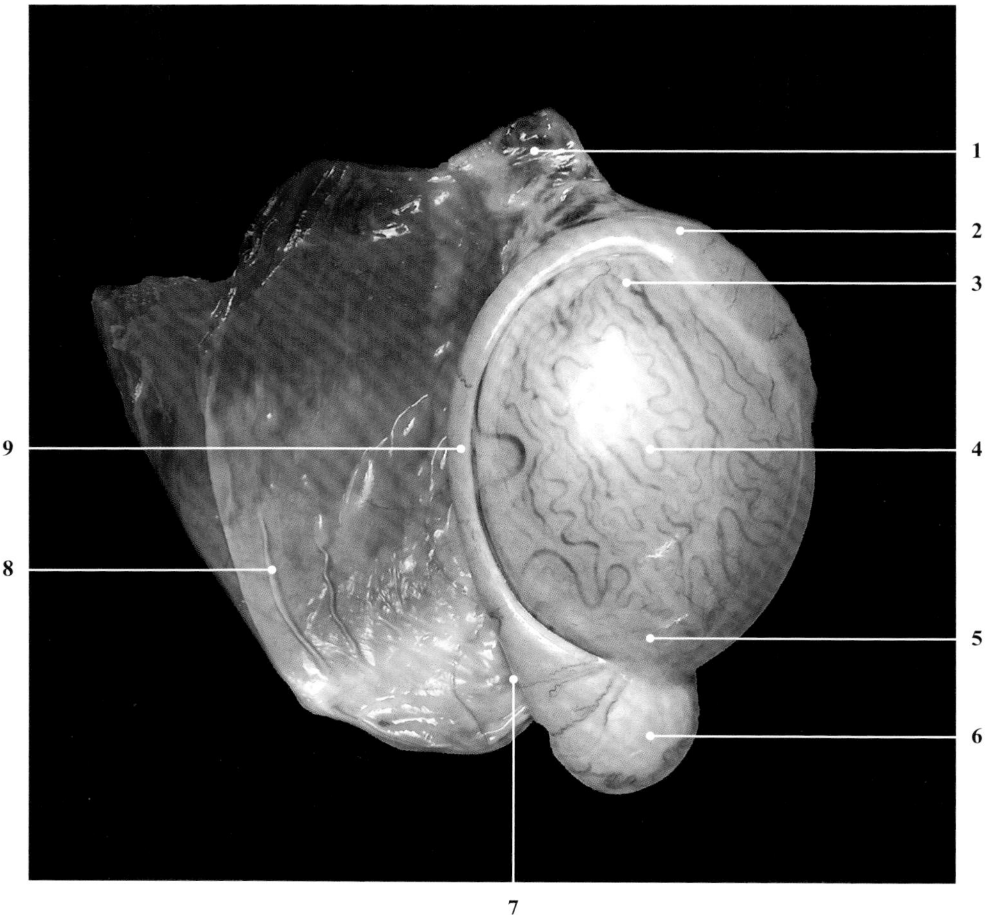

图 7-20　阴囊韧带
Figure 7-20　Scrotal ligament.

1- 精索 spermatic cord
2- 附睾头 head of epididymis
3- 睾丸头 head of testis
4- 睾丸体 body of testis
5- 睾丸尾 tail of testis
6- 附睾尾 tail of epididymis
7- 阴囊韧带 scrotal ligament
8- 阴囊壁 scrotal wall
9- 附睾体 body of epididymis

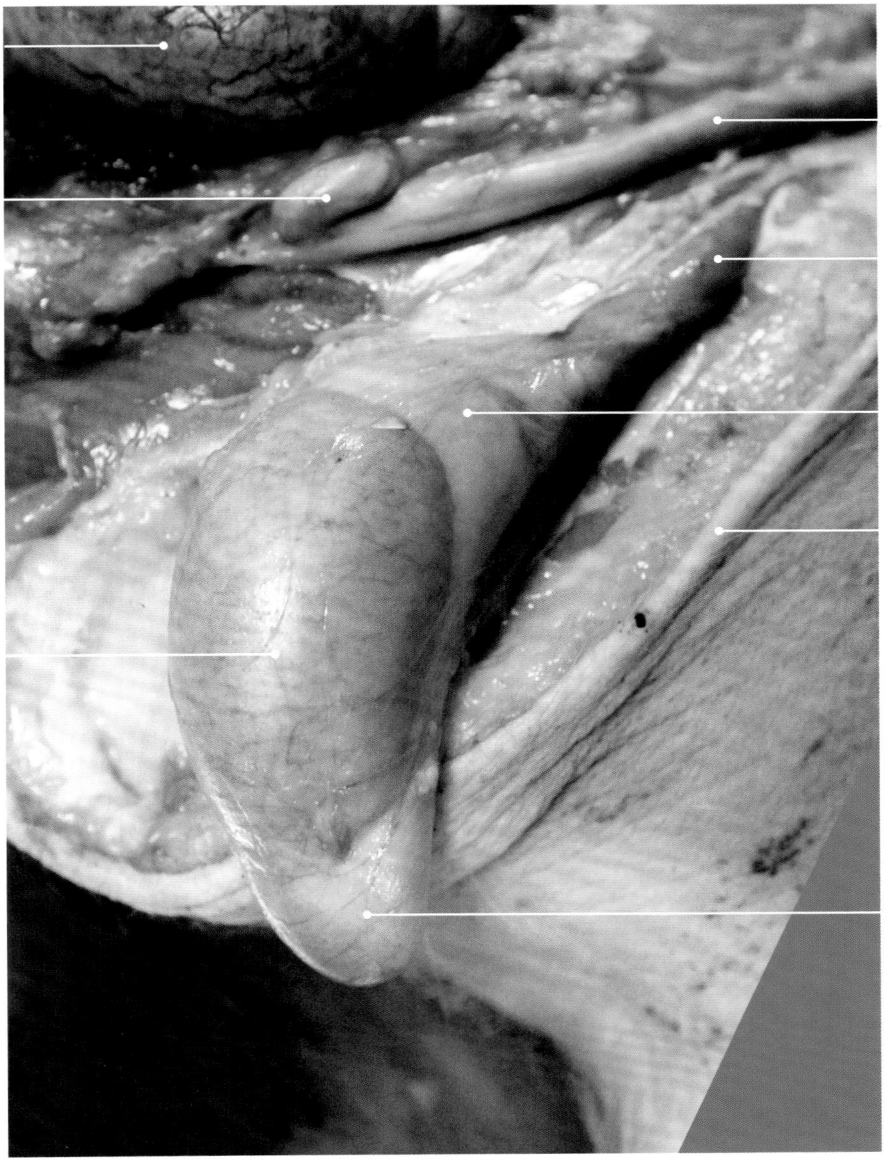

图 7-21 公猪阴囊内睾丸
Figure 7-21 Testis within the boar scrotum.

1- 阴茎 penis
2- 精索 spermatic cord
3- 附睾头 head of epididymis
4- 阴囊皮肤 scrotal skin
5- 附睾尾 tail of epididymis
6- 睾丸固有鞘膜 proper vagina tunica of testis
7- 乙状弯曲 sigmoid flexure
8- 右侧睾丸 right testis

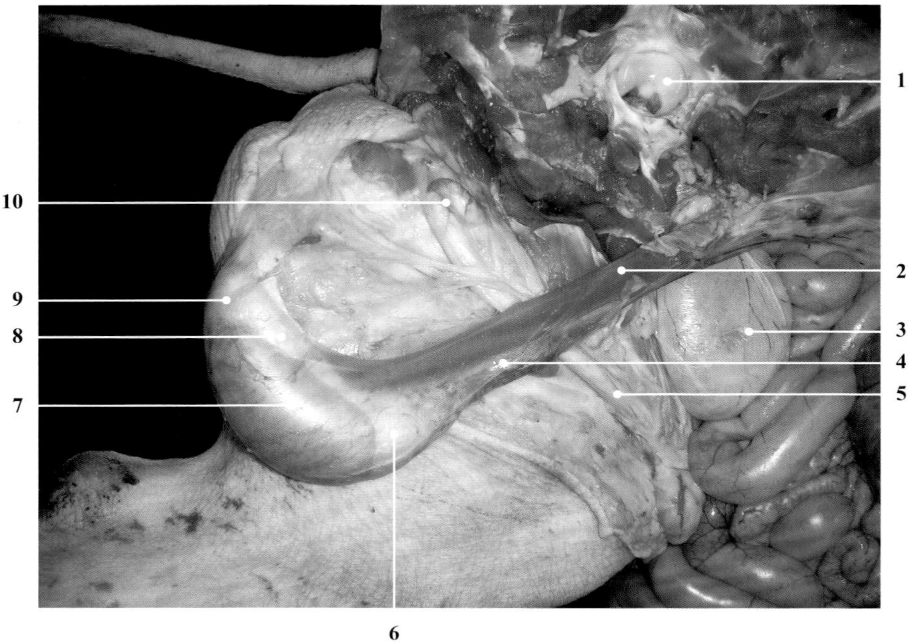

图 7-22　精索
Figure 7-22　Spermatic cord.

1- 髋臼 acetabulum
2- 提睾肌 cremaster muscle
3- 膀胱 urinary bladder
4- 精索 spermatic cord
5- 阴茎 penis
6- 附睾头 head of epididymis
7- 睾丸 testis
8- 附睾体 body of epididymis
9- 附睾尾 tail of epididymis
10- 乙状弯曲 sigmoid flexure

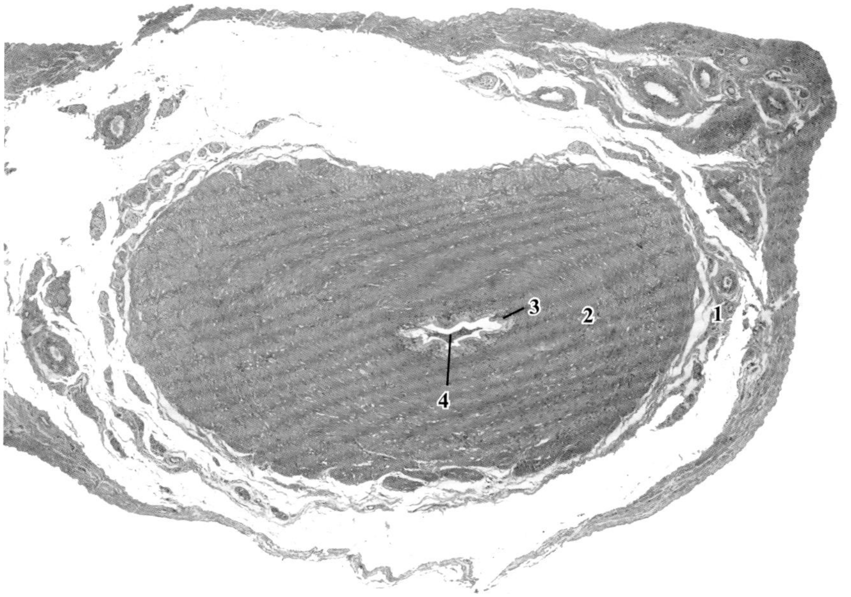

图 7-23　输精管组织切片（HE 染色，低倍镜）
Figure 7-23　Histological section of the deferent duct (HE staining, lower power).

1- 浆膜 serosa
2- 肌层 muscular layer
3- 假复层上皮 pseudostratified epithelium
4- 精子 spermatozoa

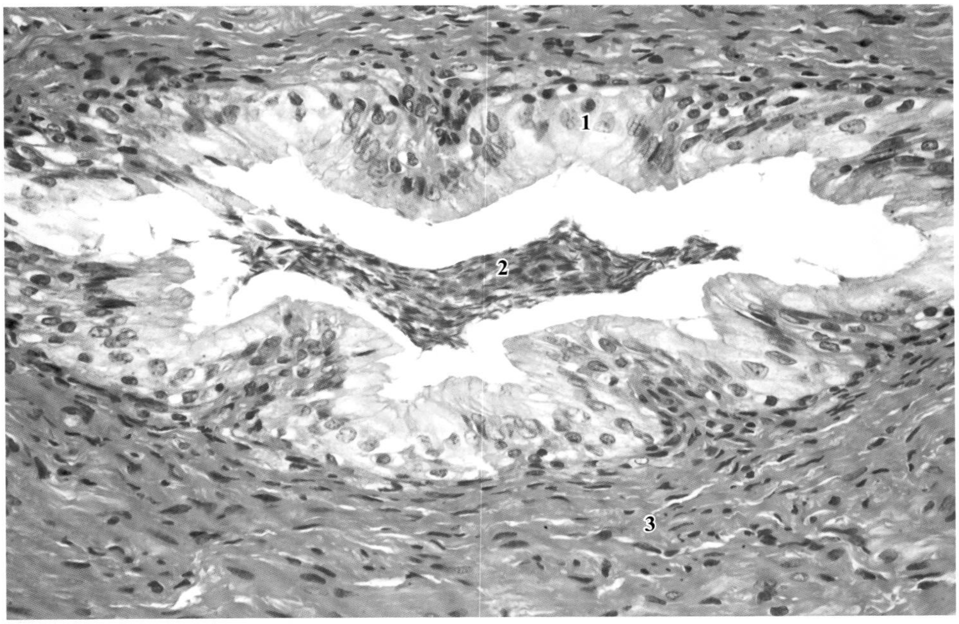

图 7-24 输精管组织切片（HE 染色，高倍镜）
Figure 7-24 Histological section of the deferent duct (HE staining, higher power).

1- 假复层上皮 pseudostratified epithelium
2- 精子 spermatozoa
3- 肌层 muscular layer

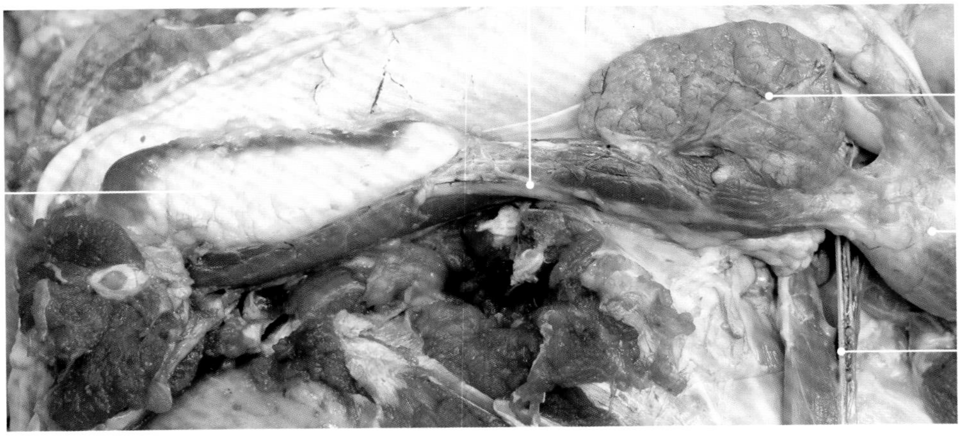

图 7-25 副性腺（左外侧观）
Figure 7-25 Accessory genital gland (left lateral view).

1- 精囊腺 vesicular gland
2- 膀胱 urinary bladder
3- 精索 spermatic cord
4- 尿道球腺 bulbourethral gland
5- 尿生殖道骨盆部 pelvis part of urogenital tract

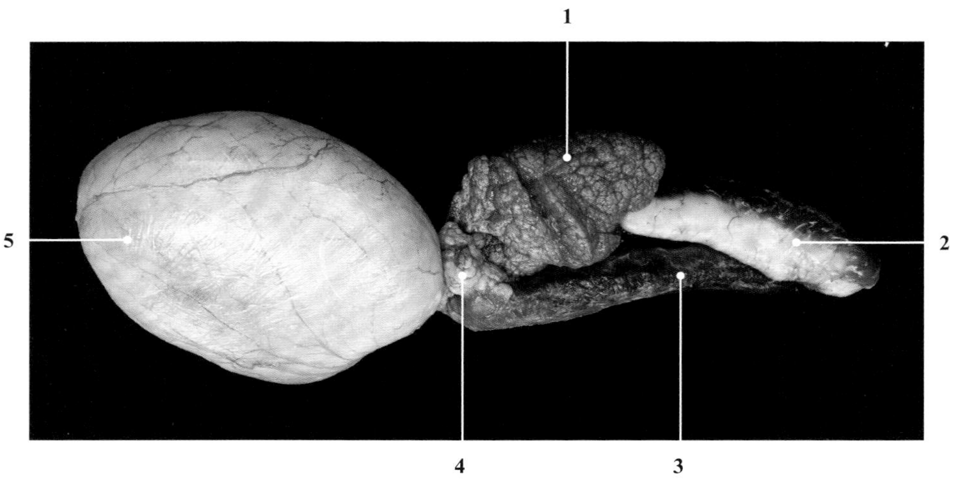

图 7-26 副性腺（右外侧观）
Figure 7-26 Accessory genital gland (right lateral view).

1- 精囊腺 vesicular gland
2- 尿道球腺 bulbourethral gland
3- 尿生殖道骨盆部 pelvis part of urogenital tract
4- 前列腺 prostate gland
5- 膀胱 urinary bladder

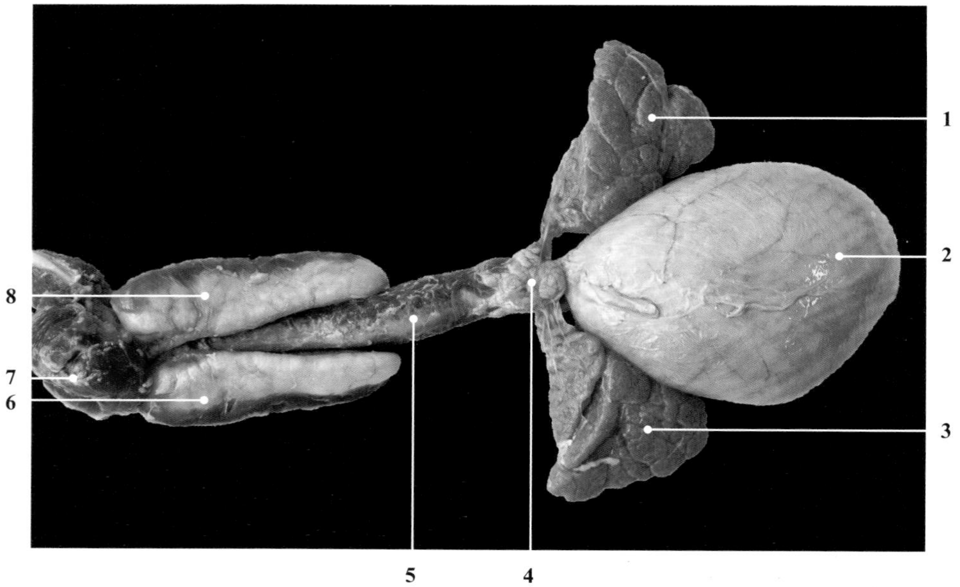

图 7-27 副性腺（背侧观）
Figure 7-27 Accessory genital gland (dorsal view).

1- 左侧精囊腺 left vesicular gland
2- 膀胱 urinary bladder
3- 右侧精囊腺 right vesicular gland
4- 前列腺 prostate gland
5- 尿生殖道骨盆部 pelvis part of urogenital tract
6- 右侧尿道球腺 right bulbourethral gland
7- 球海绵体肌 bulbocavernous muscle
8- 左侧尿道球腺 left bulbourethral gland

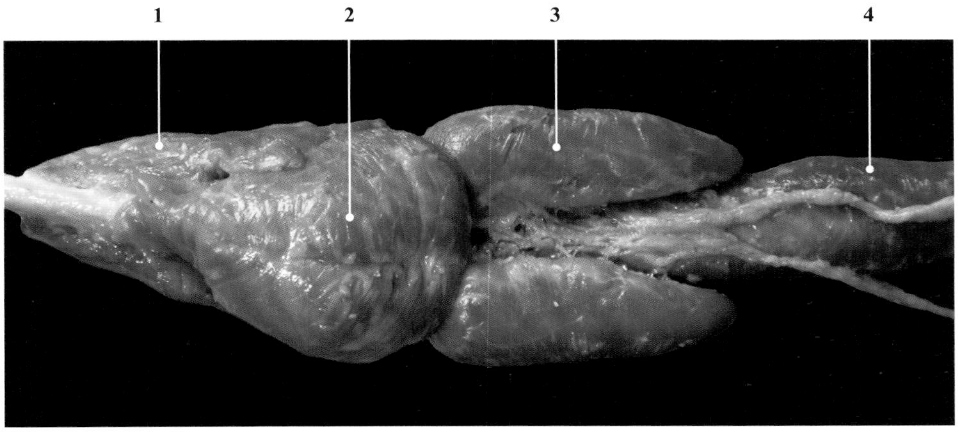

图 7-28 球海绵体肌与坐骨海绵体肌
Figure 7-28 Bulbocavernous muscle and ischiocavernosus muscle.

1- 坐骨海绵体肌 ischiocavernosus muscle
2- 球海绵体肌 bulbocavernous muscle
3- 尿道球腺 bulbourethral gland
4- 尿生殖道骨盆部 pelvis part of urogenital tract

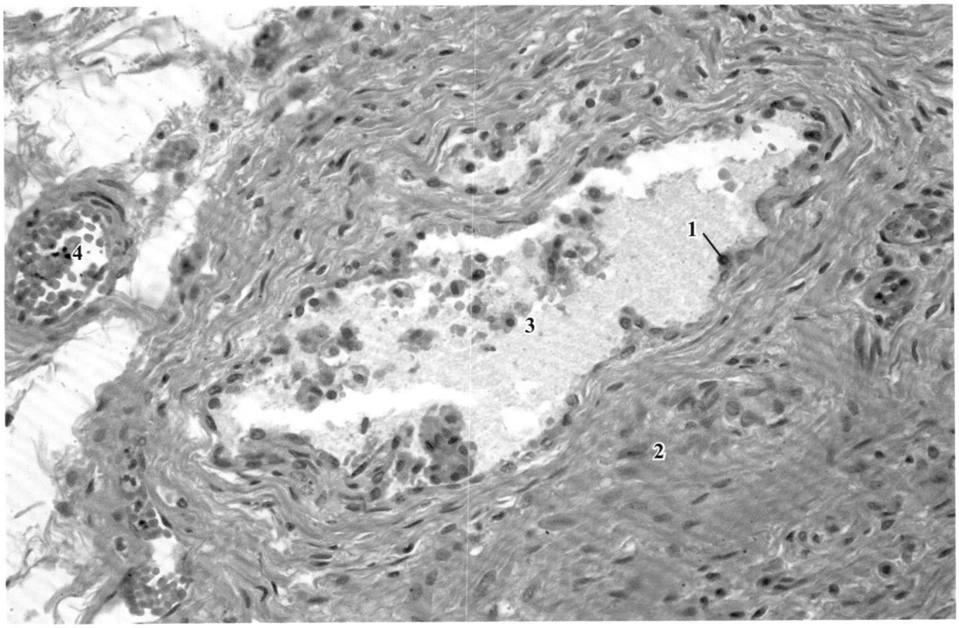

图 7-29 精囊腺组织切片（HE 染色，低倍镜）
Figure 7-29 Histological section of the vesicular gland (HE staining, lower power).

1- 基底细胞 basal cell
2- 小叶间隔 interlobular septum
3- 腺腔 lumen of gland
4- 血管 blood vessel

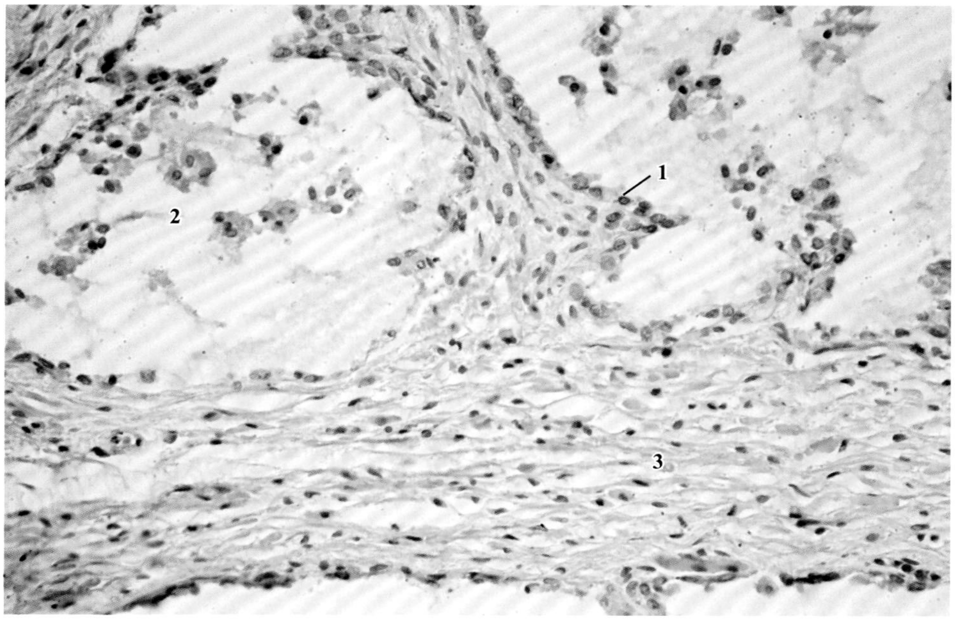

图 7-30 精囊腺组织切片（HE 染色，高倍镜）
Figure 7-30 Histological section of the vesicular gland (HE staining, higher power).

1- 基底细胞 basal cell
2- 腺腔 lumen of gland
3- 小叶间隔 interlobular septum

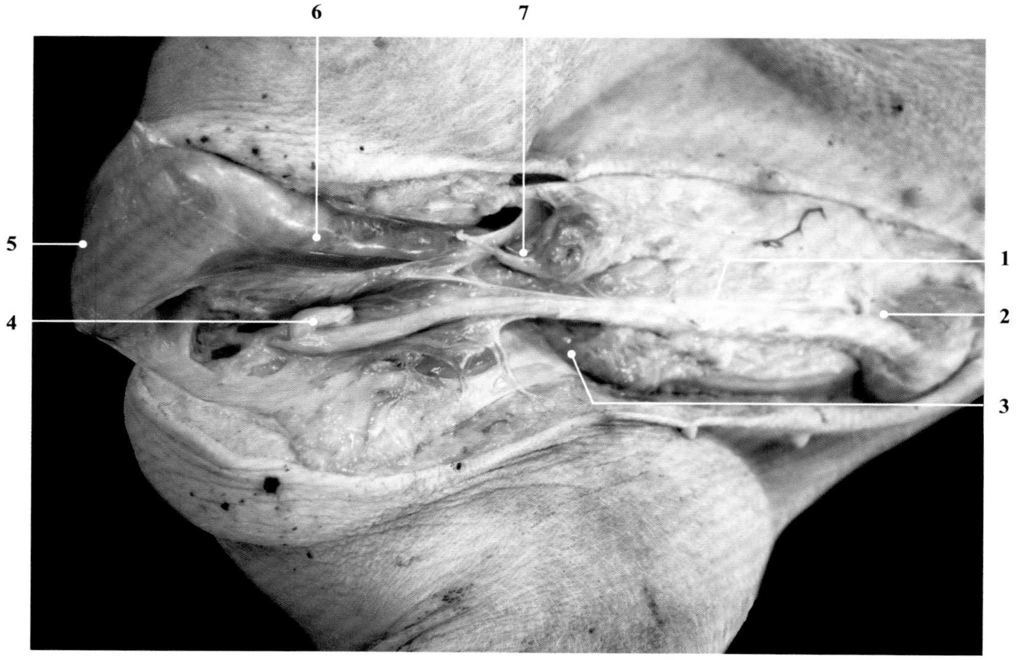

图 7-31 公猪阴茎乙状弯曲
Figure 7-31 Sigmoid flexure of the penis of a boar.

1- 阴茎 penis
2- 包皮憩室（包皮盲囊） preputial diverticulum
3- 左侧阴囊淋巴结 left scrotal lymph node
4- 乙状弯曲 sigmoid flexure
5- 睾丸 testis
6- 精索 spermatic cord
7- 右侧阴囊淋巴结 right scrotal lymph node

图 7-32 阴茎退缩肌
Figure 7-32 Retractor penis muscle.

1- 乙状弯曲 sigmoid flexure
2- 阴茎缩肌 retractor penis muscle
3- 阴茎 penis

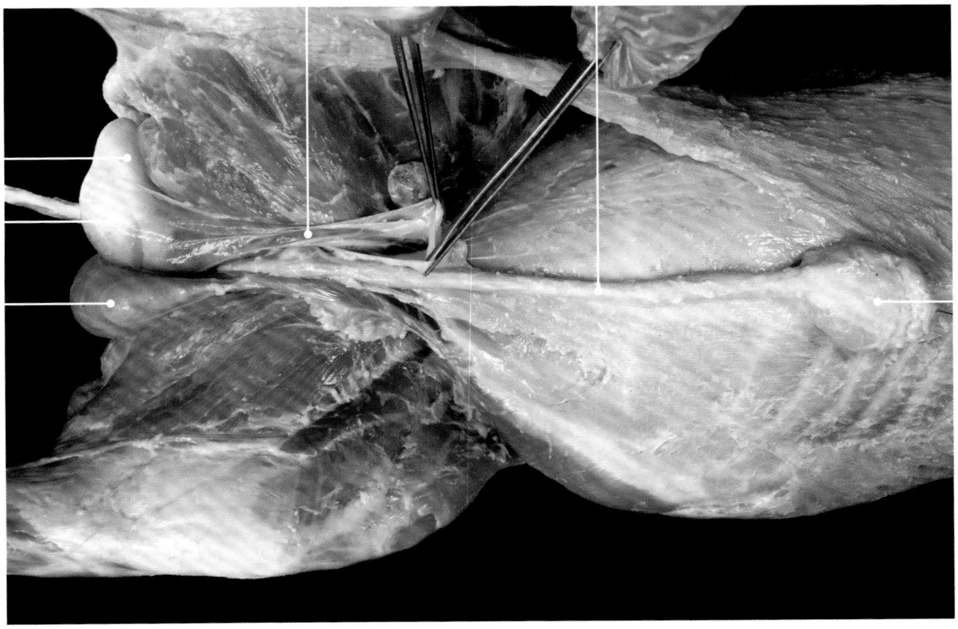

图 7-33 阴茎（腹侧观）
Figure 7-33 Penis (ventral view).

1- 包皮憩室（包皮盲囊） preputial diverticulum
2- 左侧睾丸 left testis
3- 右侧睾丸 right testis
4- 右侧附睾 right epididymis
5- 精索 spermatic cord
6- 阴茎 penis

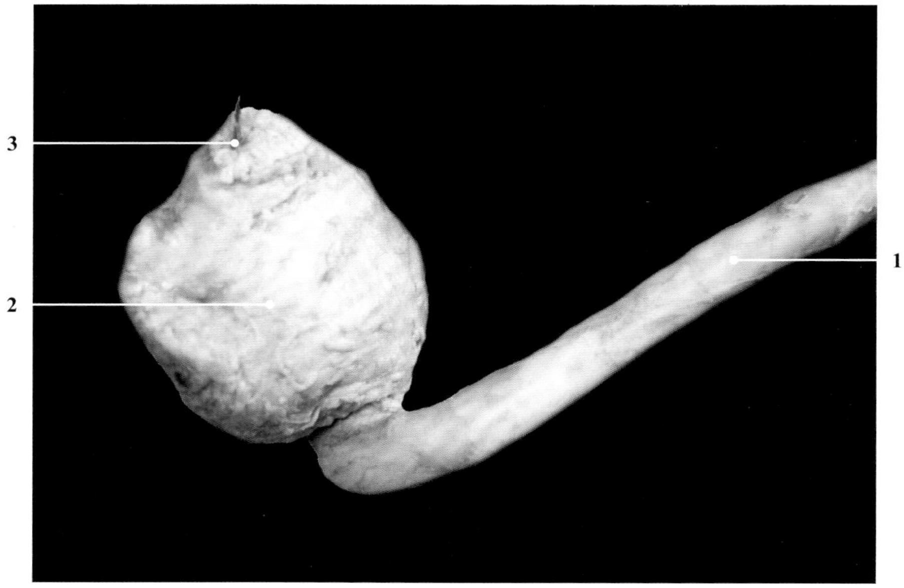

图 7-34　包皮憩室
Figure 7-34　Preputial diverticulum.

1- 阴茎 penis
2- 包皮憩室（包皮盲囊） preputial diverticulum
3- 尿道外口 external orifice of urethra

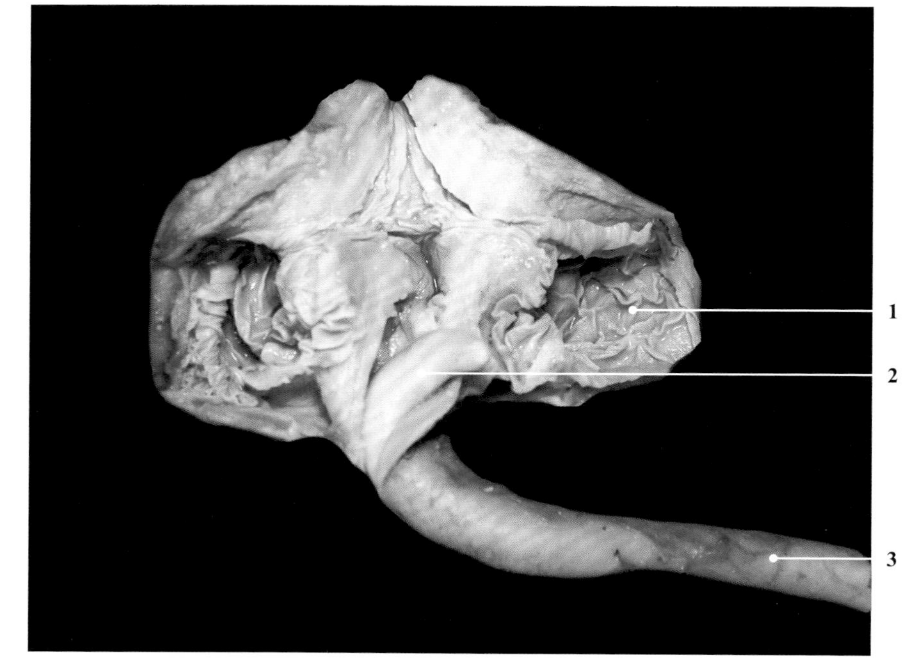

图 7-35　包皮憩室剖面
Figure 7-35　Section of preputial diverticulum.

1- 包皮憩室（包皮盲囊） preputial diverticulum
2- 阴茎头 glans of penis
3- 阴茎 penis

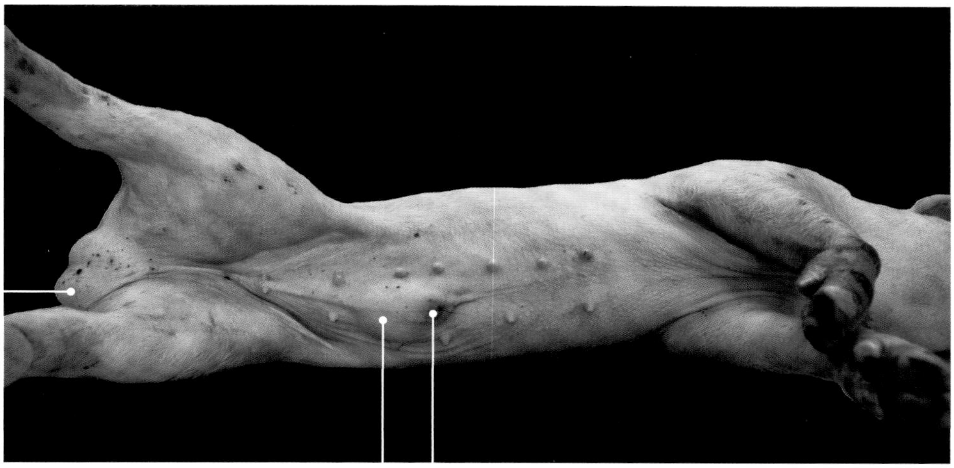

图 7-36 包皮
Figure 7-36 Prepuce.

1- 尿道外口 external orifice of urethra
2- 包皮 prepuce
3- 阴囊 scrotum

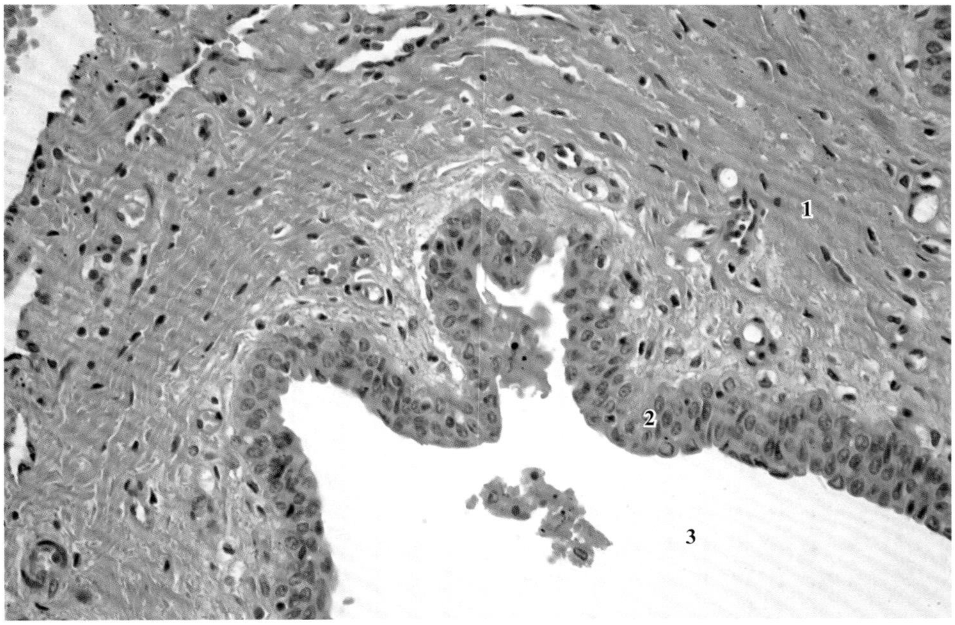

图 7-37 尿道组织切片（HE 染色，高倍镜）
Figure 7-37 Histological section of the urethra (HE staining, higher power).

1- 尿道海绵体 cavernous body of urethra
2- 复层柱状上皮 stratified columnar epithelium
3- 尿道腔 urethral lumen

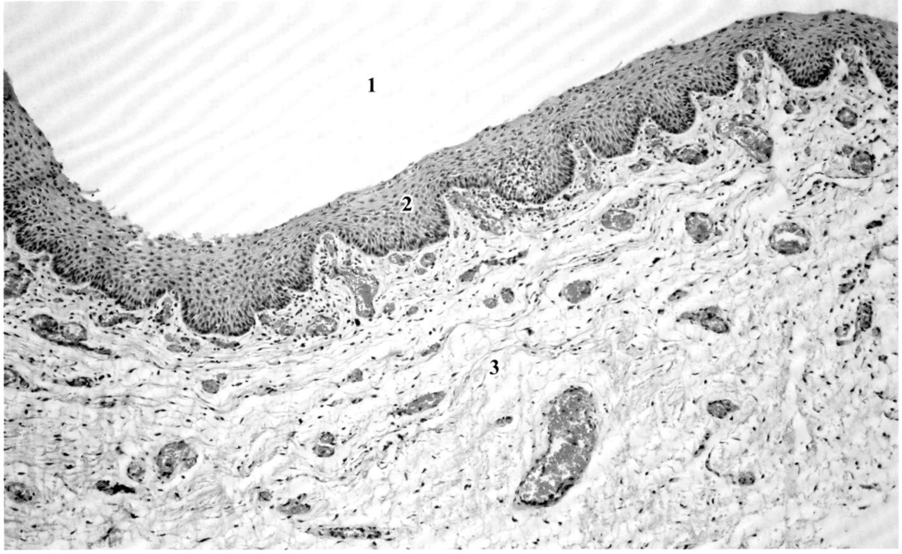

1- 尿道腔 urethral lumen
2- 复层柱状上皮 stratified columnar epithelium
3- 海绵体 corpus cavernous body (spongy body)

图 7-38　后段尿道组织切片（HE 染色，高倍镜）
Figure 7-38　Histological section of the posterior segment of a urethra (HE staining, higher power).

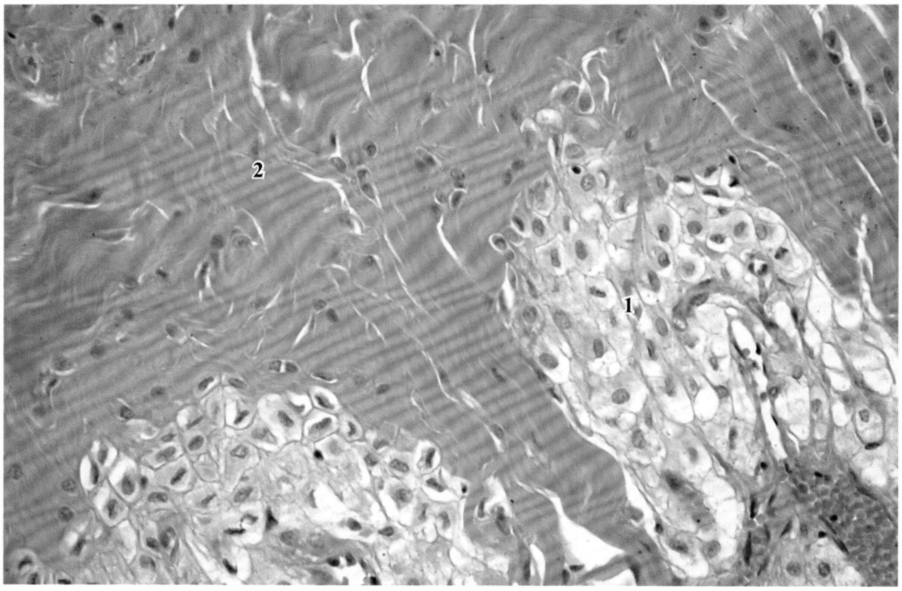

1- 阴茎海绵体 cavernous body of penis
2- 平滑肌 smooth muscle

图 7-39　阴茎海绵体组织切片（HE 染色，高倍镜）
Figure 7-39　Histological section of the cavernous body of penis (HE staining, higher power).

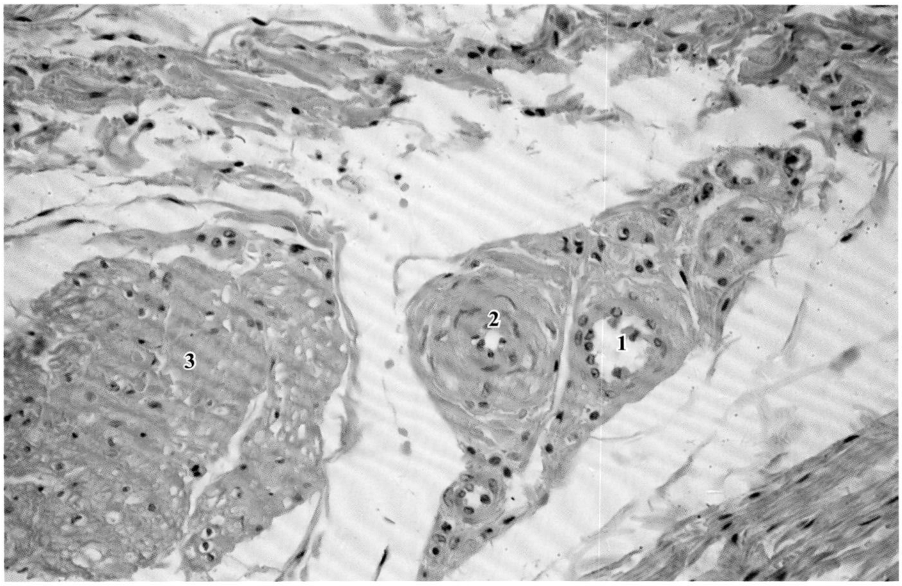

1- 小静脉 small vein
2- 小动脉 small artery
3- 海绵体 cavernous body

图 7-40 阴茎海绵体组织切片（HE 染色，高倍镜）
Figure 7-40 Histological section of the cavernous body of penis (HE staining, higher power).

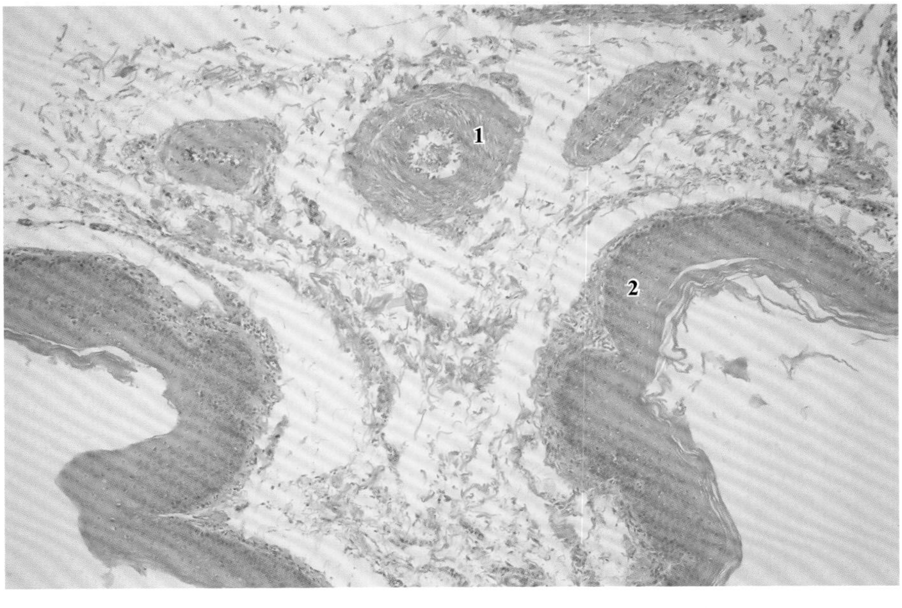

1- 小动脉 arteriole
2- 复层扁平上皮 stratified squamous epithelium

图 7-41 包皮憩室组织切片（HE 染色，低倍镜）
Figure 7-41 Histological section of the preputial diverticulum (HE staining, lower power).

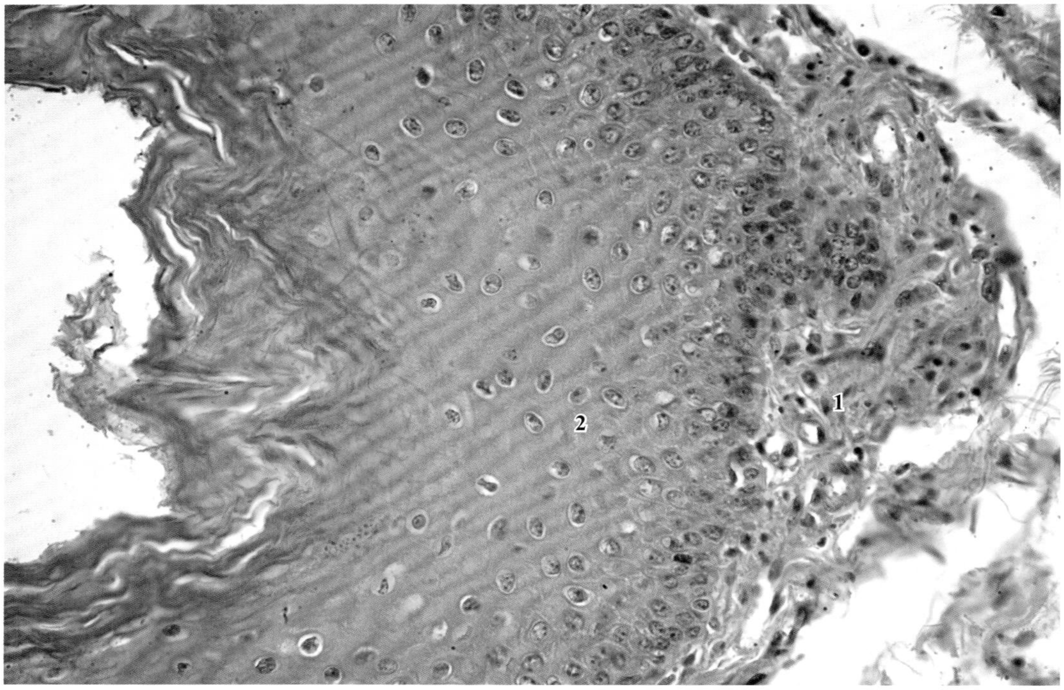

图 7-42 包皮憩室组织切片（HE 染色，高倍镜）
Figure 7-42 Histological section of the preputial diverticulum (HE staining, higher power).

1- 真皮乳头层 papillary layer of dermis
2- 复层扁平上皮 stratified squamous epithelium

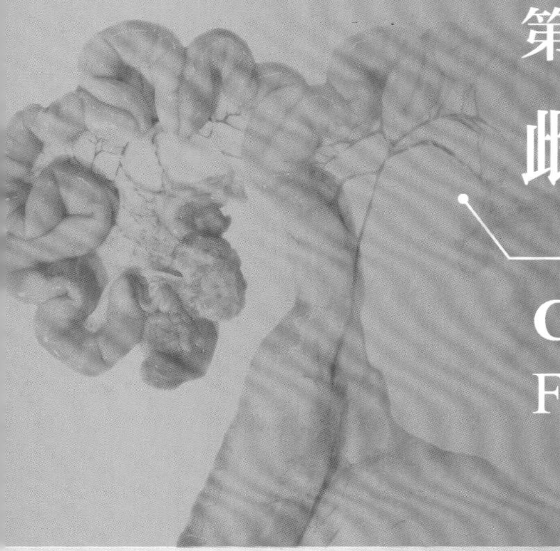

第八章

雌性生殖系统

Chapter 8
Female reproductive system

母猪生殖器官包括卵巢、输卵管、子宫、阴道、阴道前庭和阴门。

一、卵巢

卵巢（ovary）是产生卵子和分泌雌性激素的器官，其位置、形态、大小因年龄和个体不同而有较大变化。性成熟前的小母猪，卵巢较小，表面光滑，位于荐骨岬两侧稍后方，腰小肌腱附近，位置比较固定。接近性成熟时，卵巢体积增大，表面有突出的卵泡，呈桑椹状。卵巢位置稍下垂前移，位于髋结节前缘横断面的腰下部。性成熟后经产母猪卵巢变得更大，表面因有卵泡、黄体突出而呈结节状。卵巢位于髋结节前缘前方约4 cm的横断面上，包于发达的卵巢囊内。卵巢借卵巢系膜系于腰下部或骨盆腔前口处，借卵巢固有韧带与子宫角相连。卵巢固有韧带与输卵管系膜之间形成卵巢囊（ovarian bursa）。

卵巢的一般结构可分为被膜、皮质和髓质。被膜包括表面上皮（superficial epithelium）和白膜（tunica albuginea）。卵巢表面除卵巢系膜附着部外，都覆盖着一层表面上皮，或称生殖上皮（germinal epithelium）。在表面上皮的内侧是由致密结缔组织构成的白膜。卵巢的实质分为外周的皮质和中央的髓质。皮质是由基质、处于不同发育阶段的卵泡、闭锁卵泡和黄体等构成。髓质为富含弹性纤维、血管、淋巴管和神经等的疏松结缔组织。

卵泡（ovarian follicle）是由位于中央的一个卵母细胞和围绕其周围的许多卵泡细胞（follicular cell）组成。根据结构变化，卵泡分为原始卵泡（primordial follicle）、生长卵泡（growing follicle）和成熟卵泡（mature follicle）3个阶段。原始卵泡于胚胎时期开始形成，性成熟后，在促卵泡素的作用下，静止的原始卵泡开始生长发育，称为生长卵泡。根据发育阶段不同，可将其分为初级卵泡、次级卵泡和三级卵泡3个连续阶段：生长卵泡根据其有无卵泡腔分为初级卵泡和次级卵泡。三级卵泡发育到即将排卵的阶段，即为成熟卵泡。此时卵泡体积最大（5 ~ 10 mm）。猪为多胎动物，一次可排多个卵，排卵后卵泡壁和卵泡壁细胞向卵泡腔塌陷，颗粒细胞和卵泡内膜细胞增生分化，吸收血液，形成黄体（corpus luteum），黄体有一部分突出于卵巢表面，由于缺少黄体色素，黄体呈肉色。

二、输卵管

输卵管（uterine tube）是连接卵巢和子宫角之间的一对弯曲的管道，被输卵管系膜包围固定。输卵管系膜位于卵巢外侧，是连接卵巢系膜和子宫阔韧带的浆膜褶。输卵管具有收集、输送卵细胞的功能，也是卵细胞受精的场所。

输卵管的前端扩大呈漏斗状，称为输卵管漏斗（infundibulum of uterine tube），漏斗的边缘不规则，呈伞状，称输卵管伞（fimbriae of uterine tube）。漏斗中央深处有一口为输卵管腹腔口（abdominal orifice of uterine tube），与腹膜腔相通，卵细胞由此进入输卵管。输卵管的前段管径最粗，称为输卵管壶腹（ampulla of uterine tube），卵细胞常在此受精，之后进入子宫着床。后段较短，细而直，管壁较厚，称输卵管峡（isthmus of uterine tube），末端以输卵管子宫口与子宫角相连通。

输卵管的管壁由黏膜、肌层和浆膜构成。黏膜形成纵的输卵管褶，其上具有纤毛；肌层主要是环形平滑肌；浆膜包围在输卵管的外面，并形成输卵管系膜。

三、子宫

子宫（uterus）是胎儿附植和孕育的地方，为有腔的肌质器官，壁较厚，由内向外可分为内膜、肌层和浆膜3层。母猪属于双角子宫（uterus bicornis），分为子宫角（uterine horn）、子宫体（uterine body）和子宫颈（uterine cervix）3部分。子宫角为子宫的前部，较长，形成襻状弯曲类似小肠。其前端以输卵管子宫口与输卵管相通，向后延续为子宫体。子宫体位于骨盆腔内，一部分伸入腹腔内，呈圆筒状，不发达，向后延续为子宫颈。子宫颈为子宫后段的缩细部，较长，壁很厚，黏膜形成许多纵褶，内腔狭窄，称为子宫颈管（cervical canal of uterus），子宫颈管呈螺旋状，前端以子宫颈内口与子宫体相通，后端以子宫颈外口（uterine external ostium）通阴道，无子宫颈阴道突。

子宫被子宫阔韧带（broad ligament of uterus）所固定。后者为一宽厚的腹膜褶，内有丰富的结缔组织、血管、神经及淋巴管。子宫阔韧带的外侧前部有一发达的浆膜褶，称为子宫圆韧带（round ligament of uterus）。

四、阴道

阴道（vagina）是母猪的交配器官和产道，呈扁管状，位于骨盆腔内，在子宫后方，向后延接阴道前庭，其背侧与直肠相邻，腹侧与膀胱及尿道相邻。猪的阴道腔直径很大，无阴道穹隆（fundus of vagina）。

五、阴道前庭

阴道前庭（vaginal vestibule）或称尿生殖前庭（urogenital vestibulum），是母猪的交配器官和产道，也是尿液排出的经路。阴道前庭位于骨盆腔内，直肠的腹侧，其前接阴道，后端以阴门与外界相通。在阴道前庭前端腹侧壁上有一条环形黏膜褶称为阴瓣（hymen），阴瓣的后方有尿道外口（external urethral orifice），可作为前庭与阴道的分界。在阴道前庭的两侧壁有前庭大腺（greater vestibular gland）和前庭小腺（lesser vestibular gland）的开口。

六、阴门

阴门（vulva）位于肛门腹侧，由左右两阴唇（labium of vulva）构成，两阴唇间的裂缝称为阴门裂（rima vulvae）。阴唇上下两端的联合，分别称为阴唇背侧联合和阴唇腹侧联合。在腹侧联合前方有一阴蒂窝（clitoral fossa），内有细长的阴蒂（clitoris），突出于阴蒂窝的表面。

七、雌性尿道

雌性尿道较短，位于阴道腹侧，前端与膀胱颈相接，后端开口于阴道前庭起始部的腹侧壁，为尿道外口。

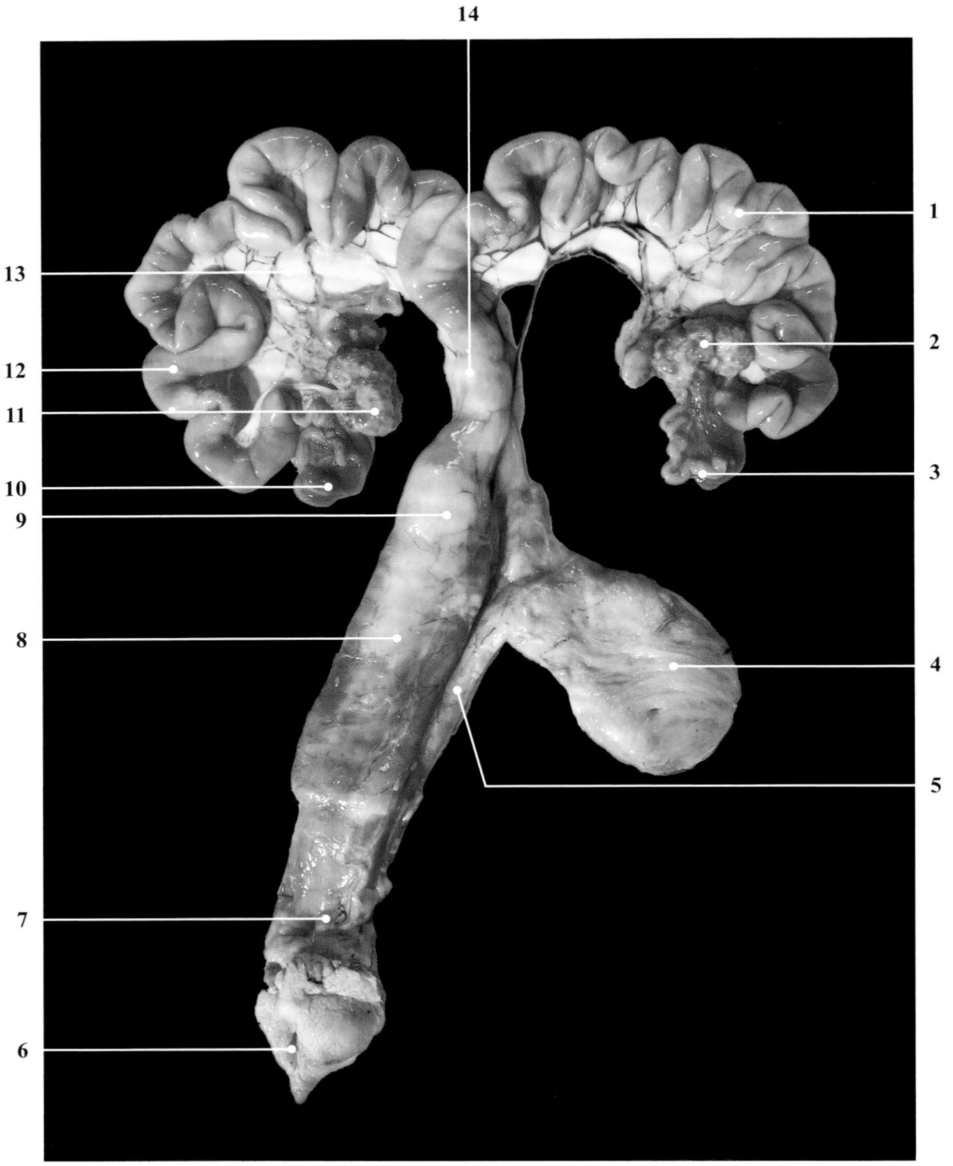

图 8-1　母猪生殖系统（背侧观）
Figure 8-1　Reproductive system of a sow (dorsal view).

1- 右侧子宫角 right uterine horn
2- 右侧卵巢 right ovary
3- 右侧输卵管 right uterine tube
4- 膀胱 urinary bladder
5- 尿道 urethra
6- 阴门 vulva
7- 阴道前庭 vaginal vestibule
8- 阴道 vagina
9- 子宫颈 uterine cervix
10- 左侧输卵管 left uterine tube
11- 左侧卵巢 left ovary
12- 左侧子宫角 left uterine horn
13- 子宫阔韧带 broad ligament of uterus
14- 子宫体 uterine body

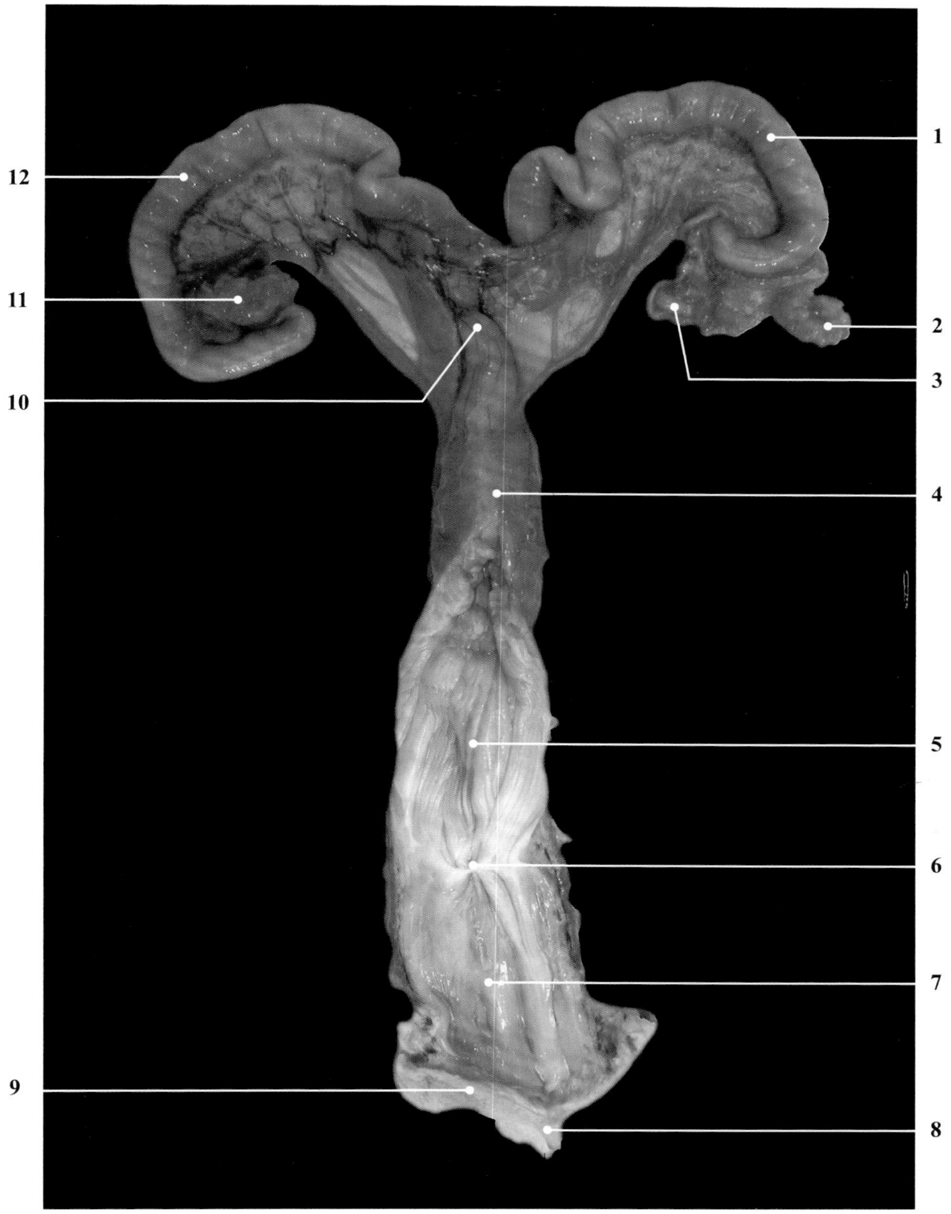

图 8-2 母猪生殖系统（背侧观，阴道和阴道前庭切开）
Figure 8-2 Reproductive system of a sow, partially opened on midline (dorsal view, opening the vagina and vaginal vestibule).

1- 右侧子宫角 right uterine horn
2- 右侧卵巢 right ovary
3- 右侧输卵管 right uterine tube
4- 子宫颈 uterine cervix
5- 阴道 vagina
6- 阴瓣 hymen
7- 阴道前庭 vaginal vestibule
8- 阴蒂 clitoris
9- 阴唇 labium of vulva
10- 子宫体 uterine body
11- 左侧卵巢 left ovary
12- 左侧子宫角 left uterine horn

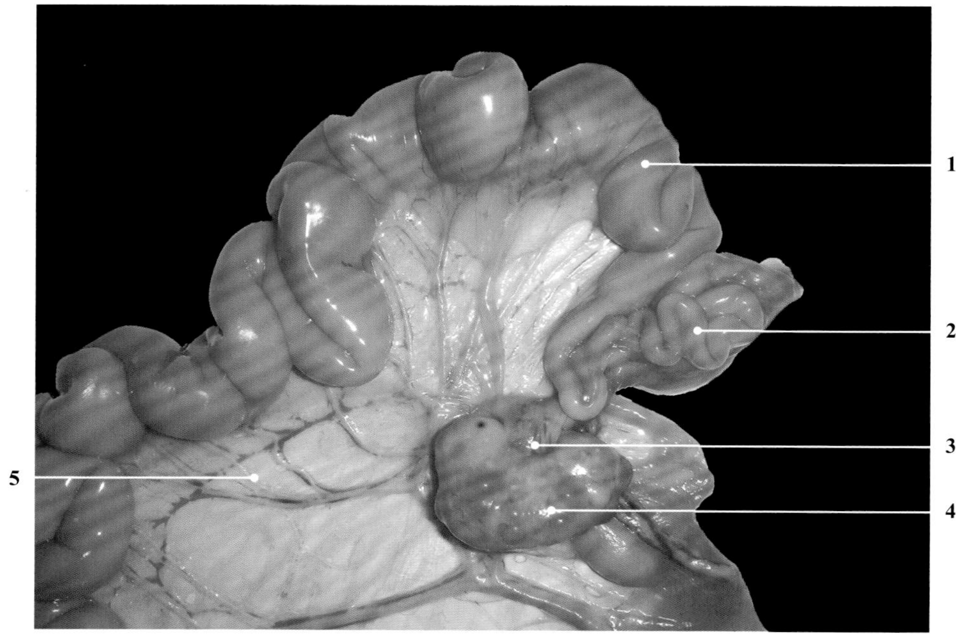

图 8-3　左侧卵巢
Figure 8-3　Left ovary.

1- 子宫角 uterine horn
2- 输卵管 uterine tube
3- 卵巢门 hilum of the ovary
4- 卵巢 ovary
5- 子宫阔韧带 broad ligament of uterus

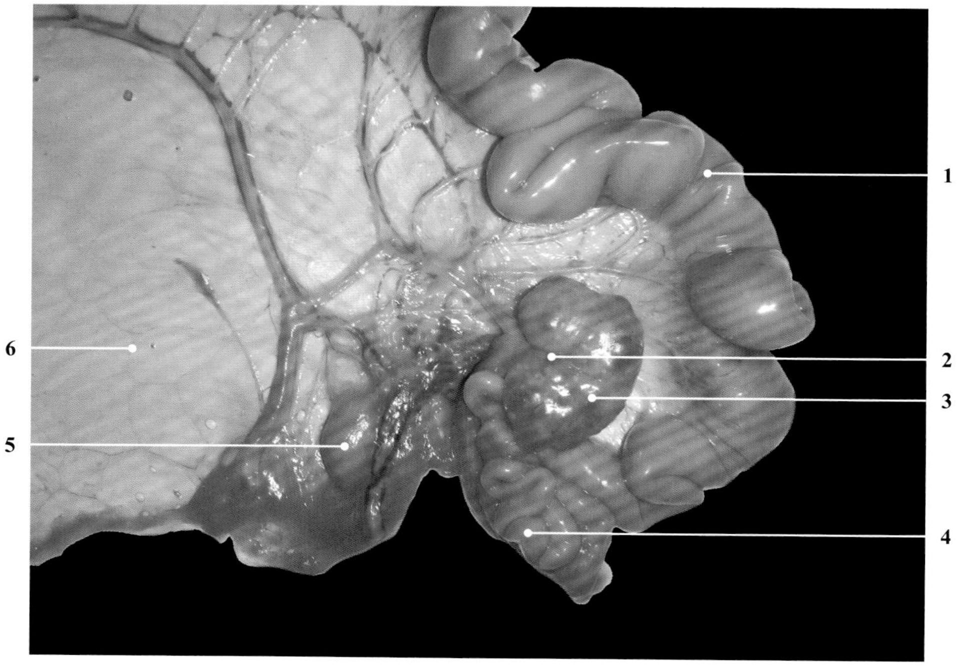

图 8-4　右侧卵巢
Figure 8-4　Right ovary.

1- 子宫角 uterine horn
2- 卵巢门 hilum of the ovary
3- 卵巢 ovary
4- 输卵管 uterine tube
5- 卵巢淋巴结 lymph node of the ovary
6- 子宫阔韧带 broad ligament of uterus

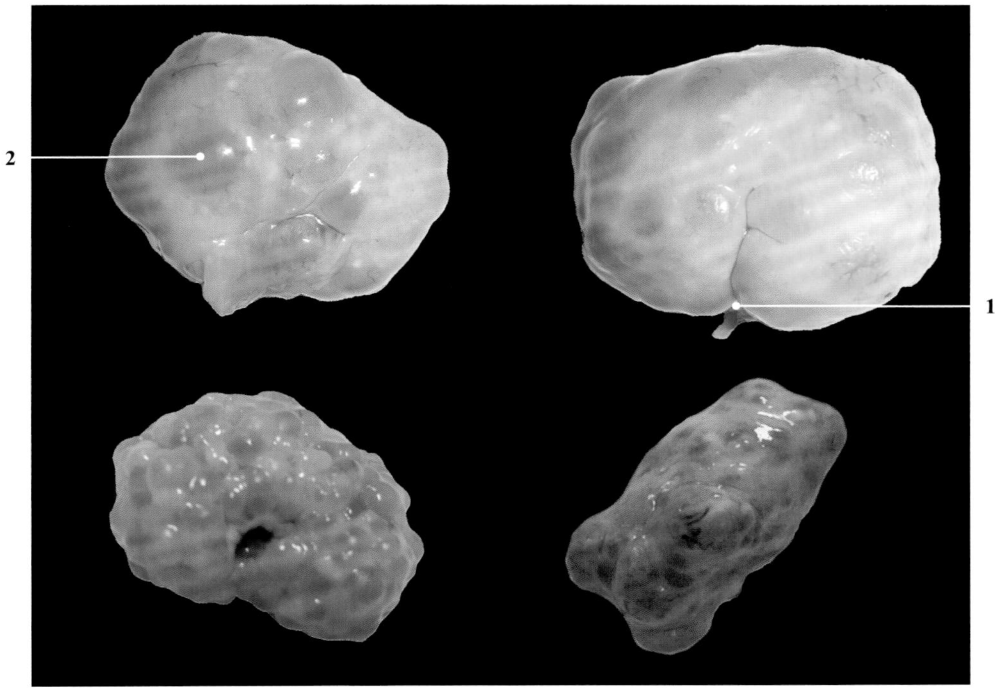

图 8-5　卵巢形态
Figure 8-5　The morphology of ovary.

1- 卵巢门 hilum of the ovary

2- 卵泡 ovarian follicle

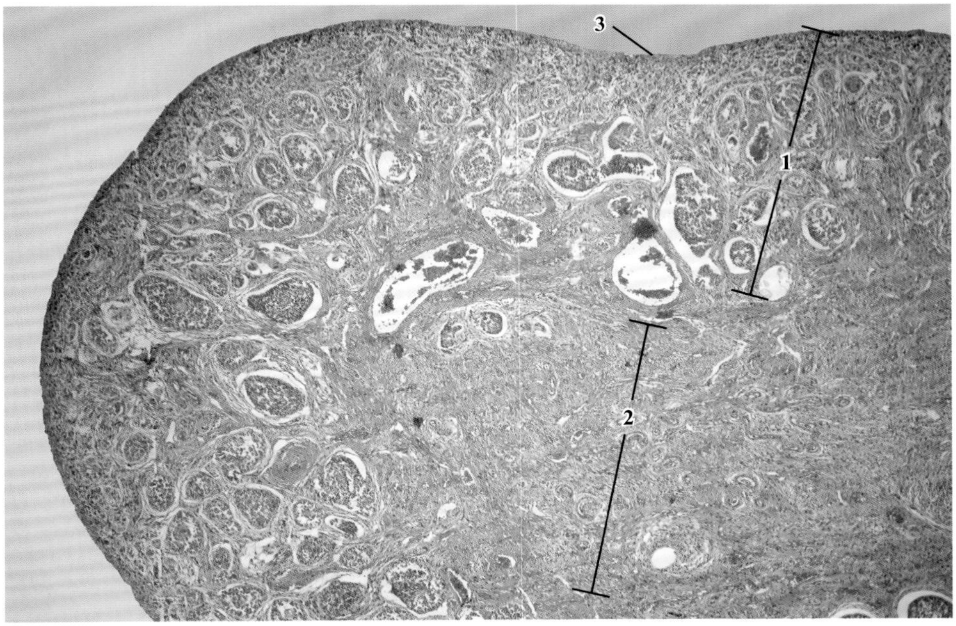

图 8-6　卵巢组织切片（HE 染色，低倍镜）
Figure 8-6　Histological section of the ovary (HE staining, lower power).

1- 皮质 cortex

2- 髓质 medulla

3- 被膜 capsule

1- 表面上皮（生殖上皮）
superficial epithelium
(germinal epithelium)
2- 白膜 tunica albuginea
3- 血管 blood vessel

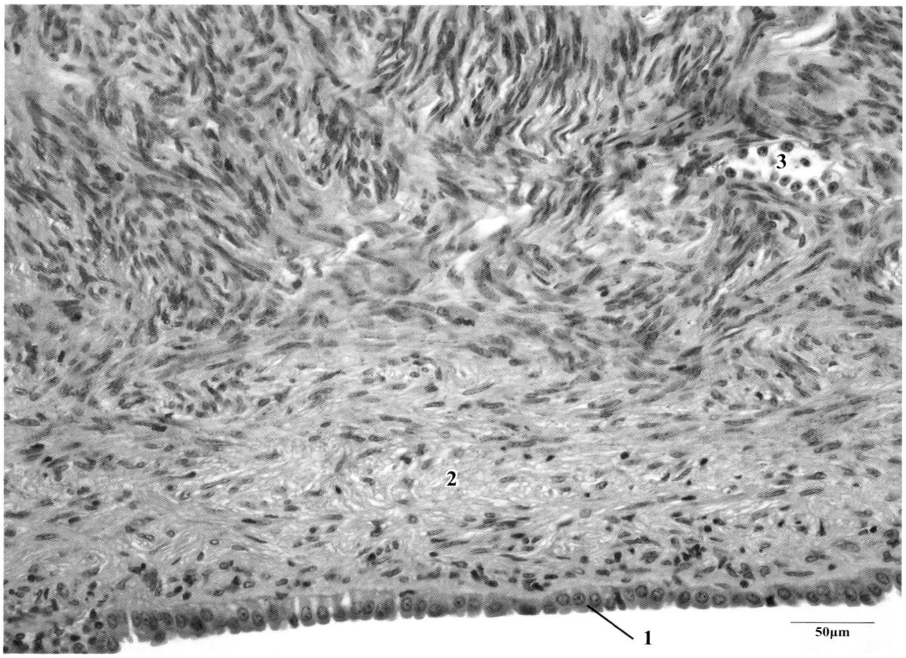

图 8-7 卵巢组织切片，示被膜（HE 染色，高倍镜）
Figure 8-7 Histological section of an ovary showing the capsule (HE staining, higher power).

1- 皮质 cortex
2- 髓质 medulla
3- 血管 blood vessel

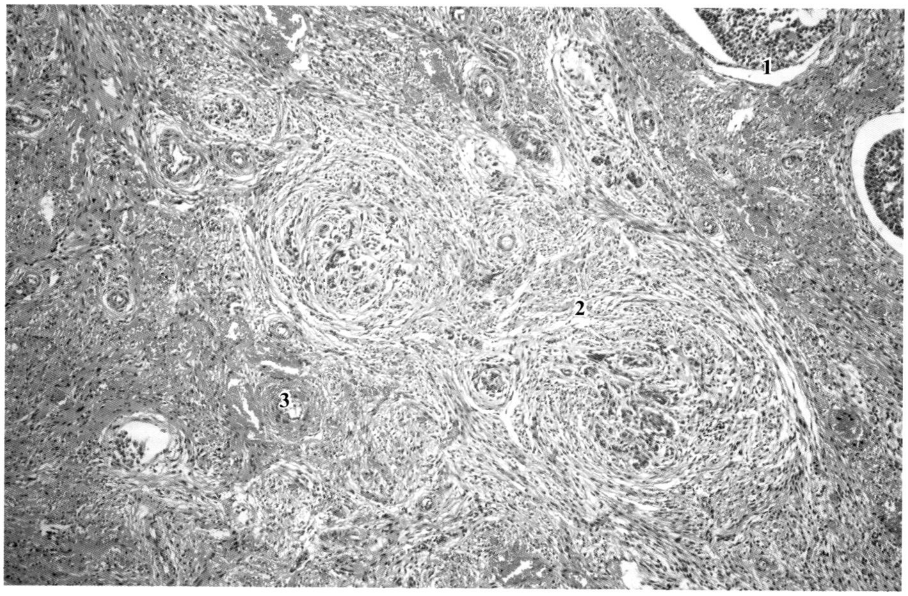

图 8-8 卵巢组织切片，示髓质（HE 染色，高倍镜）
Figure 8-8 Histological section of an ovary showing the medulla (HE staining, higher power).

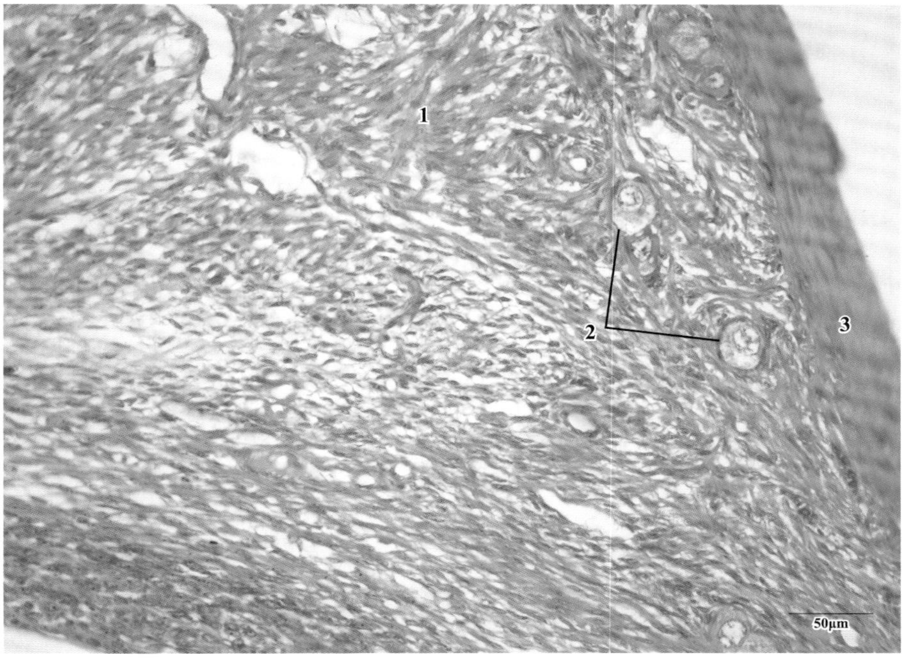

1- 间质 stroma
2- 原始卵泡 primordial follicle
3- 被膜 capsule

图 8-9　卵巢组织切片，示皮质（HE 染色，高倍镜）
Figure 8-9　Histological section of an ovary showing the cortex (HE staining, higher power).

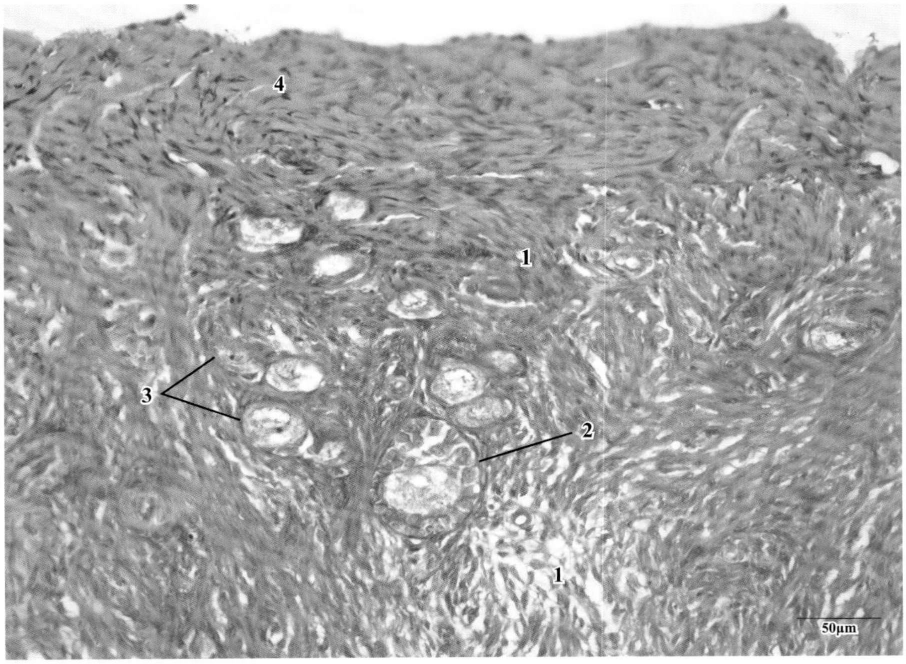

1- 间质 stroma
2- 初级卵泡 primary follicle
3- 原始卵泡 primordial follicle
4- 被膜 capsule

图 8-10　卵巢组织切片，示皮质（HE 染色，高倍镜）
Figure 8-10　Histological section of an ovary showing the cortex (HE staining, higher power).

1- 间质 stroma
2- 原始卵泡 primordial follicle
3- 初级卵泡 primary follicle
4- 初级卵母细胞 primary oocyte

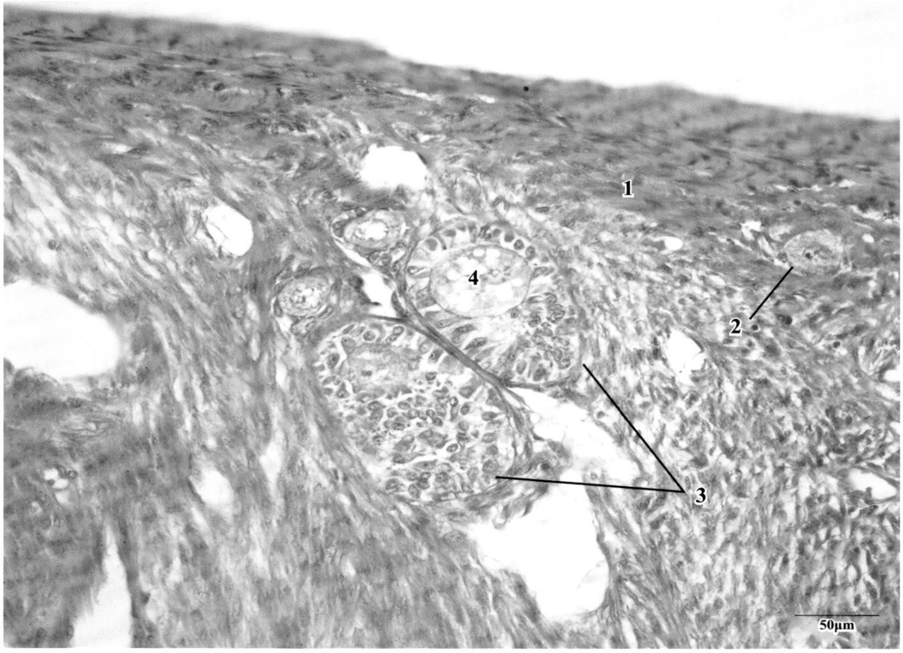

图 8-11　卵巢组织切片，示皮质（HE 染色，高倍镜）
Figure 8-11　Histological section of an ovary showing the cortex (HE staining, higher power).

1- 卵泡细胞 follicular cell
2- 卵泡膜 follicular theca
3- 初级卵母细胞的胞质 cytoplasm of primary oocyte
4- 初级卵母细胞的胞核 nucleus of primary oocyte
5- 透明带 pellucid zone

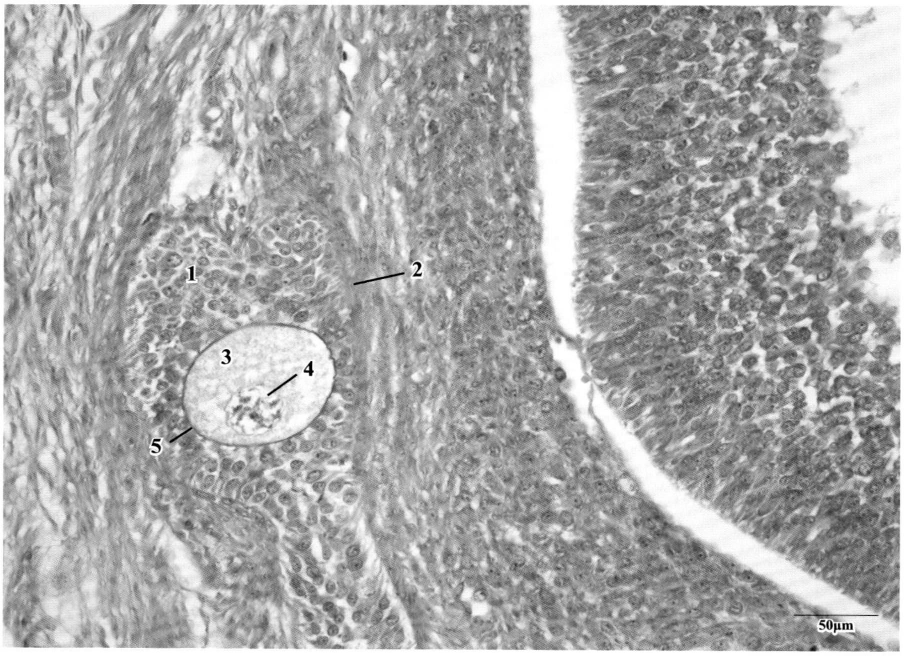

图 8-12　卵巢组织切片，示窦前卵泡（多层初级卵泡）（HE 染色，高倍镜）
Figure 8-12　Histological section of an ovary showing the preantral follicle (primary follicle with multilaminar granular cell) (HE staining, higher power).

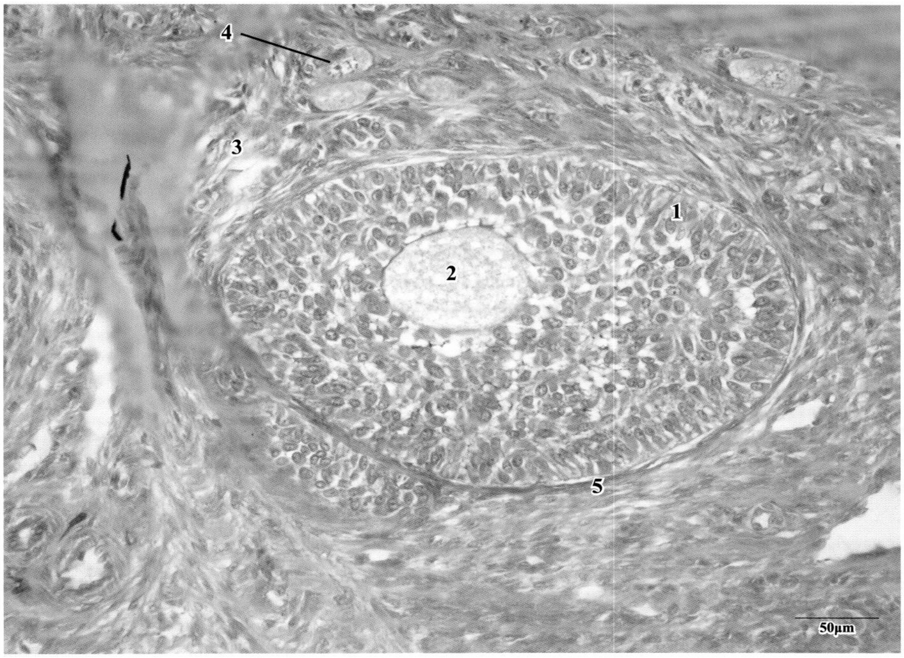

1- 卵泡细胞 follicular cell
2- 初级卵母细胞的胞质 cytoplasm of primary oocyte
3- 间质 stroma
4- 原始卵泡 primordial follicle
5- 卵泡膜 follicular theca

图 8-13　卵巢组织切片，示窦前卵泡（多层初级卵泡）（HE 染色，高倍镜）
Figure 8-13　Histological section of an ovary showing the preantral follicle (primary follicle with multilaminar granular cell) (HE staining, higher power).

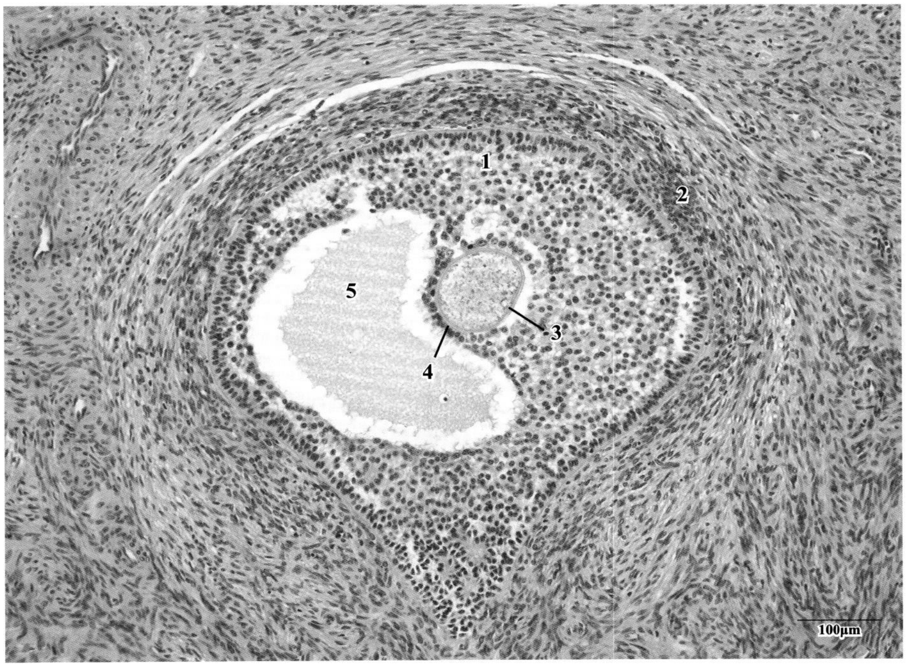

1- 颗粒细胞 granular cell
2- 卵泡膜 follicular theca
3- 透明带 pellucid zone
4- 卵丘 ovarian cumulus
5- 卵泡腔 follicular cavity

图 8-14　卵巢组织切片，示次级卵泡（幼龄窦状卵泡或囊状卵泡）（HE 染色，低倍镜）
Figure 8-14　Histological section of an ovary showing the secondary follicle (a young tertiary follicle or vesicular follicle) (HE staining, lower power).

1- 颗粒细胞 granular cell
2- 透明带 pellucid zone
3- 初级卵母细胞的细胞质 cytoplasm of primary oocyte
4- 卵泡腔 follicular cavity
5- 卵丘 ovarian cumulus
6- 颗粒膜 membrana granulosa
7- 卵泡膜内膜 internal theca of follicular theca
8- 卵泡膜外膜 external theca of follicular theca

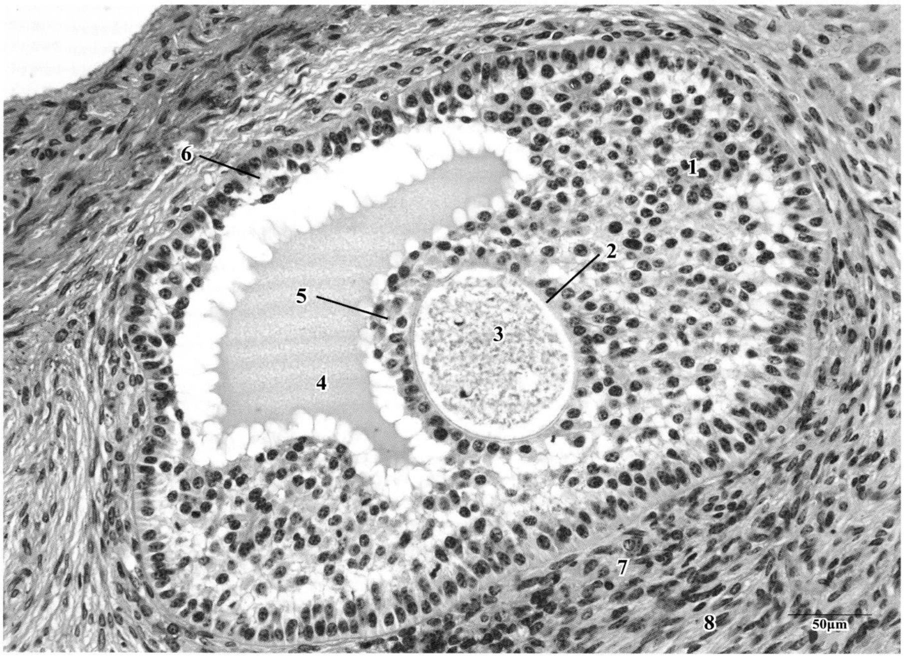

图 8-15 卵巢组织切片，示次级卵泡（幼龄窦状卵泡或囊状卵泡）（HE 染色，高倍镜）
Figure 8-15 Histological section of an ovary showing the secondary follicle (a young tertiary follicle or vesicular follicle) (HE staining, higher power).

1- 卵泡膜 follicular theca
2- 颗粒膜 membrana granulosa
3- 卵丘 ovarian cumulus
4- 卵泡腔 follicular cavity

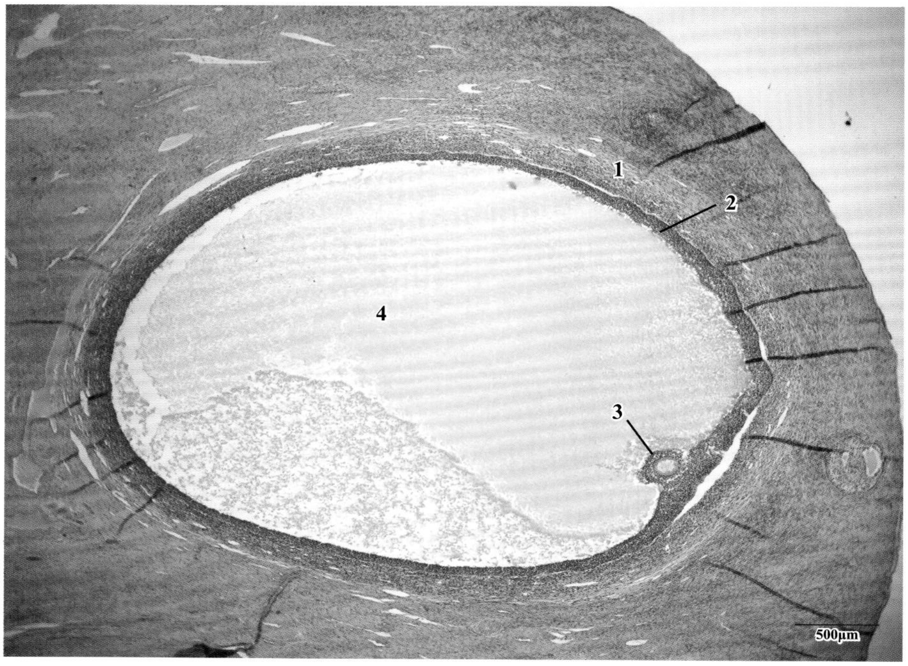

图 8-16 卵巢组织切片，示成熟卵泡（格拉夫卵泡）（HE 染色，低倍镜）
Figure 8-16 Histological section of an ovary showing the mature follicle (Graafian follicle) (HE staining, lower power).

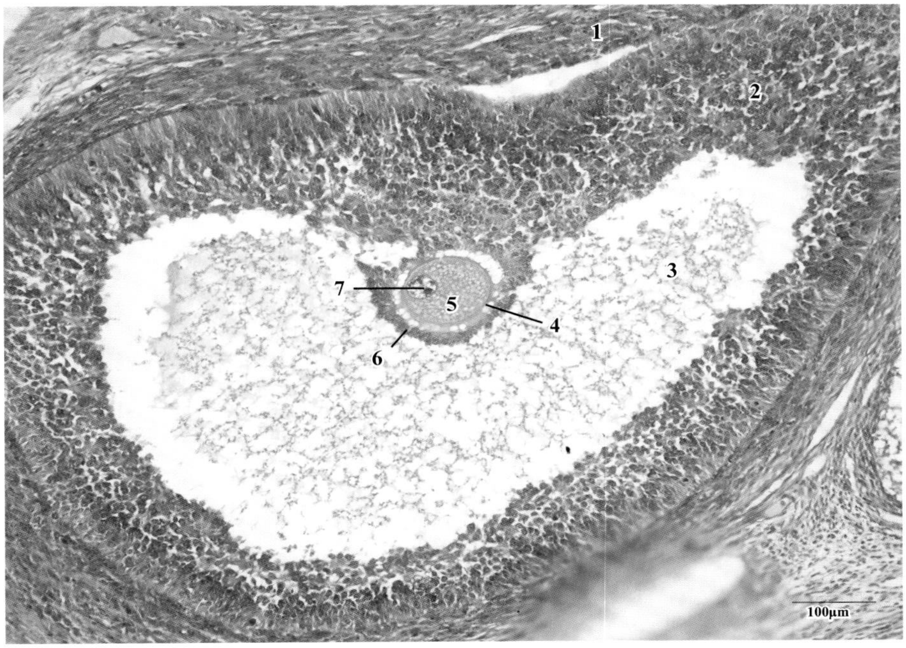

1- 卵泡膜 follicular theca
2- 颗粒细胞 granular cell
3- 卵泡腔 follicular cavity
4- 透明带 pellucid zone
5- 初级卵母细胞的胞质 cytoplasm of primary oocyte
6- 卵丘 ovarian cumulus
7- 初级卵母细胞的胞核 nucleus of primary oocyte

图 8-17　成熟卵泡组织切片，示卵丘（HE 染色，低倍镜）
Figure 8-17　Histological section of a mature follicle showing the ovarian cumulus (HE staining, lower power).

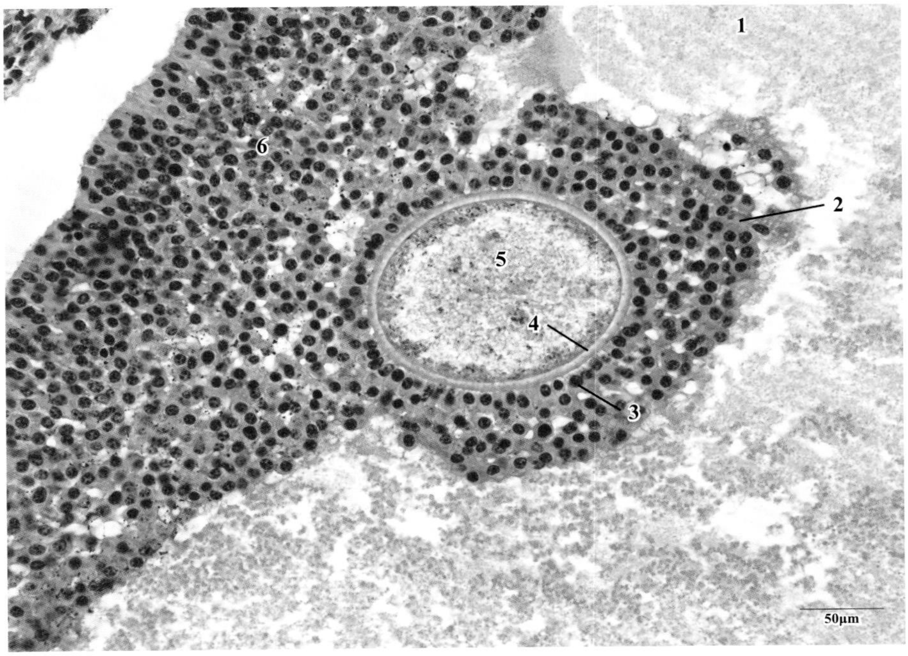

1- 卵泡腔 follicular cavity
2- 卵丘 ovarian cumulus
3- 放射冠 corona radiata
4- 透明带 pellucid zone
5- 初级卵母细胞的胞质 cytoplasm of primary oocyte
6- 颗粒细胞 granular cell

图 8-18　成熟卵泡组织切片，示卵丘（HE 染色，高倍镜）
Figure 8-18　Histological section of a mature follicle showing the ovarian cumulus (HE staining, higher power).

1- 被膜 capsule
2- 卵泡膜 follicular theca
3- 颗粒膜 membrana granulosa
4- 颗粒细胞 granular cell
5- 卵泡腔 follicular cavity
6- 次级卵泡 secondary follicle

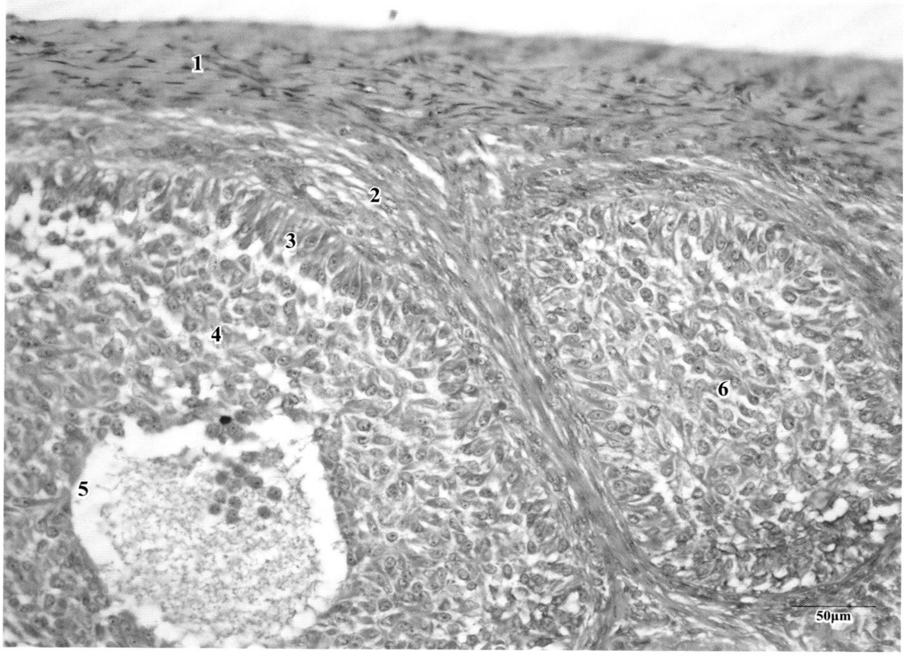

图 8-19　卵巢组织切片，示次级卵泡（幼龄窦状卵泡或囊状卵泡）（HE 染色，高倍镜）
Figure 8-19　Histological section of an ovary showing the secondary follicle (a young tertiary follicle or vesicular follicle) (HE staining, higher power).

1- 颗粒膜 membrana granulosa
2- 卵泡膜内膜 internal theca of follicular theca
3- 卵泡膜外膜 external theca of follicular theca

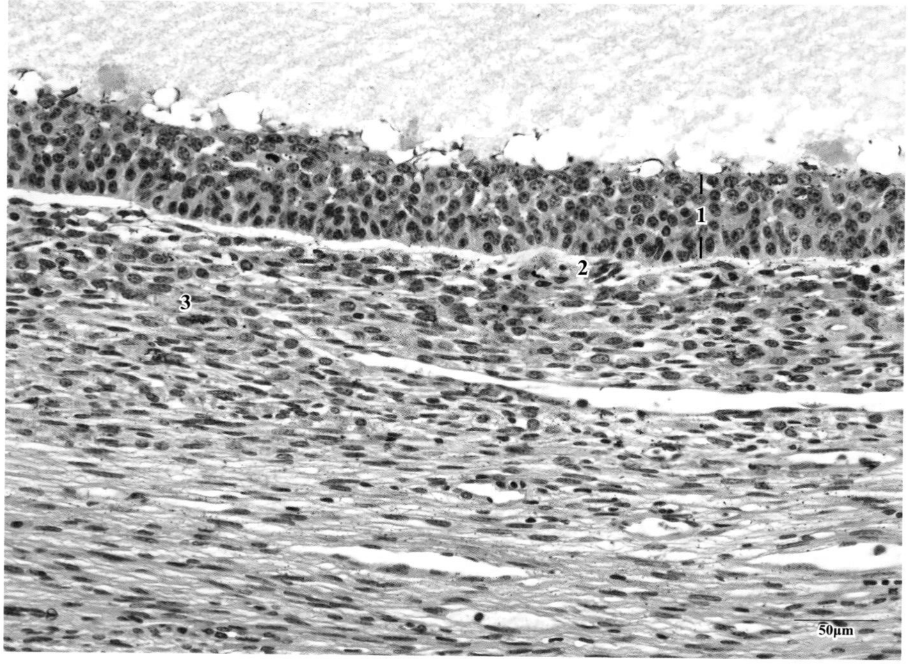

图 8-20　成熟卵泡组织切片，示卵泡膜（HE 染色，高倍镜）
Figure 8-20　Histological section of a mature follicle showing the follicular theca (HE staining, higher power).

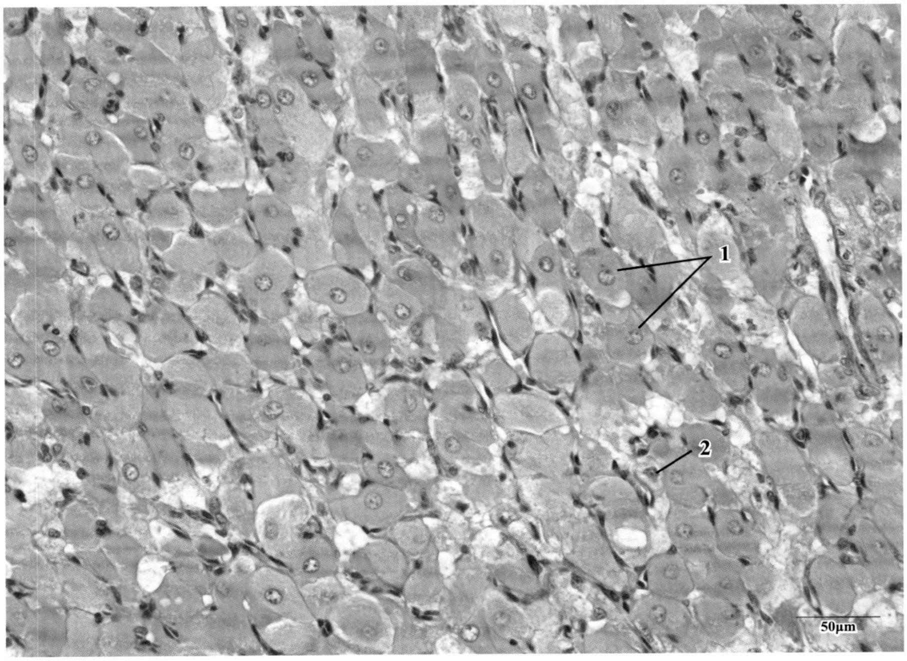

1- 颗粒黄体细胞 granulosa lutein cell

2- 膜黄体细胞 theca lutein cell

图 8-21　黄体组织切片（HE 染色，高倍镜）

Figure 8-21　Histological section of the corpus luteum (HE staining, higher power).

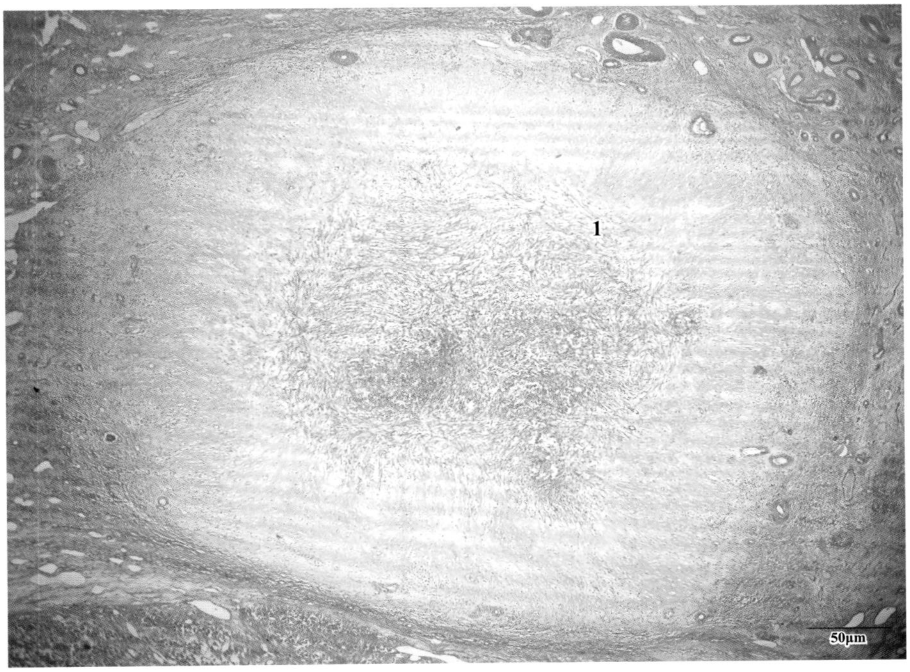

1- 白体 corpus albicans

图 8-22　白体组织切片（HE 染色，低倍镜）

Figure 8-22　Histological section of the corpus albicans (HE staining, lower power).

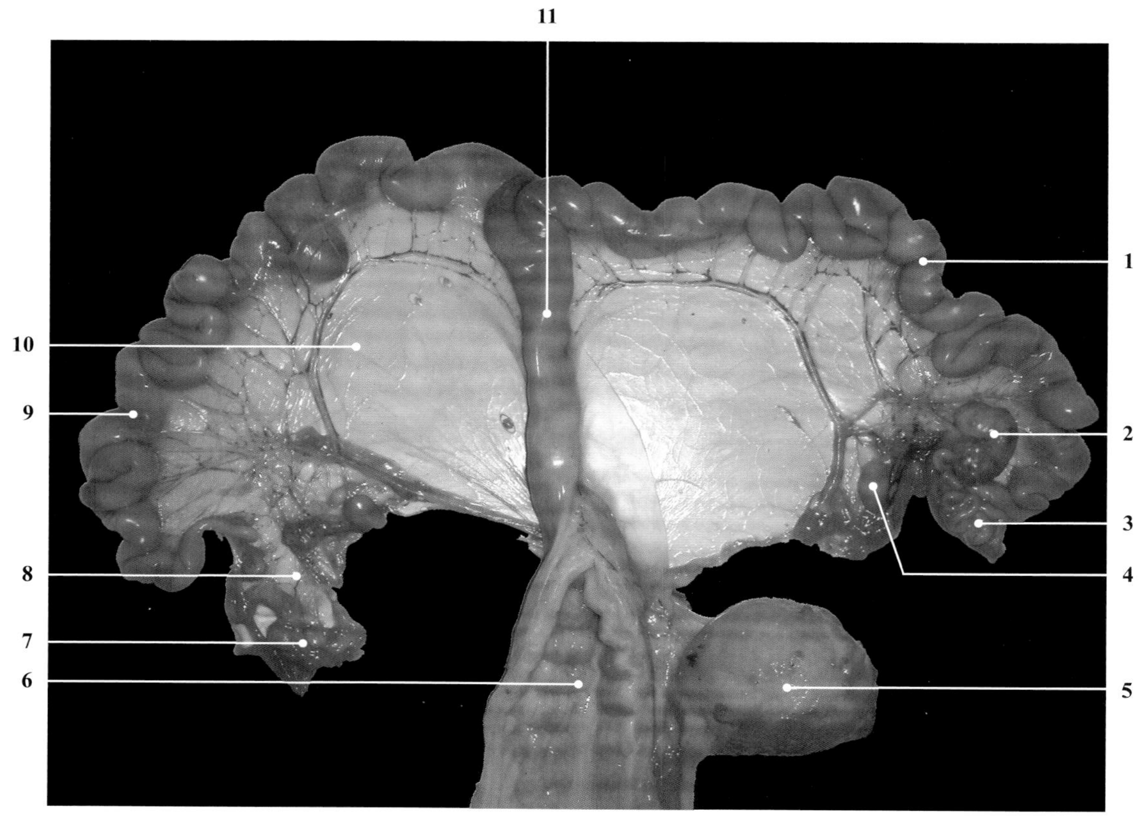

图 8-23 输卵管
Figure 8-23 Uterine tube.

1- 右侧子宫角 right uterine horn
2- 右侧卵巢 right ovary
3- 右侧输卵管 right uterine tube
4- 卵巢淋巴结 lymph node of the ovary
5- 膀胱 urinary bladder
6- 子宫颈 uterine cervix
7- 左侧输卵管 left uterine tube
8- 输卵管系膜 mesosalpinx
9- 左侧子宫角 left uterine horn
10- 子宫阔韧带 broad ligament of uterus
11- 子宫体 uterine body

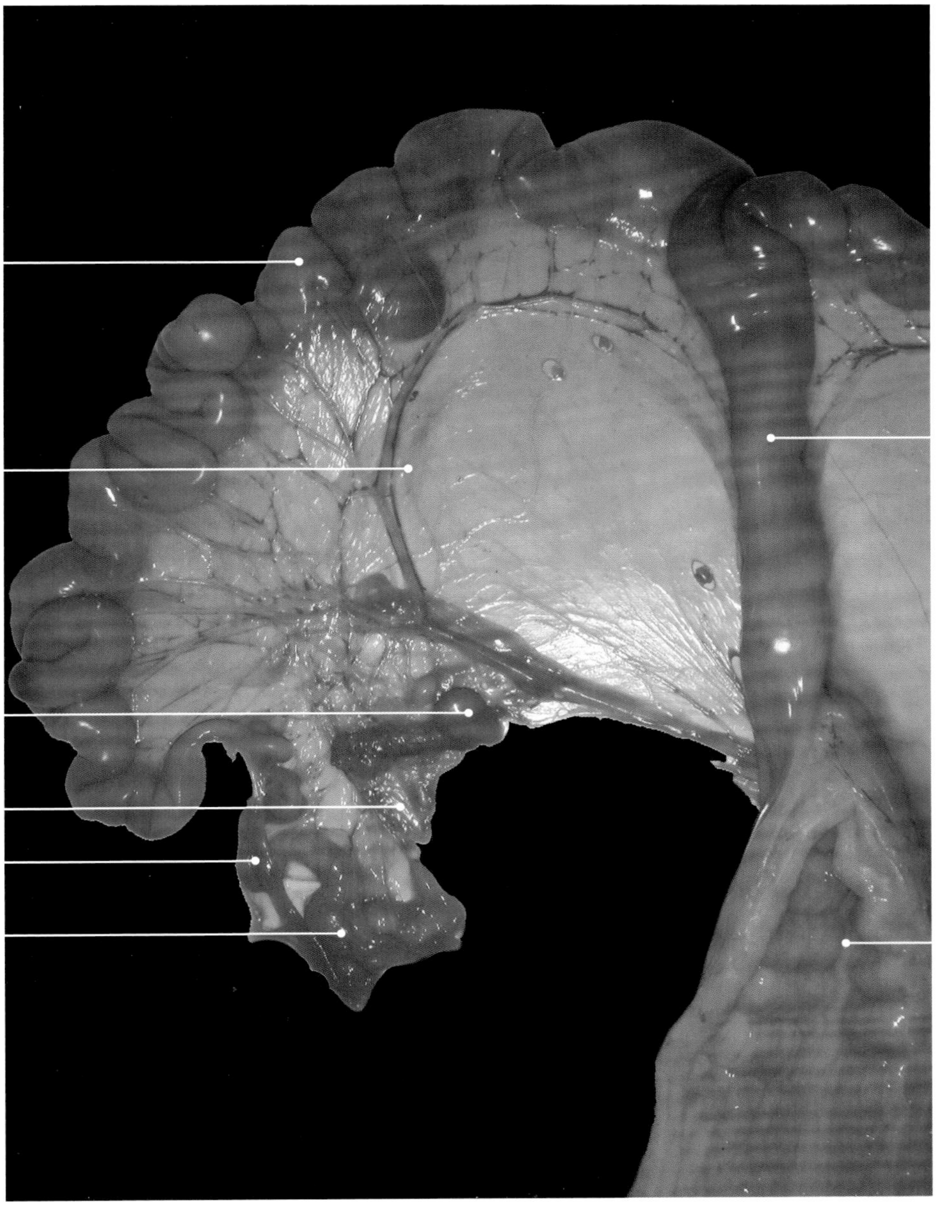

图 8-24　左侧输卵管
Figure 8-24　Left uterine tube.

1- 子宫体 uterine body
2- 子宫黏膜 uterine mucosa
3- 输卵管壶腹 ampulla of uterine tube
4- 输卵管峡部 isthmus of uterine tube
5- 输卵管伞 fimbriae of uterine tube
6- 卵巢淋巴结 lymph node of the ovary
7- 子宫阔韧带 broad ligament of uterus
8- 子宫角 uterine horn

1- 肌层 muscular layer
2- 黏膜皱襞 mucosal fold
3- 浆膜 serosa

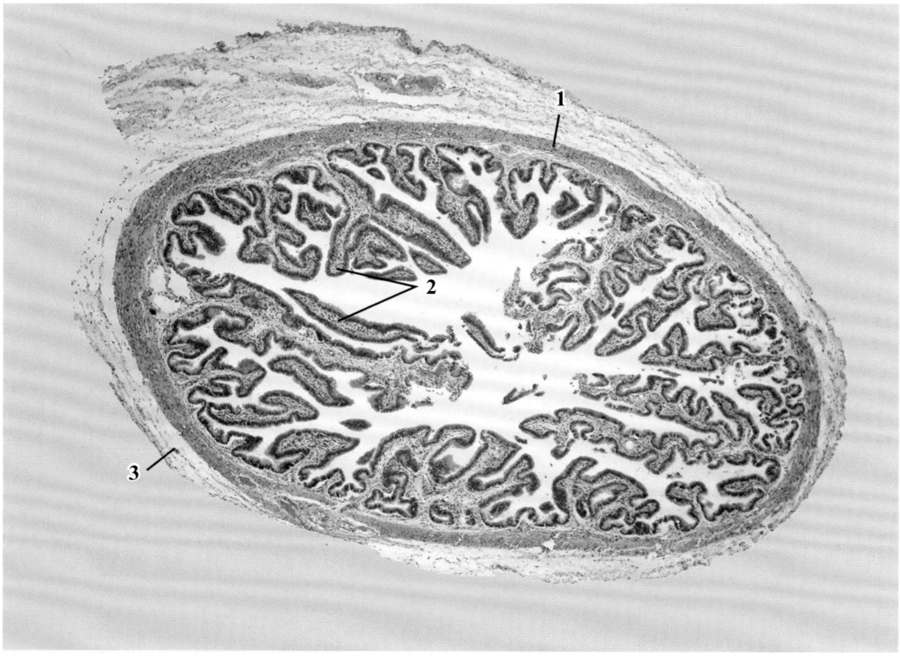

图 8-25　输卵管组织切片（HE 染色，低倍镜）
Figure 8-25　Histological section of the uterine tube (HE staining, lower power).

1- 黏膜皱襞 mucosal fold
2- 肌层 muscular layer

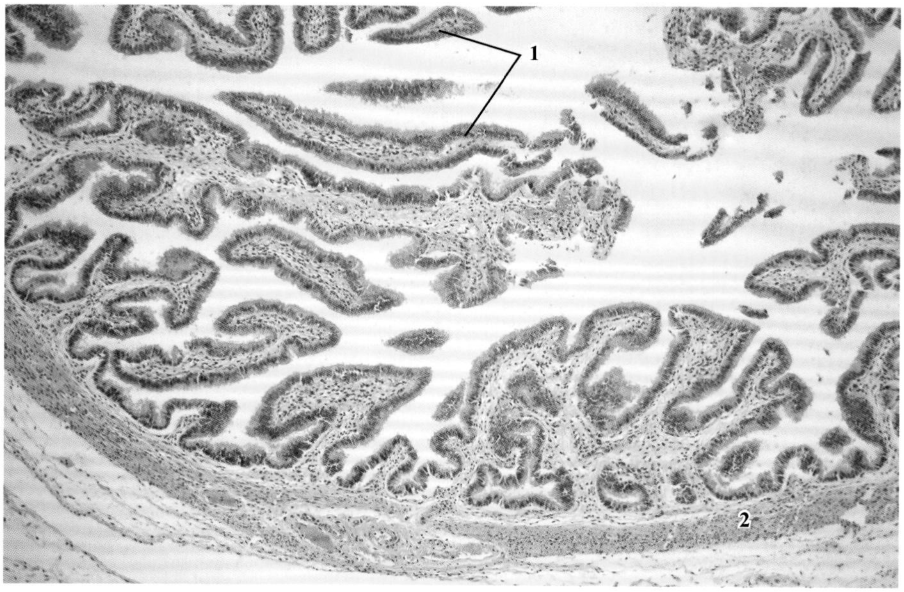

图 8-26　输卵管组织切片（HE 染色，高倍镜）
Figure 8-26　Histological section of the uterine tube (HE staining, higher power).

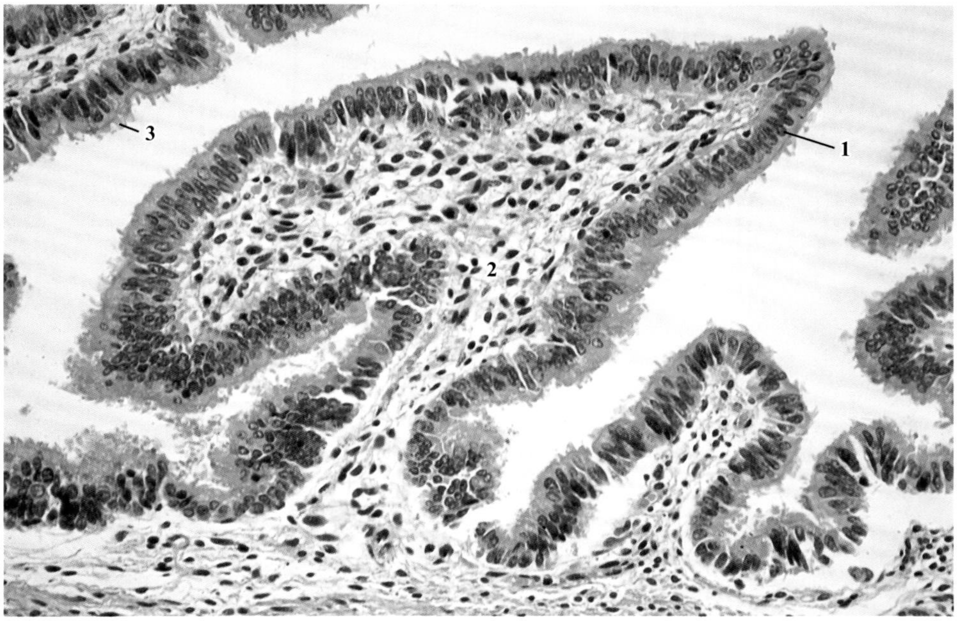

图 8-27 输卵管组织切片（HE 染色，高倍镜）
Figure 8-27 Histological section of the uterine tube (HE staining, higher power).

1- 假复层柱状上皮 pseudostratified columnar epithelium
2- 固有层 lamina propria
3- 分泌泡 secretory vacuole

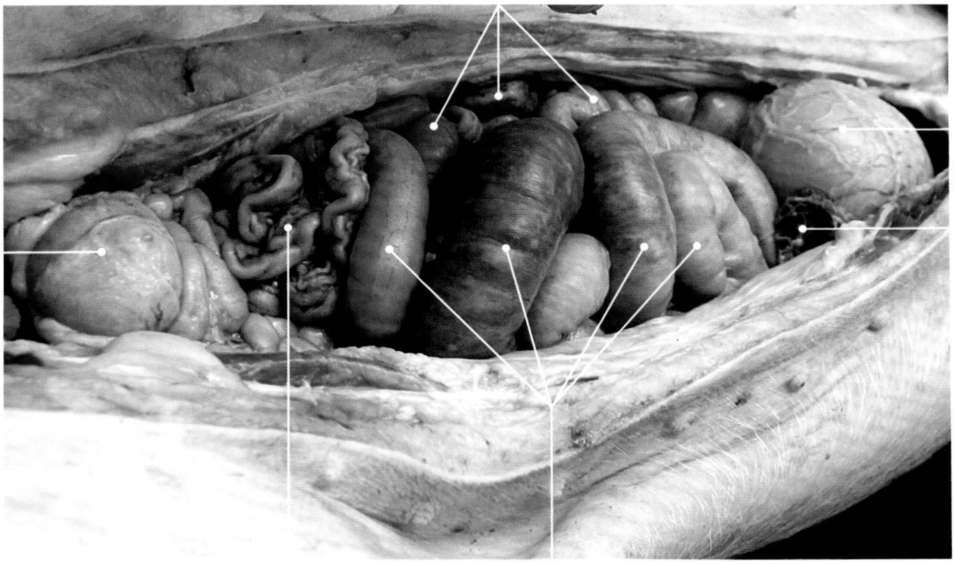

图 8-28 子宫的位置
Figure 8-28 Anatomical position of the uterus.

1- 胃 stomach
2- 脾 spleen
3- 大肠 large intestine
4- 子宫角 uterine horn
5- 膀胱 urinary bladder
6- 小肠 small intestine

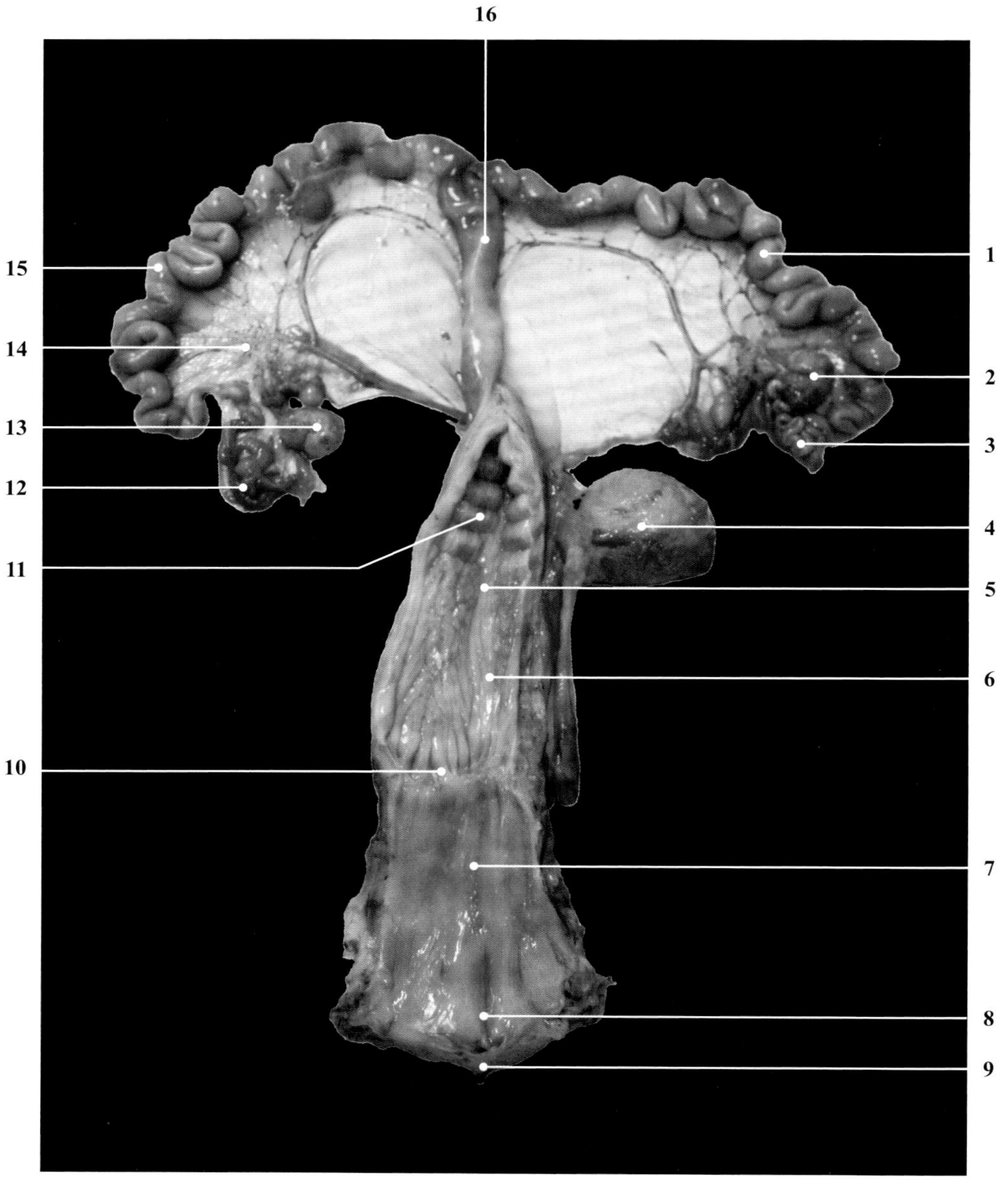

图 8-29　子宫（子宫颈切开）
Figure 8-29　The uterus, partially opened on uterine cervix.

1- 右侧子宫角 right uterine horn
2- 右侧卵巢 right ovary
3- 右侧输卵管 right uterine tube
4- 膀胱 urinary bladder
5- 子宫颈外口 external uterine orifice
6- 阴道 vagina
7- 阴道前庭 vaginal vestibule
8- 阴蒂 clitoris
9- 阴唇腹侧联合 ventral commissure of labium
10- 阴瓣 hymen
11- 子宫颈 uterine cervix
12- 左侧输卵管 left uterine tube
13- 左侧卵巢 left ovary
14- 子宫阔韧带 broad ligament of uterus
15- 左侧子宫角 left uterine horn
16- 子宫体 uterine body

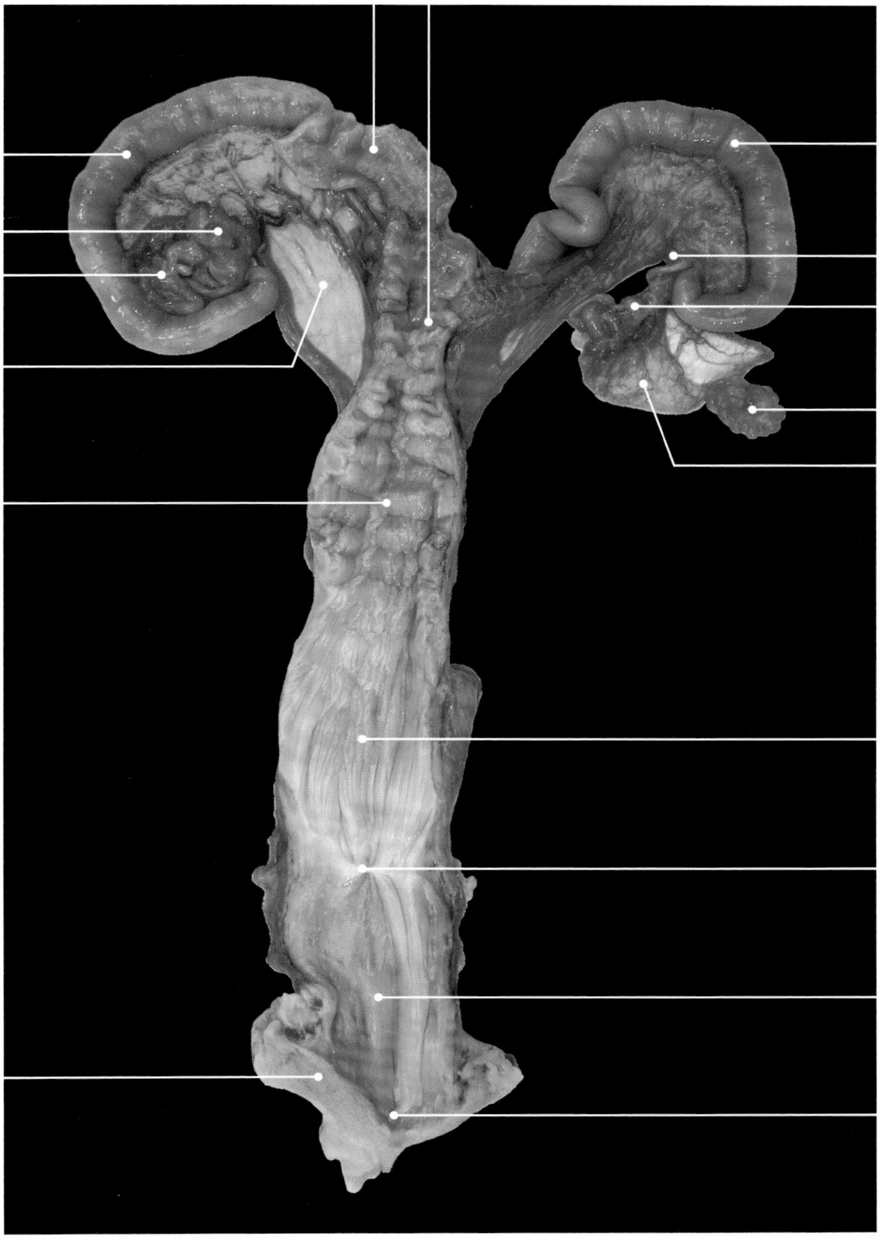

图 8-30 子宫（子宫体和子宫颈切开）
Figure 8-30 The uterus, partially opened on uterine body and cervix.

1- 右侧子宫角 right uterine horn
2- 输卵管峡 isthmus of uterine tube
3- 输卵管壶腹 ampulla of uterine tube
4- 右侧卵巢 right ovary
5- 输卵管伞 fimbriae of uterine tube
6- 阴道 vagina
7- 阴瓣 hymen
8- 阴道前庭 vaginal vestibule
9- 阴蒂 clitoris
10- 阴唇 labium of vulva
11- 子宫颈 uterine cervix
12- 子宫阔韧带 broad ligament of uterus
13- 左侧输卵管 left uterine tube
14- 左侧卵巢 left ovary
15- 左侧子宫角 left uterine horn
16- 子宫角切开 opening uterine horn
17- 子宫体切开 opening uterine body

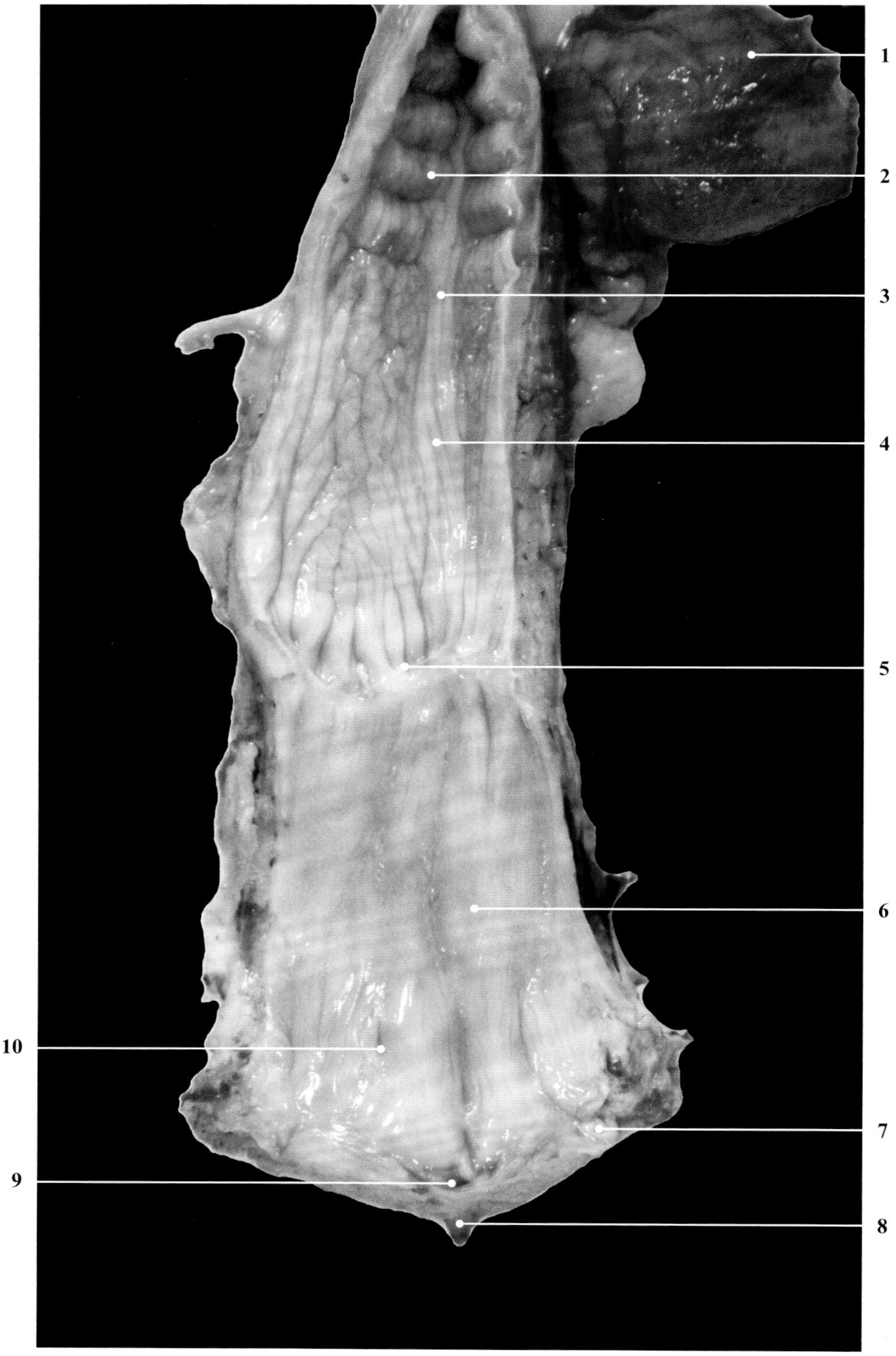

图 8-31　子宫颈
Figure 8-31　The uterine cervix.

1- 膀胱 urinary bladder
2- 子宫颈 uterine cervix
3- 子宫颈外口 external uterine orifice
4- 阴道 vagina
5- 阴瓣 hymen
6- 阴道前庭 vaginal vestibule
7- 阴唇 labium of vulva
8- 阴唇腹侧联合 ventral commissure of labium
9- 阴蒂 clitoris
10- 前庭腺开口 vestibular gland opening

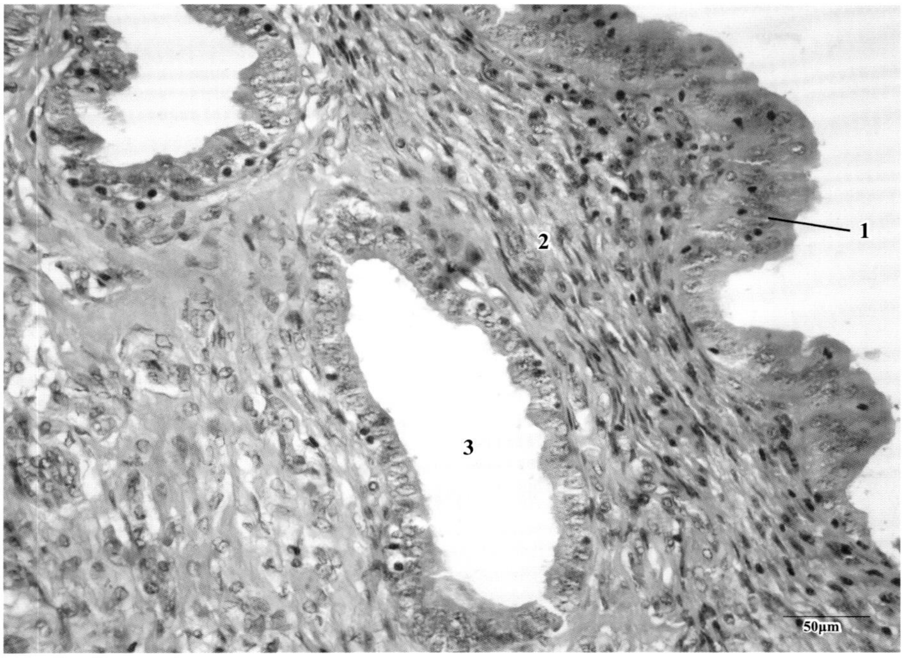

1- 假复层柱状上皮 pseudostratified columnar epithelium
2- 固有层 lamina propria
3- 子宫腺 uterine gland

图 8-32　子宫组织切片，示内膜层（HE 染色，高倍镜）
Figure 8-32　Histological section of a uterus showing the endometrium (HE staining, higher power).

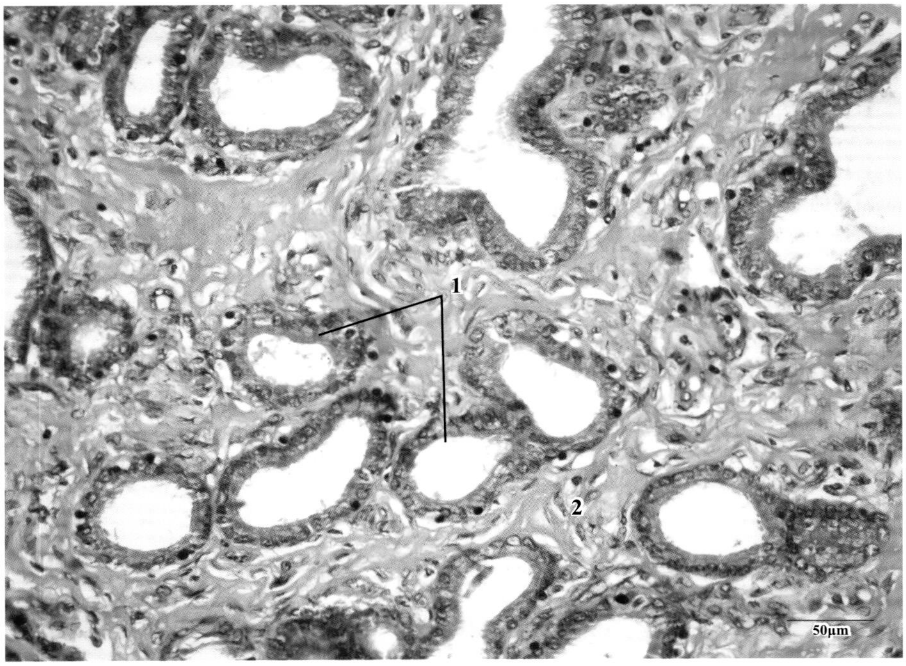

1- 子宫腺 uterine gland
2- 固有层 lamina propria

图 8-33　子宫组织切片，示子宫腺（HE 染色，高倍镜）
Figure 8-33　Histological section of a uterus showing the uterine gland (HE staining, higher power).

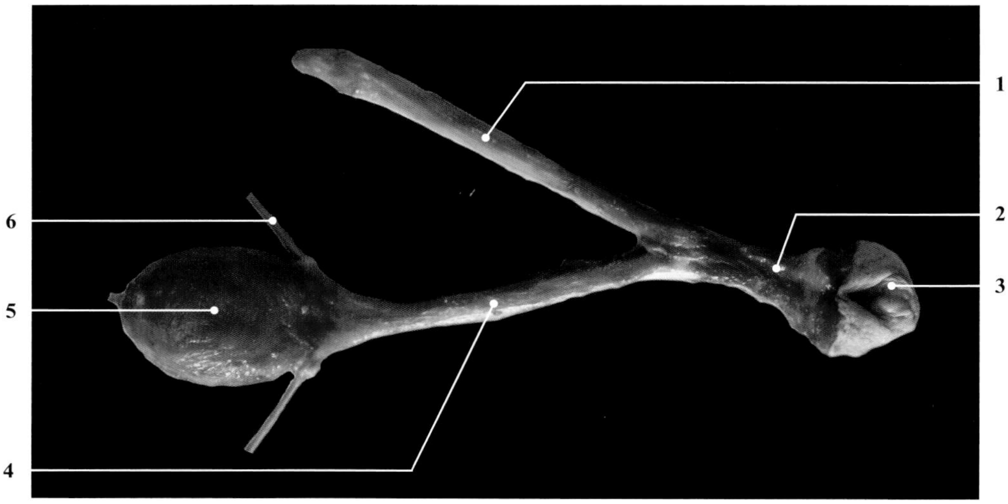

图 8-34 阴道
Figure 8-34 The vagina.

1- 阴道 vagina
2- 阴道前庭 vaginal vestibule
3- 阴门 vulva
4- 尿道 urethra
5- 膀胱 urinary bladder
6- 输尿管 ureter

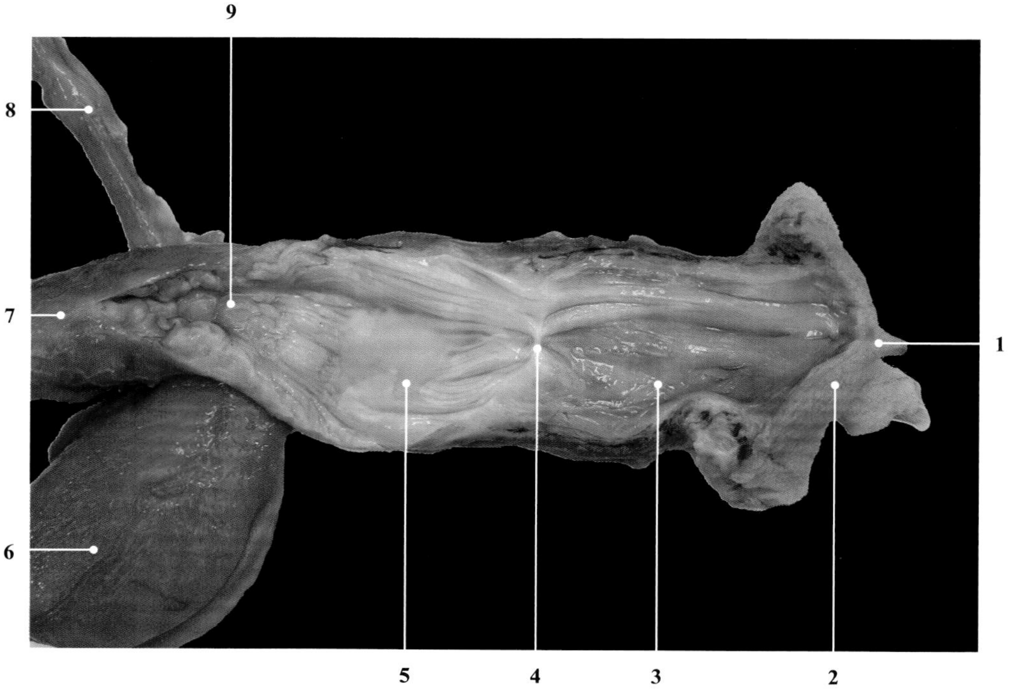

图 8-35 阴道和阴道前庭
Figure 8-35 The vagina and vaginal vestibule.

1- 阴蒂 clitoris
2- 阴唇 labium of vulva
3- 阴道前庭 vaginal vestibule
4- 阴瓣 hymen
5- 阴道 vagina
6- 膀胱 urinary bladder
7- 子宫颈 uterine cervix
8- 输尿管 ureter
9- 子宫颈外口 external uterine orifice

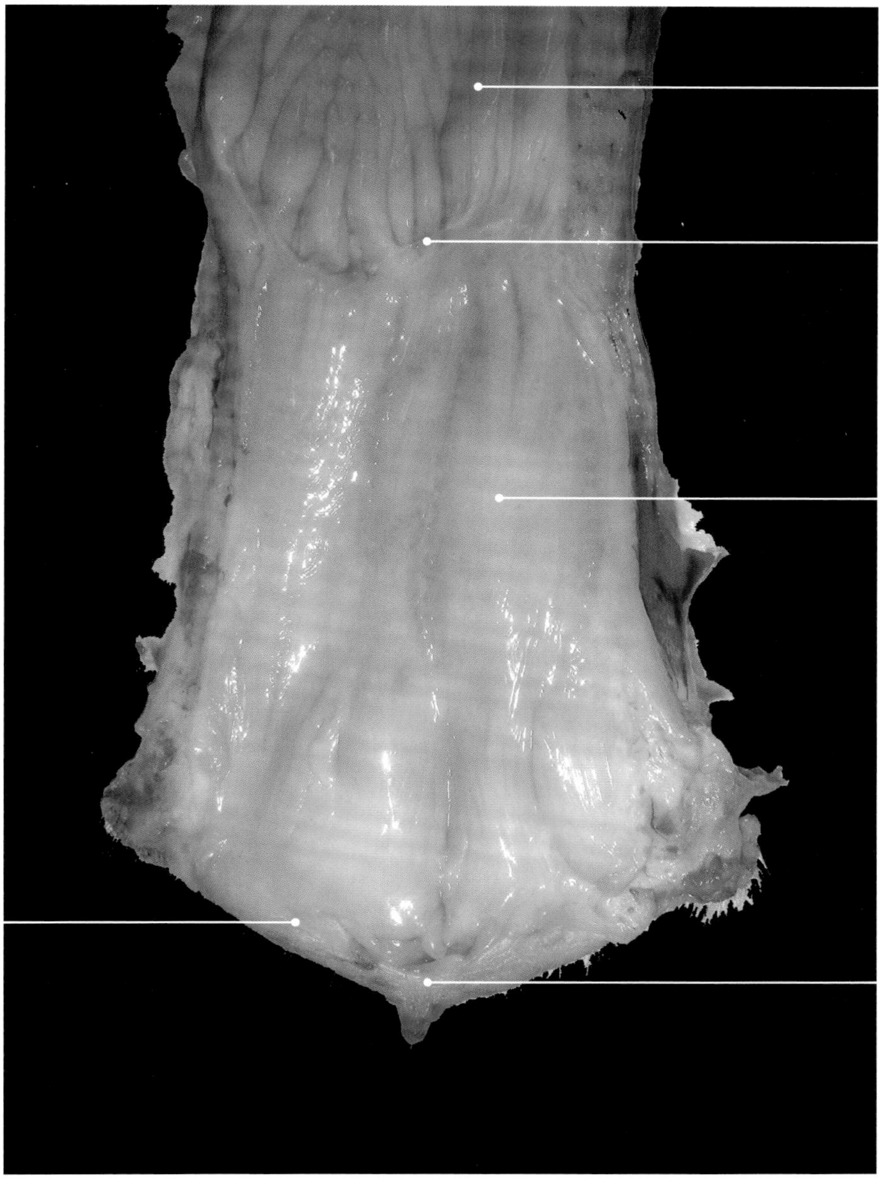

图 8-36　阴道前庭黏膜
Figure 8-36　The mocous membrane of vaginal vestibule.

1- 阴道 vagina
2- 阴瓣 hymen
3- 阴道前庭 vaginal vestibule
4- 阴蒂 clitoris
5- 阴唇 labium of vulva

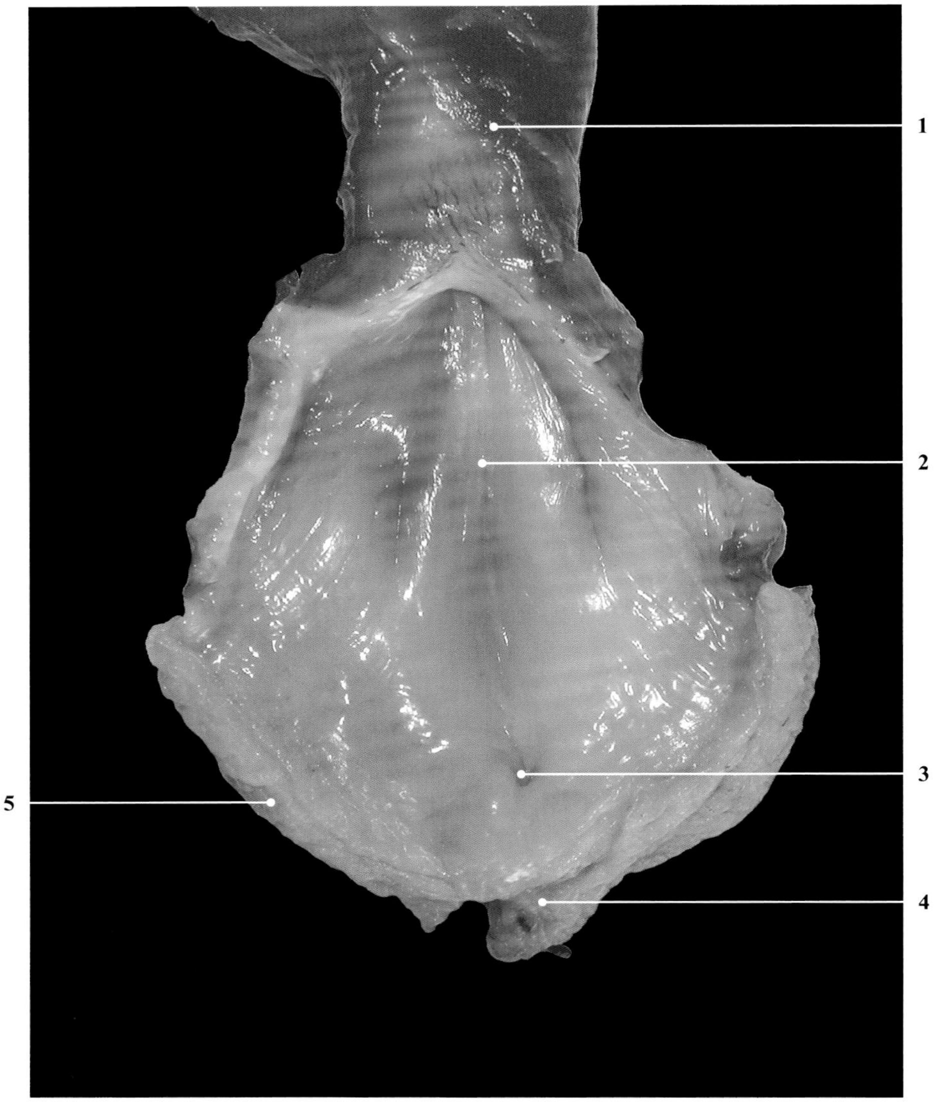

图 8-37　阴蒂窝
Figure 8-37　The fossa clitoridis.

1- 阴道 vagina
2- 阴道前庭 vaginal vestibule
3- 阴蒂窝 fossa clitoridis
4- 阴唇腹侧联合 ventral commissure of labium
5- 阴唇 labium of vulva

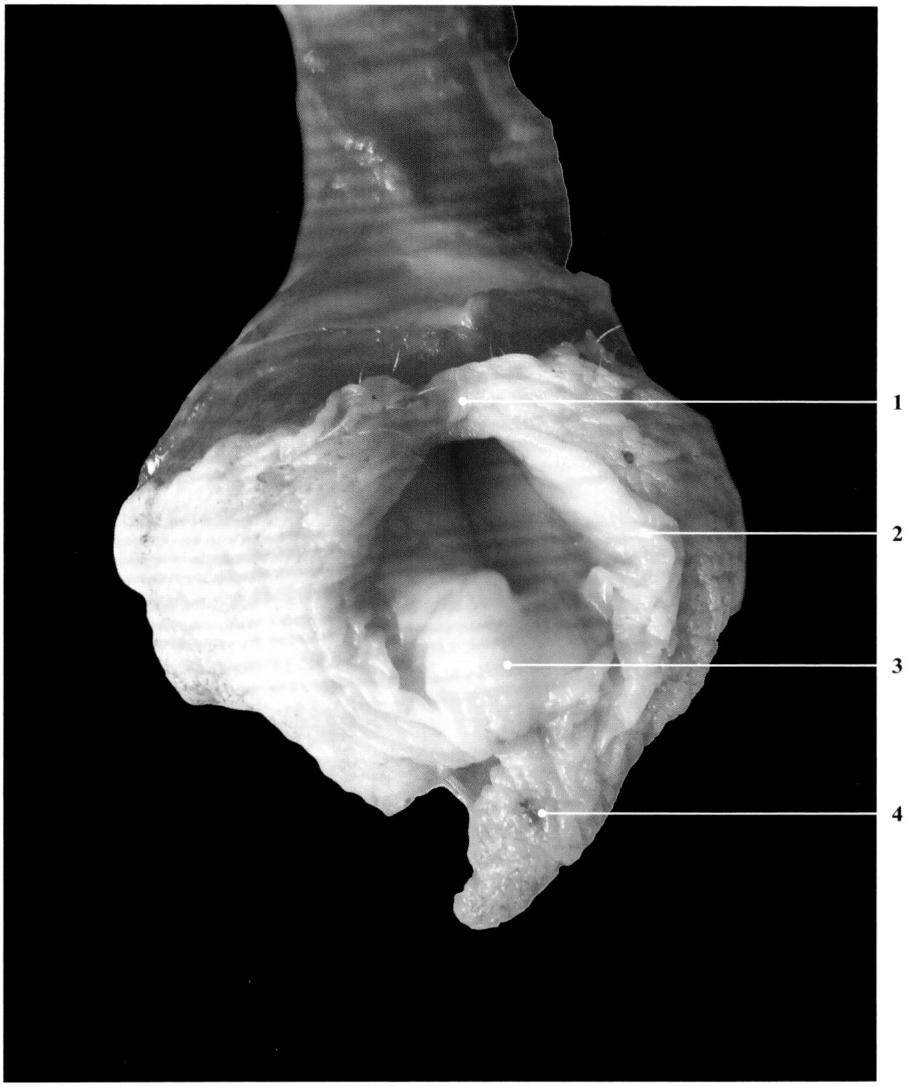

图 8-38 阴唇
Figure 8-38 The labium of vulva.

1- 阴唇背侧联合 dorsal commissure of labium
2- 阴唇 labium of vulva
3- 阴蒂 clitoris
4- 阴唇腹侧联合 ventral commissure of labium

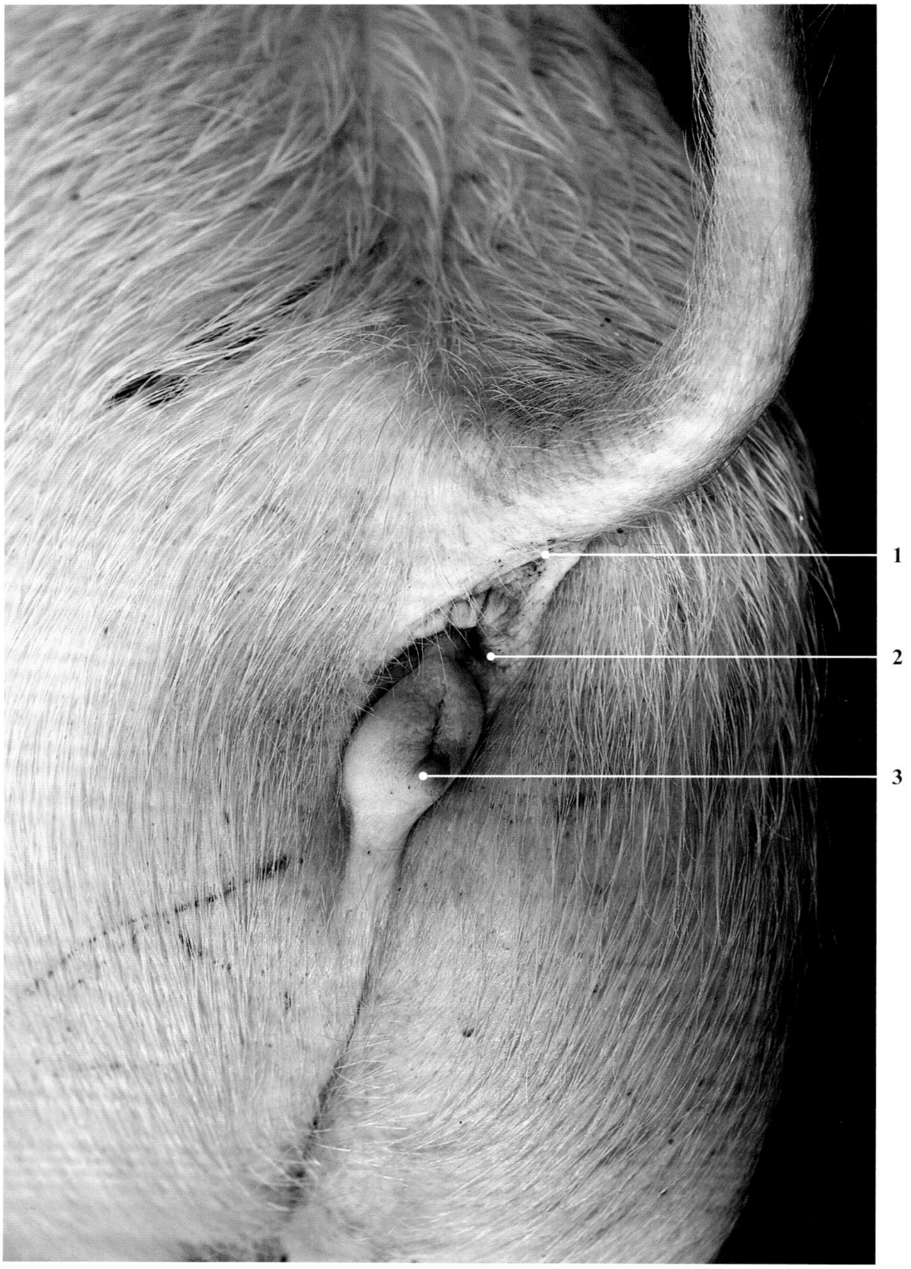

图 8-39　阴门
Figure 8-39　The vulva.

1- 尾根 root of the tail
2- 肛门 anus
3- 阴门 vulva

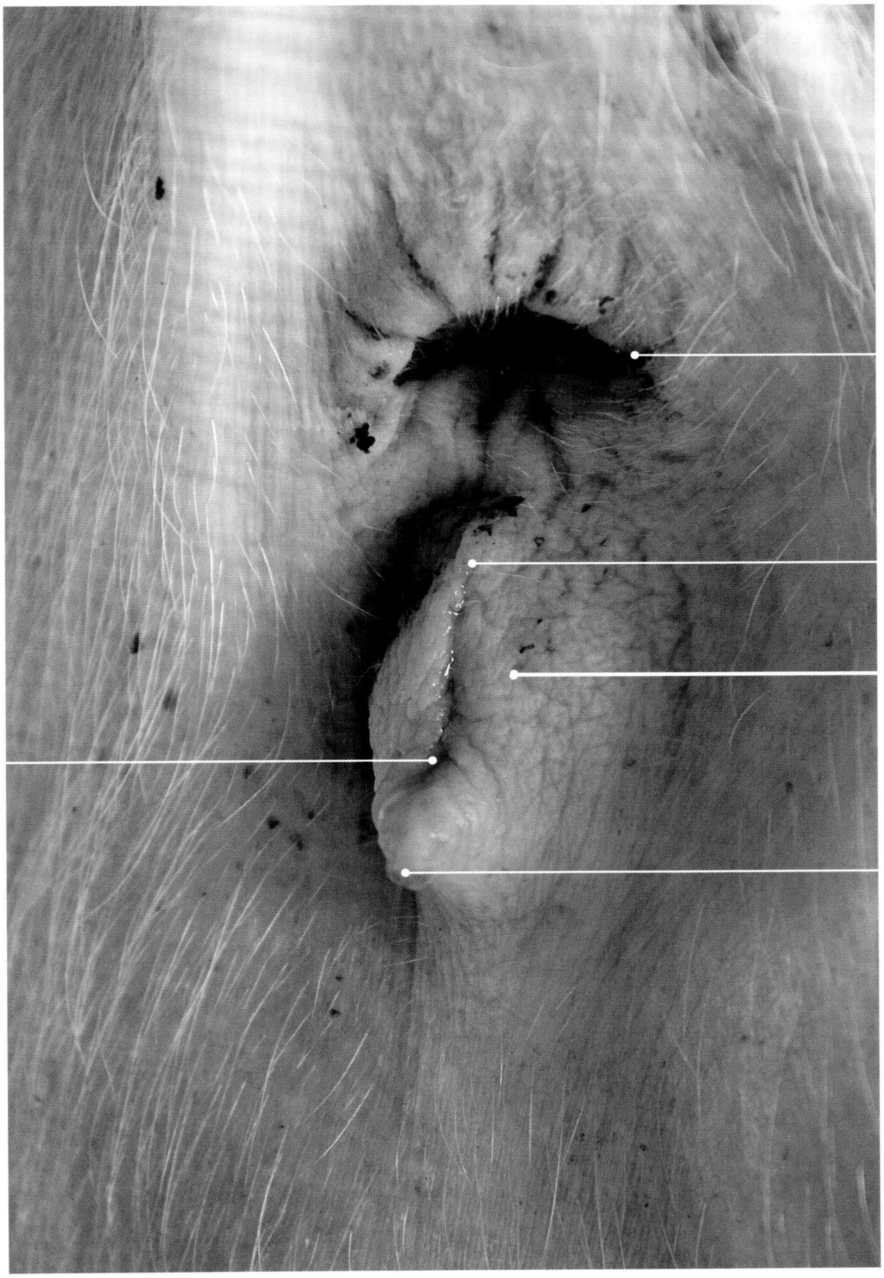

图 8-40 阴门与阴唇
Figure 8-40 The vulva and labium.

1- 肛门 anus
2- 阴唇背侧联合 dorsal commissure of labium
3- 阴唇 labium of vulva
4- 阴唇腹侧联合 ventral commissure of labium
5- 阴门 vulva

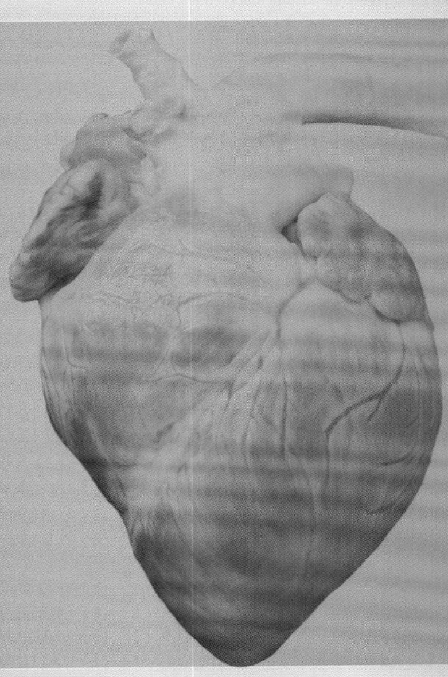

第九章

心血管系统

Chapter 9

Cardiovascular system

心血管系统（cardiovascular system）由心脏、动脉、毛细血管和静脉组成，管腔内充满血液，是血液的循环系统。

一、心脏

心脏（heart）是血液循环的动力器官。在神经体液的调节下，能够进行节律性的收缩和舒张，推动血液按一定的方向流动。

1．心脏的位置与形态

心脏为中空的肌质器官，位于纵隔（mediastinum）内，夹于左、右肺之间，外有心包（pericardium）包裹。猪的心脏较为宽厚圆钝，约在胸腔下3/5，第2至第5肋之间，略偏左侧，心尖与第7软骨和胸骨的结合处相对。心脏左侧面（心耳面）微微凸起；右侧面（心房面）凸起明显。心脏表面有3条沟，即冠状沟（coronary groove）、锥旁室间沟（paraconal interventricular groove）或称左纵沟（left longitudinal groove）、窦下室间沟（subsinuosal interventricular groove）或称右纵沟（right longitudinal groove），可作为心腔的外表分界。在冠状沟和室间沟内有营养心脏的血管，并填充有脂肪。

2．心腔的结构

心腔被纵走的房间隔（interatrial septum）和室间隔（interventricular septum）分为互不相通的左右两半，每半又分为上部的心房（cardiac atrium）和下部的心室（cardiac ventricle），同侧的心房与心室经房室口（atrioventricular orifice）相通。因此，心腔包括4个腔，分别是右心房、右心室、左心房和左心室。

（1）右心房（right atrium）：位于右心室的背侧，构成心基的右前背侧部，接纳来自前、后腔静脉和冠状窦的血液，壁薄腔大，由腔静脉窦（sinus of the venae cavae）和右心耳（right auricle）组成。腔静脉窦为体循环静脉的入口部，背侧壁及后壁分别有前、后腔静脉口，两腔静脉口之间有半月形的静脉间结节（intervenous tubercle）。在后腔静脉口腹侧的冠状窦（coronary sinus）为猪左奇静脉的入口处，窦口大并且无明显的冠状窦瓣。在后腔静脉口附近的房间隔上，有深浅不一的凹窝，称卵圆窝（oval fossa），为胚胎时期卵圆孔（foramen ovale）的遗迹。猪的卵圆窝近似圆形，内缘明显。右心耳为锥形盲囊，内有梳状肌（pectinate muscle）。

（2）右心室（right ventricle）：位于右心房腹侧，心室的右前部，不达心尖，上部有2个开口，即右房室口和肺动脉口（orifice of pulmonary trunck）。右心室接受来自右心房的静脉血，通过动脉圆锥把血液泵入肺动脉干（pulmonary trunk），进而把血运送到肺。肺动脉口为右心室的出口，纤维环上附有3片袋状的半月瓣（semilunar valve），称为肺动脉瓣（pulmonary valve）。右房室口为右心室的入口，呈卵圆形，此处由致密结缔组织构成纤维环，周缘附着有3片三角形的瓣膜，称为右房室瓣（right atrioventricular valve），或称三尖瓣（tricuspid valve）。瓣膜向下突入心室，其游离缘由腱索（chordae tendineae）与乳头肌（papillary muscle）相连。

右心室壁上有乳头肌和隔缘肉柱（septomarginal trabecula，又称心横肌）。乳头肌分为3个，一个位于前壁，两个在室间隔。猪的前乳头肌发达，尖端分出4～6个肌突通过腱索连于三尖瓣游离缘。猪的隔缘肉柱发出点均位于室间隔的上、中1/3交接部，连于心室侧壁与室间隔之间。

（3）左心房（left atrium）：位于左心室的背侧，构成心基的左后部，接受来自肺静脉的动脉血。左心房背侧壁的后部，有5 ~ 8个肺静脉口（orifice of pulmonary vein）。左心耳（left auricle）位于左心房前部，其盲端向前伸达肺干后方，腔内有梳状肌。猪的左心耳内面可见比右心耳更为发达的梳状肌。

（4）左心室（left ventricle）：位于左心房腹侧，向下伸达心尖，其心室壁比右心室壁厚。上部有2个开口，即主动脉口（aortic orifice）和左房室口。左心室接受来自肺的动脉血，并通过主动脉把血液运送到身体的绝大部分。主动脉口为左心室的出口，纤维环上附有3片袋状的半月瓣，称为主动脉瓣（aortic valve）。左房室口为左心室的入口，由致密结缔组织组成的纤维环围成，周缘有两片三角形瓣膜，称为左房室瓣（left atrioventricular valve）或二尖瓣（bicuspid valve）。猪的左心室前乳头肌较右心室前乳头肌更为发达，末端分出4 ~ 7个肌突通过腱索连于二尖瓣游离缘。

3. 心壁的构造

心壁由心外膜、心肌和心内膜构成。心外膜（epicardium）为覆盖心外表面的浆膜，即心包浆膜的脏层，外层为间皮，内层为富含血管、神经和脂肪的结缔组织。心肌（myocardium）为心壁的中层，最厚，由心肌纤维组成，被房室口纤维环分为心房肌和心室肌两个独立的肌系，因此心房和心室可分别收缩和舒张。心室壁心肌可分为内纵肌、中环肌和外斜肌三层。肌束间有丰富的结缔组织和毛细血管。心室肌纤维粗而长，有分支，横小管较多；心房肌纤维较细、无分支，横小管缺乏。心内膜（endocardium）为紧贴心肌内表面的光滑薄膜，与心底血管的内膜相连续。其深面有血管、淋巴管、神经和心脏传导系的分支。在房室口和动脉口折叠形成房室瓣、主动脉瓣和肺动脉瓣；瓣膜表面为内皮，内部为致密结缔组织。

心肌细胞（cardiac muscle cell）呈柱形，又称心肌纤维（cardiac muscle fiber）。多数细胞有分支，长度80 ~ 150 μm，直径10 ~ 30 μm，一般为单一胞核，偶见双核，胞核呈圆形，位于细胞的中央，肌浆丰富，含有大量的线粒体、糖原和少量的脂滴，并有脂褐素存在。肌纤维的纵轴末端相互连接构成网状，其连接处称为闰盘（intercalated disc）。闰盘是心肌纤维之间的界限和传递兴奋冲动的重要结构。心肌纤维肌节的构成与骨骼肌纤维类似，含有粗、细两种肌丝，具有横小管和肌浆网结构。另具有以下特点：①心肌原纤维排列不整齐，被线粒体和其他肌浆成分分隔成粗细不等的肌丝束，导致横纹不如骨骼肌明显；②横小管较粗，肌浆网稀疏，纵小管和终池不发达，横小管与终池多形成二联体结构，很难见到三联体；③闰盘常呈阶梯状，存在中间连接、桥粒以及缝隙连接等，对维持心肌纤维的同步运动和信号传导起重要作用；④心肌纤维还具有内分泌功能。

4. 心传导系统

心传导系统（conduction system of heart）由特殊的心肌纤维组成，能自发性地产生和传导兴奋，使心肌进行有规律的收缩和舒张。心传导系统包括窦房结（sinuatrial node）、房室结（atrioventricular node）、房室束（atrioventricular bundle）和浦肯野纤维（Purkinje fiber）。

5．心的血管和淋巴管

心本身的血液循环称冠状循环（coronary circulation），由冠状动脉、毛细血管和心静脉组成。

（1）冠状动脉（coronary artery）：为心的营养动脉，分左冠状动脉和右冠状动脉，分别起始于主动脉根部，经左、右心耳与肺动脉干之间穿出，沿冠状沟和室间沟走行，分支分布于心房和心室，在心肌内形成丰富的毛细血管网。

（2）心静脉（cardiac vein）：包括冠状窦及其属支、心右静脉和心最小静脉。冠状窦（coronary sinus）位于冠状沟内，经冠状窦口注入右心房，其属支有心大静脉（great cardiac vein）和心中静脉（middle cardiac vein）。心右静脉（right cardiac vein）有数支，沿右心室上行注入右心房。心最小静脉（the smallest cardiac vein）是行于心肌内的小静脉，直接开口于各心腔，或者主要是开口于右心房梳状肌之间。

心脏接受自主神经系统支配，包括交感神经和副交感神经。

6．心包

心包（pericardium）为包在心外的锥形囊，其内有少量心包液（pericardial fluid），腹侧以胸骨心包韧带（stenopericardiac ligament）附着于胸骨后部，具有保护心脏的作用。心包发炎导致心包液增多和心包增厚，通过超声检查，可检测到无回声区域，为心包积液。

二、血管

血管（blood vessel）是血液循环的管道系统。根据其结构和功能可分为动脉（artery）、静脉（vein）和毛细血管（blood capillary）。动脉是血液由心脏流向全身各部器官的血管，管壁厚且富有弹性和收缩性。静脉是全身各部位的血液回流心脏的血管，管壁薄，管腔大，有些部位的静脉具有防止血液倒流的静脉瓣结构。毛细血管是体内分布最广的血管，在器官组织内分支互相吻合成网状，管壁很薄，仅有一层内皮细胞构成。

动脉和静脉的管壁从管腔面向外可分为内膜、中膜和外膜3层。有些血管壁还分布有营养血管、神经和特殊的感受器。内膜（internal tunic）从内腔向外又分为内皮、内皮下层和内弹性膜3层。内皮（endothelium）为单层扁平上皮。内皮下层（subendothelial layer）由含有少量胶原纤维、弹性纤维的薄层结缔组织构成。某些血管的内皮下层还会存在纵行的平滑肌细胞。内弹性膜（internal elastic membrane）存在于某些动脉内皮下层，由弹性蛋白构成，膜上有孔。中膜（middle tunic）在动、静脉都存在，动脉的中膜比静脉的厚。大动脉的中膜主要由弹性膜和弹性纤维构成，其间有少许平滑肌和胶原纤维。中动脉主要是平滑肌，肌间有少许弹性纤维和胶原纤维。外膜（external tunic）由疏松结缔组织构成，静脉的外膜比动脉的厚。

1．动脉

（1）体循环（systemic circulation）的动脉包括主动脉及其各级分支。主动脉（aorta）起始于左心室的主动脉口，在肺动脉干与左、右心房之间上升，称升主动脉（ascending aorta）；出心包后向后向上呈弓状延伸至第6胸椎腹侧，称主动脉弓（aortic arch）。主动脉弓向后延续为降主动脉（descending aorta），细分为胸主动脉和腹主动脉。胸主动脉（thoracic aorta）沿胸椎腹侧向后延伸至膈，穿过膈上的主动脉裂孔（aortic foramen）后为腹主动脉（abdominal aorta），沿腰椎腹侧向后伸延，在第5～6腰椎腹侧分为左、右髂外动脉（external iliac artery），左、右髂内动脉（internal iliac artery）及荐中动脉（median sacral artery）。升主动脉起始处膨大形成主动脉球（aortic bulb），内面有3个主动脉窦（aortic sinus），由此处分出冠状动脉，供应心脏的血液。

1）从主动脉弓凸面向前分出臂头动脉干（brachiocephalic trunk），沿气管腹侧与前腔静脉之间向前延伸，至肋间隙处分出右锁骨下动脉（right subclavian artery）和双颈动脉干（bicarotid trunk）。左锁骨下动脉（left subclavian artery）直接由主动脉弓向前分出，位于臂头动脉干背侧。

双颈动脉干在胸前口处气管腹侧分为左、右颈总动脉（common carotid artery），沿颈静脉沟（jugular vein groove）深部前伸，在寰枕关节腹侧，分为枕动脉（occipital artery）、颈内动脉（internal carotid artery）和颈外动脉（external carotid artery）三大支，主要分布于头颈部。在颈总动脉分叉处或附近有颈动脉窦（carotid sinus）和颈动脉体（carotid body）。

锁骨下动脉在胸腔内分出肋颈动脉干（costocervical trunk）、胸廓内动脉（internal thoracic artery）、颈浅动脉（superficial cervical artery）和椎动脉（vertebral artery）等分支后，主干延续为腋动脉，为前肢的动脉主干。

2）胸主动脉是主动脉弓的直接延续，其侧支分为壁支和脏支。壁支为成对的肋间动脉（intercostal artery），分布于胸壁、膈及腹前部的肌肉和皮肤。脏支为支气管食管动脉（bronchoesophageal artery），分布于肺和食管等。

3）腹主动脉是胸主动脉的直接延续，其侧支分为壁支和脏支，壁支主要为成对的腰动脉（lumbar artery）；脏支有不成对的腹腔动脉（celiac artery）、肠系膜前动脉（cranial mesenteric artery）和肠系膜后动脉（caudal mesenteric artery），以及成对的肾动脉（renal artery）、睾丸动脉（testicular artery）或卵巢动脉（ovarian artery）。

4）髂内动脉为骨盆部的动脉主干，成对，其主要分支有脐动脉（umbilical artery）、髂腰动脉（iliolumbar artery）、臀前动脉（cranial gluteal artery）、前列腺动脉（prostatic artery）或阴道动脉（vaginal artery）、臀后动脉（caudal gluteal artery）、闭孔动脉（obturator artery）和阴部内动脉（internal pudendal artery），分布于荐臀部的肌肉、皮肤和骨盆腔内的器官。

5）髂外动脉是后肢的动脉主干，按部位顺次为髂外动脉、股动脉（femoral artery）、腘动脉（popliteal artery）、胫前动脉（anterior tibial artery）、足背动脉（dorsal pedal artery）和跖背侧第3动脉（dorsal metatarsal artery Ⅲ）。

6）荐中动脉为腹主动脉的延续干，沿荐骨腹侧正中向后延伸，分出荐支分布于脊髓和附近的肌肉，主干向后伸达尾椎腹侧称尾正中动脉（caudal median artery），分支分布于尾部。

（2）肺循环（pulmonary circulation）的动脉：肺动脉干（pulmonary trunk）起始于右心室的肺干口，在左、右心耳之间，主动脉左侧向上向后伸延，分为左、右肺动脉（pulmonary artery）经肺门入肺，右肺动脉在入肺前还分出一支到右肺的前叶。肺动脉在肺内随支气管反复分支，最后在肺泡周围形成毛细血管网，在此进行气体交换。肺动脉干与主动脉之间有动物导管索，或称动脉韧带（arterial ligament）相连，为胚胎期动脉导管的遗迹。

2. 静脉

体循环的静脉包括心静脉、奇静脉、前腔静脉和后腔静脉。

（1）心静脉（cardiac vein）属支有心大静脉、心中静脉、心右静脉和心小静脉。

（2）奇静脉（azygos vein）为收集大部分胸壁、气管、食管和腹壁前部血液回流的静脉主干。猪左奇静脉由属支静脉汇成后直接注入冠状窦，其属支有第1、2对腰静脉，肋腹背侧静脉，部分肋间背侧静脉，食管静脉，支气管静脉。

（3）前腔静脉（*cranial vena cava*）为收集头、颈、前肢和部分胸壁和腹壁血液回流的静脉干，其属支有颈内静脉（internal jugular vein）、颈外静脉（external jugular vein）、锁骨下静脉（subclavian vein）、肋颈静脉（costocervical vein）和胸廓内静脉（internal thoracic vein）。

（4）后腔静脉（*caudal vena cava*）为收集腹部、骨盆部、尾部及后肢血液汇流的静脉干，沿途有腰静脉（lumbar vein）、肝静脉（hepatic vein）、肾静脉（renal vein）、睾丸静脉（testicular vein）或卵巢静脉（ovarian vein）、髂总静脉（common iliac vein）等汇入。

三、血液

血液（blood）是流动在心血管内的液态结缔组织，占体重的3%～5%，由血细胞和液态的血浆构成，执行运输营养、排泄代谢产物、机体防御免疫和调节体温与渗透压等功能。

血细胞（blood cell, hemocyte）是血液中的有形成分，悬浮于血浆中，包括红细胞、白细胞和血小板。红细胞（erythrocyte, red blood cell, RBC）为双面凹的圆盘状，无胞核和细胞器，是血液中数量最多的一种细胞，（5～10）$\times 10^6$/ml。白细胞（leukocyte, white blood cell, WBC）是具有胞核和细胞器的球形细胞，体积比红细胞大，但数量少，（6～25）$\times 10^3$/ml。白细胞包括中性粒细胞（neutrophilic granulocyte, neutrophil）、嗜酸性粒细胞（eosinophilic granulocyte, eosinophil）、嗜碱性粒细胞（basophilic granulocyte, basophil）、单核细胞（monocyte）和淋巴细胞（lymphocyte）。各种白细胞的分类百分比分别为中性粒细胞占37%、嗜酸性粒细胞占4%、嗜碱性粒细胞占1%、单核细胞占5%、淋巴细胞占53%。血小板（blood platelet）是骨髓巨核细胞的碎片，没有胞核，有细胞器，为圆形或椭圆形的小体，数量为（115～425）$\times 10^3$/ml。

血浆（plasma）为淡黄色的液体，约占血液总量的50%以上，其中水约占90%，其余为有机物和无机物。血液自然凝固后，凝血块周围析出的淡黄色清亮液体称为血清（serum）。

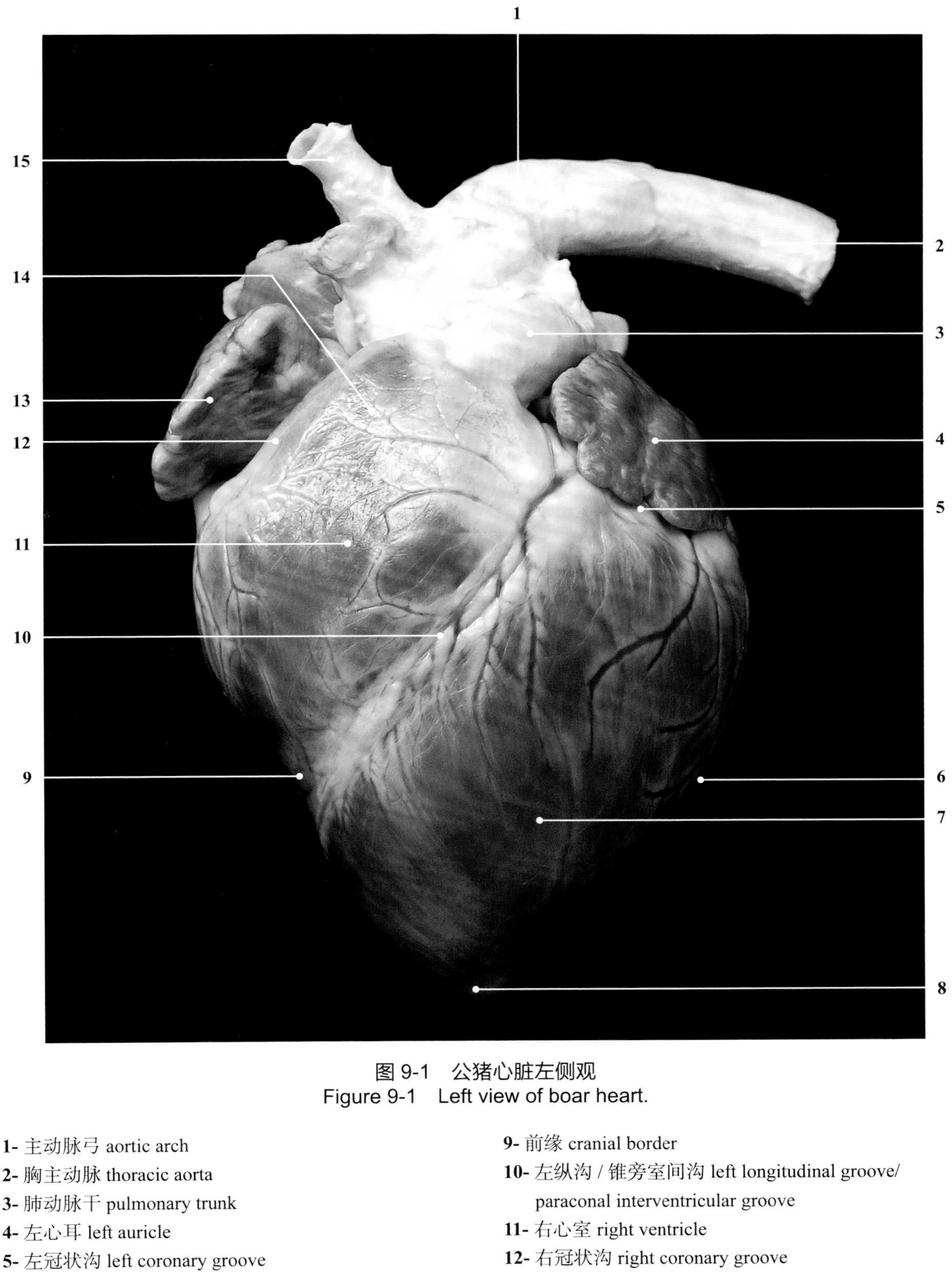

图 9-1 公猪心脏左侧观
Figure 9-1 Left view of boar heart.

1- 主动脉弓 aortic arch
2- 胸主动脉 thoracic aorta
3- 肺动脉干 pulmonary trunk
4- 左心耳 left auricle
5- 左冠状沟 left coronary groove
6- 后缘 caudal border
7- 左心室 left ventricle
8- 心尖 cardiac apex
9- 前缘 cranial border
10- 左纵沟 / 锥旁室间沟 left longitudinal groove/ paraconal interventricular groove
11- 右心室 right ventricle
12- 右冠状沟 right coronary groove
13- 右心耳 right auricle
14- 动脉圆锥 arterial cone
15- 臂头动脉干 brachiocephalic trunk

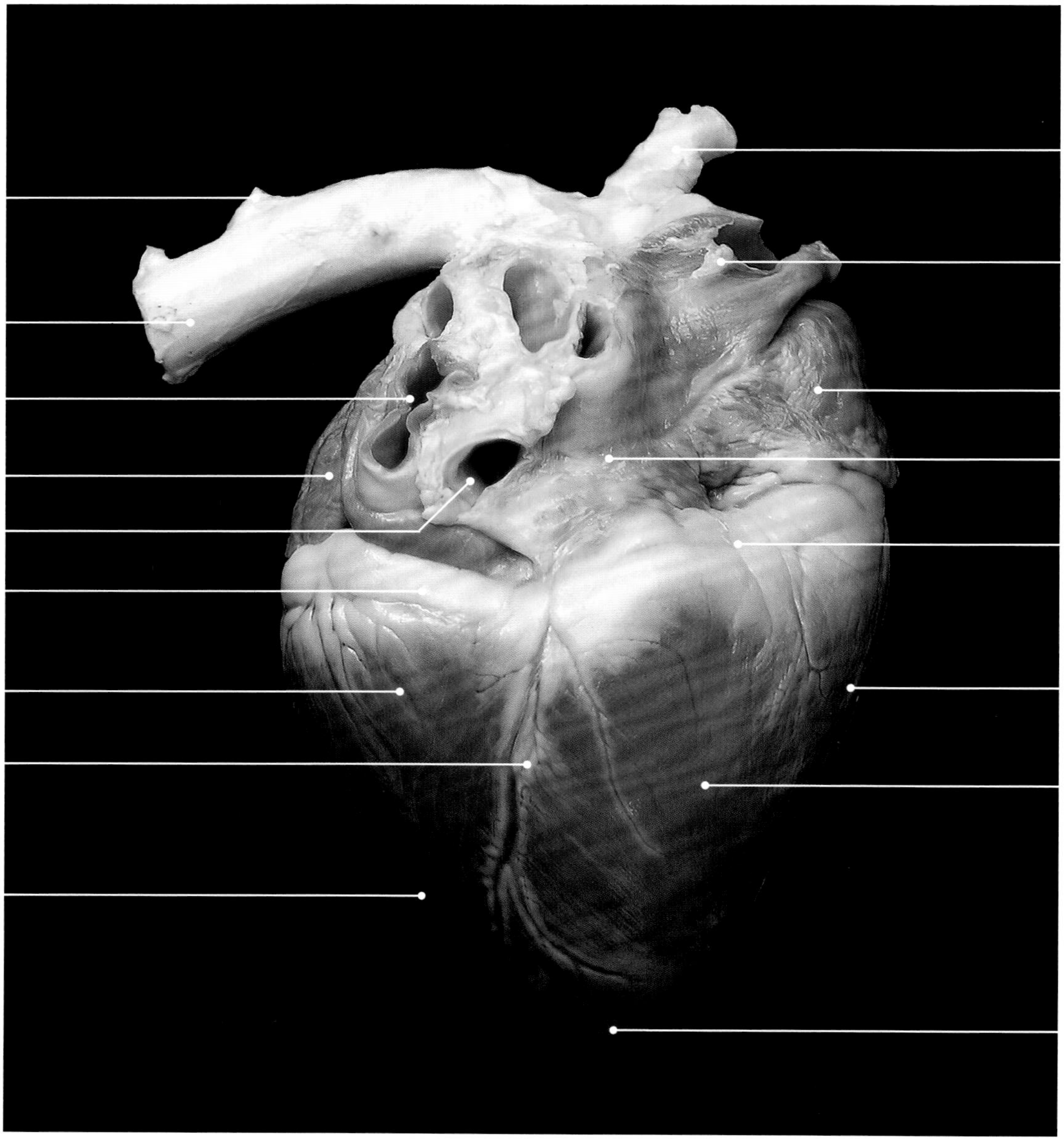

图 9-2 公猪心脏右侧观
Figure 9-2 Right view of boar heart.

1- 臂头动脉干 brachiocephalic trunk
2- 前腔静脉 *cranial vena cava*
3- 右心耳 right auricle
4- 心基 cardiac base
5- 右冠状沟 right coronary groove
6- 前缘 cranial border
7- 右心室 right ventricle
8- 心尖 cardiac apex
9- 后缘 caudal border
10- 右纵沟 / 窦下室间沟 right longitudinal groove / subsinuosal interventricular groove
11- 左心室 left ventricle
12- 左冠状沟 left coronary groove
13- 后腔静脉 *caudal vena cava*
14- 左心耳 left auricle
15- 肺静脉 pulmonary vein
16- 胸主动脉 thoracic aorta
17- 肋间动脉 intercostal artery

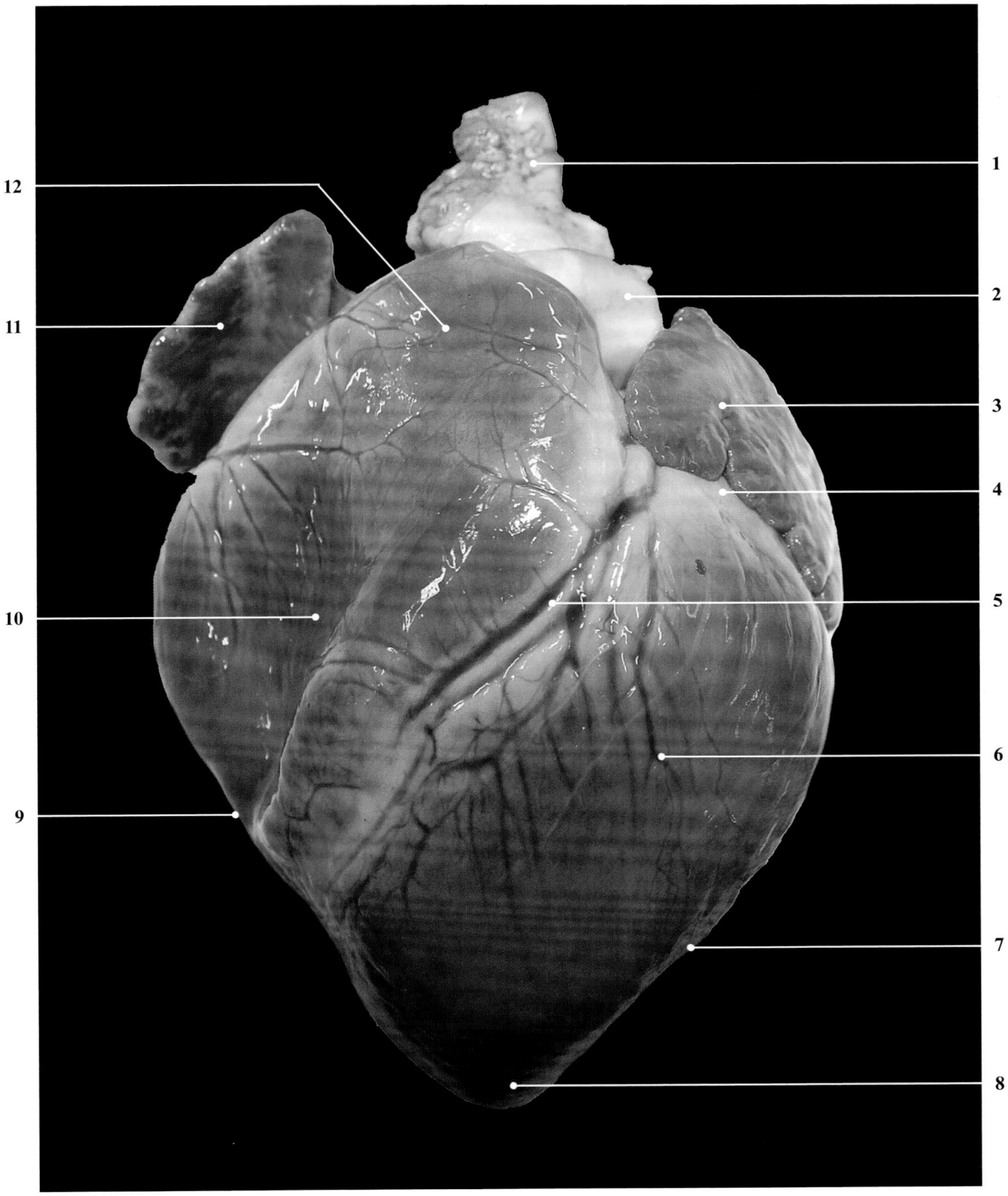

图 9-3 母猪心脏左侧观
Figure 9-3 Left view of sow heart.

1- 主动脉 aorta
2- 肺动脉干 pulmonary trunk
3- 左心耳 left auricle
4- 左冠状沟 left coronary groove
5- 左纵沟 / 锥旁室间沟 left longitudinal groove / paraconal interventricular groove
6- 左心室 left ventricle
7- 后缘 caudal border
8- 心尖 cardiac apex
9- 前缘 cranial border
10- 右心室 right ventricle
11- 右心耳 right auricle
12- 动脉圆锥 arterial cone

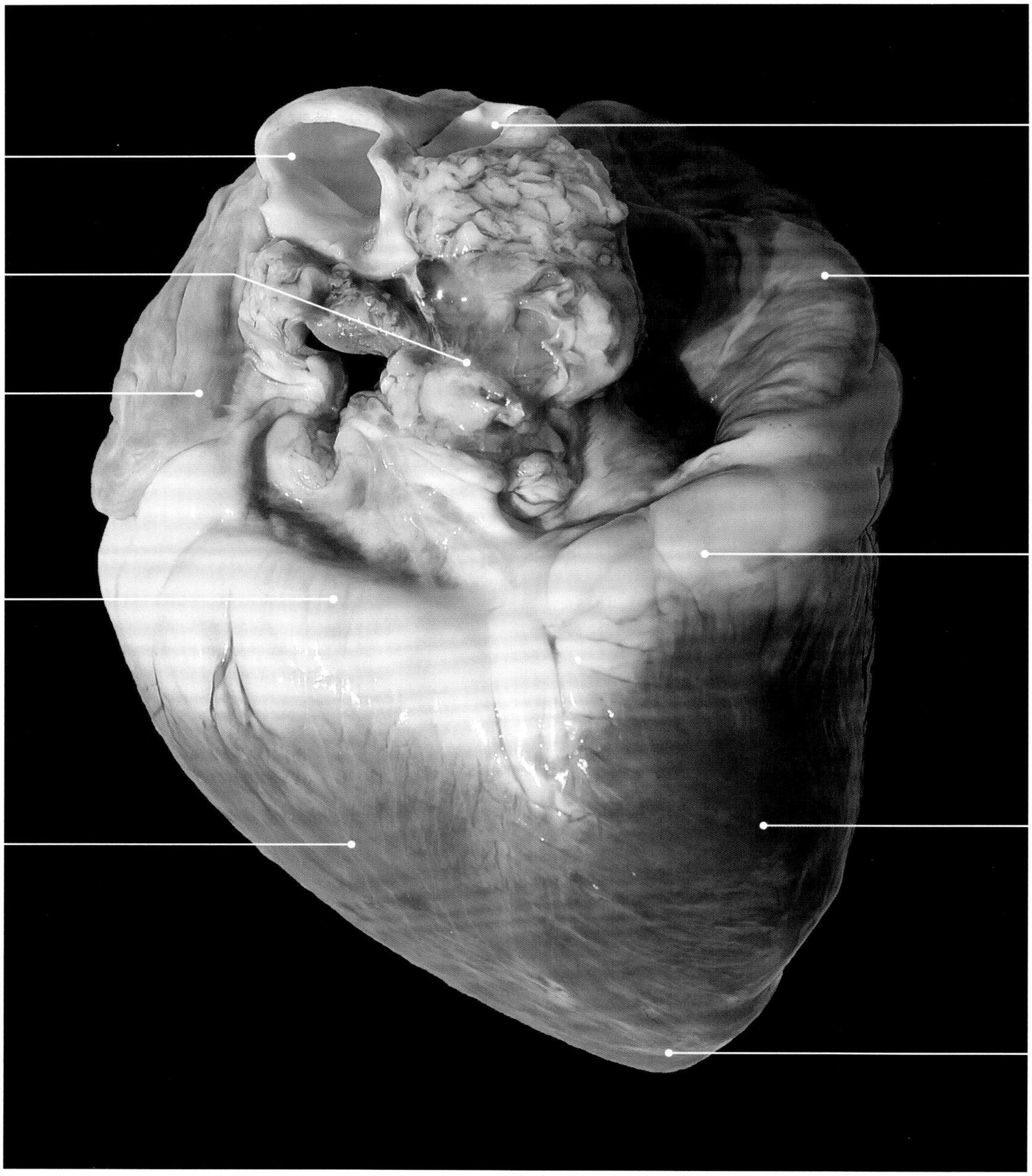

图 9-4 母猪心脏右侧观
Figure 9-4 Right view of sow heart.

1- 臂头动脉干 brachiocephalic trunk
2- 右心耳 right auricle
3- 右冠状沟 right coronary groove
4- 右心室 right ventricle
5- 心尖 cardiac apex
6- 左心室 left ventricle
7- 左冠状沟 left coronary groove
8- 左心耳 left auricle
9- 心基 cardiac base
10- 主动脉 aorta

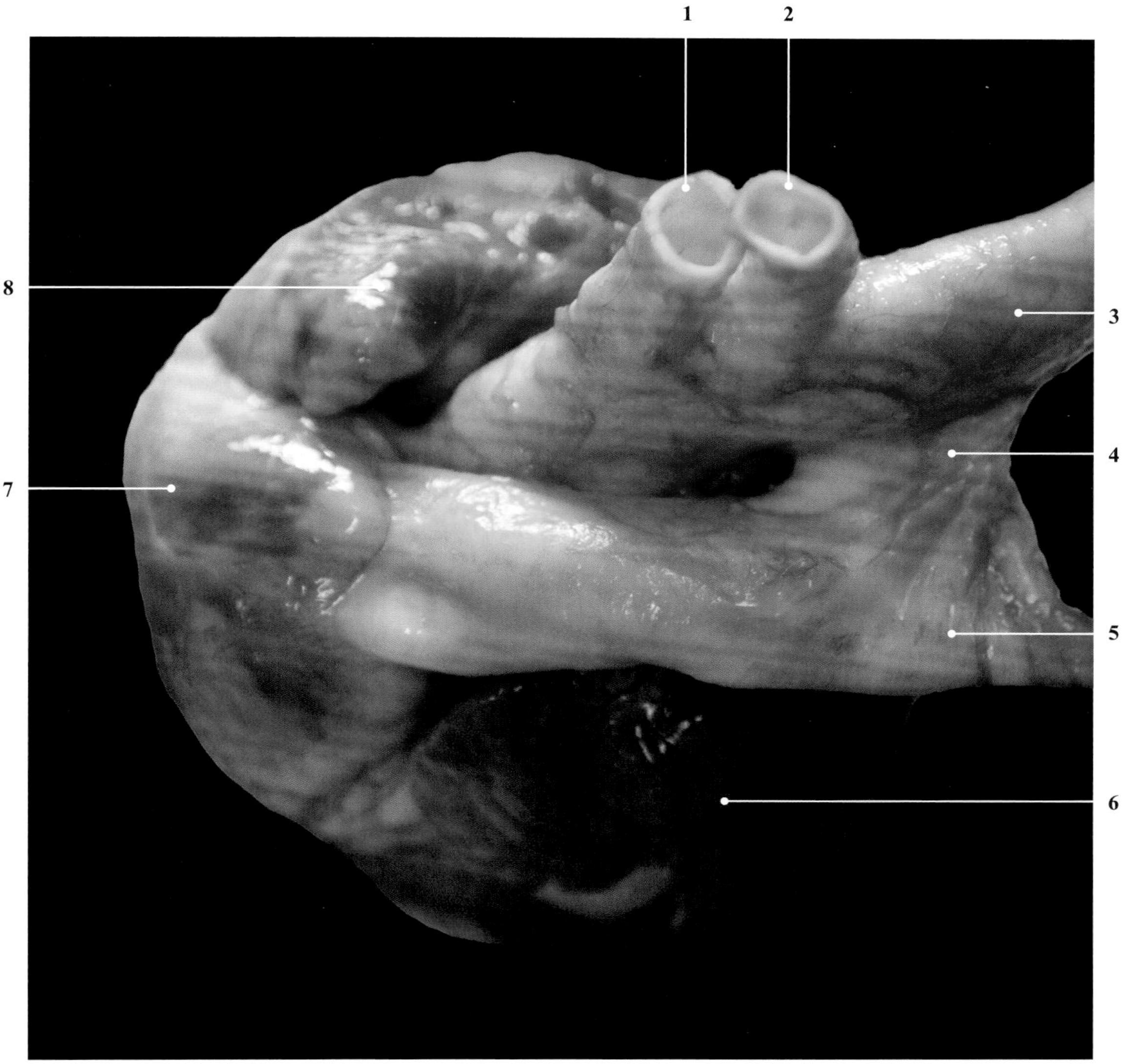

图 9-5 母猪动脉韧带
Figure 9-5 Arterial ligament of sow.

1- 臂头动脉 brachiocephalic artery
2- 左锁骨下动脉 left subclavian artery
3- 胸主动脉 thoracic aorta
4- 动脉韧带 arterial ligament
5- 肺动脉干 pulmonary trunk
6- 左心耳 left auricle
7- 动脉圆锥 arterial cone
8- 右心耳 right auricle

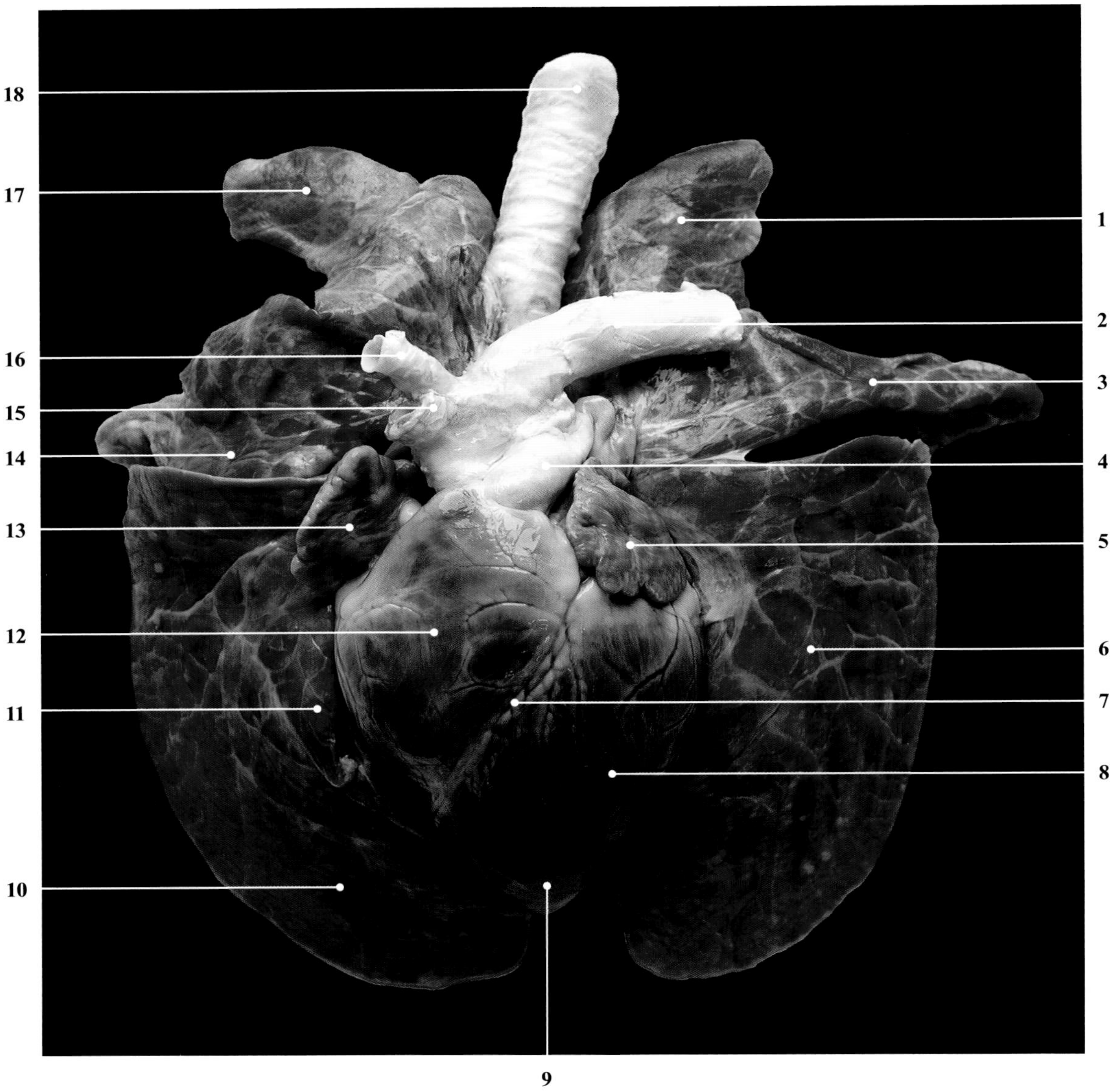

图 9-6　心脏和肺腹侧观
Figure 9-6　Ventral view of the heart and lung.

1- 左肺前叶（尖叶） left cranial lobe (apical lobe) of the lung
2- 主动脉 aorta
3- 左肺中叶（心叶） left middle lobe (cardiac lobe) of the lung
4- 肺动脉干 pulmonary trunk
5- 左心耳 left auricle
6- 左肺后叶（膈叶） left caudal lobe (diaphragmatic lobe) of the lung
7- 左纵沟（锥旁室间沟）left longitudinal groove（paraconal interventricular groove）
8- 左心室 left ventricle
9- 心尖 cardiac apex
10- 右肺后叶（膈叶） right caudal lobe (diaphragmatic lobe) of the lung
11- 副叶 accessory lobe
12- 右心室 right ventricle
13- 右心耳 right auricle
14- 右肺中叶（心叶） right middle lobe (cardiac lobe) of the lung
15- 臂头动脉 brachiocephalic artery
16- 左锁骨下动脉 left subclavian artery
17- 右肺前叶（尖叶） right cranial lobe (apical lobe) of the lung
18- 气管 trachea

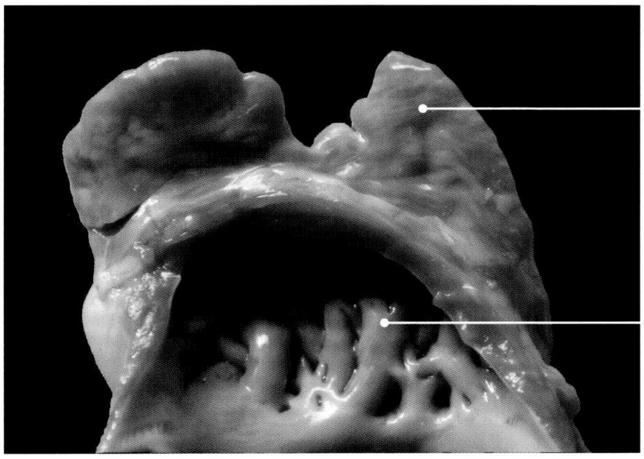

图 9-7　心耳梳状肌
Figure 9-7　Pectinate muscle of the auricle.

1- 心耳 auricle
2- 梳状肌 pectinate muscle

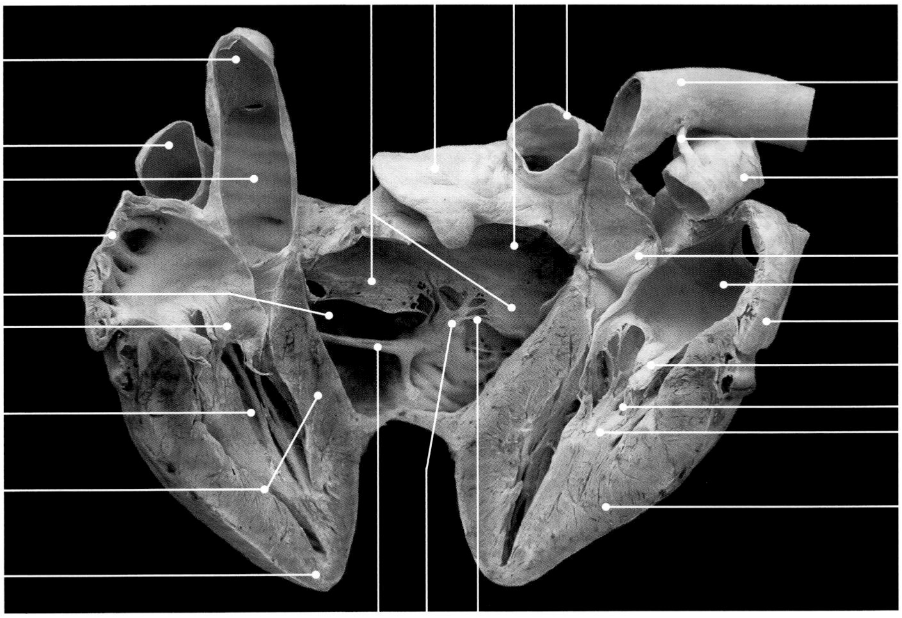

图 9-8　心腔结构（塑化标本）
Figure 9-8　Structure of the heart chamber (plastinated preparation).

1- 三尖瓣 tricuspid valve
2- 右心耳 right auricle
3- 右心房 right atrium
4- 前腔静脉 *cranial vena cava*
5- 主动脉弓 aortic arch
6- 动脉韧带 / 动脉导管索 arterial ligament
7- 肺动脉干 pulmonary trunk
8- 半月瓣 semilunar valve
9- 左心房 left atrium
10- 左心耳 left auricle
11- 二尖瓣 bicuspid valve
12- 二尖瓣的腱索 chordae tendineae of the bicuspid valve
13- 二尖瓣的乳头肌 papillary muscle of the bicuspid valve
14- 左心室壁 left ventricle wall
15- 三尖瓣的腱索 chordae tendineae of the tricuspid valve
16- 三尖瓣的乳头肌 papillary muscle of the tricuspid valve
17- 隔缘肉柱 / 心横肌 septomarginal trabecula / transverse muscle of the heart
18- 心尖 cardiac apex
19- 室间隔 interventricular septum
20- 左心室 left ventricle
21- 左房室口 left atrioventricular orifice
22- 右心室 right ventricle
23- 左心耳 left auricle
24- 主动脉 aorta
25- 臂头动脉干 brachiocephalic trunk
26- 左锁骨下动脉 left subclavian artery

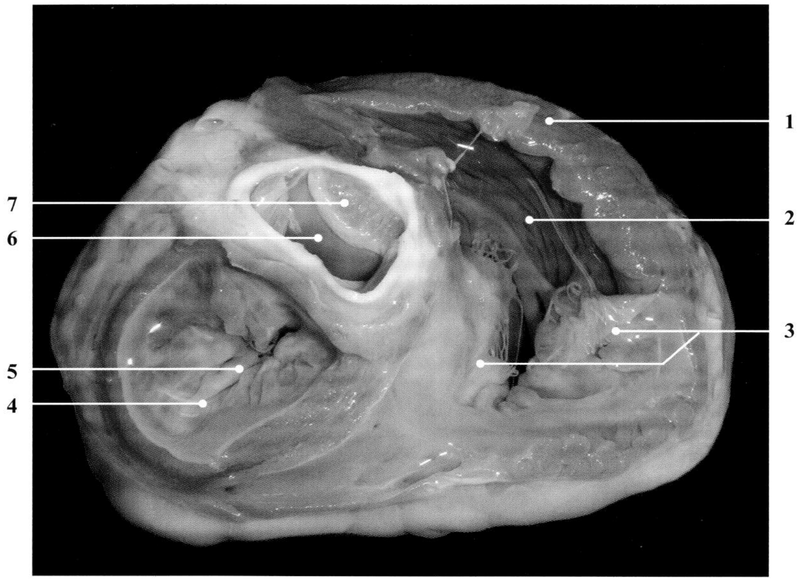

图 9-9 心瓣膜 -1
Figure 9-9 Cardiac valve-1.

1- 右心室壁 right ventricular wall
2- 右房室口 right atrioventricular orifice
3- 三尖瓣 tricuspid valve
4- 二尖瓣 bicuspid valve
5- 左房室口 left atrioventricular orifice
6- 主动脉口 aortic orifice
7- 主动脉瓣 aortic valve

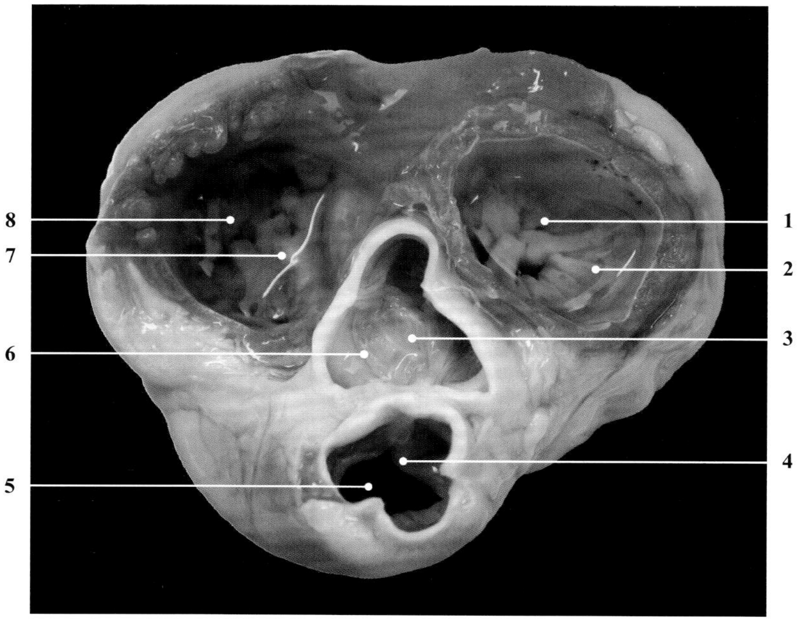

图 9-10 心瓣膜 -2
Figure 9-10 Cardiac valve-2.

1- 左房室口 left atrioventricular orifice
2- 二尖瓣 bicuspid valve
3- 主动脉瓣 aortic valve
4- 肺动脉瓣 pulmonary valve
5- 肺动脉口 orifice of pulmonary trunk
6- 主动脉口 aortic orifice
7- 三尖瓣 tricuspid valve
8- 右房室口 right atrioventricular orifice

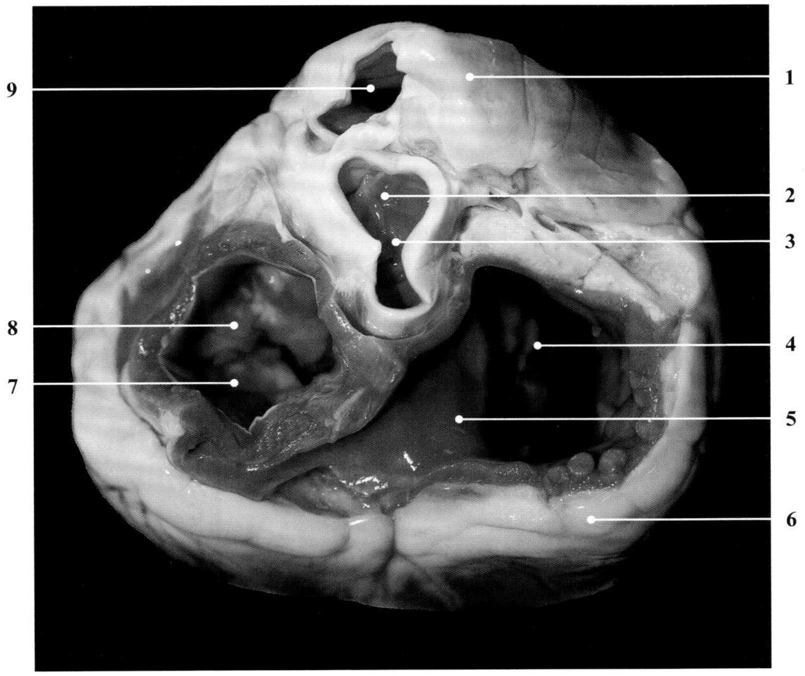

图 9-11　心瓣膜 -3
Figure 9-11　Cardiac valve-3.

1- 动脉圆锥 arterial cone
2- 主动脉瓣 aortic valve
3- 主动脉口 aortic orifice
4- 右房室口 right atrioventricular orifice
5- 三尖瓣 tricuspid valve
6- 冠状沟 coronary groove
7- 左房室口 left atrioventricular orifice
8- 二尖瓣 bicuspid valve
9- 肺动脉口 orifice of pulmonary trunk

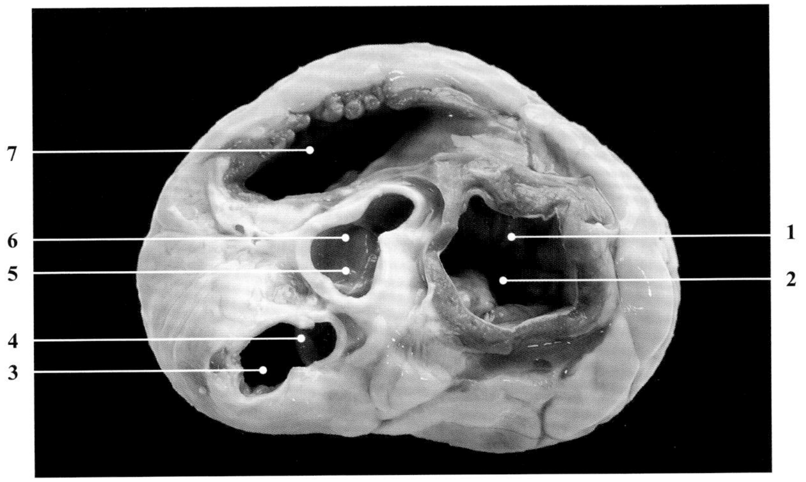

图 9-12　心瓣膜 -4
Figure 9-12　Cardiac valve-4.

1- 二尖瓣 bicuspid valve
2- 左房室口 left atrioventricular orifice
3- 肺动脉口 orifice of pulmonary trunk
4- 肺动脉瓣 pulmonary valve
5- 主动脉瓣 aortic valve
6- 主动脉口 aortic orifice
7- 右房室口 right atrioventricular orifice

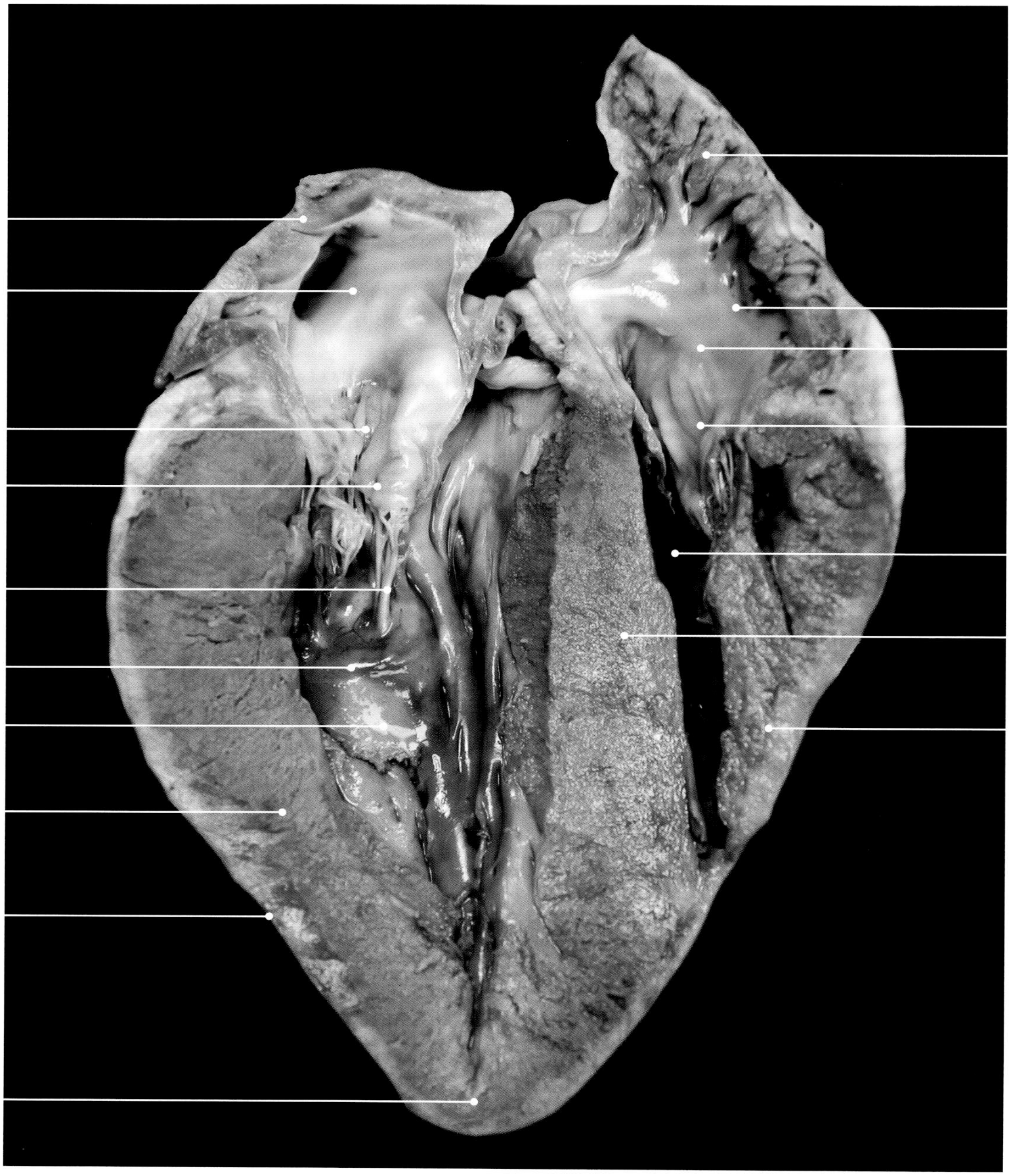

图 9-13　心剖面（固定标本）
Figure 9-13　Cross section of the heart (fixed preparation).

1- 右心耳 right auricle
2- 右心房 right atrium
3- 右房室口 right atrioventricular orifice
4- 三尖瓣 Tricuspid valve
5- 右心室 right ventricle
6- 室间隔 interventricular septum
7- 右心室壁 right ventricular wall
8- 心尖 cardiac apex
9- 心外膜 epicardium
10- 左心室壁 left ventricle wall
11- 左心室 left ventricle
12- 乳头肌 papillary muscle
13- 腱索 chordae tendineae
14- 二尖瓣 bicuspid valve
15- 左房室口 left atrioventricular orifice
16- 左心房 left atrium
17- 左心耳 left auricle

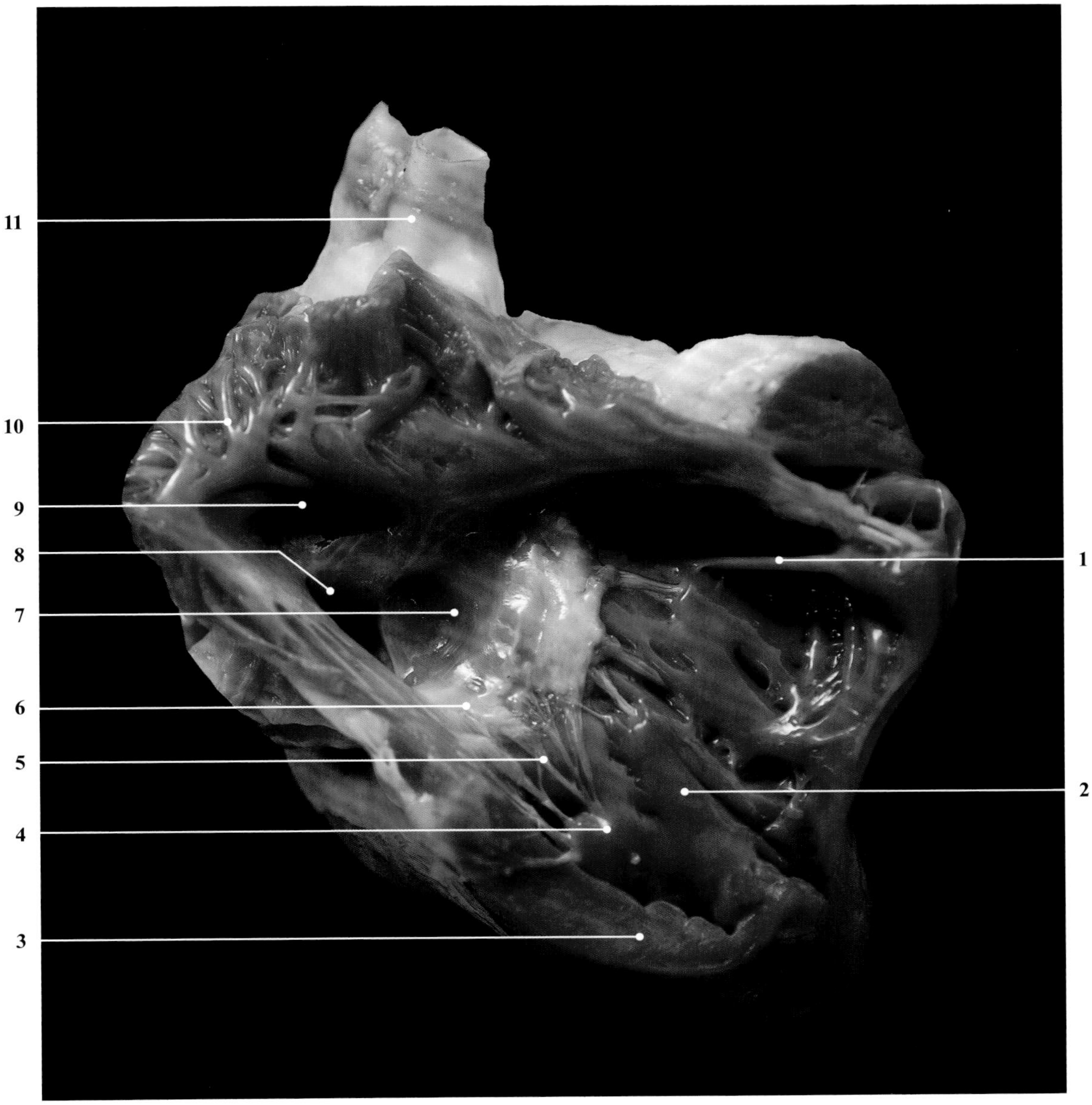

图 9-14 右心房和右心室腔面
Figure 9-14 Right atrium and ventricular cavity.

1- 隔缘肉柱 / 心横肌 septomarginal trabecula / transverse muscle of the heart
2- 室间隔 interventricular septum
3- 右心室壁 right ventricular wall
4- 乳头肌 papillary muscle
5- 腱索 chordae tendineae
6- 三尖瓣 tricuspid valve
7- 右房室口 right atrioventricular orifice
8- 冠状窦 coronary sinus
9- 卵圆窝 oval fossa
10- 梳状肌 pectinate muscle
11- 主动脉 aorta

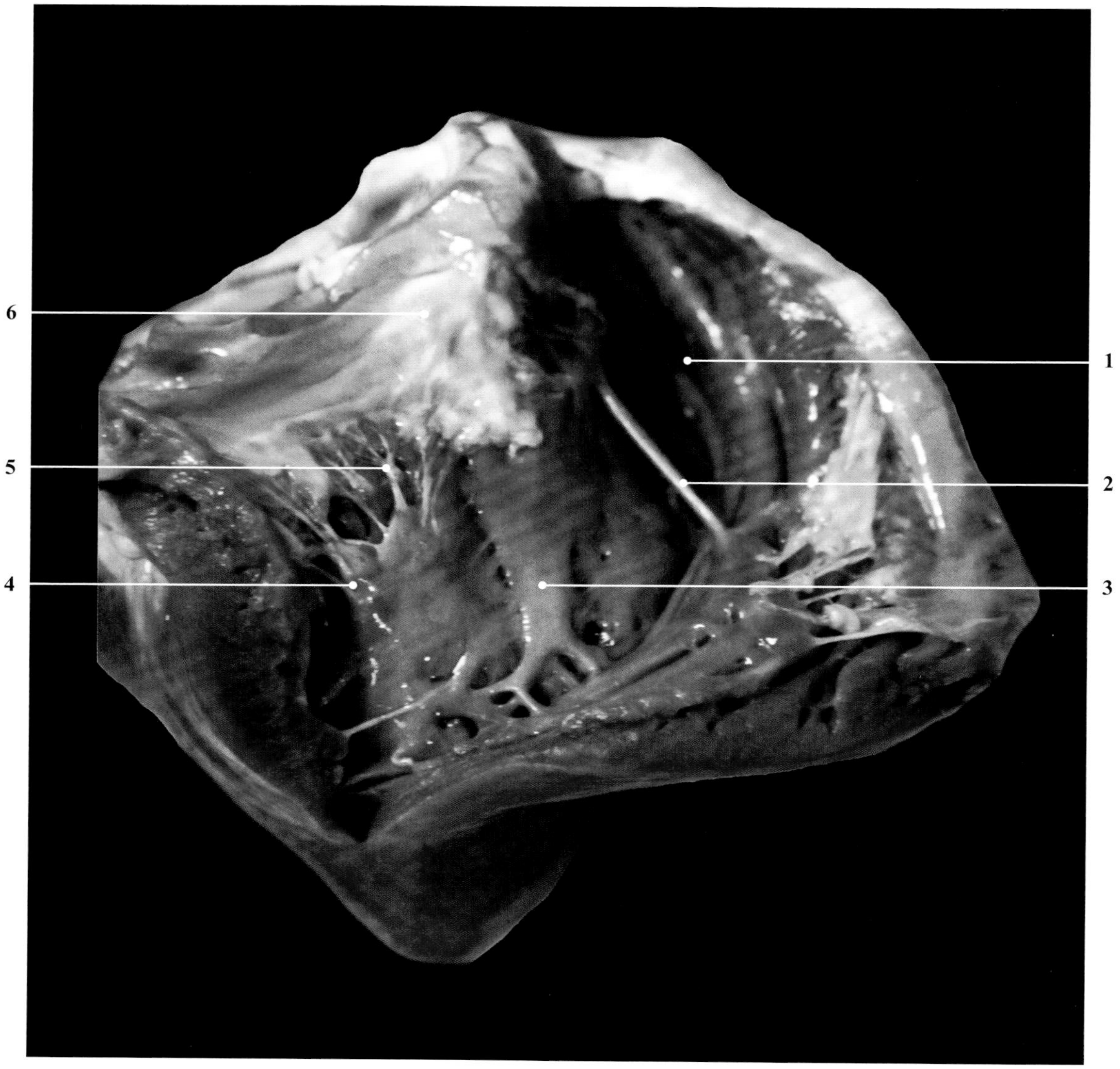

图 9-15　右心室内腔面
Figure 9-15　Right ventricular chamber.

1- 右心室 right ventricle
2- 隔缘肉柱 / 心横肌 septomarginal trabecula / transverse muscle of the heart
3- 室间隔 interventricular septum
4- 乳头肌 papillary muscle
5- 腱索 chordae tendineae
6- 三尖瓣 tricuspid valve

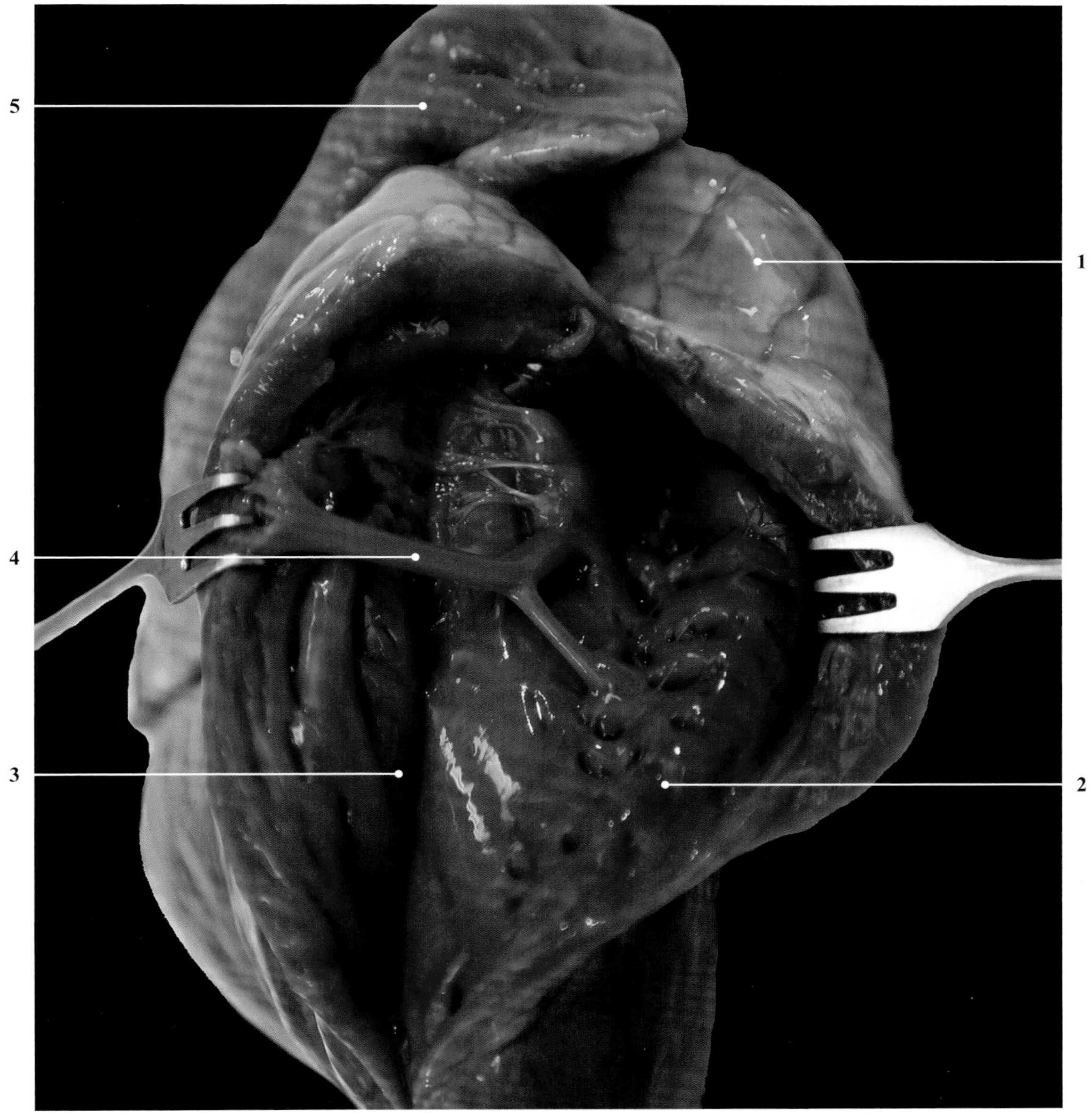

图 9-16 母猪右心室心横肌
Figure 9-16 Transverse muscle of the right ventricle in sow heart.

1- 右心耳 right auricle
2- 心内膜 endocardium
3- 右心室 right ventricle
4- 隔缘肉柱 / 心横肌 septomarginal trabecula / transverse muscle of the heart
5- 左心耳 left auricle

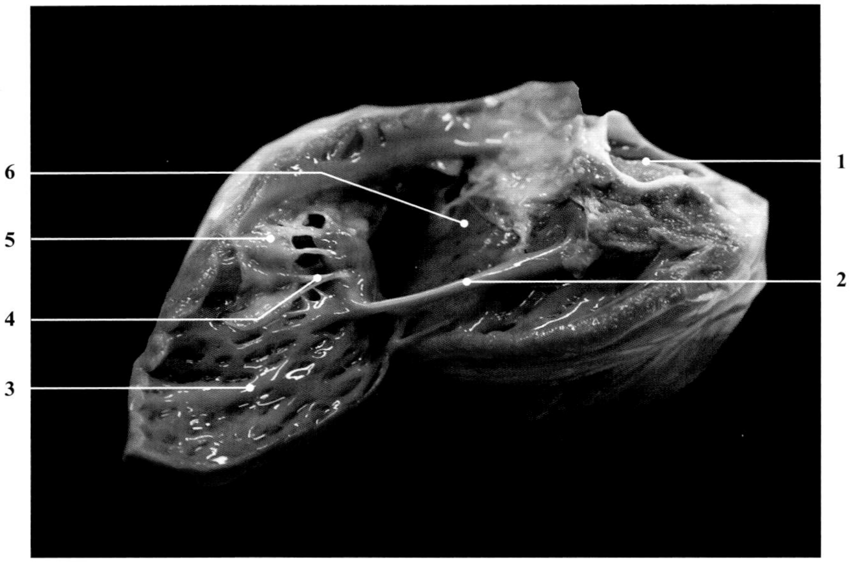

图 9-17　公猪右心室心横肌
Figure 9-17　Transverse muscle of the right ventricle in boar.

1- 主动脉口 aortic orifice
2- 隔缘肉柱 / 心横肌 septomarginal trabecula / transverse muscle of heart
3- 右心室壁 right ventricular wall
4- 腱索 chordae tendineae
5- 三尖瓣 tricuspid valve
6- 室间隔 interventricular septum

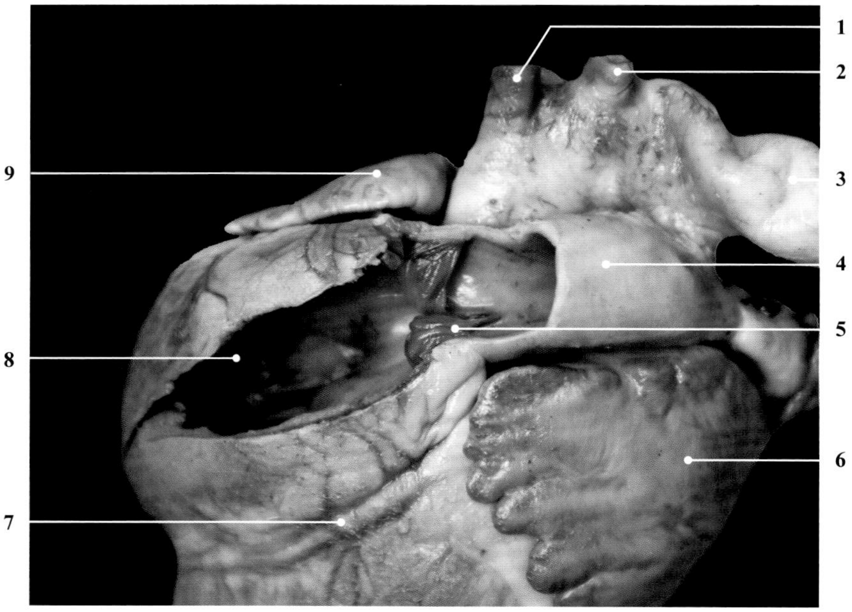

图 9-18　肺动脉和肺动脉瓣
Figure 9-18　Pulmonary artery and valve.

1- 臂头动脉干 brachiocephalic trunk
2- 左锁骨下动脉 left subclavian artery
3- 主动脉 aorta
4- 肺动脉 pulmonary artery
5- 肺动脉瓣 pulmonary valve
6- 左心耳 left auricle
7- 左纵沟 / 锥旁室间沟 left longitudinal groove / paraconal interventricular groove
8- 右心室 right ventricle
9- 右心耳 right auricle

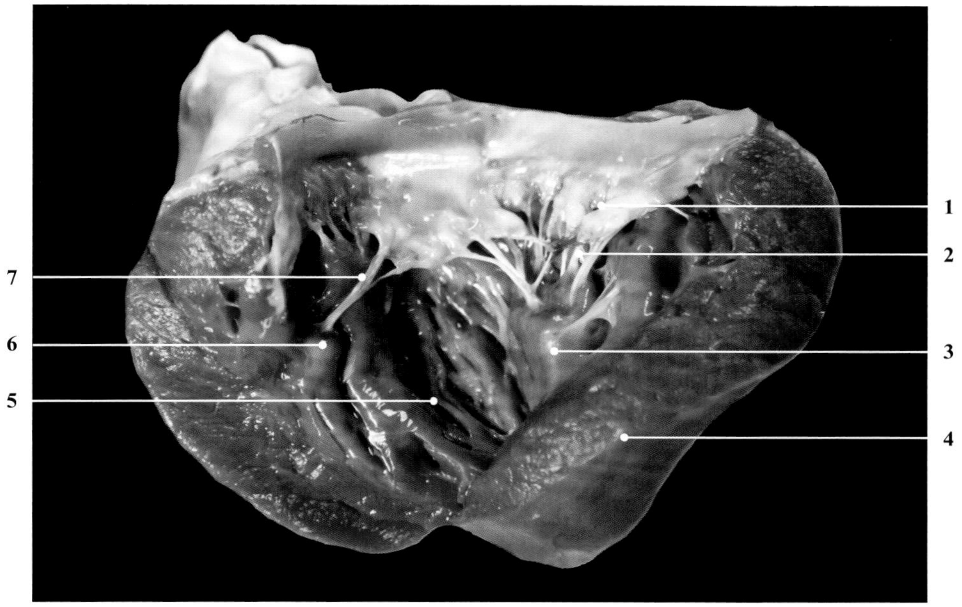

图 9-19 母猪左心室瓣膜
Figure 9-19 Cardiac valve of the left ventricle in sow.

1- 二尖瓣 bicuspid valve
2- 腱索 chordae tendineae
3- 乳头肌 papillary muscle
4- 左心室壁 left ventriclar wall
5- 左心室腔 left ventriclar chamber
6- 乳头肌 papillary muscle
7- 腱索 chordae tendineae

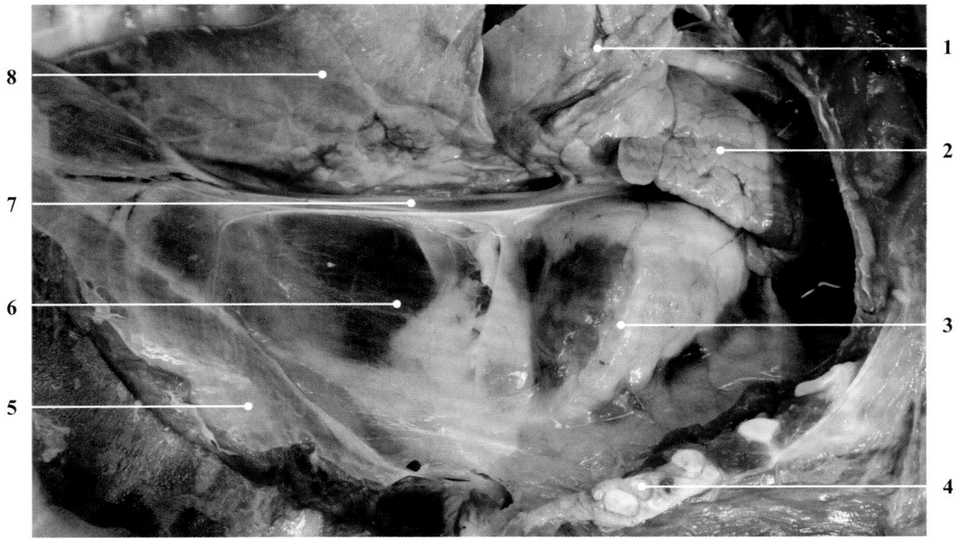

图 9-20 心包和纵隔
Figure 9-20 Pericardium and mediastinum.

1- 右肺中叶（心叶） right middle lobe (cardiac lobe) of the lung
2- 右肺前叶（尖叶） right cranial lobe (apical lobe) of the lung
3- 心包和心脏 pericardium and heart
4- 胸骨 breast bone, sternum
5- 膈 diaphragm
6- 纵隔 mediastinum
7- 后腔静脉 caudal vena cava
8- 右肺后叶（膈叶） right caudal lobe (diaphragmatic lobe) of the lung

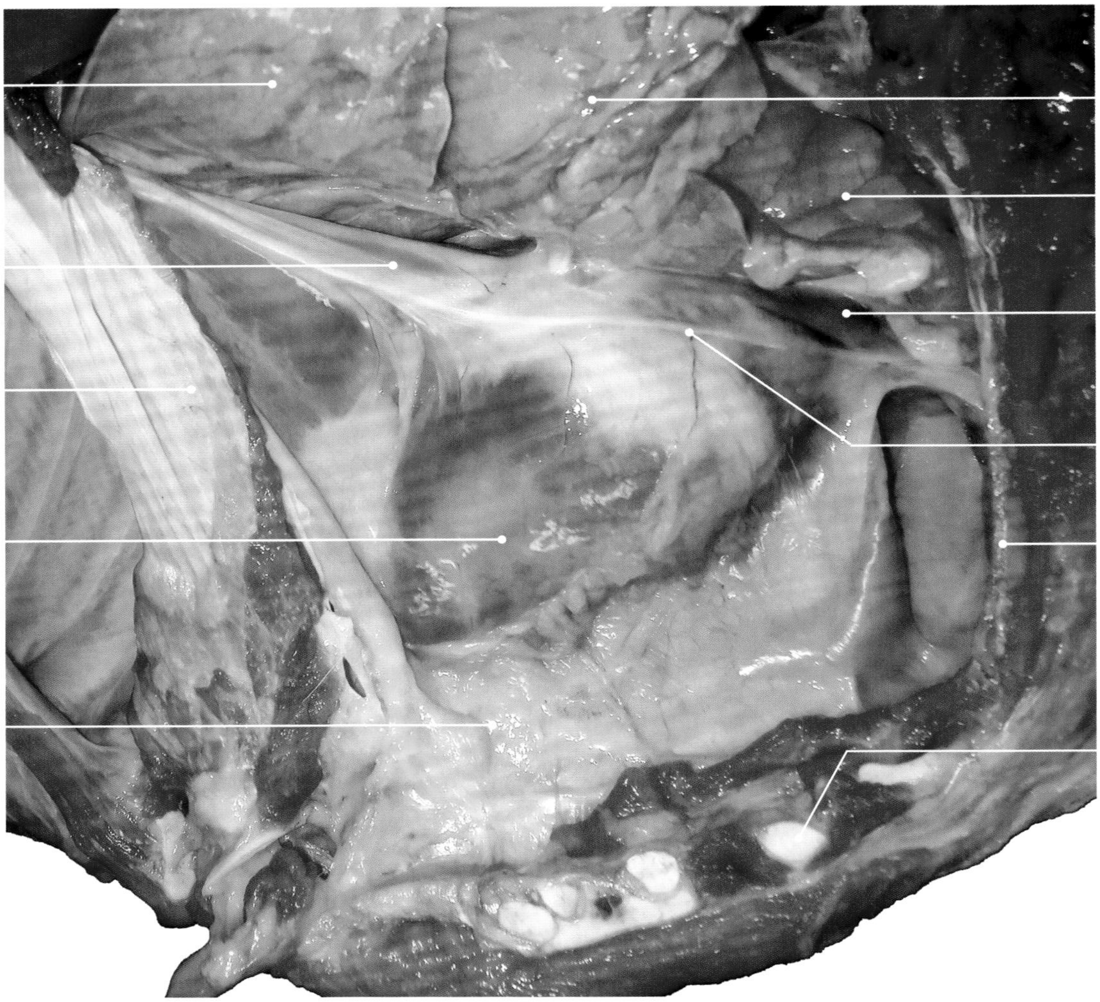

图 9-21 胸骨心包韧带
Figure 9-21 Stenopericardiac ligament.

1- 右肺中叶（心叶） right middle lobe (cardiac lobe) of the lung
2- 右肺前叶（尖叶） right cranial lobe (apical lobe) of the lung
3- 前腔静脉 *cranial vena cava*
4- 膈神经 phrenic nerve
5- 第 1 肋 lst rib
6- 胸骨 breast bone, sternum
7- 胸骨心包韧带 stenopericardiac ligament
8- 心包和心脏 pericardium and heart
9- 膈 diaphragm
10- 后腔静脉 caudal vena cava
11- 右肺后叶（膈叶） right caudal lobe (diaphragmatic lobe) of the lung

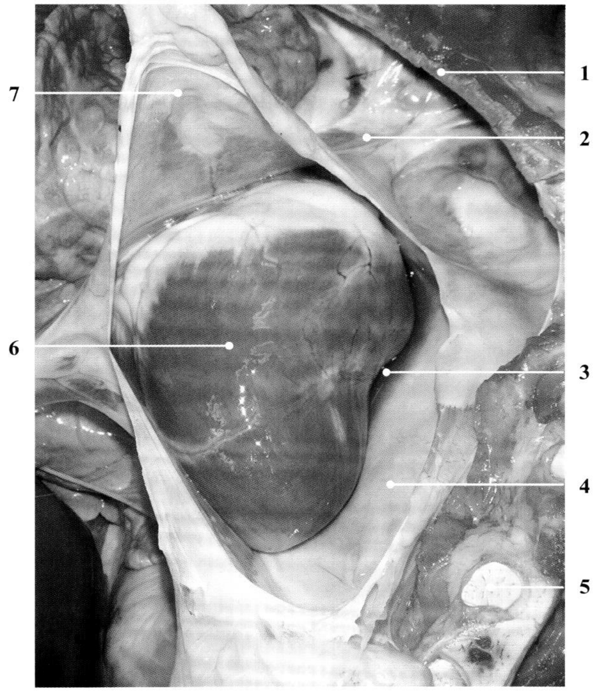

图 9-22　心包腔和心包液
Figure 9-22　Pericardial cavity and fluid.

1- 第 1 肋 lst rib
2- 前腔静脉 *cranial vena cava*
3- 心包液 pericardial fluid
4- 心包腔 pericardial cavity
5- 胸骨 breast bone, sternum
6- 心脏 heart
7- 心包 pericardium

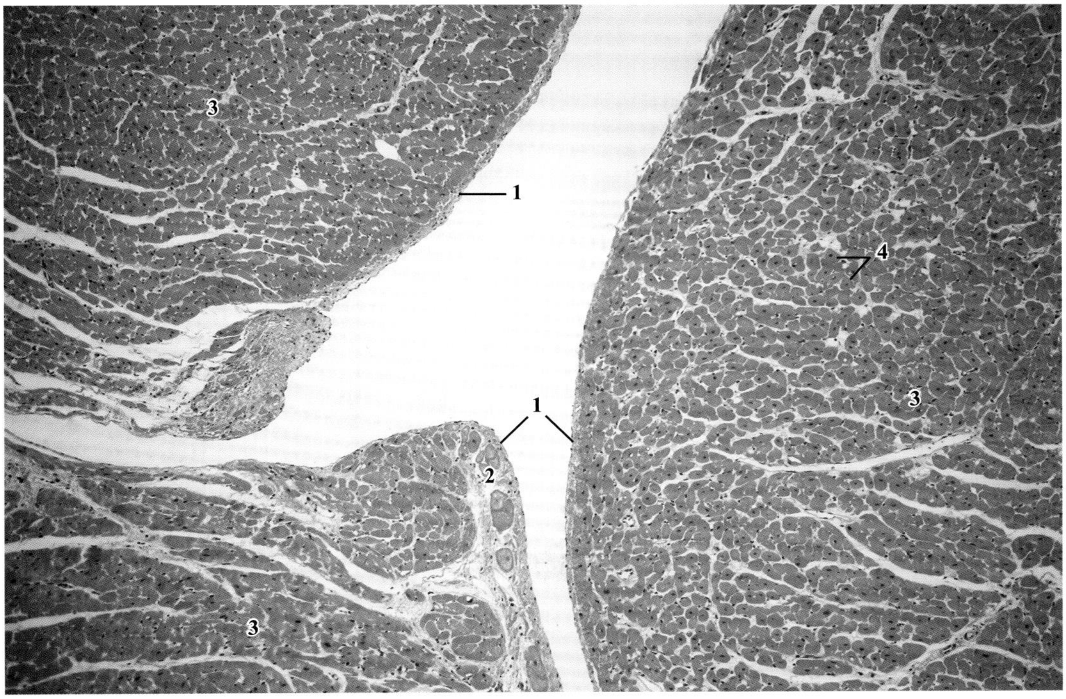

图 9-23　右心耳组织切片（横断面；HE 染色，低倍镜）
Figure 9-23　Histological section of the right auride (cross section; HE staining, lower power).

1- 心内膜 endocardium
2- 浦肯野纤维 Purkinje fiber
3- 心肌膜 myocardium
4- 心肌细胞横断面 cross section of the cardiac muscle cell

1- 肌束膜 perimysium
2- 心肌细胞核 nucleus of the cardiac muscle cell
3- 肌原纤维 myofibril
4- 心肌细胞横断面 cross section of the cardiac muscle cell
5- 毛细血管 blood capillary

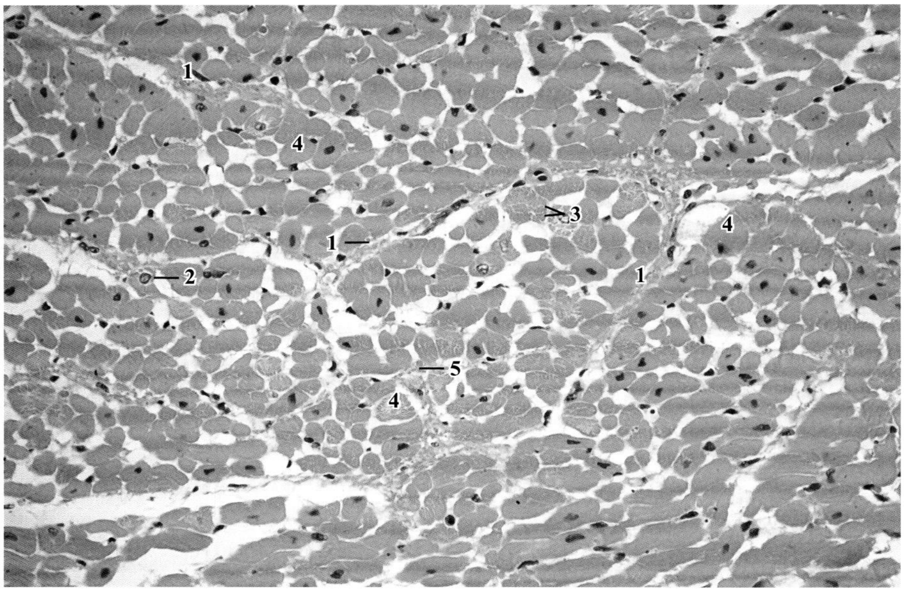

图 9-24　左心耳组织切片（横断面；HE 染色，高倍镜）
Figure 9-24　Histological section of the left auricle (cross section; HE staining; higher power).

1- 心肌细胞核 nucleus of the cardiac muscle cell
2- 闰盘 intercalated disk
3- 心肌细胞横纹 cross striation of the cardiac muscle cell
4- 成纤维细胞核 nucleus, fibroblast
5- 红细胞 red blood cell (RBC)

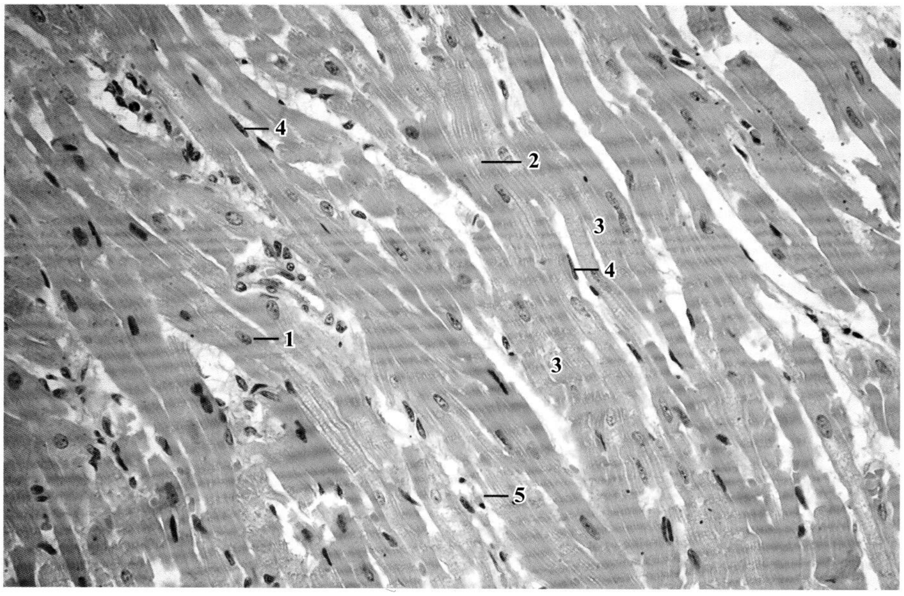

图 9-25　左心耳组织切片（纵切面；HE 染色，高倍镜）
Figure 9-25　Histological section of the left auricle (longitudinal section; HE staining, higher power).

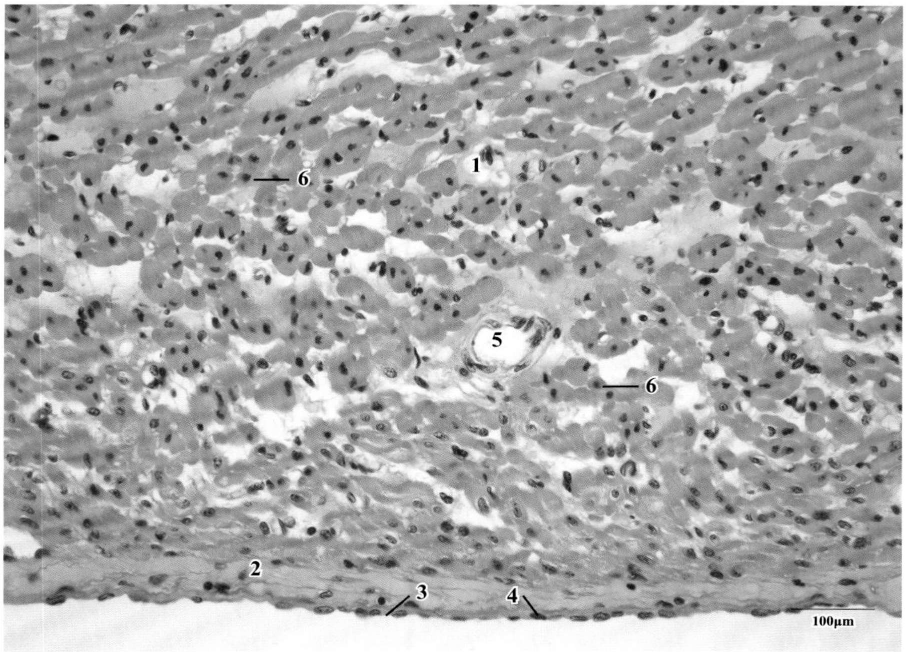

1- 心肌膜 myocardium
2- 心外膜 epicardium
3- 间皮 mesothelium
4- 间皮细胞核 nucleus of mesothelial cell
5- 静脉 vein
6- 心肌细胞横断面 cross section of the cardiac muscle cell

图 9-26 心壁组织切片（横断面；HE 染色，高倍镜）
Figure 9-26 Histological section of the cardiac wall (cross section；HE staining, higher power).

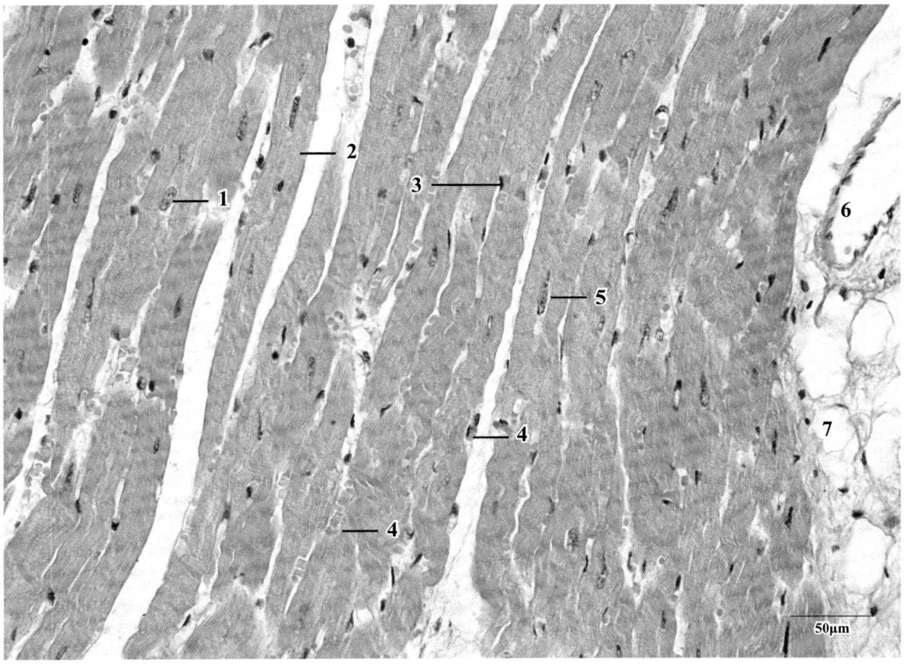

1- 心肌细胞核 nucleus of the cardiac muscle cell
2- 心肌细胞横纹 cross striation of the cardiac muscle cell
3- 内皮细胞 endothelial cell
4- 毛细血管 blood capillary
5- 成纤维细胞核 nucleus, fibroblast
6- 静脉 vein
7- 脂肪细胞 adipose cell

图 9-27 心肌组织切片（纵切面；HE 染色，高倍镜）
Figure 9-27 Histological section of the cardiac muscle (longitudinal section; HE staining, higher power).

1- 心肌细胞分支 branch of the cardiac muscle cell
2- 心肌细胞纵切面 longitudinal section of the cardiac muscle cell
3- 双核心肌细胞 binucleate cardiac muscle cell
4- 成纤维细胞核 nucleus, fibroblast

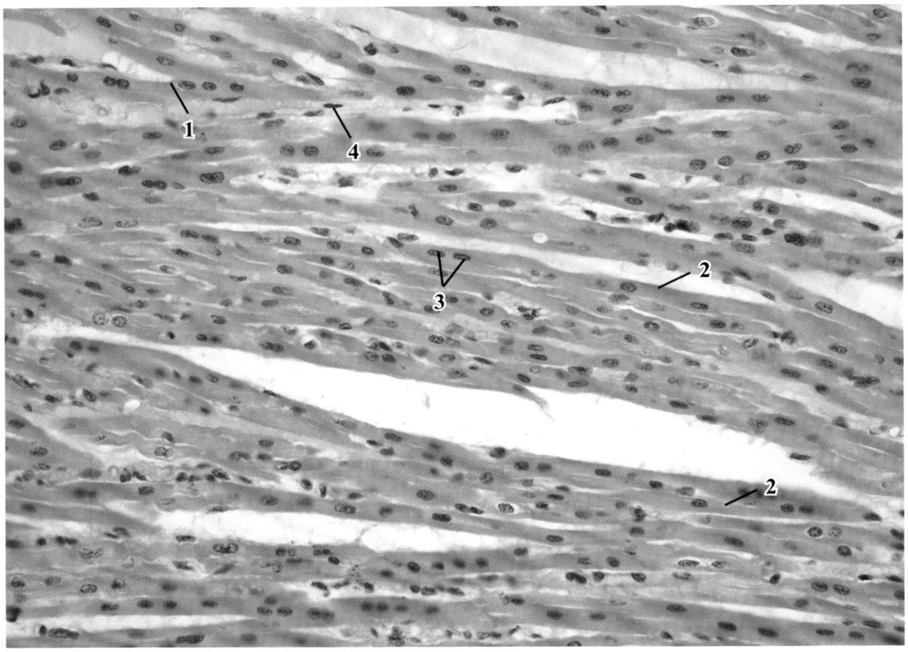

图 9-28 心肌组织切片（纵切面；HE 染色，高倍镜）
Figure 9-28 Histological section of the cardiac muscle (longitudinal section; HE staining, higher power).

1- 心肌细胞核 nucleus of the cardiac muscle cell
2- 闰盘 intercalated disk
3- 心肌细胞 cardiac muscle cell
4- 毛细血管 blood capillary
5- 内皮细胞 endothelial cell
6- 成纤维细胞核 nucleus, fibroblast

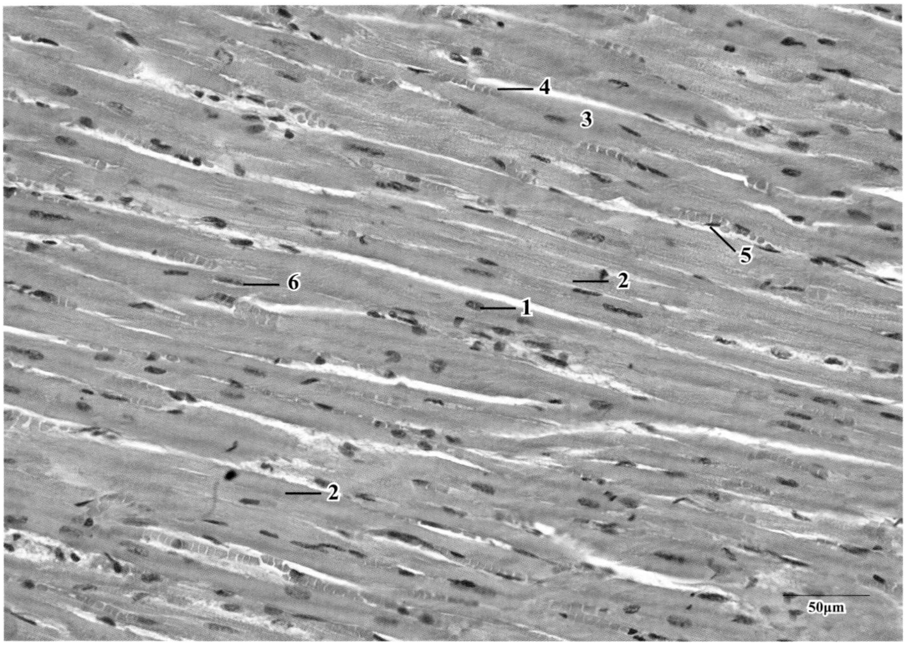

图 9-29 心肌组织切片（纵切面；HE 染色，高倍镜）
Figure 9-29 Histological section of the cardiac muscle (longitudinal section; HE staining, higher power).

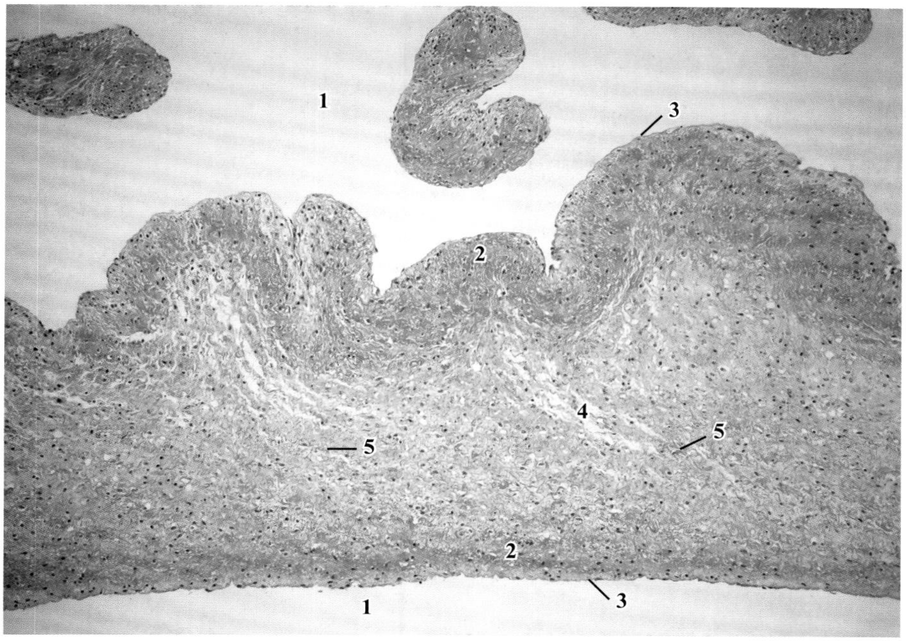

1- 心室腔 ventricular chamber
2- 心内膜 endocardium
3- 内皮 endothelium
4- 致密结缔组织 dense connective tissue
5- 弹性纤维 elastic fiber

图 9-30　二尖瓣组织切片（HE 染色，低倍镜）
Figure 9-30　Histological section of the bicuspd valve (HE staining, lower power).

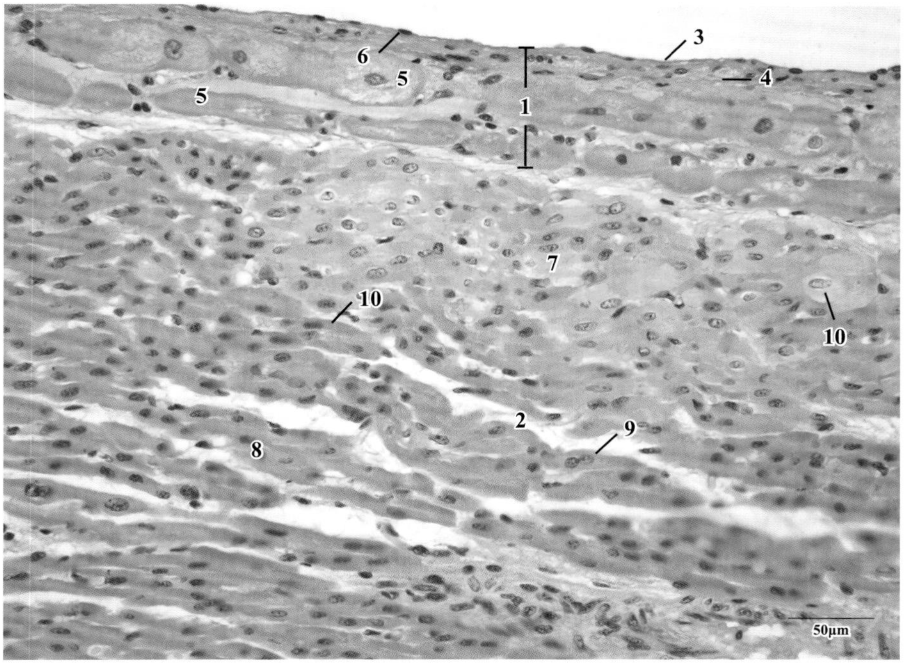

1- 心内膜 endocardium
2- 心肌膜 myocardium
3- 内皮 cardiac endothelium
4- 内皮下层 subendothelial layer
5- 浦肯野纤维 Purkinje fiber
6- 内皮细胞核 nucleus of endothelial cell
7- 心肌纤维，横断面 cardiac muscle fiber, cross section
8- 心肌纤维，纵切面 cardiac muscle fiber, longitudinal section
9- 双核心肌细胞 binucleate cardiac muscle cell
10- 心肌细胞 cardiac muscle cell

图 9-31　心壁组织切片（HE 染色，高倍镜）
Figure 9-31　Histological section of the cardiac wall (HE staining, higher power).

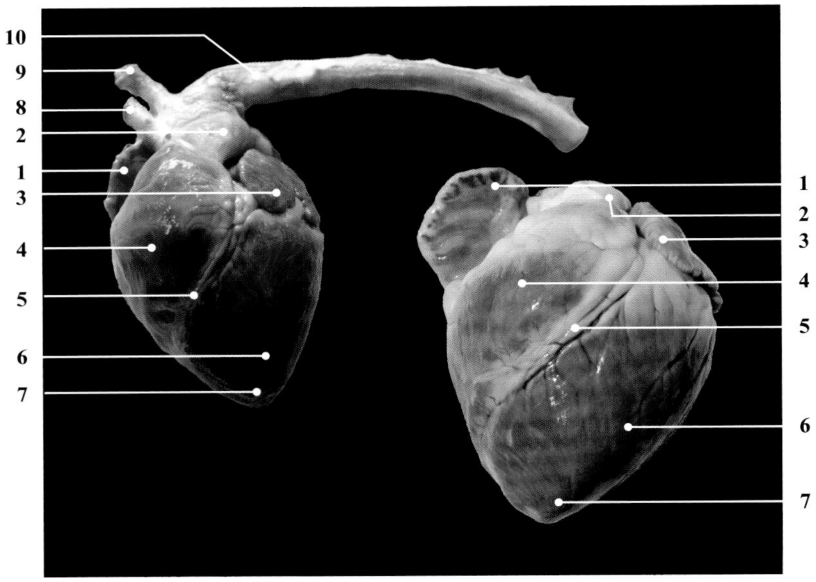

图 9-32　主动脉弓
Figure 9-32　Aortic arch.

1- 右心耳 right auricle
2- 肺动脉干 pulmonary trunk
3- 左心耳 left auricle
4- 右心室 right ventricle
5- 左纵沟 / 锥旁室间沟 left longitudinal groove / paraconal interventricular groove
6- 左心室 left ventricle
7- 心尖 cardiac apex
8- 臂头动脉 brachiocephalic artery
9- 左锁骨下动脉 left subclavian artery
10- 主动脉 aorta

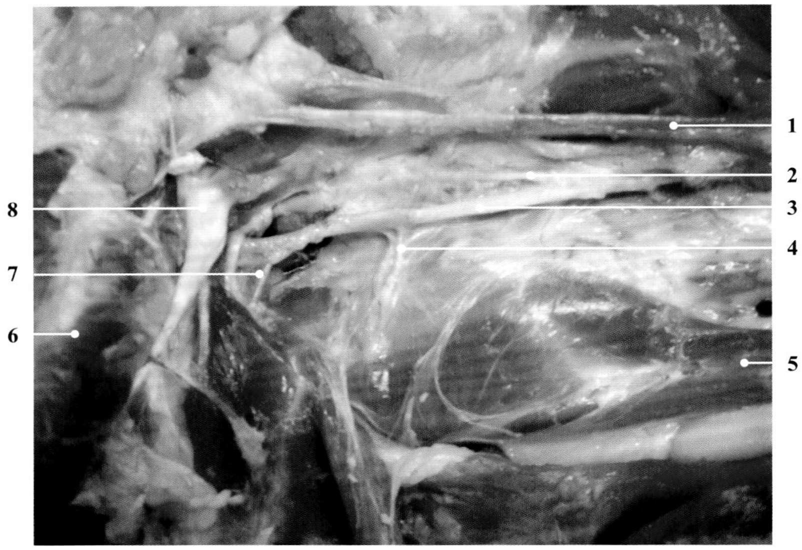

图 9-33　颈总动脉
Figure 9-33　Common carotid artery.

1- 颈静脉 jugular vein
2- 迷走交感干 vagosympathetic trunk
3- 颈总动脉 common carotid artery
4- 甲状腺动脉 thyroid artery
5- 胸骨舌骨肌 sternohyoid muscle
6- 咬肌 masseter muscle
7- 舌动脉 lingual artery
8- 舌骨 hyoid bone

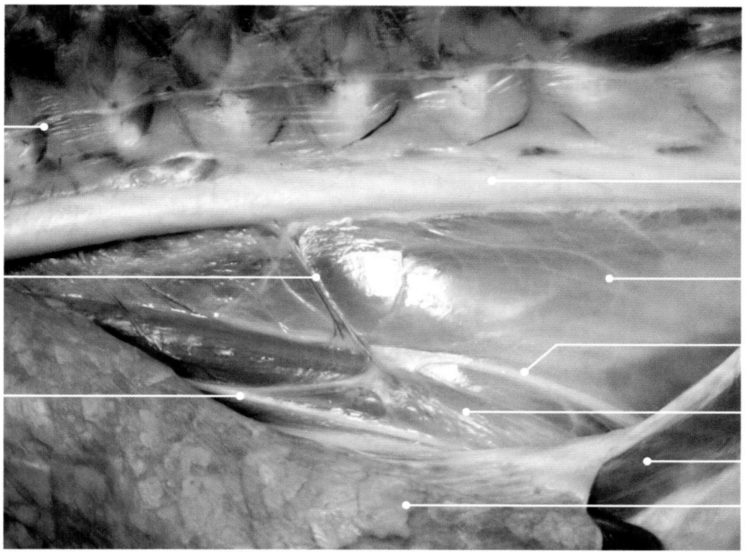

图 9-34 支气管食管动脉
Figure 9-34 Bronchoesophageal artery.

1- 胸主动脉 thoracic aorta
2- 纵隔 mediastinum
3- 迷走神经背侧干 dorsal trunk of vagus trunk
4- 食管 esophagus
5- 膈 diaphragm
6- 肺 lung
7- 迷走神经腹侧干 ventral trunk of vagus nerve
8- 支气管食管动脉 bronchoesophageal artery
9- 交感干 sympathetic trunk

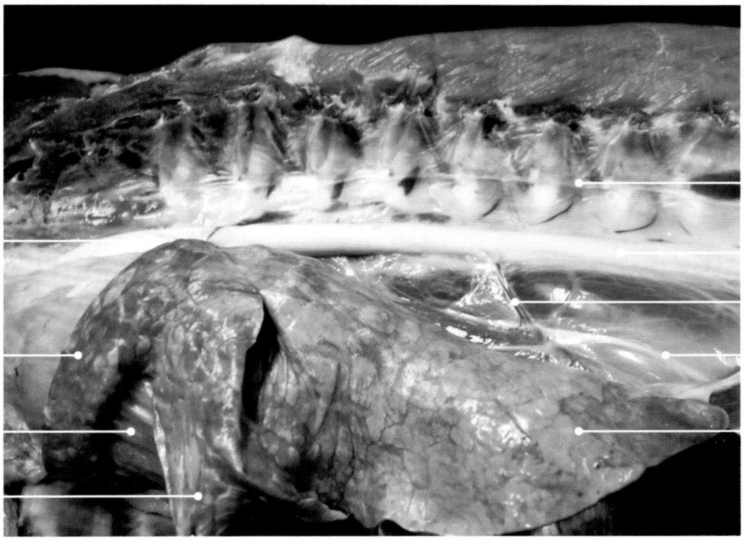

图 9-35 胸主动脉
Figure 9-35 Thoracic aorta.

1- 交感干 sympathetic trunk
2- 胸主动脉 thoracic aorta
3- 支气管食管动脉 bronchoesophageal artery
4- 迷走神经 vagus nerve
5- 左肺后叶（膈叶） left caudal lobe (diaphragmatic lobe) of the lung
6- 左肺中叶（心叶） left middle lobe (cardiac lobe) of the lung
7- 心包 pericardium
8- 左肺前叶（尖叶） left cranial lobe (apical lobe) of the lung
9- 主动脉 aorta

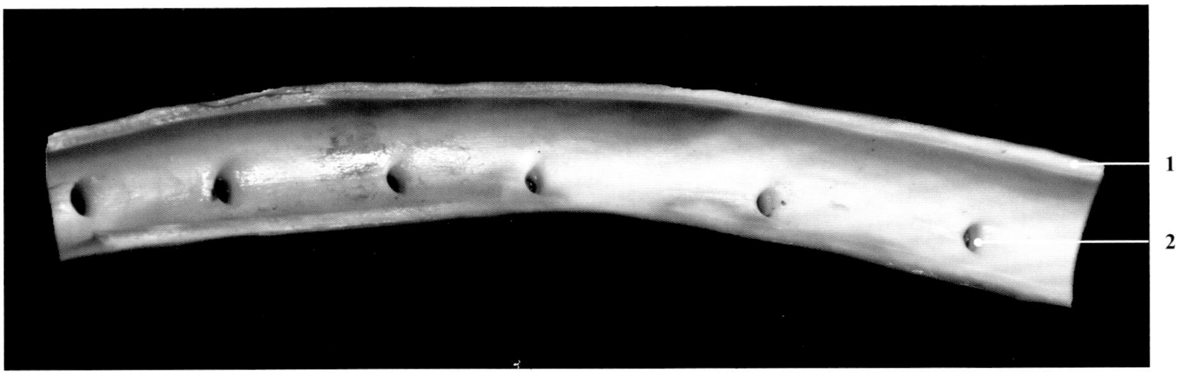

图 9-36　肋间动脉
Figure 9-36　Intercostal artery.

1- 主动脉管壁 aortic wall

2- 肋间动脉 intercostal artery

图 9-37　肋间动脉、静脉和神经
Figure 9-37　Intercostal artery, vein and nerve.

1- 肋间静脉 intercostal vein

2- 肋间动脉 intercostal artery

3- 肋骨 costal bone

4- 肋间内肌 internal intercostal muscle

5- 肋间神经 intercostal nerve

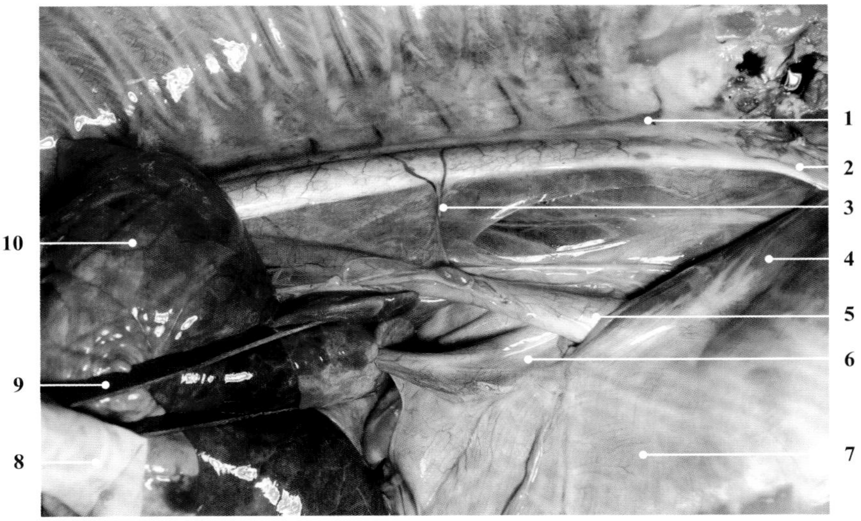

图 9-38 腔静脉裂孔、食管裂孔和主动脉裂孔
Figure 9-38 Vena caval hiatus, esophageal hiatus and aortic hiatus.

1- 奇静脉 azygos vein
2- 主动脉裂孔 aortic hiatus
3- 支气管食管动脉、静脉 bronchoesophageal artery and vein
4- 膈肉质缘 pulpa part of diaphragm
5- 食管及食管裂孔 esophagus and esophageal hiatus
6- 后腔静脉及腔静脉裂孔 caudal vena cava and vena caval hiatus
7- 膈中心腱 intermediate tendon of diaphragm
8- 手指 finger
9- 镊 tweezers
10- 肺 lung

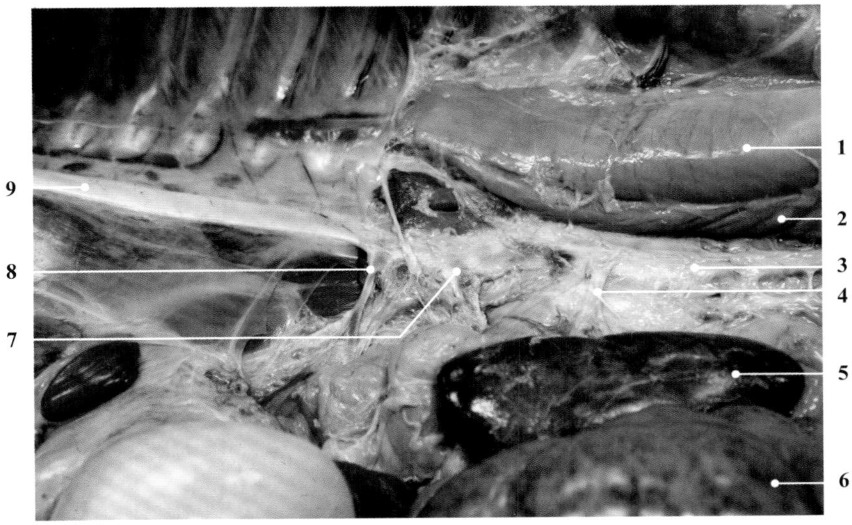

图 9-39 腹腔动脉和肠系膜前动脉
Figure 9-39 Celiac artery and cranial mesenteric artery.

1- 腰大肌 major psoas muscle
2- 腰小肌 minor psoas muscle
3- 腹主动脉 abdominal aorta
4- 肾动脉 renal artery
5- 肾 kidney
6- 大肠 large intestine
7- 肠系膜前动脉 cranial mesenteric artery
8- 腹腔动脉 celiac artery
9- 腹主动脉 abdominal aorta

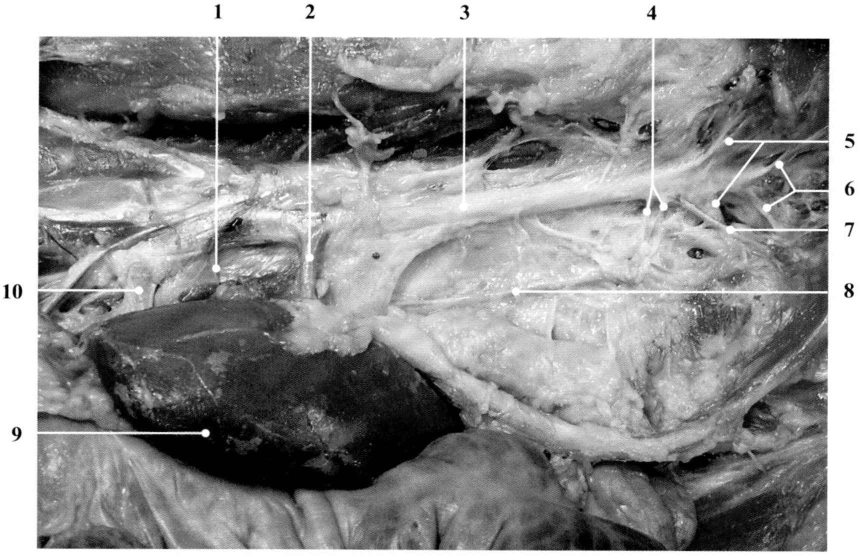

图 9-40　母猪腹主动脉分支
Figure 9-40　Branch of abdominal aorta in sow.

1- 肠系膜前动脉 cranial mesenteric artery
2- 肾动脉 renal artery
3- 腹主动脉 abdominal aorta
4- 子宫卵巢动脉 uteroovarian artery
5- 髂外动脉 external iliac artery
6- 髂内动脉 internal iliac artery
7- 肠系膜后动脉 caudal mesenteric artery
8- 输尿管 ureter
9- 肾 kidney
10- 腹腔动脉 celiac artery

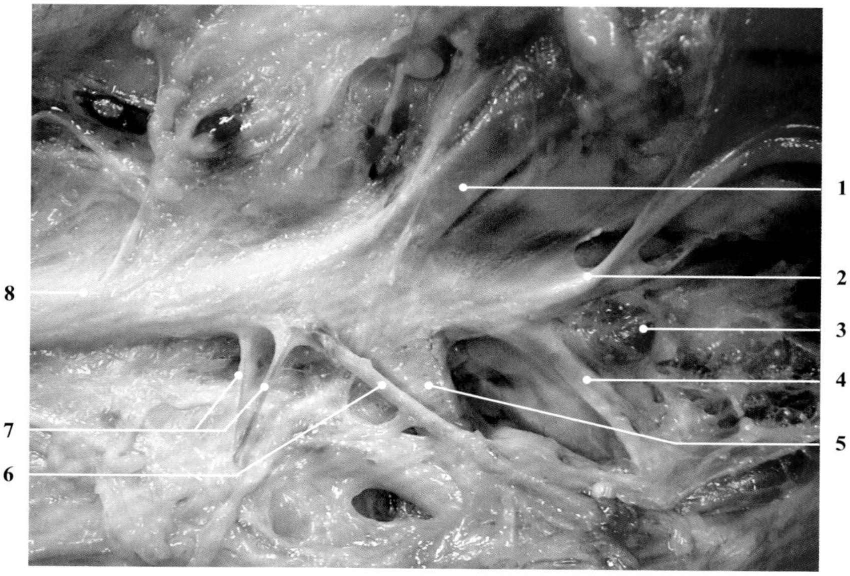

图 9-41　母猪骨盆部动脉分布
Figure 9-41　Distribution of the pelvic artery in sow.

1- 左髂外动脉 left external iliac artery
2- 左髂内动脉 left internal iliac artery
3- 荐淋巴结 sacral lymph node
4- 右髂内动脉 right internal iliac artery
5- 右髂外动脉 right external iliac artery
6- 肠系膜后动脉 caudal mesenteric artery
7- 子宫卵巢动脉 uteroovarian artery
8- 腹主动脉 abdominal aorta

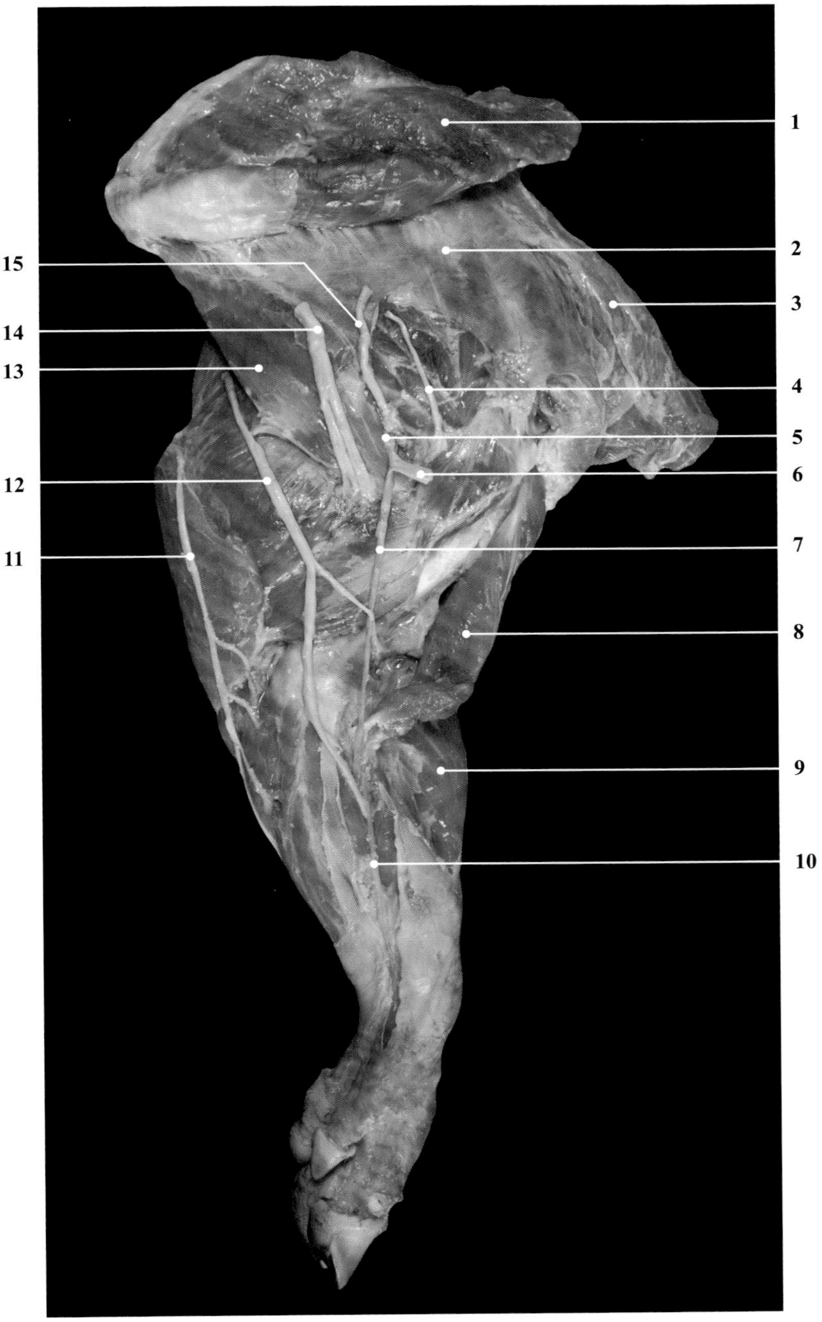

图 9-42 左前肢动脉分布（内侧观）
Figure 9-42 Distribution of the left forelimb artery (medial view).

1- 腹侧锯肌 ventral serrate muscle
2- 肩胛下肌 subscapular muscle
3- 冈上肌 supraspinatus muscle
4- 肩胛下神经 subscapular nerve
5- 肩胛下动脉 subscapular artery
6- 腋动脉 axillary artery
7- 臂动脉 brachial artery
8- 臂二头肌 biceps muscle of the forearm
9- 腕桡侧伸肌 radial extensor muscle of the carpus
10- 正中动脉 median artery
11- 尺神经 ulnar nerve
12- 正中神经 median nerve
13- 大圆肌 major teres muscle
14- 桡神经 radial nerve
15- 腋神经 axillary nerve

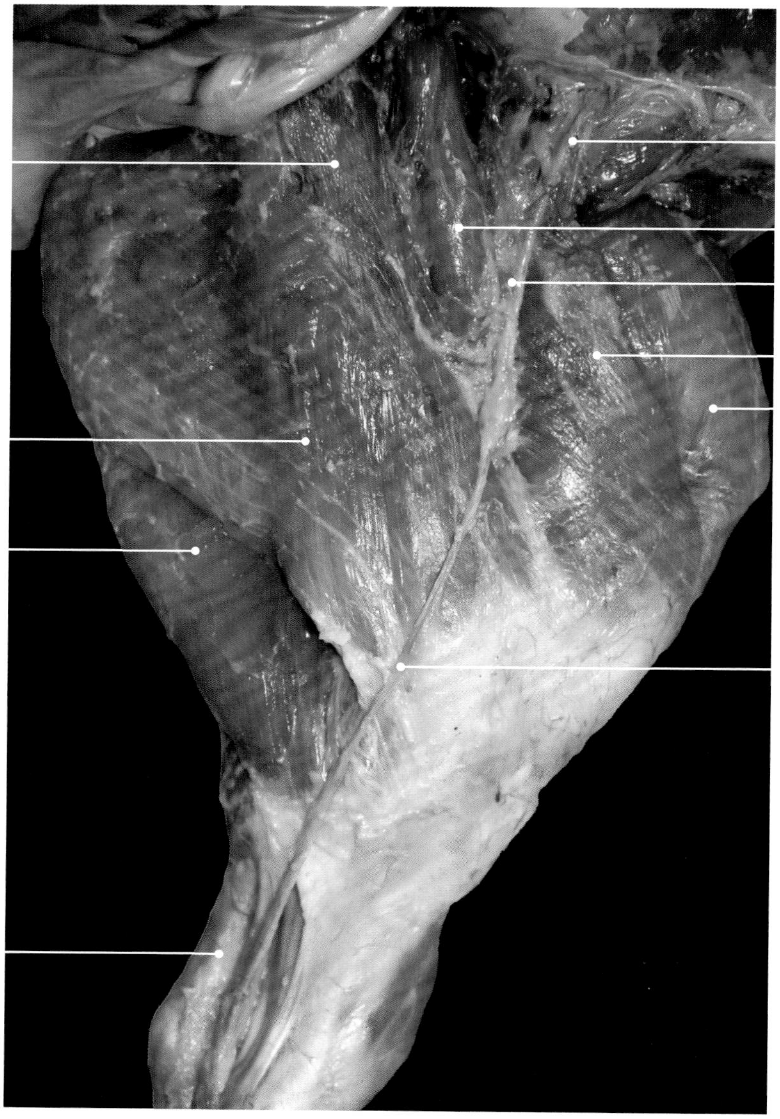

图 9-43　左后肢动脉分布（内侧观）
Figure 9-43　Distribution of the left hind limb artery (medial view).

1- 髂外动脉 external iliac artery
2- 耻骨肌 pectineal muscle
3- 股动脉 femoral artery
4- 股内侧肌 medial vastus muscle
5- 股直肌 rectus femoris muscle
6- 隐动脉 saphenous artery
7- 跟（总）腱 common calcaneal tendon
8- 半腱肌 semitendinous muscle
9- 半膜肌 semimembranous muscle
10- 内收肌 adductor muscle

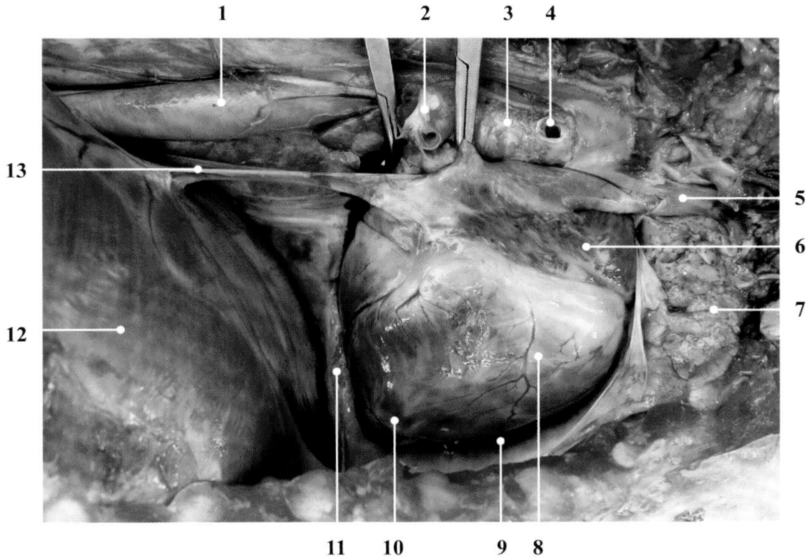

图 9-44 前、后腔静脉
Figure 9-44 Cranial and caudal vena cava.

1- 食管 esophagus
2- 支气管 bronchus
3- 气管 trachea
4- 右尖叶支气管 bronchus of right apical lobe
5- 前腔静脉 *cranial vena cava*
6- 右心房 right atrium
7- 胸腺 thymus
8- 右心室 right ventricle
9- 心包腔 pericardial cavity
10- 心尖 cardiac apex
11- 心包 pericardium
12- 膈 diaphragm
13- 后腔静脉 *caudal vena cava*

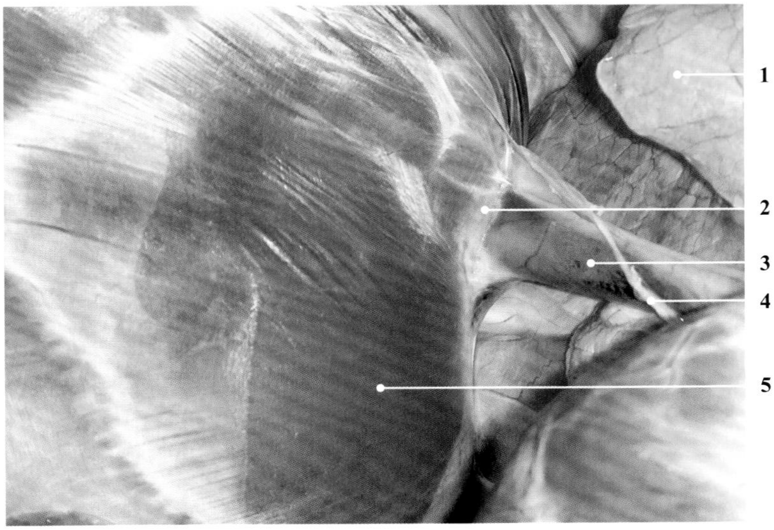

图 9-45 腔静脉裂孔
Figure 9-45 Vena caval hiatus.

1- 肺 lung
2- 腔静脉裂孔 vena caval hiatus
3- 后腔静脉 *caudal vena cava*
4- 膈神经 phrenic nerve
5- 膈 diaphragm

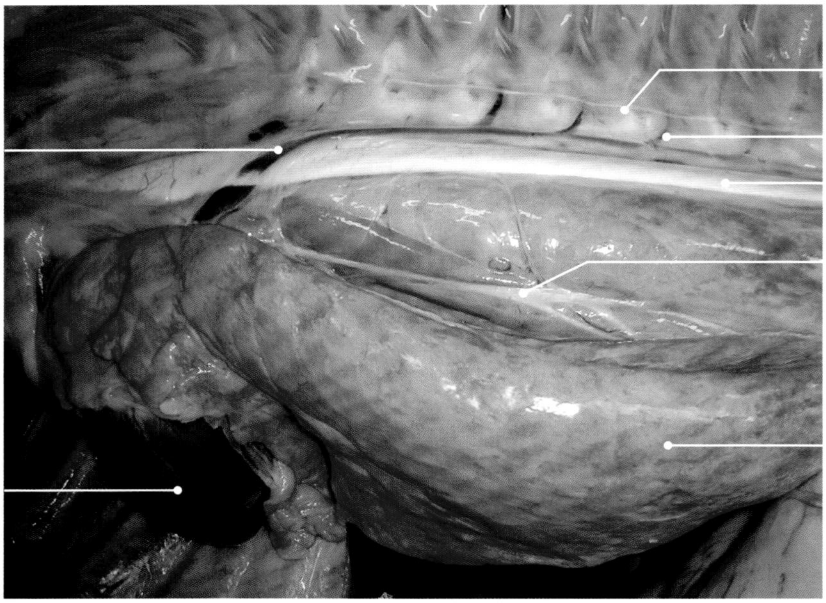

图 9-46　奇静脉
Figure 9-46　Azygos vein.

1- 交感干 sympathetic trunk
2- 肋间静脉 intercostal vein
3- 胸主动脉 thoracic aorta
4- 迷走神经 vagus nerve
5- 肺 lung
6- 心脏 heart
7- 奇静脉 azygos vein

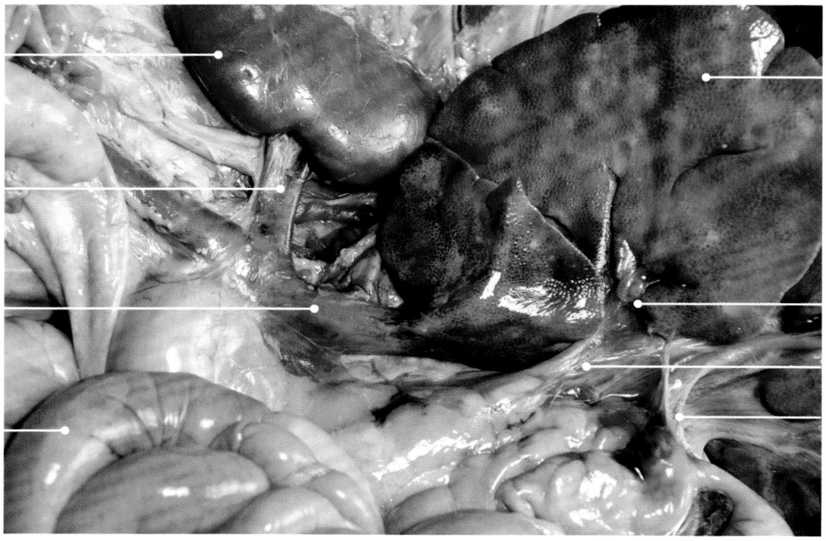

图 9-47　肝门静脉
Figure 9-47　Hepatic portal vein.

1- 肝 liver
2- 肝门 hepatic hilus
3- 肝门静脉 hepatic portal vein
4- 胆囊管 duct of gallbladder
5- 小肠 small intestine
6- 后腔静脉 *caudal vena cava*
7- 肾静脉 renal vein
8- 肾 kidney

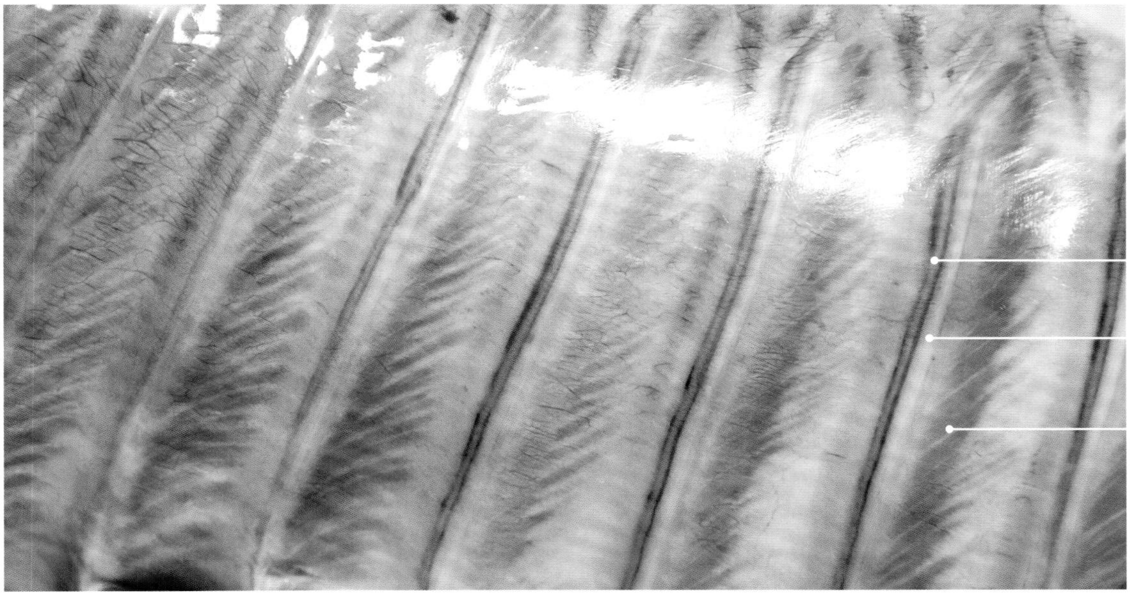

图 9-48 肋间静脉
Figure 9-48 Intercostal vein.

1- 肋间静脉 intercostal vein
2- 肋间神经 intercostal nerve
3- 肋间内肌 internal intercostal muscle

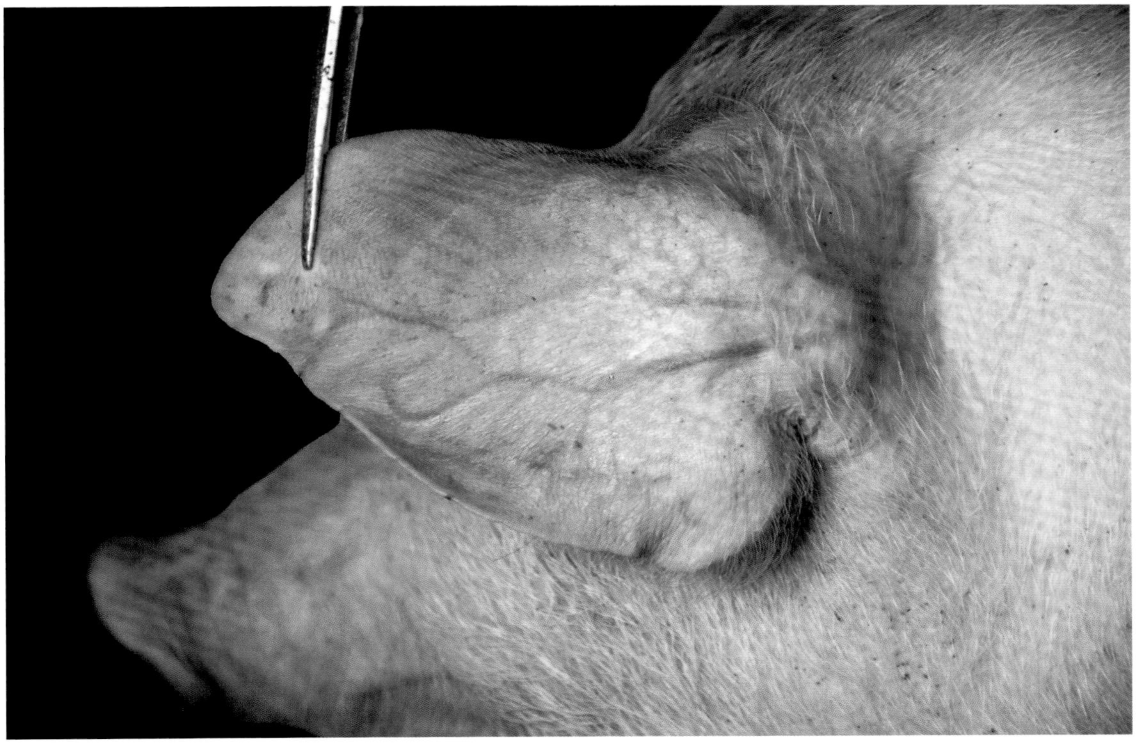

图 9-49 耳静脉
Figure 9-49 Auricular vein.

1- 动脉 artery
2- 中膜 middle tunic
3- 外弹性膜 external elastic membrane
4- 外膜 external tunic
5- 静脉 vein
6- 纵行平滑肌束 longitudinal smooth muscle bundle
7- 脂肪组织 adipose tissue

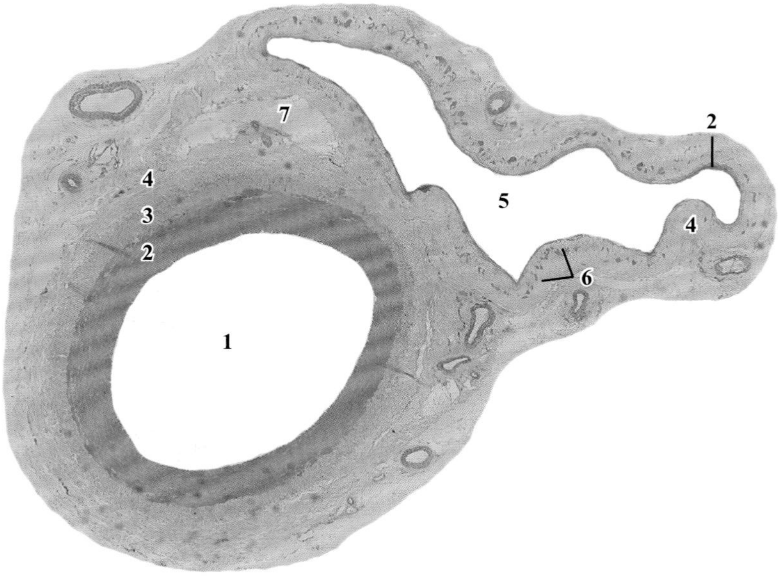

图 9-50　中型动脉和静脉组织切片（HE 染色，低倍镜）
Figure 9-50　Histological section of the medium artery and vein (HE staining, lower power).

1- 内膜 internal tunic
2- 中膜 middle tunic
3- 外膜 external tunic
4- 内弹性膜 internal elastic membrane
5- 内皮细胞核 nucleus of endothelial cell
6- 平滑肌纤维 smooth muscle fiber
7- 弹性纤维 elastic fiber
8- 外弹性膜 external elastic membrane

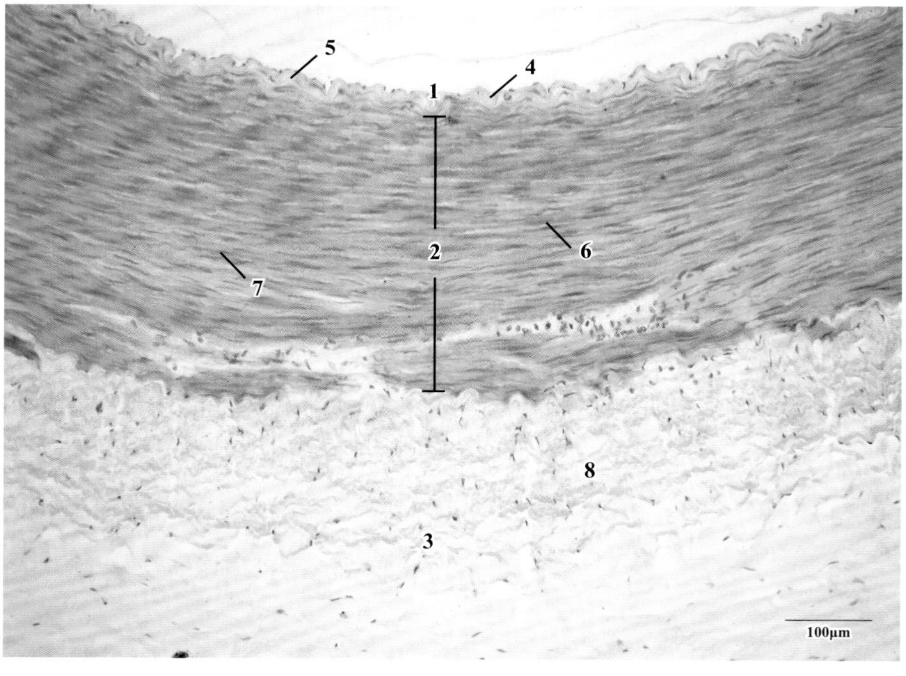

图 9-51　中型动脉组织切片（HE 染色，高倍镜）
Figure 9-51　Histological section of the medium artery (HE staining, higher power).

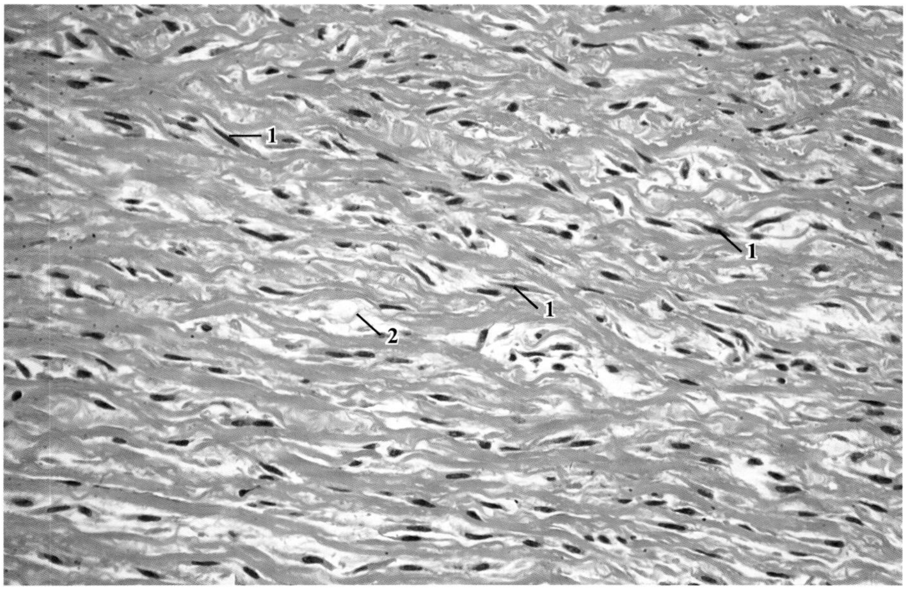

1- 平滑肌细胞核 nuclei of the small muscle cell

2- 弹性纤维 elastic fiber

图 9-52　中型动脉中膜组织切片（HE 染色，高倍镜）
Figure 9-52　Histdogical section of the middle tunic of medium artery (HE staining, higher power).

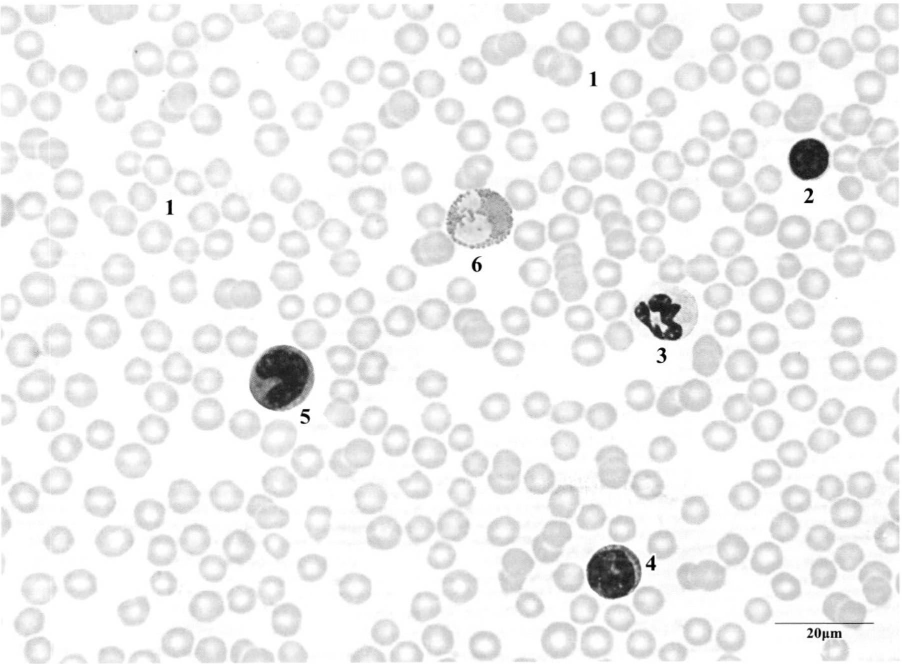

1- 红细胞 erythrocyte, red blood cell (RBC)

2- 小淋巴细胞 small lymphocyte

3- 中性粒细胞 neutrophilic granulocyte, neutrophil

4- 中淋巴细胞 medium-sized lymphocyte

5- 单核细胞 monocyte

6- 嗜酸性粒细胞 eosinophilic granulocyte, eosinophil

图 9-53　血涂片（瑞氏染色）
Figure 9-53　Blood smear (Wright's staining).

1- 红细胞 erythrocyte, red blood cell (RBC)
2- 中性粒细胞 neutrophilic granulocyte, neutrophil
3- 中淋巴细胞 medium-sized lymphocyte
4- 嗜酸性粒细胞 eosinophilic granulocyte, eosinophil

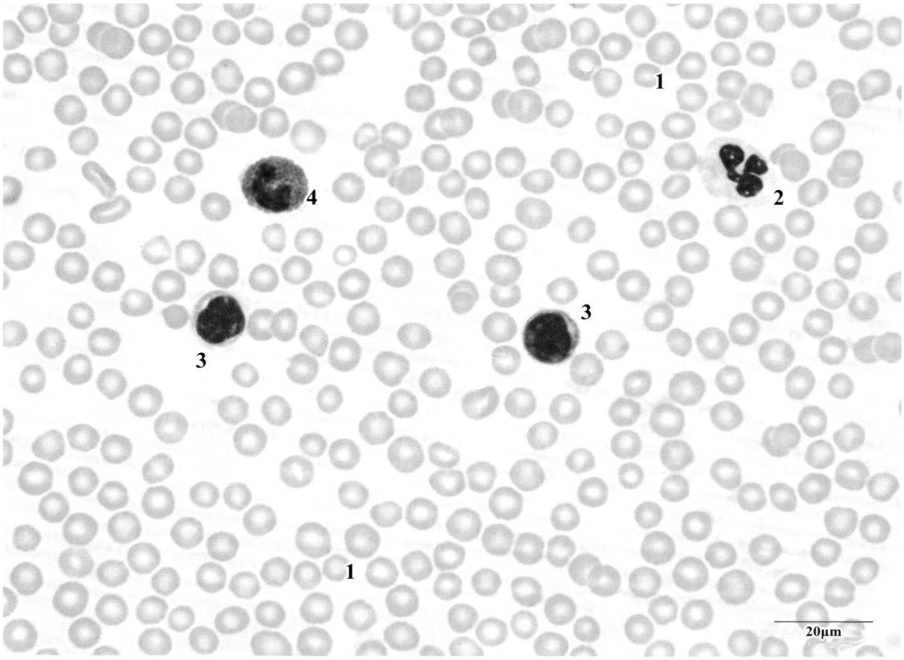

图 9-54　血涂片（瑞氏染色）
Figure 9-54　Blood smear (Wright's staining).

1- 红细胞 erythrocyte, red blood cell (RBC)
2- 嗜碱性粒细胞 basophilic granulocyte, basophil
3- 单核细胞 monocyte
4- 嗜酸性粒细胞 eosinophilic granulocyte, eosinophil
5- 中淋巴细胞 medium-sized lymphocyte

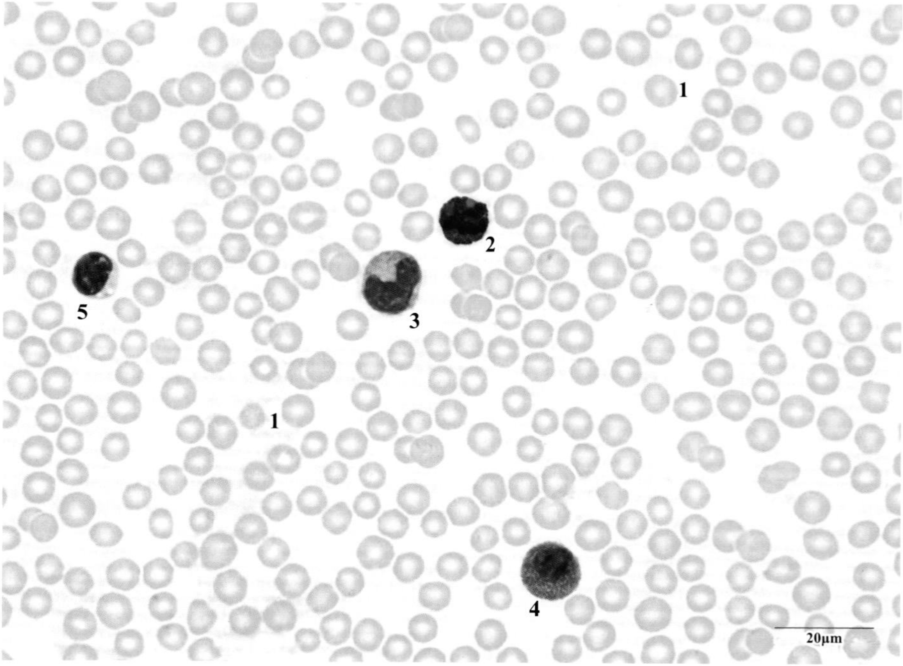

图 9-55　血涂片（瑞氏染色）
Figure 9-55　Blood smear (Wright's staining).

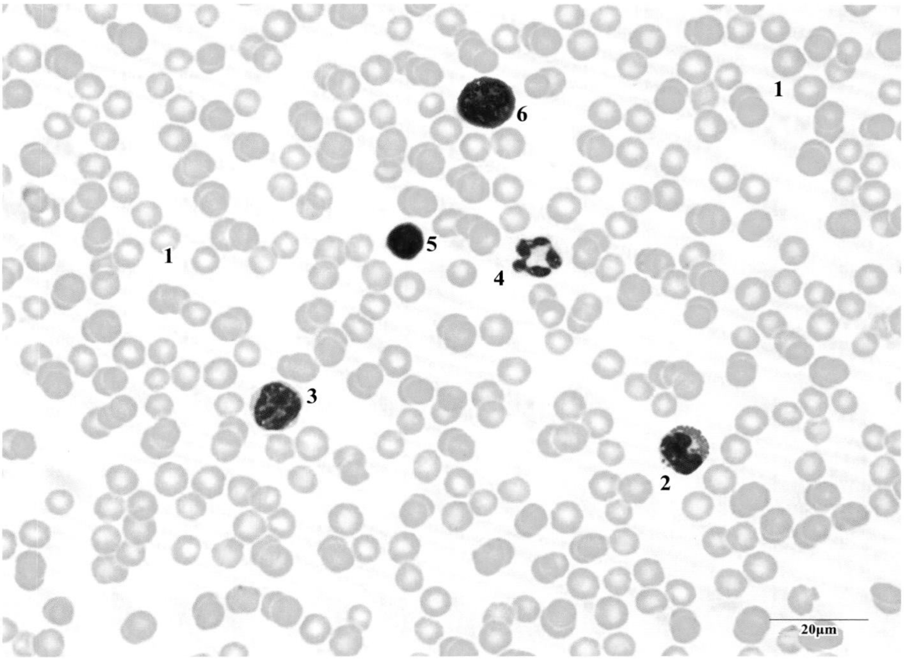

1- 红细胞 erythrocyte, red blood cell (RBC)
2- 嗜酸性粒细胞 eosinophilic granulocyte, eosinophil
3- 中淋巴细胞 medium-sized lymphocyte
4- 中性粒细胞 neutrophilic granulocyte, neutrophil
5- 小淋巴细胞 small lymphocyte
6- 嗜碱性粒细胞 basophilic granulocyte, basophil

图 9-56　血涂片（瑞氏染色）
Figure 9-56　Blood smear (Wright's staining).

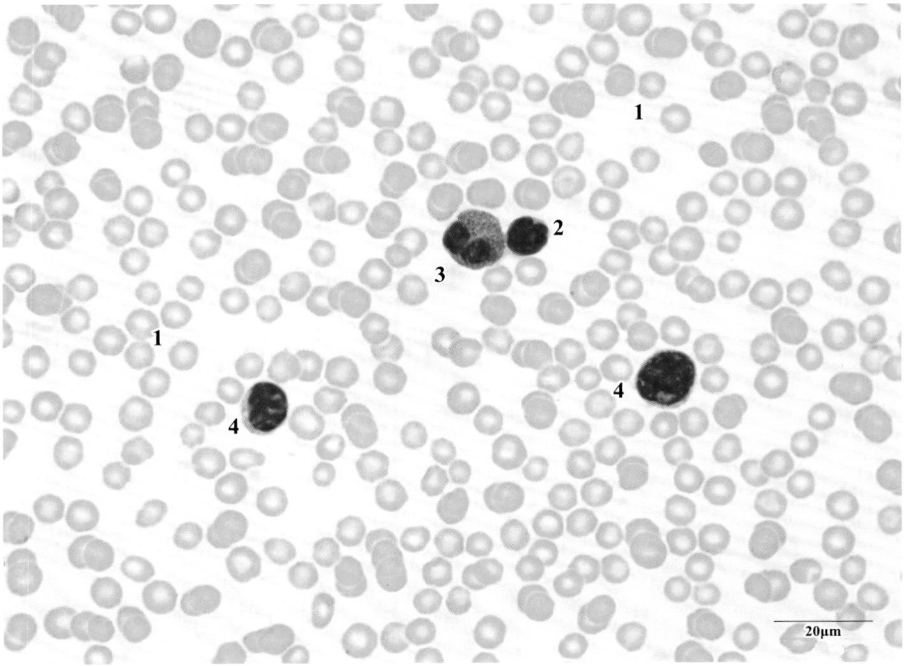

1- 红细胞 erythrocyte, red blood cell (RBC)
2- 小淋巴细胞 small lymphocyte
3- 嗜酸性粒细胞 eosinophilic granulocyte, eosinophil
4- 中淋巴细胞 medium-sized lymphocyte

图 9-57　血涂片（瑞氏染色）
Figure 9-57　Blood smear (Wright's staining).

1- 红细胞 erythrocyte, red blood cell (RBC)
2- 小淋巴细胞 small lymphocyte
3- 中淋巴细胞 medium-sized lymphocyte
4- 嗜酸性粒细胞 eosinophilic granulocyte, eosinophil
5- 中性粒细胞 neutrophilic granulocyte, neutrophil

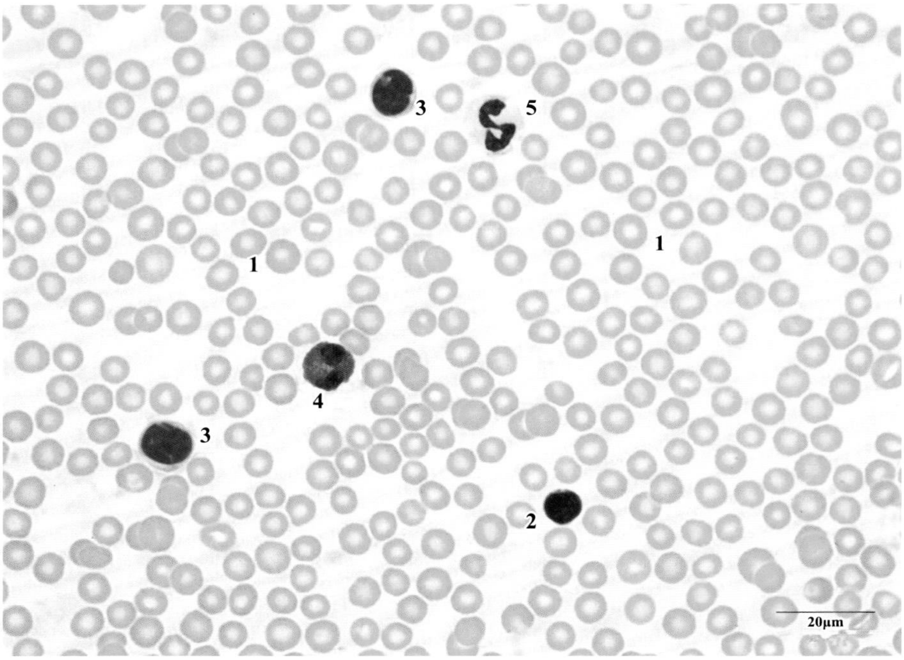

图 9-58 血涂片（瑞氏染色）
Figure 9-58 Blood smear (Wright's staining).

1- 红细胞 erythrocyte, red blood cell (RBC)
2- 嗜酸性粒细胞 eosinophilic granulocyte, eosinophil
3- 中性粒细胞 neutrophilic granulocyte, neutrophil
4- 中淋巴细胞 medium-sized lymphocyte

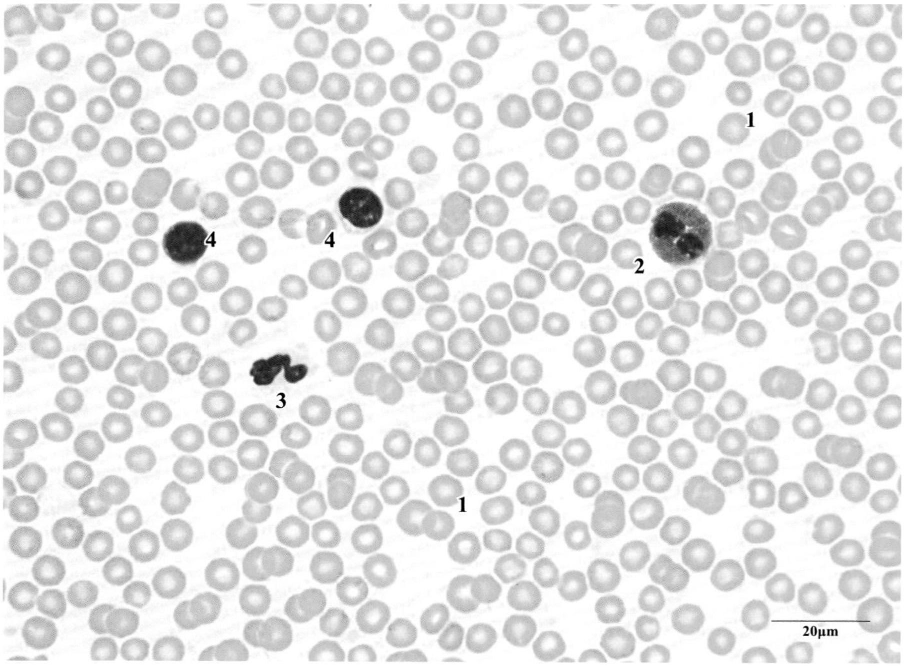

图 9-59 血涂片（瑞氏染色）
Figure 9-59 Blood smear (Wright's staining).

第十章

淋巴系统

Chapter 10

Lymphatic system

淋巴系统（lymphatic system）由淋巴管道、淋巴组织、淋巴器官和淋巴组成。淋巴组织（器官）可产生淋巴细胞（lymphocyte），参与免疫活动，因而淋巴系统是机体内主要的防卫系统。淋巴系统的免疫活动还协同神经及内分泌系统，参与机体神经体液调节，共同维持代谢平衡、生长发育和繁殖等。

一、淋巴管道

淋巴管道（lymphatic vessels）是起始于组织间隙，为淋巴液通过的管道，最后注入静脉的管道系。根据汇集顺序、管径大小及管壁薄厚，可分为毛细淋巴管、淋巴管、淋巴干和淋巴导管。

1. 毛细淋巴管（lymphatic capillary）：以盲端起始于组织间隙，起始部稍膨大；毛细淋巴管的管径较毛细血管的大，粗细不匀，彼此吻合成网；管壁只有一层内皮细胞，通常无基膜和外膜细胞，且相邻细胞以叠瓦状排列，细胞之间裂隙多而宽，因此，通透性也比毛细血管大。

当血液经动脉输送到毛细血管时，其中一部分液体经毛细血管动脉端滤出，进入组织间隙形成组织液。组织液与周围组织细胞进行物质交换后，大部分渗入毛细血管静脉端，少部分则渗入毛细淋巴管，成为淋巴（lymph）。淋巴在淋巴管内向心流动，最后注入静脉。淋巴是无色透明或微黄色的液体，由淋巴浆和淋巴细胞组成。小肠绒毛内的毛细淋巴管尚可吸收脂肪，其淋巴呈乳白色，称为乳糜（chyle）。

2. 淋巴管（lymphatic vessel）：由毛细淋巴管汇集而成，管壁较薄，管径较细，瓣膜多，管径粗细不均，常呈串珠状；在淋巴管的行程中，通常要通过一个或多个淋巴结。根据淋巴液对淋巴结的流向，淋巴管还可分成输入淋巴管和输出淋巴管。

3. 淋巴干（lymphatic trunk）：为机体一个区域内大的淋巴集合管，由淋巴管汇集而成，多与大血管伴行。主要淋巴干有气管淋巴干（tracheal lymphatic trunk）、腰淋巴干（lumbar lymphatic trunk）和内脏淋巴干（visceral lymphatic trunk），其中内脏淋巴干由肠淋巴干和腹腔淋巴干汇合形成，注入乳糜池（cisterna chyli）。

4. 淋巴导管（lymphatic duct）：为最大的淋巴集合管，由淋巴干汇集而成，包括胸导管（thoracic duct）和右淋巴导管（right lymphatic duct）。

二、淋巴组织

淋巴组织（lymphoid tissue）是富含淋巴细胞的网状组织，即在网状细胞的网眼中充满淋巴细胞，并含有少量的单核细胞、浆细胞。淋巴组织可因淋巴细胞的聚集程度和方式的不同，分为弥散淋巴组织（diffuse lymphoid tissue）和淋巴小结（lymphatic nodule）。

三、淋巴器官

淋巴器官（lymphoid organ）是以淋巴组织为主构成的实质性器官。根据发生和功能的特点，可分为中枢淋巴器官和周围淋巴器官。

1. 中枢淋巴器官（central lymphoid organ）：又称初级淋巴器官（primary lymphoid

organ），包括胸腺和骨髓。在胚胎发育过程中出现较早，其原始淋巴细胞来源于骨髓的干细胞，在胸腺内胸腺素的作用下，分化成T淋巴细胞。

胸腺（thymus）：猪为颈胸腺型，较发达，位于颈部气管两侧和胸腔前段纵隔内，呈红色或粉红色，分左、右两叶，每叶又被结缔组织分隔成许多小叶。胸腺在小猪发达，大猪则逐渐萎缩退化，到老龄时几乎全被脂肪组织代替。

胸腺为实质性器官，表面包有结缔组织性被膜（capsule）；被膜伸入其内部形成小叶间隔，把胸腺分成许多胸腺小叶（thymic lobule）。胸腺小叶由浅层的皮质和深层的髓质组成。由于小叶间隔不完整，相邻小叶的髓质常相连。

皮质（cortex）以胸腺上皮细胞构成支架，间隙内含有大量胸腺细胞和少量巨噬细胞。胸腺上皮细胞（thymic epithelial cell）又称上皮性网状细胞（epithelial reticular cell），呈星形，有突起，以桥粒连接成网。胸腺细胞（thymocyte）密集于皮质内，外周的大，幼稚，靠近髓质的较小，较成熟。进入胸腺的淋巴干细胞由浅层往深层移动，大部分在分化过程中凋亡，少数进入髓质或经皮质与髓质交界处的毛细血管后微静脉迁至周围淋巴器官或淋巴组织中。

髓质（medulla）有大量胸腺上皮细胞，少量初始T细胞、巨噬细胞、交错突细胞和肌样细胞。胸腺上皮细胞包括：①髓质上皮细胞，球形或多边形，胞体较大，能分泌胸腺激素，部分形成胸腺小体；②胸腺小体上皮细胞，呈扁平状，构成胸腺小体。胸腺小体（thymic corpuscle, Hassall's corpuscle）散在于髓质内，由胸腺上皮细胞呈同心圆排列而成。外周的细胞较幼稚，核明显，可分裂；近中心的细胞核渐退化，含较多角蛋白；中心细胞完全角化，呈强嗜酸性染色。

2. 周围淋巴器官（peripheral lymphoid organ）：也称次级淋巴器官（secondary lymphoid organ），包括淋巴结、脾、扁桃体、血淋巴结等。周围淋巴器官发育较迟，其淋巴细胞最初由中枢淋巴器官迁移而来，定居在特定区域内，在抗原的刺激下可进行分裂分化。其中T淋巴细胞形成具有特异性的免疫淋巴细胞，起细胞免疫作用；B淋巴细胞转化为能产生抗体的浆细胞，参与体液免疫反应。

（1）脾（spleen）：狭而长，呈紫红色，质地柔软。位于胃大弯的左侧，以胃韧带连接于胃大弯。脾可产生淋巴细胞和巨噬细胞，参与机体免疫活动，同时脾位于血液循环的通路上，脾内没有淋巴窦，而有大量血窦，具有造血、灭血、滤血、贮血等功能。

脾的表面包以结缔组织构成的被膜。被膜伸入脾实质内形成小梁（trabecula），小梁互相吻合形成网状支架。被膜和小梁内含有平滑肌纤维，它的收缩可调节脾内的血量。

脾的实质为脾髓，分白髓和红髓。白髓（white pulp）呈灰白色，由密集的淋巴组织围绕动脉而成，包括动脉周围淋巴鞘和脾小结。动脉周围淋巴鞘（periarterial lymphatic sheath）是围绕中央动脉周围的弥散淋巴组织，含大量 T 细胞及少量巨噬细胞与交错突细胞。脾小结（splenic nodule）或称脾小体（splenic corpuscle），位于动脉周围淋巴鞘一侧的淋巴小结，主要含B 细胞，发育良好者也有生发中心，与淋巴结的淋巴小结不一样的是脾小结内有中央动脉分支穿过。红髓（red pulp）位于白髓周围，是富含血管的弥散淋巴组织，包括脾索和脾血窦。脾索（splenic cord）富含血细胞的淋巴索；内含T细胞、B细胞、浆细胞、巨噬细胞和其他血细胞。脾血窦（splenic sinusoid）位于脾索之间，形态不规则，相连成网。窦壁由一层长杆状内皮细胞围成，细胞间隙大，基膜不完整；周围有大量巨噬细胞。脾索内血细胞穿越内皮间隙进入脾血窦。

（2）淋巴结（lymph node）：大小不一，直径从1毫米到几厘米不等。淋巴结一侧凹陷为淋巴结门，是输入淋巴管、血管及神经出入之处，另一侧隆凸，有输出淋巴管。淋巴结是位于淋巴管径路上唯一的淋巴器官，数量多，常单独或少量聚群沿血管分布。淋巴结表面包有致密结缔组织形成被膜，被膜伸入实质形成小梁，构成淋巴结内部粗大的支架。实质包括皮质和髓质。皮质含大量淋巴小结（lymphatic nodule）和小结间弥散淋巴组织。髓质含有淋巴索和淋巴窦。淋巴索为条索状淋巴组织；主要含B细胞，还含有T细胞、浆细胞、肥大细胞、巨噬细胞等。淋巴窦位于淋巴索之间，腔宽大；窦内巨噬细胞多，过滤功能强。仔猪淋巴结的皮质和髓质的位置与其他家畜的相反。淋巴小结位于深层的中央区域，而淋巴索和淋巴窦则位于周围。在成年猪的皮质和髓质则混合排列。

淋巴结的主要功能是产生淋巴细胞，过滤淋巴，清除侵入体内的细菌和异物，参与免疫反应，是机体重要的防卫器官，同时又是造血器官。局部淋巴结肿大，常反映其汇流区域有病变，尤其是主要浅在淋巴结对临床诊断和兽医卫生检疫有重要意义。

一个淋巴结或淋巴结群常位于机体的同一部位，并汇集几乎相同区域的淋巴，这个淋巴结或淋巴结群就是该区域的淋巴中心（lymphocentrum）。猪全身有18个淋巴中心。淋巴中心和淋巴结的命名，主要根据其所在部位或引流区域。全身的淋巴中心可分属于7个部位，即头部、颈部、前肢、胸腔、腹腔、腹壁与骨盆壁和后肢。

1）头部淋巴中心和淋巴结：下颌淋巴中心，有下颌淋巴结（mandibular lymph node）；腮腺淋巴中心，有腮腺淋巴结（parotid lymph node）；咽后淋巴中心，有咽后内侧淋巴结（medial retropharyngeal lymph node）和咽后外侧淋巴结（lateral retropharyngeal lymph node）。

2）颈部淋巴中心和淋巴结：颈浅淋巴中心，有颈浅淋巴结（superficial cervical lymph node），又称肩前淋巴结。猪的颈浅淋巴结分背侧和腹侧两组，背侧淋巴结相当于其他家畜的颈浅淋巴结，腹侧淋巴结则位于腮腺后缘和胸头肌之间。颈深淋巴中心，有颈深前淋巴结（cranial deep cervical lymph node）和颈深后淋巴结（caudal deep cervical lymph node）。

3）前肢淋巴中心和淋巴结：腋淋巴中心，猪只有第1肋腋淋巴结（primary costal axillary lymph node）。

4）胸腔淋巴中心和淋巴结：胸背侧淋巴中心、胸腹侧淋巴中心、纵隔淋巴中心和支气管淋巴中心。

5）腹腔内脏淋巴中心和淋巴结：腹腔淋巴中心、肠系膜前淋巴中心和肠系膜后淋巴中心。

6）腹壁和骨盆壁的淋巴中心和淋巴结：腰淋巴中心、荐髂淋巴中心、腹股沟淋巴中心和坐骨淋巴中心。

7）后肢淋巴中心和淋巴结：有腘淋巴中心。

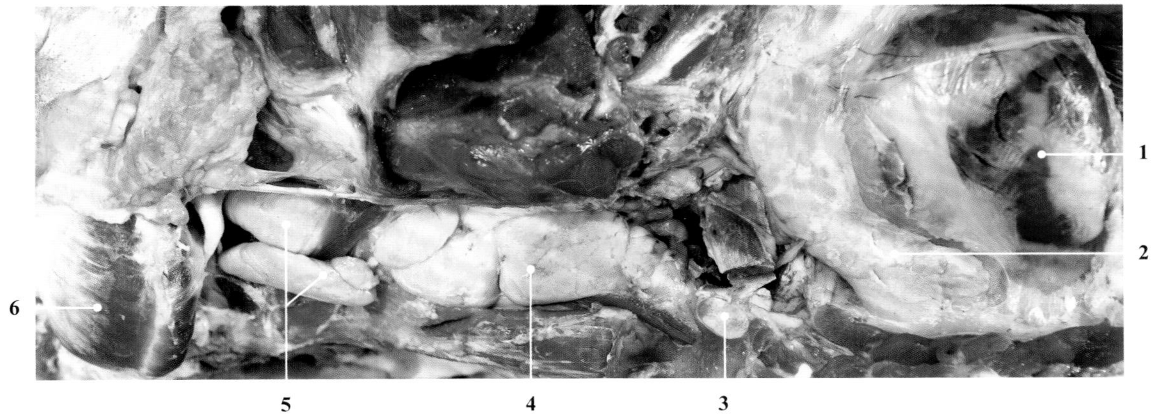

图 10-1 公猪胸腺位置（左外侧观）
Figure 10-1 Anatomical position of the thymus of a boar (left lateral view).

1- 心脏和心包 heart and pericardium
2- 胸腺胸叶 thoracic lobe of thymus
3- 胸骨柄 manubrium of sternum
4- 胸腺中间叶 intermediate lobe of thymus
5- 胸腺颈叶 cervical lobe of thymus
6- 咬肌 masseter muscle

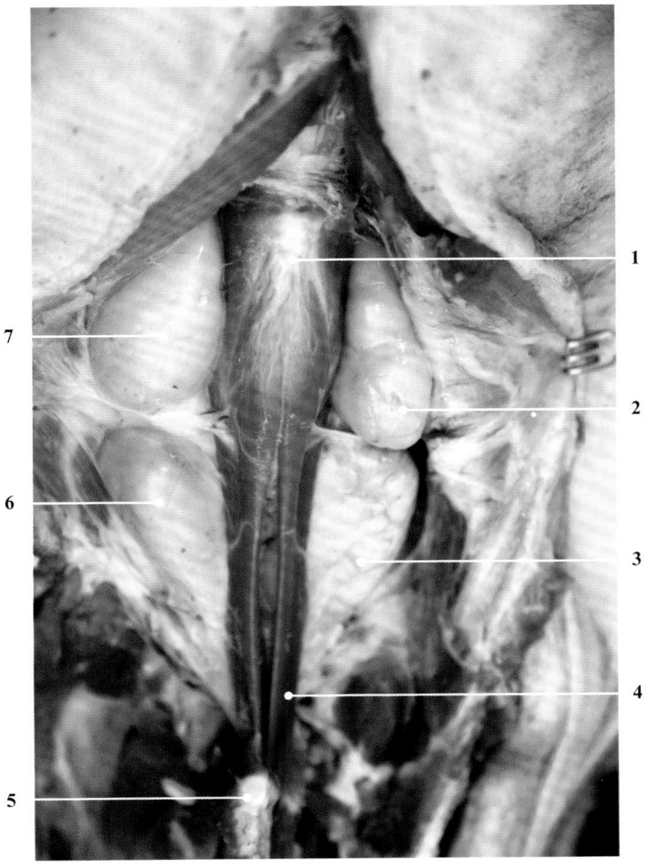

图 10-2 公猪胸腺位置（腹侧观）
Figure 10-2 Anatomical position of the thymus of a boar (ventral view).

1- 喉 larynx
2- 颈叶（左侧）cervical lobe of thymus (left side)
3- 中间叶（左侧）intermediate lobe of thymus (left side)
4- 胸骨甲状肌 sterno-thyroid muscle
5- 胸骨柄 manubrium of sternum
6- 中间叶（右侧）intermediate lobe of thymus (right side)
7- 颈叶（右侧）cervical lobe of thymus (right side)

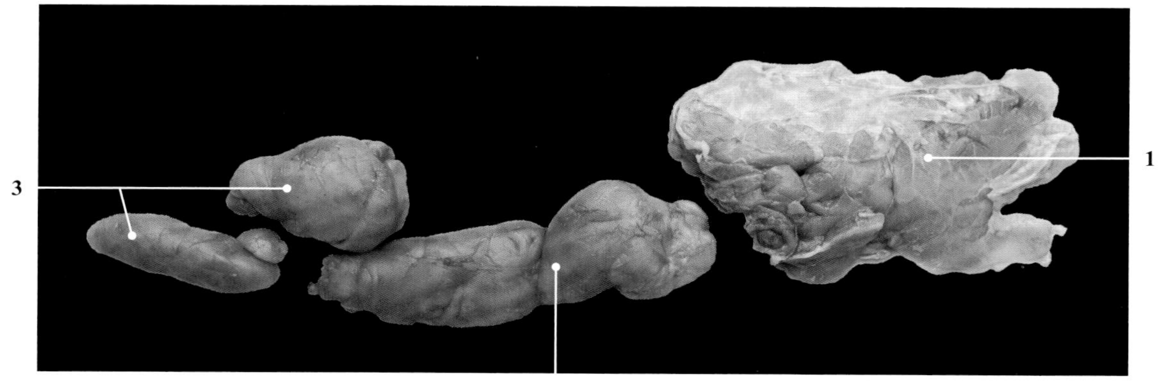

图 10-3　公猪胸腺（左侧）
Figure 10-3　Thymus of a boar (left aspect).

1- 胸叶 thoracic lobe of thymus
2- 中间叶 intermediate lobe of thymus
3- 颈叶 cervical lobe of thymus

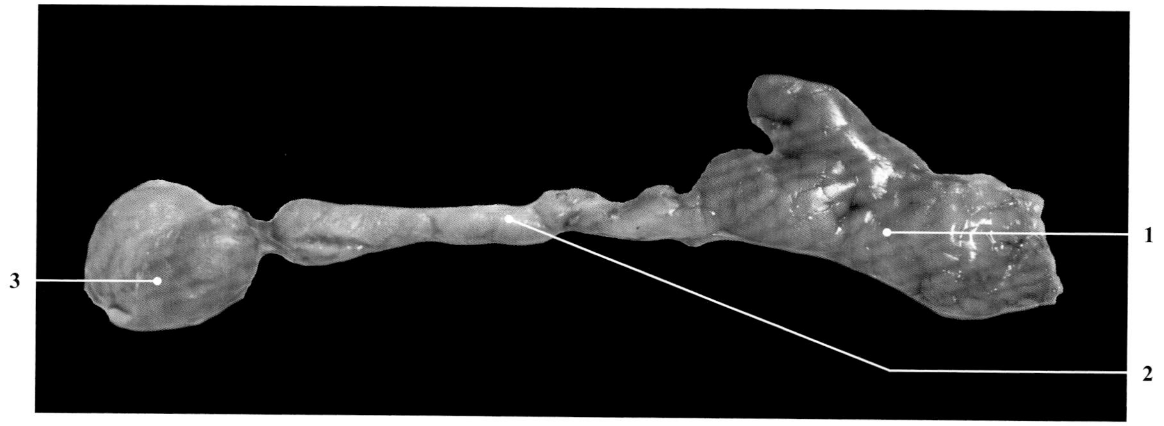

图 10-4　母猪胸腺（左侧）
Figure 10-4　Thymus of a sow (left aspect).

1- 胸叶 thoracic lobe of thymus
2- 中间叶 intermediate lobe of thymus
3- 颈叶 cervical lobe of thymus

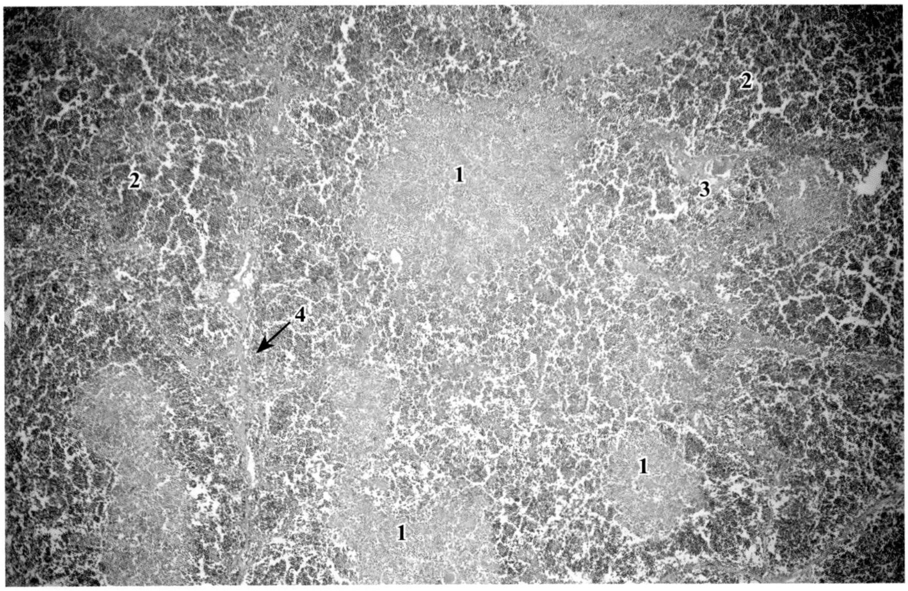

1- 胸腺髓质 thymic medulla
2- 胸腺皮质 thymic cortex
3- 静脉 vein
4- 小叶间隔 interlobular septum

图 10-5 胸腺组织切片（HE 染色，低倍镜）
Figure 10-5 Histological section of the thymus (HE staining, lower power).

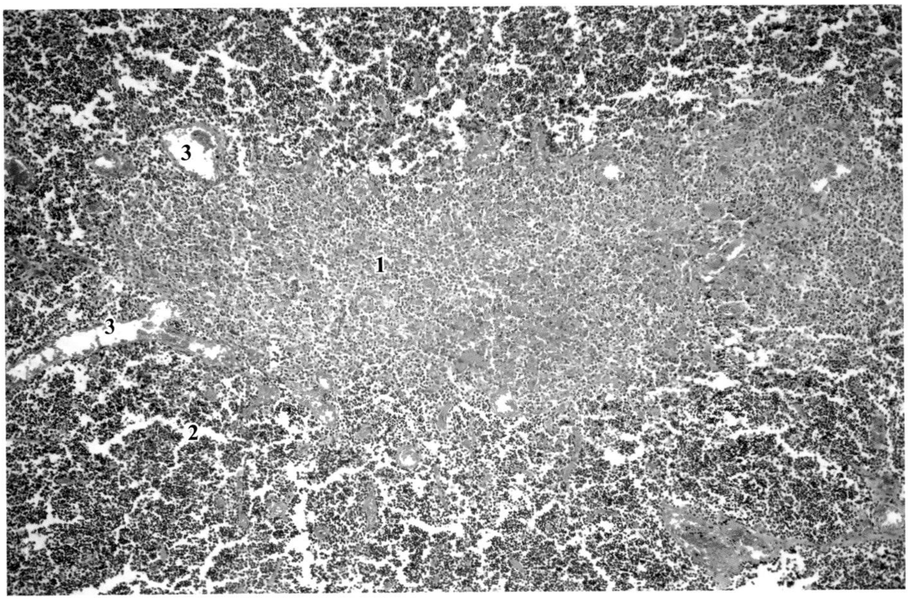

1- 胸腺髓质 thymic medulla
2- 胸腺皮质 thymic cortex
3- 静脉 vein

图 10-6 胸腺小叶组织切片（HE 染色，高倍镜）
Figure 10-6 Histological section of the thymic lobule (HE staining, higher power).

1- 静脉 vein
2- 小叶间隔 interlobular septum
3- 胸腺细胞 thymocyte
4- 胸腺上皮细胞（上皮性网状细胞）thymic epithelial cell (epithelial reticular cell)

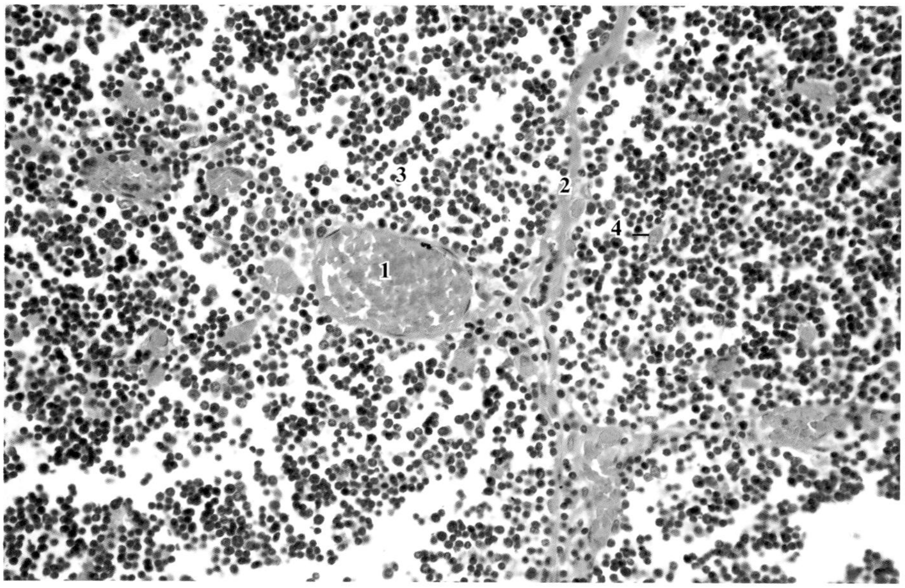

图 10-7　胸腺皮质组织切片（HE 染色，高倍镜）
Figure 10-7　Histological section of the thymic cortex (HE staining, higher power).

1- 胸腺小体 thymic corpuscle, Hassall's corpuscle
2- 静脉 vein
3- 交错突细胞 interdigitating cell
4- 髓质上皮细胞 epithelial cell of thymic medulla
5- 巨噬细胞 macrophage
6- 肌样细胞 myoid cell
7- 胸腺细胞 thymocyte

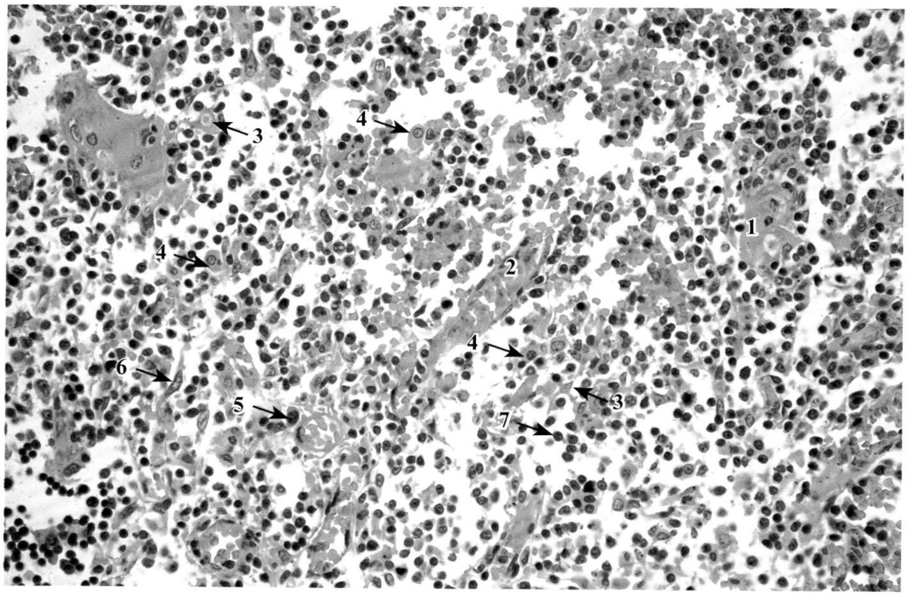

图 10-8　胸腺髓质组织切片（HE 染色，高倍镜）
Figure 10-8　Histological section of the thymic medulla (HE staining, higher power).

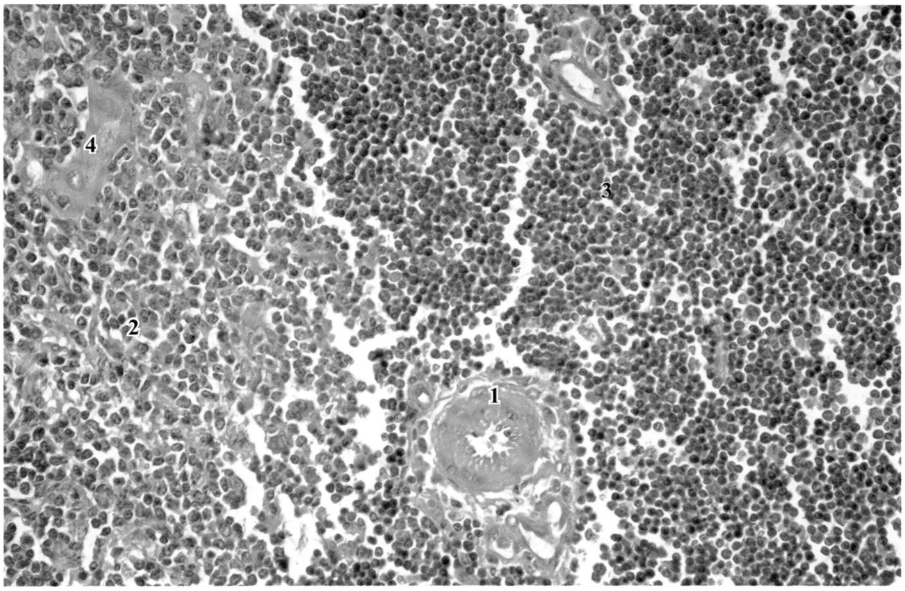

1- 动脉 artery
2- 胸腺髓质 thymic medulla
3- 胸腺皮质 thymic cortex
4- 胸腺小体 thymic corpuscle, Hassall's corpuscle

图 10-9　胸腺小叶组织切片（HE 染色，高倍镜）
Figure 10-9　Histological section of the thymic lobule (HE staining, higher power).

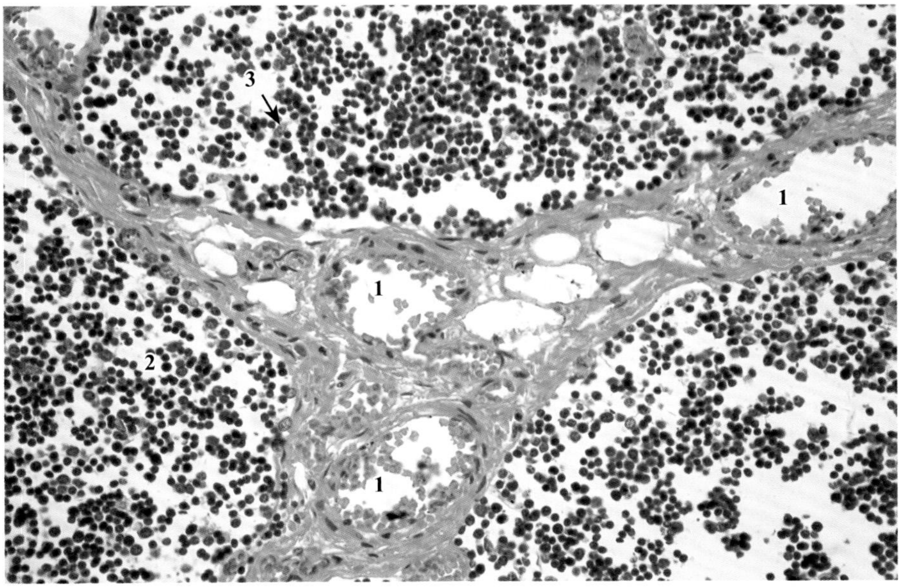

1- 血管 blood vessel
2- 胸腺细胞 thymocyte
3- 胸腺上皮细胞（上皮性网状细胞）thymic epithelial cell (epithelial reticular cell)

图 10-10　胸腺皮质组织切片（HE 染色，高倍镜）
Figure 10-10　Histological section of the thymic cortex (HE staining, higher power).

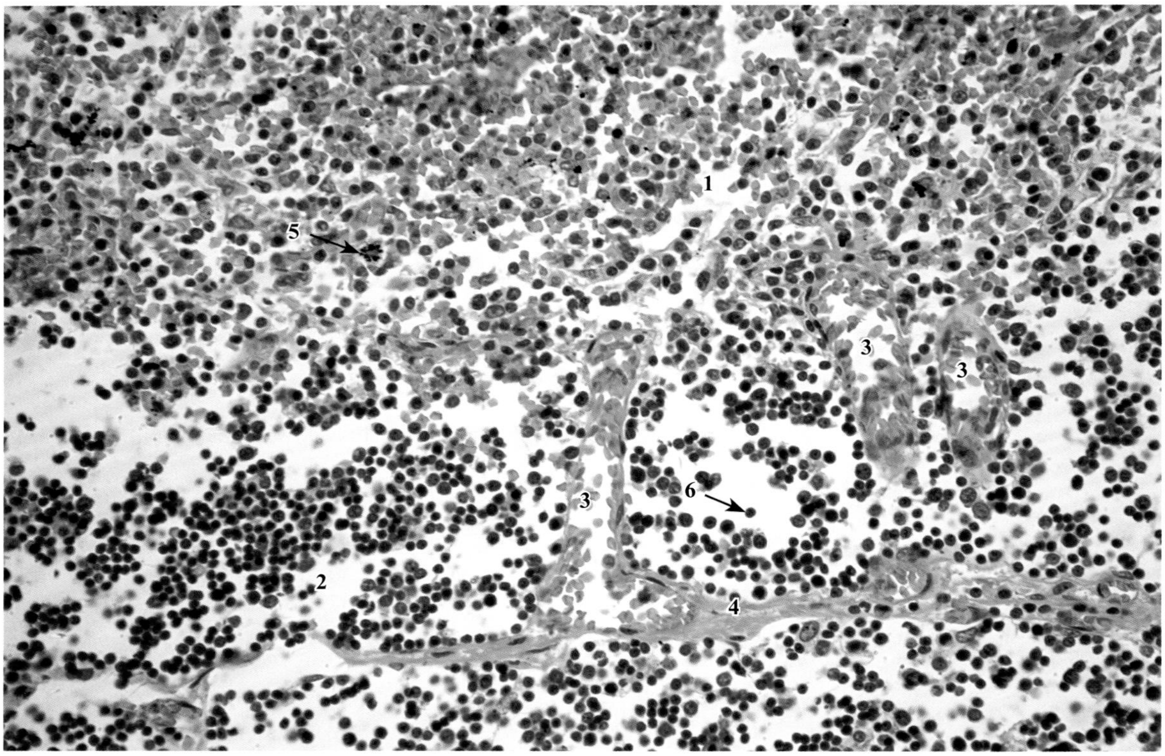

图 10-11 胸腺小叶组织切片（HE 染色，高倍镜）
Figure 10-11 Histological section of the thymic lobule (HE staining, higher power).

1- 胸腺髓质 thymic medulla
2- 胸腺皮质 thymic cortex
3- 静脉 vein
4- 小叶间隔 interlobular septum
5- 巨噬细胞 macrophage
6- 胸腺细胞 thymocyte

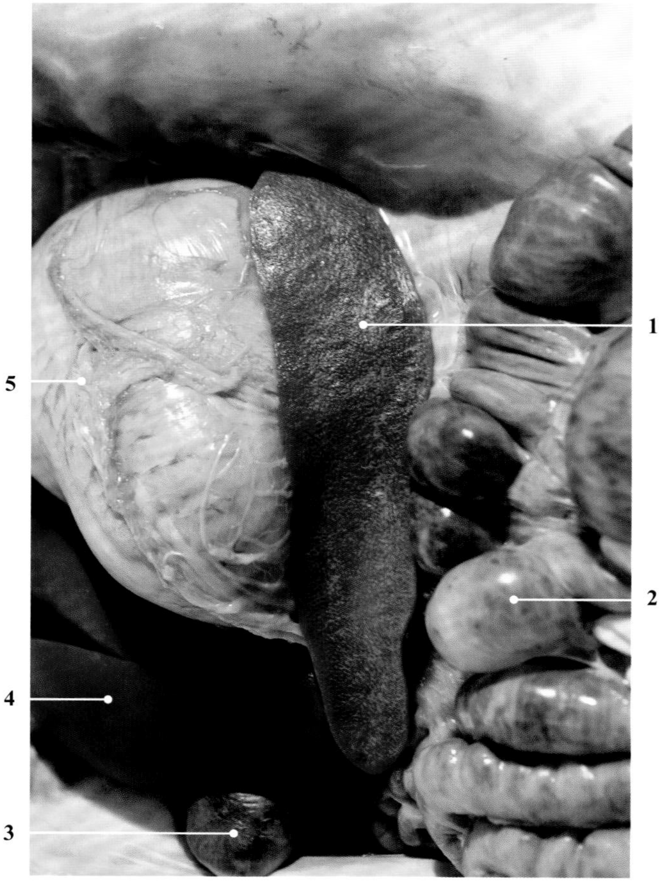

图 10-12 胃和脾的位置关系（左外侧观）
Figure 10-12 Position relation between the spleen and stomach (left lateral view).

1- 脾 spleen
2- 大肠 large intestine
3- 胆囊 gallbladder
4- 肝 liver
5- 胃 stomach

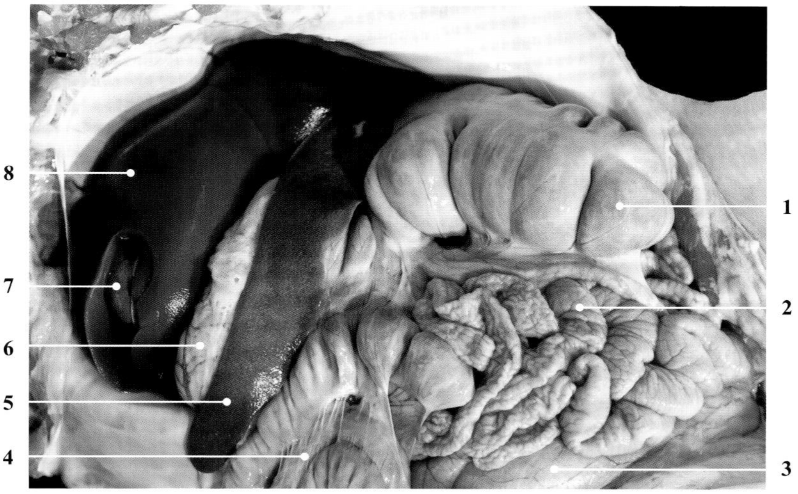

图 10-13 肝、胃、脾的位置关系（左外侧观）
Figure 10-13 Position relation between the liver, stomach and spleen (left lateral view).

1- 盲肠 caecum
2- 空肠 jejunum
3- 膀胱 urinary bladder
4- 结肠 colon
5- 脾 spleen
6- 胃 stomach
7- 胆囊 gallbladder
8- 肝 liver

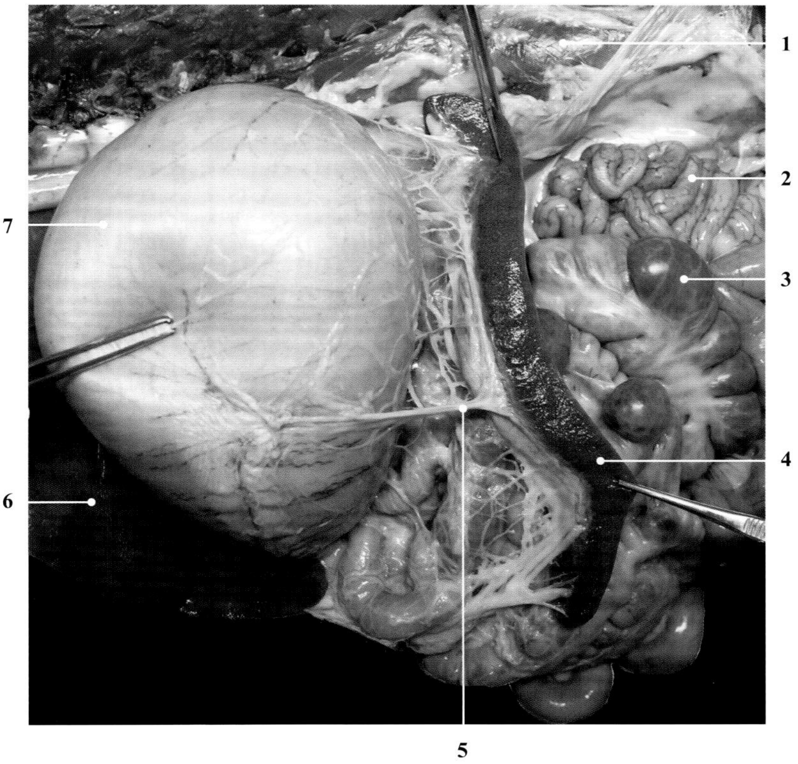

图 10-14 母猪胃脾韧带
Figure 10-14 Gastrosplenic ligament of a sow.

1- 肾 kidney
2- 小肠 small intestine
3- 大肠 large intestine
4- 脾 spleen
5- 胃脾韧带 gastrosplenic ligament
6- 肝 liver
7- 胃 stomach

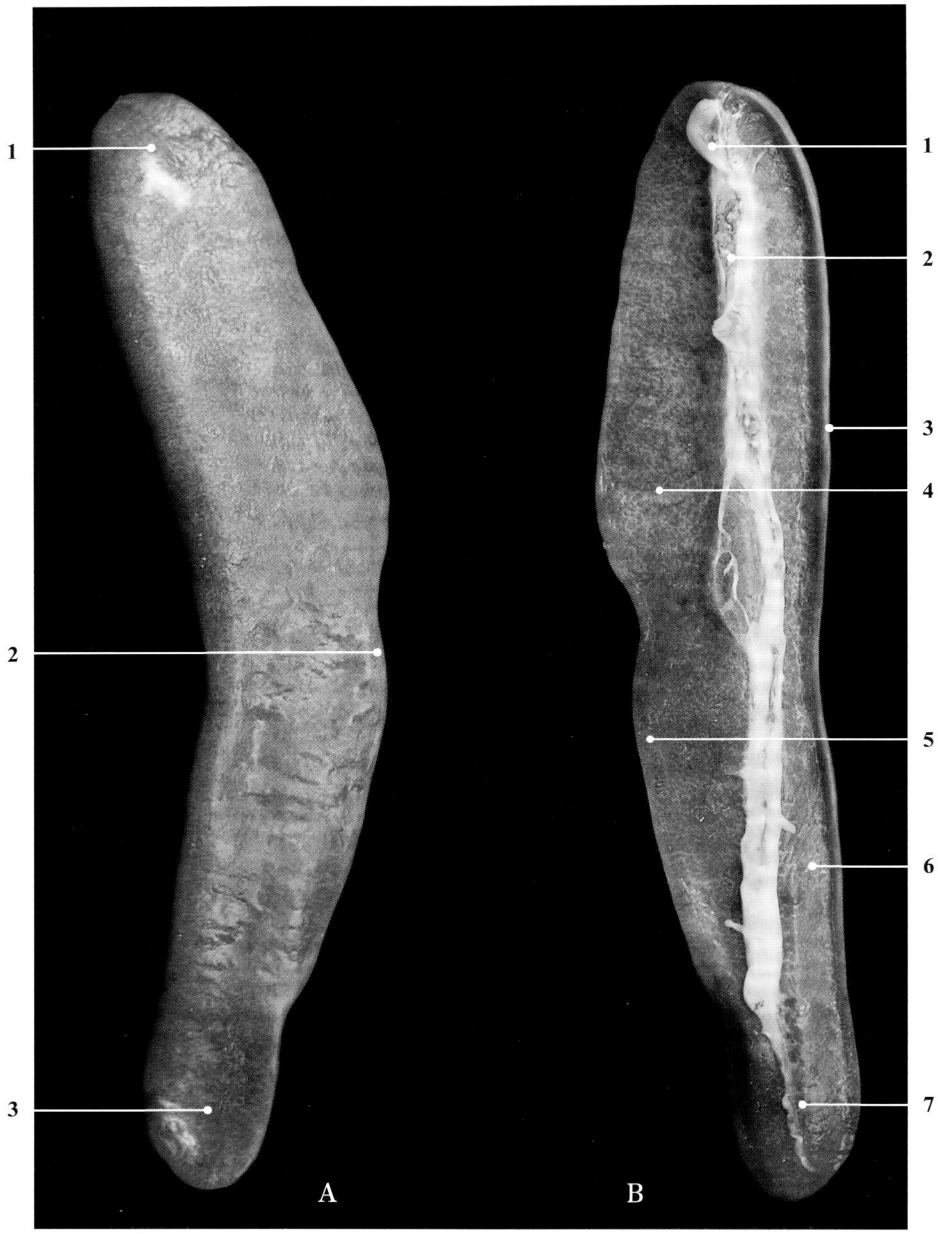

图 10-15 脾的形态
Figure 10-15 Morphology of the spleen.

A- 膈面 Diaphragmatic surface
1- 背侧端 dorsal extremity
2- 后缘 posterior border
3- 腹侧端 ventral extremity

B- 脏面 Visceral surface
1- 背侧端 dorsal extremity
2- 脾门 splenic hilum
3- 前缘 anterior border
4- 肠面 intestinal surface
5- 后缘 posterior border
6- 胃面 gastric surface
7- 腹侧端 ventral extremity

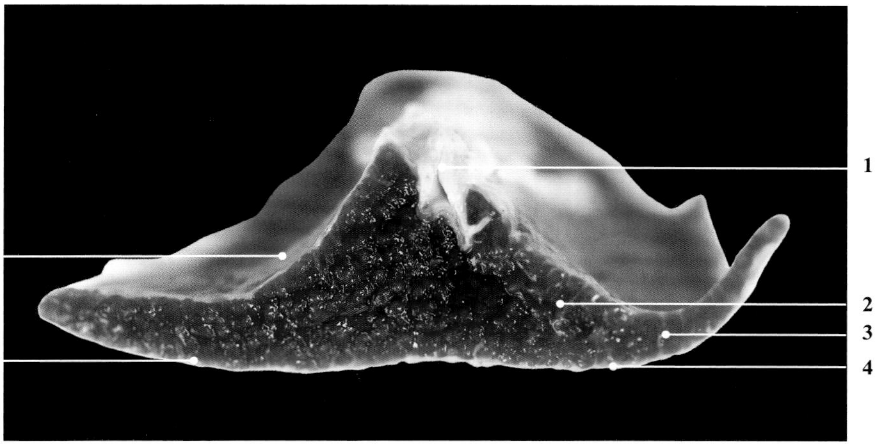

1- 脾门 splenic hilum
2- 脾实质 splenic parenchyma
3- 脾小梁 spleen trabecular
4- 被膜 capsule
5- 膈面 diaphragmatic surface
6- 脏面 visceral surface

图 10-16 脾横断面
Figure 10-16 Cross section of a spleen.

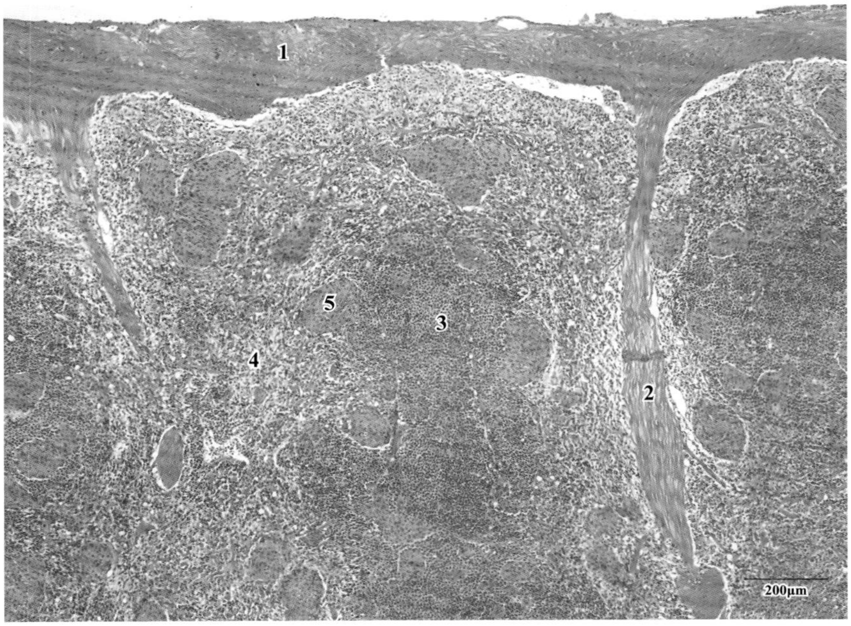

1- 被膜 capsule
2- 小梁 trabecula
3- 白髓 white pulp
4- 红髓 red pulp
5- 椭球 ellipsoid

图 10-17 脾组织切片，示被膜（HE 染色，低倍镜）
Figure 10-17 Histological section of a spleen showing the capsule (HE staining, lower power).

1- 被膜 capsule
2- 小梁 trabecula
3- 白髓 white pulp
4- 红髓 red pulp
5- 椭球 ellipsoid

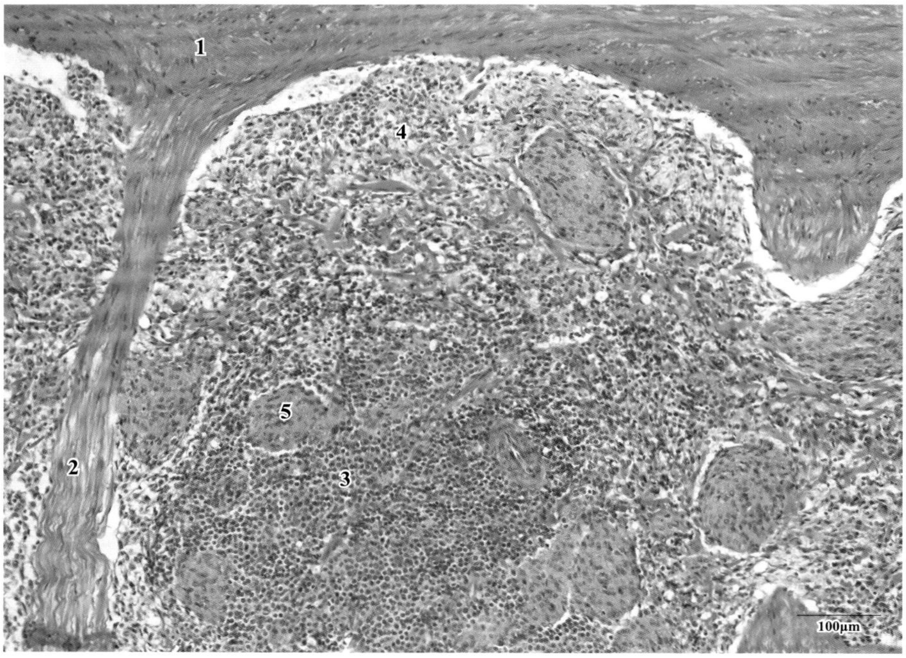

图 10-18　脾组织切片，示被膜（HE 染色，高倍镜）
Figure 10-18　Histological section of a spleen showing the capsule (HE staining, higher power).

1- 被膜 capsule
2- 椭球 ellipsoid
3- 红髓 red pulp
4- 脾索 splenic cord
5- 脾窦 splenic sinusoid
6- 小梁 trabecula

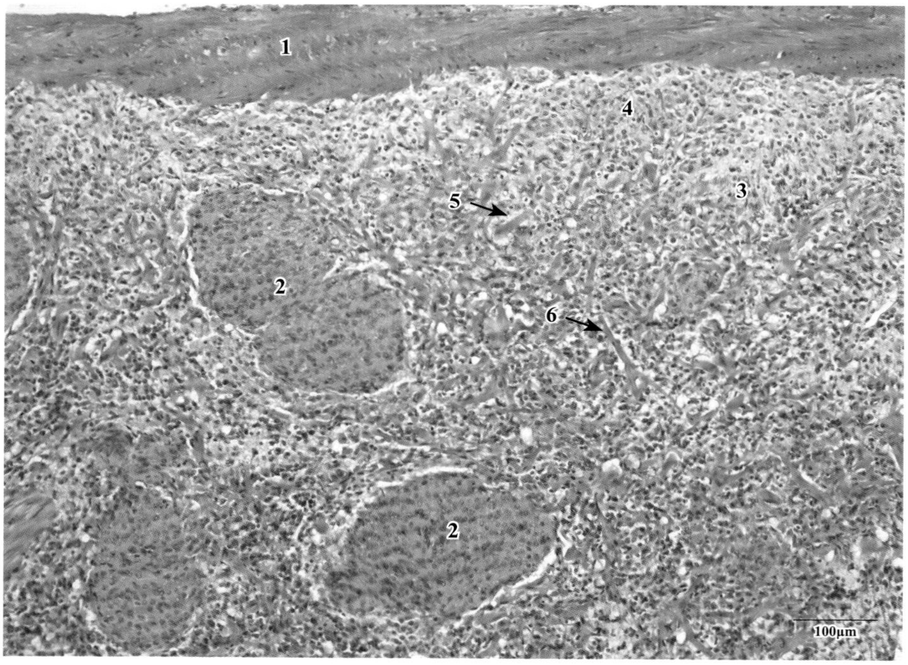

图 10-19　脾组织切片，示被膜（HE 染色，高倍镜）
Figure 10-19　Histological section of a spleen showing the capsule (HE staining, higher power).

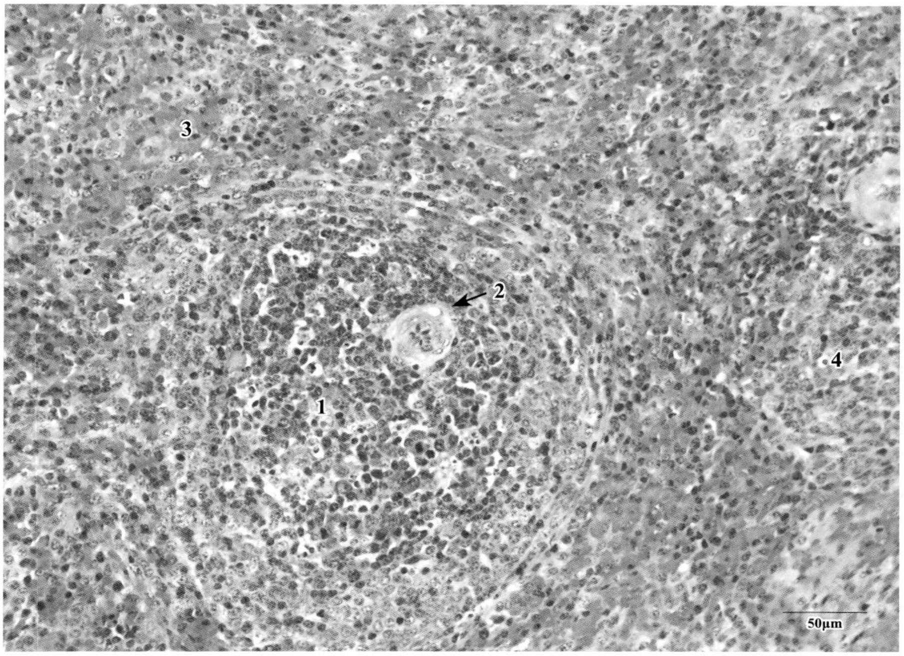

1- 脾小结 splenic nodule
2- 中央动脉 central artery
3- 红髓 red pulp
4- 白髓 white pulp

图 10-20　脾组织切片，示脾小结（HE 染色，高倍镜）
Figure 10-20　Histological section of a spleen showing the splenic nodule (HE staining, higher power).

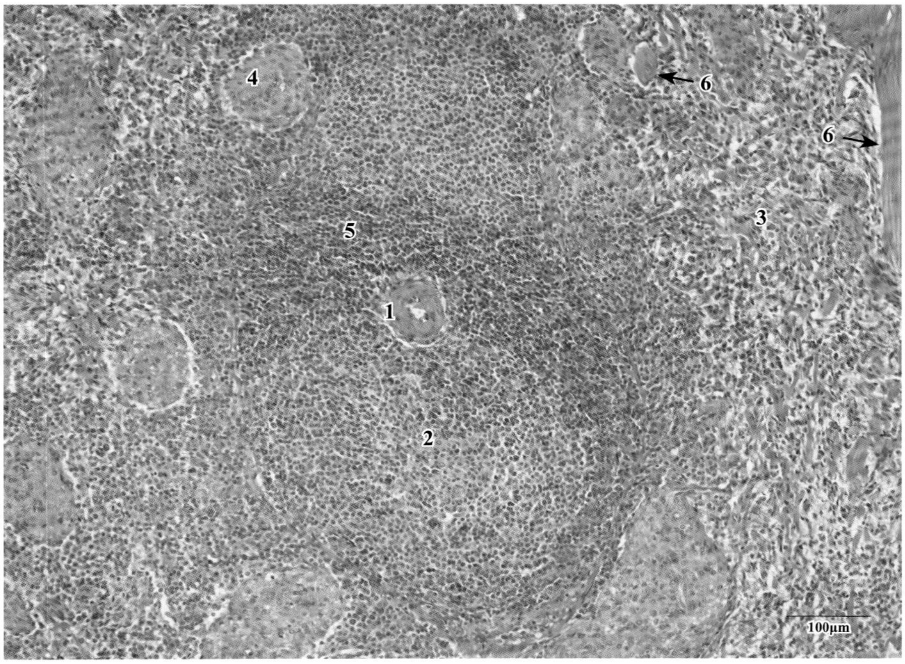

1- 中央动脉 central artery
2- 脾小结 splenic nodule
3- 红髓 red pulp
4- 椭球 ellipsoid
5- 动脉周围淋巴鞘 periarterial lymphatic sheath
6- 小梁 trabecula

图 10-21　脾组织切片，示动脉周围淋巴鞘（HE 染色，高倍镜）
Figure 10-21　Histological section of a spleen showing the periarterial lymphatic sheath (HE staining, higher power).

1- 脾小结 splenic nodule
2- 椭球 ellipsoid
3- 中央动脉 central artery
4- 红髓 red pulp
5- 白髓 white pulp
6- 小梁 trabecula
7- 动脉周围淋巴鞘 periarterial lymphatic sheath

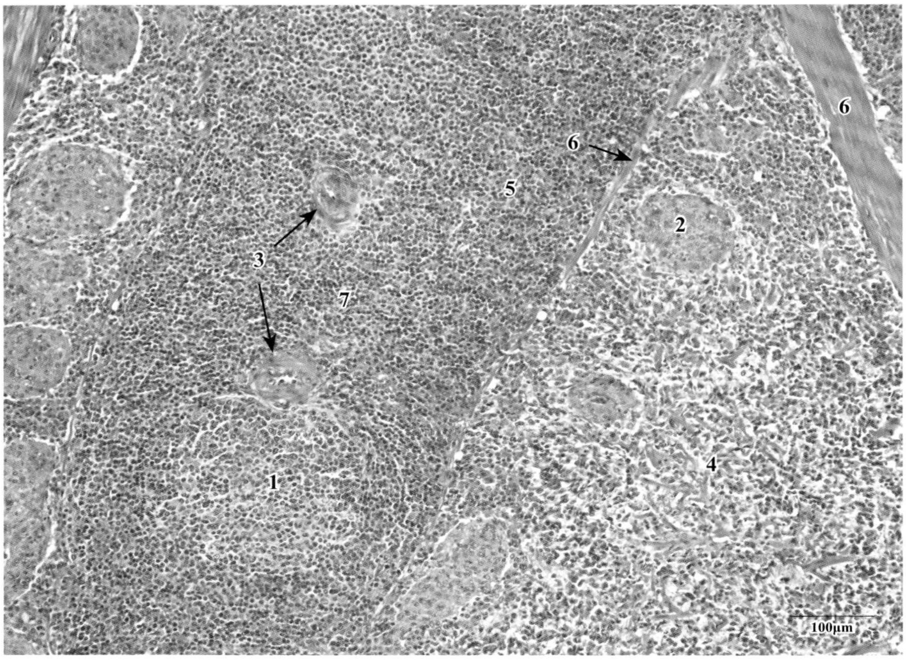

图 10-22　脾组织切片，示白髓（HE 染色，高倍镜）
Figure 10-22　Histological section of a spleen showing the white pulp (HE staining, higher power).

1- 中央动脉 central artery
2- 脾小结 splenic nodule
3- 动脉周围淋巴鞘 periarterial lymphatic sheath

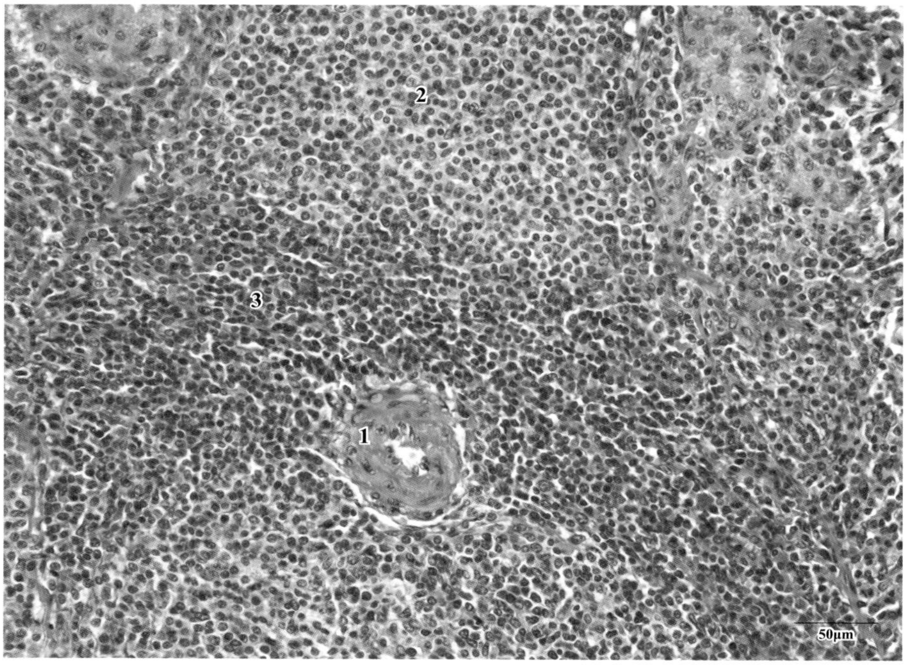

图 10-23　脾组织切片，示白髓（HE 染色，高倍镜）
Figure 10-23　Histological section of a spleen showing the white pulp (HE staining, higher power).

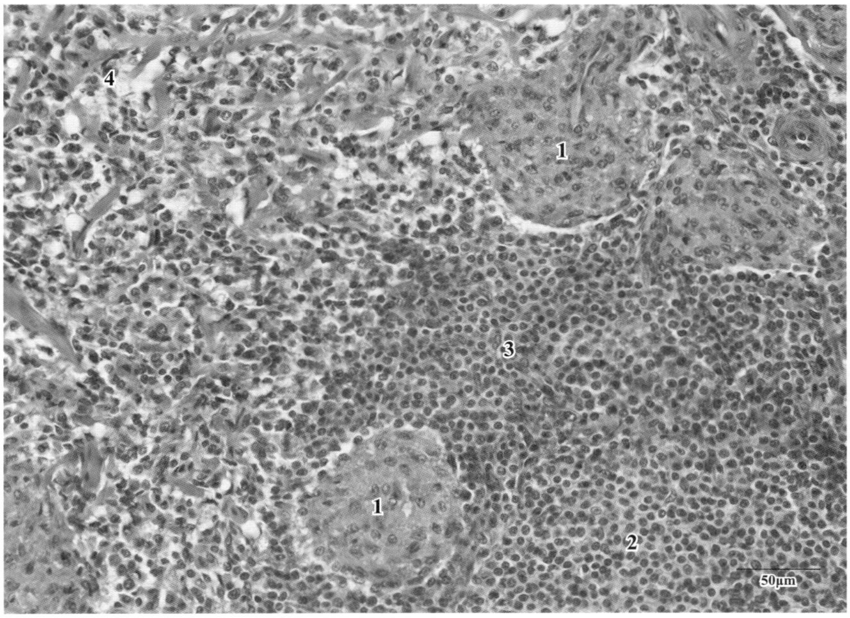

1- 椭球 ellipsoid
2- 脾小结 splenic nodule
3- 动脉周围淋巴鞘 periarterial lymphatic sheath
4- 红髓 red pulp

图 10-24 脾组织切片，示椭球（HE 染色，高倍镜）
Figure 10-24 Histological section of a spleen showing the ellipsoid (HE staining, higher power).

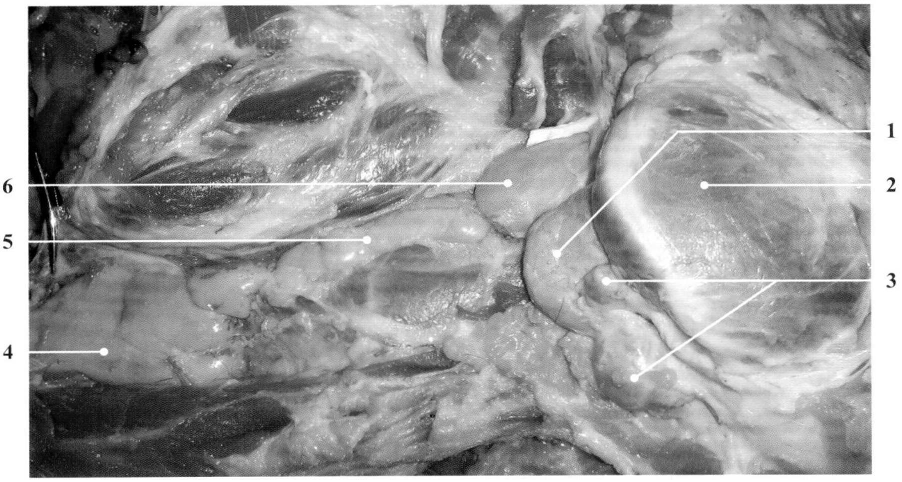

1- 颌下腺 mandibular gland
2- 咬肌 masseter muscle
3- 下颌淋巴结 mandibular lymph node
4- 胸腺胸叶 thoracic lobe of thymus
5- 胸腺中间叶 intermediate lobe of thymus
6- 胸腺颈叶 cervical lobe of thymus

图 10-25 下颌淋巴结
Figure 10-25 Mandibular lymph node.

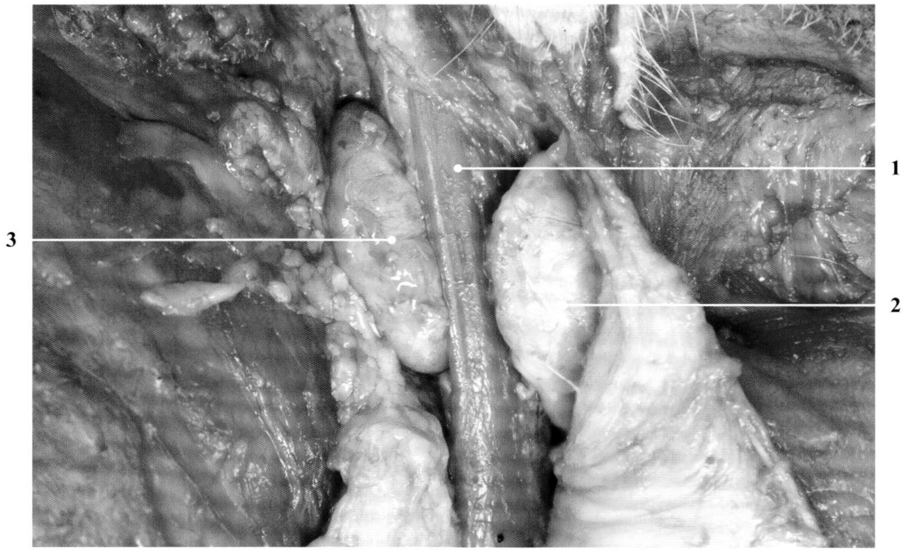

图 10-26　腮腺淋巴结和咽后外侧淋巴结
Figure 10-26　Parotid lymph node and lateral retropharyngeal lymph node.

1- 腮耳肌 parotidoauricular muscle
2- 咽后外侧淋巴结 lateral retropharyngeal lymph node
3- 腮腺淋巴结 parotid lymph node

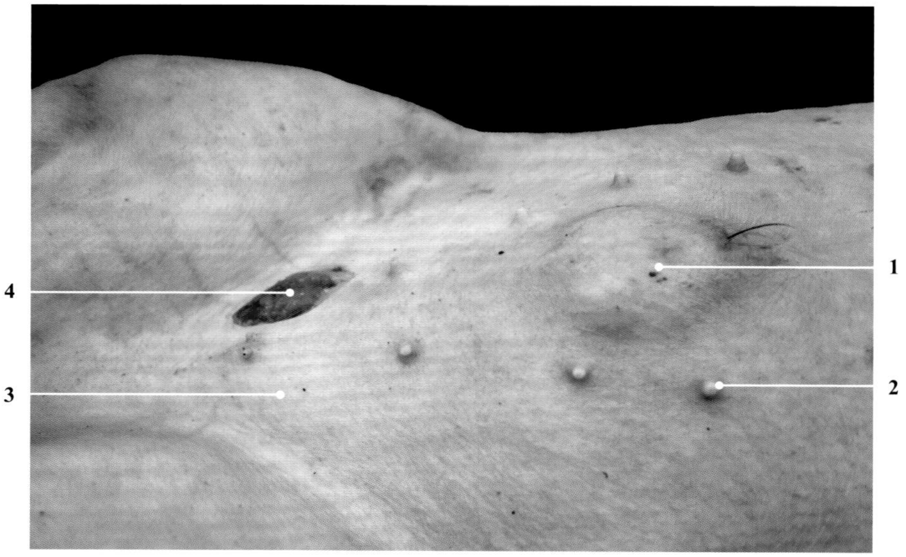

图 10-27　阴囊淋巴结 -1
Figure 10-27　Scrotal lymph node - 1.

1- 包皮憩室（包皮盲囊） preputial diverticulum
2- 乳头 teat
3- 腹壁皮肤 abdominal skin
4- 阴囊淋巴结（腹股沟浅淋巴结）scrotal lymph node (superficial inguinal lymph node)

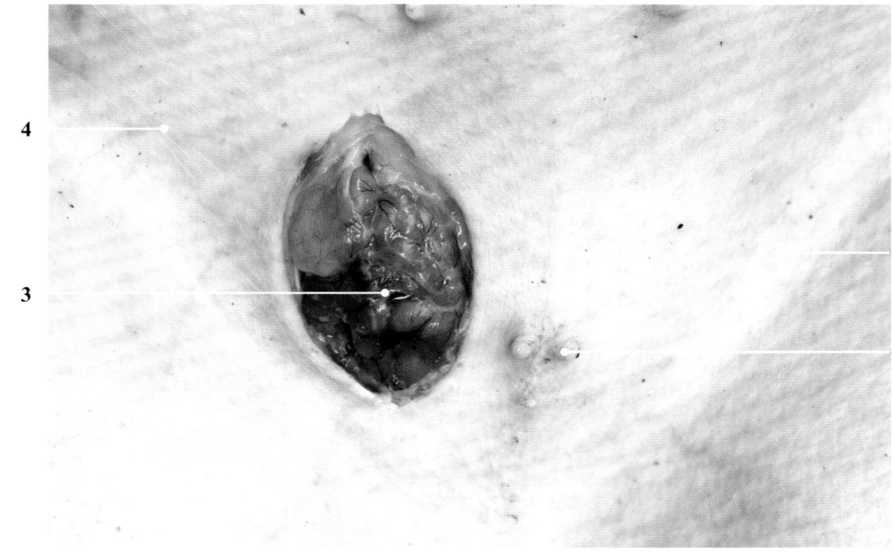

图 10-28 阴囊淋巴结 -2
Figure 10-28 Scrotal lymph node - 2.

1- 腹股沟部 inguinal region
2- 乳头 teat
3- 阴囊淋巴结（腹股沟浅淋巴结）scrotal lymph node (superfical inguinal lymph node)
4- 腹壁皮肤 abdominal skin

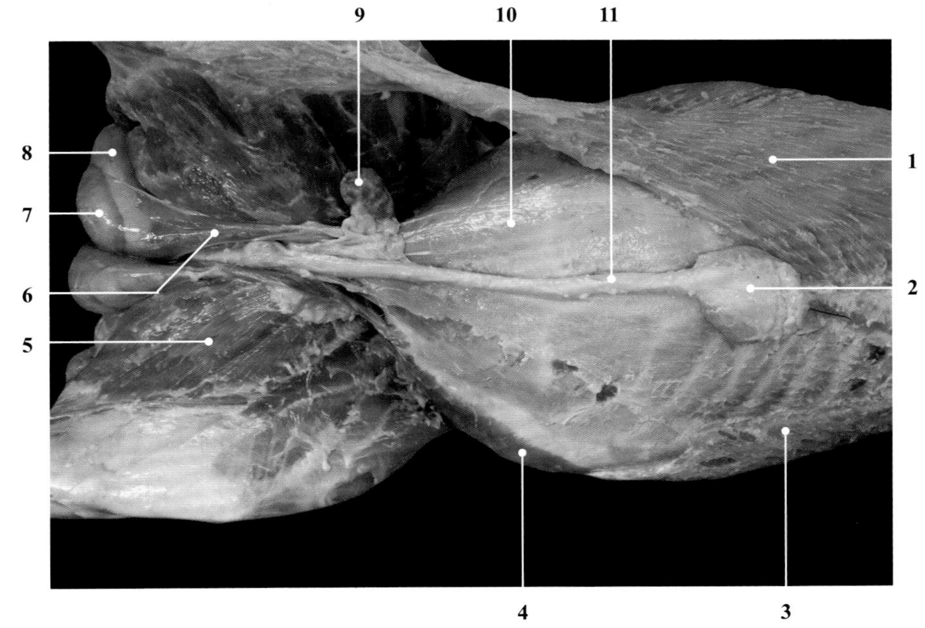

图 10-29 阴囊淋巴结 -3
Figure 10-29 Scrotal lymph node - 3.

1- 右侧腹外斜肌 right external oblique abdominal muscle
2- 包皮憩室（包皮盲囊） preputial diverticulum
3- 左侧腹外斜肌 left external oblique abdominal muscle
4- 腹内斜肌 internal oblique abdominal muscle
5- 股薄肌 gracilis muscle
6- 精索 spermatic cord
7- 睾丸 testis
8- 附睾 epididymis
9- 阴囊淋巴结（腹股沟浅淋巴结）scrotal lymph node (superficial inguinal lymph node)
10- 腹外斜肌腱膜 aponeurosis of external oblique abdominal muscle
11- 阴茎 penis

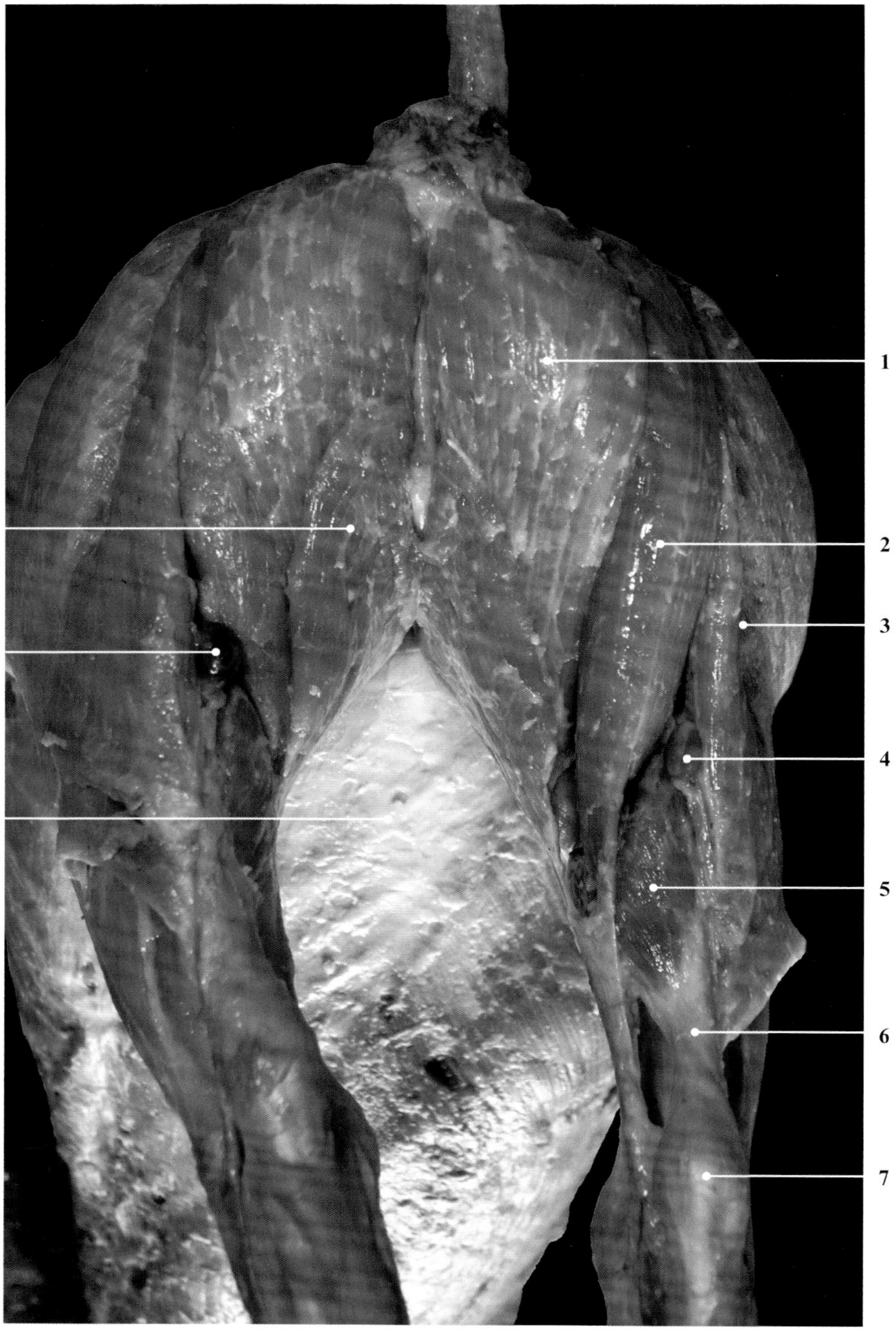

图 10-30　腘淋巴结（后面观）
Figure 10-30　Popliteal lymph node (caudal view).

1- 半膜肌 semimembranous muscle
2- 半腱肌 semitendinous muscle
3- 股二头肌 biceps muscle of the thigh
4- 右侧腘淋巴结 right popliteal lymph node
5- 腓肠肌 gastrocnemius muscle
6- 跟（总）腱 common calcaneal tendon
7- 跟结节 calcaneal tubercle
8- 腹底壁 bottom of abdomen
9- 左侧腘淋巴结 left popliteal lymph node
10- 股薄肌 gracilis muscle

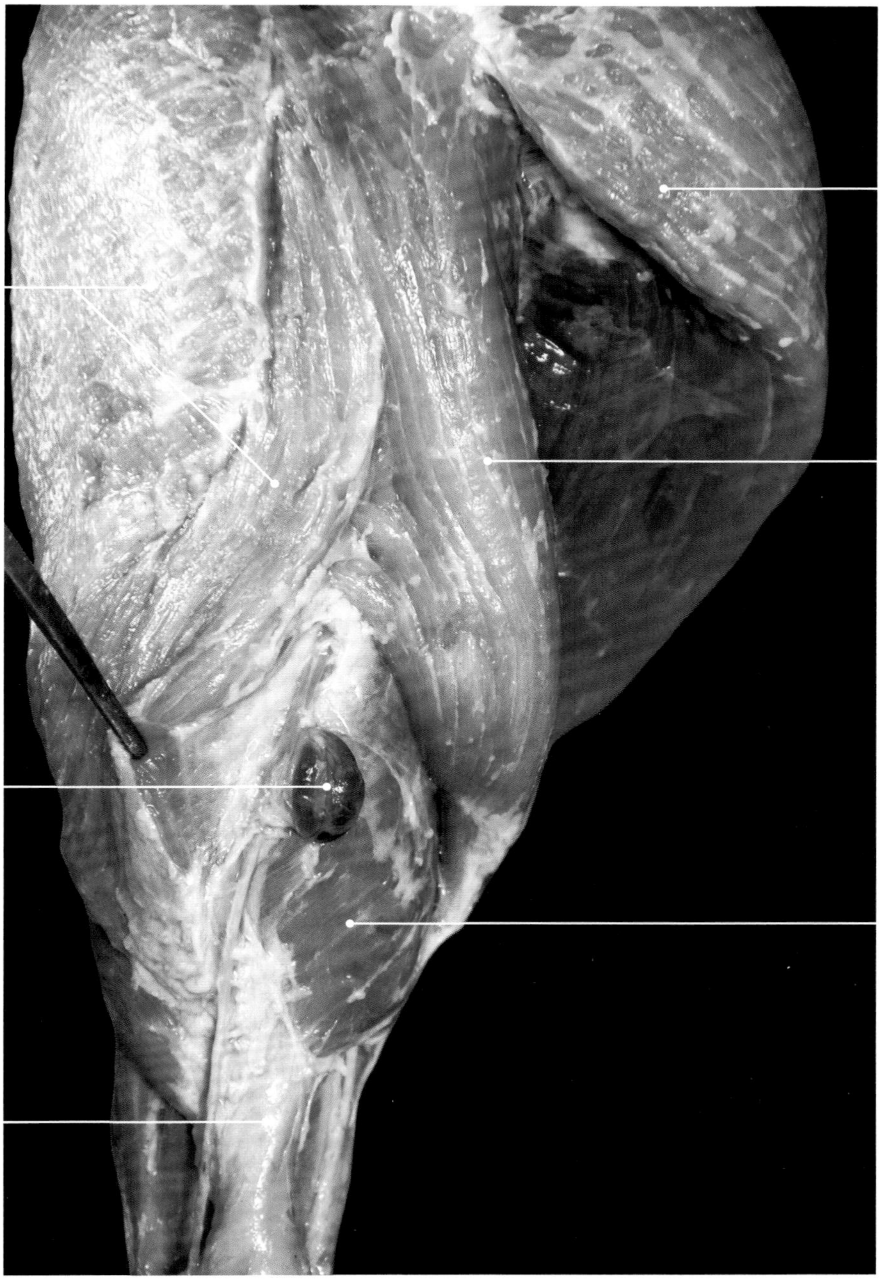

图 10-31 左后肢腘淋巴结（后面观）
Figure 10-31 Popliteal lymph node of left hindlimb (caudal view).

1- 半膜肌 semimembranous muscle
2- 半腱肌 semitendinous muscle
3- 腓肠肌 gastrocnemius muscle
4- 跟（总）腱 common calcaneal tendon
5- 腘淋巴结 popliteal lymph node
6- 股二头肌 biceps muscle of the thigh

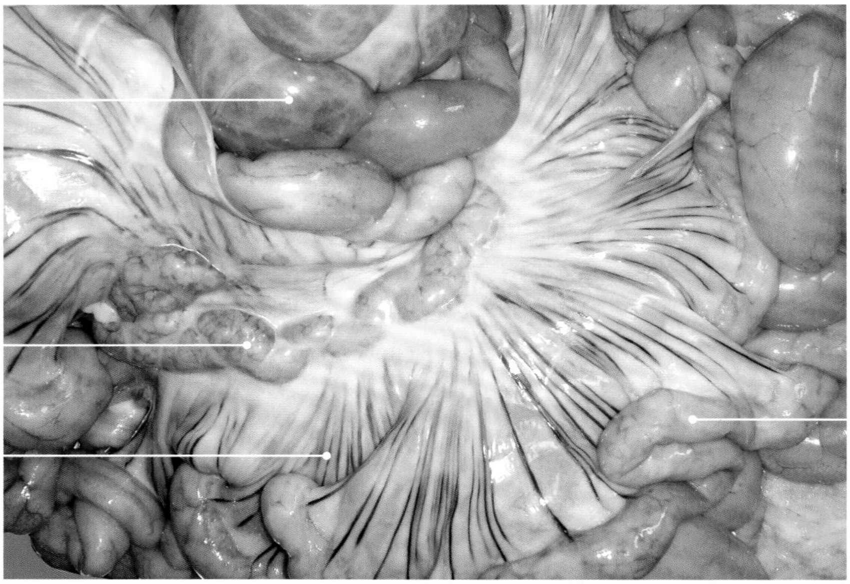

图 10-32 空肠淋巴结
Figure 10-32 Jejunal lymph nodes.

1- 空肠 jejunum
2- 肠系膜 mesentery
3- 空肠淋巴结 jejunal lymph nodes
4- 大肠 large intestine

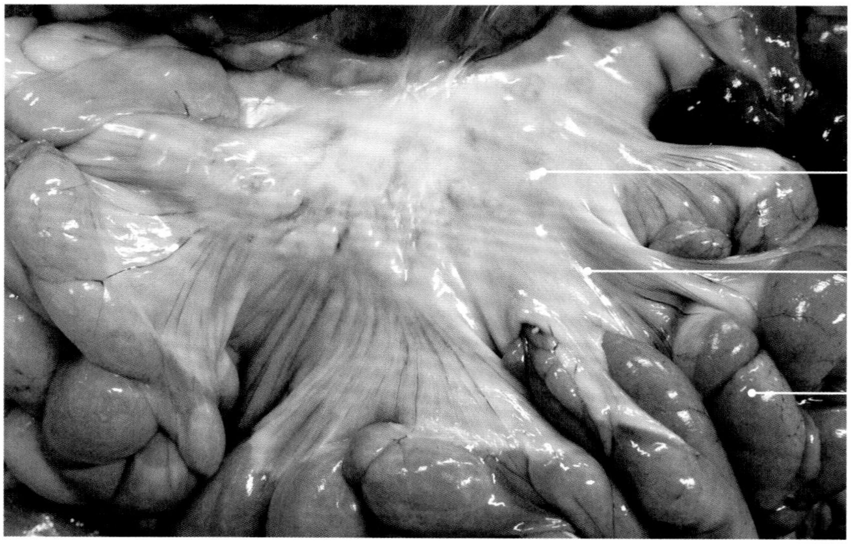

图 10-33 肠系膜淋巴结
Figure 10-33 Mesenteric lymph node.

1- 空肠淋巴结 jejunal lymph nodes
2- 肠系膜 mesentery
3- 空肠 jejunum

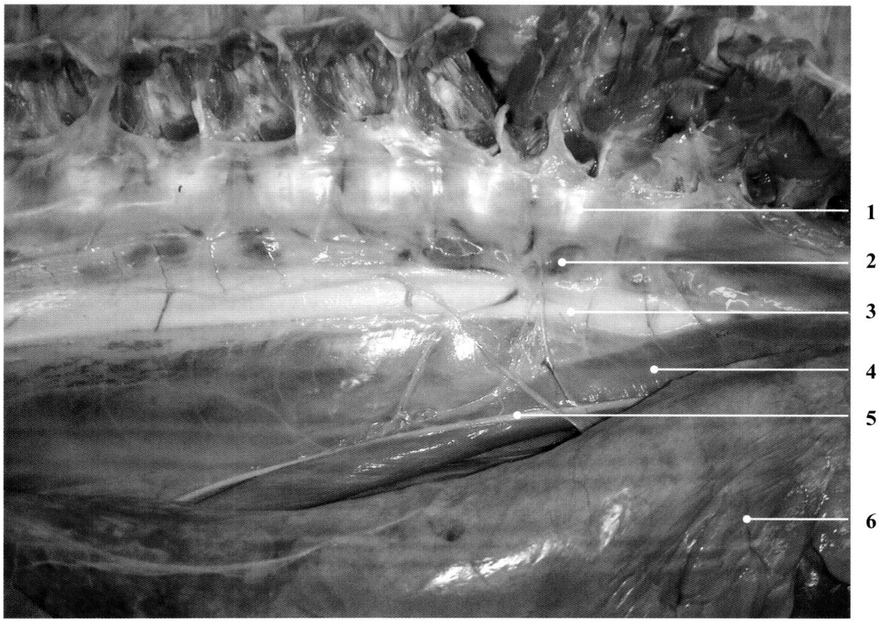

1- 胸椎 thoracic vertebrae
2- 纵隔淋巴结 mediastinal lymph node
3- 胸主动脉 thoracic aorta
4- 食管 esophagus
5- 迷走神经 vagus nerve
6- 肺 lung

图 10-34　纵隔淋巴结
Figure 10-34　Mediastinal lymph node.

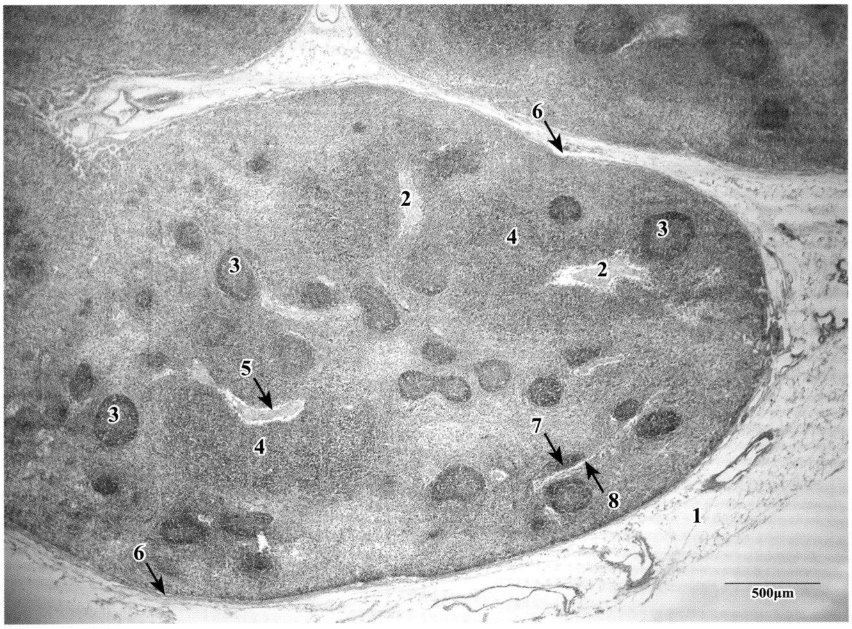

1- 被膜 capsule
2- 小梁 trabecula
3-淋巴小结 lymphatic nodule
4- 弥散淋巴组织 diffuse lymphoid tissue
5- 小梁周围淋巴窦 peritrabecular sinus
6- 被膜下淋巴窦 subcapsular sinus
7- 髓索 medullary cord
8- 髓窦 medullary sinus

图 10-35　淋巴结组织切片，示被膜（HE 染色，低倍镜）
Figure 10-35　Histological section of a lymph node showing the capsule (HE staining, lower power).

1- 被膜 capsule
2- 小梁 trabecula
3- 弥散淋巴组织 diffuse lymphatic tissue
4- 淋巴小结 lymphatic nodule
5- 髓质 medulla
6- 皮质 cortex

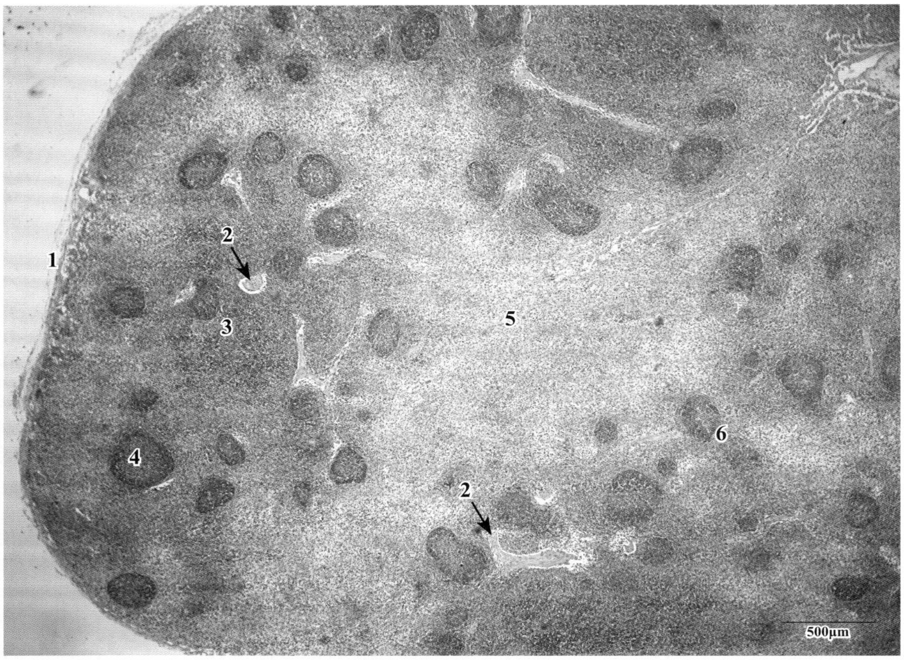

图 10-36　淋巴结组织切片，示被膜（HE 染色，低倍镜）
Figure 10-36　Histological section of a lymph node showing the capsule (HE staining, lower power).

1- 被膜 capsule
2- 被膜下窦 subcapsular sinus
3- 淋巴小结 lymphatic nodule
4- 弥散淋巴组织 diffuse lymphatic tissue
5- 小梁 trabecula
6- 小梁周围淋巴窦 peritrabecular sinus

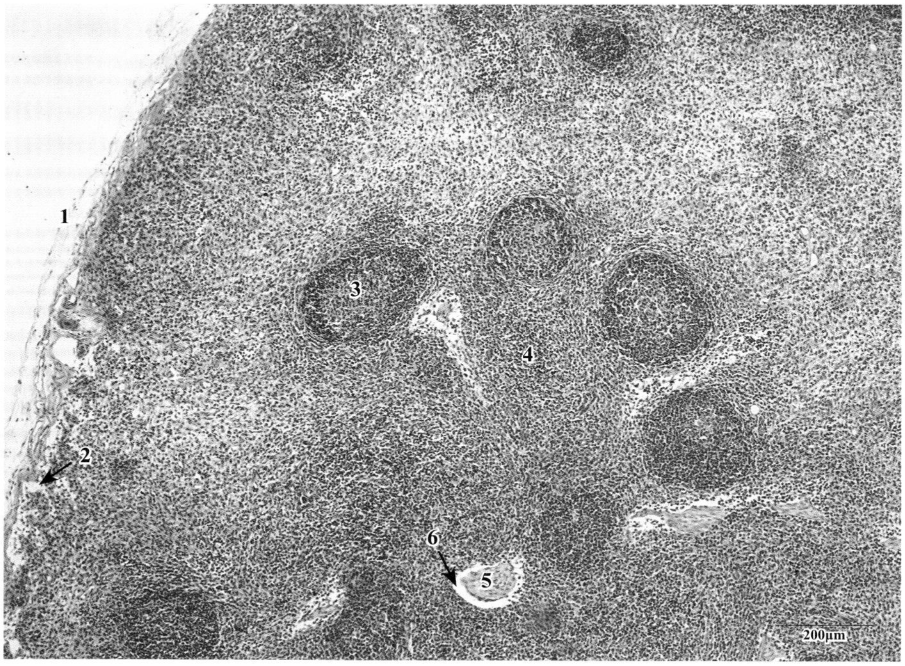

图 10-37　淋巴结组织切片，示皮质（HE 染色，低倍镜）
Figure 10-37　Histological section of a lymph node showing the cortex (HE staining, lower power).

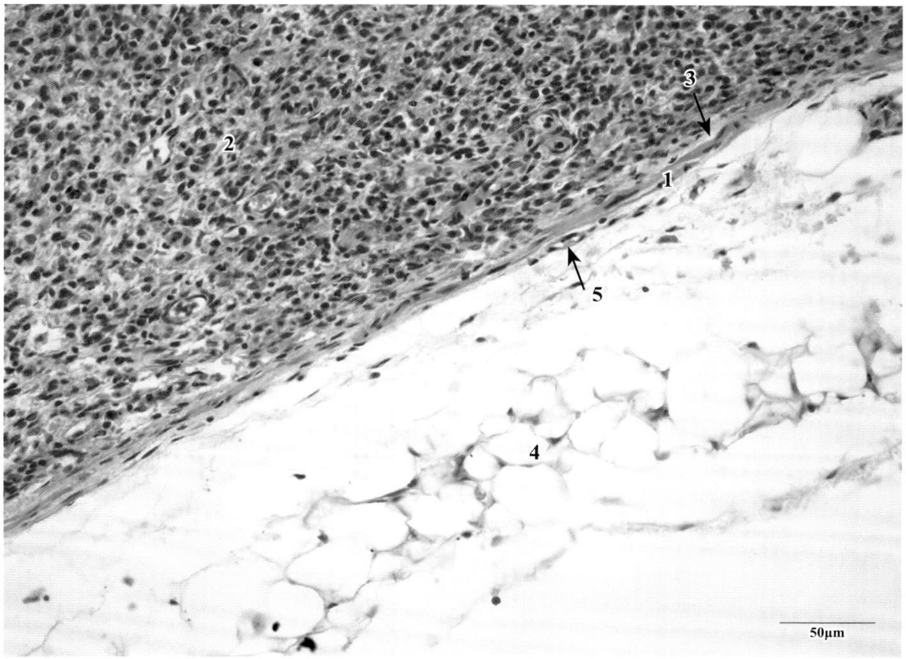

1- 被膜 capsule
2- 弥散淋巴组织 diffuse lymphatic tissue
3- 被膜下窦 subcapsular sinus
4- 脂肪组织 adipose tissue
5- 毛细血管 blood capillary

图 10-38 淋巴结组织切片，示被膜（HE 染色，高倍镜）
Figure 10-38 Histological section of a lymph node showing the capsule (HE staining, higher power).

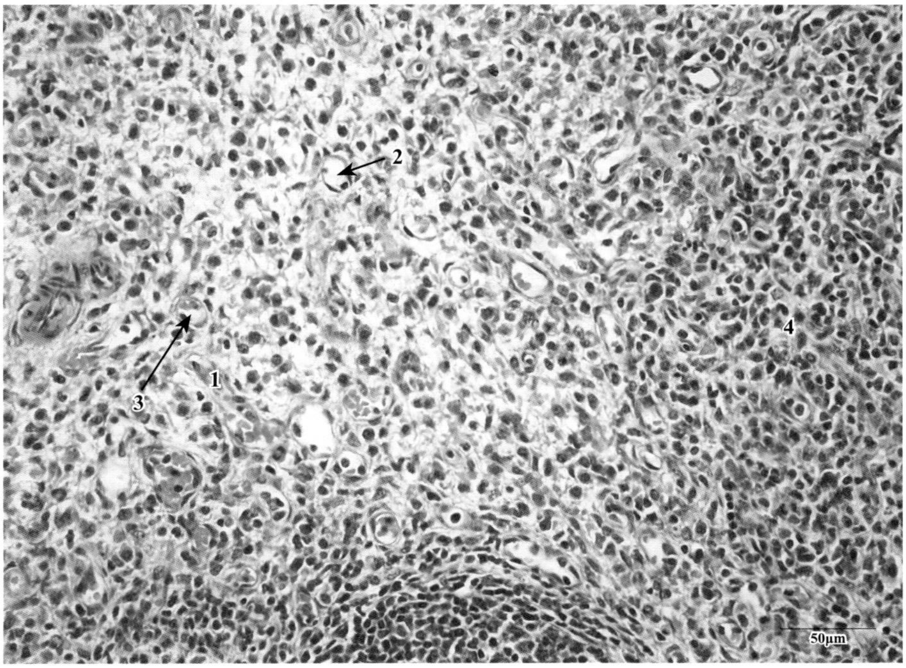

1- 髓索 medullary cord
2- 髓窦 medullary sinus
3- 毛细血管 blood capillary
4- 弥散淋巴组织 diffuse lymphatic tissue

图 10-39 淋巴结组织切片，示髓质（HE 染色，高倍镜）
Figure 10-39 Histological section of a lymph node showing the medulla (HE staining, higher power).

1- 淋巴小结生发中心 germinal center of lymphatic nodule
2- 淋巴小结暗区 dark region of lymphatic nodule
3- 小梁 trabecula

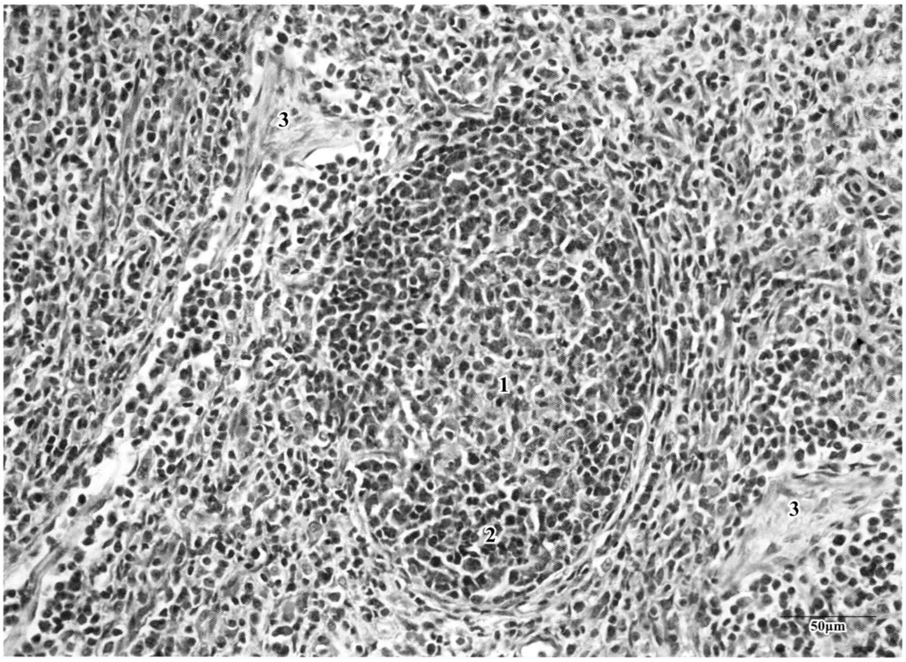

图 10-40　淋巴结组织切片，示淋巴小结（HE 染色，高倍镜）
Figure 10-40　Histological section of a lymph node showing the lymphatic nodule (HE staining, higher power).

1- 淋巴小结生发中心 germinal center of lymphatic nodule
2- 淋巴小结暗区 dark region of lymphatic nodule
3- 小梁 trabecula

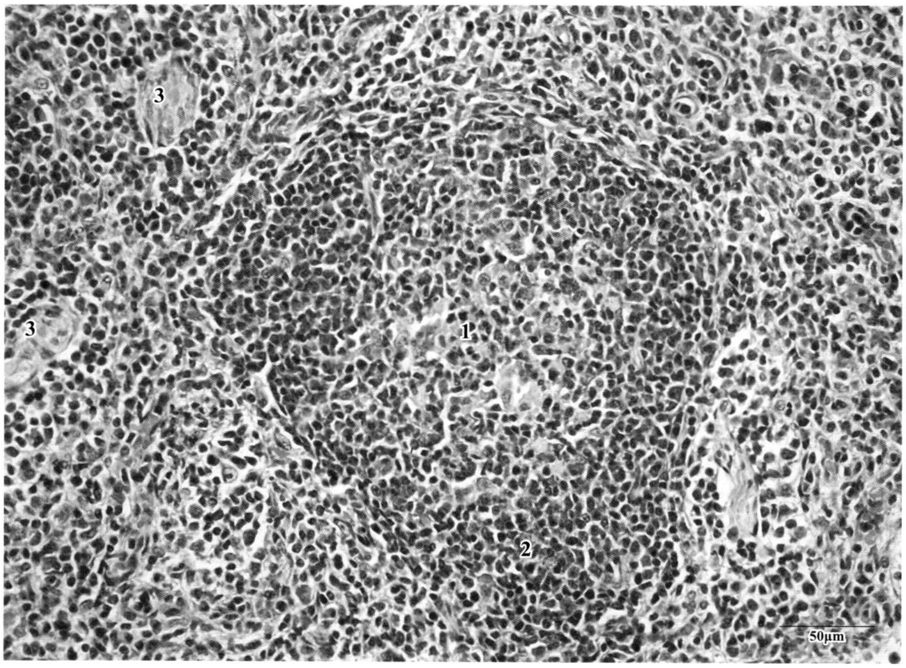

图 10-41　淋巴结组织切片，示淋巴小结（HE 染色，高倍镜）
Figure 10-41　Histological section of a lymph node showing the lymphatic nodule (HE staining, higher power).

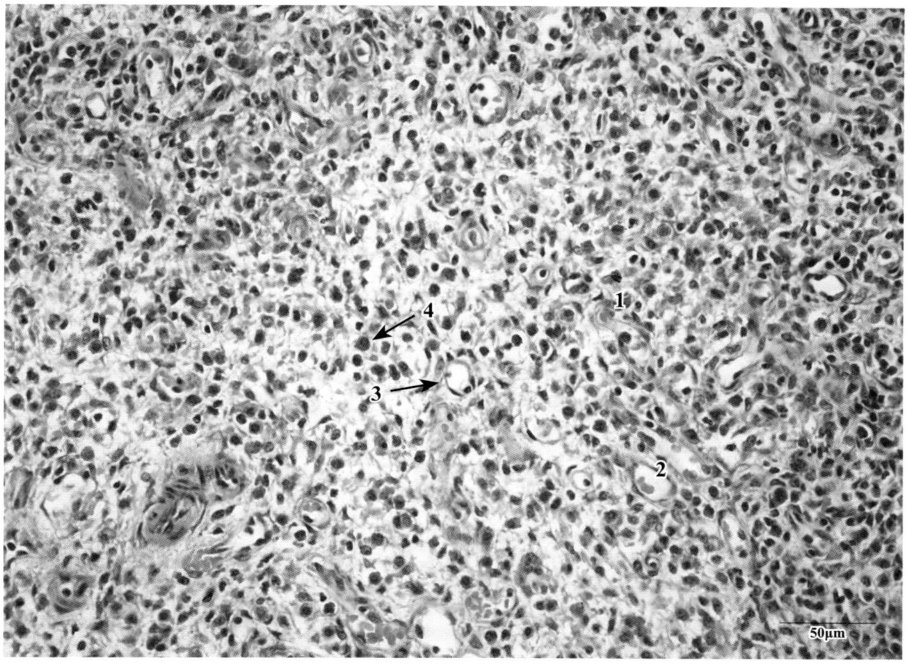

1- 髓索 medullary cord
2- 毛细血管 blood capillary
3- 髓窦 medullary sinus
4- 粒细胞 granulocyte

图 10-42 淋巴结组织切片，示髓质（HE 染色，高倍镜）
Figure 10-42 Histological section of a lymph node showing the medulla (HE staining, higher power).

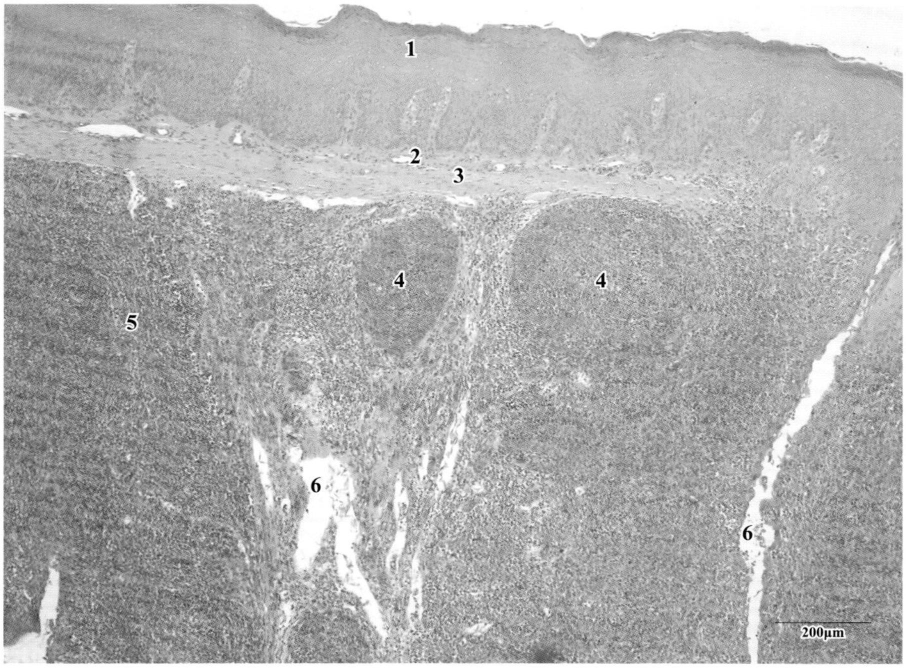

1- 复层扁平上皮 stratified squamous epithelium
2- 固有层 laminae propria
3- 肌层 muscular layer
4- 淋巴小结 lymphatic nodule
5- 弥散淋巴组织 diffuse lymphatic tissue
6- 隐窝 crypt

图 10-43 腭帆扁桃体组织切片（HE 染色，低倍镜）
Figure 10-43 Histological section of the palatine velum tonsil (HE staining, lower power).

1- 复层扁平上皮 stratified squamous epithelium
2- 固有层 laminae propria
3- 肌层 muscular layer
4- 淋巴小结 lymphatic nodule

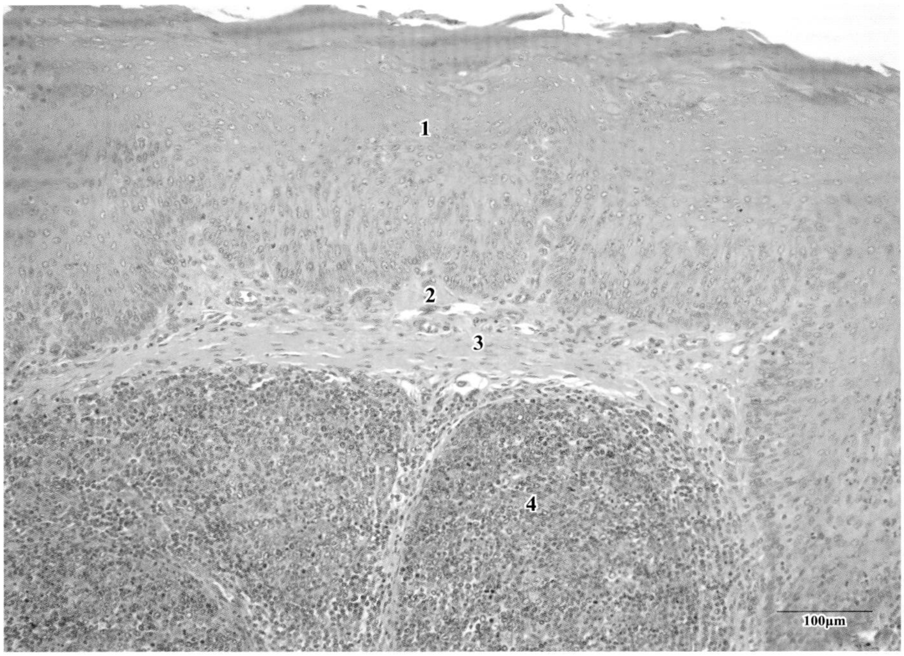

图 10-44　腭帆扁桃体组织切片（HE 染色，高倍镜）
Figure 10-44　Histological section of the palatine velum tonsil (HE staining, higher power).

1- 复层扁平上皮 stratified squamous epithelium
2- 隐窝 crypt
3- 淋巴小结 lymphatic nodule
4- 味蕾 taste bud
5- 弥散淋巴组织 diffuse lymphatic tissue
6- 静脉 vein
7- 舌腺 lingual gland
8- 横纹肌 striated muscle
9- 脂肪细胞 adipose cell

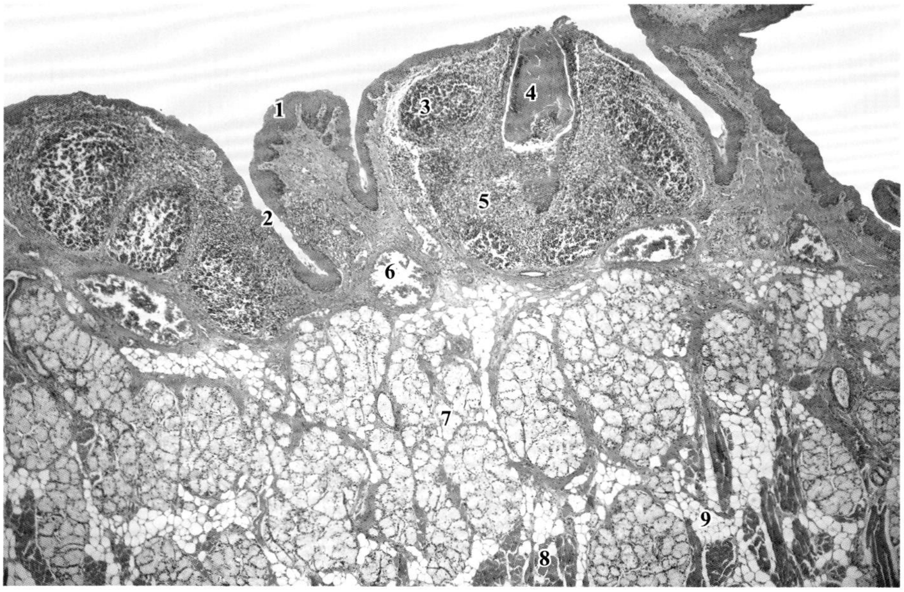

图 10-45　舌扁桃体组织切片（HE 染色，低倍镜）
Figure 10-45　Histological section of the lingual tonsil (HE staining, lower power).

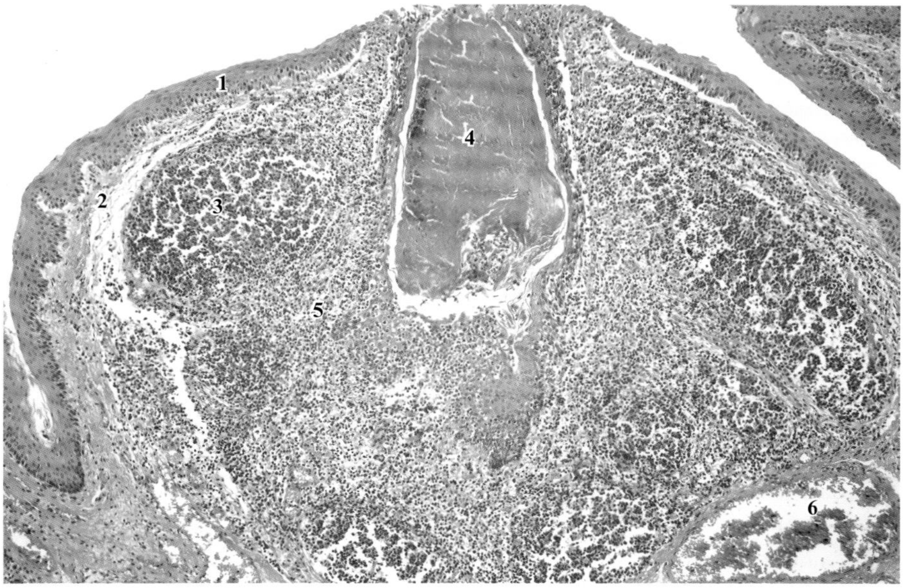

1- 复层扁平上皮 stratified squamous epithelium
2- 固有层 laminae propria
3- 淋巴小结 lymphatic nodule
4- 味蕾 taste bud
5- 弥散淋巴组织 diffuse lymphatic tissue
6- 静脉 vein

图 10-46　舌扁桃体组织切片（HE 染色，低倍镜）
Figure 10-46　Histological section of the lingual tonsil (HE staining, lower power).

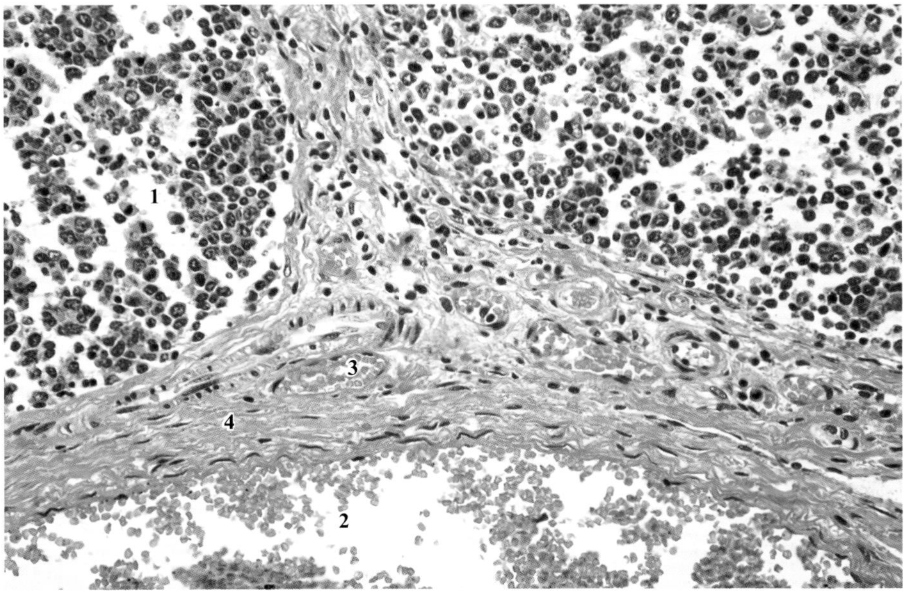

1- 淋巴细胞 lymphocyte
2- 中央静脉 central vein
3- 小静脉 small vein
4- 肌层 muscular layer

图 10-47　舌扁桃体组织切片（HE 染色，高倍镜）
Figure 10-47　Histological section of the lingual tonsil (HE staining, higher power).

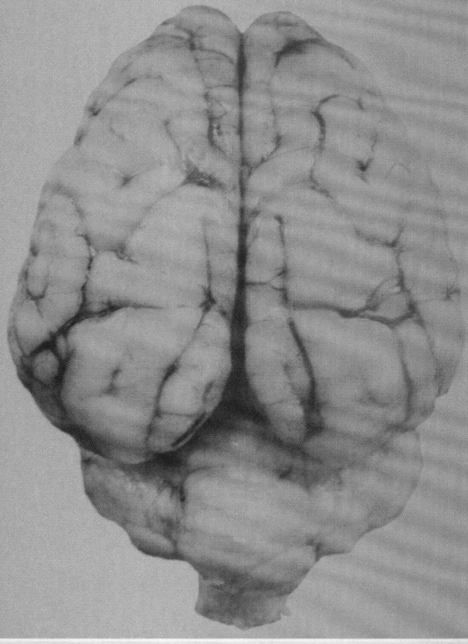

第十一章

神经系统

Chapter 11

Nervous system

神经系统（nervous system）是猪体内起主导作用的调节机构，在内分泌、免疫和感觉器官的配合下，通过对各种刺激的应答反应，调节和协调环境与机体、体内各个器官系统之间的关系，一方面使机体适应外界环境的变化，另一方面协调机体内各系统、各器官、器官内各组织的活动，使机体成为统一的整体。

神经系统的基本结构是神经组织。神经组织由神经细胞和神经胶质细胞构成。神经细胞是神经系统构造和功能的基本单位，故亦称神经元（neuron）。神经元分胞体和突起两部分。胞体也称核周体（perikaryon），是神经元的营养和代谢中心、信息整合中心和神经递质合成中心，由胞核、胞质和胞膜组成。胞质中除含有一般的细胞器外，富含尼氏体（Nissl body）和神经元纤维（neurofibril）。突起由细胞体发出，可分为树突（dendrite）和轴突（axon）两种。轴突的起始部稍凸起称轴丘（axon hillock），其内无尼氏体分布。神经元之间相互接触并发生功能联系的点，称突触（synapse）。神经胶质细胞（neuroglial cell）是神经系统的辅助成分，简称神经胶质（neuroglia）。神经纤维可划分为有髓鞘的有髓神经纤维（myelinated nerve fiber）和无髓鞘的无髓神经纤维（nonmyelinated nerve fiber）。

神经系统按其结构和功能可分为中枢神经系统（central nervous system）和周围神经系统（peripheral nervous system）。前者包括脑和脊髓，后者包括脑神经、脊神经和自主神经。

一、脊髓

脊髓（spinal cord）呈白色、背腹向稍扁的长圆柱状，位于椎管内，自枕骨大孔后缘向后伸延至第1荐椎中部，分为颈髓、胸髓、腰髓、荐髓和尾髓。脊髓有2个膨大部，位于颈髓后部和胸髓前部的称颈膨大（cervical enlargement），由其发出的脊神经形成臂神经丛，分布于前肢；位于腰荐髓间的称腰膨大（lumbar enlargement），发出的脊神经分布于骨盆腔及后肢；腰膨大之后的脊髓逐渐缩细形成圆锥状，称为脊髓圆锥（medullary cone）。自脊髓圆锥向后的细丝称为终丝（terminal filament），其中央由软膜构成，外面包裹的硬膜附着于尾椎椎体的背侧，有固定脊髓的作用。在脊髓圆锥和终丝的周围被荐神经和尾神经根所包裹，此结构称马尾（cauda equina）。

脊髓内部中央有细长纵走的中央管（central canal），前通第4脑室，后达终丝的起始部，在脊髓圆锥内扩张呈棱形状称终室（last loculus）。中央管内含脑脊髓液（cerebrospinal fluid）。在中央管的周围是“H”形的灰质（grey matter），灰质外面是白质（white matter）。灰质分背侧的背侧角（柱）和腹侧的腹侧角（柱），以及位于胸髓和腰髓段灰质的外侧和腹侧角基部的外侧角（柱）。在中央管周围连接左右侧的灰质称为灰质连合（grey commissure）。脊髓灰质由大量的神经元胞体、少量的神经纤维以及神经胶质细胞构成。在背侧柱中主要是中间神经元的胞体；腹侧柱内为运动神经元的胞体；胸腰段脊髓外侧柱为交感神经节前神经元的胞体；荐段脊髓的中间外侧柱内为副交感神经节前神经元的胞体。根据灰质内细胞形态和功能的差别，目前将其划分为10个板层（神经核）。白质可分背侧索、外侧索和腹侧索，主要由神经纤维构成，为脊髓上下传导冲动的传导径路，包括脑与脊髓之间长距离的上行（感觉性）、下行（运动性）传导束和脊髓内短距离联络性的固有束。

脊髓外面包裹3层结缔组织的脊膜（spinal meninges），由内向外依次为脊软膜（spinal pia mater）、脊蛛网膜（spinal arachnoid）和脊硬膜（spinal dura mater）。脊蛛网膜与脊软膜之间形成相当大的腔隙，称为蛛网膜下腔（subarachnoid cavity），内含脑脊髓液。脊硬膜与脊蛛网膜之间形成狭窄的硬膜下腔（subdural cavity），内含淋巴液。在脊硬膜与椎管之间有一较宽的腔隙，称为硬膜外腔（epidural cavity），内含静脉和大量脂肪，有脊神经通过。在腰荐间隙处，蛛网膜下腔间隙增大，可作为临床上抽取脑脊液或注射药物的部位。

二、脑

脑（brain, encephalon）位于颅腔内，可分为大脑、小脑和脑干。

1. 脑干（brain stem）：由后向前依次分为延髓（medulla oblongata）、脑桥（pons）、中脑（mesencephalon）和间脑（diencephalon），是脊髓向前的直接延续。脑干从前向后依次发出第3～12对脑神经，大脑、小脑、脊髓之间要通过脑干进行联系。

延髓后端在枕骨大孔处接脊髓，前端连脑桥；腹侧部位于枕骨基底部上，背侧部大部分被小脑所遮盖。脑桥位于延髓的前端，中脑的后方，小脑的腹侧。脑桥背侧面凹，为第4脑室（4th ventricle）底壁的前部。中脑位于脑桥和间脑之间，其脑室是中脑导水管（mesencephalic aqueduct），将中脑分为背侧的四叠体（quadrigeminal bodies）和腹侧的大脑脚（cerebral peduncle）。间脑位于中脑的前方，前外侧被大脑半球所遮盖；腹侧的前端为视交叉，后端为乳头体的后缘，内有第3脑室（3rd ventricle）。间脑可分为上丘脑（epithalamus）、丘脑（thalamus）、后丘脑（metathalamus）和下丘脑（hypothalamus）。

2. 小脑（cerebellum）：位于大脑后方，在延髓和脑桥的背侧。小脑的表面有许多平行的横沟和两条平行的纵沟。横沟深浅不一，浅的横沟将小脑表面分隔成小脑回，深的横沟将小脑分成许多小叶。纵沟将小脑分隔为两侧的小脑半球（cerebellar hemisphere）和中央的蚓部（vermis）。小脑腹面的两侧部有前、中、后3对小脑脚（cerebellar peduncle），分别与中脑、脑桥和延髓联系。

小脑的表面为灰质，称小脑皮质（cerebellar cortex），由外向内分为分子层、浦肯野细胞层和颗粒层；深部为白质，称小脑髓质。由于横沟的深浅不一，故髓质呈树枝状伸入小脑各叶，形成髓树（medullary arbor)，又称小脑树（cerebellar arbor）。

3. 大脑（cerebrum）：或称端脑（telencephalon），位于脑干前背侧，后端以大脑横裂（transverse cerebral fissure）与小脑分开，背侧正中的大脑纵裂（longitudinal cerebral fissure）将大脑分为左、右大脑半球（cerebral hemisphere），纵裂的底是连接两半球的横行宽纤维板，即胼胝体（corpus callosum）。每个大脑半球包括大脑皮质、白质、嗅脑和基底核。大脑半球内有侧脑室（lateral ventricle）。

大脑表层被覆一层灰质，称大脑皮质或大脑皮层（cerebral cortex），其表面凹凸不平。凹陷处为脑沟（sulcus），凸起处为脑回（gyrus），以增加大脑皮质的面积。大脑皮质的神经元胞体成层排列，由浅至深分为分子层、外颗粒层、外锥体细胞层、内颗粒层、内锥体细胞层和多形细胞层等6层。

4．脑膜和脑脊髓液：脑的外面包有3层膜，由外向内依次为脑硬膜（encephalic dura mater）、脑蛛网膜（encephalic arachnoid）和脑软膜（encephalic pia mater）。与脊髓相似，蛛网膜与软膜之间形成蛛网膜下腔，内含脑脊髓液；硬膜与蛛网膜之间形成硬膜下腔，内含淋巴液。脑室（brain ventricle）系统由侧脑室（每个大脑半球各有一个）、第3脑室、中脑导水管和第4脑室组成，其内含有脑脊髓液。脑脊髓液（cerebrospinal fluid）为无色透明液体，由侧脑室、第3脑室和第4脑室的脉络丛产生，具有营养脑、脊髓的作用，并在维持脑组织的渗透压和颅内压的相对恒定及减少外力震荡有重要作用。

三、脊神经

脊神经（spinal nerve）由脊髓发出的背侧根（感觉根）和腹侧根（运动根）在椎间孔附近聚集而成。背侧根与腹侧根汇合之前有一膨大，属感觉神经节，主要由假单极神经元（pseudounipolar neuron）的胞体聚集而成，称脊神经节（spinal ganglion）。脊神经按发出部位分为颈神经、胸神经、腰神经、荐神经和尾神经，再分支分布于躯干的体表和骨骼肌。

1．颈神经（cervical nerve）：起于颈髓，分背侧支和腹侧支。背侧支又分为内侧支和外侧支，分别穿行头半棘肌的内侧面，或头最长肌、颈最长肌和夹肌之间，最终分布于颈部背、外侧的肌肉和皮肤；腹侧支自前向后逐渐变粗，前4或5对颈神经的腹侧支小，分布于颈部腹外侧的肌肉和皮肤，后3对颈神经的腹侧支较大，参与组成臂神经丛和膈神经（phrenic nerve）。

2．胸神经（thoracic nerve）：起于胸髓，分背侧支和腹侧支。背侧支又分为内侧支和外侧支，内侧支分布于背多裂肌和棘肌等背部深层肌肉；外侧支分布于背最长肌和背髂肋肌，并从髂肋肌沟穿出后成为背皮神经，分布到背部皮肤、胸壁上方1/3部的皮肤。腹侧支称为肋间神经（intercostal nerve），主要分布于肋间肌。第1和第2胸神经的腹侧支主要参与形成臂神经丛。最后胸神经（the last thoracic nerve）的腹侧支，又称为肋腹神经（costoabdominal nerve），分布于腹部的皮肤，也分出分支到乳腺。

3．臂神经丛（brachial plexus）：由第6～8颈神经腹侧支和第1、第2胸神经腹侧支组成，主要分布于前肢的肌肉和皮肤以及部分肩带肌、胸腔和腹腔侧壁。其主要分支有肩胛上神经（suprascapular nerve）、肩胛下神经（subscapular nerve）、腋神经（axillary nerve）、胸肌神经（pectoral nerve）、肌皮神经（musculocutaneous nerve）、桡神经（radial nerve）、尺神经（ulnar nerve）和正中神经（median nerve）等。

4．腰神经（lumbar nerve）：起于腰髓，分背侧支和腹侧支。背侧支又分为内侧支和外侧支，内侧支在背腰最长肌深面分布于多裂肌等；外侧支有肌支至背腰最长肌，主干穿出背腰最长肌和臀中肌分布于腰臀部的皮肤。第1～4腰神经腹侧支形成髂腹下神经（iliohypogastric nerve）、髂腹股沟神经（ilioinguinal nerve）、生殖股神经（genitofemoral nerve）和股外侧皮神经（lateral femoral cutaneous nerve）；第4～6腰神经腹侧支参与构成腰荐神经丛。

5．荐神经（sacral nerve）：起于荐髓，分背侧支和腹侧支。背侧支经荐背侧孔出椎管，分布于臀部的皮肤以及尾根部的肌肉、皮肤。腹侧支经荐腹侧孔出椎管，第1、第2荐神经的腹侧支参与构成腰荐神经丛；第3～4对荐神经的腹侧支形成阴部神经（pudendal nerve）与直肠后神经（caudal rectal nerve）；最后一对荐神经腹侧支分布于尾的腹侧。

6. 腰荐神经丛（lumbosacral plexus）：由第4～6腰神经和第1～2荐神经腹侧支构成，位于腰荐部腹侧，其分支有股神经（femoral nerve）、坐骨神经（sciatic nerve）、闭孔神经（obturator nerve）、臀前神经（cranial gluteal nerve）和臀后神经（caudal gluteal nerve），主要分布于后肢。

7. 尾神经（coccygeal nerve）：起于尾髓，分背侧支和腹侧支，背侧支相互吻合形成尾背侧神经，伸至尾尖，腹侧支相互吻合形成尾腹侧神经伸至尾尖，分别分布于尾背、腹侧的肌肉和皮肤。

四、脑神经

脑神经（cranial nerve）自脑发出，有12对，按其与脑相连的前后顺序及其功能、分布和行程而命名。脑神经通过颅骨上的孔或裂进出颅腔，主要分布于头部和颈部等。

1. 嗅神经（olfactory nerve）：为传导嗅觉的感觉神经，起于鼻腔嗅区黏膜中的嗅细胞，其中枢突聚集成嗅丝，穿过筛板，入颅腔连接嗅球。

2. 视神经（optic nerve）：为传导视觉的感觉神经，由眼球视网膜节细胞的轴突构成，经视神经孔入颅腔，将视觉冲动传至大脑皮质。

3. 动眼神经（oculomotor nerve）：由来自中脑运动核的躯体传出纤维和副交感核的内脏传出神经组成，支配眼球和上眼睑的运动，并参与瞳孔和晶状体对光反射的调节。

4. 滑车神经（trochlear nerve）：为运动神经，起于中脑的滑车神经核，由脑干背侧发出，分布于眼球背侧斜肌，参与调节眼球的运动。

5. 三叉神经（trigeminal nerve）：为最粗大的脑神经，连于脑桥，属混合神经，由眼神经（ophthalmic nerve）、上颌神经（maxillary nerve）和下颌神经（mandibular nerve）组成。

6. 外展神经（abducent nerve）：为运动神经，起于延髓内的外展神经核，分布于眼球外直肌和眼球退缩肌，参与调节眼球的运动。

7. 面神经（facial nerve）：属混合神经，由延髓斜方体外侧发出，分布于颜面肌群，以及头部除腮腺以外的腺体，如泪腺、鼻腺和腭腺。

8. 前庭耳蜗神经（vestibulocochlear nerve）：属感觉神经，连斜方体的外侧缘，自内耳道进入耳内。传导听觉和平衡觉，分为前庭神经（vestibular nerve）和耳蜗神经（cochlear nerve）。

9. 舌咽神经（glossopharyngeal nerve）：属混合神经，自延髓的腹外侧缘发出，其根在前庭耳蜗神经根后方与迷走神经根的前面，经颈静脉孔出颅腔。主要分布于舌、咽部的肌肉和味蕾。

10. 迷走神经（vagus nerve）：为混合神经，起于延髓的腹侧面、舌咽神经根后方，是脑神经中行程最远、分布区域最广的神经，分布于咽、喉、食管、胃、肠、肝、胰、肺、心和肾等器官，调节平滑肌、心肌、腺体的活动。

11. 副神经（accessory nerve）：为运动神经，由两根组成，颅根起自延髓腹外侧缘，脊髓根由前部颈段脊髓腹侧柱发出的腹根分支组成，分布于喉、咽肌、胸头肌、斜方肌和臂头肌。

12. 舌下神经（hypoglossal nerve）：为运动神经，起自延髓的舌下神经核，自延髓腹侧下橄榄体的外侧缘发出，经舌下神经孔出颅腔，分布于舌肌和舌骨肌。

五、自主神经系统

自主神经系统（autonomic nervous system）又称植物神经系统（vegetative nervous system）或内脏神经系统（visceral nervous system），分布于内脏器官、血管和皮肤的平滑肌、心肌和腺体，分为交感神经（sympathetic nerve）和副交感神经（parasympathetic nerve）。

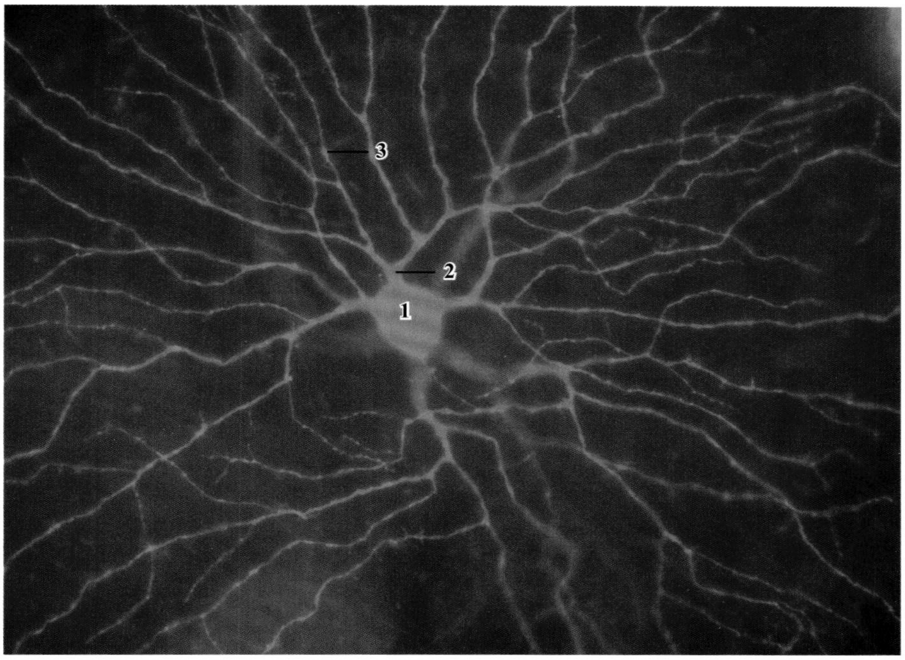

1- 胞体 soma
2- 树突 dendrite
3- 树突棘 dendritic spine

图 11-1　神经元（DiI 染色，高倍镜）
Figure 11-1　Neuron (DiI staining, higher power).

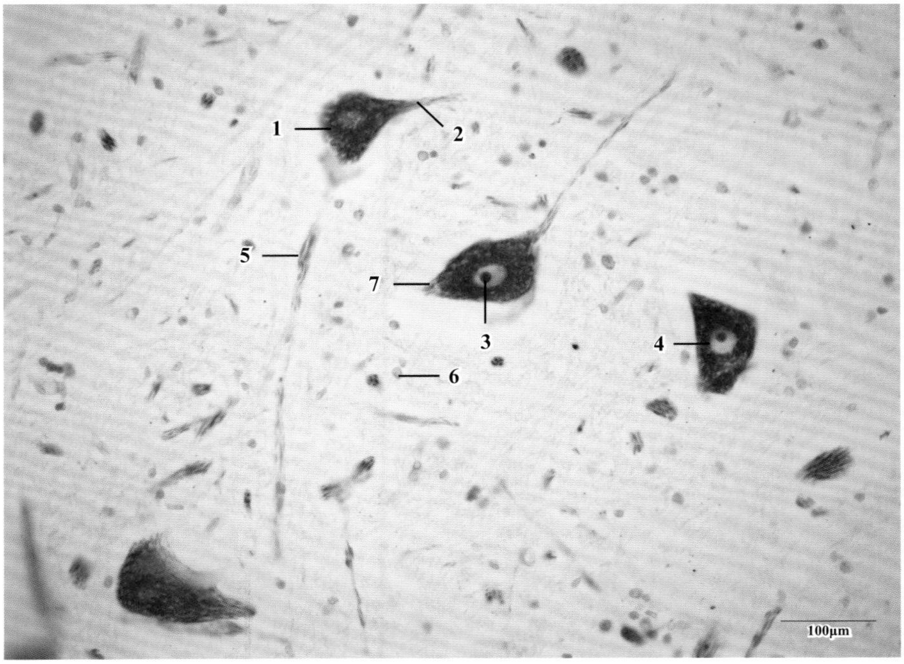

1- 尼氏体 nissl body
2- 树突 dendrite
3- 核仁 nucleolus
4- 细胞核 nucleus
5- 轴突 axon
6- 胶质细胞 glial cell
7- 轴丘 axon hillock

图 11-2　神经元（Nissl 染色，高倍镜）
Figure 11-2　Neuron (Nissl staining, higher power).

1- 轴丘 axon hillock
2- 胞体 soma
3- 树突 dendrite
4- 细胞核 nucleus
5- 核仁 nucleolus
6- 胶质细胞 glial cell

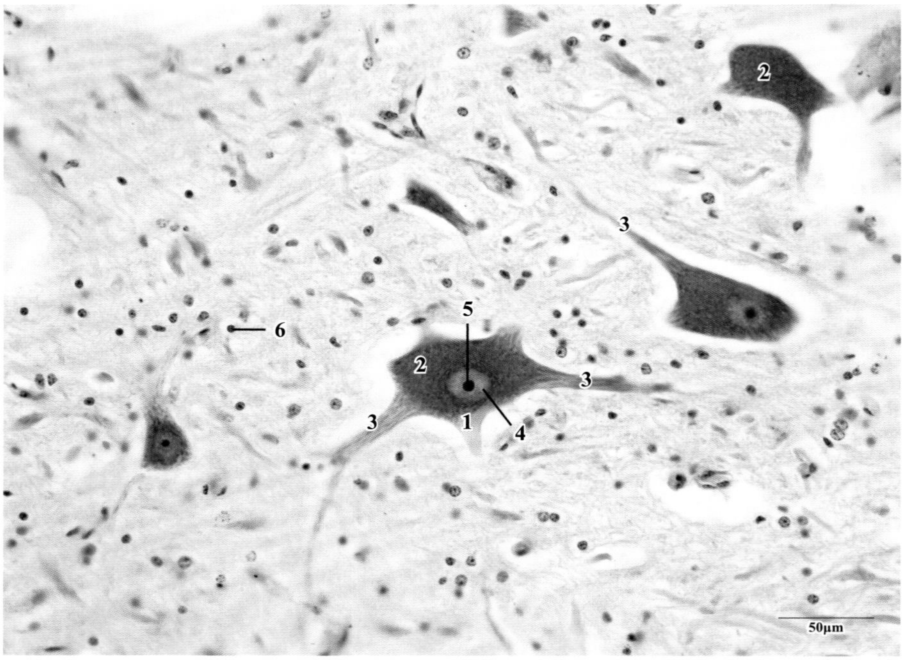

图 11-3 多级神经元（甲苯胺蓝染色，高倍镜）
Figure 11-3 Multipolar neuron (Toluidine blue staining, higher power).

1- 分子层 molecular layer
2- 浦肯野细胞 Purkinje cell
3- 颗粒层 granular layer
4- 髓质 medulla

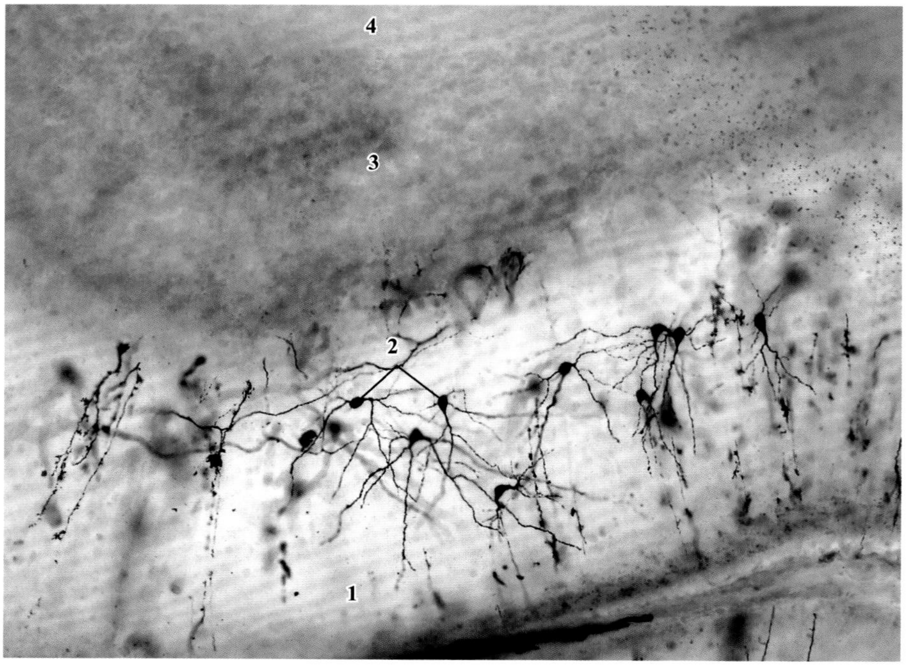

图 11-4 小脑（银染，高倍镜）
Figure 11-4 Cerebellum（silver staining, higher power）.

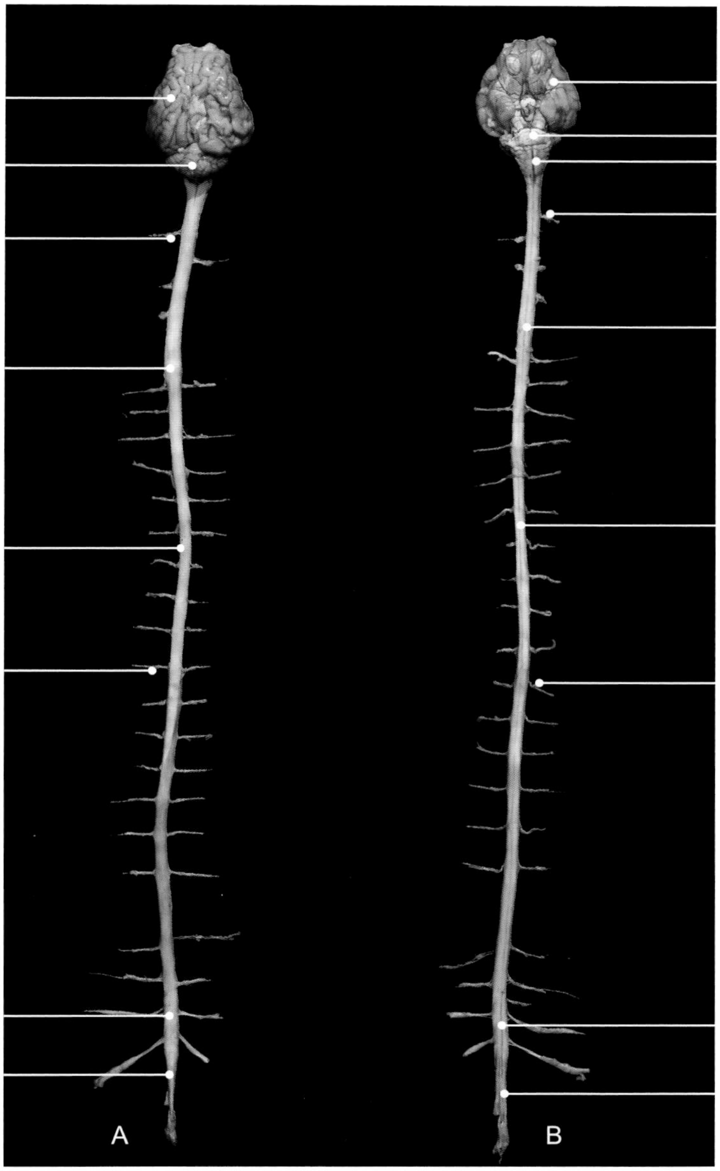

图 11-5 脑和脊髓（背面观和腹面观）
Figure 11-5 Brain and spinal cord (dorsal view and ventral view).

A- 背面观 dorsal view

1- 大脑 cerebrum
2- 小脑 cerebellum
3- 脊神经节 spinal ganglion
4- 颈膨大 cervical intumescence
5- 胸段脊髓 thoracic spinal cord
6- 脊神经根 spinal nerve root
7- 腰膨大 lumbar intumescence
8- 脊髓圆锥 medullary cone

B- 腹面观 ventral view

1- 大脑 cerebrum
2- 脑桥 pons
3- 延髓 medulla oblongata
4- 脊神经节 spinal ganglion
5- 颈膨大 cervical intumescence
6- 胸段脊髓 thoracic spinal cord
7- 脊神经根 spinal nerve root
8- 腰膨大 lumbar intumescence
9- 马尾 cauda equina

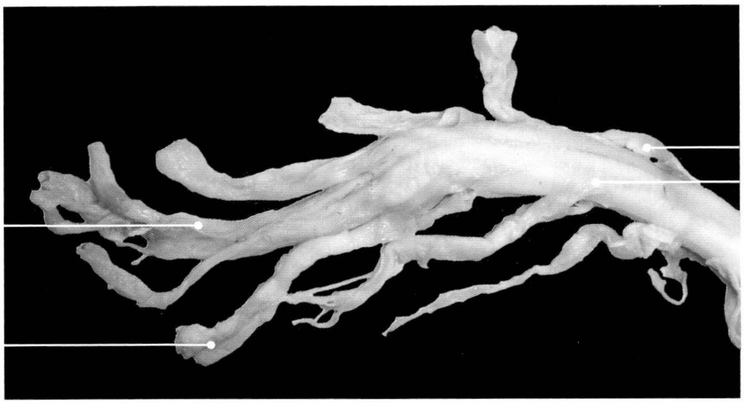

图 11-6　脊髓和马尾
Figure 11-6　Spinal cord and cauda equina.

1- 脊神经节 spinal ganglion
2- 腰膨大 lumbar intumescence
3- 脊神经根 spinal nerve root
4- 脊髓圆锥 medullary cone

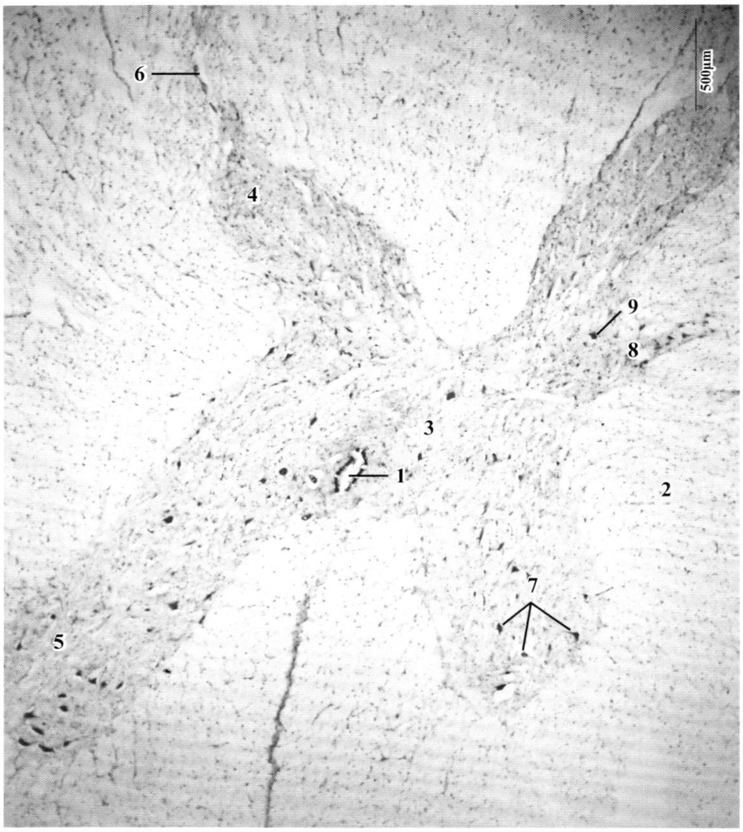

图 11-7　胸段脊髓横断面（Nissl 染色，低倍镜）
Figure 11-7　Cross section of the thoracic spinal cord （Nissl staining, lower power）.

1- 中央管 central canal
2- 白质 white matter
3- 灰质 gray matter
4- 背侧角 dorsal horn
5- 腹侧角 ventral horn
6- 背侧根 dorsal root
7- 运动神经元 motor neuron
8- 外侧角 lateral horn
9- 交感神经节前神经元 sympathetic preganglionic neuron

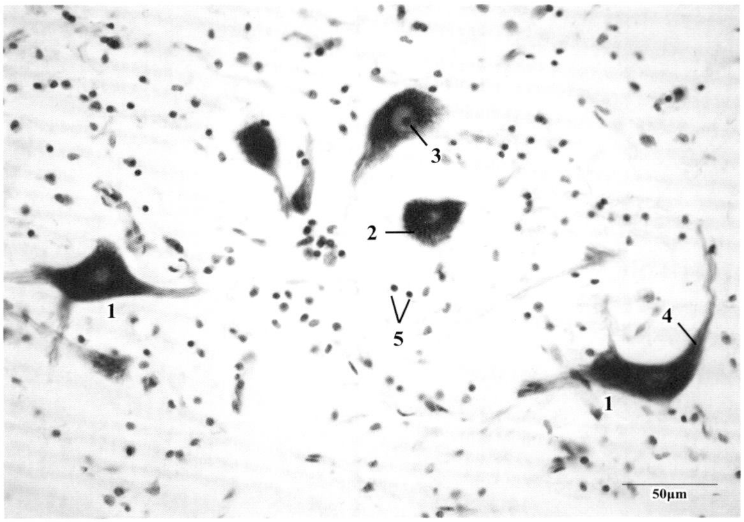

1- 多极神经元 multipolar neuron
2- 尼氏体 nissl body
3- 核仁 nucleolus
4- 树突 dendrite
5- 胶质细胞 glial cell

图 11-8 胸段脊髓神经元（Nissl 染色，高倍镜）
Figure 11-8 Neuron of the thoracic spinal cord (Nissl staining, higher power).

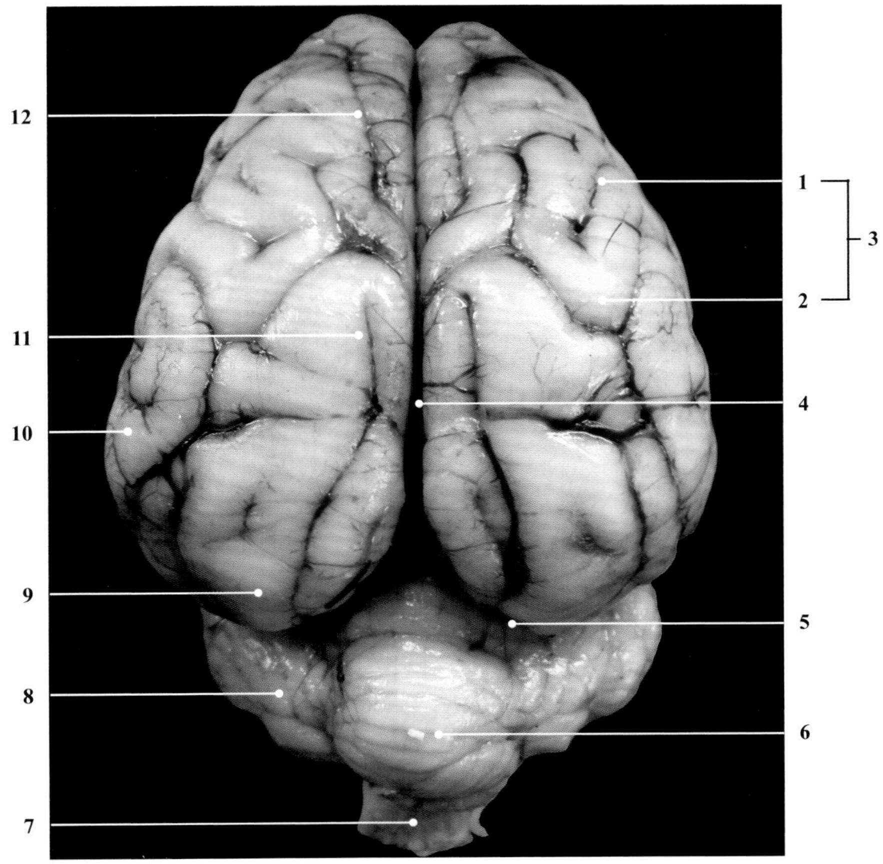

1- 脑沟 sulcus
2- 脑回 gyrus
3- 大脑皮质 cerebral cortex
4- 大脑纵裂 longitudinal cerebral fissure
5- 大脑横裂 transverse cerebral fissure
6- 小脑蚓部 vermis of cerebellum
7- 延髓 medulla oblongata
8- 小脑半球 cerebellar hemisphere
9- 枕叶 occipital lobe
10- 颞叶 temporal lobe
11- 顶叶 parietal lobe
12- 额叶 frontal lobe

图 11-9 脑的背侧面
Figure 11-9 Dorsal aspect of the brain.

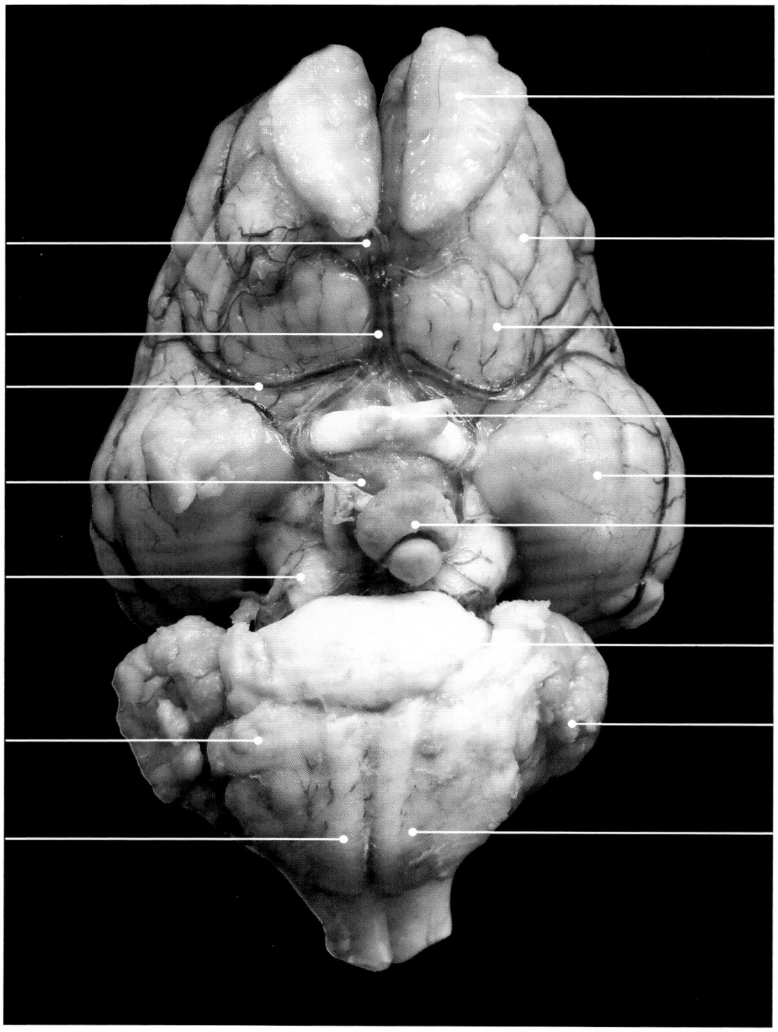

图 11-10　脑的腹侧面
Figure11-10　Ventral aspect of the brain.

1- 嗅球 olfactory bulb
2- 外侧嗅束 lateral olfactory tract
3- 嗅三角 olfactory trigone
4- 视交叉 optic chiasma
5- 梨状叶 piriform lobe
6- 垂体 hypophsis, pituitary gland
7- 脑桥 pons
8- 小脑半球 cerebellar hemisphere
9- 延髓 medulla oblongata
10- 锥体 pyramid
11- 斜方体 trapezoid body
12- 大脑脚 cerebral peduncle
13- 下丘脑 hypothalamus
14- 大脑中动脉 middle cerebral artery
15- 大脑前动脉 anterior cerebral artery
16- 内侧嗅束 medial olfactory tract

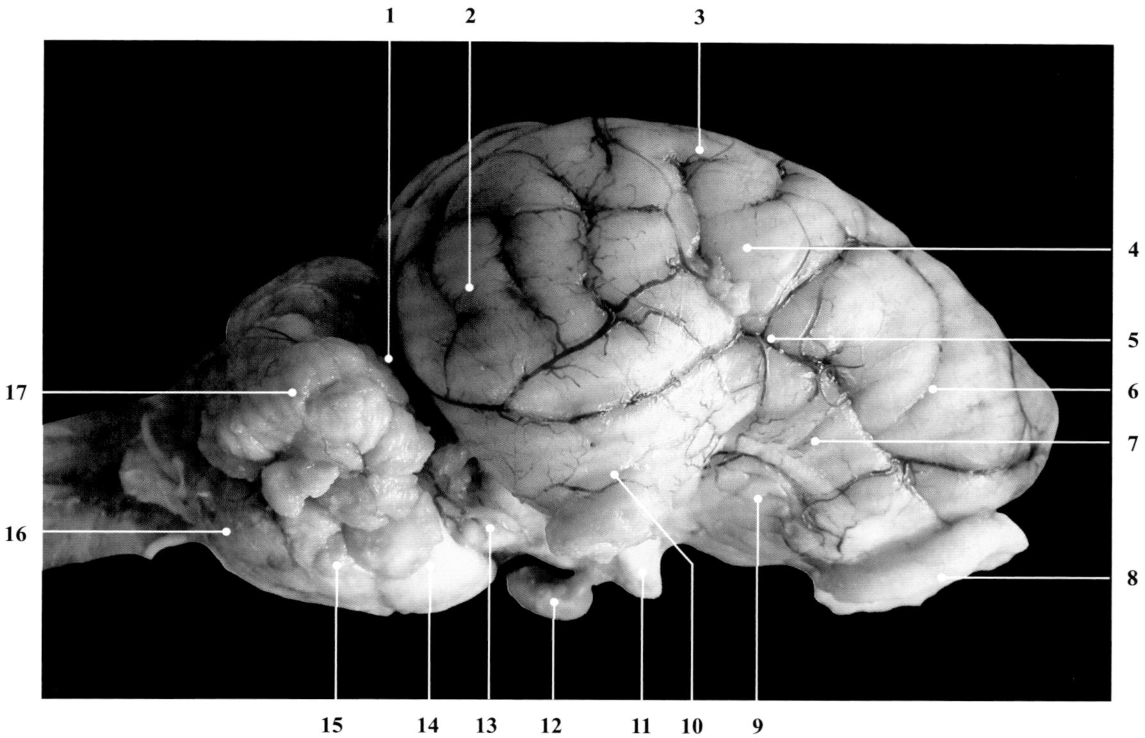

图 11-11 脑的外侧面
Figure 11-11 Lateral aspect of the brain.

1- 大脑横裂 transverse cerebral fissure
2- 枕叶 occipital lobe
3- 顶叶 parietal lobe
4- 颞叶 temporal lobe
5- 嗅沟 olfactory groove
6- 额叶 frontal lobe
7- 外侧嗅束 lateral olfactory tract
8- 嗅球 olfactory bulb
9- 嗅三角 olfactory trigone
10- 梨状叶 piriform lobe
11- 视交叉 optic chiasma
12- 垂体 hypophysis, pituitary gland
13- 大脑脚 cerebral peduncle
14- 脑桥 pons
15- 斜方体 trapezoid body
16- 延髓 medulla oblongata
17- 小脑半球 cerebellar hemisphere

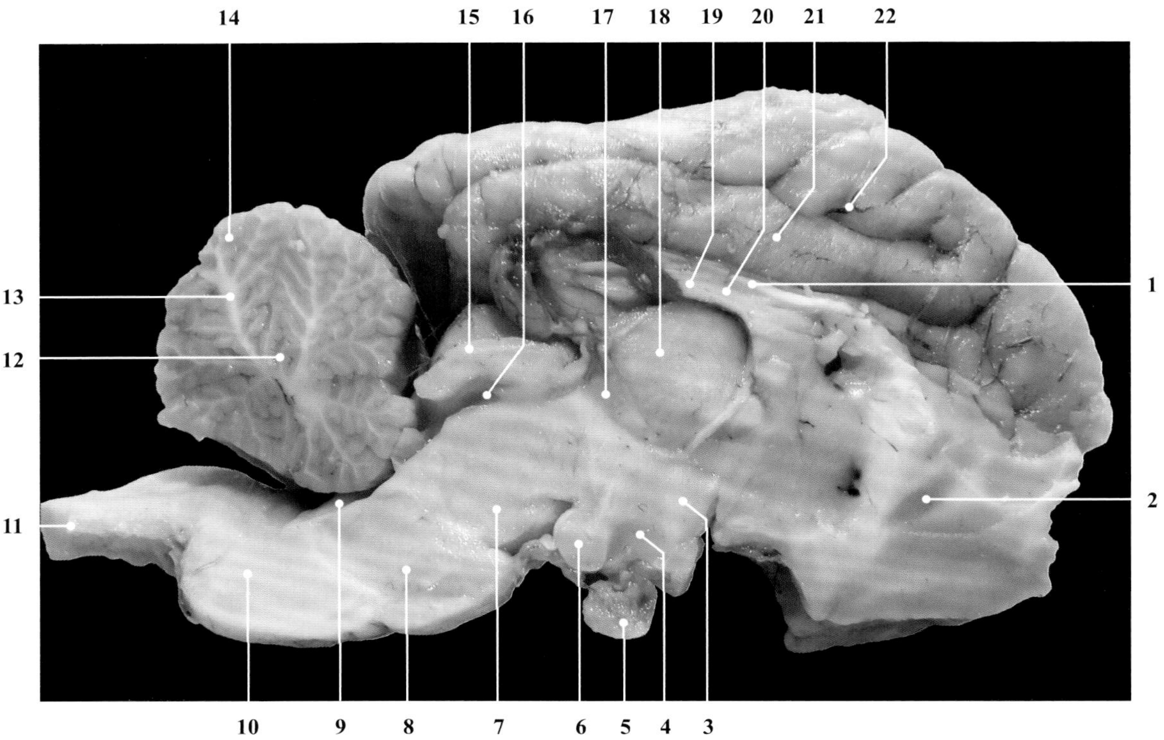

图 11-12　脑的矢状面
Figure 11-12　Median section of the brain.

1- 胼胝体 corpus callosum
2- 嗅脑 rhinencephalon
3- 下丘脑 hypothalamus
4- 漏斗 infundibulum
5- 垂体 hypophysis, pituitary gland
6- 乳头体 mamillary body
7- 大脑脚 cerebral peduncle
8- 脑桥 pons
9- 第 4 脑室 4th ventricle
10- 延髓 medulla oblongata
11- 脊髓 spinal cord
12- 小脑 cerebellum
13- 小脑树 cerebellar arbor
14- 小脑皮质 cerebellar cortex
15- 四叠体 quadrigeminal bodies
16- 中脑导水管 mesencephalic aqueduct
17- 第 3 脑室 3rd ventricle
18- 丘脑黏合部 interthalamic adhesion
19- 穹隆 fornix
20- 侧脑室 lateral ventricle
21- 扣带回 cingulate gyrus
22- 扣带沟 cingulate suleus

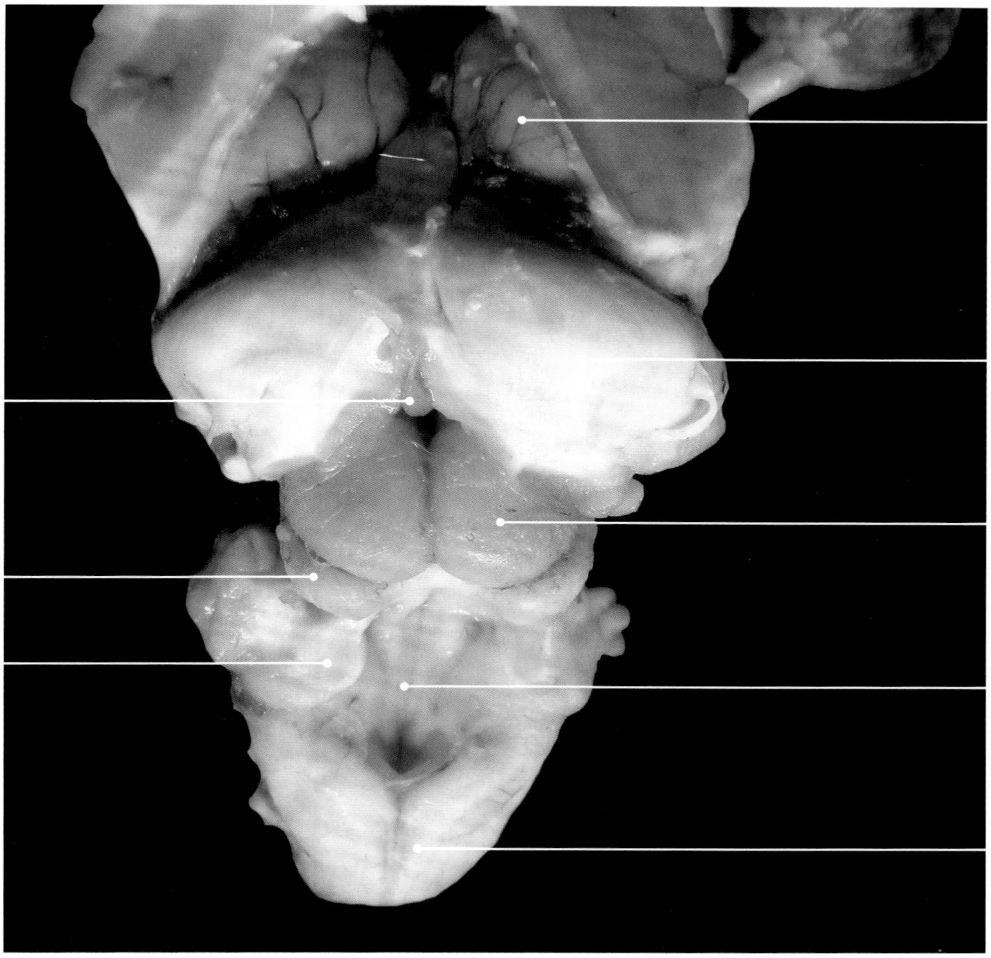

图 11-13 脑干背侧面
Figure 11-13 Dorsal aspect of the brain stem.

1- 纹状体 striatum
2- 海马 hippocampus
3- 前丘 rostral colliculus
4- 菱形窝 rhomboid fossa
5- 延髓 medulla oblongata
6- 小脑脚 cerebellar peduncle
7- 后丘 caudal colliculus
8- 松果体 pineal gland

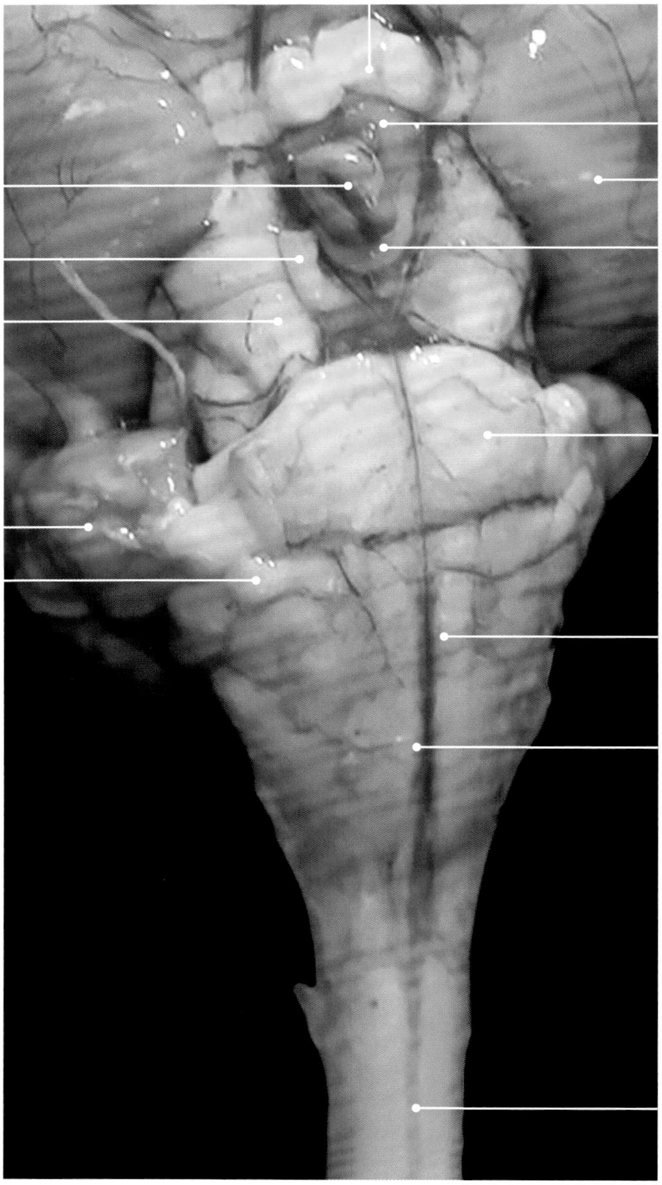

图 11-14　脑干腹侧面
Figure 11-14　Ventral aspect of the brain stem.

1- 视交叉 optic chiasma
2- 下丘脑 hypothalamus
3- 梨状叶 piriform lobe
4- 乳头体 mamillary body
5- 脑桥 pons
6- 锥体 pyramid
7- 延髓 medulla oblongata
8- 脊髓 spinal cord
9- 斜方体 trapezoid body
10- 小脑 cerebellum
11- 大脑脚 cerebral peduncle
12- 动眼神经根 root of oculomotor nerve
13- 垂体 hypophsis, pituitary gland

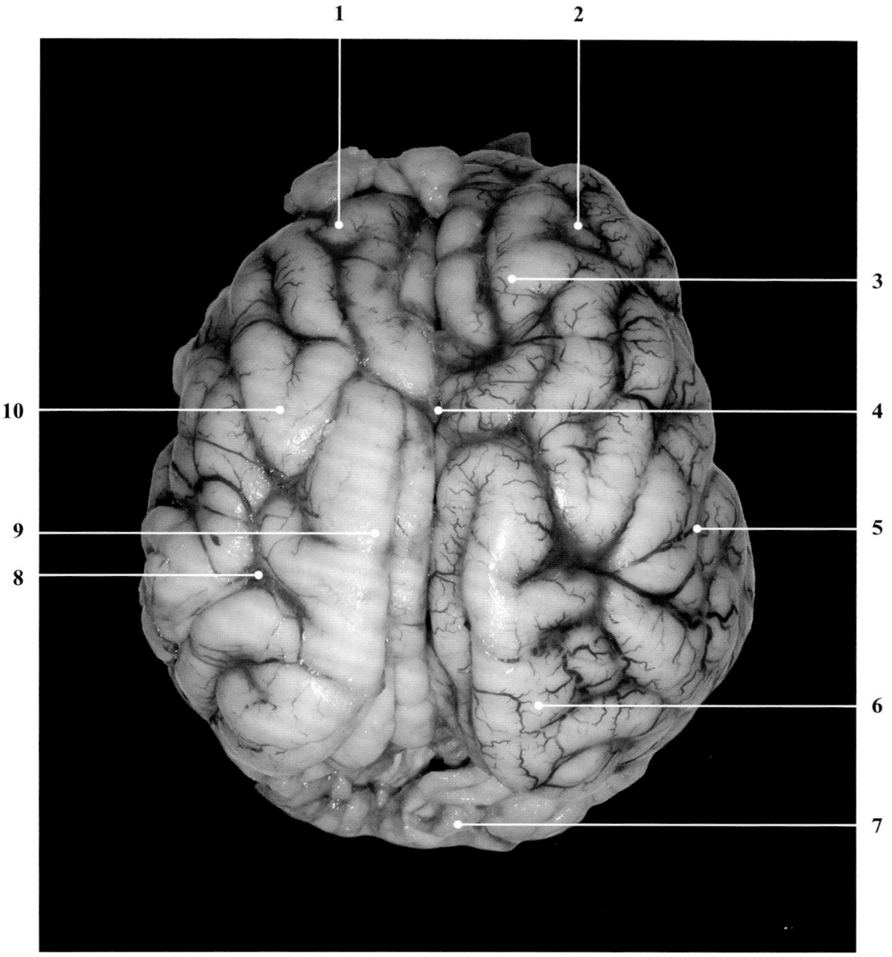

图 11-15 大脑背侧面
Figure 11-15 Dorsal aspect of the cerebrum.

1- 左侧大脑半球 left cerebral hemisphere
2- 右侧大脑半球 right cerebral hemisphere
3- 额叶 temporal lobe
4- 大脑纵裂 longitudinal cerebral fissure
5- 颞叶 frontal lobe
6- 枕叶 occipital lobe
7- 小脑 cerebellum
8- 脑沟 sulcus
9- 顶叶 parietal lobe
10- 脑回 gyrus

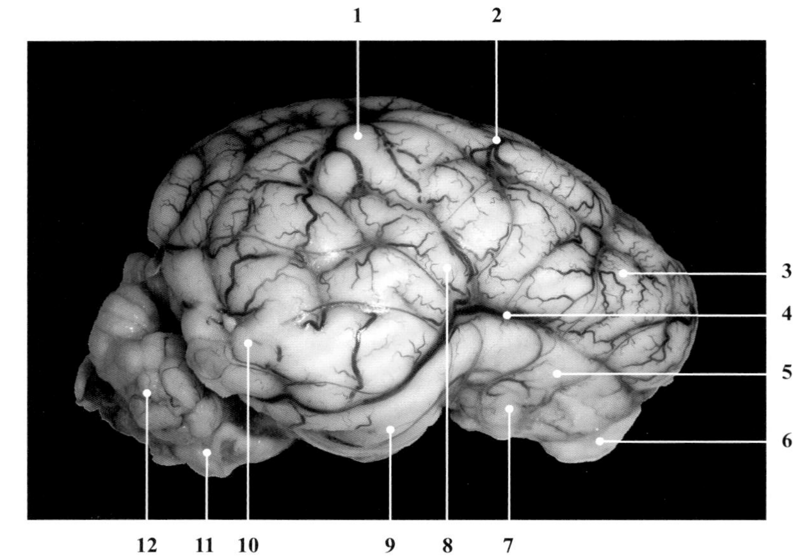

图 11-16　大脑半球外侧面
Figure 11-16　Lateral aspect of the cerebral hemisphere.

1- 顶叶 parietal lobe
2- 血管 blood vessel
3- 额叶 frontal lobe
4- 嗅沟 olfactory groove
5- 外侧嗅束 lateral olfactory tract
6- 嗅球 olfactory bulb
7- 嗅三角 olfactory trigone
8- 颞叶 temporal lobe
9- 梨状叶 piriform lobe
10- 枕叶 occipital lobe
11- 延髓 medulla oblongata
12- 小脑 cerebellum

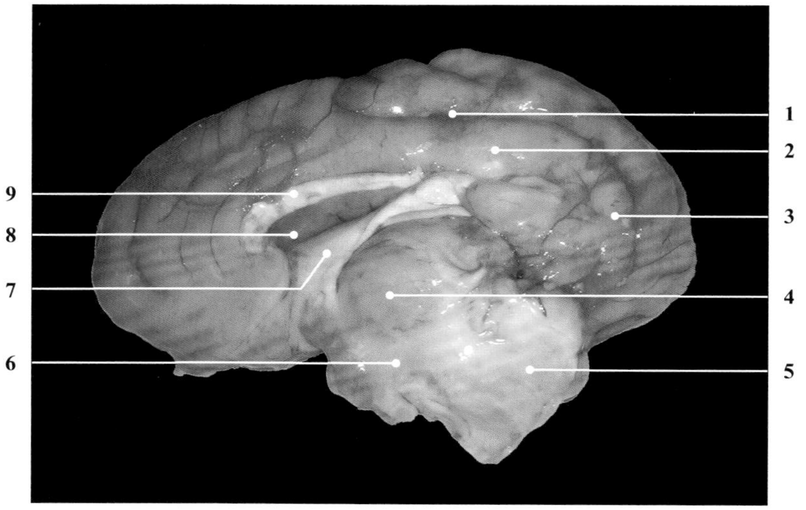

图 11-17　大脑半球内侧面
Figure 11-17　Median section of the cerebral hemisphere.

1- 扣带沟 cingulate suleus
2- 扣带回 cingulate gyrus
3- 大脑皮质 cerebral cortex
4- 丘脑黏合部 interthalamic adhesion
5- 中脑 mesencephalon
6- 下丘脑 hypothalamus
7- 穹隆 fornix
8- 侧脑室 lateral ventricle
9- 胼胝体 corpus callosum

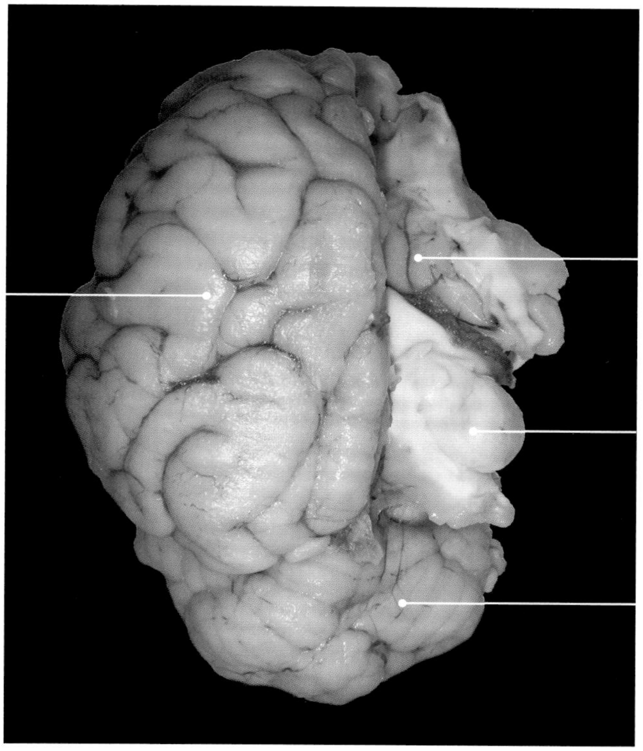

图 11-18 海马与纹状体（切除部分右侧大脑半球）
Figure11-18 Hippocampus and corpus striatum (Right cerebral hemisphere was removed).

1- 右侧纹状体 right striatum
2- 右侧海马 right hippocampus
3- 小脑 cerebellum
4- 左侧大脑半球 left cerebral hemisphere

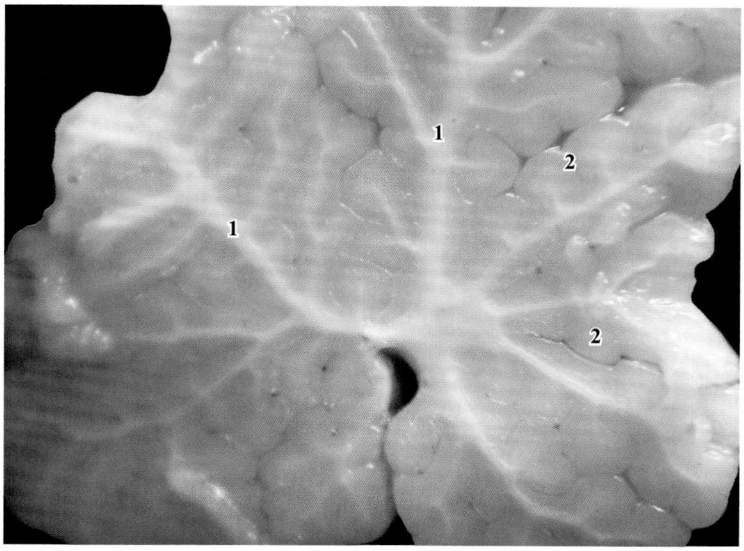

图 11-19 小脑矢状面
Figure 11-19 Sagittal section of the cerebellum.

1- 髓树 medullary arbor
2- 皮质 cortex

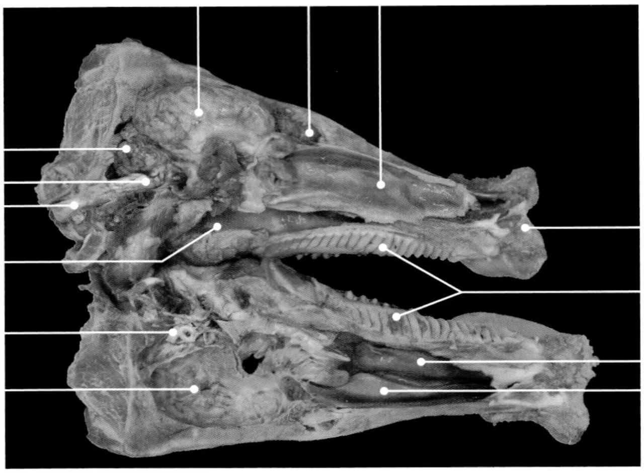

图 11-20　颅腔内脑
Figure 11-20　Brain in cranial cavity.

1- 大脑 cerebrum
2- 鼻旁窦 paranasal sinus
3- 鼻中隔 nasal septum
4- 吻突 rostral disc
5- 硬腭 hard palate
6- 下鼻甲 ventral nasal concha
7- 上鼻甲 dorsal nasal concha
8- 颅腔 cranial cavity
9- 内耳所在地 site of internal ear
10- 鼻后孔 posterior nasal apertures
11- 脊髓 spinal cord
12- 脑干 brain stem
13- 小脑 cerebellum

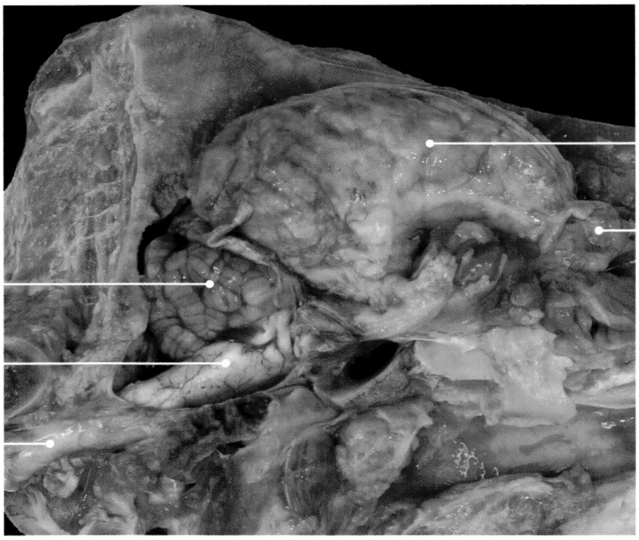

图 11-21　颅腔内脑的放大
Figure 11-21　Enlargement of the brain in cranial cavity.

1- 大脑和脑硬膜 cerebrum and encephalic dura mater
2- 嗅球 olfactory bulb
3- 脊髓 spinal cord
4- 延髓 medulla oblongata
5- 小脑 cerebellum

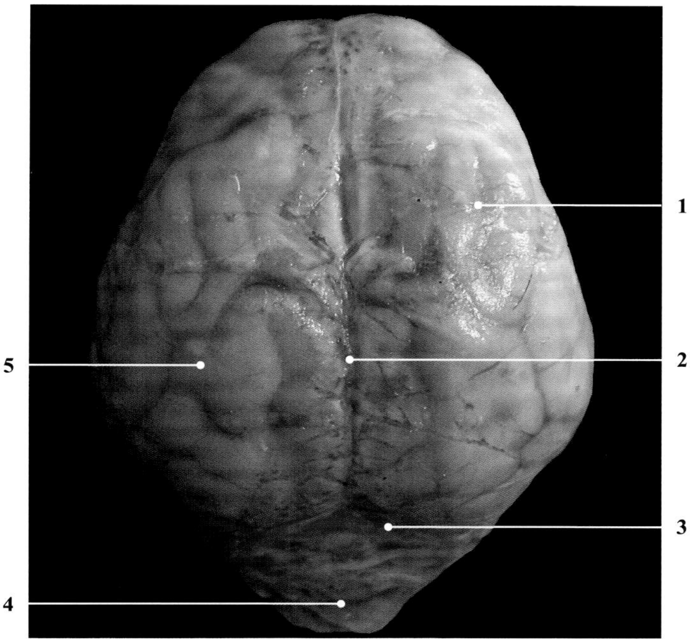

1- 脑硬膜 encephalic dura mater
2- 大脑纵裂（大脑镰） longitudinal cerebral fissure (cerebral falx)
3- 大脑横裂（小脑幕） transverse cerebral fissure (tentorium of cerebellum)
4- 小脑 cerebellum
5- 大脑 cerebrum

图 11-22 脑膜
Figure 11-22 Dura mater.

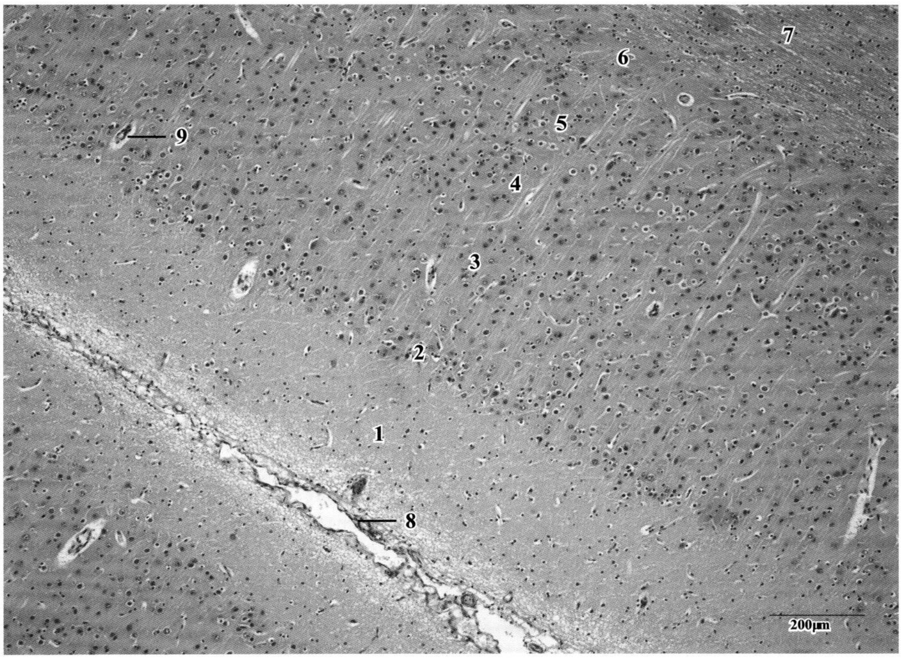

1- 分子层 molecular layer
2- 外颗粒层 external granular layer
3- 外锥体细胞层 external pyramidal layer
4- 内颗粒层 internal granular layer
5- 内锥体细胞层 internal pyramidal layer
6- 多形细胞层 multiform layer
7- 髓质 medulla
8- 脑软膜 encephalic pia mater
9- 血管 blood vessel

图 11-23 大脑皮层组织切片（HE 染色，低倍镜）
Figure11-23 Histological section of the cerebral cortex (HE staining, lower power).

1- 锥体细胞 pyramidal cell
2- 顶树突 apical dendrite
3- 锥体细胞核 nucleus of pyramidal cell
4- 颗粒细胞 granular cell
5- 梭形细胞 spindle cell

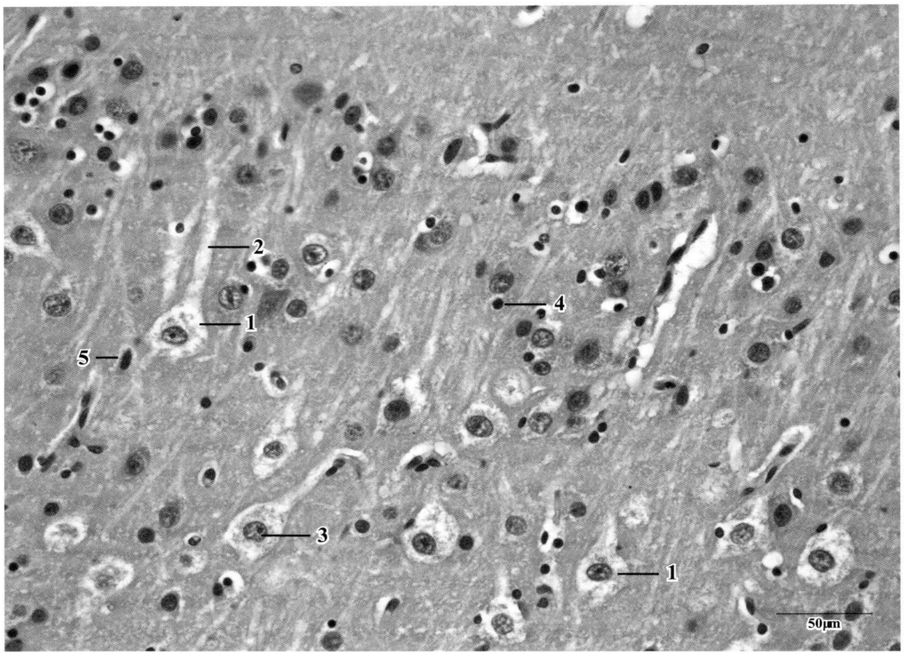

图 11-24　大脑皮层组织切片（HE 染色，高倍镜）
Figure 11-24　Histological section of the cerebral cortex (HE staining, higher power).

1- 海马沟 hippocampal sulcus
2- 齿状回颗粒层 granular layer, dentate gyrus
3- 多形细胞层 polymorphic layer
4- 海马缘 hippocampal edge
5- 分子层 molecular layer
6- 海马回 hippocampal gyrus
7- 腔隙层 stratum lacunosum
8- 辐射层 stratum radiatum
9- 锥体细胞层 pyramidal layer
10- 静脉 vein

图 11-25　海马组织切片（HE 染色，低倍镜）
Figure 11-25　Histological section of the hippocampe (HE staining, lower power).

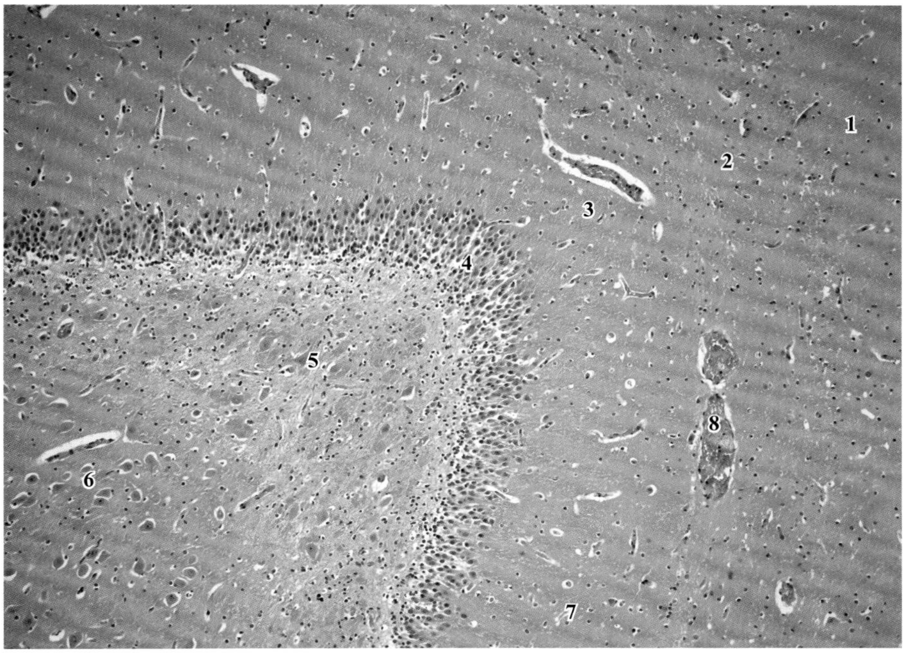

1- 锥体细胞层 pyramidal layer
2- 腔隙层 stratum lacunosum
3- 海马回 hippocampal gyrus
4- 齿状回颗粒层 granular layer, dentate gyrus
5- 多形细胞层 polymorphic layer
6- 海马缘 hippocampal edge
7- 分子层 molecular layer
8- 静脉 vein

图 11-26 海马组织切片（HE 染色，高倍镜）
Figure 11-26 Histological section of the hippocampus (HE staining, higher power).

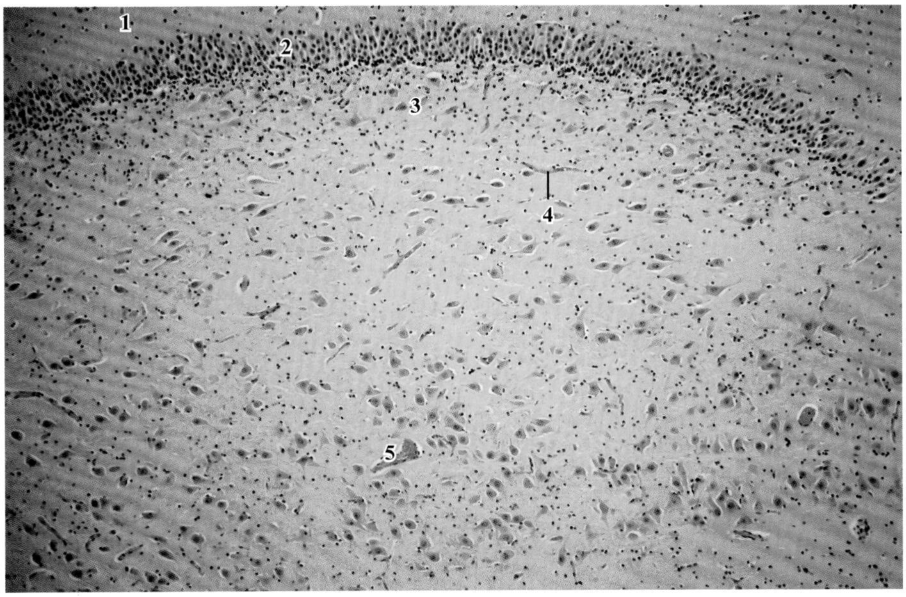

1- 齿状回分子层 molecular layer, dentate gyrus
2- 齿状回颗粒层 granular layer, dentate gyrus
3- 齿状回多形细胞层 polymorphic layer, dentate gyrus
4- 毛细血管 blood capillary
5- 静脉 vein

图 11-27 海马组织切片（HE 染色，高倍镜）
Figure 11-27 Histological section of the hippocampus (HE staining, higher power).

1- 齿状回分子层 molecular layer, dentate gyrus
2- 齿状回颗粒层 granule layer, dentate gyrus
3- 齿状回多形细胞层 polymorphic layer, dentate gyrus
4- 多形细胞 polymorphic layer
5- 颗粒细胞 granular cell
6- 胶质细胞 glial cell
7- 血管 blood vessel

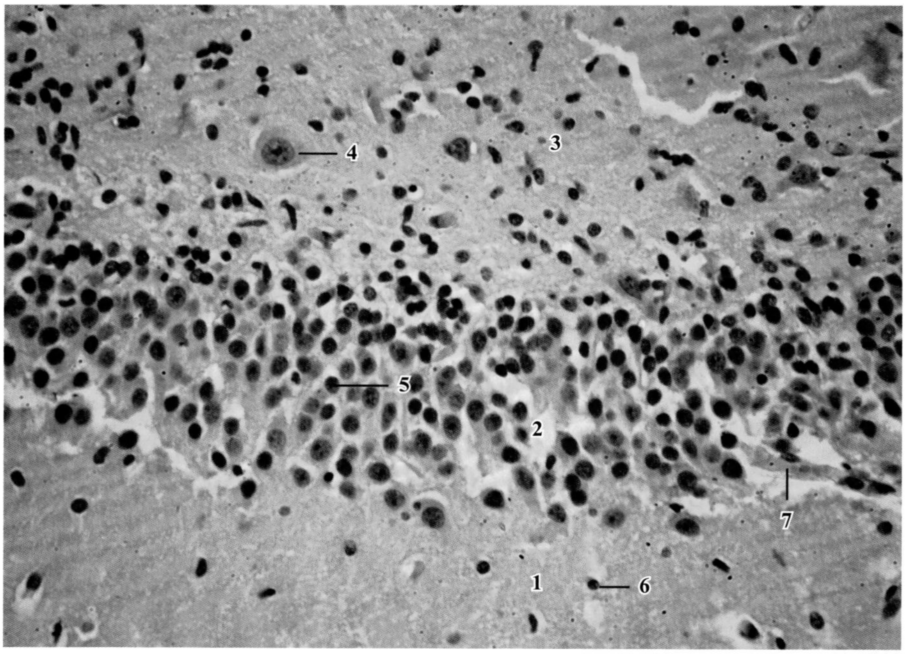

图 11-28　海马组织切片（HE 染色，高倍镜）
Figure 11-28　Histological section of the hippocampus (HE staining, higher power).

1- 脑软膜 encephalic pia mater
2- 分子层 molecular layer
3- 浦肯野细胞 Purkinje cell
4- 颗粒层 granular layer
5- 星形细胞 astrocyte

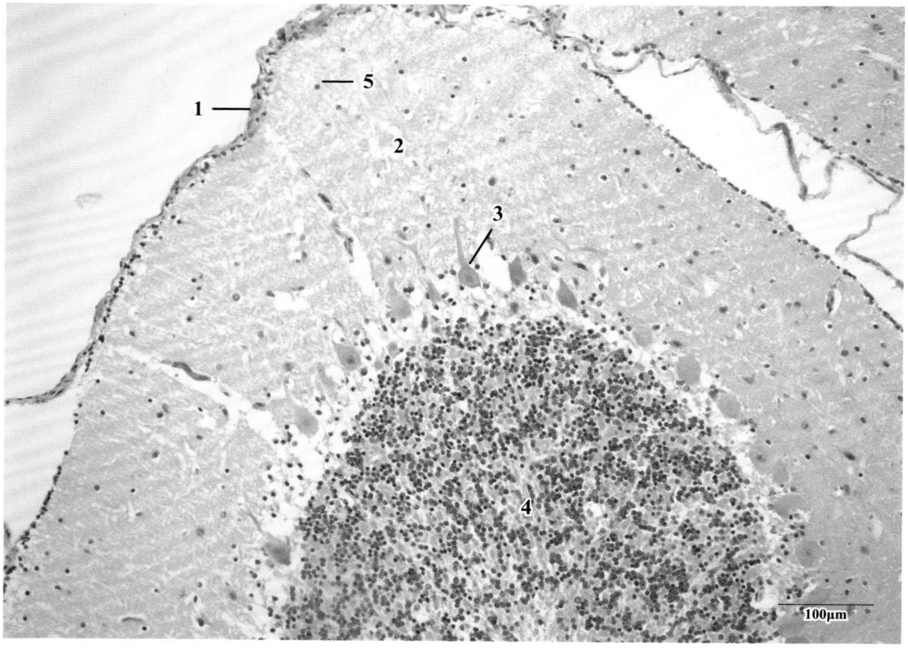

图 11-29　小脑皮质组织切片（HE 染色，高倍镜）
Figure 11-29　Histological section of the cerebellar cortex (HE staining, higher power).

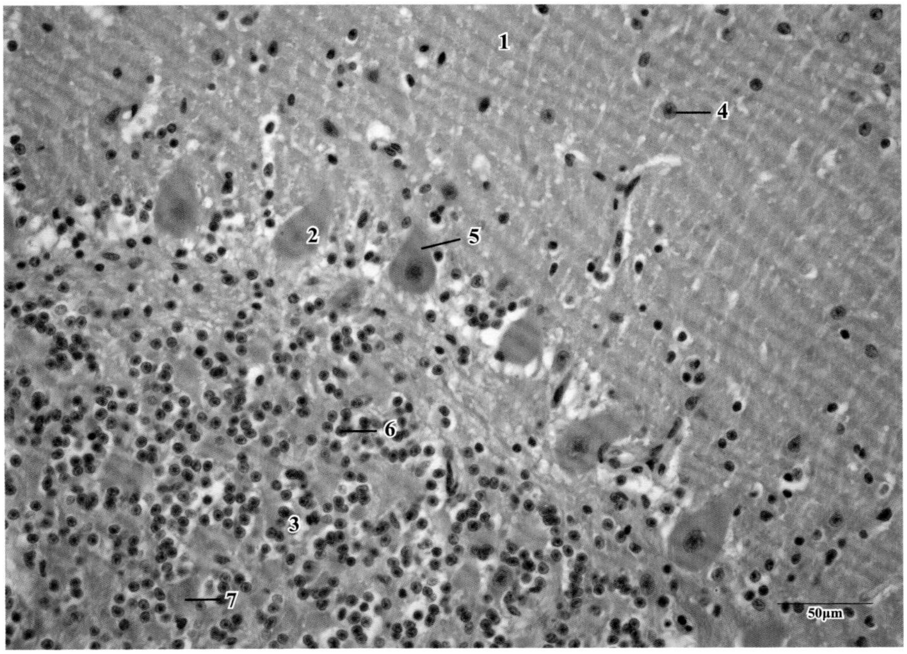

1- 分子层 molecular layer
2- 浦肯野细胞层 Purkinje cell layer
3- 颗粒层 granular layer
4- 篮状细胞 basket cell
5- 浦肯野细胞 Purkinje cell
6- 颗粒细胞 granular cell
7- 高尔基细胞 Golgi cell

图 11-30　小脑皮质组织切片（HE 染色，高倍镜）
Figure11-30　Histological section of the cerebellar cortex (HE staining, higher power).

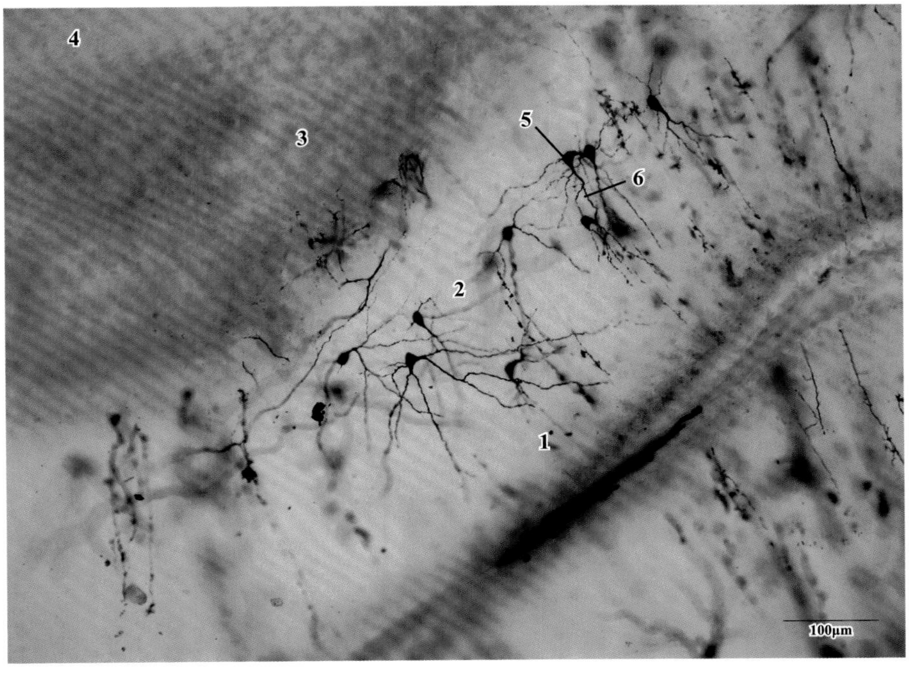

1- 分子层 molecular layer
2- 浦肯野细胞层 Purkinje cell layer
3- 颗粒层 granular layer
4- 小脑髓质 medulla of cerebellum
5- 浦肯野细胞 Purkinje cell
6- 树突 dendrite

图 11-31　小脑皮质组织切片（银染，高倍镜）
Figure 11-31　Histological section of the cerebellar cortex (Silver staining, higher power).

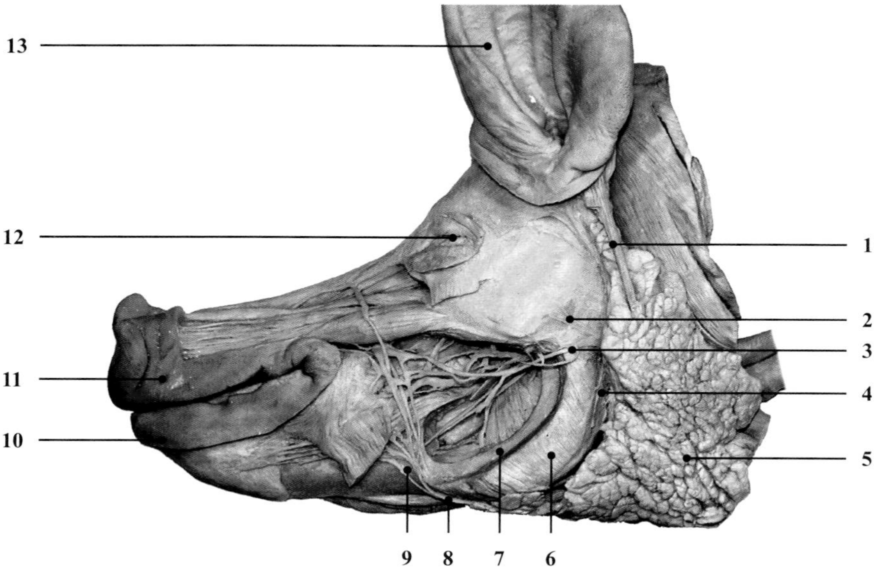

图 11-32　面神经
Figure 11-32　Facial nerve.

1- 腮耳肌 parotidoauricular muscle
2- 耳睑神经 auriculo-palpebral nerve
3- 面神经颊背侧支 dorsal buccal branch of the facial nerve
4- 面神经颈支 cervical branch of the facial nerve
5- 腮腺 parotid gland
6- 咬肌 masseter muscle
7- 下颌骨 mandible
8- 面神经颊腹侧支 ventral buccal branch of the facial nerve
9- 下唇神经 lower labial nerve
10- 下唇 lower lip
11- 上唇 upper lip
12- 眼 eye
13- 耳 ear

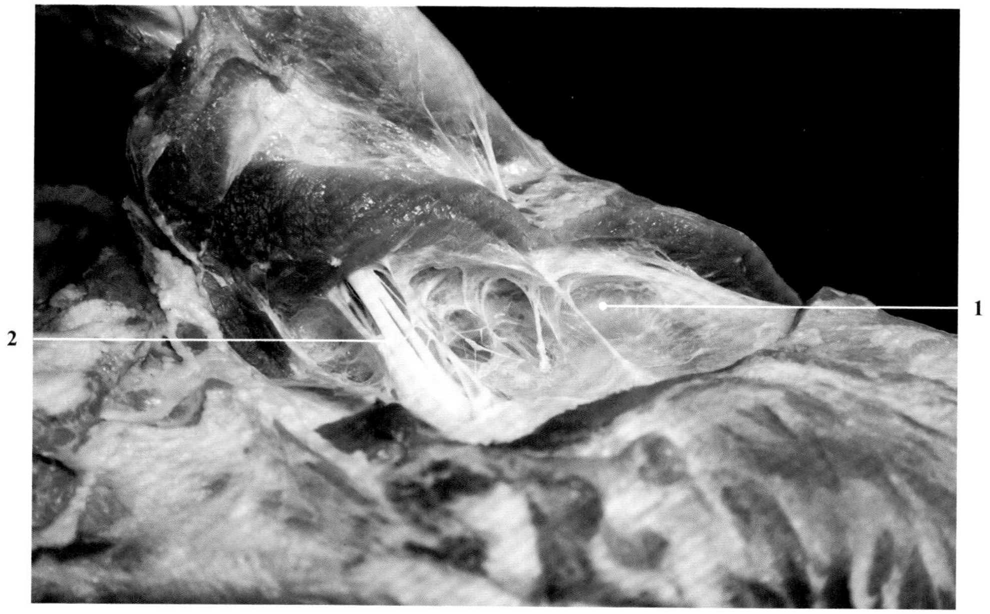

图 11-33　腋窝与臂神经丛
Figure 11-33　Axilla and brachial plexus.

1- 腋窝 axilla
2- 臂神经丛 brachial plexus

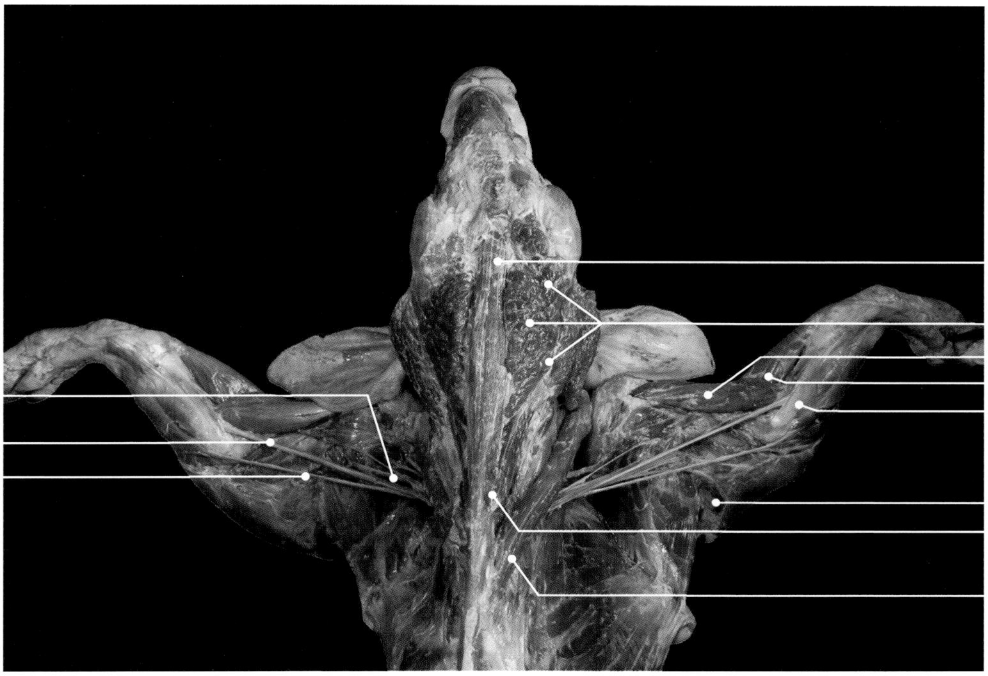

图 11-34 臂神经丛 -1
Figure 11-34 Brachial plexus-1.

1- 胸骨甲状舌骨肌 sterno-thyrohyoid muscle
2- 腮腺 parotid gland
3- 臂二头肌 biceps muscle of the forearm
4- 腕桡侧伸肌 radial extensor muscle of the carpus
5- 腕桡侧屈肌 radial flexor muscle of the carpus
6- 臂三头肌 triceps muscle of the forearm
7- 胸头肌 sternocephalic muscle
8- 斜角肌 scalene muscle
9- 尺神经 ulnar nerve
10- 正中神经 median nerve
11- 桡神经 radial nerve

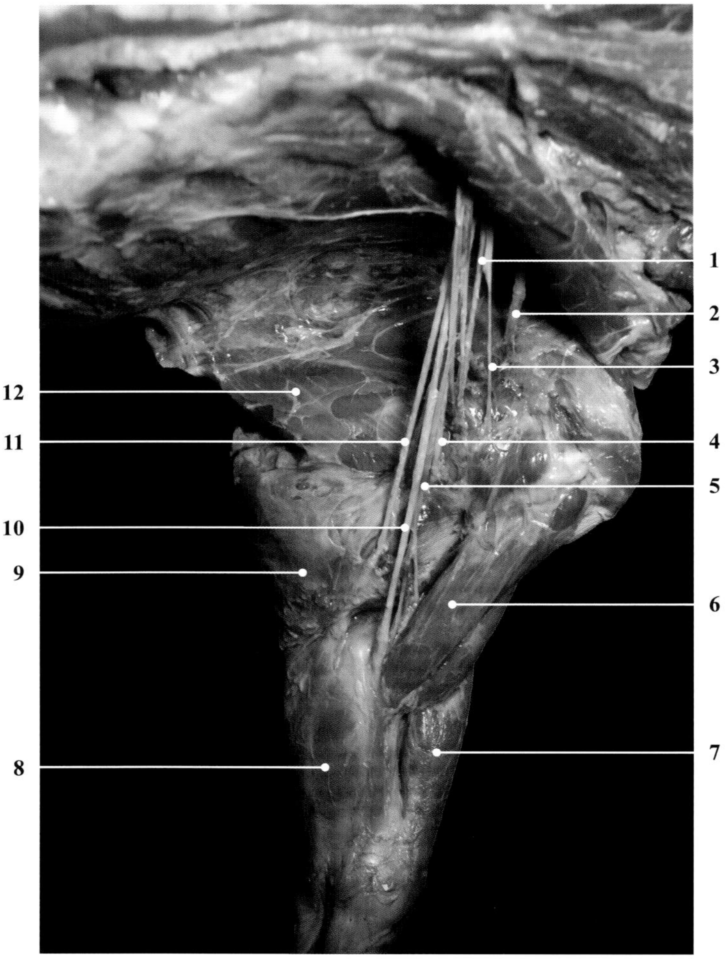

图 11-35　臂神经丛 -2（左前肢内侧面）
Figure 11-35　Brachial plexus-2 (medial aspect of left forelimb).

1- 腋神经 axillary nerve
2- 肩胛上神经 suprascapular nerve
3- 肩胛下神经 subscapular nerve
4- 桡神经 radial nerve
5- 臂动脉 brachial artery
6- 臂二头肌 biceps muscle of the forearm
7- 腕桡侧伸肌 radial extensor muscle of the carpus
8- 腕尺侧屈肌 ulnar flexor muscle of the carpus
9- 前臂筋膜张肌 tensor muscle of the antebrachial fascia
10- 正中神经 median nerve
11- 尺神经 ulnar nerve
12- 大圆肌 major teres muscle

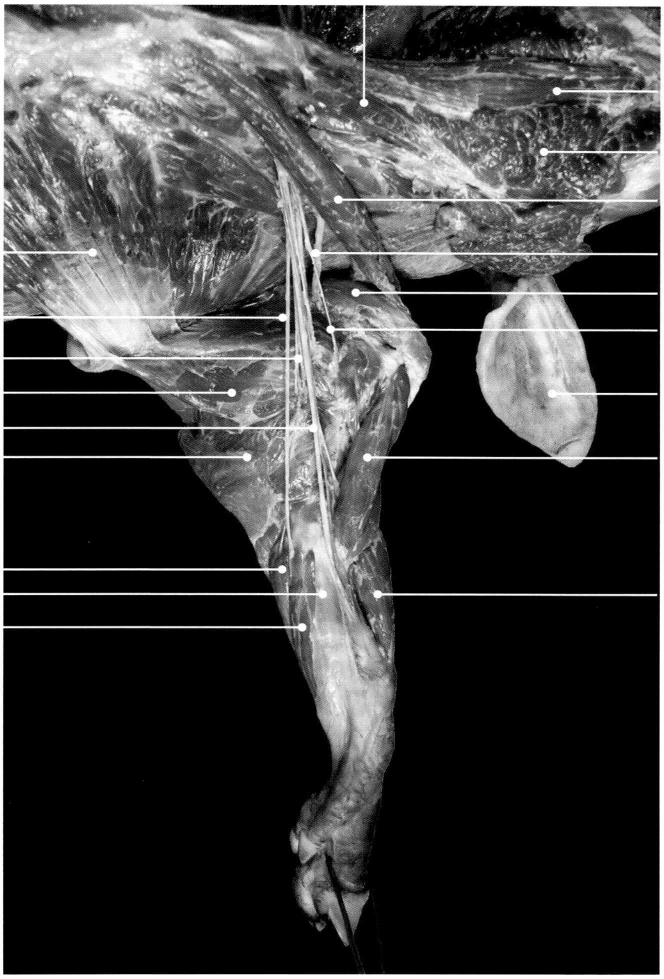

图 11-36　臂神经丛 -3（左前肢内侧面）
Figure 11-36　Brachial plexus-3 (medial aspect of left forelimb).

1- 胸头肌 sternocephalic muscle
2- 胸骨甲状舌骨肌 sterno-thyrohyoid muscle
3- 腮腺 parotid gland
4- 胸浅肌 superficial pectoral muscle
5- 肩胛上神经 suprascapular nerve
6- 冈上肌 supraspinous muscle
7- 腋神经 axillary nerve
8- 耳 ear
9- 臂二头肌 biceps muscle of the forearm
10- 腕桡侧伸肌 radial extensor muscle of the carpus
11- 腕尺侧屈肌 ulnar flexor muscle of the carpus
12- 腕桡侧屈肌 radial flexor muscle of the carpus
13- 指浅屈肌 superficial digital flexor muscle
14- 前臂筋膜张肌 tensor muscle of the antebrachial fascia
15- 正中神经 median nerve
16- 肩胛下肌 subscapular muscle
17- 桡神经 radial nerve
18- 尺神经 ulnar nerve
19- 胸腹侧锯肌 thoracic part of the ventral serrate muscle

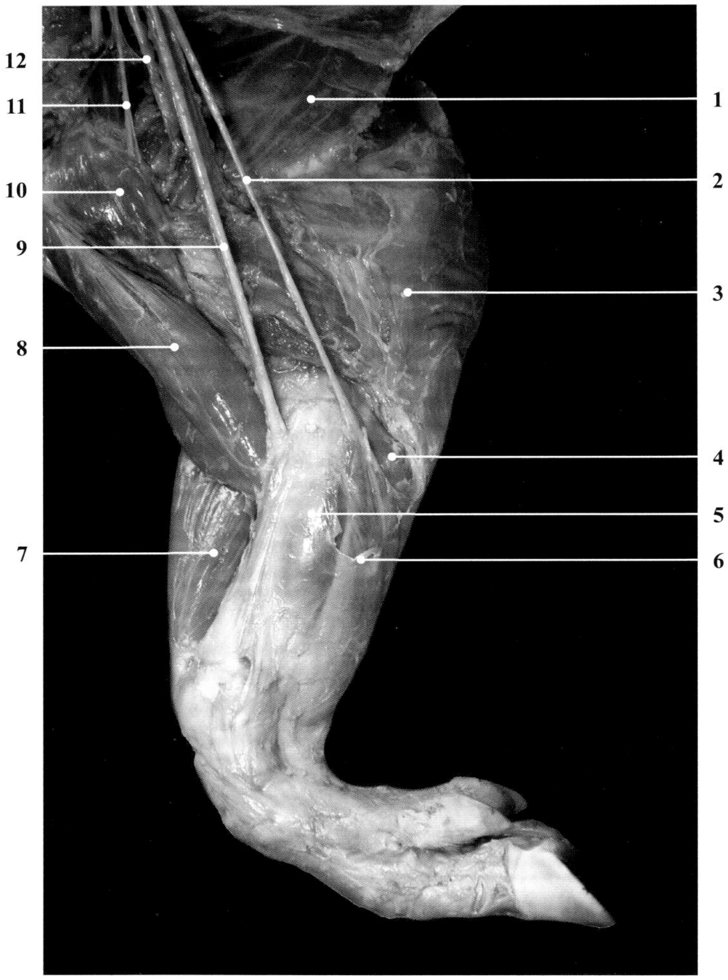

图 11-37　臂神经丛 -4（右前肢内侧面）
Figure 11-37　Brachial plexus-4 (medial aspect of right forelimb).

1- 大圆肌 major teres muscle
2- 尺神经 ulnar nerve
3- 前臂筋膜张肌 tensor muscle of the antebrachial fascia
4- 指浅屈肌 superficial digital flexor muscle
5- 腕桡侧屈肌 radial flexor muscle of the carpus
6- 腕尺侧屈肌 ulnar flexor muscle of the carpus
7- 腕桡侧伸肌 radial extensor muscle of the carpus
8- 臂二头肌 biceps muscle of the forearm
9- 正中神经 median nerve
10- 喙臂肌 coracobrachial muscle
11- 肌皮神经 musculocutaneous nerve
12- 桡神经 radial nerve

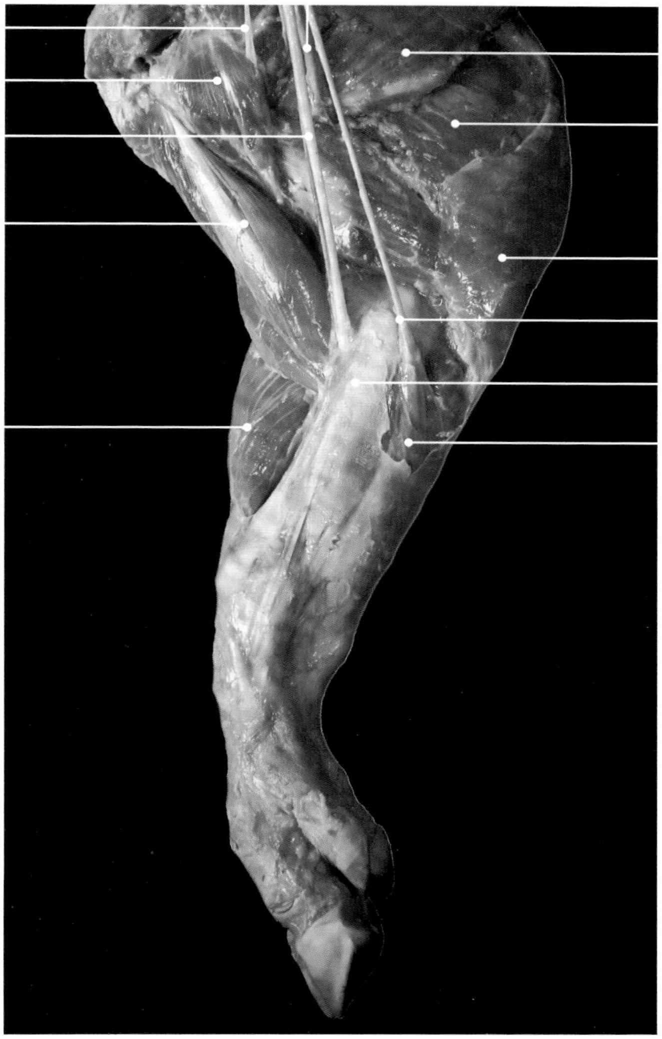

图 11-38 臂神经丛 -5（右前肢内侧面）
Figure 11-38 Brachial plexus-5 (medial aspect of right forelimb).

1- 桡神经 radial nerve
2- 大圆肌 major teres muscle
3- 臂三头肌 triceps muscle of the forearm
4- 前臂筋膜张肌 tensor muscle of the antebrachial fascia
5- 尺神经 ulnar nerve
6- 腕桡侧屈肌 radial flexor muscle of the carpus
7- 腕尺侧屈肌 ulnar flexor muscle of the carpus
8- 腕桡侧伸肌 radial extensor muscle of the carpus
9- 臂二头肌 biceps muscle of the forearm
10- 正中神经 median nerve
11- 喙臂肌 coracobrachial muscle
12- 肌皮神经 musculocutaneous nerve

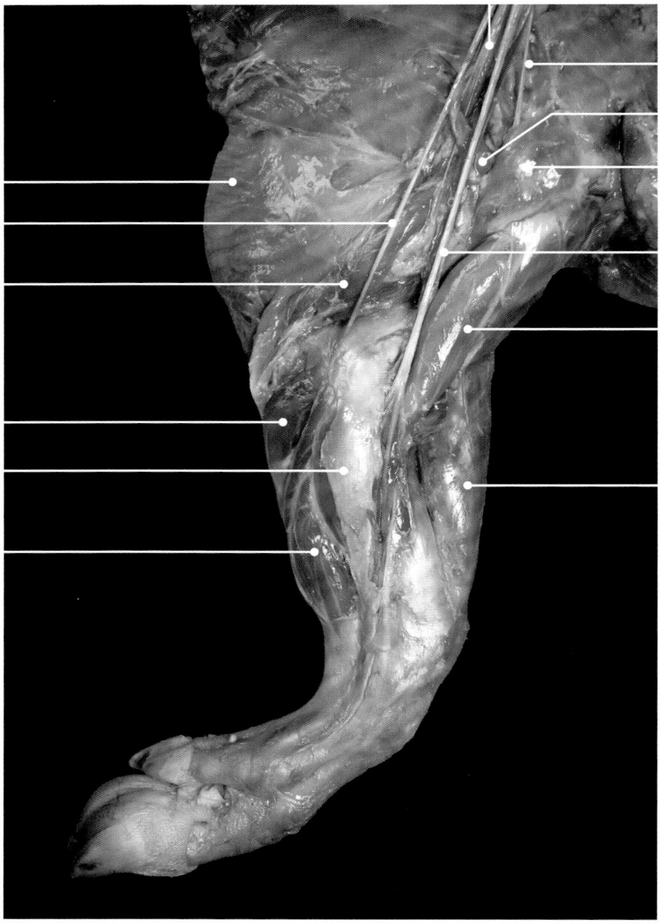

图 11-39 臂神经丛 -6（左前肢内侧面）
Figure 11-39 Brachial plexus-6 (medial aspect of left forelimb).

1- 桡神经 radial nerve
2- 肌皮神经 musculocutaneous nerve
3- 第 1 肋腋淋巴结 primary costal axillary lymph node
4- 喙臂肌 coracobrachial muscle
5- 正中神经 median nerve
6- 臂二头肌 biceps muscle of the forearm
7- 腕桡侧伸肌 radial extensor muscle of the carpus
8- 腕尺侧屈肌 ulnar flexor muscle of the carpus
9- 腕桡侧屈肌 radial flexor muscle of the carpus
10- 指浅屈肌 superficial digital flexor muscle
11- 臂三头肌内侧头 medial head, triceps muscle of the forearm
12- 尺神经 ulnar nerve
13- 前臂筋膜张肌 tensor muscle of the antebrachial fascia

图 11-40 腹壁神经
Figure 11-40 Abdominal nerve.

1- 髂腹股沟神经 ilioguinal nerve
2- 生殖股神经 genitofemoral nerve
3- 腹横肌 transverse abdominal muscle
4- 腹直肌 straight abdominal muscle
5- 肋腹神经 costoabdominal muscle
6- 髂腹下神经 iliohypogastric nerve

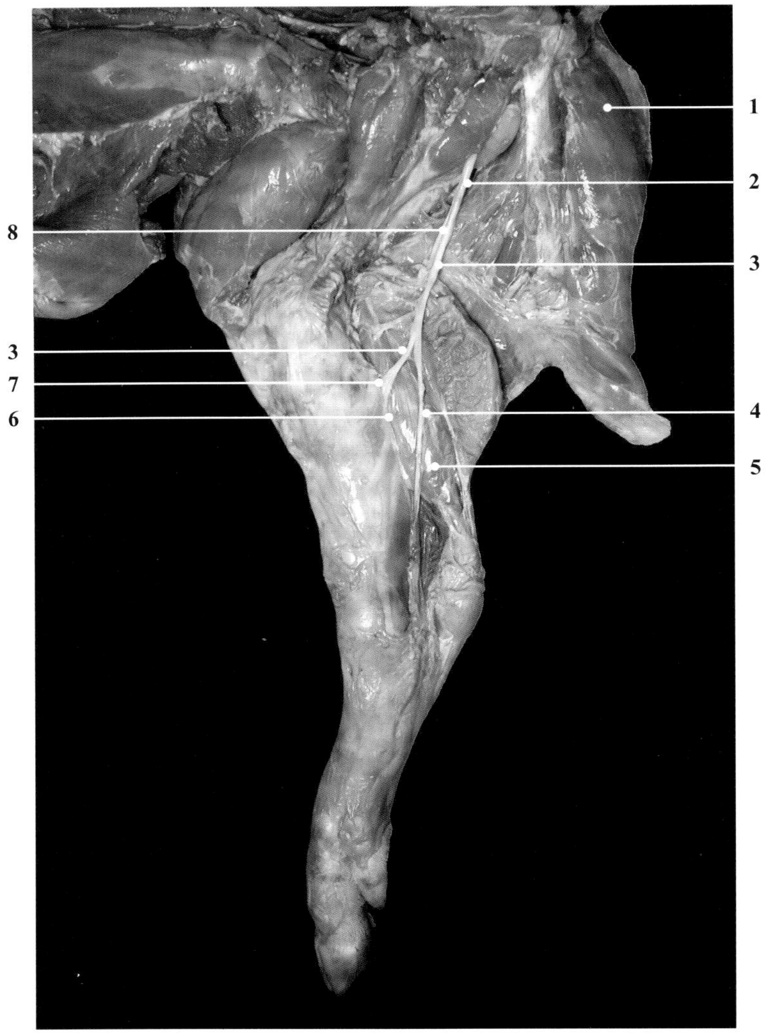

图 11-41　左后肢坐骨神经（外侧观）
Figure 11-41　Sciatic nerve of left hindlimb (lateral view).

1- 半腱肌 semitendinous muscle
2- 坐骨神经 sciatic nerve
3- 腓总神经 common peroneal nerve
4- 小腿后皮神经 caudal cutaneous sural nerve
5- 腓肠肌 gastrocnemius muscle
6- 腓浅神经 superficial fibular nerve
7- 腓深神经 deep fibular nerve
8- 胫神经 tibial nerve

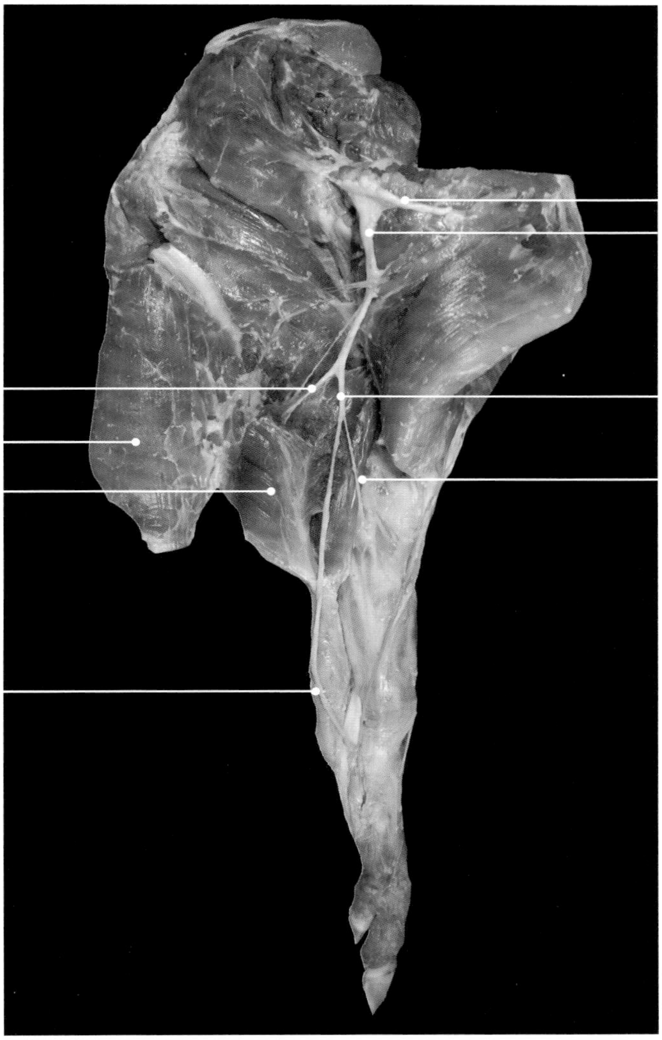

图 11-42 右后肢坐骨神经 -1 （外侧观）
Figure 11-42 Sciatic nerve of right hindlimb-1 (lateral view).

1- 臀前神经 cranial gluteal nerve
2- 坐骨神经 sciatic nerve
3- 胫神经 tibial nerve
4- 小腿外侧皮神经 lateral cutaneous sural nerve
5- 小腿后皮神经 caudal cutaneous sural nerve
6- 腓肠肌 gastrocnemius muscle
7- 半腱肌 semitendinous muscle
8- 坐骨神经肌支 muscular branch of sciatic nerve

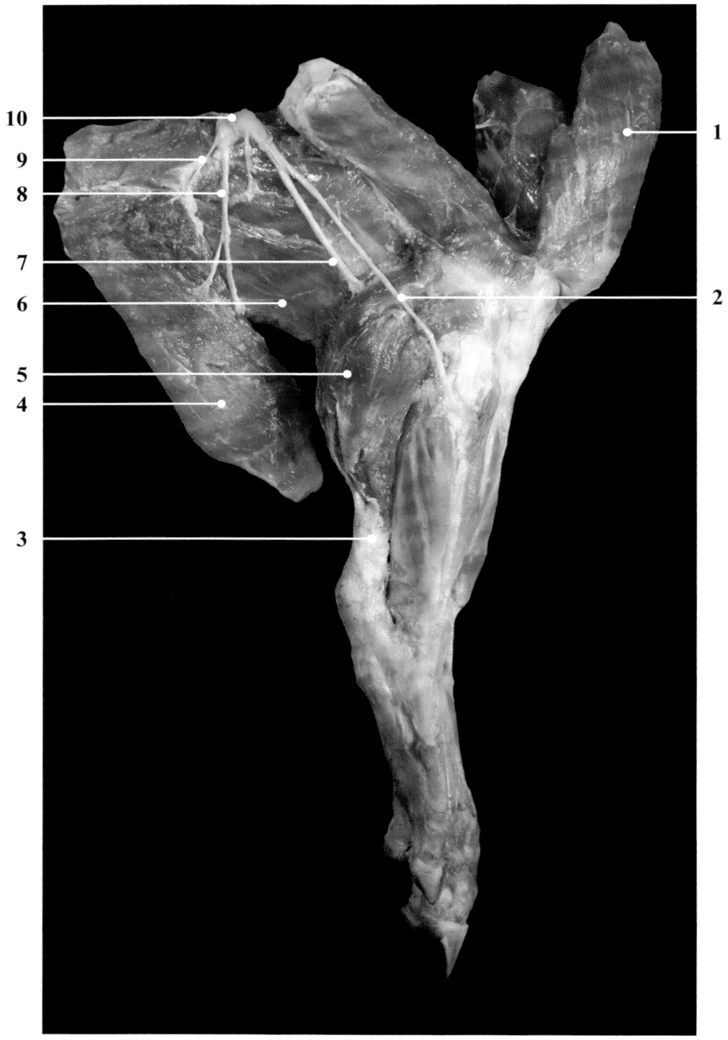

图 11-43　右后肢坐骨神经 -2（外侧观）
Figure 11-43　Sciatic nerve of right hindlimb-2 (lateral view).

1- 股四头肌 quadriceps muscle of the thigh
2- 腓总神经 common peroneal nerve
3- 跟总腱 common calcaneal tendon
4- 半腱肌 semitendinous muscle
5- 腓肠肌 gastrocnemius muscle
6- 半膜肌 semimembranous muscle
7- 胫神经 tibial nerve
8- 坐骨神经肌支 muscular branch of sciatic nerve
9- 股后皮神经 caudal femoral cutaneous nerve
10- 坐骨神经 sciatic nerve

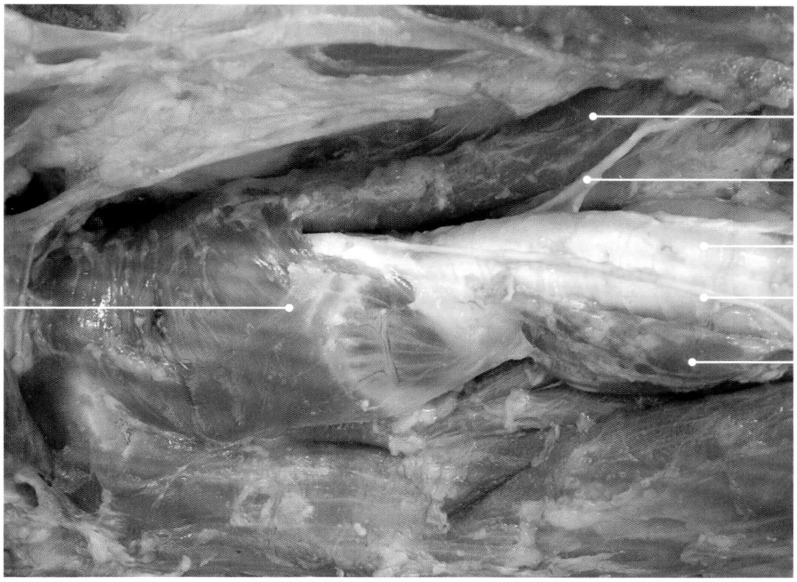

图 11-44 喉返神经（左侧观）
Figure 11-44 Recurrent laryngeal nerve (left view).

1- 食管 esophagus
2- 右侧喉返神经 right recurrent laryngeal nerve
3- 气管 trachea
4- 左侧喉返神经 left recurrent laryngeal nerve
5- 甲状腺 thyroid gland
6- 喉 larynx

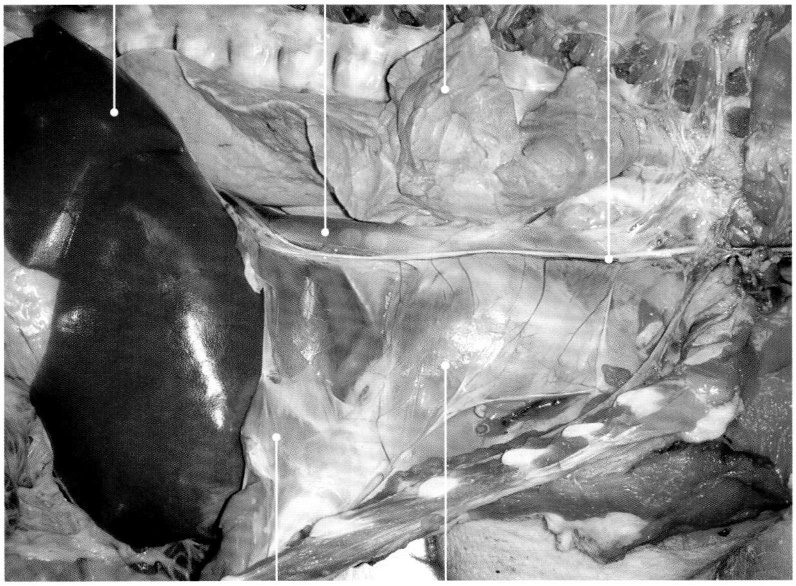

图 11-45 膈神经（右侧观）
Figure 11-45 Phrenic nerve (right view).

1- 肝 liver
2- 后腔静脉 *caudal vena cava*
3- 肺 lung
4- 膈神经 phrenic nerve
5- 心包 pericardium
6- 膈 diaphragm

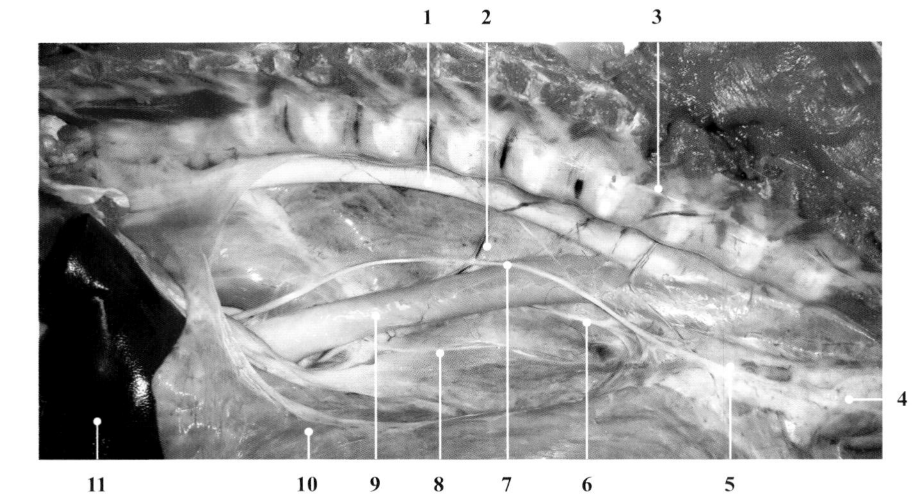

图 11-46 右侧迷走神经
Figure 11-46 Right vagus nerve.

1- 胸主动脉 thoracic aorta
2- 食管支气管动、静脉 esophageal bronchial artery and vein
3- 交感神经干 sympathetic trunk
4- 气管 trachea
5- 右侧迷走神经 right vagus nerve
6- 左侧迷走神经腹侧支 left ventral branch of vagus nerve
7- 右侧迷走神经背侧支 right dorsal branch of vagus nerve
8- 右侧迷走神经腹侧支 right ventral branch of vagus nerve
9- 食管 esophagus
10- 肺 lung
11- 肝 liver

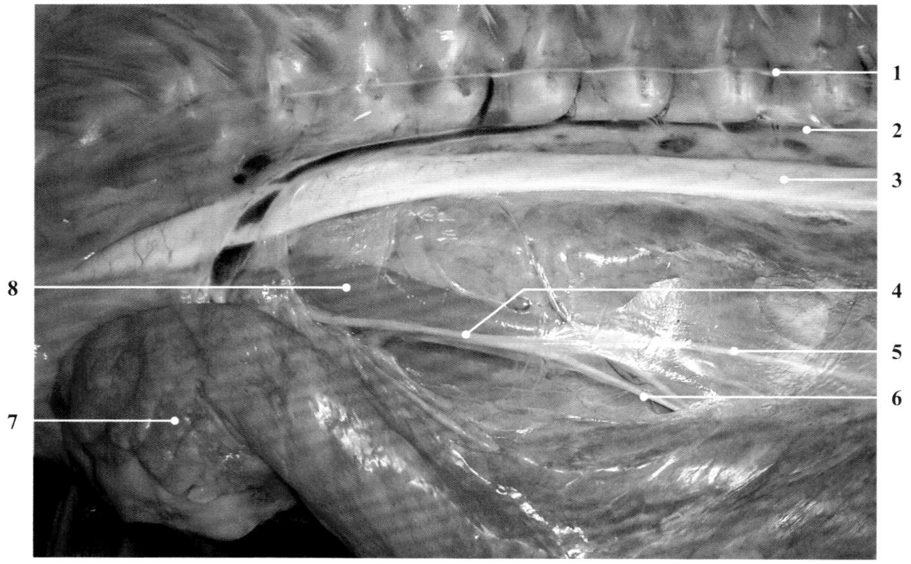

图 11-47 左侧迷走神经 -1
Figure 11-47 Left vagus nerve-1.

1- 交感干 sympathetic trunk
2- 左奇静脉 left azygos vein
3- 胸主动脉 thoracic aorta
4- 左迷走神经 left vagus nerve
5- 迷走神经背侧支 dorsal branch of the vagus nerve
6- 迷走神经腹侧支 ventral branch of the vagus nerve
7- 肺 lung
8- 食管 esophagus

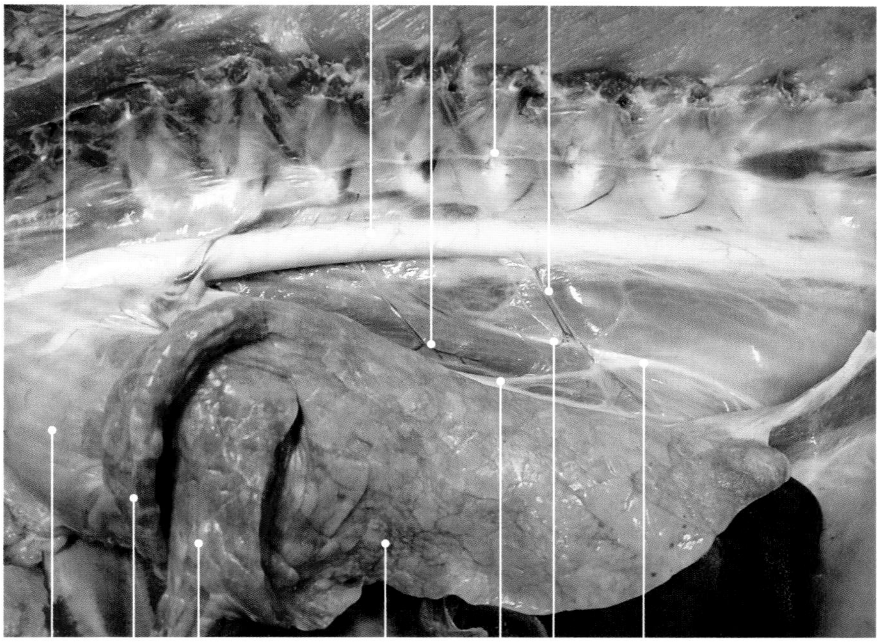

图 11-48 左侧迷走神经 -2
Figure 11-48 Left vagus nerve-2.

1- 主动脉弓 aortic arch
2- 胸主动脉 thoracic aorta
3- 食管 esophagus
4- 交感干 sympathetic trunk
5- 食管支气管动脉 esophageal bronchial artery
6- 迷走神经背侧干 dorsal trunk of vagus nerve
7- 右侧迷走神经背侧支 right dorsal branch of vagus nerve
8- 左侧迷走神经背侧支 left dorsal branch of vagus nerve
9- 左肺后叶（膈叶） left caudal lobe (diaphragmatic lobe) of the lung
10- 左肺中叶（心叶） left middle lobe (cardiac lobe) of the lung
11- 左肺前叶（尖叶） left cranial lobe (apical lobe) of the lung
12- 心脏 heart

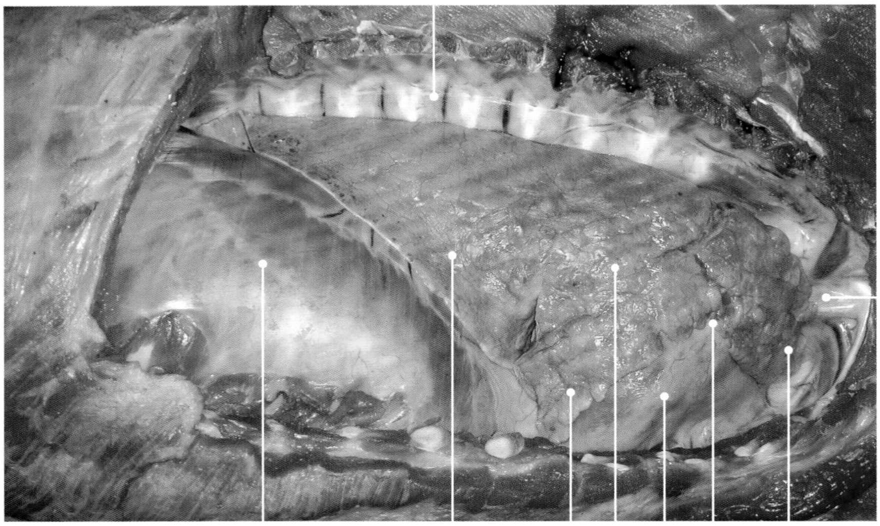

图 11-49　右侧交感干
Figure 11-49　Right sympathetic trunk.

1- 交感干 sympathetic trunk

2- 前腔静脉 *cranial vena cava*

3- 右肺前叶（尖叶） right cranial lobe (apical lobe) of the lung

4- 心切迹 cardiac notch

5- 心脏 heart

6- 肺 lung

7- 右肺中叶（心叶） right middle lobe (cardiac lobe) of the lung

8- 右肺后叶（膈叶） right caudal lobe (diaphragmatic lobe) of the lung

9- 膈 diaphragm

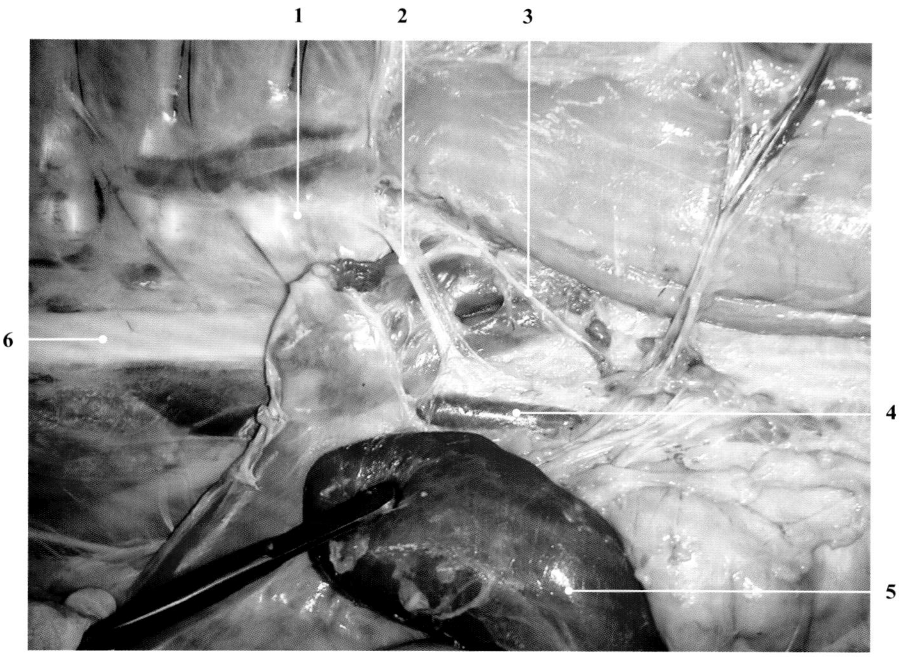

1- 交感干 sympathetic trunk
2- 内脏大神经 greater splanchnic nerve
3- 内脏小神经 lesser splanchnic nerve
4- 肾上腺 adrenal gland
5- 肾 kidney
6- 胸主动脉 thoracic aorta

图 11-50 内脏大神经与内脏小神经
Figure 11-50 Greater splanchnic nerve and lesser splanchnic nerve.

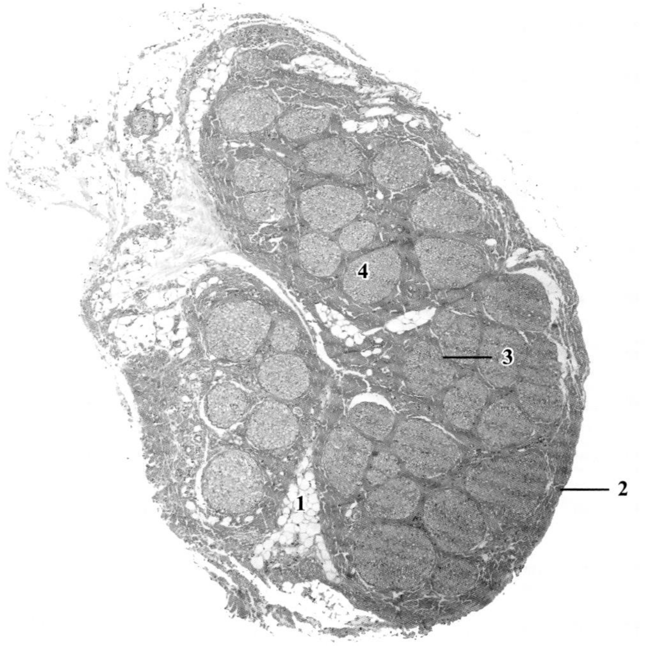

1- 脂肪组织 adipose tissue
2- 神经外膜 epineurium
3- 神经束膜 perineurium
4- 神经束 nerve tract

图 11-51 坐骨神经组织切片（HE 染色，低倍镜）
Figure 11-51 Histological section of the sciatic nerve (HE staining, lower power).

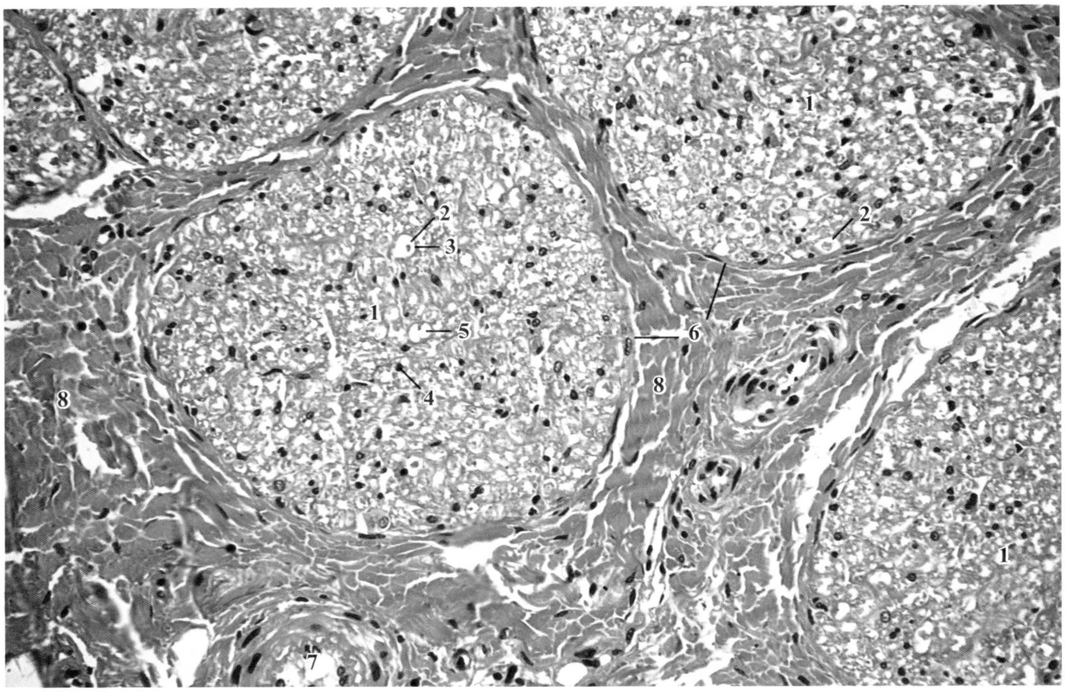

图 11-52　坐骨神经组织切片（HE 染色，高倍镜）
Figure 11-52　Histological section of the sciatic nerve (HE staining, higher power).

1- 神经束 nerve tract
2- 神经纤维 nerve fiber
3- 神经内膜 endoneurium
4- 施旺细胞核 nucleus of the Schwann cell
5- 施旺细胞胞膜沟 groove in plasma membrane of the Schwann cell
6- 神经束膜 perineurium
7- 动脉 artery
8- 平滑肌 smooth muscle

第十二章

内分泌系统

Chapter 12

Endocrine system

内分泌系统（endocrine system）由内分泌器官、内分泌组织和内分泌细胞组成。其中内分泌器官（endocrine organ）包括垂体、松果体、甲状腺、甲状旁腺和肾上腺；内分泌细胞群（内分泌组织）和细胞散在分布其他器官内的，如胰岛、黄体、肾小球旁器、消化道黏膜内的内分泌细胞等。内分泌腺（endocrine gland）是内分泌器官的主要组成部分，其与外分泌腺的主要区别是无输出导管，腺细胞的分泌物直接进入血液或淋巴，随血液循环传递到全身。内分泌腺细胞分泌的物质称为激素（hormone）。

一、垂体

垂体（hypophysis, pituitary gland）位于下丘脑的腹侧，蝶骨体上面的垂体窝内，借漏斗连于下丘脑，呈椭圆形，外包坚韧的硬脑膜。根据垂体发生和结构特点，可分为神经垂体和腺垂体2部分。

1．腺垂体（adenohypophysis）：较大，位于神经垂体的前方和两侧，后端包裹着神经垂体。腺垂体分为远侧部、中间部和结节部。远侧部位于垂体的前腹侧，中间部位于远侧部和神经垂体之间，与远侧部之间有垂体裂（pituitary crack），故习惯上以垂体裂为分界线，位于垂体裂前方的部分称垂体前叶（anterior lobe of hypophysis），位于垂体裂后方的中间部和神经部称垂体后叶（posterior lobe of hypophysis）。结节部位于垂体柄的周围。

（1）远侧部的腺细胞根据其着色的差异，分为嗜酸性细胞、嗜碱性细胞和嫌色细胞3种。嗜酸性细胞（acidophilic cell）数量较多，呈圆形或椭圆形，胞质内含嗜酸性颗粒。根据电镜下颗粒的不同又分为生长激素细胞（somatotroph）和催乳激素细胞（mammotropic cell）。嗜碱性细胞（basophilic cell）数量较少，呈椭圆形或多边形，胞质内含嗜碱性颗粒。颗粒内含糖蛋白类激素。电镜下嗜碱性细胞分为促甲状腺激素细胞（thyrotroph, TSH cell）、促性腺激素细胞（gonadotroph）和促肾上腺皮质激素细胞（corticotroph, ACTH cell）。嫌色细胞（chromophobe cell）的细胞数量多，体积小，呈圆形或多角形，胞质少，着色浅，细胞界限不清楚。电镜下，部分嫌色细胞胞质内含少量分泌颗粒，其余大多数嫌色细胞具有长的分支突起，突起伸入腺细胞之间起支持作用。

（2）结节部由嫌色细胞及嗜碱性细胞组成，可分泌黑色素细胞刺激素，促使黑色素细胞分泌增加。

（3）中间部包围着神经垂体的漏斗，此处有垂体门脉通过，含丰富的纵行毛细血管，腺细胞呈索状纵向排列于血管之间，由嫌色细胞和少量嗜色细胞组成，能分泌少量促性腺激素和促甲状腺激素。

2．神经垂体（neurohypophysis）：较小，位于腺垂体后方，由神经部和漏斗部组成。神经垂体与下丘脑直接相连，主要由来自下丘脑视上核（supraoptic nucleus）和室旁核（paraventricular nucleus）的无髓神经纤维和神经胶质细胞构成，并含有较丰富的窦状毛细血管和少量网状纤维，但可贮存下丘脑视上核和室旁核神经细胞的分泌颗粒。这些分泌颗粒能融合成光镜下可见的嗜酸性团块，为赫令体（Herring body）。视上核的神经内分泌细胞主要合成抗利尿激素（antidiuretic hormone, ADH）；室旁核的神经内分泌细胞主要合成催产素（oxytocin, OT）。神经部内的特殊分化的神经胶质细胞，为垂体细胞（pituicyte），有支持和营养神经纤维的作用，还可分泌一些化学物质以调节神经纤维的活动和激素的释放。

二、松果体

松果体（pineal body）又称松果腺（pineal gland），呈狭窄的长锥形体，红褐色，位于丘脑和四叠体之间。由于其位于第Ⅲ脑室顶，又称脑上腺（*glandula pinealis*），其一端借松果体柄与第Ⅲ脑室顶相连，第Ⅲ脑室突向柄内形成松果体隐窝。

松果体表面被以由软脑膜延续而来的结缔组织被膜，被膜随血管伸入实质内，将实质分为许多不规则小叶，小叶主要由松果体细胞（pinealocyte）、神经胶质细胞和神经纤维等组成。松果体细胞又称主细胞，光镜下成簇或成索状排列。胞体呈不规则的圆形或多角形，胞核大而圆。松果体细胞内含有丰富的5-羟色胺，在特殊酶的作用下转变为褪黑激素（melatonin）。

三、甲状腺

甲状腺（thyroid gland）位于喉后方，气管的腹侧，前端始于甲状软骨，后端尖，棕红色。甲状腺外覆有纤维囊，称甲状腺被囊（tunic of thyroid gland），此囊伸入腺组织将腺体分成大小不等的小叶，囊外包有颈深筋膜。甲状腺实质内充满大量滤泡，滤泡间的结缔组织内含有散在的滤泡旁细胞。

甲状腺滤泡由单层立方的腺上皮细胞环绕而成，中心为滤泡腔。腺上皮细胞是甲状腺激素合成和释放的部位，滤泡腔内充满均匀的胶质，是甲状腺激素复合物，也是甲状腺激素的贮存库。甲状腺主要分泌甲状腺素（thyroxine），促进机体生长发育。甲状腺的滤泡旁细胞（parafollicular cell，又称C细胞）单个镶嵌在上皮细胞之间，或散布在滤泡间的结缔组织中，分泌降钙素（calcitonin），有增强成骨细胞活性、促进骨组织钙化、使血钙降低等作用。

四、甲状旁腺

甲状旁腺（parathyroid gland）一对，较小，呈圆形，位于甲状腺外侧附近。甲状旁腺表面覆有薄层的结缔组织被膜，被膜的结缔组织携带血管、淋巴管和神经伸入腺内，形成小梁，将腺分为不完全的小叶。小叶内腺实质细胞排列成索或团状，细胞间质较多，有丰富的毛细血管。腺细胞有主细胞和嗜酸性细胞。主细胞（chief cell）分泌甲状旁腺素，以胞吐方式释放入毛细血管。甲状旁腺素的功能是调节血钙浓度，影响体内钙和磷的代谢。嗜酸性细胞（oxyphil cell）较主细胞大，数量少，单个或成群散布于主细胞之间。

五、肾上腺

肾上腺（adrenal gland）一对，呈三棱柱状体，长而窄，表面有沟，位于左、右肾的前内侧缘。左侧的肾上腺常比右侧的大而长。肾上腺表面包有致密结缔组织的被膜，含有散在的平滑肌；实质分为周围的皮质和中央的髓质。

1．皮质（cortex）：占腺体大部分，从外向内可分为球状带、束状带和网状带3部分。

（1）球状带（glomerular zone）位于被膜下方，细胞排列不规则。球状带细胞分泌盐皮质激素（醛固酮），调节电解质和水盐代谢。

（2）束状带（fasciculate zone）是球状带的延续，此层最厚。细胞较大，呈多角形，界限清楚，呈束状平行排列。胞核圆，位于中央。束状带细胞分泌糖皮质激素，调节糖、脂肪和蛋白质的代谢。

（3）网状带（reticular zone）位于皮质深层与髓质相毗连，此层最薄。细胞排列成条索状且相互吻合成网，细胞小，胞核深染，胞质弱嗜酸性。网状带细胞分泌雄激素和少量雌激素。

2. 髓质（medulla）：位于肾上腺中央，其中心有一肾上腺中央静脉，汇合皮质和髓质的血液，经肾上腺静脉离开肾上腺。髓质细胞呈团索状排列，腺细胞呈卵圆形或多角形。髓质细胞可分泌肾上腺素和去甲肾上腺素，可使小动脉收缩，心跳加快，血压升高。

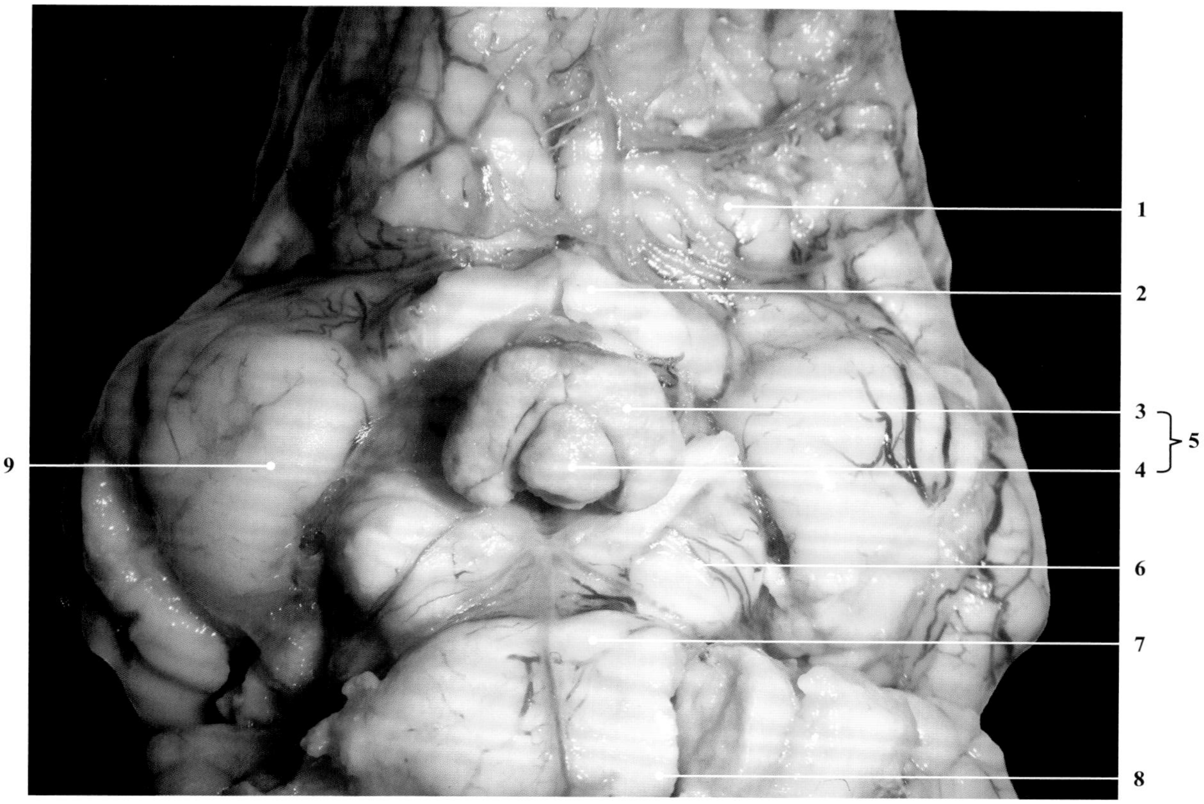

图 12-1 垂体
Figure 12-1 Hypophysis.

1- 嗅三角 olfactory trigone
2- 视交叉 optic chiasm
3- 腺垂体 adenohypophysis
4- 神经垂体 neurohypophysis
5- 垂体 hypophysis
6- 中脑大脑脚 cerebral peduncle of the mesencephalon
7- 脑桥 pons
8- 延髓 medulla oblongata
9- 梨状叶 piriform lobe

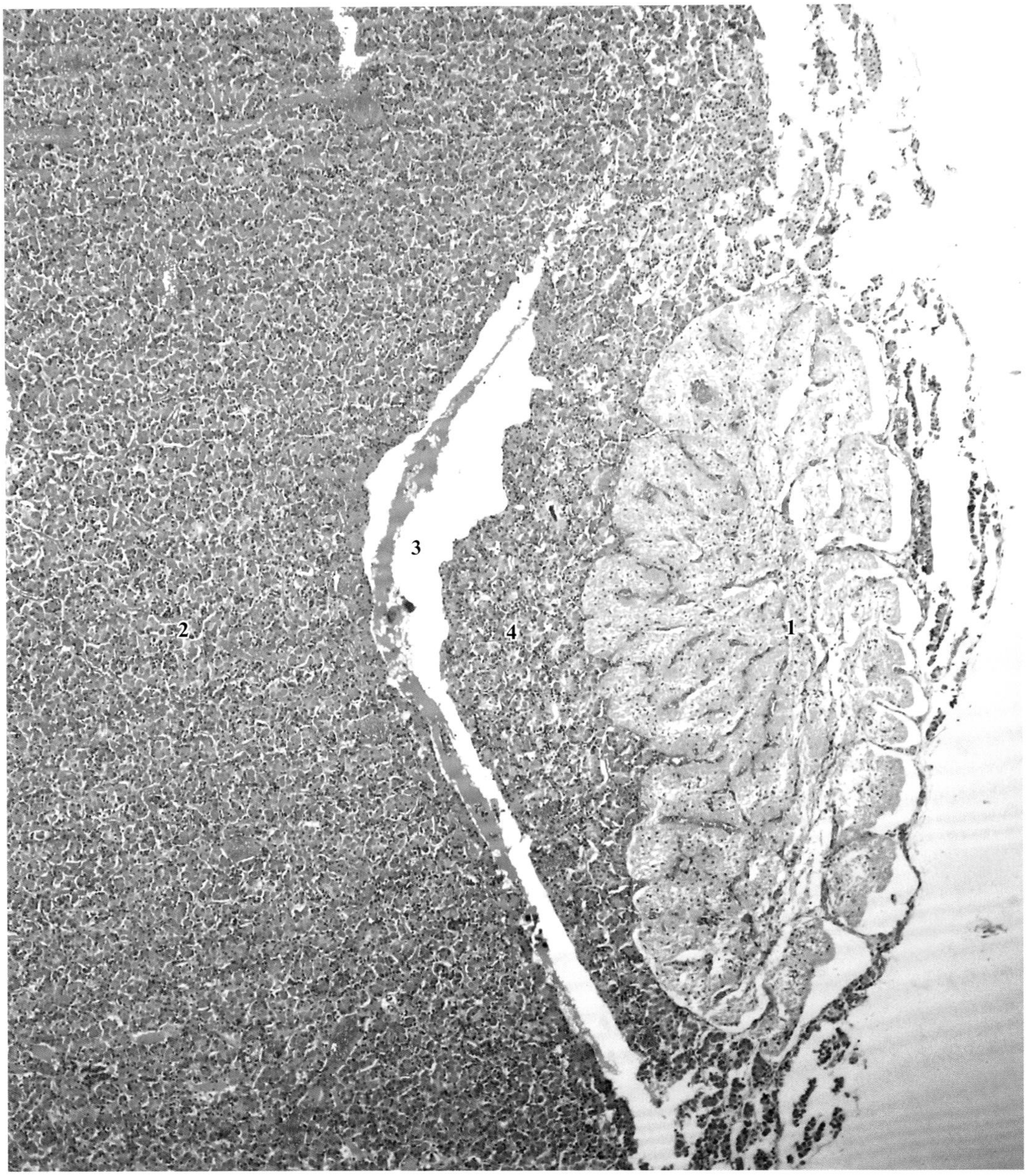

图 12-2　垂体组织切片（HE 染色，低倍镜）
Figure 12-2　Histological section of the hypophysis (HE staining, lower power).

1- 神经垂体神经部 pars nervosa, neurohypophysis
2- 腺垂体远侧部 pars distalis, adenohypophysis
3- 垂体裂 pituitary crack
4- 腺垂体中间部 pars intermedia, adenohypophysis

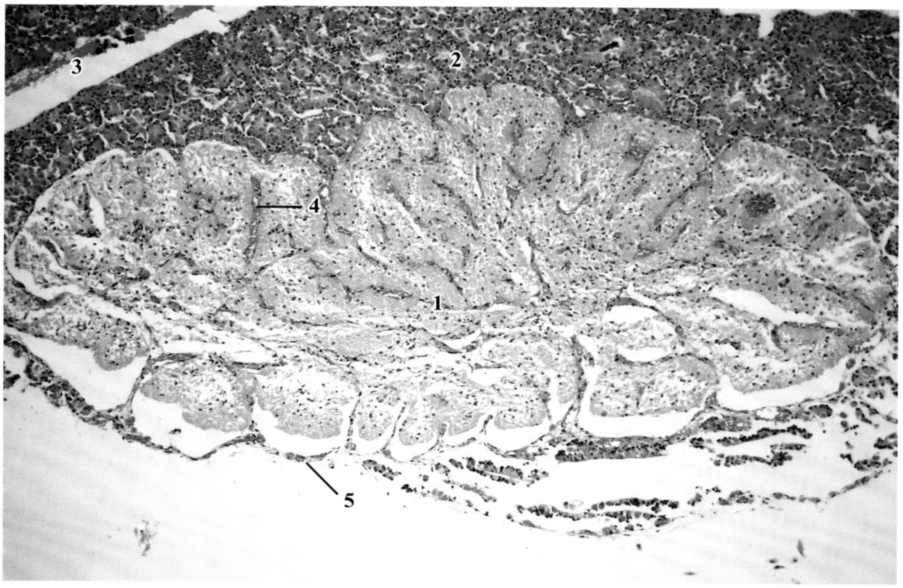

1- 神经垂体神经部 pars nervosa, neurohypophysis
2- 腺垂体中间部 pars intermedia, adenohypophysis
3- 垂体裂 pituitary crack
4- 血管 blood vessl
5- 被膜 capsule

图 12-3 神经垂体组织切片（HE 染色，高倍镜）
Figure 12-3 Histological section of the neurohypophysis (HE staining, higher power).

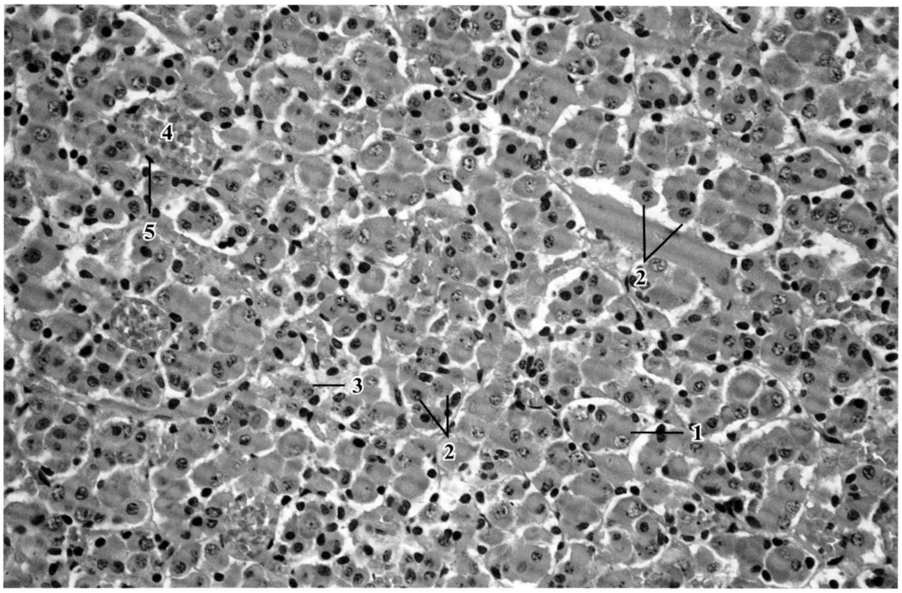

1- 嗜碱性细胞 basophilic cell
2- 嗜酸性细胞 acidophilic cell
3- 嫌色细胞 chromophobe cell
4- 血管 blood vessel
5- 内皮细胞 endothelial cell

图 12-4 腺垂体组织切片（HE 染色，高倍镜）
Figure 12-4 Histological section of the adenohypophysis (HE staining, higher power).

1- 嗜酸性细胞
acidophilic cell
2- 嗜碱性细胞
basophilic cell
3- 嫌色细胞
chromophobe cell
4- 静脉 vein

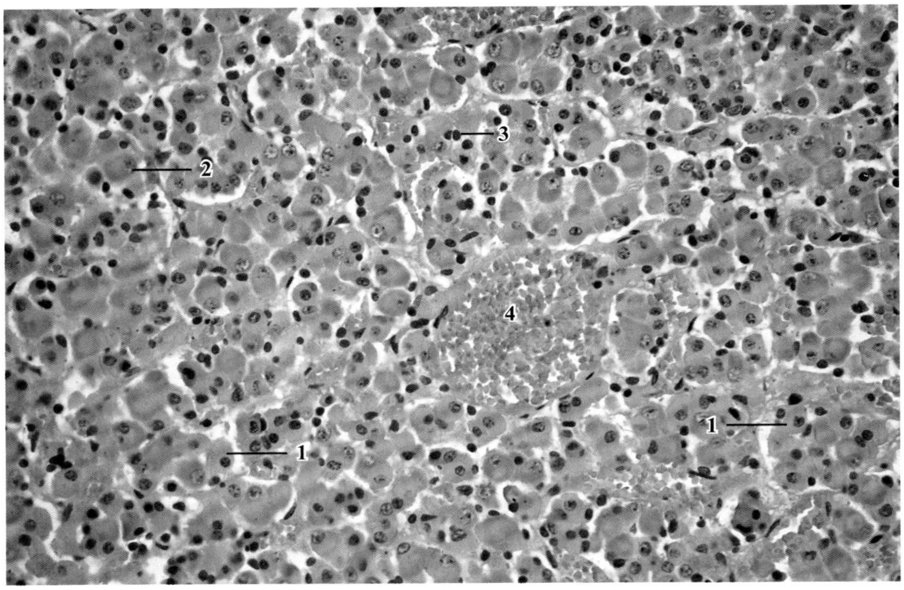

图 12-5 腺垂体组织切片（HE 染色，高倍镜）
Figure 12-5 Histological section of the adenohypophysis (HE staining, higher power).

1- 神经垂体神经部
pars nervosa,
neurohypophysis
2- 腺垂体中间部
pars intermedia,
adenohypophysis
3- 嗜酸性细胞
acidophilic cell
4- 嫌色细胞
chromophobe cell
5- 嗜碱性细胞
basophilic cell
6- 垂体细胞（神经胶质细胞）pituicyte (neuroglial cell)
7- 静脉 vein

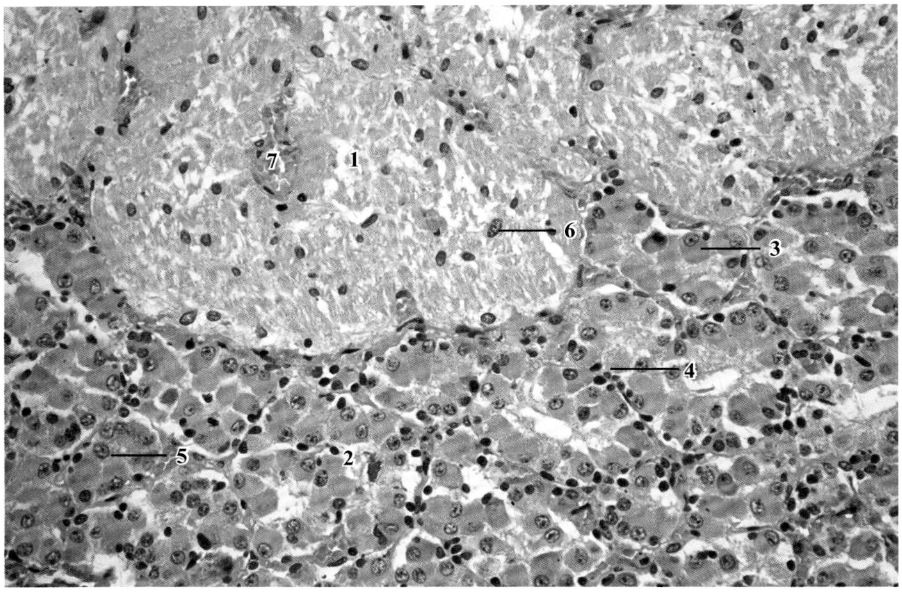

图 12-6 垂体组织切片（HE 染色，高倍镜）
Figure 12-6 Histological section of the hypophysis (HE staining, higher power).

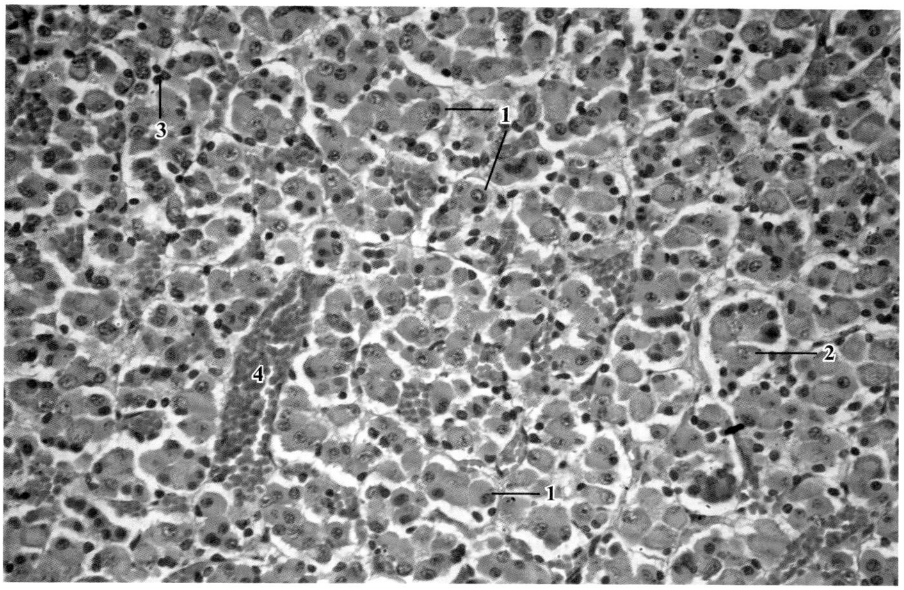

1- 嗜酸性细胞 acidophilic cell
2- 嗜碱性细胞 basophilic cell
3- 嫌色细胞 chromophobe cell
4- 静脉 vein

图 12-7 腺垂体组织切片（HE 染色，高倍镜）
Figure 12-7 Histological section of the adenohypophysis (HE staining, higher power).

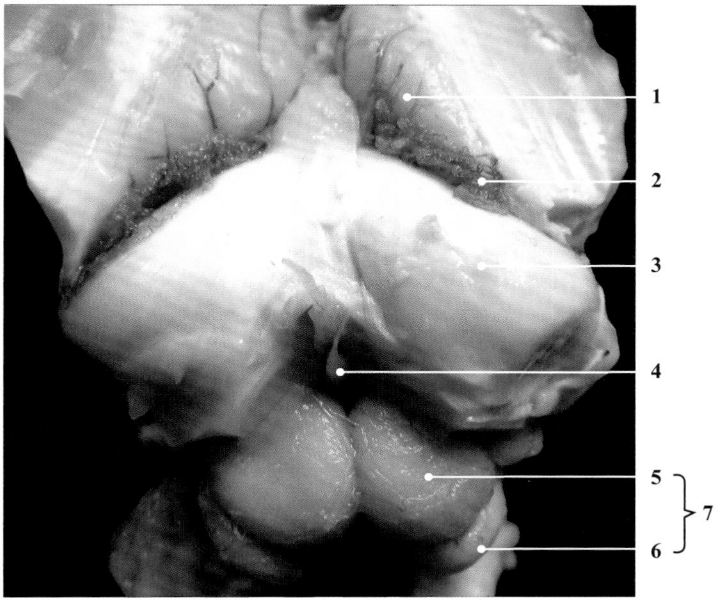

图 12-8 松果体位置
Figure 12-8 Location of the pineal gland.

1- 纹状体 corpus striatum
2- 脉络丛 choroid plexus
3- 海马 hippocampus
4- 松果体 pineal gland
5- 前丘 rostral colliculus
6- 后丘 caudal colliculus
7- 四叠体 quadrigeminal bodies

1- 松果体 pineal gland
2- 血管 blood vessel

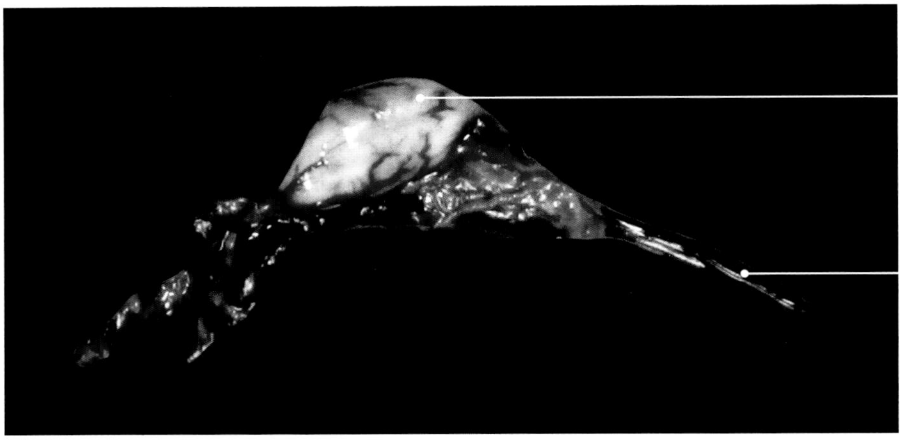

图 12-9　松果体
Figure 12-9　Pineal gland.

1- 松果体实质 pineal parenchyma
2- 松果体柄 pineal stalk
3- 被膜 capsule
4- 血管 blood vessel
5- 松果体小叶 pineal lobule

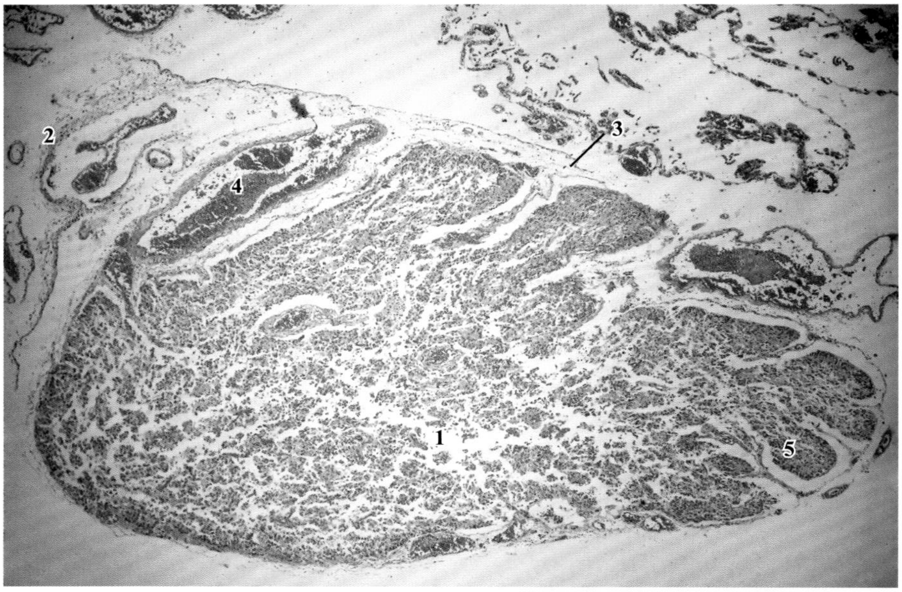

图 12-10　松果体组织切片（HE 染色，低倍镜）
Figure 12-10　Histological section of the pineal gland (HE staining, lower power).

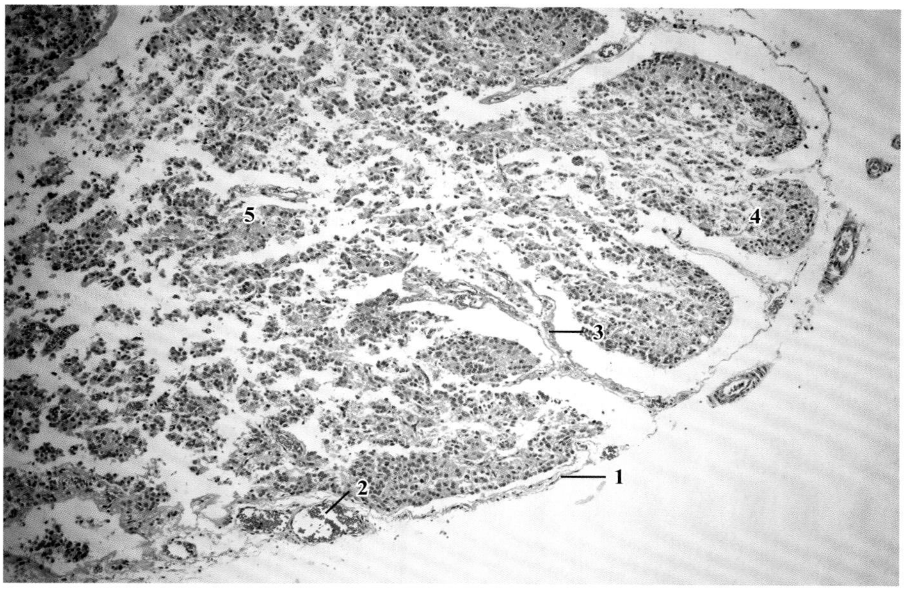

1- 被膜 capsule
2- 血管 blood vessel
3- 结缔组织 connective tissue
4- 松果体小叶 pineal lobule
5- 松果体实质 pineal parenchyma

图 12-11 松果体组织切片（HE 染色，高倍镜）
Figure 12-11 Histological section of the pineal gland (HE staining, higher power).

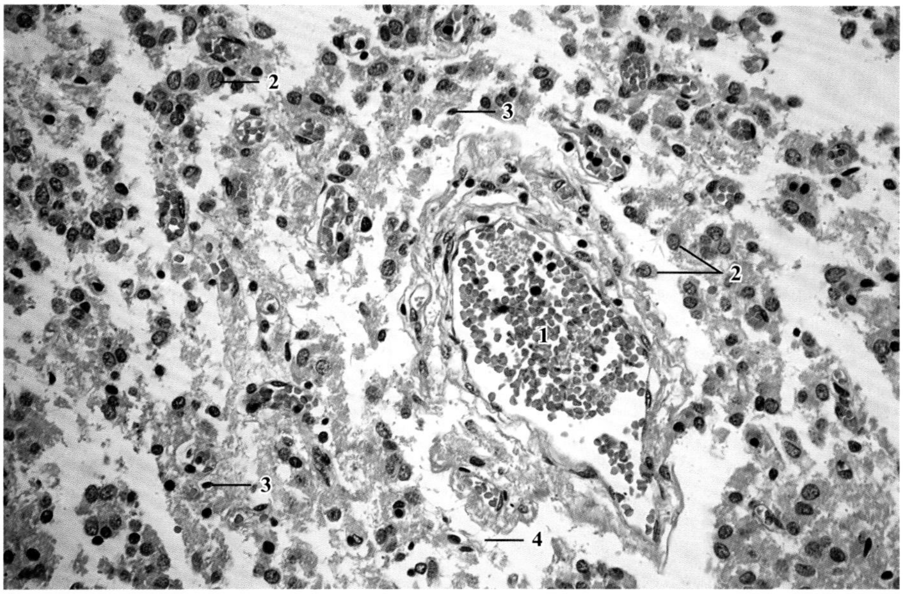

1- 血管 blood vessel
2- 松果体细胞 pinealocyte
3- 星形胶质细胞 astroglial cell
4- 神经胶质细胞纤维 neuroglial cell fiber

图 12-12 松果体组织切片（HE 染色，高倍镜）
Figure 12-12 Histological section of the pineal gland (HE staining, higher power).

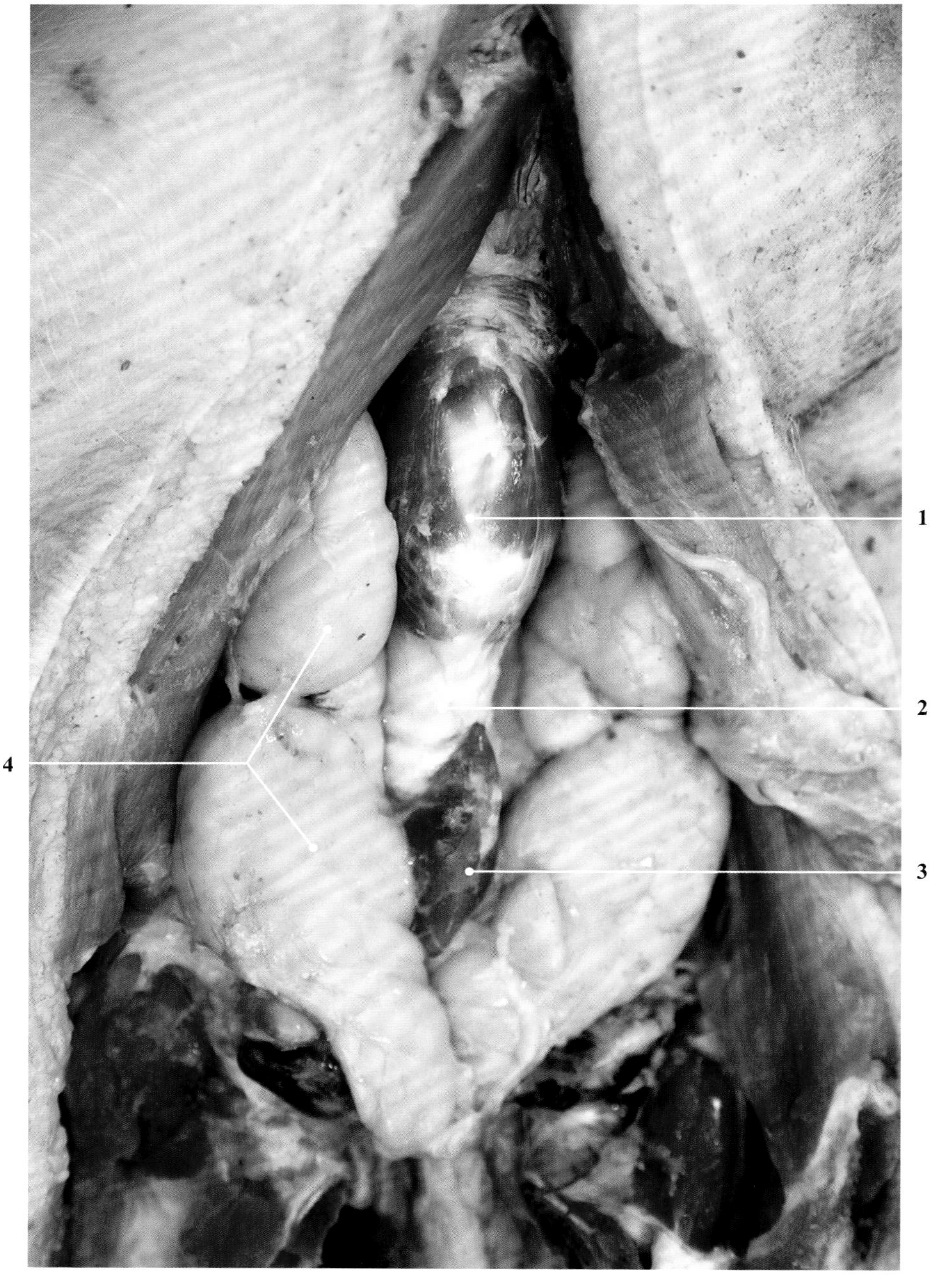

图 12-13 甲状腺位置（腹侧观）
Figure 12-13 Location of the thyroid gland (ventral view).

1- 喉 larynx
2- 气管 trachea
3- 甲状腺 thyroid gland
4- 胸腺 thymus

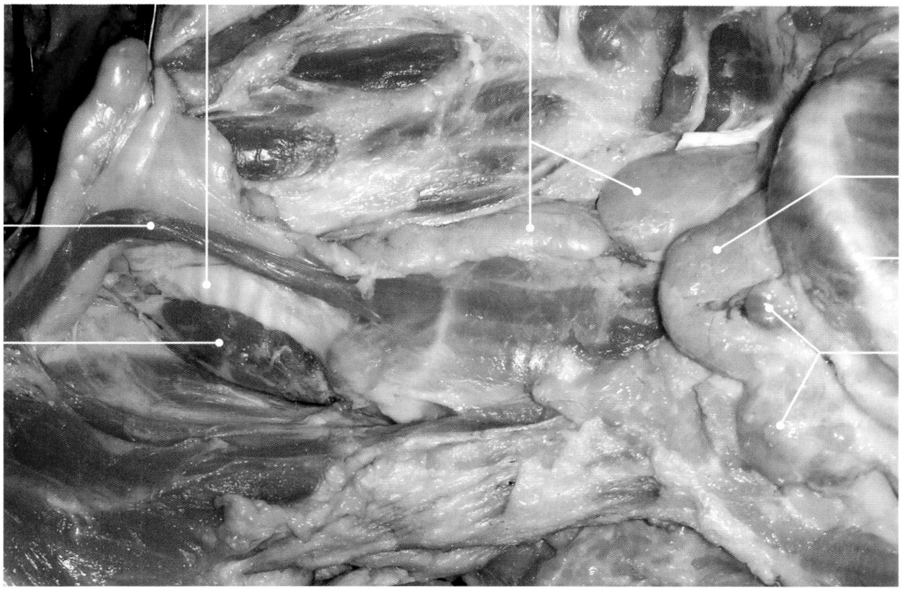

图 12-14 母猪甲状腺位置（右外侧观）
Figure 12-14 Location of the thyroid gland of a sow (right lateral view).

1- 颌下腺 mandibular gland
2- 下颌骨 mandible
3- 下颌淋巴结 mandibular lymph node
4- 甲状腺 thyroid gland
5- 胸骨甲状舌骨肌 sterno-thyrohyoid muscle
6- 气管 trachea
7- 胸腺 thymus

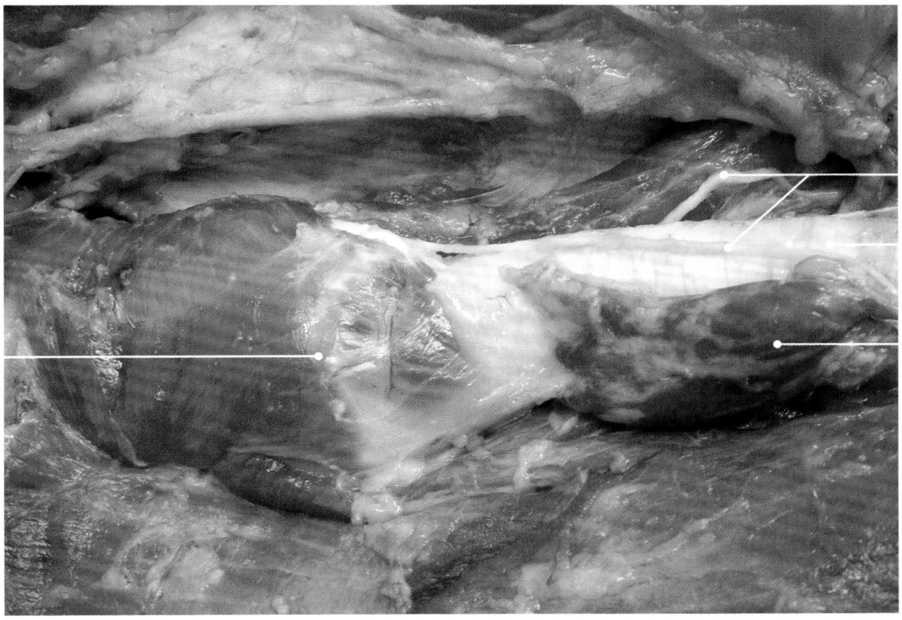

图 12-15 公猪甲状腺位置（左外侧观）
Figure 12-15 Location of the thyroid gland of a boar (left lateral view).

1- 喉返神经 recurrent laryngeal nerve
2- 气管 trachea
3- 甲状腺 thyroid gland
4- 喉 larynx

5 6

1

4

2

3

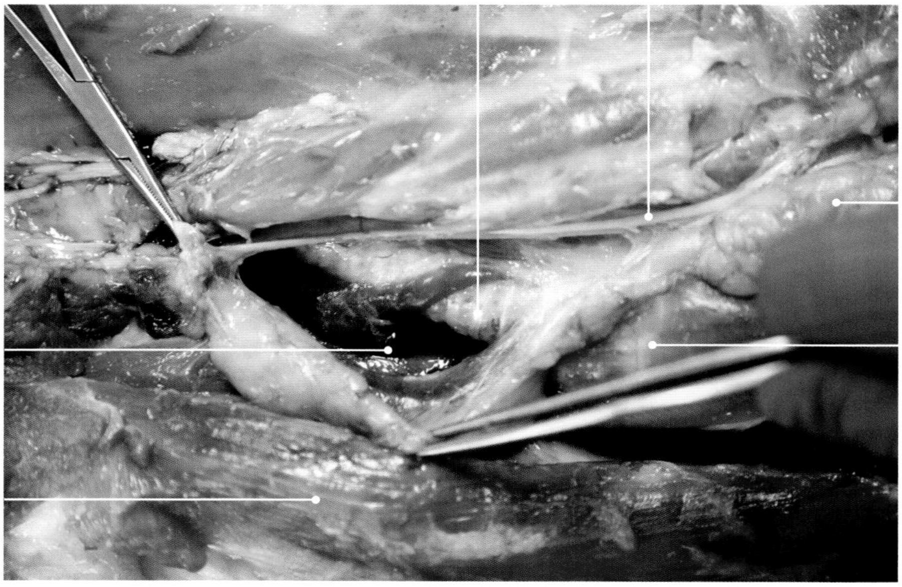

图 12-16 公猪甲状腺位置（右外侧观）
Figure 12-16 Location of the thyroid gland of a boar (right lateral view).

1- 胸腺 thymus
2- 喉 larynx
3- 胸头肌 sternocephalic muscle
4- 甲状腺 thyroid gland
5- 气管 trachea
6- 迷走交感干 vagosympathetic trunk

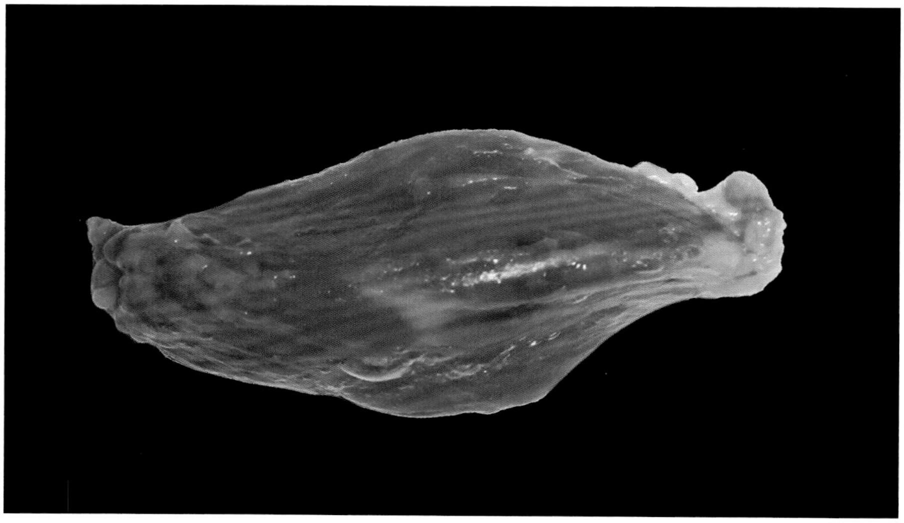

图 12-17 母猪甲状腺背侧面
Figure 12-17 Dorsal aspect of the thyroid gland of a sow.

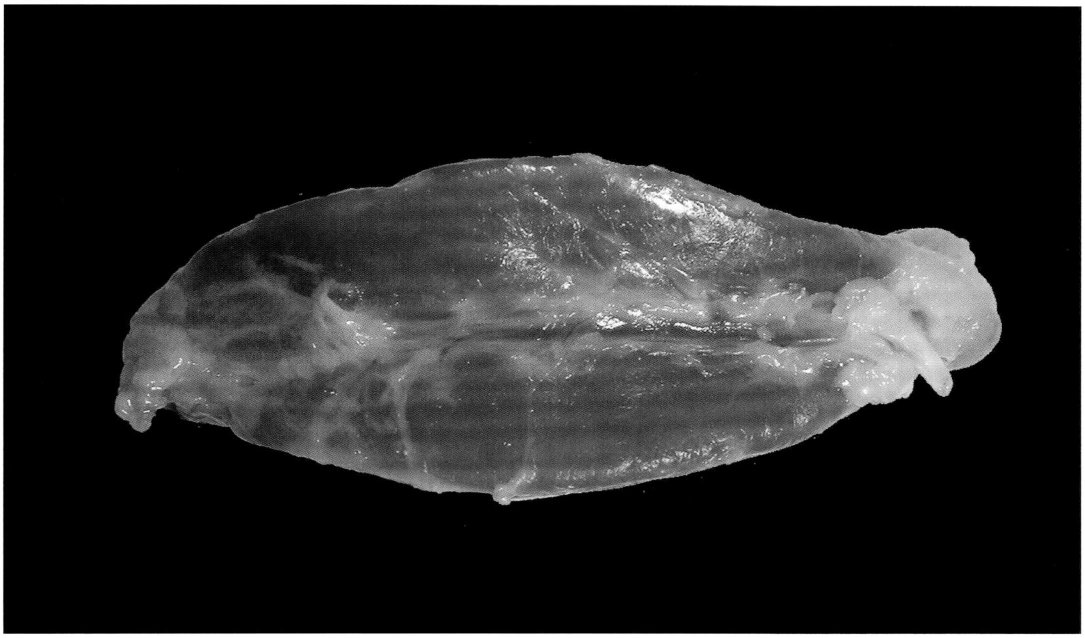

图 12-18 母猪甲状腺腹侧面
Figure 12-18 Ventral aspect of the thyroid gland of a sow.

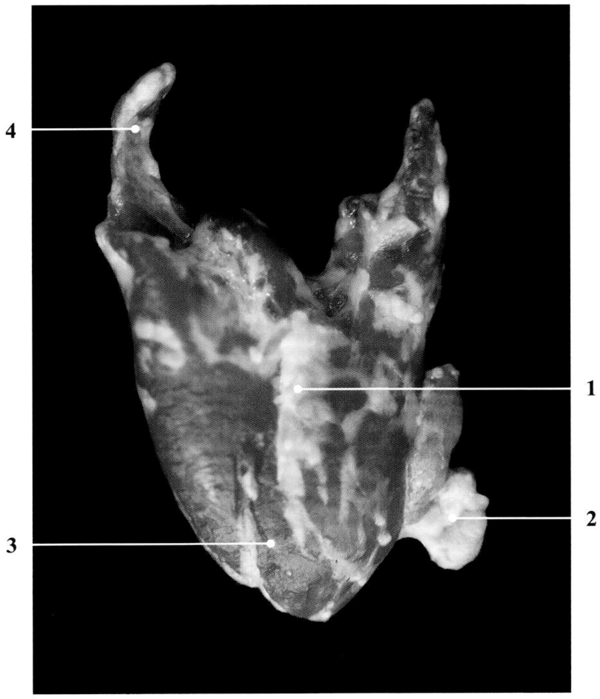

图 12-19 公猪甲状腺
Figure 12-19 Thyroid gland of the boar.

1- 甲状腺 thyroid gland
2- 甲状旁腺 parathyroid gland
3- 甲状腺后部 caudal pars of the thyroid gland
4- 甲状腺前部 cranial pars of the thyroid gland

1- 滤泡 follicle
2- 滤泡细胞 follicular cell
3- 胶质 colloid
4- 内皮细胞 endothelial cell
5- 滤泡旁细胞 parafollicular cell

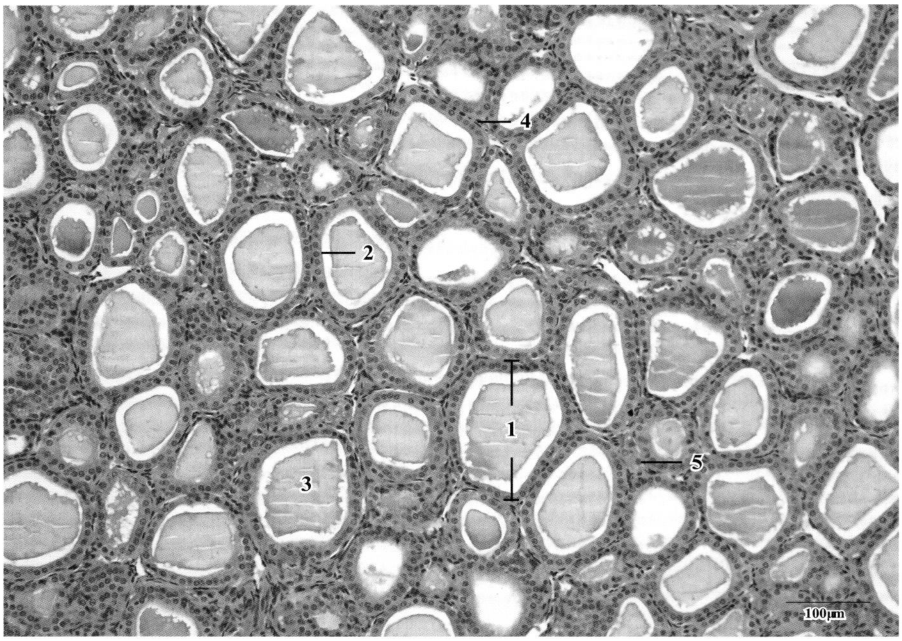

图 12-20　甲状腺滤泡组织切片（HE 染色，低倍镜）
Figure 12-20　Histological section of the thyroid follicles (HE staining, lower power).

1- 胶质 colloid
2- 滤泡细胞 follicular cell
3- 内皮细胞 endothelial cell
4- 滤泡旁细胞 parafollicular cell

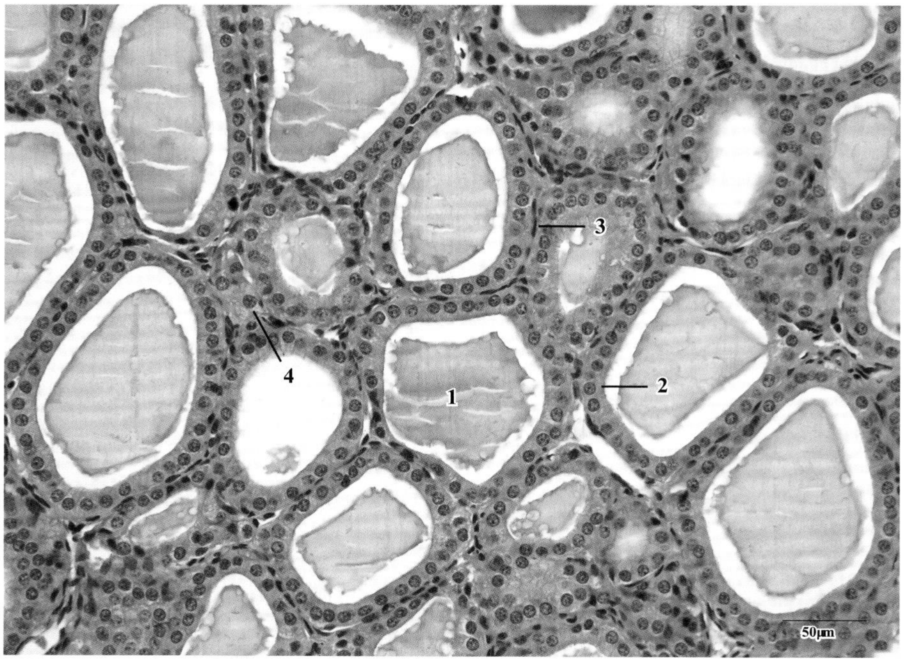

图 12-21　甲状腺滤泡组织切片（HE 染色，高倍镜）
Figure 12-21　Histological section of the thyroid follicles (HE staining, higher power).

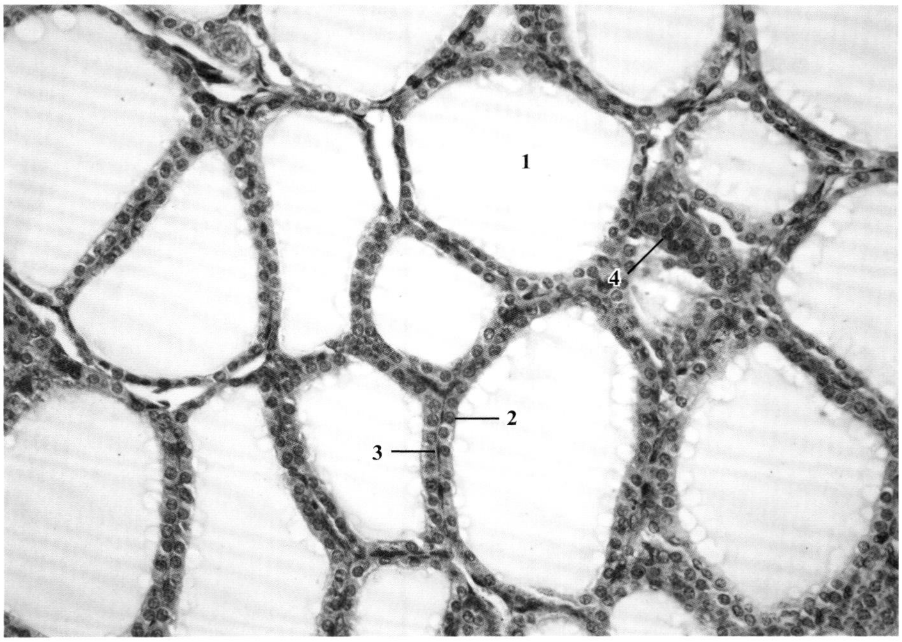

1- 滤泡 follicle
2- 滤泡细胞 follicular cell
3- 内皮细胞 endothelial cell
4- 滤泡旁细胞群 parafollicular cell group

图 12-22　甲状腺滤泡组织切片（HE 染色，低倍镜）
Figure 12-22　Histological section of the thyroid follicles (HE staining, lower power).

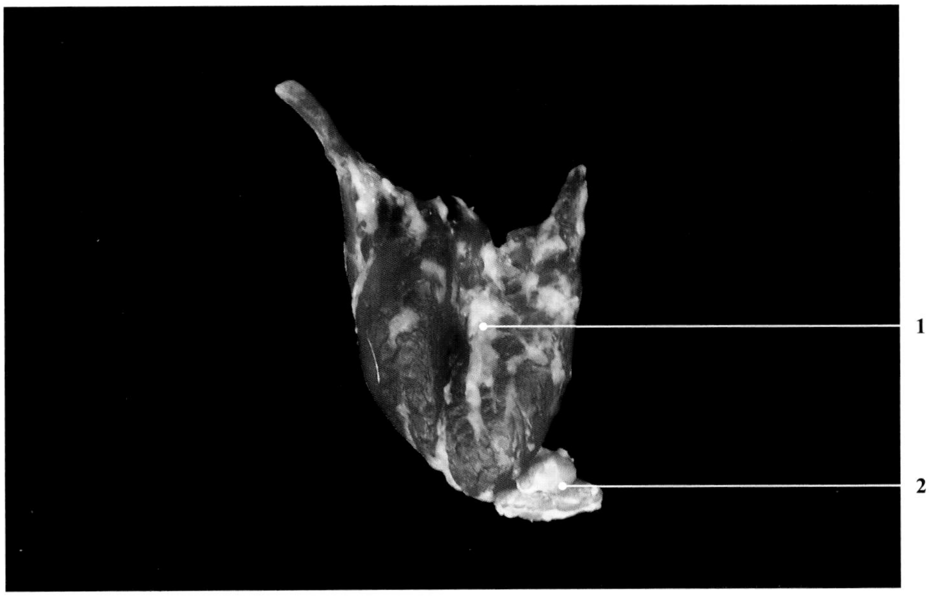

1- 甲状腺 thyroid gland
2- 甲状旁腺 parathyroid gland

图 12-23　公猪甲状旁腺
Figure 12-23　Parathyroid gland of the boar.

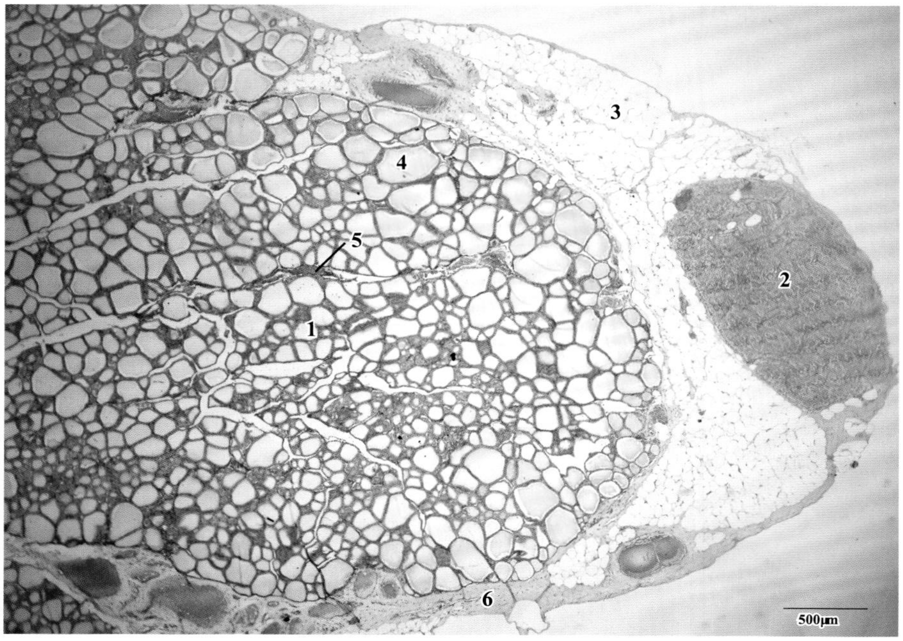

图 12-24　甲状旁腺组织切片（HE 染色，低倍镜）
Figure 12-24　Histological section of the parathyroid gland (HE staining, lower power).

1- 甲状腺 thyroid gland
2- 甲状旁腺 parathyroid gland
3- 脂肪组织 adipose tissue
4- 滤泡 follicle
5- 血管 blood vessel
6- 被膜 capsule

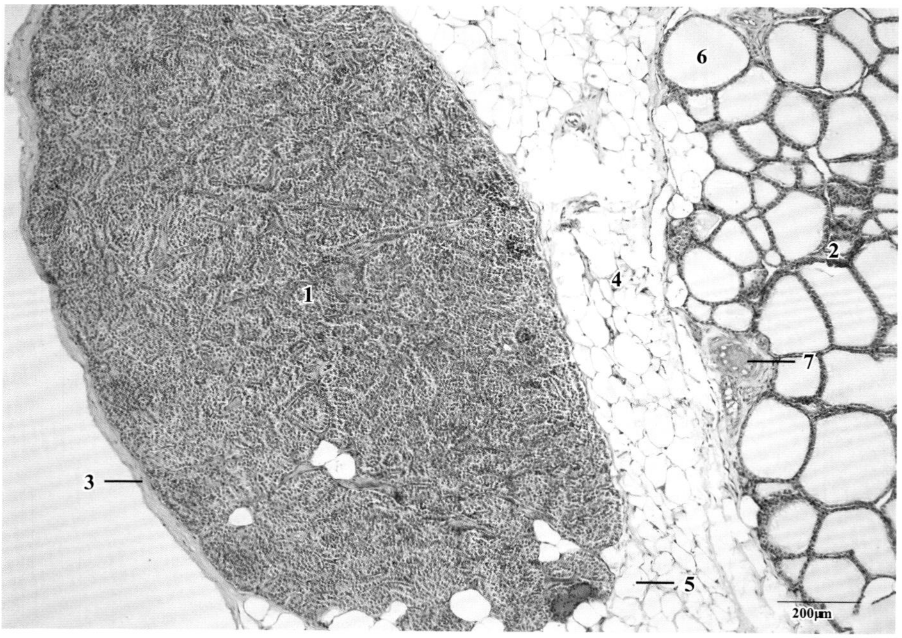

图 12-25　甲状旁腺组织切片（HE 染色，高倍镜）
Figure 12-25　Histological section of the parathyroid gland (HE staining, higher power).

1- 甲状旁腺 parathyroid gland
2- 甲状腺 thyroid gland
3- 被膜 capsule
4- 脂肪组织 adipose tissue
5- 脂肪细胞 adipocyte
6- 滤泡 follicle
7- 血管 blood vessel

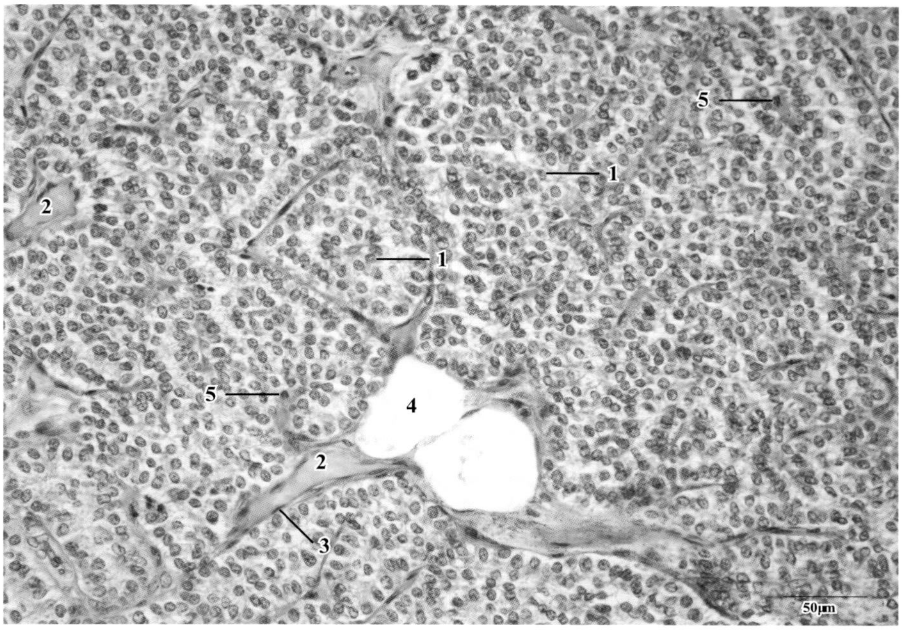

图 12-26 甲状旁腺组织切片（HE 染色，高倍镜）
Figure 12-26 Histological section of the parathyroid gland (HE staining, higher power).

1- 主细胞 chief cell
2- 静脉 vein
3- 内皮细胞 endothelial cell
4- 脂肪组织 adipose tissue
5- 嗜酸性细胞 oxyphil cell

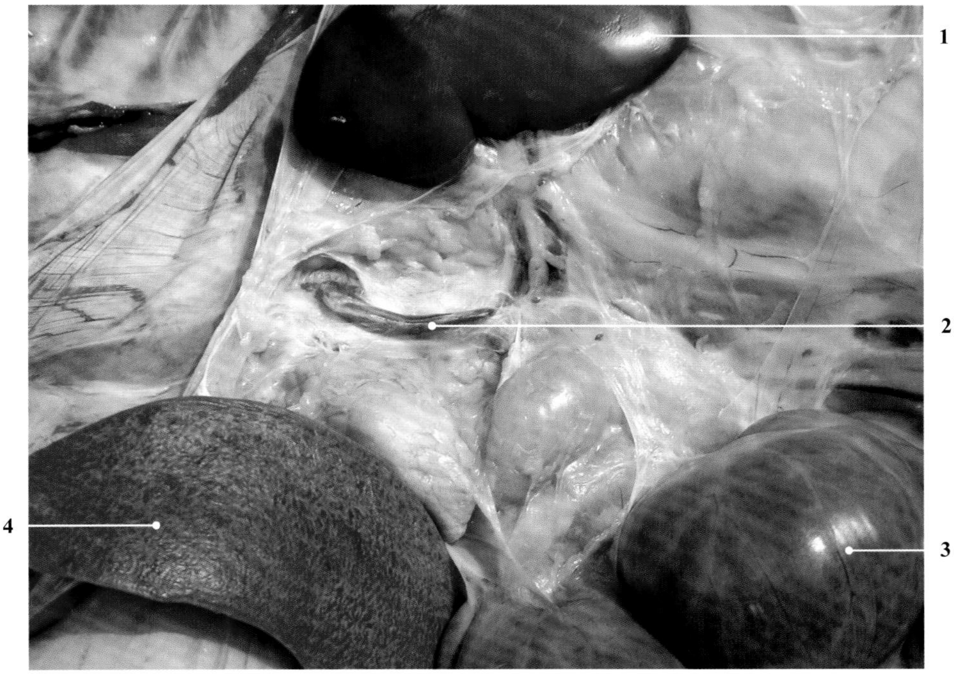

图 12-27 肾上腺位置（左外侧观）
Figure 12-27 Location of the adrenal gland (left lateral view).

1- 肾 kidney
2- 左侧肾上腺 left adrenal gland
3- 大肠 large intestine
4- 脾 spleen

图 12-28　右侧肾上腺位置
Figure 12-28　Location of the right adrenal gland.

1- 被膜 capsule
2- 肾 kidney
3- 肾门 renal hilum
4- 右侧肾上腺 right adrenal gland

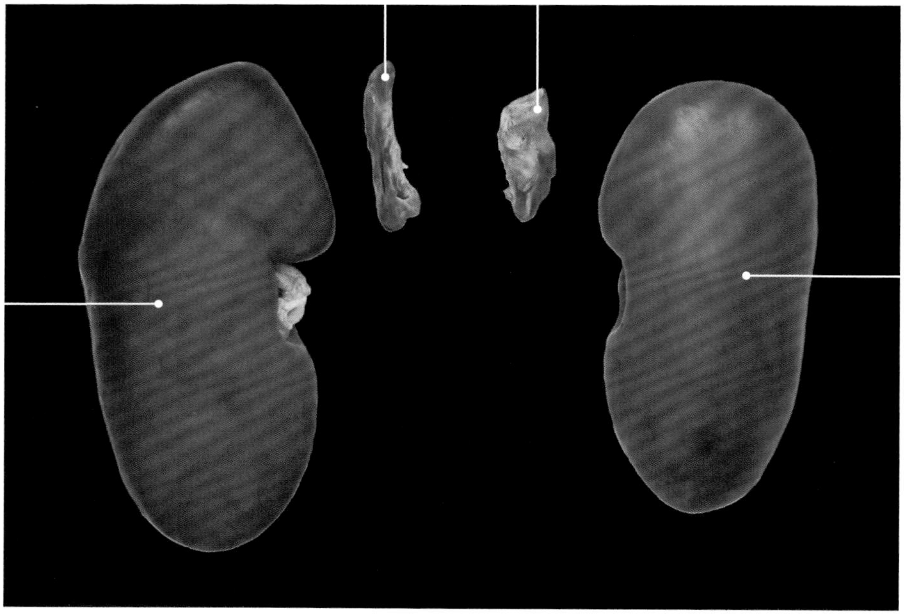

图 12-29　肾上腺与肾的位置关系
Figure 12-29　Position relationship between the adrenal gland and kidney.

1- 右肾 right kidney
2- 左肾 left kidney
3- 左侧肾上腺 left adrenal gland
4- 右侧肾上腺 right adrenal gland

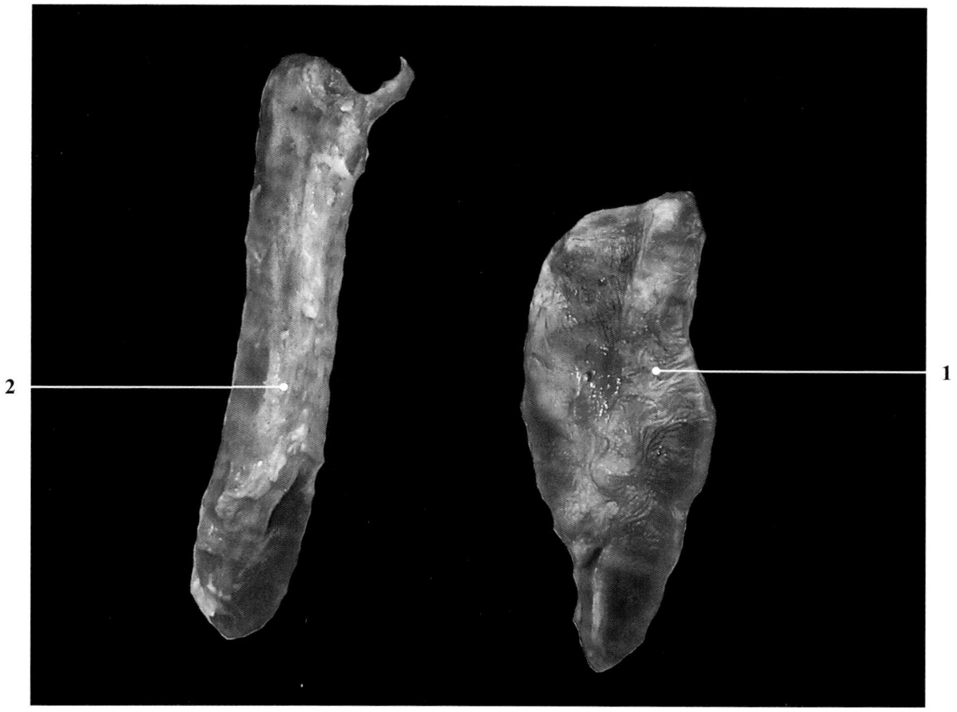

图 12-30　肾上腺形态
Figure 12-30　Morphology of the adrenal gland.

1- 右侧肾上腺 right adrenal gland　　**2-** 左侧肾上腺 left adrenal gland

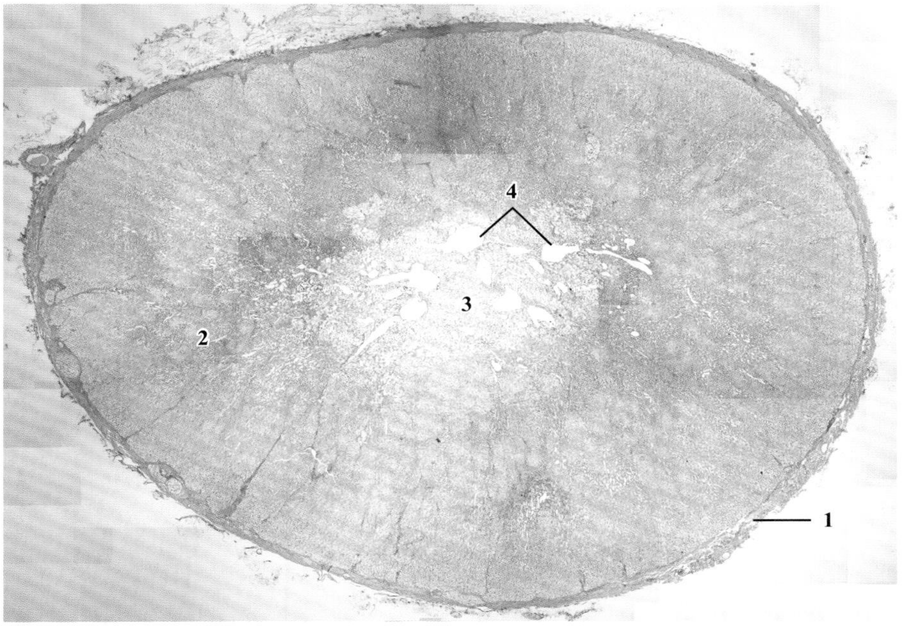

图 12-31　肾上腺组织切片（HE 染色，低倍镜）
Figure 12-31　Histological section of the adrenal gland (HE staining, lower power).

1- 被膜 capsule　　**3-** 髓质 medulla
2- 皮质 cortex　　**4-** 静脉 vein

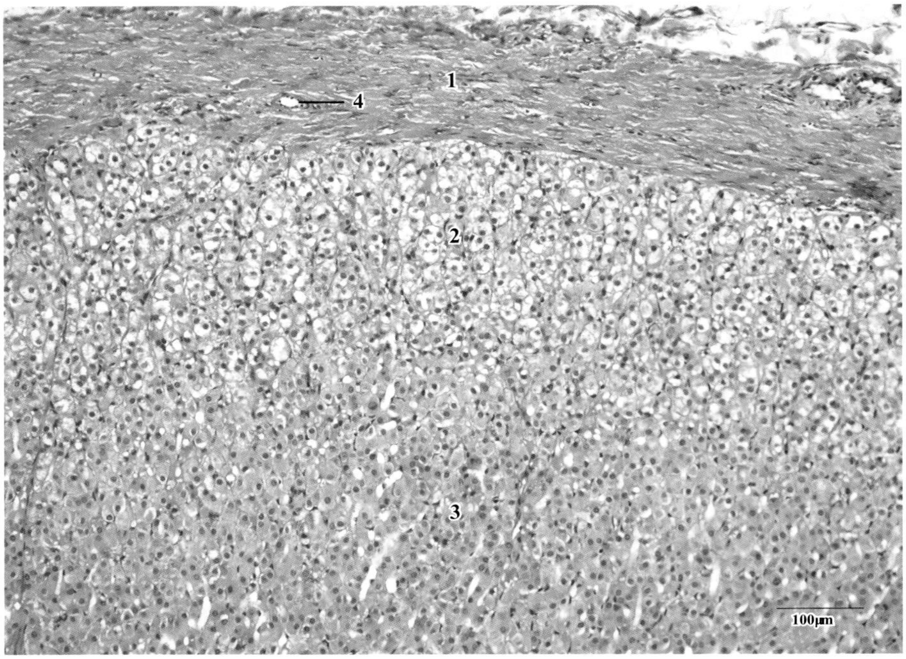

1- 被膜 capsule
2- 球状带 glomerular zone
3- 束状带 fasciculate zone
4- 血管 blood vessel

图 12-32　肾上腺组织切片示被膜（HE 染色，高倍镜）
Figure 12-32　Histological section of a adrenal gland showing the capsule (HE staining, higher power).

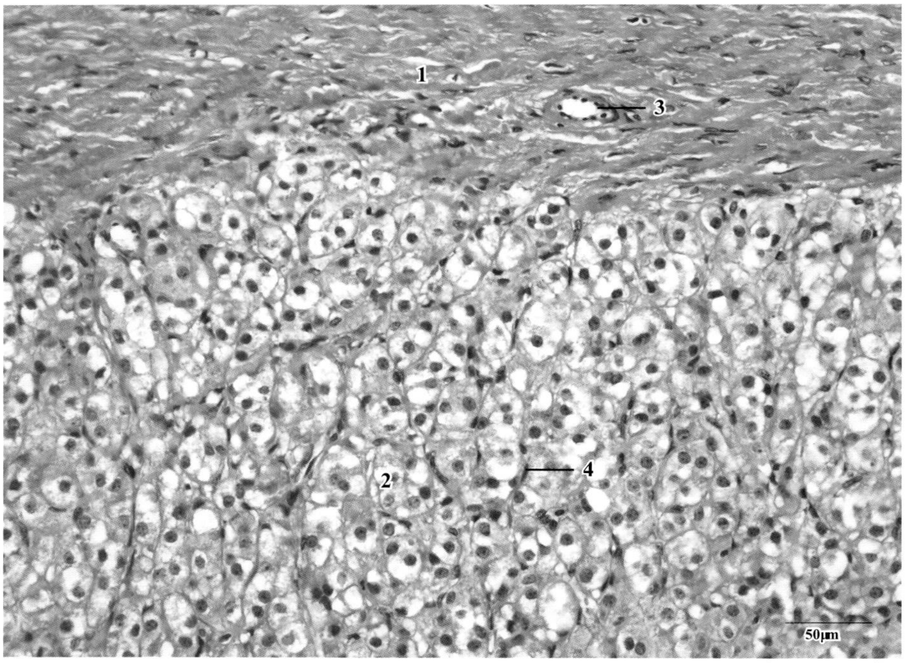

1- 被膜 capsule
2- 球状带 glomerular zone
3- 血管 blood vessel
4- 内皮细胞 endothelial cell

图 12-33　肾上腺组织切片示球状带（HE 染色，高倍镜）
Figure 12-33　Histological section of a adrenal gland showing the glomerular zone (HE staining, higher power).

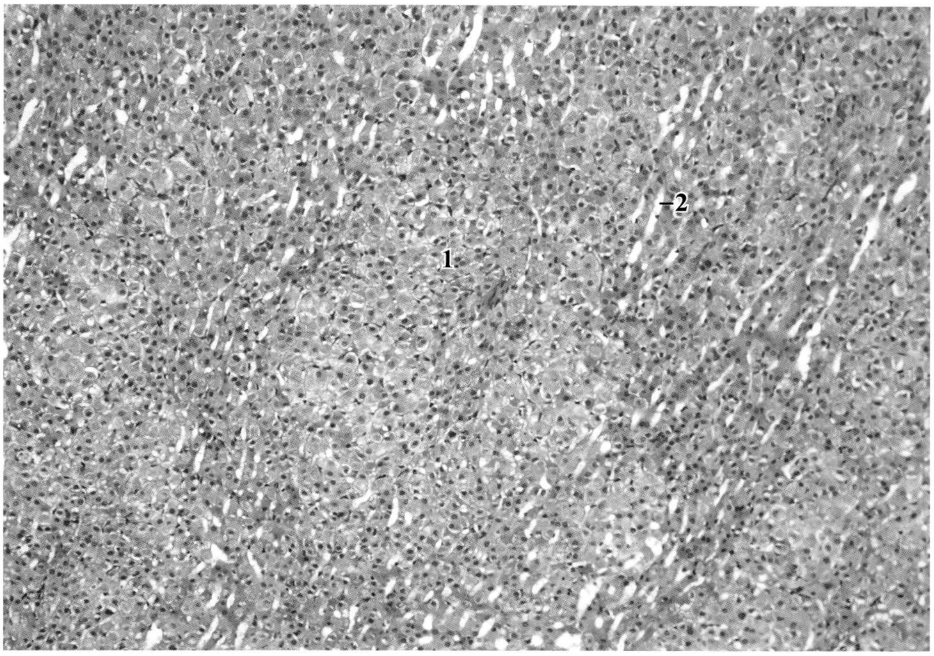

图 12-34　肾上腺皮质部束状带组织切片（HE 染色，高倍镜）
Figure 12-34　Histological section of the fasciculate zone of adrenal gland (HE staining, higher power).

1- 束状带 fasciculate zone　　**2-** 血窦 blood sinus

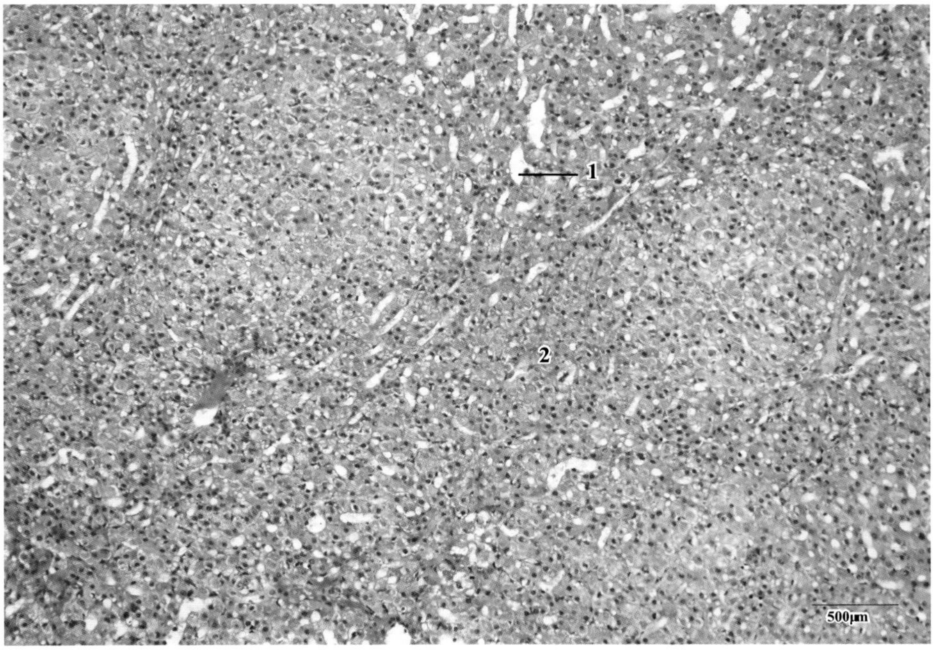

图 12-35　肾上腺皮质部束状带组织切片（HE 染色，高倍镜）
Figure 12-35　Histological section of the fasciculate zone of adrenal gland (HE staining, higher power).

1- 血窦 blood sinus　　**2-** 束状带 fasciculate zone

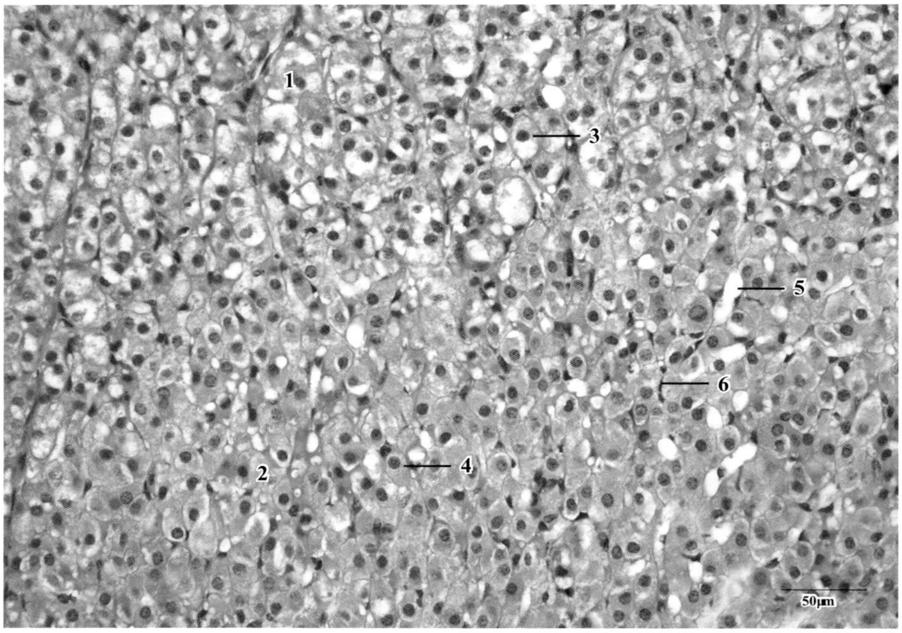

图 12-36　肾上腺皮质部球状带和束状带组织切片（HE 染色，高倍镜）
Figure 12-36　Histological section of the glomerular and fasciculate zones of adrenal gland (HE staining, higher power).

1- 球状带 glomerular zone
2- 束状带 fasciculate zone
3- 球状带细胞 glomerular zone cell
4- 束状带细胞 fasciculate zone cell
5- 血窦 blood sinus
6- 内皮细胞 endothelial cell

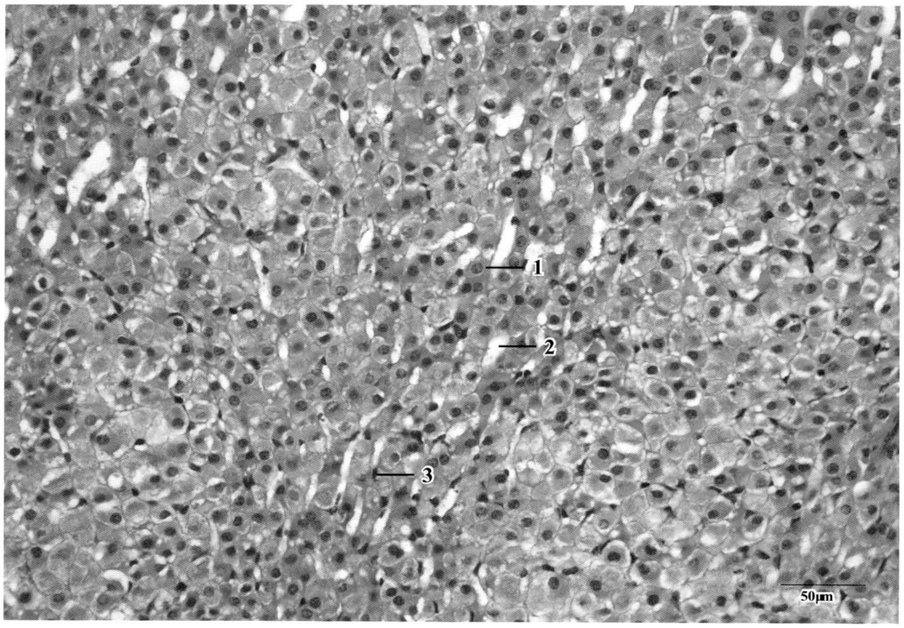

图 12-37　肾上腺皮质部网状带组织切片（HE 染色，高倍镜）
Figure 12-37　Histological section of the reticular zone of adrenal gland (HE staining, higher power).

1- 网状带细胞 reticular zone cell
2- 血窦 blood sinus
3- 内皮细胞 endothelial cell

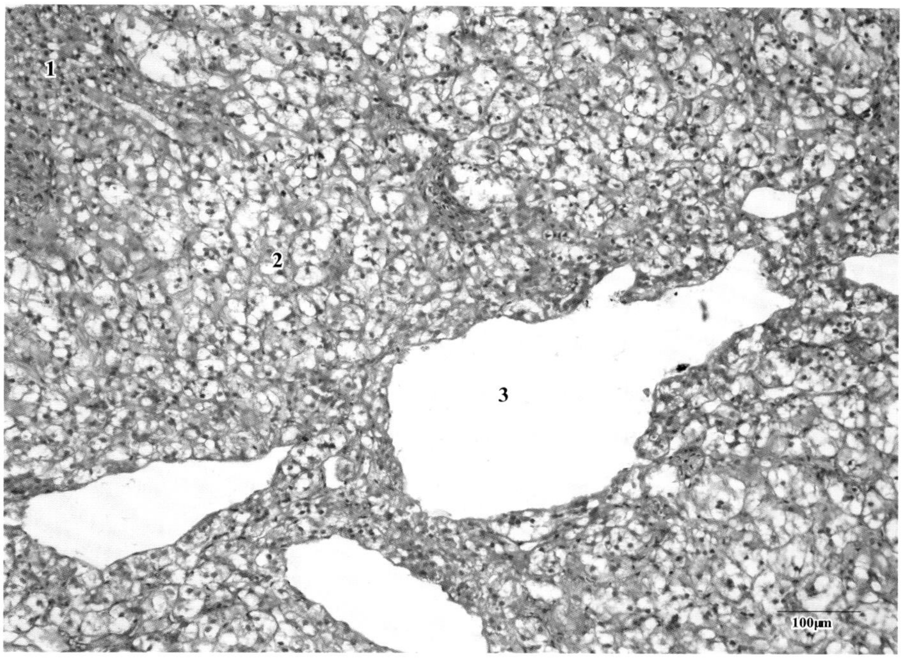

图 12-38　肾上腺髓质组织切片（HE 染色，高倍镜）
Figure 12-38　Histological section of the adrenal medulla (HE staining, higher power).

1- 网状带 reticular zone
2- 髓质 medulla
3- 静脉 vein

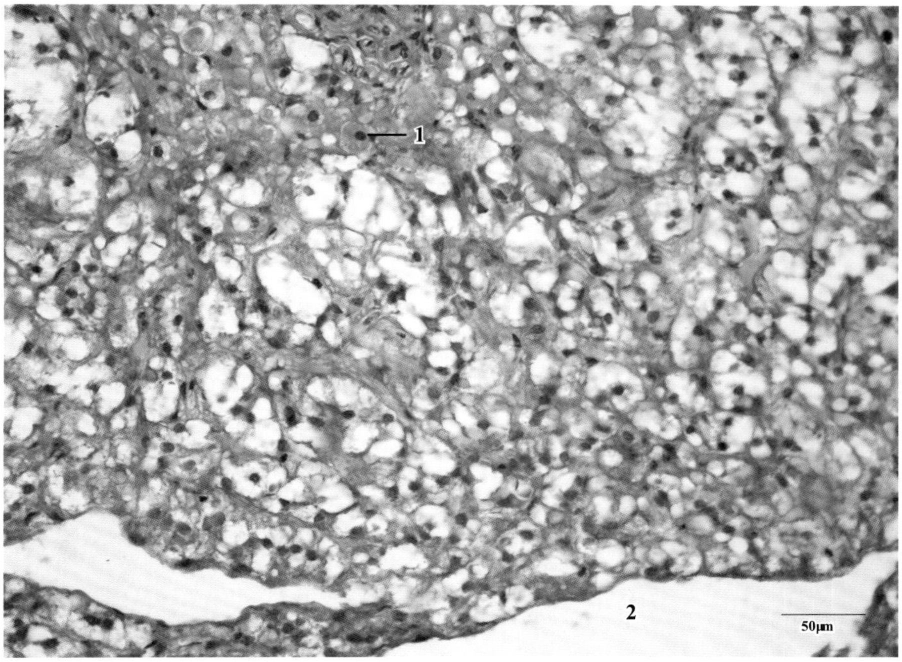

图 12-39　肾上腺髓质组织切片（HE 染色，高倍镜）
Figure 12-39　Histological section of the adrenal medulla (HE staining, higher power).

1- 交感神经节细胞 sympathetic ganglion cell
2- 静脉 vein

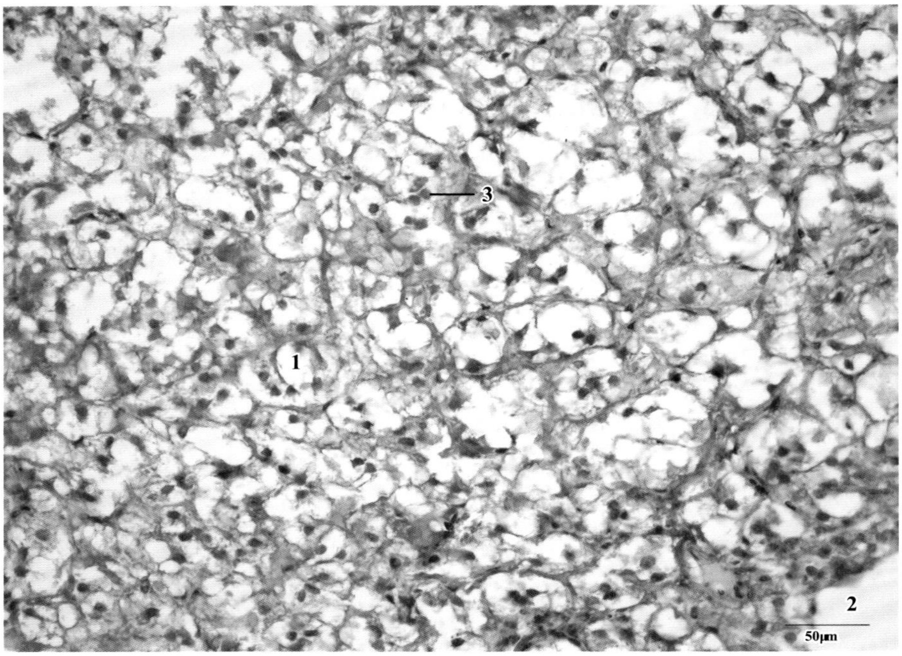

图 12-40 肾上腺髓质组织切片（HE 染色，高倍镜）
Figure 12-40 Histological section of the adrenal medulla (HE staining, higher power).

1- 髓质 medulla
2- 静脉 vein
3- 髓细胞 medullary cell

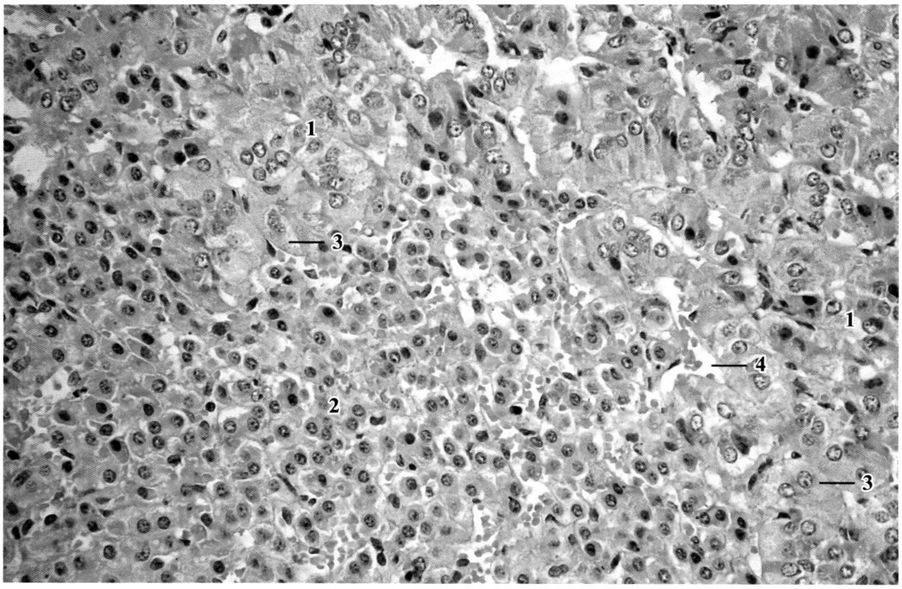

图 12-41 肾上腺网状带和髓质组织切片（HE 染色，高倍镜）
Figure 12-41 Histological section of the reticular zone and medulla of adrenal gland (HE staining, higher power).

1- 网状带 reticular zone
2- 髓质 medulla
3- 网状带细胞 reticular zone cell
4- 血窦 blood sinus

第十三章

感觉器官

Chapter 13

Sense organ

感觉器官（sense organ）是由感受器和附属器构成。感受器是感觉神经末梢的特殊装置，广泛分布于身体各器官和组织内。感受器的功能是接受机体内、外环境各种不同的刺激，并将刺激转变为神经冲动，经过感觉神经传入中枢神经，最后到达大脑皮质，产生相应的感觉。

一、视觉器官

视觉器官（visual organ）又称眼（eye），由眼球和附属器官构成。能感受光的刺激，经视神经传到中枢而产生视觉。

1. 眼球（eyeball）：是视觉器官的主要部分，位于眼眶内，后端有视神经与脑相连。眼球近似球形，由眼球壁和眼球内容物组成。

（1）眼球壁：从外向内由纤维膜、血管膜和视网膜3层构成。

1）纤维膜（fibrous membrane, fibrous tunic）又叫白膜，由致密结缔组织构成，厚而坚韧，具有保护眼球内容物的作用。位于眼球壁外层，分为后部的巩膜和前部的角膜两部分。巩膜（sclera）占纤维膜的后4/5，具有保护眼球和维持眼球形状的作用。巩膜前部与角膜相连接的地方为角膜巩膜缘（corneoscleral junction），呈环状，其深面有巩膜静脉窦，是眼房水流出的通道，有调节眼压的作用。如果眼房水不能顺利流出，眼压就会增加，导致青光眼。巩膜的后腹侧、视神经纤维穿出的部位有巩膜筛板（cribriform plate of sclera）。角膜（cornea）占纤维膜的前1/5，无色透明，具有折光作用。角膜前面隆凸，后面凹陷。周缘较厚，中部薄，嵌入巩膜中。角膜主要由平行排列的胶原纤维组成，呈板层状，内无血管（其营养由角膜周围的血管供给），但分布有丰富的感觉神经末梢，故感觉灵敏。角膜发炎时应及时治疗，否则会造成角膜混浊，影响视力。

2）血管膜（vascular tunic）是眼球壁的中层，位于纤维膜与视网膜之间，富有血管和色素细胞，具有输送营养和吸收眼内分散光线的作用，并形成暗的环境，有利于视网膜对光色的感应。血管膜由后向前分为脉络膜、睫状体和虹膜三部分。脉络膜（choroid）呈暗褐色，衬于巩膜的内面，但没有照膜。睫状体（ciliary body）位于脉络膜和虹膜之间，是血管膜中部的增厚部分，呈环带形围于晶状体周围（宽约1 cm），形成睫状环。睫状环可分为内部的睫状突和外部的睫状肌。睫状突是睫状体内表面许多呈放射状排列的皱褶，约有100个。睫状肌由平滑肌构成，位于睫状体的外部，肌纤维起于角膜与巩膜连接处，向后止于睫状环。在注视近或远距离的物体时，睫状肌能调节晶状体的形状变化。虹膜（iris）位于血管膜的前部，在晶状体之前，呈圆盘状。虹膜的颜色因色素细胞多少和分布不同而有差异，一般为灰褐色或黄褐色。虹膜的周缘连于睫状体，其中央有一孔以透过光线，称为瞳孔（pupil）。

3）视网膜（retina）位于眼球壁内层，分为视部和盲部。视部具有感光作用，衬于脉络膜的内面，且与其紧密相连，薄而柔软。平时略呈淡红色，死后混浊，变为灰白色，易于从脉络膜上脱落。盲部位于睫状体和虹膜的内面，很薄，无感光作用，外层为色素上皮，内层无神经元。

视网膜视部属神经组织，具有很强的感光作用。主要由色素上皮细胞、视锥细胞、视杆细胞、双极细胞、水平细胞、无长突细胞、Müller细胞和视网膜神经节细胞构成的10层结构，由外至内分别为色素上皮层（pigment epithelium layer）、视锥和视杆细胞层（cones and rods layer）、外界膜（external limiting membrane）、外颗粒层（outer nuclear layer）、外网状层（outer plexiform layer）、内颗粒层（internal nuclear layer）、内网状层（inner plexiform layer）、神经节细胞层（ganglion cell layer）、神经纤维层（nerve fiber layer）和内界膜（inner limiting membrane）。

（2）眼球内容物：是眼球内一些无色透明的折光结构，包括晶状体、眼房水和玻璃体。其作用是与角膜一起组成眼的折光系统，将通过眼球的光线经过折射，使焦点集中在视网膜上，形成影像。

1）晶状体（lens）位于虹膜与玻璃体之间，富有弹性，外面包有一弹性囊。晶状体周缘借着睫状小带连于睫状突。睫状肌的收缩和弛缓，可以改变睫状小带对晶状体的拉力，从而改变晶状体的凸度，以调节视力。

2）眼房和眼房水。眼房（eye-chamber）是位于角膜后面和晶状体前面之间的空隙，被虹膜分为眼前房和眼后房，两者以瞳孔相交通。眼房水（aqueous humor）为眼房里的无色透明液体，由睫状突和虹膜产生，然后在眼前房的周缘渗入巩膜静脉窦而至眼静脉。

3）玻璃体（vitreous body）位于晶状体与视网膜之间，是无色透明的半流动状胶体，外包一层很薄的透明膜，为玻璃体膜。玻璃体除有折光作用外，还有支持视网膜的作用。

2. 眼的附属器官（adnexa of eye）：有眼睑、泪器、眼球肌和眶骨膜等，它们对眼球有保护、运动和支持作用。

（1）眼睑（eyelid）：为覆盖于眼球前方的皮肤褶，分为上眼睑和下眼睑。上眼睑和下眼睑间形成眼裂（rima oculi）。眼睑的外面为皮肤，中间主要为眼轮匝肌，内面衬着一薄层湿润而富有血管的膜，为睑结膜（palpebral conjunctiva）。睑结膜还折转覆盖在眼球巩膜的前部，为球结膜（bulbar conjunctiva）。睑结膜与球结膜共同称为眼结膜。当眼睑闭合时，结膜合成一完整的结膜囊（conjunctival sac）。眼睑缘长有睫毛。

第3眼睑（third eyelid），又称瞬膜（palpebra tertius），是位于内眼角的半月状结膜皱褶，褶内有三角形软骨板，为弹性软骨。

（2）泪器（lacrimal apparatus）：由泪腺和泪道组成。泪腺（lacrimal gland）位于额骨眶上突的基部、眼球的背外侧，以数条输出管开口于上眼睑结膜。泪腺为管泡腺，分泌黏液型泪液，有湿润和清洁结膜及角膜的作用。泪道（lacrimal passage）是泪液排出的通道，分泪管和鼻泪管两段。猪无泪囊，鼻泪管开口于下鼻道后部。

（3）眼球肌（muscles of eyeball）：是一些使眼球灵活运动的横纹肌，位于眶骨膜内，均起始于视神经孔周围的眼眶壁，止于眼球巩膜，包括眼球退缩肌（retractor muscles of eyeball）、眼球直肌（straight muscles of eyeball）、眼球斜肌（oblique muscles of eyeball）和上睑提肌（levator muscle of the upper eyelid）。

（4）眶骨膜（orbital periosteum）：为眼眶内衬一层致密坚韧的圆锥状纤维鞘，又称眼鞘，包围着眼球、眼肌、眼的血管和神经及泪腺。它源于骨膜，其内、外间隙中充填着大量脂肪，与眼眶和眶骨膜一起构成眼的保护器官。

二、位听器官

位听器官包括位觉器官（position sense organ）和听觉器官（auditory organ）两部分。这两部分功能虽然不同，而结构上难以分开。位听器官由外耳、中耳和内耳三部分构成。外耳收集声波，中耳传导声波，内耳中有听觉感受器和位置觉感受器。

1．外耳（external ear）：包括耳郭、外耳道和鼓膜三部分。

（1）耳郭（auricle）：以耳郭软骨为基础，内、外均覆有皮肤。耳郭背面隆凸称为耳背，与耳背相对应的凹面称为耳舟。耳郭前、后缘向上汇合形成耳尖，耳郭下部叫耳根，在腮腺深部连于外耳道。

（2）外耳道（external acoustic meatus）：是从耳郭基部到鼓膜的通道，外口大、内口小，内口朝向中耳。由软骨性外耳道和骨性外耳道两部分构成。外侧部是软骨性外耳道，其上部与耳郭软骨相接，下部固着于骨性外耳道的外口。内侧部是骨性外耳道即颞骨的外耳道，呈漏斗状，内面衬有皮肤。在软骨管部的皮肤含有皮脂腺和耵聍腺（ceruminous gland）。后者为变异的汗腺，分泌耳蜡，又称耵聍。

（3）鼓膜（tympanic membrane）：位于外耳道底部，是外耳和中耳的分界。鼓膜分3层，外层为表皮层，来自外耳道皮肤；中层为纤维层，由致密胶质纤维构成；内层为黏膜层，为鼓室黏膜的延续部分。

2．中耳（middle ear）：由鼓室、听小骨和咽鼓管组成。

（1）鼓室（tympanic cavity）：是颞骨里一个含有空气的骨腔，内面被覆黏膜。鼓室的外侧壁是鼓膜，与外耳道隔开；内侧壁为骨质壁或迷路壁，与内耳为界。在内侧壁上有一隆起称为岬（promontory），岬的前方有前庭窗，被镫骨底及环状韧带封闭；岬的后方有蜗窗（fenestra cochleae），被第2鼓膜所封闭。鼓室的前下方有孔通咽鼓管。

（2）听小骨（auditory ossicle）：位于鼓室内，共有3块，由外向内依次为锤骨（malleus）、砧骨（incus）和镫骨（stapes）。它们彼此以关节连成一个骨链，一端以锤骨柄附着于鼓膜，另一端以镫骨底的环状韧带附着于前庭窗。鼓膜接受声波而振动，再经此骨链将声波传递到内耳。

（3）咽鼓管（pharyngotympanic tube）：又称耳咽管（auditory tube），为连接于咽和鼓室之间一个沟状管道，起自鼓室而开口于咽腔。

3．内耳（internal ear）：又称迷路（因结构复杂而得名），位于岩颞骨岩部内，分为骨迷路和膜迷路两部分。它们是盘曲于鼓室内侧骨质内的骨管，在骨管内套有膜管。骨管称骨迷路；膜管称膜迷路。膜迷路内充满内淋巴，在膜迷路与骨迷路之间充满外淋巴，它们起着传递声波刺激和感受位置变动刺激的作用。

（1）骨迷路（osseous labyrinth）：位于鼓室内侧的骨质内，由前庭、3个骨质半规管和耳蜗三部分构成。

前庭（vestibule）为位于骨迷路中部较为扩大的空腔，呈球形，向前下方与耳蜗相通，向后上方与骨半规管相通。前庭的外侧壁(即鼓室的内侧壁)上有前庭窗和蜗窗；前庭的内侧壁是构成内耳道底的部分，壁上有前庭嵴，嵴的前方有一球囊隐窝（spherical recess）；后方有一椭圆囊隐窝（elliptical recess）；后下方有一前庭小管内口。

骨半规管（canales semicirculares ossei）位于前庭的后上方，由3个彼此互相垂直的半环形骨管组成，按其位置分别为前半规管、后半规管和外半规管。每个半规管的一端膨大，为骨壶腹（osseous ampulla），另一端为骨脚（bony crura）。

耳蜗（cochlea）位于前庭的前下方，由一耳蜗螺旋管围绕蜗轴（由骨松质构成）盘旋数圈而成，呈圆锥形。管的起端与前庭相通，盲端终止于蜗顶。沿蜗轴向螺旋管内发出骨螺旋板，将螺旋管不完全地分隔为前庭阶和鼓室阶两部分。

（2）膜迷路（membranous labyrinth）：为套于骨迷路内，互相通连的膜性囊和管（由纤维组织构成，内面衬有单层上皮），形状与骨迷路相似，由椭圆囊、球囊、膜半规管和耳蜗管组成。

椭圆囊（utriculus）位于前庭的椭圆隐窝内，与3个膜半规管相通。

球囊（sacculus）位于前庭的球状隐窝内，一端与椭圆囊相通，另一端与耳蜗管相通。

膜半规管（semicircular duct）套于骨半规管内，与骨半规管的形状一致，膜壶腹和膜脚均开口于椭圆囊。在椭圆囊、球囊和膜半规管壶腹的壁上，均有一增厚的部分，分别形成椭圆囊斑（macula utriculi）、球囊斑（macula sacculi）和壶腹嵴（crista ampullaris）。

耳蜗管（cochlear duct）位于耳蜗螺旋管内，与耳蜗螺旋管的形状一致。一端与球囊相通连，另一端终止于蜗顶。在耳蜗管的基底膜上有感觉上皮的隆起，称为螺旋器（spiral organ），又称柯蒂器（organ of Corti），为听觉感受器，声波经一系列途径传到耳蜗后，由耳蜗管内的螺旋器将其转化为神经冲动，再经前庭耳蜗神经的耳蜗支传到脑，而产生听觉。

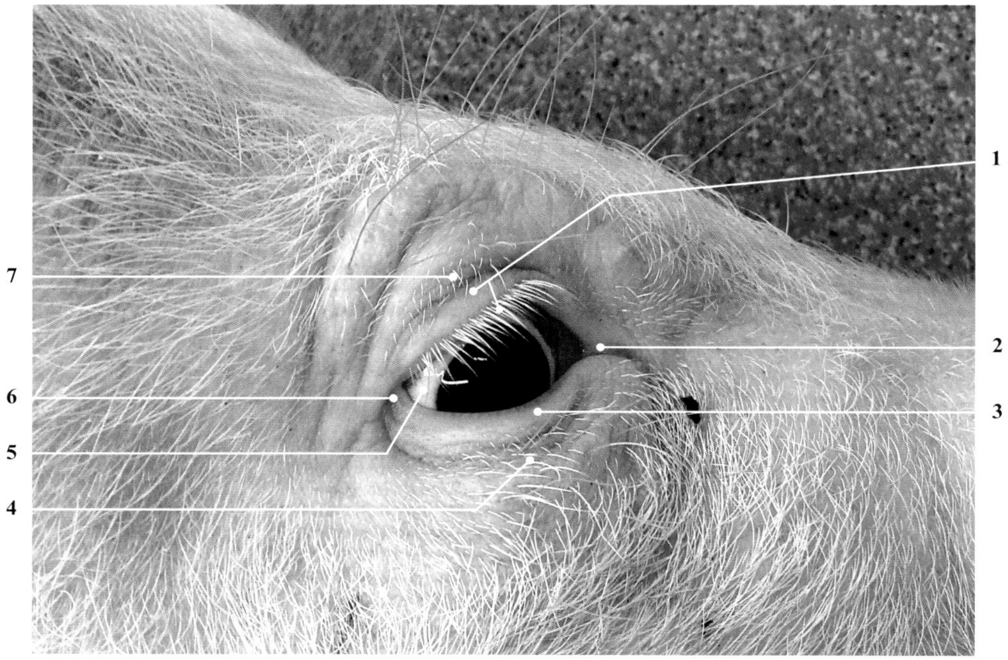

图 13-1 公猪右侧眼
Figure 13-1 Right eye of a boar.

1- 上眼睑及睫毛 superior eyelid and eyelash
2- 内眼角 medial canthus
3- 下眼睑 inferior eyelid
4- 眶下触毛 infraorbital tactile hair
5- 眼球 eye ball
6- 外眼角 external canthus
7- 眶上触毛 supraorbital tactile hair

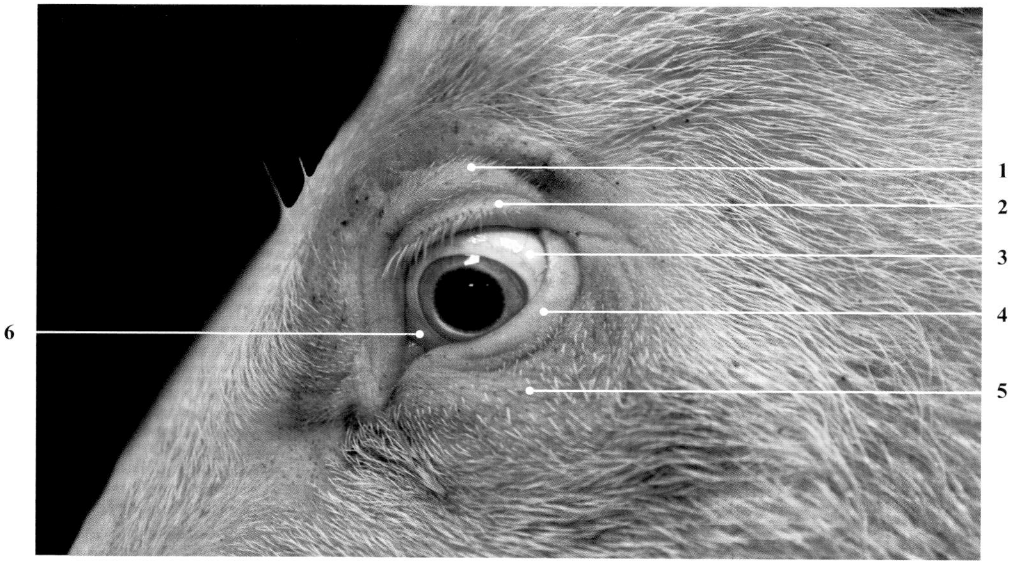

图 13-2 母猪左侧眼
Figure 13-2 Left eye of a sow.

1- 眶上触毛 supraorbital tactile hair
2- 上眼睑及睫毛 superior eyelid and eyelash
3- 眼球 eye ball
4- 下眼睑 inferior eyelid
5- 眶下触毛 infraorbital tactile hair
6- 第 3 眼睑（瞬膜） third eyelid (palpebra tertius)

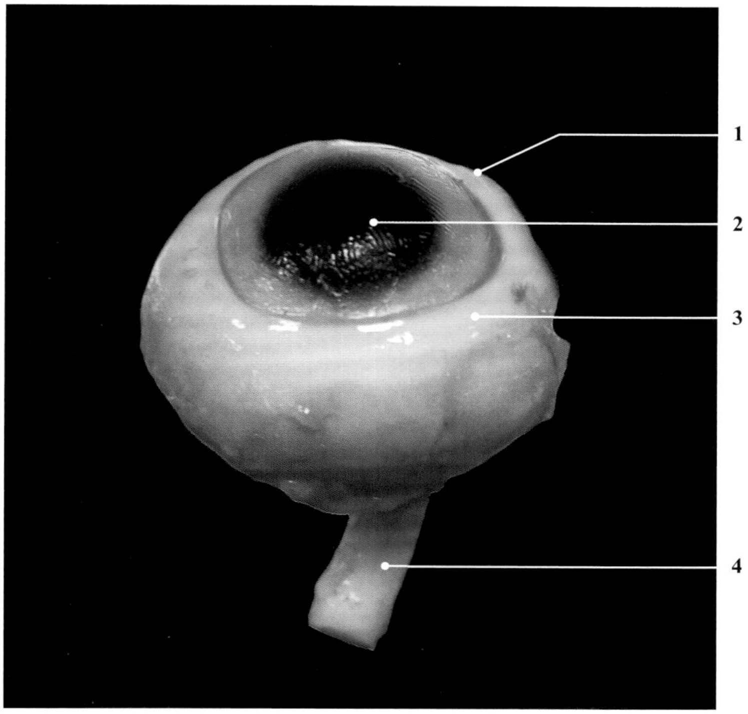

图 13-3 母猪眼球（正面观）
Figure 13-3 Eyeball of a sow (anterior view).

1- 角膜巩膜缘 corneoscleral junction
2- 角膜 cornea
3- 巩膜 sclera
4- 视神经 optic nerve

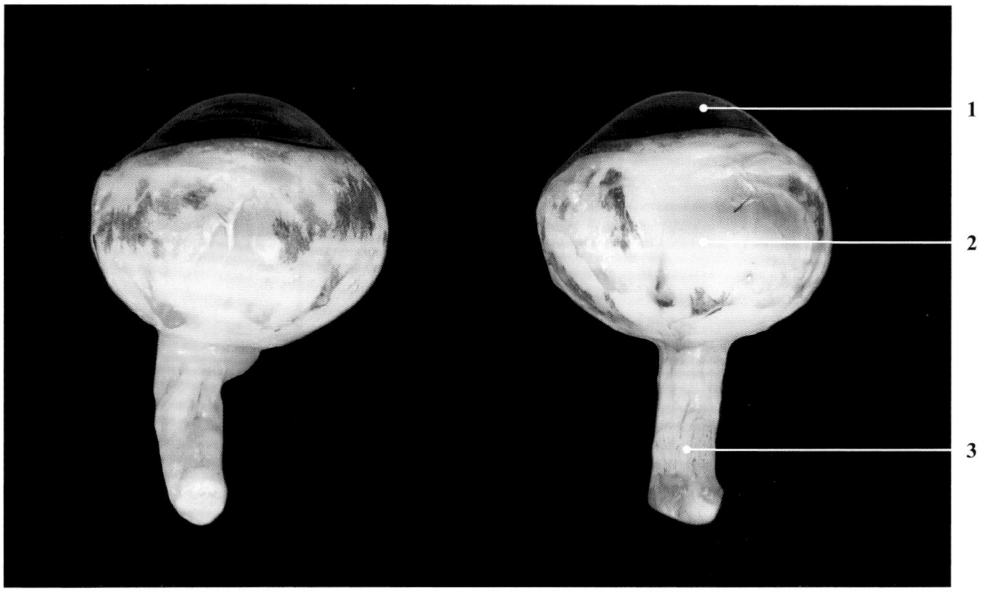

图 13-4 眼球和视神经（侧面观）
Figure 13-4 Eyeball and optic nerve (lateral view).

1- 角膜 cornea
2- 巩膜 sclera
3- 视神经 optic nerve

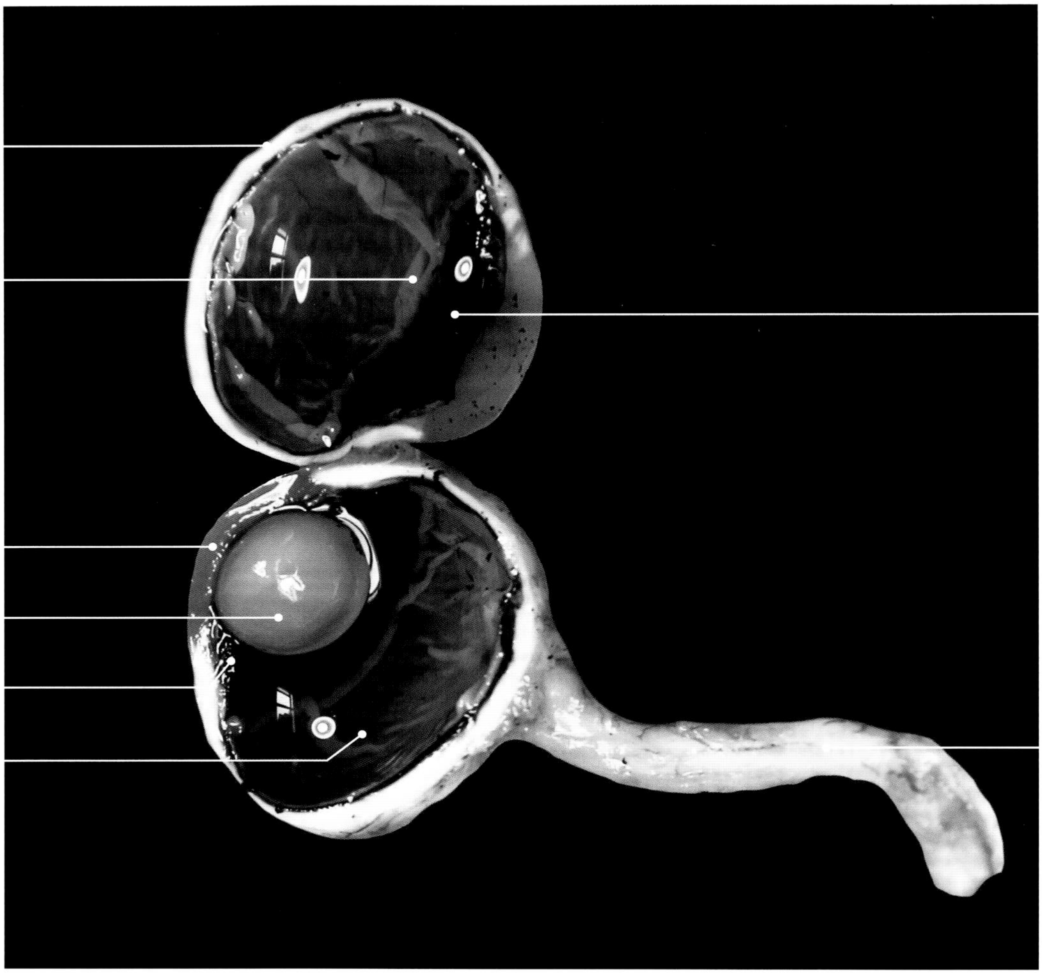

图 13-5 眼球（剖面）
Figure 13-5 Eyeball (cross section).

1- 脉络膜 choroid
2- 视神经 optic nerve
3- 视网膜 retina
4- 睫状体 ciliary body
5- 晶状体 lens
6- 虹膜 iris
7- 巩膜 sclera

图 13-6 母猪玻璃体（剖面）
Figure 13-6 Vitreous body of a sow (cross section).

1- 视网膜 retina
2- 巩膜 sclera
3- 玻璃体 vitreous body
4- 睫状体 ciliary body
5- 晶状体 lens

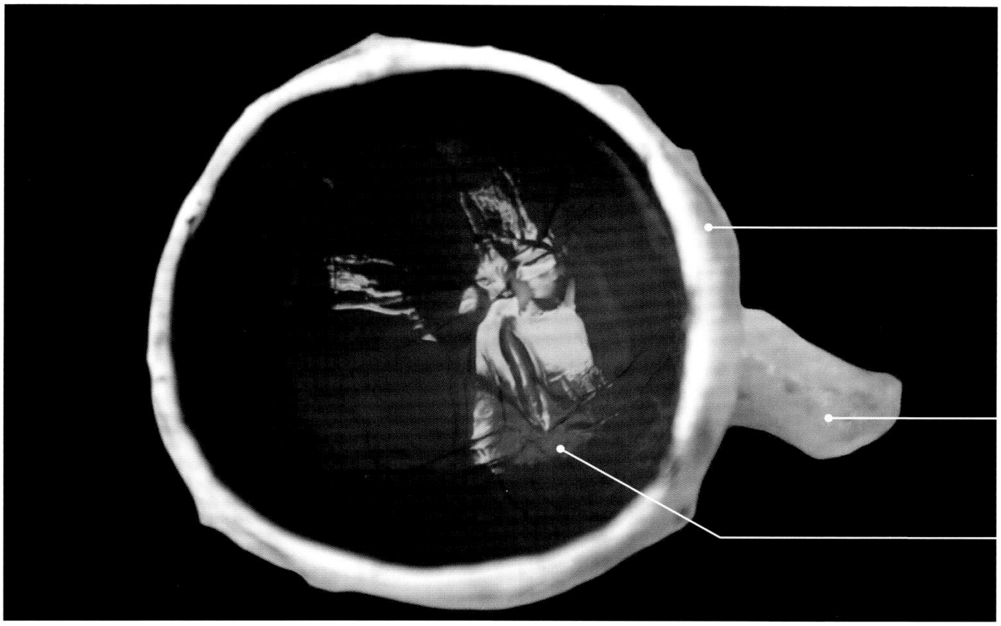

图 13-7 母猪脉络膜（剖面）
Figure 13-7 Choroid of a sow (cross section).

1- 巩膜 sclera
2- 视神经 optic nerve
3- 脉络膜 choroid

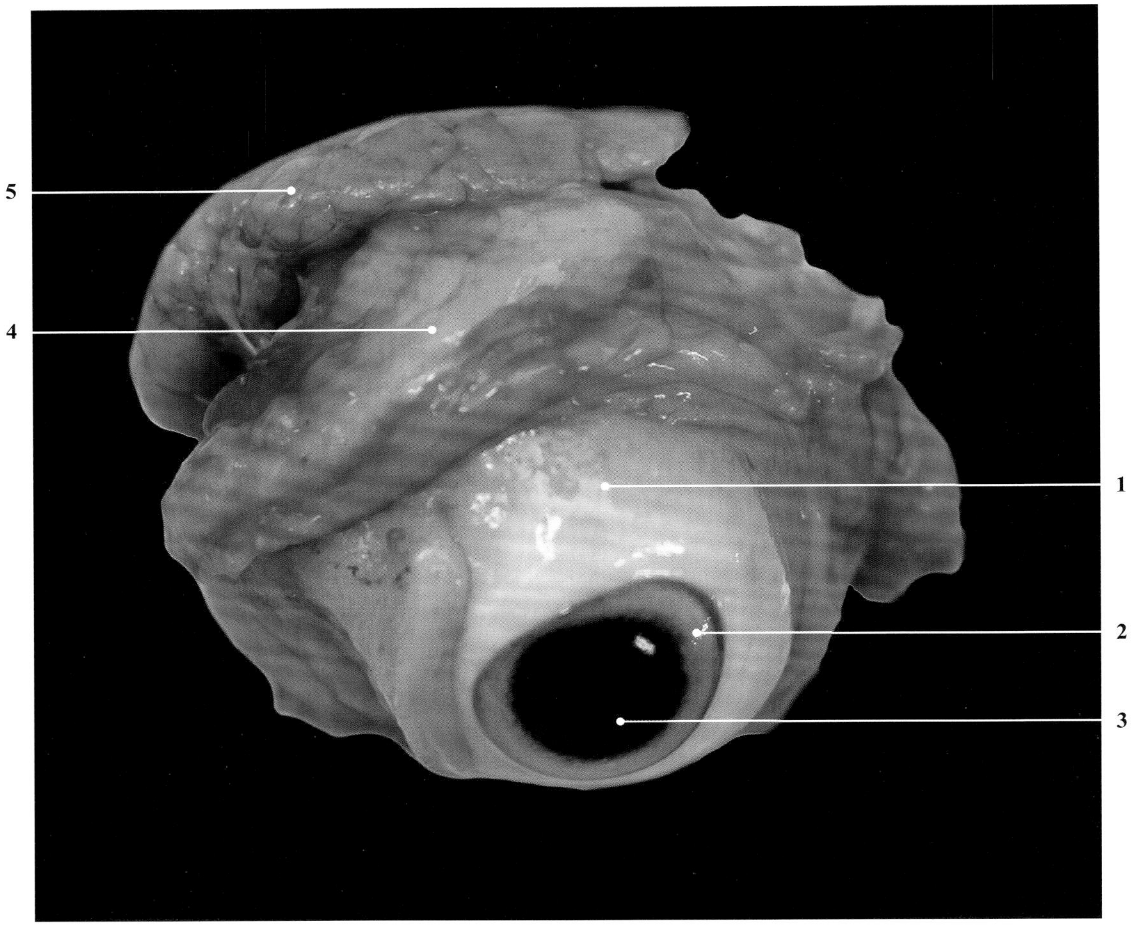

图 13-8　泪腺
Figure 13-8　Lacrimal gland.

1- 巩膜 sclera
2- 虹膜 iris
3- 角膜与瞳孔 cornea and pupil
4- 脂肪 fat
5- 泪腺 lacrimal gland

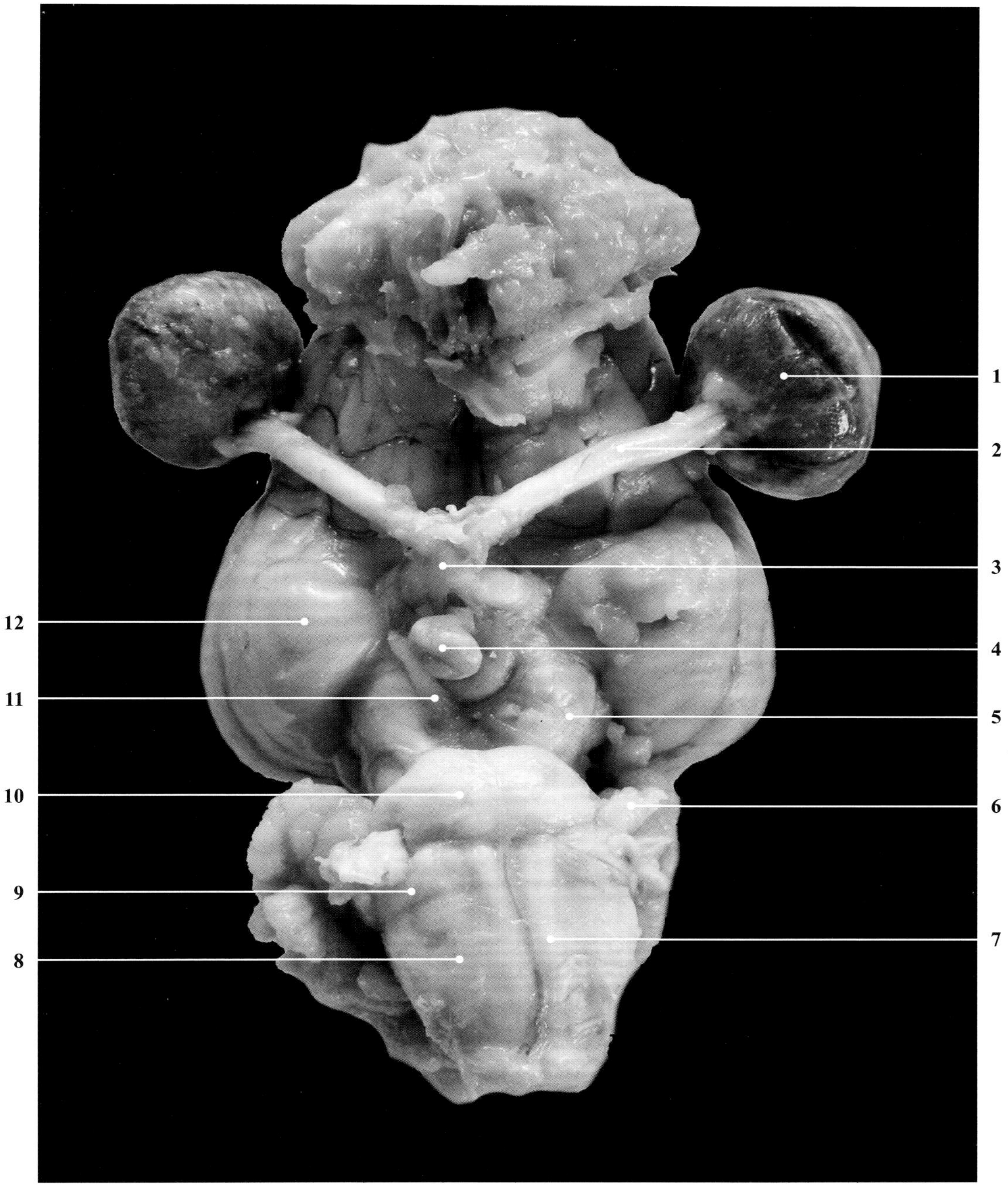

图 13-9　视神经与视交叉
Figure 13-9　Optic nerve and optic chiasma.

1- 眼球 eye ball
2- 视神经 optic nerve
3- 视交叉 optic chiasma
4- 垂体 pituitary gland
5- 大脑脚 cerebral peduncle
6- 三叉神经根 roots of trigeminal nerve
7- 锥体 pyramid
8- 延髓 medulla oblongata
9- 斜方体 trapezoid body
10- 脑桥 pons
11- 动眼神经根 root of oculomotor nerve
12- 梨状叶 piriform lobe

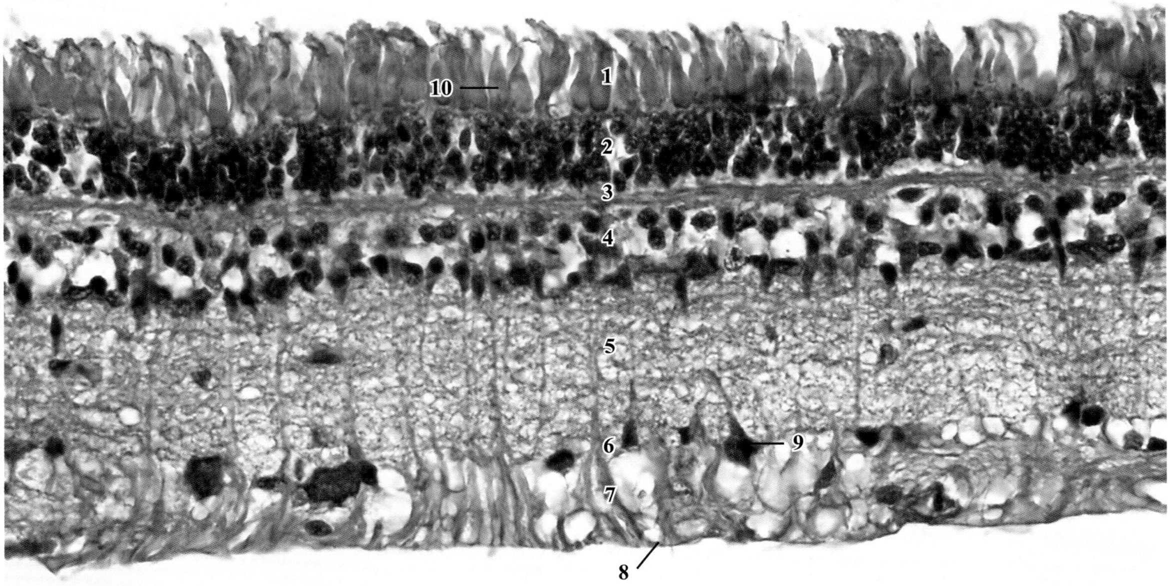

图 13-10　视网膜组织切片（HE 染色，高倍镜）
Figure 13-10　Histological section of a retina (HE staining, high power).

1- 视锥和视杆细胞层 cones and rods layer
2- 外颗粒层 outer nuclear layer
3- 外网状层 outer plexiform layer
4- 内颗粒层 inner nuclear layer
5- 内网状层 inner plexiform layer
6- 神经节细胞层 ganglion cell layer
7- 神经纤维层 nerve fiber layer
8- 内界膜 inner limiting membrane
9- 视网膜神经节细胞 retinal ganglion cell
10- 视锥细胞 cone cell

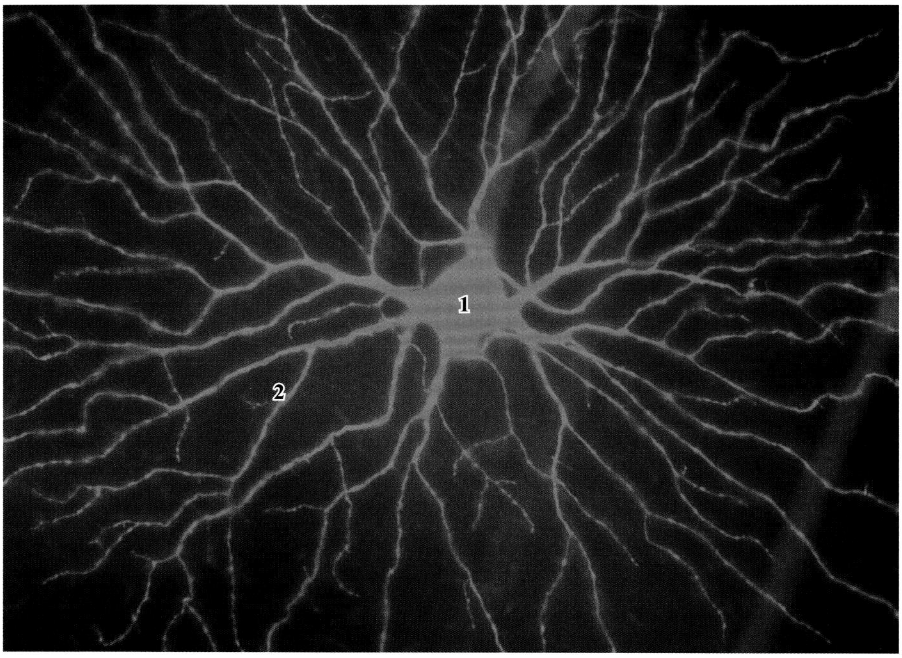

1- 胞体 soma
2- 树突 dendrite

图 13-11　α 亚型视网膜神经节细胞（DiI 标记）
Figure 13-11　Retinal ganglion cell, α subtype (DiI labeled).

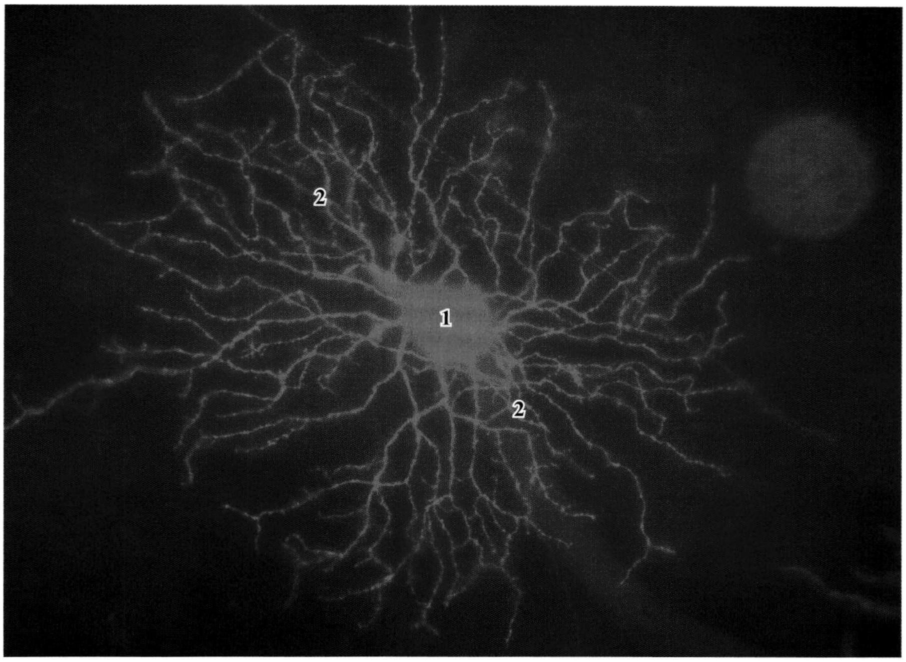

1- 胞体 soma
2- 树突 dendrite

图 13-12 β 亚型视网膜神经节细胞（DiI 标记）
Figure 13-12 Retinal ganglion cell, β subtype (DiI labeled).

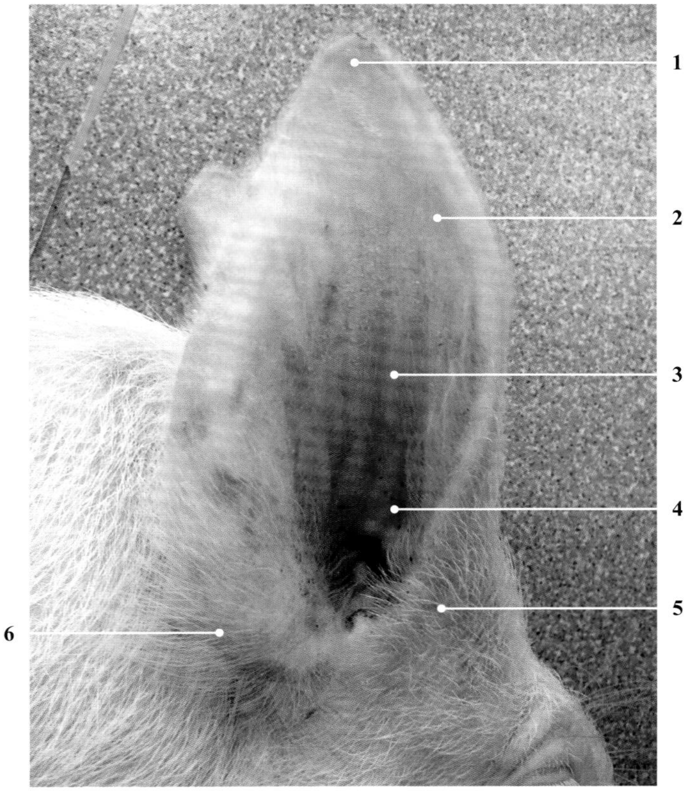

图 13-13 外耳（正面观）
Figure 13-13 External ear (anterior view).

1- 耳尖 apex of ear
2- 耳郭 auricle
3- 对耳轮 anthelix
4- 外耳道 external auditory meatus
5- 耳根 root of ear
6- 耳毛 tragi

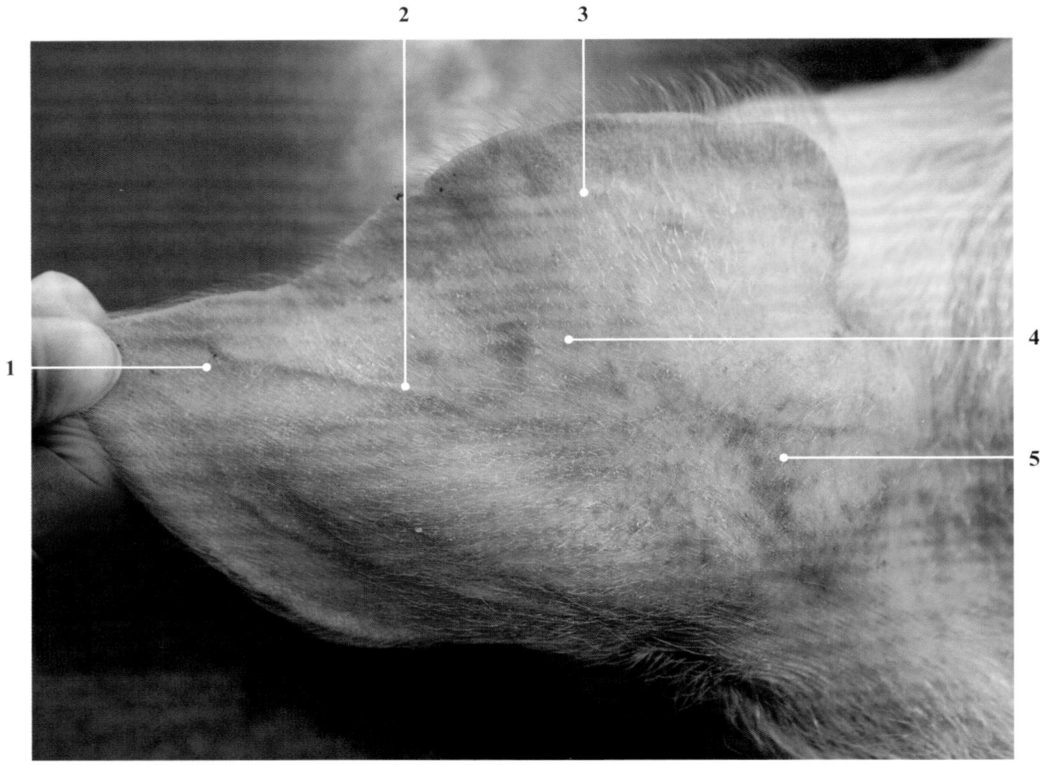

图 13-14　外耳（背侧观）
Figure 13-14　External ear (dorsal view).

1- 耳尖 apex of ear
2- 耳大静脉 great auricular vein
3- 耳前静脉 anterior auricular vein
4- 耳背 back of ear
5- 耳根 root of ear

参考文献

1. 陈耀星. 动物解剖学彩色图谱［M］. 北京：中国农业大学出版社, 2013.
2. 陈耀星. 畜禽解剖学［M］. 第3版. 北京：中国农业大学出版社, 2010.
3. 陈耀星，动物局部解剖学［M］. 第2版. 北京：中国农业大学出版社, 2010.
4. Horst Erich König, Yans-Goorg Liebich. 家畜兽医解剖学教程与彩色图谱［M］. 第3版. 陈耀星，刘为民译. 北京：中国农业大学出版社，2009.
5. 董常生. 家畜解剖学［M］. 第5版. 北京：中国农业出版社，2015.
6. William J. Bacha, Jr, Linda M. Bacha. 兽医组织学彩色图谱. 陈耀星译. 北京：中国农业大学出版社，2007.
7. 彭克美，等. 动物组织学及胚胎学［M］. 北京：高等教育出版社，2009.
8. 张立教，秦鹏春，段莫超，等. 猪的解剖组织［M］. 第2版. 北京：科学出版社，1984.
9. 林辉. 猪解剖图谱［M］. 北京：农业出版社，1992.
10. 李德雪，尹昕. 动物组织学彩色图谱［M］. 长春：吉林科学技术出版社，1995.
11. Sack WO. Horowitz / Kramer Atlas of Musculoskeletal Anatomy of the Pig［M］. New York: Veterinary Textbooks, Ithaca, 1982.
12. Colville T, Bassert JM. Clinical Anatomy and Physiology for Veterinary Technicians［M］. 2nd ed. Canada: Mosby, Inc., 2008.
13. Dyce KM, Sack WO, Wending CJG. Textbook of Veterinary Anatomy［M］. 4th ed. Canada: W B Saunders Company, 2010.
14. Popesko P. Atlas of topographical anatomy of the domestic animals［M］. London: WB Saunders Company Philadelphia, 1985.
15. World Association of Veterinary Anatomists［M］. 4th ed. Belgium: Nomina Anatomica Veterinaria, 1992.

中英索引

Ⅰ型肺泡细胞 type Ⅰ alveolar cell 238
Ⅱ型肺泡细胞 type Ⅱ alveolar cell 238
α亚型视网膜神经节细胞 retinal ganglion cell, α subtype 478
β亚型视网膜神经节细胞 retinal ganglion cell, β subtype 479

B

白膜 tunica albuginea 268,274,278,298,304
白髓 white pulp 371,381,382,383,384
白体 corpus albicans 311
白细胞 leukocyte，white blood cell，WBC 332
白质 white matter 400,408
半棘肌 semispinal muscle 89
半腱肌 semitendinous muscle 83,85,102,103,104,105,106,108,119,360,388,389,432,433,434
瓣膜 cardiac valve 347
半膜肌 semimembranous muscle 83,102,103,104,105,106,107,108,109,360,388,389,434
半月瓣 semilunar valve 328,339
鲍曼腔 Bowman’s space 254,255
包皮 prepuce 247,265,270,293
包皮口 preputial opening 270
包皮憩室（包皮盲囊）preputial diverticulum 84,87,246,247,270,271,272,290,292,295,296,386,387
胞体 soma 201,226,405,406,478,479
鲍曼囊 Bowman’s capsule 241
杯状细胞 goblet cell 116,171,173,176,177,178,180,181,182,184,185,207,224,237
背部 back 4,5,7
背侧端 dorsal extremity 380
背侧根 dorsal root 408
背侧弓 dorsal arch 43,44
背侧角 dorsal horn 408
背侧结节 dorsal tubercle 43,44
背侧缘 dorsal border 61
背阔肌 broadest muscle of the back 82,84,85,86,87,88,90,93,95,97,387
背腰最长肌 dorsal-lumbus longest muscle 81,84,85,86,88,89,90,104
被毛 clothing hair 3,121,203
被膜 capsule 250,252,253,303,305,309,371,381,382,391,392,393,447,450,451,458,460,461,462
被膜下窦 subcapsular sinus 391,392,393
被囊细胞 capsular cell 226
被皮系统 integumentary system 2
贲门 cardia 149,152,153,154,155
贲门部 cardiac part 149,150,151,152
贲门腺 cardiac gland 116,160,161
贲门腺区 region of cardiac gland 153,154,155,
鼻 nose 206
鼻唇提肌 nasolabial levator muscle 81
鼻骨 nasal bone 29,35,36,38,39,139
鼻颌切迹 nasomaxillary notch 39
鼻后孔 posterior nasal apertures 37,144,210,212,418
鼻甲 nasal conchae 206,210
鼻甲骨 turbinal bone 29
鼻孔 nostril 121,122,206,209
鼻旁窦 paranasal sinus 206,210,418
鼻腔 nasal cavity 114,206
鼻软骨 nasal cartilage 206
鼻中隔 nasal septum 211,212,418
比目鱼肌 soleus muscle 83,10
闭孔 obturator foramen 69,70
闭孔动脉 obturator artery 83,108,331
闭孔神经 obturator nerve 403

闭孔外肌 external obturator muscle 83
壁细胞 parietal cell 116,158,159,160
臂部 brachial region 4,5
臂动脉 brachial artery 98,359,426
臂二头肌 biceps muscle of the forearm 82,98,101,359,425,426,427,428,429,430
臂骨 bone of arm 30
臂肌 brachial muscle 82,93,95,96,98,100
臂肌沟 bravhial muscle groove 63
臂三头肌 triceps muscle of the forearm 82,84,85,90,101,425,429
臂三头肌内侧头 medial head of triceps muscle of the forearm 93,95,96,98,100,430
臂三头肌长头 long head of triceps muscle of the forearm 93,95,96,98,100
臂神经丛 brachial plexus 402,424,425,426,427,428,429
臂头动脉 brachiocephalic artery 337,338,354
臂头动脉干 brachiocephalic trunk 331,333,334,336,339,346
臂头肌 brachinocephalic muscle 82,84,87,88,93,94,95,97,100,119
扁桃体 tonsil 115
变移上皮 transitional epithelium 258,259,262,263,298
表面上皮 superficial epithelium 304
表皮 epidermis 2,13,19,24
髌骨 patella 31,75
玻璃体 vitreous body 469,475,474
勃起组织 erectile tissue 270
布伦纳腺 Brunner's gland 117,172

C

侧部 lateral part 55
侧脑室 lateral ventricle 401,412,416
肠面 intestinal surface 380
肠腔 intestinal lumen 174,175,183
肠系膜 mesentery 166,390
肠系膜后动脉 caudal mesenteric artery 331,358
肠系膜淋巴结 mesenteric lymph node 166,390
肠系膜前动脉 cranial mesenteric artery 331,357,358
肠腺 intestinal gland 183,184,185
肠隐窝 intestinal crypt 116,174,175,176,177,180
尘细胞 dust cell 208,238
成熟卵泡 mature follicle 298,309,310
成纤维细胞 fibroblast 195,218,225,226
成纤维细胞核 nucleus,fibroblast 350,351,352
尺骨 ulna 31,58,60,68
尺骨茎突 styloid process of ulna 65,66
尺骨体（干）body of ulna 65,66
尺神经 ulnar nerve 98,359,402,425,426,427,428,429,430
尺腕骨 ulnar carpal bone 67,68
齿 tooth 114
齿槽间缘 interalveolar margin 41
齿槽缘 alveolar margin 41
齿突 dens 45,47
齿状回多形细胞层 polymorphic layer, dentate gyrus 421
齿状回分子层 molecular layer, dentate gyrus 421,422
齿状回颗粒层 granular layer,dentate gyrus 420,421,422
耻骨 pubis 31,70
耻骨后支 caudal branch of pubis 69
耻骨肌 pectineal muscle 83,108,360
耻骨联合 pubic symphysis 69
耻骨前支 cranial branch of pubis 69
耻骨梳 pecten of pubis 69
耻骨体 body of pubis 69
出球小动脉 efferent arteriole 255
初级精母细胞 primary spermatocyte 268,275,276
初级淋巴器官 primary lymphoid organ 370
初级卵母细胞 primary oocyte 306
初级卵母细胞的胞核 nucleus of primary oocyte 306,307,308,309
初级卵泡 primary follicle 305,306
穿通管 perfoating canal 33
垂体 hypophsis, pituitary gland 410,411,412,414,442,445,446,448
垂体后叶 posterior lobe of hypophysis 442
垂体裂 hypophyseal cleft 442,446,447
垂体前叶 anterior lobe of hypophysis 442

垂体细胞（神经胶质细胞）pituicyte (neuroglial cell) 442,448
垂直肌 vertical muscle 130,131,132
锤骨 malleus 470
唇 lip 114
次级精母细胞 secondary spermatocyte 268,275
次级淋巴器官 secondary lymphoid organ 371
次级卵泡 secondary follicle 309
粗隆沟 tuberositial groove 74
粗面内质网 rough endoplasmic reticulum 196,197,202
促甲状腺激素细胞 thyrotroph, TSH cell 442
促肾上腺皮质激素细胞 corticotroph, ACTH cell 442
促性腺激素细胞 gonadotroph 442
催产素 oxytocin,OT 442
催乳激素细胞 mammotropic cell 442

D

大肠 large intestine 117,165,190,199,248,249,315,357,378,379,390,459
大结节 greater tubercle 64
大结节后部 caudal part of greater tubercle 63
大结节前部 cranial part of greater tubercle 63
大脑 cerebrum 144,210,401,407,415,418,419
大脑半球 cerebral hemisphere 401,415,416,417
大脑横裂 transverse cerebral fissure 401,409,411,419
大脑脚 cerebral peduncle 401,410,411,412,414,445,476
大脑镰 cerebral falx 419
大脑皮质（大脑皮层）cerebral cortex 401,409,416
大脑前动脉 anterior cerebral artery 410
大脑中动脉 middle cerebral artery 410
大脑纵裂 longitudinal cerebral fissure 401,409,415,419
大网膜 greater omentum 151,152,156,157,164,190
大圆肌 major teres muscle 82,97,98,359,426,428,429
大转子 greater trochanter 71,72
大转子后部 caudal part of greater trochanter 72
大转子前部 cranial part of greater trochanter 72
单层立方上皮 simple cuboidal epithelium 193,195
单层柱状上皮 simple columnar epithelium 158,171,180
单核肝细胞 hepatocyte with single nucleus 192,193
单核细胞 monocyte 332,365,366
胆囊 gallbladder 117,163,164,165,186,188,189,190,191,199,378,379
胆囊管 cystic duct 188,189,190,191,199,362
导管 duct 137,148,172
导管口 opening of duct 19
镫骨 stapes 470
低倍镜 lower power 179
第1颈椎（寰椎）1st cervical vertebra (atlas) 43,44
第1肋 1st rib 58,60,348,349
第1肋骨 1th costal bone 59
第1肋软骨 1st costal cartilage 56
第1尾椎 1st coccygeal vertebra 55
第1腰椎 1st lumbar vertebra 53
第10肋 10th rib 58
第10胸椎 10th thoracic vertebra 50
第1肋腋淋巴结 primary costal axillary lymph node 372,430
第2跗骨 2nd tarsal bone 77
第2颈椎（枢椎）2nd cervical vertebra (axis) 45,46,47
第2肋软骨 2nd costal cartilage 56,57
第2腕骨 2nd carpal bone 67,68
第2掌骨 2nd metacarpal bone 68
第2指 2nd finger(dewclaw) 98
第2趾 2nd toe (hoof) 103
第2跖骨 2nd metatarsal bone 77
第3腓骨肌 3rd fibular muscle 83,102,103,106,107,108,119
第3跗骨 3rd tarsal bone 76,77
第3脑室 3rd ventricle 401,412
第3腕骨 3rd carpal bone 67,68
第3胸椎 3rd thoracic vertebra 50,51,52,54
第3眼睑 third eyelid 469,472
第3掌骨 3rd metacarpal bone 68
第3跖骨 3rd metatarsal bone 77

第3指 3rd finger (hoof) 98,99
第3指近指节骨（系骨）3rd proximal phalanx (os compedale) 68
第3指远指节骨（蹄骨）3rd distal phalanx (coffin bone) 68
第3指中指节骨（冠骨）3rd middle phalanx (os coronale) 68
第3趾 3rd toe (hoof) 103
第3趾近趾节骨（系骨）3rd proximal phalanx (os compedale) 77
第3趾远趾节骨（蹄骨）3rd distal phalanx (coffin bone) 77
第3趾中趾节骨（冠骨）3rd middle phalanx (os coronale) 77
第4跗骨 4th tarsal bone 76,77
第4颈椎 4th cervical vertebra 42,48,49
第4腕骨 4th carpal bone 67,68
第4胸椎 4th thoracic vertebra 50
第4掌骨 4th metacarpal bone 68
第4跖骨 4th metatarsal bone 77
第4脑室 4th ventricle 401,412
第4指 4th finger(hoof) 96,99
第4趾 4th toe (hoof) 103,107
第4指伸肌 4th digital extensor muscle 96
第4指伸肌腱 4th digital extensor tendon 95
第5肋软骨 5th costal cartilage 56,57
第5掌骨 5th metacarpal bone 68
第5指 5th finger (dewclaw) 96,98,99
第5趾 5th toe (dewclaw) 103,107
第5跖骨 5th metatarsal bone 77
第5指伸肌 5th digital extensor muscle 96
第5指伸肌腱 5th digital extensor tendon 95
第7颈椎 7th cervical vertebra 42
第7颈椎棘突 spinous process of the 7th cervical vertebra 42
第8 肋骨 8th costal bone 59
蝶窦 sphenoidal sinus 206
蝶骨 sphenoid bone 28,37
蝶骨体 basisphenoid 39
耵聍腺 ceruminous gland 3,470
顶骨 parietal bone 28,35,36,38,39,40,138
顶嵴 parietal crest 36,138
顶间骨 interparietal bone 28
顶树突 apical dendrite 420
顶叶 parietal lobe 409,411,415,416
动脉 artery 224,235,237,254,258,330,364,377,440
动脉韧带（动脉导管索）arterial ligament 332,337,339
动脉圆锥 arterial cone 333,335,337,341
动脉周围淋巴鞘 periarterial lymphatic sheath 371,383,384,385
动眼神经 oculomotor nerve 403
动眼神经根 root of oculomotor nerve 414,476
窦房结 sinuatrial node 329
窦下室间沟 subsinuosal interventricular groove 328,334
端脑 telencephalon 401
对耳轮 anthelix 479
多级神经元 multipolar neuron 406,409
多形细胞 polymorphic layer 422
多形细胞层 polymorphic layer 419,420,421

E

额窦 frontal sinus 39,144,206,211,212
额骨 frontal bone 28,34,35,36,38,39,138
额骨颧突 zygomatic process of frontal bone 35,138
额皮肌 cutaneous muscle of forehead 86
额叶 frontal lobe 409,411,415,416
腭垂 uvula 114
腭大孔 greater palatine foramen 37
腭帆扁桃体 tonsil of palatine velum 114,115,125,126,129,395,396
腭缝 palatine raphe 114,123,124
腭骨 palatine bone 29,39
腭骨垂直板 perpendicular plate of palatine bone 37
腭骨水平板 horizontal plate of palatine bone 37
腭裂 palatine fissure 37
腭小孔 lesser palatine foramen 37
腭褶 palatine fold 114,123,124
腭正中缝 median palatine suture 37
耳 ear 119,120,121,140,209,210,424,427
耳背 back of ear 480

耳部 aural region 4,5,6,7
耳大静脉 great auricular vein 480
耳根 root of ear 479,480
耳郭 auricle 470,479
耳尖 apex of ear 479,480
耳睑神经 auriculo-palpebral nerve 424
耳静脉 auricular vein 363
耳毛 tragi 479
耳前静脉 anterior auricular vein 480
耳蜗 cochlea 471
耳蜗管 cochlear duct 471
耳蜗神经 cochlear nerve 403
耳咽管 auditory tube 470
耳状面 auricular articular surface 55
二腹肌 digastric muscle 81
二尖瓣 bicuspid valve 329,339,340,341,342,347,353
二尖瓣的腱索 chordae tendineae of the bicuspid valve 339
二尖瓣的乳头肌 papillary muscle of the bicuspid valve 339
二头肌粗隆 bicipital tuberosity 72
二头肌沟 bicipital groove 64

F

方叶 quadrate hepatic lobe 186,188
房室瓣 atrioventricular valve 328,329
房室口 atrioventricular orifice 328,339,340,341,342,343
房间隔 interatrial septum 328
房室结 atrioventricular node 329
房室束 atrioventricular bundle 329
放射冠 corona radiate 309
飞节 hock 32
腓肠肌 gastrocnemius muscle 83,103,106,107,108,109,111,388,389,432,433,434
腓骨 fibula 31,34,70,73,74,75,77
腓骨头 head of fibula 73,74
腓骨长肌 long fibular muscle 83,102,103,106,107
腓浅神经 superficial fibular nerve 432
腓深神经 deep fibular nerve 432
腓总神经 common peroneal nerve 432,434
肺 lung 119,120,166,190,191,199,207,230,231,233,234,235,236,338,355,357,362,363,391,435,436,438
肺动脉 pulmonary artery 331,346
肺动脉瓣 pulmonary valve 328,340,341,346
肺动脉干 pulmonary trunk 328,331,333,335,337,338,339,354
肺动脉口 orifice of pulmonary trunck 328,340,341
肺段支气管 segmental bronchus 208
肺钝缘（背侧缘）blunt edge (dorsal margin) of the lung 228
肺静脉 pulmonary vein 334
肺静脉口 orifice of pulmonary vein 329
肺巨噬细胞 pulmonary macrophage 208
肺门 hilum of the lung 207
肺泡 pulmonary alveolus 208,236,237,238
肺泡隔 alveolar septum 236,237
肺泡管 alveolar duct 208,236
肺泡巨噬细胞 pulmonary alveolar macrophage 238
肺泡囊 alveolar sac 208,236,237
肺锐缘（腹侧缘）sharp edge (ventral margin) of the lung 228
肺实质 pulmonary parenchyma 235
肺小叶 pulmonary lobule 208
肺胸膜 pulmonary pleura 207
肺循环 pulmonary circulation 331
肺叶支气管 lobar bronchus 207,235
分泌单位 secretory unit 172
分泌导管 excretory duct 20,21,23,218
分泌泡 secretory vacuole 315
分泌小管 secretory tubulus 20
分子层 molecular laye 406,419,420,421,422,423
冯 · 埃布纳腺 von Ebner gland 134
缝匠肌 sartorius muscle 83,108,109
跗部 tarsal region 4,5,7
跗骨 tarsal bone 31,70,76
跗关节 tarsal joint 32,85,102
辐射层 stratum radiatum 420
附睾 epididymis 90,247,268,271,272,277,278,279,291,387

附睾体 body of epididymis 268,273,277,278,279,284,286
附睾头 head of epididymis 268,273,274,277,279,280,284,285,286
附睾尾 tail of epididymis 268,273,274,277,278,279,281,282,284,285,286
附睾尾韧带（阴囊韧带）scrotal ligament 268,269,284
复层扁平上皮 stratified squamous epithelium 131,132,133,135,136,145,148,218,295,296,395,396,397
复层柱状上皮 stratified columnar epithelium 264,293,294
副交感神经 parasympathetic nerve 404
副神经 accessory nerve 403
副腕骨 accessory carpal bone 67
副性腺 accessory genital gland 287,288
副叶 accessory lobe 219,220,229,230,231,232,233,234,338
腹白线 abdominal linea alba 81
腹壁 abdominal wall 156
腹壁肌 abdominal wall muscles 90,91
腹壁皮肤 abdominal skin 386,387
腹壁神经 abdominal nerve 431
腹部 abdomen 4,5,6,7
腹部乳房 abdomen breast 10
腹侧端 ventral extremity 380
腹侧弓 ventral arch 44
腹侧嵴 ventral crest 52,54
腹侧角 ventral horn 408
腹侧结节 ventral tubercle 43,44
腹侧锯肌 ventral serrate muscle 82,97,98,359
腹底壁 bottom of the abdomen wall 259,388
腹股沟部 inguinal region 387
腹股沟管 inguinal canal 82,90
腹股沟管皮下环 external ring of inguinal canal 90
腹股沟浅淋巴结 superficial inguinal lymph node 386,387
腹横肌 transverse abdominal muscle 81,431
腹膜 peritoneum 170
腹膜壁层 parietal peritoneum 170
腹膜腔 peritoneal cavity 170
腹内斜肌 internal oblique abdominal muscle 81,84,87,90,103,108,387
腹腔动脉 celiac artery 331,357,358
腹腔内脏器官 abdominal visceral organ 163,164
腹外斜肌 external oblique abdominal muscle 81,84,85,86,87,88,90,104,387
腹外斜肌腱膜 aponeurosis of external oblique abdominal muscle 387
腹直肌 straight abdominal muscle 81,85,90,431
腹主动脉 abdominal aorta 249,331,357,358

G

盖细胞 tectorial cell 262
肝 liver 90,117,119,120,147,156,163,164,165,166,186,187,188,190,191,192,196,197,199,362,378,379,435,436
肝板 hepatic plate 117,191,192,193,194
肝静脉 hepatic vein 186,187,332
肝门 hepatic porta 186,188,189,362
肝门静脉 hepatic portal vein 190,362
肝尾叶 caudate hepatic lobe 188,189,199
肝细胞 hepatocyte 117,195,197
肝细胞核 nucleus of hepatocyte 193,196
肝小叶 hepatic lobule 117,189,192,193,194,195
肝右内叶 right medial hepatic lobe 186,187,188,189,199
肝右外叶 right lateral hepatic lobe 186,187,188,199
肝圆韧带 round ligament of liver 186,187,188,190,191
肝左内叶 left medial hepatic lobe 186,187,188
肝左外叶 left lateral hepatic lobe 186,187,188
感觉器官 sense organ 468
冈上肌 supraspinatus muscle 82,84,85,93,95,96,97,98,359,427
冈上窝 supraspinous fossa 61
冈下肌 infraspinous muscle 82,84,86,93,96
冈下肌止点区 area of insertion for infraspinatus muscle 63
冈下窝 infraspinous fossa 61
肛门 anus 117,203,260,282,283,324,325
高倍镜 high power 179
高尔基复合体 Golgi complex 202

高尔基细胞 Golgi cell 423
睾丸 testis 84,87,90,119,120,247,268,271,272,273,276,285,286,290,291,387
睾丸动脉 testicular artery 331
睾丸附睾缘 margo epididymidis of testis 274,277
睾丸间质细胞 Leydig cell 268,275,276
睾丸静脉 testicular vein 332
睾丸输出小管 eferent ductule of testis 280
睾丸体 body of testis 273,277,284
睾丸网 testicular network 268
睾丸尾 tail of testis 273,277,284
睾丸小隔 septula testis 268,275,276
睾丸游离缘 margo liber of testis 274,277
睾丸纵隔 mediastinum testis 268,274,278
隔缘肉柱 septomarginal trabecula 328,339,343,344,345
膈 diaphragm 81,90,119,120,156,163,165,227,347,348,355,361,435,438,
膈脚 crus of diaphragm 248
膈面 diaphragmatic surface 380,381
膈肉质缘 pulpa part of diaphragm 357
膈神经 phrenic nerve 348,361,402,435
膈中心腱 central tendon of diaphragm 248,357
根鞘 root sheath 3
跟（总）腱 common calcaneal tendon 83,85,102,103,360,388,389,434
跟骨 calcaneus 34,70,76,77
跟结节 calcaneal tuberosity 77,107,108,120,388
弓间隙 interarcuate space 55
肱骨 humerus 30,34,58,60
肱骨干（体）shaft of humerus 63,64
肱骨嵴 humerus crest 63
肱骨颈 neck of humerus 63,64,
肱骨髁 humeral condyle 63,64
肱骨头 head of humerus 63,64
巩膜 sclera 468,473,474,475,476
巩膜筛板 cribriform plate of sclera 468
股薄肌 gracilis muscle 83,87,90,387,388
股部 thigh region 4,5
股动脉 femoral artery 331,360
股二头肌 biceps muscle of the thigh 38,83,84,85,86,102,103,105,106,388
股方肌 quadrate muscle of the thigh 83
股骨 femoral bone,femur 31,34,70,75,119
股骨滑车 trochlea of femur 71,75
股骨内侧观 medial view of the femur 72
股骨前面观 cranial view of the femur 71
股骨滑车粗隆 trochlea tuberosity of femur 71
股骨颈 neck of femur 71
股骨内侧髁 medial condyle of femur 75
股骨体（干）shaft of femur 71,72
股骨头 head of femur 70,71,72
股骨头窝 fovea of femoral head 71,72
股后肌群 posterior thigh muscle groups 106
股后皮神经 caudal femoral cutaneous nerve 434
股内侧肌 medial vastus muscle 83,108,109,360
股内侧肌群 interfemus muscle groups 109
股神经 femoral nerve 403
股四头肌 quadriceps muscle of the thigh 83,84,85,90,102,103,434
股外侧肌 lateral vastus muscle 83,103,107
股外侧皮神经 lateral femoral cutaneous nerve 402
股直肌 rectus femoris muscle 83,107,109,360
股中间肌 intermedial vastus muscle 83
骨半规管 canales semicirculares ossei 471
骨单位 osteon 33
骨骼 skeleton 34
骨骼肌 skeletal muscle 80,132
骨壶腹 osseous ampulla 471
骨间板 interstitial lalmella 33
骨间肌 interosseus 83
骨间隙 interosseous space 70,73,74
骨脚 bony crura 471
骨迷路 osseous labyrinth 470
骨密质 compact bone 32
骨膜 periosteum 32
骨盆 pelvis 31
骨盆后口 caudal opening of the pelvis 70
骨盆联合 pelvic symphysis 108
骨松质 spongy bone 32
骨髓 bone marrow 33
骨质 bone substance 32

骨组织 bone tissue,osseous tissue 32
鼓膜 tympanic membrane 470
鼓泡 tympanic bulla 37,39
鼓舌骨 tympanohyoid 41,42
鼓室 tympanic cavity 470
固有层 lamina propri 130,131,132,133,135,136,145,146,147,148,157,158,160,161,171,172,173,175,176,177,180,181,182,184,185,218,224,225,226,236,237,258,259,262,263,264,315,319,395,396,397
固有层初级乳头 primary papilla of lamina propria 131
固有层次级乳头 secondary papilla of lamina propria 131
固有层的血管 blood vessel within lamina propria 158
固有层乳头 papilla of lamina propria 132,133
固有韧带 proper ligament 268
固有鞘膜 proper vagina tunica 268,285
关节凹 articular fovea 44
关节面 articular sarface 65
关节窝 articular fossa 43
关节盂（肩臼）glenoid cavity (glenoid fossa) 61,62,63
管腔 lumens 275,276
冠关节 coronal joint 32
冠状动脉 coronary artery 330
冠状窦 coronary sinus 328,330,343
冠状沟 coronary groove 328,333,334,336,335 341
冠状突 coronoid process 41
冠状循环 coronary circulation 330
腘动脉 popliteal artery 331
腘肌 popliteus muscle 83
腘肌面 popliteal surface 72
腘淋巴结 popliteal lymph node 105,388,389
腘切迹 popliteal notch 73

H

哈塞耳代小体 Hassall's corpuscle 371
海马 hippocampus 413,417,420,421,422,449
海马沟 hippocampal sulcus 420
海马回 hippocampal gyrus 420,421
海马缘 hippocampal edge 420,421
海绵体 cavernous body 294,295
汗腺 sweat gland 3,12,13,15,16,18,19,20
核仁 nucleolus 196,405,406,409
核周体 perikaryon 400
颌前骨 premaxillare bone 29
颌下腺 mandibular gland 115,120,141,142,143,385,452
赫令体 Herring body 442
黑素细胞 melanocyte 14
恒齿 permanent tooth 138,139
横行肌 transversal muscle 133
横结肠 transverse colon 117
横突 transverse process 46,47,50,51,52,53,54,56
横突管 transverse canal 30
横突后支 caudal branch of transverse process 48,49
横突孔 transverse foramen 30,45,46,48,49
横突前支 cranial branch of transverse process 48
横纹 cross striation 80,111
横纹肌 striated muscle 396
横线 transverse line 55
红髓 red pulp 371,381,382,383,384,385
红细胞 erythrocyte, red blood cell (RBC) 24,110,193,238,332,350,365,366,367,368
虹膜 iris 468,474,476
喉 larynx 141,144,206,210,213,217,219,220,373,435,452,453,454
喉返神经 recurrent laryngeal nerve 435,453
喉口 aperture of larynx 125,144,213,214,215,216
喉腔 laryngeal cavity 206,213,216
喉软骨 laryngeal cartilage 206,215
后背侧锯肌 caudal dorsal serrate muscle 81,84,89,90
后端 caudal extremity 250,251
后关节凹 caudal articular fovea 44
后关节突 caudal articular process 45,46,47,48,49,51,53,54,55
后角 caudal angle 61,62
后肋窝 caudal costal fossa 51,52
后腔静脉 *caudal vena cava* 90,190,191,199,332,334,347,348,351,361,362,435

后腔静脉裂孔 postcaval vein hiatus 81
后丘 caudal colliculus 413,449
后丘脑 metathalamus 401
后缘 caudal border 61,62,333,334,335,380
后肢 hindlimb 2
呼吸系统 respiratory system 206
呼吸性细支气管 respiratory bronchiole 208,236
壶腹嵴 crista ampullaris 471
滑车 trochlea 73,74
滑车结节 trochlear tubercle 75
滑车切迹 trochlear notch 65,66
滑车神经 trochlear nerve 403
滑膜囊 synovial bursa 80
环甲肌 cricothyreoid muscle 214
环状软骨 cricoid cartilage 206,213,215,216,217
寰椎 atlas 30,42
寰椎翼 wing of atlas 43,44
换毛 molting 3
黄体 corpus luteum 298,311
灰质 grey matter 400,408
灰质连合 grey commissure 400
回肠 ileum 116,147,162,167,179,181,182,183
回肠黏膜 ileal mucosa 179
回盲韧带 ileocaecal ligament 167,179
会厌 epiglottis 213
会厌软骨 epiglottic cartilage 206,213,214,215,217
喙臂肌 coracobrachial muscle 82,98,428,429,430
混合腺 mixed gland 145,226
混合腺泡 mixed acinus 218

J

肌层 muscular layer 131,147,157,174,180,222,258,259,286,287,314,395,396,397
肌腹 muscle belly 80
肌腱 muscle tendon 80
肌节 sarcomere 80
肌内膜 endomysium 80,110,111
肌皮神经 musculocutaneous nerve 402,428,429,430
肌上皮细胞核 nucleus of the myoepithelial cell 20
肌束膜 perimysium 80,110,111,350
肌外膜 epimysium 80
肌细胞核 nucleus of the muscle cell 110,111
肌纤维 muscle fiber 110,111
肌线 muscular line 73
肌样细胞 myoid cell 376
肌样细胞的胞核 nucleus of myoid cell 275,276
肌原纤维 myofibril 350
基底层 basal layer 2,13,14,15,25
基底纹 basal striation 142
基底细胞 basal cell 13,14,20,21,22,224,280,289,290
基膜 basal lamina (basement membrane) 176,178,224
激素 hormone 442
棘层 spinous cell layer 2,13,14,15,25
棘突 spinous process 45,46,47,48,49,50,51,52,53,54
棘细胞 spinous cell 14
集合小管 collecting tubule 242
脊膜 spinal meninges 401
脊软膜 spinal pia mater 401
脊神经 spinal nerve 402
脊神经根 spinal nerve root 407,408
脊神经节 spinal ganglion 402,407,408
脊髓 spinal cord 144,210,400,407,408,412,414,418
脊髓圆锥 medullary cone 400,407,408
脊硬膜 spinal dura mater 401
脊蛛网膜 spinal arachnoid 401
脊柱 vertebral column 30
夹肌 splenius muscle 81,89
颊 cheek 114,140
颊齿 cheek tooth 114
颊肌 buccinator muscle 81,84,88
甲状旁腺 parathyroid gland 443 ,455,457,458,459
甲状软骨 thyroid cartilage 206,213,214,215,217
甲状舌骨 thyrohyoid 42
甲状舌骨软骨 cartilage of the thyrohyoid bone 42
甲状腺 thyroid gland 435,443,452,453,454,455,457,458
甲状腺被囊 tunic of thyroid gland 443
甲状腺动脉 thyroid artery 354
甲状腺后部 caudal pars of the thyroid gland 455
甲状腺滤泡 thyroid follicle 456,457
甲状腺前部 cranial pars of the thyroid gland 455
甲状腺素 thyroxine 443

岬 promontory 470
假单极神经元 pseudounipolar neuron 402
假复层上皮 pseudostratified epithelium 279,280,281,282,286,287,280
假复层纤毛柱状上皮 pseadostratified ciliated columnar epithelium 223,224,226,236,237
假复层柱状上皮 pseudostratified columnar epithelium 315,319
间脑 diencephalon 401
间皮 mesothelium 351
间皮细胞核 nucleus of mesothelial cell 351
间质 stroma 305,306,307
间质细胞 interstitial cell 268,275,276
间质组织 interstitial tissue 275
肩臂皮肌 cutaneous omobrachial muscle 81
肩带部肌 shoulder girdle muscles 94
肩关节 humeral joint 32
肩胛部 shoulder joint 4,5,7
肩胛冈 spine of scapula 61
肩胛冈结节 tuberosity of scapular spine 61
肩胛骨 scapula 30,34,58,60,119
肩胛骨关节盂 glenoid cavity of the scapula 63
肩胛横突肌 omotransverse muscle 82,93,95,97
肩胛结节 scapular tuber 61,63
肩胛颈 neck of scapula 61
肩胛切迹 scapular notch 61,62
肩胛上神经 suprascapular nerve 402,426,427
肩胛舌骨肌 omohyoid muscle 81
肩胛下动脉 subscapular artery 359
肩胛下肌 subscapular muscle 82,97,98,359,427
肩胛下神经 subscapular nerve 98,359,402,426
肩胛下窝 subscapular fossa 62
睑结膜 palpebral conjunctiva 469
荐背侧孔 dorsal sacral foramen 55
荐部 sacral part 6,7
荐腹侧孔 ventral sacral foramen 55
荐骨 sacrum 30,55,70
荐骨岬 promontory of sacrum 55
荐骨翼 wing of sacrum 55
荐结节 sacral tuber 34,69,70
荐淋巴结 sacral lymph node 358
荐髂关节 sacroiliac joint 32
荐神经 sacral nerve 402
荐臀部 sacral-gluteal region 4
荐外侧嵴 lateral sacral crest 55
荐中动脉 median sacral artery 331
荐椎 sacral vertebrae 30
剑胸软骨结合 xiphoid sternum synchondrosis 57
剑状软骨 xiphoid cartilage 56,57,58,60,163
腱鞘 tendon sheath 80
腱索 chordae tendineae 328,339,342,343,344,345,347
浆半月 serous de milune 143
浆膜 serosa 174,175,180,286,314
浆液腺泡 serous acinus 135,142,143,160,161,172,218,225
降钙素 calcitonin 443
降结肠 decending colon 117
降主动脉 descending aorta 330
交错突细胞 interdigitating cell 376
交感干 sympathetic trunk 262,355,436,437,438,439
交感神经 sympathetic nerve 404
交感神经节前神经元 sympathetic preganglionic neuron 408
交感神经节细胞 sympathetic ganglion cell 465
胶原纤维 collagenous fiber 195,238
胶原纤维束 collagen fiber bundle 15,17,18,22,23,218
胶质 colloid 456
胶质细胞 glial cell 405,406,409,422
角化的复层扁平上皮 keratinized stratified squamous epithelium 130
角膜 cornea 468,473,476
角膜巩膜缘 corneoscleral junction 468,473
角舌骨 ceratohyoid 42
角质层 horny layer 2,13,14,15,24,25
角质细胞 keratinocyte 13,14
尖叶支气管 apical lobar bronchus 235,361
节段性支气管 segmental bronchus,tertiary bronchus 235
结肠 colon 117,120,147,156,162,163,164,167,179,184,379
结肠壁 colonic wall 183
结肠黏膜 colonic mucosa 179,183

结肠襻 colon loop 166,168,169
结肠腺 colonic gland 185
结缔组织 connective tissue 451
结节间沟 intertubercular sulcus 64
结膜囊 conjunctival sac 469
睫毛 eyelash 472
睫状体 ciliary body 468,474,475
筋膜 fascia 80
近侧前臂间隙 proximal space of forearm 65,66
近端小管 proximal tubule 241,254
近端小管曲部 proximal convoluted tubule 241
近指节间关节 proximal interphalangeal joint 32
茎舌骨肌 stylohyoid muscle 81
茎突舌骨 stylohyoid 41,42
晶状体 lens 469,474,475
精阜 seminal hillock 269
精囊腺 vesicular gland 246,247,269,271,272,287,288,289,290
精索 spermatic cord 90,247,269,273,274,277,278,284,285,286,287,290,291,387
精原细胞 spermatogonium 268,275,276
精子 spermatozoa 268,275,276,279,281,286,287
精子发生 spermatogenesis 268
精子细胞 spermatid 268,275,276
颈部 cervical part 4,5,6,7
颈部毛流（旋涡毛流）cervical flumina pilorum 8,9
颈动脉窦 carotid sinus 331
颈动脉体 carotid body 331
颈多裂肌 cervical multifidus muscle 81
颈静脉 jugular vein 354
颈静脉沟 jugular vein groove 331
颈菱形肌 cervical part of rhomboid muscle 95,97
颈内动脉 internal carotid artery 331
颈内静脉 internal jugular vein 332
颈黏液细胞 mucus neck cell 116
颈黏液腺细胞 mucous neck cell 159,160
颈膨大 cervical enlargement 400,407
颈皮肌 cutaneous muscle of neck 81
颈浅动脉 superficial cervical artery 331
颈浅淋巴结（肩前淋巴结）superficial cervical lymph node 372
颈深后淋巴结 caudal deep cervical lymph node 372
颈神经 cervical nerve 402
颈外动脉 external carotid artery 331
颈外静脉 external jugular vein 332
颈斜方肌 cervical part of trapezius muscle 84,85,86,93,95
颈长肌 long muscle of neck 81
颈椎 cervical vertebrae 30,34,42
颈总动脉 common carotid artery 331,354
胫骨 tibia 31,34,70,73,74,75,77
胫骨粗隆 tibial tuberosity 74
胫骨嵴 tibial crest 74
胫骨内髁 medial condyle of tibia 73,74
胫骨前肌 cranial tibial muscle 83
胫骨外髁 lateral condyle of tibia 73,74
胫前动脉 anterior tibial artery 331
胫神经 tibial nerve 432,433,434
静脉 vein 218,224,226,235,236,237,253,330,351,364,375,376,378,396,397,420,421,448,449,459,461,465,466
静脉间结节 intervenous tubercle 328
臼齿 molar 114,123,124,129
巨噬细胞 macrophage 218,376,378
距骨 talus 76,77
锯肌面 face for serrate muscle 62
菌状乳头 fungiform papilla 114,125,126,127

K

抗利尿激素 antidiuretic hormone,ADH 442
柯蒂器 organ of Corti 471
颏结节 mental tubercle 138,139
颏孔 mental foramen 35,41
颗粒层 granular layer 2,13,14,15,25,406,422,423
颗粒黄体细胞 granulosa lutein cell 311
颗粒膜 membrane granulosa 308,309,310
颗粒细胞 granular cell 14,15,306,307,308,309,420,422,423
髁间隆起 intercondylar eminence 75
髁间区 intercondyloid area 73
髁间窝 intercondylar fossa 72
髁旁突 paracondylar process 39,40

髁状突 condylar process 41
克拉拉细胞 Clara cell 208
空肠 jejunum 116,147,156,162,165,166,168,174,175,176,177,178,179,190,191,199,379,390
空肠壁 jejunal wall 173
空肠淋巴结 jejunal lymph nodes 390
空肠黏膜 jejunal mucosa 173
空肠系膜 mesojejunum 166
空泡细胞 vacuolated cell 20,21
口裂 oral fissure 114,122
口轮匝肌 orbicular muscle of the mouth 81
口腔 oral cavity 114,119,120,123,211,212
口腔底 basis cavum oris 129
扣带沟 cingulate suleus 412,416
枯否细胞 Kuffer's cell 117,192,193,197
髋骨 hip bone 31,69
髋关节 hip joint 32
髋结节 coxal tuberosity 69,70
髋臼 acetabulum 69,286
眶骨膜 orbital periosteum 469
眶上触毛 supraorbital tactile hair 472
眶上孔 supraorbital foramen 36
眶窝 orbital fossa 35,38
眶下触毛 infraorbital tactile hair 472
眶下孔 infraorbital foramen 35,36,38,138,139
阔筋膜张肌 tensor muscle of the fascia lata 83,84,85,90,102

L

拉克囊 Rathke's pouch 446,447
篮状细胞 basket cell 423
朗格汉斯岛 pancreatic islet,islet of Langerhans 118
肋 rib 30,59,70
肋（骨）沟 costal groove 59
肋（骨）角 angle of rib 59
肋（骨）结节 costal tubercle 59
肋（骨）颈 neck of rib 59
肋（骨）小头 head of rib 59
肋腹神经 costoabdominal muscle 402,431
肋骨 costal bone 30,34,60,89,356
肋骨干 shaft of rib 59
肋间动脉 intercostal artery 331,334,356
肋间肌 intercostal muscles 89
肋间静脉 intercostal vein 356,362,363
肋间内肌 internal intercostal muscle 81,89,92,356,363,402
肋间外肌 external intercostal muscle 81,89
肋间隙 intercostal space 58
肋颈动脉干 costocervical trunk 331
肋颈静脉 costocervical vein 332
肋软骨 costal cartilage 30,58,59,60
肋胸关节 costosternal joint 32
泪道 lacrimal passage 469
泪骨 lacrimal bone 35,36,38
泪孔 lacrimal foramen 38,139
泪器 lacrimal apparatus 469
泪腺 lacrimal gland 469,476
梨状叶 piriform lobe 410,411,414,416,445,476
梨状隐窝 piriform recess 115
犁骨 vomer 29,37
李氏隐窝 crypt of Lieberkühn 116,174,176,177
粒细胞 granulocyte 395
裂孔 slit pore 241
裂孔膜 slit membrane 241
淋巴 lymph 370
淋巴导管 lymphatic duct 370
淋巴干 lymphatic trunk 370
淋巴管 lymphatic vessel 192,370
淋巴管道 lymphatic vessels 370
淋巴结 lymph node 372,391,392,393,394,395
淋巴器官 lymphoid organ 370
淋巴系统 lymphatic system 370
淋巴细胞 lymphocyte 146,181,237,238,332,370,397
淋巴小结 lymphatic nodule 145,174,175,180,370,372,391,392,395,396,397
淋巴小结暗区 dark region of lymphatic nodule 394
淋巴小结生发中心 germinal center of lymphatic nodule 394
淋巴中心 lymphocentrum 372
淋巴组织 lymphoid tissue 370
菱形肌 rhomboid muscle 82
菱形窝 rhomboid fossa 413

漏斗 infundibulum 412
颅部 cranial part 2,4,5,6,7
颅骨 cranial bone 28,211,212
颅腔 cranial cavity 39,211,212,418
卵巢 ovary 245,260,298,300,301,302,303,304,305,306,307,308,310,312,316,317
卵巢动脉 ovarian artery 331,332
卵巢淋巴结 lymph node of the ovary 302,312,313
卵巢门 hilum of the ovary 302,303
卵巢囊 ovarian bursa 298
卵泡 ovarian follicle 298,303
卵泡膜 follicular theca 306,307,308,309
卵泡膜内膜 internal theca of follicular theca 308,310
卵泡膜外膜 external theca of follicular theca 308,310
卵泡腔 follicular cavity 307,308,309
卵泡细胞 follicular cell 298,306,307
卵丘 ovarian cumulus 307,308,309
卵圆孔 foramen ovale 328
卵圆窝 oval fossa 328,343
轮廓乳头 vallate papilla 114,125,126,127,129,132
螺旋器 spiral organ filtration membnane 241,471
滤过膜 filtration membrane 241
滤过屏障 filtration barrier 241
滤泡 follicle 456,457,458
滤泡旁细胞 parafollicular cell 443,456
滤泡旁细胞群 parafollicular cell group 457
滤泡细胞 follicular cell 456,457

M

马尾 cauda equine 400,407,408
脉络丛 choroid plexus 449
脉络膜 choroid 468,474,475
蔓状丛 pampiniform plexus 278
盲肠 caecum 117,120,147,162,163,164,166,167,179,379
盲肠尖 apex of caecum 167
盲肠淋巴结 caecal lymph nodes 167
盲肠黏膜 caecal mucosa 179
毛 hair 3,12
毛干 hair shaft 3,13
毛根 hair root 3,12,13,16,17,18,19
毛流 flumina pilcorum 9
毛囊 hair follicle 3,12,13,15,17,18,22
毛皮质 hair cortex 13,16,17,18,21
毛球 hair bulb 3,12,13,15,16,19
毛乳头 hair papilla 3,12,15,16,19
毛髓质 hair medulla 13,15,17,21
毛细淋巴管 lymphatic capillary 370
毛细血管 blood capillary 14,18,110,111,132,146,202,218,226,330,350,351,352,393,395,421
毛小皮 hair cuticle 15,16,17,18
酶原颗粒 zymogen granule 202
门管区 portal area 117
门静脉环 annulus portae 118
弥散淋巴组织 diffuse lymphatic tissue 146,370,391,392,393,395,396,397
迷走交感干 vagosympathetic trunk 354,454
迷走神经 vagus nerve 355,362,391,403,436,437
迷走神经背侧支 dorsal branch of vagus nerve 436,437
迷走神经腹侧支 ventral branch of vagus nerve 436
迷走神经背侧干 dorsal trunk of vagus nerve 355,437
迷走神经腹侧干 ventral trunk of the vagus nerve 355
泌尿系统 urinary system 240,244
泌尿生殖器官 urogenital organs 245,246,247
泌尿生殖系统 urogenital system 271
泌乳期乳房 lactating breast 10
泌酸细胞 oxyntic cell 116
面部 facial part 2,4,5,6,7
面骨 facial bone 29
面嵴 facial crest 35
面皮肌 cutaneous muscle of face 80
面神经 facial nerve 403,424
面神经颊背侧支 dorsal buccal branch of the facial nerve 424
面神经颊腹侧支 ventral buccal branch of the facial nerve 424
面神经颈支 cervical branch of the facial nerve 424
膜半规管 semicircular duct 471
膜黄体细胞 theca lutein cell 311
膜迷路 membranous labyrinth 471

拇长外展肌 abductor pollicis longus muscle 82

N

囊隐窝 elliptical recess 470
囊肿 cyst sac 186
脑 brain,encephalon 401,407,409,410,411,412
脑干 brain stem 401,413,414,418
脑沟 sulcus 401,409,415
脑回 gyrus 401,409,415
脑脊髓液 cerebrospinal fluid 400,402
脑膜 dura mater 419
脑桥 pons 401,407,410,411,412,414,445
脑软膜 encephalic pia mater 402,419,422
脑上腺 *glandula pinealis* 443
脑神经 cranial nerve 403
脑室 brain ventricle 402
脑硬膜 encephalic dura mater 402,418,419
脑蛛网膜 encephalic arachnoid 402
内侧冠突 medial coronoid process 65
内侧髁 medial condyle 72,73,74
内侧髁间结节 medial intercondyloid tubercle 73
内侧上髁 medial epicondyle 71,72
内侧嗅束 medial olfactory tract 410
内侧缘 medial border 250,251
内弹性膜 internal elastic membrane 195,330,364
内耳 internal ear 470
内耳道 internal acoustic meatus 39
内分泌器官 endocrine organ 442
内分泌系统 endocrine system 442
内分泌腺 endocrine gland 442
内根鞘 inner root sheath 15,16,17,18,19,21
内根鞘小皮 cuticle of internal root sheath 16,17,18
内环骨板 inner circumferential lamella 33
内环肌层 inner circular stratum 171,174,175,419,469,478
内膜 internal tunic 330,364
内皮 endothelium 330,353
内皮细胞 endothelial cell 110,111,192,193,218,264,351,352,447,456,457,459,462,464
内皮细胞核 nucleus of endothelial cell 353,364
内皮下层 subendothelial layer 330,353
内收肌 adductor muscle 83,108,109,360
内网状层 inner plexiform layer 469,478
内眼角 medial canthus 472
内脏大神经 greater splanchnic nerve 439
内脏淋巴干 visceral lymphatic trunk 370
内脏器官 visceral organs 119,120
内脏神经系统 visceral nervous system 404
内脏小神经 lesser splanchnic nerve 439
内锥体细胞层 internal pyramidal layer 419
尼氏体 Nissl body 400,405,409
黏膜 mucosa 174,175,183,184,221
黏膜固有层 mucosal lamina propria 223
黏膜肌层 lamina muscularis mucosae 157,158,160,171,172,175,176,181,182,183,184,185
黏膜上皮 mucosal epithelium 135,146,147,174,175,223
黏膜下层 submucosa 147,157,158,171,174,175,180,183,184,218,222,223,226
黏膜下腺 submucosal gland 117
黏膜皱襞 mucosal fold 314
黏液囊 mucous bursa 80
黏液腺 mucous gland 148
黏液腺泡 mucous acinus (复数acini) 132,135,137,160,161,172,218,225
尿道 urethra 242,244,245,257,258,261,263,264,293,294,300,320
尿道骨盆部 pelvis part of urethra 246,247
尿道海绵体 cavernous body of urethra 293
尿道肌 urethral muscle 269
尿道腔 urethral lumen 293,294
尿道球 urethral bulb 269
尿道球腺 bulbourethral gland 246,247,269,271,272,287,288,289
尿道峡 urethral isthmus 269
尿道外口 external urethral orifice 261,264,265,269,292,293,299
尿生殖道 urogenital tract 269
尿生殖道骨盆部 pelvis part of urogenital tract 272,287,288,289
尿生殖前庭 urogenital vestibulum 299
镊 tweezers 357
颞骨 temporal bone 29,37,40,139

颞骨颧突 zygomatic process of temporal bone 35,36,38,138
颞骨岩部 petrous part of temporal bone 39
颞肌 temporal muscle 81
颞窝 temporal fossa 36,138
颞下颌关节 temporomandibular joint 32
颞叶 temporal lobe 409,411,415,416

P

派尔结 Peyer's patch 117
膀胱 urinary bladder 119,120,163,165,166,242,244,245,246,247,257,258,259,261,263,271,272,286,287,288,300,312,315,316,318,320,379
膀胱壁 bladder wall 261
膀胱顶（膀胱尖）apex of bladder 242,257,258,260
膀胱颈 neck of bladder 242,257,258,260,261
膀胱黏膜 mucosa of bladder 261
膀胱上皮 bladder epitheliam 262,263
膀胱体 body of bladder 242,257,258,260
膀胱圆韧带 ligamentum teres vesicae 242
泡心细胞 centroacinar cell 200,201,202,203
皮肤 skin 2,19,121
皮肤腺 cutaneous gland 3
皮肌 cutaneous muscle 80,88
皮下脂肪 subcutaneous fat 13,24
皮下组织 subcutaneous tissue 2,12
皮脂腺 sebaceous gland 3,12,13,19,20,21,22
皮脂腺导管 duct of sebaceous gland 19
皮脂腺分泌导管 excretory duct of the sebaceous gland 12
皮质 cortex 251,303,304,371,392,417,443,461
皮质迷路 cortical labyrinth 253
皮质肾单位 cortical nephron 240
脾 spleen 120,156,164,166,248,249,315,371,378,379,380,382,383,384,385,459
脾窦 splenic sinusoid 382
脾门 splenic hilum 380,381
脾实质 splenic parenchyma 381
脾索 splenic cord 371,382
脾小结 splenic nodule 371,383,384,385
脾小梁 spleen trabecular 381
脾小体 splenic corpuscle 371
脾血窦 splenic sinusoid 372
胼胝体 corpus callosum 401,412,416
平滑肌 smooth muscle 193,195,223,226,236,237,279,280,281,282,294,440
平滑肌细胞核 nucleus of the smooth muscle cell 110,237, 365
平滑肌纤维 smooth muscle fiber 364
破裂孔 foramen lacerum 37
浦肯野细胞 Purkinje cell 406,422,423
浦肯野细胞层 Purkinje cell layer 423
浦肯野纤维 Purkinje fiber 329,349,353

Q

奇静脉 azygos vein 332,357,362
脐部 umbilical region 4,5
脐动脉 umbilical artery 331
鬐甲部 withers 4,5,6,7
气管 trachea 144,207,210,213,214,215,216,217,219,220,221,222,223,224,230,231,232,233,234,235,338,361,435,436,452,453,454
气管分叉处 divaricate site of trachea 235
气管肌 tracheal muscle 221,222
气管淋巴干 tracheal lymphatic trunk 370
气管黏膜固有层 mucosal lamina propria of the trachea 222
气管黏膜上皮 mucosal epithelium of the trachea 222
气管腔 tracheal cavity 221,222
气管软骨 tracheal cartilage 213,221,225
气管神经节 tracheal ganglion 226
气管外膜 adventitia of trachea 222
气管腺 tracheal gland 223,224
气-血屏障 blood-air barrier 208
髂腹股沟神经 ilioinguinal nerve 402,431
髂腹下神经 iliohypogastric nerve 402,431
髂骨 ilium 31,34,69,70
髂骨体 body of ilium 69
髂骨翼 wing of ilium 69
髂肌 iliac muscle 83

髂嵴 iliac crest 69
髂肋肌 iliocostal muscle 81,88,89
髂肋肌沟 iliocostal muscle sulcus 88,89
髂内动脉 interalilial arten 331,358
髂外动脉 external iliac artery 331,358,360
髂腰动脉 iliolumbar artery 331
髂腰肌 iliopsoas 83
髂总静脉 common iliac vein 332
前背侧锯肌 cranial dorsal serrate muscle 81
前臂部 antebrachial region 4,5,7
前臂骨 skeleton of forearm 31,34
前臂间隙 space of forearm 58,60
前臂筋膜张肌 tensor muscle of the antebrachial fascia 82,93,95,96,97,98,426,427,428,429,430
前端 cranial extremity 250,251
前关节突 cranial articular process 45,47,48,49,50,52,53,54,55
前角 cranial angle 61,62
前臼齿 premolar 114,123
前肋窝 cranial costal fossa 50,52
前列腺 prostate gland 269,288
前列腺动脉 prostatic artery 331
前腔静脉 *cranial vena cava* 332,334,339,348,349,361,438
前丘 rostral colliculus 413,449
前庭 vestibule 470
前庭襞 vestibular fold 213
前庭大腺 greater vestibular gland 299
前庭耳蜗神经 vestibulocochlear nerve 403
前庭神经 vestibular nerve 403
前庭腺开口 vestibular gland opening 318
前庭小腺 lesser vestibular gland 299
前缘 cranial border 61,62,333,334,335,380
前肢 forelimb 2,6,121
浅表肾单位 superfacial nephron 240
浅筋膜 superficial fascia 2,80
腔静脉窦 sinus of the venae cavae 328
腔静脉裂孔 vena caval hiatus 357,361
腔隙层 stratum lacunosum 420,421
切齿 incisor 114,123,124,140
切齿骨 incisive bone 29,35,38,39,139
切齿骨鼻突 nasal process of the incisive bone 36
切齿骨腭突 palatine process of the incisive bone 37
切齿骨骨体 body of the incisive bone 36,37
切齿乳头 incisive papilla 123,124
切迹 notch 55
穹隆 fornix 412,416
丘脑 thalamus 401
丘脑黏合部 interthalamic adhesion 412,416
球海绵体肌 bulbocavernous muscle 269,288,289
球节 felock joint 32
球结膜 bulbar conjunctiva 469
球囊 sacculus 471
球囊斑 macula saccule 471
球囊隐窝 spherical recess 470
球内系膜 intraglomerular 241
球内系膜细胞 intraglomerular mesangial cell 241
球旁复合体 juxtaglomerular complex 242
球旁细胞 juxtaglomerular cell 242,255
球外系膜细胞 extraglomerular mesangial cell 242,254
球状带 glomerular zone 444,462,464
球状带细胞 glomerular zone cell 464
屈肌间肌 interflexor muscle 83
躯干 trunk 2
躯干皮肌 cutaneous muscle of trunk 81
曲精小管 contorted seminiferous tubule 268
颧弓 zygomatic arch 36,37
颧骨 zygomatic bone 29,35,36,38,138,139
犬齿 canine 36,37,38,41,114,138,139
犬齿肌 canine muscle 81,88

R

桡骨 radius 31,58,60,68,97
桡骨凹 radial foveae 65,66
桡骨滑车 radial trochlea 65,66
桡骨茎突 styloid process of radius 66
桡骨颈 neck of radius 66
桡骨内侧粗隆 medial eminence of radius 65
桡骨体（干）body of radius 65,66
桡骨外侧粗隆 lateral eminence of the radius 66
桡骨窝 radial fossa 63
桡骨远端（横嵴）distal extremity of radius (transverse

ridge) 65,66
桡神经 radial nerve 98,120,359,402,425,426,427,428,429
桡腕骨 radial carpal bone 67,68
韧带窝 ligament fossa 63,64
绒毛 villus 171,176,181
溶酶体 lysosome 196,197
肉蹄 dermis of the hoof 3
肉质缘 pulpa part of diaphragm 90
乳房 mammae, breast 3,10
乳糜 chyle 370
乳糜池 cisterna chyli 370
乳头 nipple 10,11,265,386,387
乳头层 papillary layer 2
乳头肌 papillary muscle 328,339,342,343,344,347,412,414
乳突 mamilloarticular process 54
乳腺 mammary gland 3
入球小动脉 afferent arteriole 255
软腭 soft palate 144,210,211
软骨 costal cartilage 30
软骨基质 cartilage matrix 225
软骨膜 perichondrium 223,224,225,226
软骨囊 cartilage capsule 225
软骨片 cartilage plate 236
软骨细胞 chondrocyte 223,224,225
软骨陷窝 cartilage lacuna 224,225
闰管 intercalated duct 142,143,200,201,202,329,350,353

S

腮耳肌 parotido auricular muscle 88,386,424
腮腺 parotid gland 84,87,88,94,115,119,140,142,424,425,427
腮腺淋巴结 parotid lymph node 372,386
塞尔托利细胞（支持细胞）Sertoli cell (sustentacular cell) 275,276
三叉神经 trigeminal nerve 403
三叉神经根 roots of trigeminal nerve 476
三尖瓣 tricuspid valve 328,339,340,341,342,343,344,345
三角肌 deltoid muscle 82,85,93,95,96,111
三角肌粗隆 deltoid tuberosity 63
三角肌线 deltoid line 63
三角韧带 triangular ligament 188,189
色素上皮层 pigment epithelium layer 469
筛窦 ethmoid sinus 206
筛骨 ethmoid bone 28
筛骨板 ethmoidal plate 39
上鼻道 dorsal nasal meatus 206
上鼻甲 dorsal nasal concha 144,212,418
上鼻甲骨 dorsal turbinal bone 39
上唇 upper lip 123,124,424
上唇降肌 depressor muscle of the upper lip 88
上唇提肌 levator muscle of the upper lip 81,84,88
上颌窦 maxillary sinus 206
上颌骨 maxillary bone 29,34,35,36,38,39,138
上颌骨腭突 palatine process of the maxillary bone 37
上颌神经 maxillary nerve 403
上睑提肌 levator muscle of the upper eyelid 469
上臼齿 upper molar 138,139
上皮内淋巴细胞 intraepithelial lymphocyte 171,176,177,178,182
上皮性网状细胞 epithelial reticular cell 371,376,377
上前臼齿 upper premolar 138,139
上切齿 upper incisors 138,139
上丘脑 epithalamus 401
上犬齿 upper canine 139
上犬齿窝 upper canine fossa 138
上舌骨 epihyoid 41,42
上眼睑 superior eyelid 472
杓状软骨 arytenoid cartilage 206,213,214,215,216,217
舌 tongue 114,123,125,126,127,128,140,144,210
舌背 dorsum of tongue 114
舌扁桃体 lingual tonsil 114,115,396,397
舌动脉 lingual artery 354
舌根 root of tongue (lingual root) 125,126,127,128,129,132
舌骨 hyoid bone 29,41,42,354
舌肌 muscles of tongue 128
舌尖 apex of tongue (lingual apex) 125,126,127,128,

129
舌黏膜 mucous membrane of the tongue 128
舌乳头 lingual papilla 114,131
舌体 body of tongue (lingual body) 125,126,127,128,129
舌系带 lingual frenum 129
舌下神经 hypoglossal nerve 403
舌下神经孔 hypoglossal foramen 37
舌下腺 sublingual gland 115,128
舌腺 lingual gland 132,396
舌咽神经 glossopharyngeal nerve 403
伸肌沟 extensor groove 74
深筋膜 deep fascia 80
神经 nerve 226
神经垂体 neurohypophysis 442,445
神经垂体神经部 pars nervosa,neurohypophysis 446,447,448
神经胶质 neuroglia 400
神经胶质细胞 neuroglial cell 400
神经胶质细胞纤维 neuroglial cell fiber 451
神经节 ganglion 201
神经节细胞层 ganglion cell layer 469,478
神经膜细胞 neurolemmal cell 226
神经内膜 endoneurium 440
神经束 nerve tract 439,440
神经束膜 perineurium 226,439,440
神经外膜 epineurium 439
神经系统 nervous system 400
神经细胞核 nucleus of the neuron 226
神经纤维 nerve fiber 440
神经纤维层 nerve fiber layer 469,478
神经元 neuron 132,133,400,405
神经元突起 process of the neuron 226
神经元纤维 neurofibril 400
肾 kidney 119,190,191,199,240,244,245,246,247,248,249,250,251,252,257,260,271,357,358,362,379,439,460
肾被膜 kidney capsule 250
肾单位 nephron 240
肾动脉 renal artery 249,253,331,357,358
肾窦 renal sinus 240,252
肾后端 caudal extremity of the kidney 250
肾静脉 renal vein 249,253,332,362
肾门 renal hilum 240,250,251,252,257,460
肾内侧缘 medial border of the kidney 250
肾皮质 renal cortex 240,252,257
肾前端 cranial extremity of the kidney 250
肾乳头 renal papillae 240,252,257
肾上腺 adrenal gland 245,249,250,260,439,443,459,460,461,462,463,464,465,466
肾实质 renal parenchyma 250
肾髓质 renal medulla 240,252,256,257
肾小管 renal tubule 241,256
肾小囊 renal capsule 241,254,255
肾小囊腔 capsular space 254,255
肾小球 renal glomerulus 241
肾小体 renal corpuscle 240,253,255
肾盂 renal pelvis 252,257
肾盏 renal calices 252,257
肾脂囊 *renal adipose capsule* 240,248
肾柱 renal column 252
肾铸型 kidney cast 253
升结肠 ascending colon 117
升主动脉 ascending aorta 330
生精上皮 spermatogenic epithelium 275
生精细胞 spermatogenic cell 268,276
生精小管 seminiferous tubule 268,275,276
生长激素细胞 somatotroph 442
生长卵泡 growing follicle 298
生殖股神经 genitofemoral nerve 402,431
生殖系统 reproductive system 272,300,301
生殖上皮 germinal epithelium 298,304
声带 vocal cords 206,207,213,216
声门裂 rima glottidis 215
施万细胞 Schwann cell 226
施旺细胞胞膜沟 groove in plasma membrane of the Schwann cell 440
施旺细胞核 nucleus of the Schwann cell 440
十二指肠 duodenum 116,147,150,151,152,153,154,155,162,165,170,171,172,190,191,199
十二指肠壁 duodenal wall 170
十二指肠后曲 caudal duodenal flexure 162
十二指肠黏膜 duodenal mucosa 170

十二指肠乳头 duodenal papilla 170
十二指肠腺 duodenal gland 117,172
实质 parenchyma 251,274,278,450
食管 esophagus 115,144,147,149,150,151,152,153,154,155,162,210,213,214,224,355,357,361,391,435,436,437
食管口 pharyngeal opening of the esophagus 144
食管裂孔 esophageal hiatus 81,357
食管黏膜固有层 mucosal lamina propria of the esophagus 222
食管黏膜上皮 mucosal epithelium of the esophagus 222
食管腔 esophageal lumen 147,148,222
食管外膜 adventitia of the esophagus 222
食管腺 esophageal gland 115,147,148
食管咽口 pharynx entrance of the esophagus 213
食管支气管静脉 esophageal bronchial vein 436
食管支气管动脉 esophageal bronchial artery 436,437
视交叉 optic chiasma 410,411,445,476
视觉器官 visual organ 468
视上核 supraoptic nucleus 442
视神经 optic nerve 403,473,474,475,476
视神经孔 optic canal, optic foramen 38
视网膜 retina 468,474,475
视网膜神经节细胞 retinal ganglion cell 478
视杆细胞层 rods layer 469,478
视锥细胞 cone cell 478
视锥细胞层 cones layer 469,478
室间隔 interventricular septum 328,339,342,343,344,346
室旁核 paraventricular nucleus 442
嗜碱性粒细胞 basophilic granulocyte,basophil 332,366,367
嗜碱性细胞 basophilic cell 442,447,448,449
嗜酸性粒细胞 eosinophilic granulocyte,eosinophil 332,365,366,367,368
嗜酸性细胞 acidophilic cell 442,443,447,448,449,459
嗜伊红细胞 eosinophil 177,184,185
嗜银细胞 argyrophilic cell 185
手指 finger 357
枢椎 axis 30,42
梳状肌 pectinate muscle 328,339,343
疏松结缔组织 loose connective tissue 279,280,281,282
输精管 deferent duct 247,268,271,272,273,278,279,286,287,300,301,312,316
输卵管 uterine tube 298,302,312,314,315
输卵管腹腔口 abdominal orifice of uterine tube 298
输卵管壶腹 ampulla of uterine tube 298,313,317
输卵管漏斗 infundibulum of uterine tube 298
输卵管伞 fimbriae of uterine tube 260,298,313,317
输卵管系膜 mesosalpinx 312
输卵管峡 isthmus of uterine tube 298,313,317
输尿管 ureter 242,244,245,246,247,248,250,252,253,257,258,259,260,271,320,358
输尿管管腔 ureteral lumen 258
输尿管末段 final segment of the ureter 258
束状带 fasciculate zone 444,462,463,464
束状带细胞 fasciculate zone cell 464
树突 dendrite 400,405,406,409,423,478,479
树突棘 dendritic spine 405
竖毛肌 arrector muscle 3,12
刷细胞 brush cell 207
双核肝细胞 binucleate hepatocyte 192,193
双核细胞 binucleate cell 262
双核心肌细胞 binucleate cardiac muscle cell 352,353
双角子宫 uterus bicornis 299
双颈动脉干 bicarotid trunk 331
瞬膜 palpebra tertius 469,472
丝状乳头 filiform papilla 114,126,127
四叠体 quadrigeminal bodies 401,412,449
松果体 pineal gland 413,443,449,450,451
松果体柄 pineal stalk 450
松果体细胞 pinealocyte 443,451
松果体小叶 pineal lobule 450,451
松果腺 pineal gland 443
髓窦 medullary sinus 391,393,395
髓放线 medullary ray 253
髓旁肾单位 juxtamedullary nephron 240
髓树 medullary arbor 401,417
髓索 medullary cord 391,393,395

髓细胞 medullary cell 466
髓质 medulla 303,304,371,392,406,419,444,461,465,466
髓质上皮细胞 epithelial cell of thymic medulla 376
梭形细胞 spindle cell 420
锁骨下动脉 subclavian artery 331,337,338,339,346,354
锁骨下静脉 subclavian vein 332

T

弹性纤维 elastic fiber 218,223,224,225,238,353,364,365
提睾肌 cremasteric muscle 90,286
蹄 hoof 3,11,15,23
蹄壁 wall of hoof 11
蹄底 sole of hoof 11
蹄关节 coffin joint 32
蹄球 bulb of hoof 11
蹄匣 hoof capsule 3
体循环 systemic circulation 330
听觉器官 auditory organ 470
听小骨 auditory ossicle 470
同源细胞群 isogenous group 224,225
瞳孔 pupil 468,476
头半棘肌 semispinal muscle of the head 81
头部 head 2,88,210,211,212
头骨 skull 28,35,36,37,38,39,40,78
头肌 biceps muscle of the thigh 104
头长肌 long muscle of head 81
透明层 clear layer 2,15,24,25
透明带 pellucid zone 306,307,308,309
透明软骨 hyaline cartilage 222,223,224,225
透明软骨片 hyaline cartilage plate 226
突触 synapse 400
褪黑激素 melatonin 443
臀部 coxal region 5,7
臀部肌 rump muscles 104
臀股二头肌 glutaeofemorales biceps muscle 83
臀后动脉 caudal gluteal artery 331
臀后神经 caudal gluteal nerve 403
臀前动脉 cranial gluteal artery 331
臀前神经 cranial gluteal nerve 403,433
臀浅肌 superficial gluteal muscle 83,85,86,102,103,104,119
臀深肌 deep gluteal muscle 83
臀线 gluteal line 69
臀中肌 middle gluteal muscle 83,85,86,88,102,103,104,107,120
椭球 ellipsoid 381,382,383,384,385
椭圆囊 utriculus 471
椭圆囊斑 macula utriculi 471
唾液腺 salivary gland 115

W

外鼻 external nose 206
外侧冠突 lateral coronoid process 66
外侧踝 lateral malleolus 72,73,74
外侧角 lateral horn 408
外侧髁间结节 lateral intercondyloid tubercle 73
外侧上髁 lateral epicondyle 71
外侧上髁嵴 lateral supracondylar crest 63
外侧嗅束 lateral olfactory tract 410,411,416
外侧缘 lateral border 250,251
外弹性膜 external elastic membrane 364
外耳 external ear 470,479,480
外耳道 external auditory meatus 38,470,479
外根鞘 outer root sheath 15,16,17,18,19,21
外环骨板 outer circumferential lamella 33
外界膜 external limiting membrane 469
外颗粒层 external granular layer 419,469,478
外膜 adventitia 147,221,223,258,330,364
外网状层 outer plexiform layer 469,478
外眼角 external canthus 472
外展神经 abducent nerve 403
外锥体细胞层 external pyramidal layer 419
外纵肌层 outer longitudinal stratum 171,174,175
腕部 carpal region 4,5,7
腕尺侧屈肌 ulnar flexor muscle of the carpus 82,97,98,101,426,427,428,429
腕骨 carpal bone 31,34,60
腕关节 carpal joint 8,32,95,96,97,98,325
腕桡侧屈肌 radial flexor muscle of the carpus 82,93,97,

98,100,101,425,427,428,429,430
腕桡侧伸肌 radial extensor muscle of the carpus 82,85,87,93,95,96,97,98,99,100,101,119,359,425,426,427,428,429,430
腕外侧屈肌 lateral flexor muscle of the carpus 82,95,96,100,119
腕腺 carpal gland 3,12,22,23
腕斜伸肌 extensor carpiobliquus muscle 82
网状层 reticular layer 2
网状带 reticular zone 444,464,465,466
微动脉 arteriole 18,110
微静脉 venule 132
微绒毛 microvillus 280,281,282
尾 tail 203
尾部 tail region 4,7
尾根 root of the tail 203,282,283,324
尾骨 coccygeal bone 70
尾神经 coccygeal nerve 403
尾叶 caudate hepatic lobe 186
尾正中动脉 caudal median artery 331
尾椎 coccygeal vertebrae 30,34,56
卫星细胞 satellite cell 201,226
未分化细胞 undifferentiated cell 180
位觉器官 position sense organ 470
味沟 gustatory furrow 132,135
味蕾 taste bud 114,131,132,135,396,397
味腺 taste gland 134
胃 stomach 116,120,147,149,150,151,152,153,154,155,156,162,163,164,165,166,190,191,199,248,315,378,379
胃壁面 diaphragmatic surface of stomach 156
胃肠道 gastrointestinal tract 162
胃大弯 greater curvature of stomach 149,150,151,152,156,165
胃底部 fundus of stomach 149,150,151,152,157,158,159,160
胃底腺 fundic gland 116,157,158
胃底腺基部 base of fundic gland 157,158
胃底腺区 region of fundic gland 153,154,155
胃酶原细胞 zymogenic cell 116
胃面 gastric surface 380
胃黏膜面 the mucosal surface of stomach 153,154,155
胃脾韧带 gastrosplenic ligament 379
胃憩室 gastric diverticulum 116,150,151,152,156,162
胃体 body of stomach 149,150,151,152
胃腺 gastric gland 116
胃小凹 gastric pit 116,157,158
胃小弯 lesser curvature of stomach 149,150,152
纹状管 striated duct 142,143
纹状体 striatum 116,171,176,178,182,413,417,449
纹状缘 striated border 181
吻骨 rostral bone 29,35,36,38,211,212
吻镜 rostral plate 114,206
吻突 rostral disc, snout 4,5,6,7,114,119,120,121,124,144,206,209,210,211,212,418
蜗窗 fenestra cochleae 470
无毛皮肤 nonhairy skin 24,25
无髓神经纤维 nonmyelinated nerve fiber 400
无腺部 non-glandular part 153,154,155

X

吸收细胞 absorptive cell 116,176
吸收细胞核 nucleus of the absorptive cell 177,178
膝盖骨 kneecap 31,75
膝关节 stifle joint 32
膝上窝 suprapatellar fossa 71
细胞分裂 cell division 225
细胞核 nucleus 136,143,184,185,197,405,406
细胞连接 cell junction 196,203
细段 thin segment 241,254,256
细支气管 bronchiole 208,236,237
细支气管管腔 lumen of the bronchiole 237
下鼻道 ventral nasal meatus 206
下鼻甲 ventral nasal concha 144,212,418
下鼻甲骨 ventral turbinal bone 39
下唇 lower lip 424
下唇降肌 depressor muscle of the lower lip 81
下唇神经 lower labial nerve 424
下颌骨 mandible 29,34,35,41,44,139,210,211,212,424,453
下颌骨角 angle of mandible 138

下颌骨切迹 mandible notch 41
下颌骨体臼齿部 molar part of mandible body 41
下颌骨体切齿部 incisive part of mandible body 41
下颌角 angle of mandible 41
下颌淋巴结 mandibular lymph node 142,372,385,453
下颌舌骨肌 mylohyoid muscle 81
下颌神经 mandibular nerve 403
下颌窝 mandibular fossa 37
下臼齿 lower molar 138,139
下前臼齿 lower premolar 138,139
下切齿 lower incisors 138,139
下丘脑 hypothalamus 401,410,414,416
下犬齿 lower canine 138,139
下眼睑 inferior eyelid 472
纤毛细胞 ciliated cell 207
纤维膜 fibrous membrane,fibrous tunic 468
纤维囊 fibrous capsule 240
嫌色细胞 chromophobe cell 442,447,448,449
线粒体 mitochondrion 196,197,329
腺垂体 adenohypophysis 442,445,447,448,449
腺垂体远侧部 pars distalis,adenohypophysis 446
腺垂体中间部 pars intermedia,adenohypophysis 446,447,448
腺泡 acinus 23
腺泡腔 acinus cavity 23
腺腔 lumen of gland 203,289,290
腺细胞 gland cell 20,22,23
腺小叶 glandular lobular 23
项嵴 nuchal crest 40
项结节 nuchal tubercle 28,40
项韧带 nuchal ligament 32
消化管 digestive tract 114,147
消化系统 digestive system 114
消化腺 digestive gland 114
小肠 small intestine 116,119,157,163,164,165,249,259,362,379
小肠绒毛 intestinal villi 116,174,175,180
小肠腺 small intestinal gland 116,172,173,174,175,180,181,182
小动脉 small antery 157,200,254,295
小结节 less tubercle 64
小静脉 small vein 110,157,200,295,397
小颗粒细胞 small granule cell 207
小梁 trabecula 371,381,382,383,384,391,392,394
小梁周围淋巴窦 peritrabecular sinus 391,392
小淋巴细胞 small lymphocyte 365,367,368
小脑 cerebellum 144,210,401,406,407,412,414,415,416,417,418,419
小脑半球 cerebellar hemisphere 401,409,410,411
小脑脚 cerebellar peduncle 401,413
小脑幕 tentorium of cerebellum 419
小脑皮质 cerebellar cortex 401,412,422,423
小脑树 cerebellar arbor 401,412
小脑髓质 medulla of cerebellum 423
小脑蚓部 vermis of cerebellum 409
小腿部 crural region 4,5,7
小腿骨 skeleton of leg 31
小腿后皮神经 caudal cutaneous sural nerve 432,433
小腿外侧皮神经 lateral cutaneous sural nerve 433
小叶间胆管 interlobular bile duct 191,192,193,194,195
小叶间导管 interlobular duct 200
小叶间动脉 interlobular artery 191,193,194,195
小叶间隔 interlobular septum 289,290,375,376,378
小叶间结缔组织 interlobular connective tissue 137,148,172,191,192,194,200
小叶间静脉 interlobular vein 191,192,193,194,195
小叶内结缔组织 innerlobular connective tissue 23
小圆肌 minor teres muscle 82
小支气管 small bronchiole 235
小转子 lesser trochanter 71,72
斜方肌 trapezius muscle 82,119,120
斜方体 trapezoid body 410,411,414,476
斜角肌 scalene muscle 81,425
心瓣膜 cardiac valve 340,341
心包 pericardium 227,328,330,347,348,349,355,361,373,435
心包腔 pericardial cavity 349,361
心包液 pericardial fluid 330,349
心壁 cardiac wall 351,353
心传导系统 conduction system of heart 329
心大静脉 great cardiac vein 330

心耳 auricle 328,333,334,335,336,337,338,339,342,345,346,354
心耳梳状肌 pectinate muscle of the auricle 339
心房 cardiac atrium 328,329,339,341,342,343,361
心横肌 transverse muscle of the heart 339,343,344,345
心肌 myocardium 329,351,352
心肌膜 myocardium 349,351,353
心肌细胞 cardiac muscle cell 329,352,353
心肌细胞分支 branch of the cardiac muscle cell 352
心肌细胞核 nucleus of the cardiac muscle cell 350,351,352
心肌细胞横纹 cross striation of the cardiac muscle cell 350,351
心肌纤维 cardiac muscle fiber 329,353
心基 cardiac base 334,336
心尖 cardiac apex 333,334,335,336,338,339,342,354,361
心静脉 cardiac vein 332,330
心内膜 endocardium 329,345,349,353
心腔 heart chamber 339
心切迹 cardiac notch 230,231,232,233,438
心室 cardiac ventricle 328
心室壁 ventricular wall 339,340,342,343,345,347
心室腔 ventricular chamber 353
心外膜 epicardium 329,342,351
心血管系统 cardiovascular system 328
心右静脉 right cardiac vein 330
心脏 heart 119,120,166,190,227,228,229,328,333,335,336,338,342,347,348,349,362,373,437,438
心中静脉 middle cardiac vein 330
心最小静脉 the smallest cardiac vein 330
新生毛皮质 cortex of new hair 19
星形胶质细胞 astroglial cell 451
星形细胞 astrocyte 422
胸（廓）前口 cranial opening of the thoracic cage 60
胸壁 thoracic wall 90
胸部 thoracic part 4,5,6,7
胸部乳房 chest breast 10
胸导管 thoracic duct 370
胸段脊髓 thoracic spinal cord 407,408,409
胸腹部乳房 thoracoabdominal breast 10
胸腹侧锯肌 thoracic part of the ventral serrate muscle 427
胸骨 breast bone, sternum 30,56,57,58,60,347,348,349
胸骨柄 manubrium of sternum 373
胸骨柄软骨 manubrian cartilage 56,57
胸骨嵴 sternal crest 57
胸骨甲状肌 sternothyroid muscle 81,373
胸骨甲状舌骨肌 sternothyrohyoid muscle 81,87,94,141,425,427,453
胸骨软骨结合 manubriogladiolar junction 56,57
胸骨舌骨肌 sternohyoid muscle 81,354
胸骨体 body of sternum 56,57
胸骨心包韧带 stenopericardiac ligament 330,348
胸横肌 transverse pectoral muscle 87,94
胸肌 pectoral muscle 82,84
胸肌神经 pectoral nerve 402
胸降肌 descending pectoral muscle 87,94
胸廓后口 caudal opening of the thoracic cage 60
胸廓内动脉 internal thoracic artery 331
胸廓内静脉 internal thoracic vein 332
胸菱形肌 thoracic part of the trapezius muscle 97
胸浅肌 superficial pectoral muscle 97,427
胸腔 troracic cavity 227
胸深肌 deep pectoral muscle 85,87,90,94,95,97
胸神经 thoracic nerve 402
胸头肌 sternocephalic muscle 81,85,87,94,100,425,427,454
胸腺 thymus 141,361,371,373,374,375,452,453,454
胸腺位置 anatomical position of the thymus 373
胸腺颈叶 cervical lobe of thymus 373,374,385
胸腺皮质 thymic cortex 375,376,377,378
胸腺上皮细胞 thymic epithelial cell 371,376,377
胸腺髓质 thymic medulla 375,376,377,378
胸腺细胞 thymocyte 371,376,377,378
胸腺小体 thymic corpuscle, Hassall's corpuscle 371,376,377
胸腺小叶 thymic lobule 371,375,377,378
胸腺胸叶 thoracic lobe of thymus 373,374,385
胸腺中间叶 intermediate lobe of thymus 373,385
胸斜方肌 thoracic part of trapezius muscle 84,85,86,88,90,95

胸主动脉 thoracic aorta 330,333,334,337,355,362,391,436,437,439
胸椎 thoracic vertebrae 30,34,50,58,60,391
胸椎棘突 spinous process of thoracic vertebra 50
嗅沟 olfactory groove 411,416
嗅脑 rhinencephalon 412
嗅球 olfactory bulb 410,411,416,418
嗅三角 olfactory trigone 410,411,416,445
嗅神经 olfactory nerve 403
悬韧带 suspensory ligament 83
悬蹄 dewclaw 3,11,12
血窦 blood sinus 117,191,192,193,194,195,463,464,466
血管 blood vessel 133,135,136,137,146,159,201,238,259,262,264,281,289,304,330,377,416,419,422,447,450,451,458,462
血管弓 hemal arch 56
血管极 vascular pole 254,255
血管膜 vascular tunic 468
血管球 glomerulus 241,254,255
血管球基膜 glomerular basement membrane 241
血管突 hemal process 56
血管系膜 mesangium 241
血浆 plasma 332
血清 serum 332
血涂片 blood smear 365,366,367,368
血细胞 blood cell, hemocyte 192,332
血小板 blood platelet 332
血液 blood 332

Y

咽 pharynx 115,123,144,145,146,210,213
咽扁桃体 pharyngeal tonsil 115
咽鼓管 pharyngotympanic tube 470
咽鼓管扁桃体 tubal tonsil 115
咽后内侧淋巴结 medial retropharyngeal lymph node 372
咽后外侧淋巴结 lateral retropharyngeal lymph node 372,386
咽后隐窝 pharyngeal recess 115,144,213
咽峡 isthmus of the fauces 144
延髓 medulla oblongata 401,407,409,410,411,412,413,414,416,418,445,476
眼 eye 5,119,120,121,122,209,424,468,472
眼部 ocular region 6
眼的附属器官 adnexa of eye 469
眼房 eye-chamber 469
眼房水 aqueous humor 469
眼睑 eyelid 469
眼睫毛 eye lash 122
眼裂 rima oculi 469
眼球 eye ball 468,472,473,474,476
眼球肌 muscles of eyeball 469
眼球退缩肌 retractor muscles of eyeball 469
眼球斜肌 oblique muscles of eyeball 469
眼球直肌 straight muscles of eyeball 469
眼神经 ophthalmic nerve 403
眼窝 orbit 138
腰部 lumbar part 4,5,6,7
腰大肌 major psoas muscle 81,92,357
腰动脉 lumbar artery 331
腰方肌 lumbar quadrate muscle 81
腰肌 psoas muscle 92
腰荐神经丛 lumbosacral plexus 403
腰静脉 lumbar vein 332
腰淋巴干 lumbar lymphatic trunk 370
腰膨大 lumbar intumescence 400,407,408
腰神经 lumbar nerve 402
腰小肌 minor psoas muscle 81,92,357
腰椎 lumbar vertebrae 5,30,34,53
咬肌 masseter muscle 81,84,85,88,110,119,140,141,354,373,385,424
咬肌窝 masseteric fossa 41
叶状乳头 foliate papilla 114,126,127,129
腋动脉 axillary artery 98,359
腋神经 axillary nerve 402,426,427
腋窝 axilla 424
胰 pancreas 117,147,162,170,190,191,198,199,200,201,202,203,248,249
胰岛 pancreatic islet (islet of Langerhans) 201
胰岛B细胞（胰岛素细胞）B cell (insulin cell) 202
胰管 pancreatic duct 118

胰环 pancreatic ring 118,198
胰体 body of the pancreas 198
胰腺细胞 acinar cell of the pancreas 200,201,202
胰腺细胞核 nucleus of the acinar cell 202
胰右叶 right pancreatic lobe 198
胰左叶 left pancreatic lobe 198
乙状弯曲 sigmoid flexure 271,272,285,286,290,291
翼骨 pterygoid bone 29,37,39
翼肌 pterygoideus 43
翼孔 alar foramen 43
翼窝 alar fossa 43
阴瓣 hymen 299,301,316,317,318,320,321
阴部内动脉 internal pudendal artery 331
阴部神经 pudendal nerve 402
阴唇 labia 203,257,263,299,301,317,318,320,321,322, 323,325
阴唇背侧联合 dorsal commissure of labium 203, 323,325
阴唇腹侧联合 ventral commissure of labium 203, 264,316,318,322,323,325
阴道 vagina 245,257,260,263,264,299,300,301,316,317, 318,320,321,322
阴道动脉 vaginal artery 331
阴道前庭 vaginal vestibule 245,257,260,261,263,264, 299,300,301,316,317,318,320,321,322
阴道穹隆 fundus of vagina 299
阴蒂 clitoris 261,264,299,301,316,317,318,320,321,323
阴蒂窝 clitoral fossa 299,322
阴茎 penis 87,90,246,247,270,271,272,285,286,290, 292,291,292,387
阴茎根 root of penis 270
阴茎海绵体 cavernous body of penis 270,294,295
阴茎脚 crus of penis 246,247
阴茎球 penis bulb 247,269
阴茎缩肌 retractor penis muscle 246,270,271,272,291
阴茎体 body of penis 270
阴茎头 glans of penis 270,292
阴茎乙状弯曲 sigmoid flexure of the penis 247,270, 290
阴门 vulva 203,244,245,260,264,299,300,320,324,325
阴门裂 rima vulvae 299
阴囊 scrotum 269,282,283,293
阴囊壁 scrotal wall 284
阴囊淋巴结 scrotal lymph node 290,386,387
阴囊皮肤 scrotal skin 285
阴囊中缝 scrotal raphe 282,283
蚓部 vermis 401
隐动脉 saphenous artery 360
隐窝 crypt 395,396
鹰嘴 olecranon 58,60,65,66
鹰嘴结节 olecranal tuber 65,66
鹰嘴窝 olecranon fossa 63,64
硬腭 hard palate 114,124,144,210,211,212,418
硬膜外腔 epidural cavity 401
硬膜下腔 subdural cavity 401
幽门 pylorus 149,152,154,155,165
幽门部 pyloric part 149,150,151,152
幽门腺 pyloric gland 116,161
幽门腺区 region of pyloric gland 153,154,155
幽门圆枕 torus pyloricus 116,154,155
有毛皮肤 hairy skin 12,13,14,21,22
有髓神经纤维 myelinated nerve fiber 400
右肺后叶（膈叶）right caudal lobe (diaphragmatic lobe) of the lung 219,220,227,228,229,230,231,232, 233,234,338,347,348,438
右肺前叶（尖叶）right cranial lobe (apical lobe) of the lung 219,220,227,228,229,230,231,232,233, 234,347,348,438
右心室 right ventricle 328,333,334,335,336,338,339, 342,344,345,346,354,361
右肺中叶（心叶）right middle lobe (cardiac lobe) of the lung 219,220,227,228,229,230,231,232,233, 234,338,347,348,438
右纵沟 right longitudinal groove 328,334
盂上结节 supraglenoid tubercle 61,63
原始卵泡 primordial follicle 298,305,306,307
远侧前臂间隙 distal space of forearm 65,66
远端小管 distal tubule 241,254,255
远直小管 distal straight tubule 256
远指节间关节 distal interphalangeal joint 32
运动神经元 motor neuron 408
运动系统 locomotor system 80

Z

脏面 visceral surface 380,381
掌部 metacarpal region 4,5,7
掌骨 metacarpal bone 31,34
掌指关节 metacarpophalangeal joint 32
真皮 dermis 2,12,13,14,15,23,25
真皮根鞘 dermis,root sheath 15,16,17,18,19,21
真皮乳头层 dermis,papillary layer 13,15,19,24,25,296
真皮网状层 dermis,reticular layer 13,15,17,19,24
真皮网状层的胶原纤维束 collagen fiber bundlem dermis reticular layer 21
砧骨 incus 470
枕动脉 occipital artery 331
枕骨 occipital bone 28,35,36,37,38,39,138
枕骨大孔 foramen magnum 37,39,40
枕骨基部 basioccipital bone 37,39,40
枕骨颈静脉突 jugular process of the occipital bone 39,40
枕骨髁 occipital condyle 37,40
枕骨鳞部 squamous part of occipital bone 40
枕颌肌 occipitomandibular muscle 81
枕嵴 occipital crest 36,40
枕叶 occipital lobe 409,411,415,416
正中动脉 median artery 98,359,402,425,426,427,428,429,430
支持细胞 sustentacular cell 161,275,276
支气管 bronchus 207,233,234,236,361,476
支气管食管静脉 bronchoesophageal vein 357
支气管食管动脉 bronchoesophageal artery 331,355
支气管树 bronchial tree 208,235
脂肪细胞 adipocyte 23,24,223,225,226,237,258,351,396,458
脂肪细胞核 nucleus of adipocyte 23,24,223
脂肪组织 adipose tissue 132,135,364,393,439,458,459
直肠 rectum 117,147,162,185,260
直肠后神经 caudal rectal nerve 402
直肠壶腹 rectal ampulla 117
直精小管 straight seminiferous tubule 268
直小静脉 straight venule 256
植物神经系统 vegetative nervous system 404
跖背侧第3动脉 dorsal metatarsal artery Ⅲ 331
跖部 metatarsal region 4,5,7
跖骨 metatarsal bone 32,34
指部 digital region 4,5,7
指骨 digital bone 31,34
指关节 finger joint 32
指浅屈肌 superficial digital flexor muscle 83,95,96,97,98,100,101,427,428,430
指深屈肌 deep digital flexor muscle 83,93
指深屈肌尺骨头 ulnar head of deep digital flexor muscle 98,100,101
指深屈肌肱骨头 humeral head of deep digital flexor muscle 97,98,101
指深屈肌桡骨头 radial head of deep digital flexor muscle 98
指外侧伸肌 lateral digital extensor muscle 82,93,96
指外侧伸肌（第4指伸肌）lateral digital extensor muscle to 4th digitorum 95,100
指外侧伸肌（第5指伸肌）lateral digital extensor muscle to 5th digitorum 95,100
指外侧伸肌腱 lateral digital extensor tendon 99,100
指总伸肌 common digital extensor muscle 82,85,95,96,100,119
指总伸肌腱 common digital extensor tendon 95,99,100
趾部 digital region 4,5,7
趾骨 digital bone 32,34
趾浅屈肌 superficial digital flexor muscle 83
趾深屈肌 deep digital flexor muscle 83,103,108
趾外侧伸肌 lateral digital extensor muscle 83,102,103,106
趾长伸肌 long digital extensor muscle 83
致密斑 macula densa 242,254,255
致密结缔组织 dense connective tissue 353
中鼻道 middle nasal meatus 206
中耳 middle ear 470
中间腕骨 intermediate carpal bone 67,68
中间叶 intermediate lobe of thymus 373,374
中淋巴细胞 medium-sized lymphocyte 365,366,367,368

中膜 middle tunic 330,364,365
中脑 mesencephalon 401,416
中脑导水管 mesencephalic aqueduct 401,412
中枢淋巴器官 central lymphoid organ 370
中枢神经系统 central nervous system 400
中心腱 central tendon 81,90
中型静脉 medium vein 364
中型动脉 medium artery 364,365
中性粒细胞 neutrophilic granulocyte, neutrophil 332,365,366,367,368
中央动脉 central artery 383
中央跗骨 central tarsal bone 76,77
中央管 central canal 400,408
中央静脉 central vein 117,191,192,193,194,397
中央乳糜管 central lacteal 117,175,176,178,180
终末细支气管 terminal bronchiole 208,236
终室 last loculus 400
周围淋巴器官 peripheral lymphoid organ 371
周围神经系统 peripheral nervous system 400
轴丘 axon hillock 400,405,406
轴突 axon 400,405
肘部 elbow region 4
肘关节 elbow joint 32
肘肌 anconeus muscle 82
肘突 anconeal process 65,66
肘窝 cubital fossa 63,64
蛛网膜下腔 subarachnoid cavity 401
主动脉 aorta 330,335,336,338,339,343,346,354,355
主动脉瓣 aortic valve 329,340,341
主动脉窦 aortic sinus 331
主动脉弓 aortic arch 330,333,339,354,437
主动脉管壁 aortic wall 356
主动脉口 aortic orifice 329,340,341,345
主动脉裂孔 aortic foramen 81,331,357
主动脉球 aortic bulb 331
主蹄 principal hoof 3,11,12
主细胞 chief cell 116,159,160,443,459
主支气管 primary bronchus 235
转子窝 trochanteric fossa 72
椎动脉 vertebral artery 331
椎弓 vertebral arch 48,49,50,51,54
椎骨 vertebrae 29
椎关节 costovertebral joint 32
椎后切迹 caudal vertebral notch 54
椎间孔 intervertebral foramen 50,53
椎间盘 intervertebral disc 50
椎孔 vertebral foramen 44,45,46,48,49,50,51,54
椎体 vertebral body 47,50,51,52,53
椎头 vertebral head 48,50,52,54,55
椎外侧孔 lateral vertebral foramen 43
椎窝 vertebral fossa 46,49,51,52,53,54
锥旁室间沟 paraconal interventricular groove 328,333,335,338,346,354
锥体 pyramid 410,414,476
锥体细胞 pyramidal cell 420
锥体细胞层 pyramidal layer 420,421
锥体细胞核 nucleus of pyramidal cell 420
锥状乳头 conical papilla 114,125,126,127,129
孖肌 gemellus 83
滋养孔 nutrient foramen 72
子宫 uterus 299,315,316,317,319
子宫角 uterine horn 245,260,299,300,301,302,312,313,315,316,317
子宫角切开 opening uterine horn 317
子宫颈 uterine cervix 245,260,263,299,300,301,312,316,317,318,320
子宫颈管 cervical canal of uterus 299
子宫颈外口 external uterine orifice 299,316,318,310,320
子宫阔韧带 broad ligament of uterus 245,299,300,302,312,313,316,317
子宫卵巢动脉 uteroovarian artery 358
子宫黏膜 uterine mucosa 313
子宫体 uterine body 131,245,260,299,300,301,312,316
子宫体切开 opening uterine body 317
子宫腺 uterine gland 319
子宫圆韧带 round ligament of uterus 299
籽骨 sesamoid bone 31,32
自主神经系统 autonomic nervous system 404
纵隔 mediastinum 228,328,347,355
纵隔淋巴结 mediastinal lymph node 391

纵隔面 mediastinal surface 228
纵行肌 longitudinal muscle 130,131
纵行平滑肌束 longitudinal smooth muscle bundle 364
足背动脉 dorsal pedal artery 331
足细胞 podocyte 241,254,255
最后胸神经 the last thoracic nerve 402
左肺后叶（膈叶）left caudal lobe (diaphragmatic lobe) of the lung 219,220,228,229,230,231,232,233,234,338,355,437
左奇静脉 left azygos vein 436
左肺前叶（尖叶）left cranial lobe (apical lobe) of the lung 219,220,228,229,230,231,232,233,234,338,355,437
左心耳 left auricle 329,334,335,336,337,338,339,342,345,346,350,354
左心室 left ventricle 329,333,334,335,336,338,339,342,354
左心室腔 left ventriclar chamber 347
左腰大肌 left major psoas muscle 92
左腰小肌 left minor psoas muscle 92
左支气管 left bronchus 230,233,234
左肺中叶（心叶）left middle lobe (cardiac lobe) of the lung 219,220,228,229,230,231,232,233,234,338,355,437
左纵沟 left longitudinal groove 328,333,335,338,346,354
坐骨 ischium 31,34,69,70
坐骨板 plate of ischium 69
坐骨大切迹 greater sciatic notch 69
坐骨弓 ischial arch 69
坐骨海绵体肌 ischiocavernosus muscle 289
坐骨棘 ischial spine 69
坐骨结节 ischial tuberosity 69,70
坐骨联合 ischiatic symphysis 69
坐骨神经 sciatic nerve 120,403,432,433,434
坐骨神经肌支 muscular branch of sciatic nerve 433,434,439,440
坐骨体 body of ischium 69
坐骨小切迹 lesser sciatic notch 69
坐骨支 ramus of ischium 69

英中索引

1st cervical vertebra (atlas) 第1颈椎（寰椎）43,44
1st coccygeal vertebra 第1尾椎 55
1st costal cartilage 第1肋软骨 56
1st lumbar vertebra 第1腰椎 53
1st rib 第1肋 58,60,348,349
1th costal bone 第1肋骨 59
2nd carpal bone 第2腕骨 67,68
2nd cervical vertebra (axis) 第2颈椎（枢椎）45,46,47
2nd costal cartilage 第2肋软骨 56,57
2nd finger(dewclaw) 第2指 98
2nd metacarpal bone 第2掌骨 68
2nd metatarsal bone 第2跖骨 77
2nd tarsal bone 第2跗骨 77
2nd toe (hoof) 第2趾 103
3rd carpal bone 第3腕骨 67,68
3rd distal phalanx (coffin bone) 第3指远指节骨（蹄骨）68
3rd distal phalanx (coffin bone) 第3趾远趾节骨（蹄骨）77
3rd fibular muscle 第3腓骨肌 83,102,103,106,107,108,119
3rd finger (hoof) 第3指 98,99
3rd metacarpal bone 第3掌骨 68
3rd metatarsal bone 第3跖骨 77
3rd middle phalanx (os coronale) 第3指中指节骨（冠骨）68
3rd middle phalanx (os coronale) 第3趾中趾节骨（冠骨）77
3rd proximal phalanx (os compedale) 第3指近指节骨（系骨）68
3rd proximal phalanx (os compedale) 第3趾近趾节骨（系骨）77
3rd tarsal bone 第3跗骨 76,77
3rd thoracic vertebra 第3胸椎 50,51,52,54
3rd toe (hoof) 第3趾 103
3rd ventricle 第3脑室 401,412
4th carpal bone 第4腕骨 67,68
4th cervical vertebra 第4颈椎 42,48,49
4th digital extensor muscle 第4指伸肌 96
4th digital extensor tendon 第4指伸肌腱 95
4th finger(hoof) 第4指 96,99
4th metacarpal bone 第4掌骨 68
4th metatarsal bone 第4跖骨 77
4th tarsal bone 第4跗骨 76,77
4th thoracic vertebra 第4胸椎 50
4th toe (hoof) 第4趾 103,107
4th ventricle 第4脑室 401,412
5th costal cartilage 第5肋软骨 56,57
5th digital extensor muscle 第5指伸肌 96
5th digital extensor tendon 第5指伸肌腱 95
5th finger (dewclaw) 第5指 96,98,99
5th metacarpal bone 第5掌骨 68
5th metatarsal bone 第5跖骨 77
5th toe (dewclaw) 第5趾 103,107
7th cervical vertebra 第7颈椎 42
8th costal bone 第8肋骨 59
10th rib 第10肋 58
10th thoracic vertebra 第10胸椎 50

A

abdomen breast 腹部乳房 10
abdomen 腹部 4,5,6,7
abdominal aorta 腹主动脉 249,331,357,358
abdominal linea alba 腹白线 81

abdominal nerve 腹壁神经 431
abdominal orifice of uterine tube 输卵管腹腔口 298
abdominal skin 腹壁皮肤 386,387
abdominal visceral organ 腹腔内脏器官 163,164
abdominal wall muscles 腹壁肌 90,91
abdominal wall 腹壁 156
abducent nerve 外展神经 403
abductor pollicis longus muscle 拇长外展肌 82
absorptive cell 吸收细胞 116,176
accessory carpal bone 副腕骨 167
accessory genital gland 副性腺 287,288
accessory lobe 副叶 219,220,229,230,231,232,233,234,338
accessory nerve 副神经 403
acetabulum 髋臼 69,286
acidophilic cell 嗜酸性细胞 442,443,447,448,449,459
acinar cell of the pancreas 胰腺细胞 200,201,202
acinus cavity 腺泡腔 23
acinus 腺泡 23
adductor muscle 内收肌 83,108,109,360
adenohypophysis 腺垂体 442,445,447,448,449
adipocyte 脂肪细胞 23,24,223,225,226,237,258,351,396,458
adipose tissue 脂肪组织 132,135,364,393,439,458,459
adnexa of eye 眼的附属器官 469
adrenal gland 肾上腺 245,249,250,260,439,443,459,460,461,462,463,464,465,466
adventitia of the esophagus 食管外膜 222
adventitia of trachea 气管外膜 222
adventitia 外膜 147,221,223,258,330,364
afferent arteriole 入球小动脉 255
alar foramen 翼孔 43
alar fossa 翼窝 43
alveolar duct 肺泡管 208,236
alveolar margin 齿槽缘 41
alveolar sac 肺泡囊 208,236,237
alveolar septum 肺泡隔 236,237
ampulla of uterine tube 输卵管壶腹 298,313,317
anatomical position of the thymus 胸腺位置 373
anconeal process 肘突 65,66
anconeus muscle 肘肌 82
angle of mandible 下颌骨角 138
angle of mandible 下颌角 41
angle of rib 肋（骨）角 59
annulus portae 门静脉环 118
antebrachial region 前臂部 4,5,7
anterior auricular vein 耳前静脉 480
anterior cerebral artery 大脑前动脉 410
anterior lobe of hypophysis 垂体前叶 442
anterior tibial artery 胫前动脉 331
anthelix 对耳轮 479
antidiuretic hormone,ADH 抗利尿激素 442
anus 肛门 117,203,260,282,283,324,325
aorta 主动脉 330,335,336,338,339,343,346,354,355
aortic arch 主动脉弓 330,333,339,354,437
aortic bulb 主动脉球 331
aortic foramen 主动脉裂孔 81,331,357
aortic orifice 主动脉口 329,340,341,345
aortic sinus 主动脉窦 331
aortic valve 主动脉瓣 329,340,341
aortic wall 主动脉管壁 356
aperture of larynx 喉口 125,144,213,214,215,216
apex of bladder 膀胱顶（膀胱尖）242,257,258,260
apex of caecum 盲肠尖 167
apex of ear 耳尖 479,480
apex of tongue (lingual apex) 舌尖 125,126,127,128,129
apical dendrite 顶树突 420
apical lobar bronchus 尖叶支气管 235,361
aponeurosis of external oblique abdominal muscle 腹外斜肌腱膜 387
aqueous humor 眼房水 469
area of insertion for infraspinatus muscle 冈下肌止点区 63
argyrophilic cell 嗜银细胞 185
arrector muscle 竖毛肌 3,12
arterial cone 动脉圆锥 333,335,337,341
arterial ligament 动脉韧带（动脉导管索）332,337,339

arteriole 微动脉 18,110
artery 动脉 224,235,237,254,258,330,364,377,440
articular fossa 关节窝 43
articular fovea 关节凹 44
articular sarface 关节面 65
arytenoid cartilage 杓状软骨 206,213,214,215,216,217
ascending aorta 升主动脉 330
ascending colon 升结肠 117
astrocyte 星形细胞 422
astroglial cell 星形胶质细胞 451
atlas 寰椎 30,42
atrioventricular bundle 房室束 329
atrioventricular node 房室结 329
atrioventricular orifice 房室口 328,339,340,341,342,343
atrioventricular valve 房室瓣 328,329
auditory organ 听觉器官 470
auditory ossicle 听小骨 470
auditory tube 耳咽管 470
aural region 耳部 4,5,6,7
auricle 耳郭 470,479
auricle 心耳 328,333,334,335,336,337,338,339,342,345,346,354
auricular articular surface 耳状面 55
auricular vein 耳静脉 363
auriculo-palpebral nerve 耳睑神经 424
autonomic nervous system 自主神经系统 404
axilla 腋窝 424
axillary artery 腋动脉 98,359
axillary nerve 腋神经 402,426,427
axis 枢椎 30,42
axon hillock 轴丘 400,405,406
axon 轴突 400,405
azygos vein 奇静脉 332,357,362

B

B cell (insulin cell) 胰岛B细胞（胰岛素细胞）202
back of ear 耳背 480
back 背部 4,5,7
basal cell 基底细胞 13,14,20,21,22,224,280,289,290
basal layer 基底层 2,13,14,15,25
basal lamina (basement membrane) 基膜 176,178,224
basal striation 基底纹 142
base of fundic gland 胃底腺基部 157,158
basioccipital bone 枕骨基部 37,39,40
basis cavum oris 口腔底 129
basisphenoid 蝶骨体 39
basket cell 篮状细胞 423
basophilic cell 嗜碱性细胞 442,447,448,449
basophilic granulocyte,basophil 嗜碱性粒细胞 332,366,367
bicarotid trunk 双颈动脉干 331
biceps muscle of the forearm 臂二头肌 82,98,101,359,425,426,427,428,429,430
biceps muscle of the thigh 股二头肌 38,83,84,85,86,102,103,105,106,388
biceps muscle of the thigh 头肌 104
bicipital groove 二头肌沟 64
bicipital tuberosity 二头肌粗隆 72
bicuspid valve 二尖瓣 329,339,340,341,342,347,353
binucleate cardiac muscle cell 双核心肌细胞 352,353
binucleate cell 双核细胞 262
binucleate hepatocyte 双核肝细胞 192,193
bladder epitheliam 膀胱上皮 262,263
bladder wall 膀胱壁 261
blood capillary 毛细血管 14,18,110,111,132,146,202,218,226,330,350,351,352,393,395,421
blood cell, hemocyte 血细胞 192,332
blood platelet 血小板 332
blood sinus 血窦 117,191,192,193,194,195,463，464,466
blood smear 血涂片 365,366,367,368
blood vessel within lamina propria 固有层的血管 158
blood vessel 血管 133,135,136,137,146,159,201,238,259,262,264,281,289,304,330,377,416,419,422,447,450,451,458,462
blood 血液 332

blood-air barrier 气-血屏障 208
blunt edge (dorsal margin) of the lung 肺钝缘（背侧缘）228
body of bladder 膀胱体 242,257,258,260
body of epididymis 附睾体 268,273,277,278,279,284,286
body of ilium 髂骨体 69
body of ischium 坐骨体 69
body of penis 阴茎体 270
body of pubis 耻骨体 69
body of radius 桡骨体（干）65,66
body of sternum 胸骨体 56,57
body of stomach 胃体 149,150,151,152
body of testis 睾丸体 273,277,284
body of the incisive bone 切齿骨骨体 36,37
body of the pancreas 胰体 198
body of tongue (lingual body) 舌体 125,126,127,128,129
body of ulna 尺骨体（干）65,66
bone marrow 骨髓 33
bone of arm 臂骨 30
bone substance 骨质 32
bone tissue,osseous tissue 骨组织 32
bony crura 骨脚 471
bottom of the abdomen wall 腹底壁 259,388
Bowman’s capsule 鲍曼囊 241
Bowman’s space 鲍曼腔 254,255
brachial artery 臂动脉 98,359,426
brachial muscle 臂肌 82,93,95,96,98,100
brachial plexus 臂神经丛 402,424,425,426,427,428,429
brachial region 臂部 4,5
brachinocephalic muscle 臂头肌 82,84,87,88,93,94,95,97,100,119
brachiocephalic artery 臂头动脉 337,338,354
brachiocephalic trunk 臂头动脉干 331,333,334,336,339,346
brain stem 脑干 401,413,414,418
brain ventricle 脑室 402
brain,encephalon 脑 401,407,409,410,412,411
branch of the cardiac muscle cell 心肌细胞分支 352
bravhial muscle groove 臂肌沟 63
breast bone, sternum 胸骨 30,56,57,58,60,347,348,349
broad ligament of uterus 子宫阔韧带 245,299,300,302,312,313,316,317
broadest muscle of the back 背阔肌 82,84,85,86,87,88,90,93,95,97,387
bronchial tree 支气管树 208,235
bronchiole 细支气管 208,236,237
bronchoesophageal artery 支气管食管动脉 331,355
bronchoesophageal vein 支气管食管静脉 357
bronchus 支气管 207,233,234,236,361,476
Brunner’s gland 布伦纳腺 117,172
brush cell 刷细胞 207
buccinator muscle 颊肌 81,84,88
bulb of hoof 蹄球 11
bulbar conjunctiva 球结膜 469
bulbocavernous muscle 球海绵体肌 269,288,289
bulbourethral gland 尿道球腺 246,247,269,271,272,287,288,289

C

caecal mucosa 盲肠黏膜 179
caecal lymph nodes 盲肠淋巴结 167
caecum 盲肠 117,120,147,162,163,164,166,167,179,379
calcaneal tuberosity 跟结节 77,107,108,120,388
calcaneus 跟骨 34,70,76,77
calcitonin 降钙素 443
canales semicirculares ossei 骨半规管 471
canine muscle 犬齿肌 81,88
canine 犬齿 36,37,38,41,114
capsular cell 被囊细胞 226
capsular space 肾小囊腔 254,255
capsule 被膜 250,252,253,303,305,309,371,381,382,391,392,393,447,450,451,458,460,461,462
cardia 贲门 149,152,153,154,155
cardiac apex 心尖 333,334,335,336,338,339,342,354,361
cardiac atrium 心房 328,329,339,341,342,343,361

cardiac base 心基 334,336
cardiac gland 贲门腺 116,160,161
cardiac muscle cell 心肌细胞 329,352,353
cardiac muscle fiber 心肌纤维 353
cardiac notch 心切迹 230,231,232,233,438
cardiac part 贲门部 149,150,151,152
cardiac valve 瓣膜 347
cardiac valve 心瓣膜 340,341
cardiac vein 心静脉 332,330
cardiac ventricle 心室 328
cardiac wall 心壁 351,353
cardiovascular system 心血管系统 328
carotid body 颈动脉体 331
carotid sinus 颈动脉窦 331
carpal bone 腕骨 31,34,60
carpal gland 腕腺 3,12,22,23
carpal joint 腕关节 8,32,95,96,97,98,325
carpal region 腕部 4,5,7
cartilage capsule 软骨囊 225
cartilage lacuna 软骨陷窝 224,225
cartilage matrix 软骨基质 225
cartilage of the thyrohyoid bone 甲状舌骨软骨 42
cartilage plate 软骨片 236
cauda equine 马尾 400,407,408
caudal angle 后角 61,62
caudal articular fovea 后关节凹 44
caudal articular process 后关节突 45,46,47,48,49,51,53,54,55
caudal border 后缘 61,62,333,334,335,380
caudal branch of pubis 耻骨后支 69
caudal branch of transverse process 横突后支 48,49
caudal colliculus 后丘 413,449
caudal costal fossa 后肋窝 51,52
caudal cutaneous sural nerve 小腿后皮神经 432,433
caudal deep cervical lymph node 颈深后淋巴结 372
caudal dorsal serrate muscle 后背侧锯肌 81,84,89,90
caudal duodenal flexure 十二指肠后曲 162
caudal extremity of the kidney 肾后端 250
caudal extremity 后端 250,251
caudal femoral cutaneous nerve 股后皮神经 434
caudal gluteal artery 臀后动脉 331
caudal gluteal nerve 臀后神经 403
caudal median artery 尾正中动脉 331
caudal mesenteric artery 肠系膜后动脉 331,358
caudal opening of the pelvis 骨盆后口 70
caudal opening of the thoracic cage 胸廓后口 60
caudal pars of the thyroid gland 甲状腺后部 455
caudal part of greater trochanter 大转子后部 72
caudal part of greater tubercle 大结节后部 63
caudal rectal nerve 直肠后神经 402
caudal vena cava 后腔静脉 90,190,191,199,332,334,347,348,351,361,362,435
caudal vertebral notch 椎后切迹 54
caudate hepatic lobe 肝尾叶 188,189,199
caudate hepatic lobe 尾叶 186
cavernous body of penis 阴茎海绵体 270,294,295
cavernous body of urethra 尿道海绵体 293
cavernous body 海绵体 294,295
celiac artery 腹腔动脉 331,357,358
cell division 细胞分裂 225
cell junction 细胞连接 196,203
central artery 中央动脉 383
central canal 中央管 400,408
central lacteal 中央乳糜管 117,175,176,178,180
central lymphoid organ 中枢淋巴器官 370
central nervous system 中枢神经系统 400
central tarsal bone 中央跗骨 76,77
central tendon of diaphragm 膈中心腱 248,357
central tendon 中心腱 81,90
central vein 中央静脉 117,191,192,193,194,397
centroacinar cell 泡心细胞 200,201,202,203
ceratohyoid 角舌骨 42
cerebellar arbor 小脑树 401,412
cerebellar cortex 小脑皮质 401,412,422,423
cerebellar hemisphere 小脑半球 401,409,410,411
cerebellar peduncle 小脑脚 401,413
cerebellum 小脑 144,210,401,406,407,412,414,415,416,417,418,419

cerebral cortex 大脑皮质（大脑皮层） 401,409,416
cerebral falx 大脑镰 419
cerebral hemisphere 大脑半球 401,415,416,417
cerebral peduncle 大脑脚 401,410,411,412,414,445,476
cerebrospinal fluid 脑脊髓液 400,402
cerebrum 大脑 144,210,401,407,415,418,419
ceruminous gland 耵聍腺 3,470
cervical branch of the facial nerve 面神经颈支 424
cervical canal of uterus 子宫颈管 299
cervical enlargement 颈膨大 400,407
cervical flumina pilorum 颈部毛流（旋涡毛流）8,9
cervical lobe of thymus 胸腺颈叶 373,374,385
cervical multifidus muscle 颈多裂肌 81
cervical nerve 颈神经 402
cervical part of rhomboid muscle 颈菱形肌 95,97
cervical part of trapezius muscle 颈斜方肌 84,85,86,93,95
cervical part 颈部 4,5,6,7
cervical vertebrae 颈椎 30,34,42
cheek tooth 颊齿 114
cheek 颊 114,140
chest breast 胸部乳房 10
chief cell 主细胞 116,159,160,443,459
chondrocyte 软骨细胞 223,224,225
chordae tendineae of the bicuspid valve 二尖瓣的腱索 339
chordae tendineae 腱索 328,339,342,343,344,345,347
choroid plexus 脉络丛 449
choroid 脉络膜 468,474,475
chromophobe cell 嫌色细胞 442,447,448,449
chyle 乳糜 370
ciliary body 睫状体 468,474,475
ciliated cell 纤毛细胞 207
cingulate suleus 扣带沟 412,416
cisterna chyli 乳糜池 370
Clara cell 克拉拉细胞 208
clear layer 透明层 2,15,24,25
clitoral fossa 阴蒂窝 299,322
clitoris 阴蒂 261,264,299,301,316,317,318,320,321,323
clothing hair 被毛 3,121,203
coccygeal nerve 尾神经 403
coccygeal bone 尾骨 70
coccygeal vertebrae 尾椎 30,34,56
cochlea 耳蜗 471
cochlear duct 耳蜗管 471
cochlear nerve 耳蜗神经 403
coffin joint 蹄关节 32
collagen fiber bundle 胶原纤维束 15,17,18,22,23,218
collagen fiber bundlem dermis reticular layer 真皮网状层的胶原纤维束 21
collagenous fiber 胶原纤维 195,238
collecting tubule 集合小管 242
colloid 胶质 456
colon loop 结肠襻 166,168,169
colon 结肠 117,120,147,156,162,163,164,167,179,184,379
colonic gland 结肠腺 185
colonic mucosa 结肠黏膜 179,183
colonic wall 结肠壁 183
common calcaneal tendon 跟（总）腱 83,85,102,103,360,388,389,434
common carotid artery 颈总动脉 331,354
common digital extensor muscle 指总伸肌 82,85,95,96,100,119
common digital extensor tendon 指总伸肌腱 95,99,100
common iliac vein 髂总静脉 332
common peroneal nerve 腓总神经 432,434
compact bone 骨密质 32
conduction system of heart 心传导系统 329
condylar process 髁状突 41
cone cell 视锥细胞 478
cones layer 视锥细胞层 469,478
conical papilla 锥状乳头 114,125,126,127,129
conjunctival sac 结膜囊 469
connective tissue 结缔组织 451
contorted seminiferous tubule 曲精小管 268

coracobrachial muscle 喙臂肌 82,98,428,429,430
cornea 角膜 468,473,476
corneoscleral junction 角膜巩膜缘 468,473
corona radiate 放射冠 309
coronal joint 冠关节 32
coronary artery 冠状动脉 330
coronary circulation 冠状循环 330
coronary groove 冠状沟 328,333,334,336,335 341
coronary sinus 冠状窦 328,330,343
coronoid process 冠状突 41
corpus albicans 白体 311
corpus callosum 胼胝体 401,412,416
corpus luteum 黄体 298,311
cortex of new hair 新生毛皮质 19
cortex 皮质 251,303,304,371,392,417,443,461
cortical labyrinth 皮质迷路 253
cortical nephron 皮质肾单位 240
corticotroph, ACTH cell 促肾上腺皮质激素细胞 442
costal bone 肋骨 30,34,60,89,356
costal cartilage 肋软骨 30,58,59,60
costal cartilage 软骨 30
costal groove 肋（骨）沟 59
costal tubercle 肋（骨）结节 59
costoabdominal muscle 肋腹神经 402,431
costocervical trunk 肋颈动脉干 331
costocervical vein 肋颈静脉 332
costosternal joint 肋胸关节 32
costovertebral joint 椎关节 32
coxal region 臀部 5,7
coxal tuberosity 髋结节 69,70
cranial angle 前角 61,62
cranial articular process 前关节突 45,47,48,49,50,52,53,54,55
cranial bone 颅骨 28,211,212
cranial border 前缘 61,62,333,334,335,380
cranial branch of pubis 耻骨前支 69
cranial branch of transverse process 横突前支 48
cranial cavity 颅腔 39,211,212,418
cranial costal fossa 前肋窝 50,52
cranial dorsal serrate muscle 前背侧锯肌 81
cranial extremity of the kidney 肾前端 250
cranial extremity 前端 250,251
cranial gluteal artery 臀前动脉 331
cranial gluteal nerve 臀前神经 403,433
cranial mesenteric artery 肠系膜前动脉 331,357,358
cranial nerve 脑神经 403
cranial opening of the thoracic cage 胸（廓）前口 60
cranial pars of the thyroid gland 甲状腺前部 455
cranial part of greater trochanter 大转子前部 72
cranial part of greater tubercle 大结节前部 63
cranial part 颅部 2,4,5,6,7
cranial tibial muscle 胫骨前肌 83
cranial vena cava 前腔静脉 332,334,339,348,349,361,438
cremasteric muscle 提睾肌 90,286
cribriform plate of sclera 巩膜筛板 468
cricoid cartilage 环状软骨 206,213,215,216,217
cricothyreoid muscle 环甲肌 214
crista ampullaris 壶腹嵴 471
cross striation of the cardiac muscle cell 心肌细胞横纹 350,351
cross striation 横纹 80,111
crural region 小腿部 4,5,7
crus of diaphragm 膈脚 248
crus of penis 阴茎脚 246,247
crypt of Lieberkühn 李氏隐窝 116,174,176,177
crypt 隐窝 395,396
cubital fossa 肘窝 63,64
cutaneous gland 皮肤腺 3
cutaneous muscle of face 面皮肌 80
cutaneous muscle of forehead 额皮肌 86
cutaneous muscle of neck 颈皮肌 81
cutaneous muscle of trunk 躯干皮肌 81
cutaneous muscle 皮肌 80,88
cutaneous omobrachial muscle 肩臂皮肌 81
cuticle of internal root sheath 内根鞘小皮 16,17,18
cyst sac 囊肿 186
cystic duct 胆囊管 188,189,190,191,199,362

D

dark region of lymphatic nodule 淋巴小结暗区 394
decending colon 降结肠 117
deep digital flexor muscle 指深屈肌 83,93
deep digital flexor muscle 趾深屈肌 83,103,108
deep fascia 深筋膜 80
deep fibular nerve 腓深神经 432
deep gluteal muscle 臀深肌 83
deep pectoral muscle 胸深肌 85,87,90,94,95,97
deferent duct 输精管 247,268,271,272,273,278,279,286,287,300,301,312,316
deltoid line 三角肌线 63
deltoid muscle 三角肌 82,85,93,95,96,111
deltoid tuberosity 三角肌粗隆 63
dendrite 树突 400,405,406,409,423,478,479
dendritic spine 树突棘 405
dens 齿突 45,47
dense connective tissue 致密结缔组织 353
depressor muscle of the lower lip 下唇降肌 81
depressor muscle of the upper lip 上唇降肌 88
dermis of the hoof 肉蹄 3
dermis 真皮 2,12,13,14,15,23,25
dermis,papillary layer 真皮乳头层 13,15,19,24,25,296
dermis,reticular layer 真皮网状层 13,15,17,19,24
dermis,root sheath 真皮根鞘 15,16,17,18,19,21
descending aorta 降主动脉 330
descending pectoral muscle 胸降肌 87,94
dewclaw 悬蹄 3,11,12
diaphragm 膈 81,90,119,120,156,163,165,227,347,348,355,361,435,438,
diaphragmatic surface of stomach 胃壁面 156
diaphragmatic surface 膈面 380,381
diencephalon 间脑 401
diffuse lymphatic tissue 弥散淋巴组织 146,370,391,392,393,395,396,397
digastric muscle 二腹肌 81
digestive gland 消化腺 114
digestive system 消化系统 114
digestive tract 消化管 114,147
digital bone 指骨，趾骨 31,32,34
digital region 指部，趾部 4,5,7
distal extremity of radius (transverse ridge) 桡骨远端（横嵴）65,66
distal interphalangeal joint 远指节间关节 32
distal space of forearm 远侧前臂间隙 65,66
distal straight tubule 远直小管 256
distal tubule 远端小管 241,254,255
divaricate site of trachea 气管分叉处 235
dorsal arch 背侧弓 43,44
dorsal border 背侧缘 61
dorsal branch of the vagus nerve 迷走神经背侧支 436,437
dorsal buccal branch of the facial nerve 面神经颊背侧支 424
dorsal commissure of labium 阴唇背侧联合 203,323,325
dorsal extremity 背侧端 380
dorsal horn 背侧角 408
dorsal metatarsal artery Ⅲ 跖背侧第3动脉 331
dorsal nasal concha 上鼻甲 144,212,418
dorsal nasal meatus 上鼻道 206
dorsal pedal artery 足背动脉 331
dorsal root 背侧根 408
dorsal sacral foramen 荐背侧孔 55
dorsal trunk of vagus nerve 迷走神经背侧干 355,437
dorsal tubercle 背侧结节 43,44
dorsal turbinal bone 上鼻甲骨 39
dorsal-lumbus longest muscle 背腰最长肌 81,84,85,86,88,89,90,104
dorsum of tongue 舌背 114
duct of sebaceous gland 皮脂腺导管 19
duct 导管 137,148,172
duodenal gland 十二指肠腺 117,172
duodenal mucosa 十二指肠黏膜 170
duodenal papilla 十二指肠乳头 170
duodenal wall 十二指肠壁 170
duodenum 十二指肠 116,147,150,151,152,153,154,155,162,165,170,171,172,190,191,199
dura mater 脑膜 419
dust cell 尘细胞 208,238

E

ear 耳 119,120,121,140,209,210,424,427
eferent ductule of testis 睾丸输出小管 280
efferent arteriole 出球小动脉 255
elastic fiber 弹性纤维 218,223,224,225,238,353,364,365
elbow joint 肘关节 32
elbow region 肘部 4
ellipsoid 椭球 381,382,383,384,385
elliptical recess 囊隐窝 470
encephalic arachnoid 脑蛛网膜 402
encephalic dura mater 脑硬膜 402,418,419
encephalic pia mater 脑软膜 402,419,422
endocardium 心内膜 329,345,349,353
endocrine gland 内分泌腺 442
endocrine organ 内分泌器官 442
endocrine system 内分泌系统 442
endomysium 肌内膜 80,110,111
endoneurium 神经内膜 440
endothelial cell 内皮细胞 110,111,192,193,218,264,351,352,447,456,457,459,462,464
endothelium 内皮 330,353
eosinophil 嗜伊红细胞 177,184,185
eosinophilic granulocyte,eosinophil 嗜酸性粒细胞 332,365,366,367,368
epicardium 心外膜 329,342,351
epidermis 表皮 2,13,19,24
epididymis 附睾 90,247,268,271,272,277,278,279,291,387
epidural cavity 硬膜外腔 401
epiglottic cartilage 会厌软骨 206,213,214,215,217
epiglottis 会厌 213
epihyoid 上舌骨 41,42
epimysium 肌外膜 80
epineurium 神经外膜 439
epithalamus 上丘脑 401
epithelial cell of thymic medulla 髓质上皮细胞 376
epithelial reticular cell 上皮性网状细胞 371,376,377
erectile tissue 勃起组织 270
erythrocyte, red blood cell (RBC) 红细胞 24,110,193,238,332,350,365,366,367,368
esophageal bronchial artery 食管支气管动脉 436,437
esophageal bronchial vein 食管支气管静脉 436
esophageal gland 食管腺 115,147,148
esophageal hiatus 食管裂孔 81,357
esophageal lumen 食管腔 147,148,222
esophagus 食管 115,144,147,149,150,151,152,153,154,155,162,210,213,214,224,355,357,361,391,435,436,437
ethmoid bone 筛骨 28
ethmoid sinus 筛窦 206
ethmoidal plate 筛骨板 39
excretory duct of the sebaceous gland 皮脂腺分泌导管 12
excretory duct 分泌导管 20,21,23,218
extensor carpiobliquus muscle 腕斜伸肌 82
extensor groove 伸肌沟 74
external auditory meatus 外耳道 38,470,479
external canthus 外眼角 472
external carotid artery 颈外动脉 331
external ear 外耳 470,479,480
external elastic membrane 外弹性膜 364
external granular layer 外颗粒层 419,469,478
external iliac artery 髂外动脉 331,358,360
external intercostal muscle 肋间外肌 81,89
external jugular vein 颈外静脉 332
external limiting membrane 外界膜 469
external nose 外鼻 206
external oblique abdominal muscle 腹外斜肌 81,84,85,86,87,88,90,104,387
external obturator muscle 闭孔外肌 83
external pyramidal layer 外锥体细胞层 419
external ring of inguinal canal 腹股沟管皮下环 90
external theca of follicular theca 卵泡膜外膜 308,310
external urethral orifice 尿道外口 261,264,265,269,292,293,299
external uterine orifice 子宫颈外口 299,316,318,310,320
extraglomerular mesangial cell 球外系膜细胞 242,254

eye ball 眼球 468,472,473,474,476
eye lash 眼睫毛 122
eye 眼 5,119,120,121,122,209,424,468,472
eye-chamber 眼房 469
eyelash 睫毛 472
eyelid 眼睑 469

F

face for serrate muscle 锯肌面 62
facial bone 面骨 29
facial crest 面嵴 35
facial nerve 面神经 403,424
facial part 面部 2,4,5,6,7
fascia 筋膜 80
fasciculate zone cell 束状带细胞 464
fasciculate zone 束状带 444,462,463,464
felock joint 球节 32
femoral artery 股动脉 331,360
femoral nerve 股神经 403
femoral bone, femur 股骨 31,34,70,75,119
fenestra cochleae 蜗窗 470
fibroblast 成纤维细胞 195,218,225,226
fibrous capsule 纤维囊 240
fibrous membrane,fibrous tunic 纤维膜 468
fibula 腓骨 31,34,70,73,74,75,77
filiform papilla 丝状乳头 114,126,127
filtration barrier 滤过屏障 241
filtration membrane 滤过膜 241
fimbriae of uterine tube 输卵管伞 260,298,313,317
final segment of the ureter 输尿管末段 258
finger joint 指关节 32
finger 手指 357
flumina pilcorum 毛流 9
foliate papilla 叶状乳头 114,126,127,129
follicle 滤泡 456,457,458
follicular cavity 卵泡腔 307,308,309
follicular cell 卵泡细胞 298,306,307
follicular cell 滤泡细胞 456,457
follicular theca 卵泡膜 306,307,308,309
foramen lacerum 破裂孔 37
foramen magnum 枕骨大孔 37,39,40
foramen ovale 卵圆孔 328
forelimb 前肢 2,6,121
fornix 穹隆 412,416
fovea of femoral head 股骨头窝 71,72
frontal bone 额骨 28,34,35,36,39,138
frontal lobe 额叶 409,411,415,416
frontal sinus 额窦 39,144,206,211,212
fundic gland 胃底腺 116,157,158
fundus of stomach 胃底部 149,150,151,152,157,158,159,160
fundus of vagina 阴道穹隆 299
fungiform papilla 菌状乳头 114,125,126,127

G

gallbladder 胆囊 117,163,164,165,186,188,189,190,191,199,378,379
ganglion cell layer 神经节细胞层 469,478
ganglion 神经节 201
gastric diverticulum 胃憩室 116,150,151,152,156,162
gastric gland 胃腺 116
gastric pit 胃小凹 116,157,158
gastric surface 胃面 380
gastrocnemius muscle 腓肠肌 83,103,106,107,108,109,111,388,389,432,433,434
gastrointestinal tract 胃肠道 162
gastrosplenic ligament 胃脾韧带 379
gemellus 孖肌 83
genitofemoral nerve 生殖股神经 402,431
germinal center of lymphatic nodule 淋巴小结生发中心 394
germinal epithelium 生殖上皮 298,304
gland cell 腺细胞 20,22,23
glandula pinealis 脑上腺 443
glandular lobular 腺小叶 23
glans of penis 阴茎头 270,292
glenoid cavity (glenoid fossa) 关节盂（肩臼）61,62,63
glenoid cavity of the scapula 肩胛骨关节盂 63
glial cell 胶质细胞 405,406,409,422
glomerular basement membrane 血管球基膜 241

glomerular zone cell 球状带细胞 464
glomerular zone 球状带 444,462,464
glomerulus 血管球 241,254,255
glossopharyngeal nerve 舌咽神经 403
glutaeofemorales biceps muscle 臀股二头肌 83
gluteal line 臀线 69
goblet cell 杯状细胞 116,171,173,176,177,178,180,181,182,184,185,207,224,237
Golgi cell 高尔基细胞 423
Golgi complex 高尔基复合体 202
gonadotroph 促性腺激素细胞 442
gracilis muscle 股薄肌 83,87,90,387,388
granular cell 颗粒细胞 14,15,306,307,308,309,420,422,423
granular layer 颗粒层 2,13,14,15,25,406,422,423
granular layer,dentate gyrus 齿状回颗粒层 420,421,422
granulocyte 粒细胞 395
granulosa lutein cell 颗粒黄体细胞 311
great auricular vein 耳大静脉 480
great cardiac vein 心大静脉 330
greater curvature of stomach 胃大弯 149,150,151,152,156,165
greater omentum 大网膜 151,152,156,157,164,190
greater palatine foramen 腭大孔 37
greater sciatic notch 坐骨大切迹 69
greater splanchnic nerve 内脏大神经 439
greater trochanter 大转子 71,72
greater tubercle 大结节 64
greater vestibular gland 前庭大腺 299
grey commissure 灰质连合 400
grey matter 灰质 400,408
groove in plasma membrane of the Schwann cell 施旺细胞胞膜沟 440
growing follicle 生长卵泡 298
gustatory furrow 味沟 132,135
gyrus 脑回 401,409,415

H

hair bulb 毛球 3,12,13,15,16,19
hair cortex 毛皮质 13,16,17,18,21
hair cuticle 毛小皮 15,16,17,18
hair follicle 毛囊 3,12,13,15,17,18,22
hair medulla 毛髓质 13,15,17,21
hair papilla 毛乳头 3,12,15,16,19
hair root 毛根 3,12,13,16,17,18,19
hair shaft 毛干 3,13
hair 毛 3,12
hairy skin 有毛皮肤 12,13,14,21,22
hard palate 硬腭 114,124,144,210,211,212,418
Hassall's corpuscle 胸腺小体 371,376,377
head of epididymis 附睾头 268,273,274,277,279,280,284,285,286
head of femur 股骨头 70,71,72
head of fibula 腓骨头 73,74
head of humerus 肱骨头 63,64
head of rib 肋（骨）小头 59
head 头部 2,88,210,211,212
heart chamber 心腔 339
heart 心脏 119,120,166,190,227,228,229,328,333,335,336,338,342,347,348,349,362,373,437,438
hemal arch 血管弓 56
hemal process 血管突 56
hepatic lobule 肝小叶 117,189,192,193,194,195
hepatic plate 肝板 117,191,192,193,194
hepatic porta 肝门 186,188,189,362
hepatic portal vein 肝门静脉 190,362
hepatic vein 肝静脉 186,187,332
hepatocyte with single nucleus 单核肝细胞 192,193
hepatocyte 肝细胞 117,195,197
Herring body 赫令体 442
high power 高倍镜 179
hilum of the ovary 卵巢门 302,303
hilum of the lung 肺门 207
hindlimb 后肢 2
hip bone 髋骨 31,69
hip joint 髋关节 32
hippocampal edge 海马缘 420,421
hippocampal gyrus 海马回 420,421
hippocampal sulcus 海马沟 420
hippocampus 海马 413,417,420,421,422,449

hock 飞节 32
hoof capsule 蹄匣 3
hoof 蹄 3,11,15,23
horizontal plate of palatine bone 腭骨水平板 37
hormone 激素 442
horny layer 角质层 2,13,14,15,24,25
humeral condyle 肱骨髁 63,64
humeral head of deep digital flexor muscle 指深屈肌肱骨头 97,98,101
humeral joint 肩关节 32
humerus crest 肱骨嵴 63
humerus 肱骨 30,34,58,60
hyaline cartilage plate 透明软骨片 226
hyaline cartilage 透明软骨 222,223,224,225
hymen 阴瓣 299,301,316,317,318,320,321
hyoid bone 舌骨 29,41,42,354
hypoglossal foramen 舌下神经孔 37
hypoglossal nerve 舌下神经 403
hypophsis, pituitary gland 垂体 410,411,412,414,442,445,446,448
hypophyseal cleft 垂体裂 442,446,447
hypothalamus 下丘脑 401,410,414,416

I

ileal mucosa 回肠黏膜 179
ileocaecal ligament 回盲韧带 167,179
ileum 回肠 116,147,162,167,179,181,182,183
iliac crest 髂嵴 69
iliac muscle 髂肌 83
iliocostal muscle sulcus 髂肋肌沟 88,89
iliocostal muscle 髂肋肌 81,88,89
iliohypogastric nerve 髂腹下神经 402,431
ilioinguinal nerve 髂腹股沟神经 402,431
iliolumbar artery 髂腰动脉 331
iliopsoas 髂腰肌 83
ilium 髂骨 31,34,69,70
incisive bone 切齿骨 29,35,38,39,139
incisive papilla 切齿乳头 123,124
incisive part of mandible body 下颌骨体切齿部 41
incisor 切齿 114,123,124,140
incus 砧骨 470
inferior eyelid 下眼睑 472
infraorbital foramen 眶下孔 35,36,38,138,139
infraorbital tactile hair 眶下触毛 472
infraspinous fossa 冈下窝 61
infraspinous muscle 冈下肌 82,84,86,93,96
infundibulum of uterine tube 输卵管漏斗 298
infundibulum 漏斗 412
inguinal canal 腹股沟管 82,90
inguinal region 腹股沟部 387
inner circular stratum 内环肌层 171,174,175,419,469,478
inner circumferential lamella 内环骨板 33
inner plexiform layer 内网状层 469,478
inner root sheath 内根鞘 15,16,17,18,19,21
innerlobular connective tissue 小叶内结缔组织 23
integumentary system 被皮系统 2
interalilial arten 髂内动脉 331,358
interalveolar margin 齿槽间缘 41
interarcuate space 弓间隙 55
interatrial septum 房间隔 328
intercalated duct 闰管 142,143,200,201,202,329,350,353
intercondylar eminence 髁间隆起 75
intercondylar fossa 髁间窝 72
intercondyloid area 髁间区 73
intercostal artery 肋间动脉 331,334,356
intercostal muscles 肋间肌 89
intercostal space 肋间隙 58
intercostal vein 肋间静脉 356,362,363
interdigitating cell 交错突细胞 376
interfemus muscle groups 股内侧肌群 109
interflexor muscle 屈肌间肌 83
interlobular artery 小叶间动脉 191,193,194,195
interlobular bile duct 小叶间胆管 191,192,193,194,195
interlobular connective tissue 小叶间结缔组织 137,148,172,191,192,194,200
interlobular duct 小叶间导管 200
interlobular septum 小叶间隔 289,290,375,376,378
interlobular vein 小叶间静脉 191,192,193,194,195
intermedial vastus muscle 股中间肌 83

intermediate carpal bone 中间腕骨 67,68
intermediate lobe of thymus 胸腺中间叶 373,385
intermediate lobe of thymus 中间叶 373,374
internal acoustic meatus 内耳道 39
internal carotid artery 颈内动脉 331
internal ear 内耳 470
internal elastic membrane 内弹性膜 195,330,364
internal intercostal muscle 肋间内肌 81,89,92,356, 363,402
internal jugular vein 颈内静脉 332
internal oblique abdominal muscle 腹内斜肌 81,84, 87,90,103,108,387
internal pudendal artery 阴部内动脉 331
internal pyramidal layer 内锥体细胞层 419
internal theca of follicular theca 卵泡膜内膜 308, 310
internal thoracic artery 胸廓内动脉 331
internal thoracic vein 胸廓内静脉 332
internal tunic 内膜 330,364
interosseous space 骨间隙 70,73,74
interosseus 骨间肌 83
interparietal bone 顶间骨 28
interstitial sell 间质细胞 268,275,276
interstitial lalmella 骨间板 33
interstitial tissue 间质组织 275
interthalamic adhesion 丘脑黏合部 412,416
intertubercular sulcus 结节间沟 64
intervenous tubercle 静脉间结节 328
interventricular septum 室间隔 328,339,342,343,344, 346
intervertebral disc 椎间盘 50
intervertebral foramen 椎间孔 50,53
intestinal crypt 肠隐窝 116,174,175,176,180
intestinal gland 肠腺 183,184,185
intestinal lumen 肠腔 174,175,183
intestinal surface 肠面 380
intestinal villus 肠绒毛 116,174,175,180
intraepithelial lymphocyte 上皮内淋巴细胞 171, 176,177,178,182
intraglomerular mesangial cell 球内系膜细胞 241
intraglomerular 球内系膜 241
iris 虹膜 468,474,476
ischial arch 坐骨弓 69
ischial spine 坐骨棘 69
ischial tuberosity 坐骨结节 69,70
ischiatic symphysis 坐骨联合 69
ischiocavernosus muscle 坐骨海绵体肌 289
ischium 坐骨 31,34,69,70
isogenous group 同源细胞群 224,225
isthmus of the fauces 咽峡 144
isthmus of uterine tube 输卵管峡 298,313,317

J

jejunal lymph nodes 空肠淋巴结 390
jejunal mucosa 空肠黏膜 173
jejunal wall 空肠壁 173
jejunum 空肠 116,147,156,162,165,166,168,174, 175,176,177,178,179,190,191,199,379,390
jugular process of the occipital bone 枕骨颈静脉突 39,40
jugular vein groove 颈静脉沟 331
jugular vein 颈静脉 354
juxtaglomerular cell 球旁细胞 242,255
juxtaglomerular complex 球旁复合体 242
juxtamedullary nephron 髓旁肾单位 240

K

keratinized stratified squamous epithelium 角化的复层扁平上皮 130
keratinocyte 角质细胞 13,14
kidney capsule 肾被膜 250
kidney cast 肾铸型 253
kidney 肾 119,190,191,199,240,244,245,246,247,248, 249,250,251,252,257,260,271,357,358,362,379,439, 460
kneecap 膝盖骨 31,75
Kuffer’s cell 枯否细胞 117,192,193,197

L

labia 阴唇 203,257,263,299,301,317,318,320,321, 322,323,325
lacrimal apparatus 泪器 469

lacrimal bone 泪骨 35,36,38
lacrimal foramen 泪孔 38,139
lacrimal gland 泪腺 469,476
lacrimal passage 泪道 469
lactating breast 泌乳期乳房 10
lamina muscularis mucosae 黏膜肌层 157,158,160,171,172,175,176,181,182,183,184,185
lamina propri 固有层 130,131,132,133,135,136,145,146,147,148,157,158,160,161,171,172,173,175,176,177,180,181,182,184,185,218,224,225,226,236,237,258,259,262,263,264,315,319,395,396,397
large intestine 大肠 117,165,190,199,248,249,315,357,378,379,390,459
laryngeal cartilage 喉软骨 206,215
laryngeal cavity 喉腔 206,213,216
larynx 喉 141,144,206,210,213,217,219,220,373,435,452,453,454
last loculus 终室 400
lateral border 外侧缘 250,251
lateral condyle of tibia 胫骨外髁 73,74
lateral coronoid process 外侧冠突 66
lateral cutaneous sural nerve 小腿外侧皮神经 433
lateral digital extensor muscle to 4th digitorum 指外侧伸肌（第4指伸肌）95,100
lateral digital extensor muscle to 5th digitorum 指外侧伸肌（第5指伸肌）95,100
lateral digital extensor muscle 指（趾）外侧伸肌 82,83,93,96,102,103,106
lateral digital extensor tendon 指外侧伸肌腱 99,100
lateral eminence of the radius 桡骨外侧粗隆 66
lateral epicondyle 外侧上髁 71
lateral femoral cutaneous nerve 股外侧皮神经 402
lateral flexor muscle of the carpus 腕外侧屈肌 82,95,96,100,119
lateral intercondyloid tubercle 外侧髁间结节 73
lateral malleolus 外侧踝 72,73,74
lateral olfactory tract 外侧嗅束 410,411,416
lateral part 侧部 55
lateral retropharyngeal lymph node 咽后外侧淋巴结 372,386
lateral sacral crest 荐外侧嵴 55
lateral supracondylar crest 外侧上髁嵴 63
lateral vastus muscle 股外侧肌 83,103,107
lateral ventricle 侧脑室 401,412,416
lateral vertebral foramen 椎外侧孔 43
lateral horn 外侧角 408
left auricle 左心耳 329,334,335,336,337,338,339,342,345,346,350,354
left azygos vein 左奇静脉 436
left bronchus 左支气管 230,233,234
left caudal lobe (diaphragmatic lobe) of the lung 左肺后叶（膈叶）219,220,228,229,230,231,232,233,234,338,355,437
left cranial lobe (apical lobe) of the lung 左肺前叶（尖叶）219,220,228,229,230,231,232,233,234,338,355,437
left lateral hepatic lobe 肝左外叶 186,187,188
left longitudinal groove 左纵沟 328,338,333,335,346,354
left major psoas muscle 左腰大肌 92
left medial hepatic lobe 肝左内叶 186,187,188
left middle lobe (cardiac lobe) of the lung 左肺中叶（心叶）219,220,228,229,230,231,232,233,234,338,355,437
left minor psoas muscle 左腰小肌 92
left pancreatic lobe 胰左叶 198
left ventriclar chamber 左心室腔 347
left ventricle 左心室 329,333,334,335,336,338,339,342,354
lens 晶状体 469,474,475
less tubercle 小结节 64
lesser curvature of stomach 胃小弯 149,150,152
lesser palatine foramen 腭小孔 37
lesser sciatic notch 坐骨小切迹 69
lesser splanchnic nerve 内脏小神经 439
lesser trochanter 小转子 71,72
lesser vestibular gland 前庭小腺 299
leukocyte, white blood cell, WBC 白细胞 332
levator muscle of the upper eyelid 上睑提肌 469
levator muscle of the upper lip 上唇提肌 81,84,88

Leydig cell 睾丸间质细胞 268,275,276
ligament fossa 韧带窝 63,64
ligamentum teres vesicae 膀胱圆韧带 242
lingual artery 舌动脉 354
lingual frenum 舌系带 129
lingual gland 舌腺 132,396
lingual papilla 舌乳头 114,131
lingual tonsil 舌扁桃体 114,115,396,397
lip 唇 114
liver 肝 90,117,119,120,147,156,163,164,165,166,186,187,188,190,191,192,196,197,199,362,378,379,435,436
lobar bronchus 肺叶支气管 207,235
locomotor system 运动系统 80
long digital extensor muscle 趾长伸肌 83
long fibular muscle 腓骨长肌 83,102,103,106,107
long head of triceps muscle of the forearm 臂三头肌长头 93,95,96,98,100
long muscle of head 头长肌 81
long muscle of neck 颈长肌 81
longitudinal cerebral fissure 大脑纵裂 401,409,415,419
longitudinal muscle 纵行肌 130,131
longitudinal smooth muscle bundle 纵行平滑肌束 364
loose connective tissue 疏松结缔组织 279,280,281,282
lower canine 下犬齿 138,139
lower incisors 下切齿 138,139
lower labial nerve 下唇神经 424
lower lip 下唇 424
lower molar 下臼齿 138,139
lower power 低倍镜 179
lower premolar 下前臼齿 138,139
lumbar artery 腰动脉 331
lumbar intumescence 腰膨大 400,407,408
lumbar lymphatic trunk 腰淋巴干 370
lumbar nerve 腰神经 402
lumbar part 腰部 4,5,6,7
lumbar quadrate muscle 腰方肌 81
lumbar vein 腰静脉 332
lumbar vertebrae 腰椎 5,30,34,53
lumbosacral plexus 腰荐神经丛 403
lumen of gland 腺腔 203,289,290
lumen of the bronchiole 细支气管管腔 237
lumens 管腔 275,276
lung 肺 119,120,166,190,191,199,207,230,231,233,234,235,236,338,355,357,362,363,391,435,436,438
lymph node of the ovary 卵巢淋巴结 302,312,313
lymph node 淋巴结 372,391,392,393,394,395
lymph 淋巴 370
lymphatic capillary 毛细淋巴管 370
lymphatic duct 淋巴导管 370
lymphatic nodule 淋巴小结 145,174,175,180,370,372,391,392,395,396,397
lymphatic system 淋巴系统 370
lymphatic trunk 淋巴干 370
lymphatic vessel 淋巴管 192,370
lymphatic vessels 淋巴管道 370
lymphocentrum 淋巴中心 372
lymphocyte 淋巴细胞 146,181,237,238,332,370,397
lymphoid organ 淋巴器官 370
lymphoid tissue 淋巴组织 370
lysosome 溶酶体 196,197

M

macrophage 巨噬细胞 218,376,378
macula densa 致密斑 242,254,255
macula saccule 球囊斑 471
macula utriculi 椭圆囊斑 471
major psoas muscle 腰大肌 81,92,357
major teres muscle 大圆肌 82,97,98,359,426,428,429
malleus 锤骨 470
mamilloarticular process 乳突 54
mammae, breast 乳房 3,10
mammary gland 乳腺 3
mammotropic cell 催乳激素细胞 442
mandible notch 下颌骨切迹 41
mandible 下颌骨 29,34,35,41,44,139,210,211,212,424,453

mandibular fossa 下颌窝 37
mandibular gland 颌下腺 115,120,141,142,143,385,452
mandibular lymph node 下颌淋巴结 142,372,385,453
mandibular nerve 下颌神经 403
manubrian cartilage 胸骨柄软骨 56,57
manubriogladiolar junction 胸骨软骨结合 56,57
manubrium of sternum 胸骨柄 373
margo epididymidis of testis 睾丸附睾缘 274,277
margo liber of testis 睾丸游离缘 274,277
masseter muscle 咬肌 81,84,85,88,110,119,140,141,354,373,385,424
masseteric fossa 咬肌窝 41
mature follicle 成熟卵泡 298,309,310
maxillary bone 上颌骨 29,34,35,36,38,39,138
maxillary nerve 上颌神经 403
maxillary sinus 上颌窦 206
medial border of the kidney 肾内侧缘 250
medial border 内侧缘 250,251
medial canthus 内眼角 472
medial condyle of femur 股骨内侧髁 75
medial condyle of tibia 胫骨内髁 73,74
medial condyle 内侧髁 72,73,74
medial coronoid process 内侧冠突 65
medial eminence of radius 桡骨内侧粗隆 65
medial epicondyle 内侧上髁 71,72
medial head of triceps muscle of the forearm 臂三头肌内侧头 93,95,96,98,100,430
medial intercondyloid tubercle 内侧髁间结节 73
medial olfactory tract 内侧嗅束 410
medial retropharyngeal lymph node 咽后内侧淋巴结 372
medial vastus muscle 股内侧肌 83,108,109,360
medial view of the femur 股骨内侧观 72
median artery 正中动脉 98,359,402,425,426,427,428,429,430
median palatine suture 腭正中缝 37
median sacral artery 荐中动脉 331
mediastinal lymph node 纵隔淋巴结 391
mediastinal surface 纵隔面 228
mediastinum testis 睾丸纵隔 268,274,278
mediastinum 纵隔 228,328,347,355
medium artery 中型动脉 364,365
medium vein 中型静脉 364
medium-sized lymphocyte 中淋巴细胞 365,366,367,368
medulla oblongata 延髓 401,407,409,410,411,412,413,414,416,418,445,476
medulla of cerebellum 小脑髓质 423
medulla 髓质 303,304,371,392,406,419,444,461,465,466
medullary arbor 髓树 401,417
medullary cell 髓细胞 466
medullary cone 脊髓圆锥 400,407,408
medullary cord 髓索 391,393,395
medullary ray 髓放线 253
medullary sinus 髓窦 391,393,395
melanocyte 黑素细胞 14
melatonin 褪黑激素 443
membrane granulosa 颗粒膜 308,309,310
membranous labyrinth 膜迷路 471
mental foramen 颏孔 35,41
mental tubercle 颏结节 138,139
mesangium 血管系膜 241
mesencephalic aqueduct 中脑导水管 401,412
mesencephalon 中脑 401,416
mesenteric lymph node 肠系膜淋巴结 166,390
mesentery 肠系膜 166,390
mesojejunum 空肠系膜 166
mesosalpinx 输卵管系膜 312
mesothelium 间皮 351
metacarpal bone 掌骨 31,34
metacarpal region 掌部 4,5,7
metacarpophalangeal joint 掌指关节 32
metatarsal bone 跖骨 32,34
metatarsal region 跖部 4,5,7
metathalamus 后丘脑 401
microvillus 微绒毛 280,281,282
middle cardiac vein 心中静脉 330
middle cerebral artery 大脑中动脉 410
middle ear 中耳 470

middle gluteal muscle 臀中肌 83,85,86,88,102,103,104,107,120
middle nasal meatus 中鼻道 206
middle tunic 中膜 330,364,365
minor psoas muscle 腰小肌 81,92,357
minor teres muscle 小圆肌 82
mitochondrion 线粒体 196,197,329
mixed acinus 混合腺泡 218
mixed gland 混合腺 145,226
molar 臼齿 114,123,124,129
molar part of mandible body 下颌骨体臼齿部 41
molecular laye 分子层 406,419,420,421,422,423
molecular layer, dentate gyrus 齿状回分子层 421,422
molting 换毛 3
monocyte 单核细胞 332,365,366
motor neuron 运动神经元 408
mucosa of bladder 膀胱黏膜 261
mucosa 黏膜 174,175,183,184,221
mucosal epithelium of the esophagus 食管黏膜上皮 222
mucosal epithelium of the trachea 气管黏膜上皮 222
mucosal epithelium 黏膜上皮 135,146,147,174,175,223
mucosal fold 黏膜皱襞 314
mucosal lamina propria of the esophagus 食管黏膜固有层 222
mucosal lamina propria of the trachea 222 气管黏膜固有层
mucosal lamina propria 黏膜固有层 223
mucous acinus (复数acini) 黏液腺泡 132,135,137,160,161,172,218,225
mucous bursa 黏液囊 80
mucous gland 黏液腺 148
mucous membrane of the tongue 舌黏膜 128
mucous neck cell 颈黏液腺细胞 159,160
mucus neck cell 颈黏液细胞 116
multipolar neuron 多级神经元 406,409
muscle belly 肌腹 80
muscle fiber 肌纤维 110,111
muscle tendon 肌腱 80
muscles of eyeball 眼球肌 469
muscles of tongue 舌肌 128
muscular branch of sciatic nerve 坐骨神经肌支 433,434,439,440
muscular layer 肌层 131,147,157,174,180,222,258,259,286,287,314,395,396,397
muscular line 肌线 73
musculocutaneous nerve 肌皮神经 402,428,429,430
myelinated nerve fiber 有髓神经纤维 400
mylohyoid muscle 下颌舌骨肌 81
myocardium 心肌 329,351,352
myocardium 心肌膜 349,351,353
myofibril 肌原纤维 350
myoid cell 肌样细胞 376

N

nasal bone 鼻骨 29,35,36,38,39,139
nasal cartilage 鼻软骨 206
nasal cavity 鼻腔 114,206
nasal conchae 鼻甲 206,210
nasal process of the incisive bone 切齿骨鼻突 36
nasal septum 鼻中隔 211,212,418
nasolabial levator muscle 鼻唇提肌 81
nasomaxillary notch 鼻颌切迹 39
neck of bladder 膀胱颈 242,257,258,260,261
neck of femur 股骨颈 71
neck of humerus 肱骨颈 63,64,
neck of radius 桡骨颈 66
neck of rib 肋（骨）颈 59
neck of scapula 肩胛颈 61
nephron 肾单位 240
nerve fiber layer 神经纤维层 478
nerve fiber 神经纤维 440
nerve tract 神经束 439,440
nerve 神经 226
nervous system 神经系统 400
neurofibril 神经元纤维 400
neuroglia 神经胶质 400
neuroglial cell fiber 神经胶质细胞纤维 451
neuroglial cell 神经胶质细胞 400
neurohypophysis 神经垂体 442,445

neurolemmal cell 神经膜细胞 226
neuron 神经元 132,133,400,405
neutrophilic granulocyte, neutrophil 中性粒细胞 332, 365,366,367,368
nipple 乳头 10,11,265,386,387
Nissl body 尼氏体 400,405,409
non-glandular part 无腺部 153,154,155
nonhairy skin 无毛皮肤 24,25
nonmyelinated nerve fiber 无髓神经纤维 400
nose 鼻 206
nostril 鼻孔 121,122,206,209
notch 切迹 55
nuchal crest 项嵴 40
nuchal ligament 项韧带 32
nuchal tubercle 项结节 28,40
nucleolus 核仁 196,405,406,409
nucleus of adipocyte 脂肪细胞核 23,24,223
nucleus of endothelial cell 内皮细胞核 353,364
nucleus of hepatocyte 肝细胞核 193,196
nucleus of mesothelial cell 间皮细胞核 351
nucleus of myoid cell 肌样细胞的胞核 275,276
nucleus of primary oocyte 初级卵母细胞的胞核 306,307,308,309
nucleus of pyramidal cell 锥体细胞核 420
nucleus of the absorptive cell 吸收细胞核 177,178
nucleus of the acinar cell 胰腺细胞核 202
nucleus of the cardiac muscle cell 心肌细胞核 350, 351,352
nucleus of the muscle cell 肌细胞核 110,111
nucleus of the myoepithelial cell 肌上皮细胞核 20
nucleus of the neuron 神经细胞核 226
nucleus of the Schwann cell 施旺细胞核 440
nucleus of the smooth muscle cell 平滑肌细胞核 110,237,365
nucleus 细胞核 136,143,184,185,197,405,406
nucleus,fibroblast 成纤维细胞核 350,351,352
nutrient foramen 滋养孔 72

O

oblique muscles of eyeball 眼球斜肌 469
obturator artery 闭孔动脉 83,108,331
obturator foramen 闭孔 69,70
obturator nerve 闭孔神经 403
occipital artery 枕动脉 331
occipital bone 枕骨 28,35,36,37,38,39,138
occipital condyle 枕骨髁 37,40
occipital crest 枕嵴 36,40
occipital lobe 枕叶 409,411,415,416
occipitomandibular muscle 枕颌肌 81
ocular region 眼部 6
oculomotor nerve 动眼神经 403
olecranal tuber 鹰嘴结节 65,66
olecranon fossa 鹰嘴窝 63,64
olecranon 鹰嘴 58,60,65,66
olfactory bulb 嗅球 410,411,416,418
olfactory groove 嗅沟 411,416
olfactory nerve 嗅神经 403
olfactory trigone 嗅三角 410,411,416,445
omohyoid muscle 肩胛舌骨肌 81
omotransverse muscle 肩胛横突肌 82,93,95,97
opening of duct 导管口 19
opening uterine body 子宫体切开 317
opening uterine horn 子宫角切开 317
ophthalmic nerve 眼神经 403
optic canal, optic foramen 视神经孔 38
optic chiasma 视交叉 410,411,445,476
optic nerve 视神经 403,473,474,475,476
oral cavity 口腔 114,119,120,123,211,212
oral fissure 口裂 114,122
orbicular muscle of the mouth 口轮匝肌 81
orbit 眼窝 138
orbital fossa 眶窝 35,38
orbital periosteum 眶骨膜 469
organ of Corti 柯蒂器 471
orifice of pulmonary trunck 肺动脉口 328,340,341
orifice of pulmonary vein 肺静脉口 329
osseous ampulla 骨壶腹 471
osseous labyrinth 骨迷路 470
osteon 骨单位 33
outer circumferential lamella 外环骨板 33
outer longitudinal stratum 外纵肌层 171,174,175
outer plexiform layer 外网状层 469,478

outer root sheath 外根鞘 15,16,17,18,19,21
oval fossa 卵圆窝 328,343
ovarian artery 卵巢动脉 331,332
ovarian bursa 卵巢囊 298
ovarian cumulus 卵丘 307,308,309
ovarian follicle 卵泡 298,303
ovary 卵巢 245,260,298,300,301,302,303,304,305,306,307,308,310,312,316,317
oxyntic cell 泌酸细胞 116
oxytocin,OT 催产素 442

P

palatine bone 腭骨 29,39
palatine fissure 腭裂 37
palatine fold 腭褶 114,123,124
palatine process of the incisive bone 切齿骨腭突 37
palatine process of the maxillary bone 上颌骨腭突 37
palatine raphe 腭缝 114,123,124
palpebra tertius 瞬膜 469,472
palpebral conjunctiva 睑结膜 469
pampiniform plexus 蔓状丛 278
pancreas 胰 117,147,162,170,190,191,198,199,200,201,202,203,248,249
pancreatic duct 胰管 118
pancreatic islet (islet of Langerhans) 胰岛 201
pancreatic islet, islet of Langerhans 朗格汉斯岛 118
pancreatic ring 胰环 118,198
papilla of lamina propria 固有层乳头 132,133
papillary layer 乳头层 2
papillary muscle of the bicuspid valve 二尖瓣的乳头肌 339
papillary muscle 乳头肌 328,339,342,343,344,347,412,414
paraconal interventricular groove 锥旁室间沟 328,333,335,338,346,354
paracondylar process 髁旁突 39,40
parafollicular cell group 滤泡旁细胞群 457
parafollicular cell 滤泡旁细胞 443,456
paranasal sinus 鼻旁窦 206,210,418
parasympathetic nerve 副交感神经 404
parathyroid gland 甲状旁腺 443 ,455,457,458,459
paraventricular nucleus 室旁核 442
parenchyma 实质 251,274,278,450
parietal bone 顶骨 28,35,36,38,39,40,138
parietal cell 壁细胞 116,158,159,160
parietal crest 顶嵴 36,138
parietal lobe 顶叶 409,411,415,416
parietal peritoneum 腹膜壁层 170
parotidoauricular muscle 腮耳肌 88,386,424
parotid gland 腮腺 84,87,88,94,115,119,140,142,424,425,427
parotid lymph node 腮腺淋巴结 372,386
pars distalis,adenohypophysis 腺垂体远侧部 446
pars intermedia,adenohypophysis 腺垂体中间部 446,447,448
pars nervosa,neurohypophysis 神经垂体神经部 446,447,448
patella 髌骨 31,75
pecten of pubis 耻骨梳 69
pectinate muscle of the auricle 心耳梳状肌 339
pectinate muscle 梳状肌 328,339,343
pectineal muscle 耻骨肌 83,108,360
pectoral muscle 胸肌 82,84
pectoral nerve 胸肌神经 402
pellucid zone 透明带 306,307,308,309
pelvic symphysis 骨盆联合 108
pelvis part of urethra 尿道骨盆部 246,247
pelvis part of urogenital tract 尿生殖道骨盆部 272,287,288,289
pelvis 骨盆 31
penis bulb 阴茎球 247,269
penis 阴茎 87,90,246,247,270,271,272,285,286,290,292,291,292,387
perfoating canal 穿通管 33
periarterial lymphatic sheath 动脉周围淋巴鞘 371,383,384,385
pericardial cavity 心包腔 349,361
pericardial fluid 心包液 330,349
pericardium 心包 227,328,330,347,348,349,355,361,373,435
perichondrium 软骨膜 223,224,225,226
perikaryon 核周体 400

perimysium 肌束膜 80,110,111,350
perineurium 神经束膜 226,439,440
periosteum 骨膜 32
peripheral lymphoid organ 周围淋巴器官 371
peripheral nervous system 周围神经系统 400
peritoneal cavity 腹膜腔 170
peritoneum 腹膜 170
peritrabecular sinus 小梁周围淋巴窦 391,392
permanent tooth 恒齿 138,139
perpendicular plate of palatine bone 腭骨垂直板 37
petrous part of temporal bone 颞骨岩部 39
Peyer's patch 派尔结 117
pharyngeal opening of the esophagus 食管口 144
pharyngeal recess 咽后隐窝 115,144,213
pharyngeal tonsil 咽扁桃体 115
pharyngotympanic tube 咽鼓管 470
pharynx entrance of the esophagus 食管咽口 213
pharynx 咽 115,123,144,145,146,210,213
phrenic nerve 膈神经 348,361,402,435
pigment epithelium layer 色素上皮层 469
pineal gland 松果体 413,443,449,450,451
pineal gland 松果腺 443
pineal lobule 松果体小叶 450,451
pineal stalk 松果体柄 450
pinealocyte 松果体细胞 443,451
piriform lobe 梨状叶 410,411,414,416,445,476
piriform recess 梨状隐窝 115
pituicyte (neuroglial cell) 垂体细胞（神经胶质细胞）442,448
plasma 血浆 332
plate of ischium 坐骨板 69
podocyte 足细胞 241,254,255
polymorphic layer 多形细胞 422
polymorphic layer 多形细胞层 419,420,421
polymorphic layer, dentate gyrus 齿状回多形细胞层 421
pons 脑桥 401,407,410,411,412,414,445
popliteal artery 腘动脉 331
popliteal lymph node 腘淋巴结 105,388,389
popliteal notch 腘切迹 73
popliteal surface 腘肌面 72
popliteus muscle 腘肌 83
portal area 门管区 117
position sense organ 位觉器官 470
postcaval vein hiatus 后腔静脉裂孔 81
posterior lobe of hypophysis 垂体后叶 442
posterior nasal apertures 鼻后孔 37,144,210,212,418
posterior thigh muscle groups 股后肌群 106
premaxillare bone 颌前骨 29
premolar 前臼齿 114,123
prepuce 包皮 247,265,270,293
preputial diverticulum 包皮憩室（包皮盲囊）84,87,246,247,270,271,272,290,292,295,296,386,387
preputial opening 包皮口 270
primary bronchus 主支气管 235
primary costal axillary lymph node 第1肋腋淋巴结 372,430
primary follicle 初级卵泡 305,306
primary lymphoid organ 初级淋巴器官 370
primary oocyte 初级卵母细胞 306
primary papilla of lamina propria 固有层初级乳头 131
primary spermatocyte 初级精母细胞 268,275,276
primordial follicle 原始卵泡 298,305,306,307
principal hoof 主蹄 3,11,12
process of the neuron 神经元突起 226
promontory of sacrum 荐骨岬 55
promontory 岬 470
proper ligament 固有韧带 268
proper vagina tunica 固有鞘膜 268,285
prostate gland 前列腺 269,288
prostatic artery 前列腺动脉 331
proximal convoluted tubule 近端小管曲部 241
proximal interphalangeal joint 近指节间关节 32
proximal space of forearm 近侧前臂间隙 65,66
proximal tubule 近端小管 241,254
pseadostratified ciliated columnar epithelium 假复层纤毛柱状上皮 223,224,226,236,237
pseudostratified columnar epithelium 假复层柱状上皮 315,319
pseudostratified epithelium 假复层上皮 279,280,281,282,286,287,280

pseudounipolar neuron 假单极神经元 402
psoas muscle 腰肌 92
pterygoid bone 翼骨 29,37,39
pterygoideus 翼肌 43
pubic symphysis 耻骨联合 69
pubis 耻骨 31,70
pudendal nerve 阴部神经 402
pulmonary alveolar macrophage 肺泡巨噬细胞 238
pulmonary alveolus 肺泡 208,236,237,238
pulmonary artery 肺动脉 331,346
pulmonary circulation 肺循环 331
pulmonary lobule 肺小叶 208
pulmonary macrophage 肺巨噬细胞 208
pulmonary parenchyma 肺实质 235
pulmonary pleura 肺胸膜 207
pulmonary trunk 肺动脉干 328,331,333,335,337,338,339,354
pulmonary valve 肺动脉瓣 328,340,341,346
pulmonary vein 肺静脉 334
pulpa part of diaphragm 膈肉质缘 357
pulpa part of diaphragm 肉质缘 90
pupil 瞳孔 468,476
Purkinje cell layer 浦肯野细胞层 423
Purkinje cell 浦肯野细胞 406,422,423
Purkinje fiber 浦肯野纤维 329,349,353
pyloric gland 幽门腺 116,161
pyloric part 幽门部 149,150,151,152
pylorus 幽门 149,152,154,155,165
pyramid 锥体 410,414,476
pyramidal cell 锥体细胞 420
pyramidal layer 锥体细胞层 420,421

Q

quadrate hepatic lobe 方叶 186,188
quadrate muscle of the thigh股方肌 83
quadriceps muscle of the thigh 股四头肌 83,84,85,90,102,103,434
quadrigeminal bodies 四叠体 401,412,449

R

radial carpal bone 桡腕骨 67,68
radial extensor muscle of the carpus 腕桡侧伸肌 82,85,87,93,95,96,97,98,99,100,101,119,359,425,426,427,428,429,430
radial flexor muscle of the carpus 腕桡侧屈肌 82,93,97,98,100,101,425,427,428,429,430
radial fossa 桡骨窝 63
radial foveae 桡骨凹 65,66
radial head of deep digital flexor muscle 指深屈肌桡骨头 98
radial nerve 桡神经 98,120,359,402,425,426,427,428,429
radial trochlea 桡骨滑车 65,66
radius 桡骨 31,58,60,68,97
ramus of ischium 坐骨支 69
Rathke's pouch 拉克囊 446,447
rectal ampulla 直肠壶腹 117
rectum 直肠 117,147,162,185,260
rectus femoris muscle 股直肌 83,107,109,360
recurrent laryngeal nerve 喉返神经 435,453
red pulp 红髓 371,381,382,383,384,385
region of cardiac gland 贲门腺区 153,154,155,
region of fundic gland 胃底腺区 153,154,155
region of pyloric gland 幽门腺区 153,154,155
renal adipose capsule 肾脂囊 240,248
renal artery 肾动脉 249,253,331,357,358
renal calices 肾盏 252,257
renal capsule 肾小囊 241,254,255
renal column 肾柱 252
renal corpuscle 肾小体 240,253,255
renal cortex 肾皮质 240,252,257
renal glomerulus 肾小球 241
renal hilum 肾门 240,250,251,252,257,460
renal medulla 肾髓质 240,252,256,257
renal papillae 肾乳头 240,252,257
renal parenchyma 肾实质 250
renal pelvis 肾盂 252,257
renal sinus 肾窦 240,252
renal tubule 肾小管 241,256
renal vein 肾静脉 249,253,332,362
reproductive system 生殖系统 272,300,301
respiratory bronchiole 呼吸性细支气管 208,236

respiratory system 呼吸系统 206
reticular layer 网状层 2
reticular zone 网状带 444,464,465,466
retina 视网膜 468,474,475
retinal ganglion cell 视网膜神经节细胞 478
retinal ganglion cell, α subtype α 亚型视网膜神经节细胞 478
retinal ganglion cell, β subtype β 亚型视网膜神经节细胞 479
retractor muscles of eyeball 眼球退缩肌 469
retractor penis muscle 阴茎缩肌 246,270,271,272,291
rhinencephalon 嗅脑 412
rhomboid fossa 菱形窝 413
rhomboid muscle 菱形肌 82
rib 肋 30,59,70
right cardiac vein 心右静脉 330
right caudal lobe (diaphragmatic lobe) of the lung 右肺后叶（膈叶）219,220,227,228,229,230,231,232,233,234,338,347,348,438
right cranial lobe (apical lobe) of the lung 右肺前叶（尖叶）219,220,227,228,229,230,231,232,233,234,347,348,438
right lateral hepatic lobe 肝右外叶 186,187,188,199
right longitudinal groove 右纵沟 328,334
right medial hepatic lobe 肝右内叶 186,187,188,189,199
right middle lobe (cardiac lobe) of the lung 右肺中叶（心叶）219,220,227,228,229,230,231,232,233,234,338,347,348,438
right pancreatic lobe 胰右叶 198
right ventricle 右心室 328,333,334,335,336,338,339,342,344,345,346,354,361
rima glottidis 声门裂 215
rima oculi 眼裂 469
rima vulvae 阴门裂 299
rods layer 视杆细胞层 469,478
root of ear 耳根 479,480
root of oculomotor nerve 动眼神经根 414,476
root of penis 阴茎根 270
root of the tail 尾根 203,282,283,324
root of tongue (lingual root) 舌根 125,126,127,128,129,132
root sheath 根鞘 3
roots of trigeminal nerve 三叉神经根 476
rostral bone 吻骨 29,35,36,38,211,212
rostral colliculus 前丘 413,449
rostral disc, snout 吻突 4,5,6,7,114,119,120,121,124,144,206,209,210,211,212,418
rostral plate 吻镜 114,206
rough endoplasmic reticulum 粗面内质网 196,197,202
round ligament of liver 肝圆韧带 186,187,188,190,191
round ligament of uterus 子宫圆韧带 299
rump muscles 臀部肌 104

S

sacculus 球囊 471
sacral lymph node 荐淋巴结 358
sacral nerve 荐神经 402
sacral part 荐部 6,7
sacral tuber 荐结节 34,69,70
sacral vertebrae 荐椎 30
sacral-gluteal region 荐臀部 4
sacroiliac joint 荐髂关节 32
sacrum 荐骨 30,55,70
salivary gland 唾液腺 115
saphenous artery 隐动脉 360
sarcomere 肌节 80
sartorius muscle 缝匠肌 83,108,109
satellite cell 卫星细胞 201,226
scalene muscle 斜角肌 81,425
scapula 肩胛骨 30,34,58,60,119
scapular notch 肩胛切迹 61,62
scapular tuber 肩胛结节 61,63
Schwann cell 施万细胞 226
sciatic nerve 坐骨神经 120,403,432,433,434
sclera 巩膜 468,473,474,475,476
scrotal ligament 附睾尾韧带（阴囊韧带）268,269,284
scrotal lymph node 阴囊淋巴结 290,386,387

scrotal raphe 阴囊中缝 282,283
scrotal skin 阴囊皮肤 285
scrotal wall 阴囊壁 284
scrotum 阴囊 269,282,283,293
sebaceous gland 皮脂腺 3,12,13,19,20,21,22
secondary follicle 次级卵泡 309
secondary lymphoid organ 次级淋巴器官 371
secondary papilla of lamina propria 固有层次级乳头 131
secondary spermatocyte 次级精母细胞 268,275
secretory tubulus 分泌小管 20
secretory unit 分泌单位 172
secretory vacuole 分泌泡 315
segmental bronchus 肺段支气管 208
segmental bronchus,tertiary bronchus 节段性支气管 235
semicircular duct 膜半规管 471
semilunar valve 半月瓣 328,339
semimembranous muscle 半膜肌 83,102,103,104,105,106,107,108,109,360,388,389,434
seminal hillock 精阜 269
seminiferous tubule 生精小管 268,275,276
semispinal muscle of the head 头半棘肌 81
semispinal muscle 半棘肌 89
semitendinous muscle 半腱肌 83,85,102,103,104,105,106,108,119,360,388,389,432,433,434
sense organ 感觉器官 468
septomarginal trabecula 隔缘肉柱 328,339,343,344,345
septula testis 睾丸小隔 268
serosa 浆膜 174,175,180,286,314
serous acinus 浆液腺泡 135,142,143,160,161,172,218,225
serous demilune 浆半月 143
Sertoli cell (sustentacular cell) 塞尔托利细胞（支持细胞）275,276
serum 血清 332
sesamoid bone 籽骨 31,32
shaft of femur 股骨体（干）71,72
shaft of humerus 肱骨干（体）63,64
shaft of rib 肋骨干 59
sharp edge (ventral margin) of the lung 肺锐缘（腹侧缘）228
shoulder girdle muscles 肩带部肌 94
shoulder joint 肩胛部 4,5,7
sigmoid flexure of the penis 阴茎乙状弯曲 247,270,290
sigmoid flexure 乙状弯曲 271,272,285,286,290,291
simple columnar epithelium 单层柱状上皮 158,171,180
simple cuboidal epithelium 单层立方上皮 193,195
sinuatrial node 窦房结 329
sinus of the venae cavae 腔静脉窦 328
skeletal muscle 骨骼肌 80,132
skeleton of forearm 前臂骨 31,34
skeleton of leg 小腿骨 31
skeleton 骨骼 34
skin 皮肤 2,19,121
skull 头骨 28,35,36,37,38,39,40,78
slit membrane 裂孔膜 241
slit pore 裂孔 241
small artery 小动脉 157,200,254,295
small bronchiole 小支气管 235
small granule cell 小颗粒细胞 207
small intestinal gland 小肠腺 116,172,173,174,175,180,181,182
small intestine 小肠 116,119,157,163,164,165,249,259,362,379
small lymphocyte 小淋巴细胞 365,367,368
small vein 小静脉 110,157,200,295,397
smooth muscle fiber 平滑肌纤维 364
smooth muscle 平滑肌 193,195,223,226,236,237,279,280,281,282,294,440
soft palate 软腭 114,144,210,211
sole of hoof 蹄底 11
soleus muscle 比目鱼肌 83,10
soma 胞体 201,226,405,406,478,479
somatotroph 生长激素细胞 442
space of forearm 前臂间隙 58,60
spermatic cord 精索 90,247,269,273,274,277,278,284,285,286,287,290,291,387

spermatid 精子细胞 268,275,276
spermatogenesis 精子发生 268
spermatogenic cell 生精细胞 268,276
spermatogenic epithelium 生精上皮 275
spermatogonium 精原细胞 268,275,276
spermatozoa 精子 268,275,276,279,281,286,287
sphenoid bone 蝶骨 28,37
sphenoidal sinus 蝶窦 206
spherical recess 球囊隐窝 470
spinal arachnoid 脊蛛网膜 401
spinal cord 脊髓 144,210,400,407,408,412,414,418
spinal dura mater 脊硬膜 401
spinal ganglion 脊神经节 402,407,408
spinal meninges 脊膜 401
spinal nerve root 脊神经根 407,408
spinal nerve 脊神经 402
spinal pia mater 脊软膜 401
spindle cell 梭形细胞 420
spine of scapula 肩胛冈 61
spinous cell layer 棘层 2,13,14,15
spinous cell 棘细胞 14
spinous process of the 7th cervical vertebra 第7颈椎棘突 42
spinous process of thoracic vertebra 胸椎棘突 50
spinous process 棘突 45,46,47,48,49,50,51,52,53,54
spiral organ filtration membnane 螺旋器 241,471
spleen trabecular 脾小梁 381
spleen 脾 120,147,156,164,166,248,249,315,371,378,379,380,382,383,384,385,459
splenic cord 脾索 371,382
splenic corpuscle 脾小体 371
splenic hilum 脾门 380,381
splenic nodule 脾小结 371,383,384,385
splenic parenchyma 脾实质 381
splenic sinusoid 脾窦 382
splenic sinusoid 脾血窦 372
splenius muscle 夹肌 81,89
spongy bone 骨松质 32
squamous part of occipital bone 枕骨鳞部 40
stapes 镫骨 470
stenopericardiac ligament 胸骨心包韧带 330,348
sternal crest 胸骨嵴 57
sternocephalic muscle 胸头肌 81,85,87,94,100,425,427,454
sternohyoid muscle 胸骨舌骨肌 81,354
sternothyrohyoid muscle 胸骨甲状舌骨肌 81,87,94,141,425,427,453
sternothyroid muscle 胸骨甲状肌 81,373
stifle joint 膝关节 32
stomach 胃 116,120,147,149,150,151,152,153,154,155,156,162,163,164,165,166,190,191,199,248,315,378,379
straight abdominal muscle 腹直肌 81,85,90,431
straight muscles of eyeball 眼球直肌 469
straight seminiferous tubule 直精小管 268
straight venule 直小静脉 256
stratified columnar epithelium 复层柱状上皮 264,293,294
stratified squamous epithelium 复层扁平上皮 131,132,133,135,136,145,148,218,295,296,395,396,397
stratum lacunosum 腔隙层 420,421
stratum radiatum 辐射层 420
striated border 纹状缘 181
striated duct 纹状管 142,143
striated muscle 横纹肌 396
striatum 纹状体 116,171,176,178,182,413,417,449
stroma 间质 305,306,307
stylohyoid muscle 茎舌骨肌 81
stylohyoid 茎突舌骨 41,42
styloid process of radius 桡骨茎突 66
styloid process of ulna 尺骨茎突 65,66
subarachnoid cavity 蛛网膜下腔 401
subcapsular sinus 被膜下窦 391,392,393
subclavian artery 锁骨下动脉 331,337,338,339,346,354
subclavian vein 锁骨下静脉 332
subcutaneous fat 皮下脂肪 13,24
subcutaneous tissue 皮下组织 2,12
subdural cavity 硬膜下腔 401
subendothelial layer 内皮下层 330,353

sublingual gland 舌下腺 115,128
submucosa 黏膜下层 147,157,158,171,174,175,180,183,184,218,222,223,226
submucosal gland 黏膜下腺 117
subscapular artery 肩胛下动脉 359
subscapular fossa 肩胛下窝 62
subscapular muscle 肩胛下肌 82,97,98,359,427
subscapular nerve 肩胛下神经 98,359,402,426
subsinuosal interventricular groove 窦下室间沟 328,334
sulcus 脑沟 401,409,415
superfacial nephron 浅表肾单位 240
superficial cervical artery 颈浅动脉 331
superficial cervical lymph node 颈浅淋巴结（肩前淋巴结）372
superficial digital flexor muscle 指（趾）浅屈肌 83,95,96,97,98,100,101,427,428,430
superficial epithelium 表面上皮 304
superficial fascia 浅筋膜 2,80
superficial fibular nerve 腓浅神经 432
superficial gluteal muscle 臀浅肌 83,85,86,102,103,104,119
superficial inguinal lymph node 腹股沟浅淋巴结 386
superficial pectoral muscle 胸浅肌 97,427
superior eyelid 上眼睑 472
supraglenoid tubercle 盂上结节 61,63
supraoptic nucleus 视上核 442
supraorbital foramen 眶上孔 36
supraorbital tactile hair 眶上触毛 472
suprapatellar fossa 膝上窝 71
suprascapular nerve 肩胛上神经 402,426,427
supraspinatus muscle 冈上肌 82,84,85,93,95,96,97,98,359,427
supraspinous fossa 冈上窝 61
suspensory ligament 悬韧带 83
sustentacular cell 支持细胞 161,275,276
sweat gland 汗腺 3,12,13,15,16,18,19,20
sympathetic ganglion cell 交感神经节细胞 465
sympathetic nerve 交感神经 404
sympathetic preganglionic neuron 交感神经节前神经元 408
sympathetic trunk 交感干 262,355,436,437,438,439
synapse 突触 400
synovial bursa 滑膜囊 80
systemic circulation 体循环 330

T

tail of epididymis 附睾尾 268,273,274,277,278,279,281,282,284,285,286
tail of testis 睾丸尾 273,277,284
tail region 尾部 4,7
tail 尾 203
talus 距骨 76,77
tarsal bone 跗骨 31,70,76
tarsal joint 跗关节 32,85,102
tarsal region 跗部 4,5,7
taste bud 味蕾 114,131,132,135,396,397
taste gland 味腺 134
tectorial cell 盖细胞 262
telencephalon 端脑 401
temporal bone 颞骨 29,37,40,139
temporal fossa 颞窝 36,138
temporal lobe 颞叶 409,411,415,416
temporal muscle 颞肌 81
temporomandibular joint 颞下颌关节 32
tendon sheath 腱鞘 80
tensor muscle of the antebrachial fascia 82,93,95,96,97,98,426,427,428,429,430 前臂筋膜张肌
tensor muscle of the fascia lata 阔筋膜张肌 83,84,85,90,102
tentorium of cerebellum 小脑幕 419
terminal bronchiole 终末细支气管 208,236
testicular network 睾丸网 268
testicular artery 睾丸动脉 331
testicular vein 睾丸静脉 332
testis 睾丸 84,87,90,119,120,247,268,271,272,273,276,285,286,290,291,387
thalamus 丘脑 401
the last thoracic nerve 最后胸神经 402
the mucosal surface of stomach 胃黏膜面 153,154,155

the smallest cardiac vein 心最小静脉 330
theca lutein cell 膜黄体细胞 311
thigh region 4,5 股部
thin segment 细段 241,254,256
third eyelid 第3眼睑 469,472
thoracic aorta 胸主动脉 330,333,334,337,355,362,391,436,437,439
thoracic duct 胸导管 370
thoracic lobe of thymus 胸腺胸叶 373,374,385
thoracic nerve 胸神经 402
thoracic part of the trapezius muscle 胸菱形肌 97
thoracic part of the ventral serrate muscle 胸腹侧锯肌 427
thoracic part of trapezius muscle 胸斜方肌 84,85,86,88,90,95
thoracic part 胸部 4,5,6,7
thoracic spinal cord 胸段脊髓 407,408,409
thoracic vertebrae 胸椎 30,34,50,58,60,391
thoracic wall 胸壁 90
thoracoabdominal breast 胸腹部乳房 10
thymic corpuscle 胸腺小体 371,376,377
thymic cortex 胸腺皮质 375,376,377,378
thymic epithelial cell 胸腺上皮细胞 371,376,377
thymic lobule 胸腺小叶 371,375,377,378
thymic medulla 胸腺髓质 375,376,377,378
thymocyte 胸腺细胞 371,376,377,378
thymus 胸腺 141,361,371,373,374,375,452,453,454
thyrohyoid 甲状舌骨 42
thyroid artery 甲状腺动脉 354
thyroid cartilage 甲状软骨 206,213,214,215,217
thyroid follicle 甲状腺滤泡 456,457
thyroid gland 甲状腺 435,443,452,453,454,455,457,458
thyrotroph, TSH cell 促甲状腺激素细胞 442
thyroxine 甲状腺素 443
tibia 胫骨 31,34,70,73,74,75,77
tibial crest 胫骨嵴 74
tibial nerve 胫神经 432,433,434
tibial tuberosity 胫骨粗隆 74
tongue 舌 114,123,125,126,127,128,140,144,210
tonsil of palatine velum 腭帆扁桃体 114,115,125,126,129,395,396
tonsil 扁桃体 115
tooth 齿 114
torus pyloricus 幽门圆枕 116,154,155
trabecula 小梁 371,381,382,383,384,391,392,394
trachea 气管 144,207,210,213,214,215,216,217,219,220,221,222,223,224,230,231,232,233,234,235,338,361,435,436,452,453,454
tracheal cartilage 气管软骨 213,221,225
tracheal cavity 气管腔 221,222
tracheal ganglion 气管神经节 226
tracheal gland 气管腺 223,224
tracheal lymphatic trunk 气管淋巴干 370
tracheal muscle 气管肌 221,222
tragi 耳毛 479
transitional epithelium 变移上皮 258,259,262,263,298
transversal muscle 横行肌 133
transverse abdominal muscle 腹横肌 81,431
transverse canal 横突管 30
transverse cerebral fissure 大脑横裂 401,409,411,419
transverse colon 横结肠 117
transverse foramen 横突孔 30,45,46,48,49
transverse line 横线 55
transverse muscle of the heart 心横肌 339,343,344,345
transverse pectoral muscle 胸横肌 87,94
transverse process 横突 46,47,50,51,52,53,54,56
trapezius muscle 斜方肌 82,119,120
trapezoid body 斜方体 410,411,414,476
triangular ligament 三角韧带 188,189
triceps muscle of the forearm 臂三头肌 82,84,85,90,101 ,425,429
tricuspid valve 三尖瓣 328,339,340,341,342,343,344,345
trigeminal nerve 三叉神经 403
trochanteric fossa 转子窝 72
trochlea of femur 股骨滑车 71,75
trochlea tuberosity of femur 股骨滑车粗隆 71
trochlea 滑车 73,74

trochlear nerve 滑车神经 403
trochlear notch 滑车切迹 65,66
trochlear tubercle 滑车结节 75
troracic cavity 胸腔 227
trunk 躯干 2
tubal tonsil 咽鼓管扁桃体 115
tuberositial groove 粗隆沟 74
tuberosity of scapular spine 肩胛冈结节 61
tunia media of the arteriole 微动脉中膜 110
tunic of thyroid gland 甲状腺被囊 443
tunica albuginea 白膜 268,274,278,298,304
turbinal bone 鼻甲骨 29
tweezers 镊 357
tympanic bulla 鼓泡 37,39
tympanic cavity 鼓室 470
tympanic membrane 鼓膜 470
tympanohyoid 鼓舌骨 41,42
type Ⅰ alveolar cell Ⅰ型肺泡细胞 238
type Ⅱ alveolar cell Ⅱ型肺泡细胞 238
ulna 尺骨 31,58,60,68
ulnar carpal bone 尺腕骨 67,68
ulnar flexor muscle of the carpus 腕尺侧屈肌 82,97,98,101,426,427,428,429
ulnar head of deep digital flexor muscle 指深屈肌尺骨头 98,100,101
ulnar nerve 尺神经 98,359,402,425,426,427,428,429,430
umbilical artery 脐动脉 331
umbilical region 脐部 4,5
undifferentiated cell 未分化细胞 180
upper canine fossa 上犬齿窝 138
upper canine 上犬齿 139
upper incisors 上切齿 138,139
upper lip 上唇 123,124,424
upper molar 上臼齿 138,139
upper premolar 上前臼齿 138,139
ureter 输尿管 242,244,245,246,247,248,250,252,253,257,258,259,260,271,320,358
ureteral lumen 输尿管管腔 258
urethra 尿道 242,244,245,257,258,261,263,264,293,294,300,320
urethral bulb 尿道球 269
urethral isthmus 尿道峡 269
urethral lumen 尿道腔 293,294
urethral muscle 尿道肌 269
urinary bladder 膀胱 119,120,163,165,166,242,244,245,246,247,257,258,259,261,263,271,272,286,287,288,300,312,315,316,318,320,379
urinary system 泌尿系统 240,244
urogenital organs 泌尿生殖器官 245,246,247
urogenital system 泌尿生殖系统 271
urogenital tract 尿生殖道 269
urogenital vestibulum 尿生殖前庭 299
uterine body 子宫体 131,245,260,299,300,301,312,316
uterine cervix 子宫颈 245,260,263,299,300,301,312,316,317,318,320
uterine gland 子宫腺 319
uterine horn 子宫角 245,260,299,300,301,302,312,313,315,316,317
uterine mucosa 子宫黏膜 313
uterine tube 输卵管 298,302,312,314,315
uteroovarian artery 子宫卵巢动脉 358
uterus 子宫 299,315,316,317,319
uterus bicornis 双角子宫 299
utriculus 椭圆囊 471
uvula 腭垂 114
vacuolated cell 空泡细胞 20,21
vagina 阴道 245,257,260,263,264,299,300,301,316,317,318,320,321,322
vaginal artery 阴道动脉 331
vaginal vestibule 阴道前庭 245,257,260,261,263,264,299,300,301,316,317,318,320,321,322
vagosympathetic trunk 迷走交感干 354,454
vagus nerve 迷走神经 355,362,391,403,436,437
vallate papilla 轮廓乳头 114,125,126,127,129,132
vascular pole 血管极 254,255
vascular tunic 血管膜 468
vegetative nervous system 植物神经系统 404
vein 静脉 218,224,226,235,236,237,253,330,351,364,375,376,378,396,397,420,421,448,449,459,461,465,466

vena caval hiatus 腔静脉裂孔 357,361
ventral arch 腹侧弓 44
ventral branch of vagus nerve 迷走神经腹侧支 436
ventral buccal branch of the facial nerve 面神经颊腹侧支 424
ventral commissure of labium 阴唇腹侧联合 203,264,316,318,322,323,325
ventral crest 腹侧嵴 52,54
ventral extremity 腹侧端 380
ventral horn 腹侧角 408
ventral nasal concha 下鼻甲 144,212,418
ventral nasal meatus 下鼻道 206
ventral sacral foramen 荐腹侧孔 55
ventral serrate muscle 腹侧锯肌 82,97,98,359
ventral trunk of the vagus nerve 迷走神经腹侧干 355
ventral tubercle 腹侧结节 43,44
ventral turbinal bone 下鼻甲骨 39
ventricular chamber 心室腔 353
ventricular wall 心室壁 339,340,342,343,345,347
venule 微静脉 132
vermis of cerebellum 小脑蚓部 409
vermis 蚓部 401
vertebrae 椎骨 29
vertebral arch 椎弓 48,49,50,51,54
vertebral artery 椎动脉 331
vertebral body 椎体 47,50,51,52,53
vertebral column 脊柱 30
vertebral foramen 椎孔 44,45,46,48,49,50,51,54
vertebral fossa 椎窝 46,49,51,52,53,54
vertebral head 椎头 48,50,52,54,55
vertical muscle 垂直肌 130,131,132
vesicular gland 精囊腺 246,247,269,271,272,287,288,289,290
vestibular fold 前庭襞 213
vestibular gland opening 前庭腺开口 318
vestibular nerve 前庭神经 403
vestibule 前庭 470
vestibulocochlear nerve 前庭耳蜗神经 403
villus 绒毛 171,176,181
visceral lymphatic trunk 内脏淋巴干 370
visceral nervous system 内脏神经系统 404
visceral organs 内脏器官 119,120
visceral surface 脏面 380,381
visual organ 视觉器官 468
vitreous body 玻璃体 469,475,474
vocal cords 声带 206,207,213,216
vomer 犁骨 29,37
von Ebner gland 冯 · 埃布纳腺 134
vulva 阴门 203,244,245,260,264,299,300,320,324,325

W

wall of hoof 蹄壁 11
white matter 白质 400,408
white pulp 白髓 371,381,382,383,384
wing of atlas 寰椎翼 43,44
wing of ilium 髂骨翼 69
wing of sacrum 荐骨翼 55
withers 鬐甲部 4,5,6,7

X

xiphoid cartilage 剑状软骨 56,57,58,60,163
xiphoid sternum synchondrosis 剑胸软骨结合 57

Z

zygomatic arch 颧弓 36,37
zygomatic bone 颧骨 29,35,36,38,138,139
zygomatic process of frontal bone 额骨颧突 35,138
zygomatic process of temporal bone 颞骨颧突 35,36,38,138
zymogen granule 酶原颗粒 202
zymogenic cell 胃酶原细胞 116